FUNDAMENTALS OF NUCLEAR MODELS
Foundational Models

FUNDAMENTALS OF NUCLEAR MODELS
Foundational Models

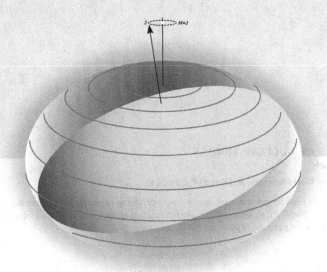

David J Rowe
University of Toronto, Canada

John L Wood
Georgia Institute of Technology, USA

 World Scientific

NEW JERSEY · LONDON · SINGAPORE · BEIJING · SHANGHAI · HONG KONG · TAIPEI · CHENNAI

Published by

World Scientific Publishing Co. Pte. Ltd.

5 Toh Tuck Link, Singapore 596224

USA office: 27 Warren Street, Suite 401-402, Hackensack, NJ 07601

UK office: 57 Shelton Street, Covent Garden, London WC2H 9HE

British Library Cataloguing-in-Publication Data
A catalogue record for this book is available from the British Library.

FUNDAMENTALS OF NUCLEAR MODELS
Foundational Models

ISBN-13 978-981-256-955-4
ISBN-10 981-256-955-3
ISBN-13 978-981-256-956-1 (pbk)
ISBN-10 981-256-956-1 (pbk)

Printed in Singapore.

To

Una and Elizabeth

Preface

"Everything should be made as simple as possible, but not simpler."

[Albert Einstein]

In the language of Philip Anderson,[1] nuclear physics is an *emergent phenomenon*. In other words, no body of theoretical physicists, however smart they might be, could have predicted the existence of a nucleus, let alone its properties, from first principles (even if such first principles were known). Nuclear physics is an experimental science. Nevertheless, the theoretical interpretation of what is observed and the construction of logically consistent predictive theories are essential for guiding the course of experiment and for developing an understanding of the subject.

Physics progresses by systematic experimental observations and the construction of models to interpret them. In the normal progression of physics research, the compilation of experimental data runs in parallel with the development of theoretical models in a mutually constructive way. However, sometimes the natural interpretation of a body of data can be at serious odds with the theoretical perspectives of the day. As more data are accumulated, the interpretation may be adjusted and brought into line with an evolving underlying theory. It can also happen that a body of data continues to conflict with conventional wisdom and is far too pervasive to be ignored. Explaining the data may then result in a *paradigm shift*[2] in the theoretical description of the system concerned.

The modern era of nuclear physics was initiated in the middle of the last century by such a paradigm shift. Because of the strong short-range interactions between nucleons, there had previously been little hope that nuclei could be understood at a microscopic level. Thus, radioactivity and macroscopic properties, such as sizes and masses, were described in terms of a so-called *compound nucleus* model and a *liquid drop* model. However, a simple independent-particle shell model proved to be so extraordinarily successful at interpreting a huge body of experimental data that theorists were obliged to seek an explanation of how a system of strongly-interacting nucleons could behave like weakly-interacting particles in a central field.

---

[1] Anderson P.W. (1972), *Science* **177**, 393.

[2] Kuhn T.S. (1996), *The Structure of Scientific Revolutions* (University of Chicago Press, Chicago), third edn.

The subsequent many-nucleon theory of the nucleus emerged from the need to come to terms with the shell-model perspective.

The development of the shell model and an (albeit incomplete) understanding of how it could apply, rapidly led to its acceptance as the underlying theory of nuclear structure in terms of interacting neutrons and protons. This shell-model example illustrates the importance of pursuing alternative interpretations of observed phenomena in spite of conflicts with generally accepted beliefs. One should be aware that a model can be successful in fitting a body of data for reasons quite different from those envisaged in its formulation. This can happen, for example, because different physical models may have common, or simply related, mathematical structures. Thus, fitting a model to a data set may be testing the mathematical structure of the model but not necessarily its physical interpretation. It should also be remembered that a given set of physical observables will, generally, have many representations in quantum mechanics. And, while the states of different representations may be 100% orthogonal to one another, their properties may be indistinguishable by a finite set of measurements taken with limited accuracy. Thus, to extract a convincing physical interpretation of the many models of nuclear structure, it is important to consider how these models fit into the larger overall perspective of the nucleus as a system of interacting neutrons and protons.

Expressing a successful model in microscopic terms is an essential component of understanding its significance. If all attempts to derive such an expression fail, then one must either suspect the interpretation of the model or the validity of the underlying microscopic theory. Whatever the answer, one will undoubtedly emerge from the exercise of rationalising the seemingly incompatible perspectives with a better understanding of the physics of what is going on. *Discovering that a model is successful is not nearly as important as understanding why it is successful. Identifying the limitations of a model are equally important.* Thus, a primary objective of this book is to discover the extent to which the various insightful models of nuclear structure can be embedded as submodels of the shell model. In the end, *the main criteria for acceptance of a theory of nuclear structure are consistency and elegance.*

Many models of nuclear structure phenomena are formulated as submodels of the shell model. Elliott's SU(3) model of nuclear rotations and the BCS theory of pair coupling are notable examples. In contrast, the collective models were initially formulated as liquid drop models and viewed as continuum limits of a many-nucleon system. However, the observation of rotational bands in light nuclei showed that collective states also occur for relatively small numbers of nucleons. In the unified model, the collective models were given a microscopic interpretation in terms of nucleons moving in a mean field with collective properties.

The unified model largely avoids the necessity of requiring the large nucleon-number limit. It is exceptionally insightful and has been deployed in numerous creative ways, e.g., to determine the kinds of shell model configurations necessary to explain collective phenomena. Its deficiency is that it is limited by the constraints of a mean-field approximation to a highly restricted subspace of shell model states. In parallel with the mean-field methods, an algebraic approach, based on dynamical symmetries, which relates models with shell-model coupling schemes, has been developed. Fortunately, the mean-field and algebraic approaches are complementary and can even be used in combination. Both are discussed in this book.

In the algebraic approach, the observables of a successful model are expressed in terms of a Lie algebra of observables called a *spectrum generating algebra*. One then seeks a microscopic expression of these basic observables in terms of nucleon observables (i.e., position and momentum coordinates and spins). Finding such an expression can be a very non-trivial task. It may be that the model is only valid in a limiting situation, such as a limit in which the nucleon number is large. Or it may be that the model is an approximation to a more general model which does have a microscopic expression. One can then attempt to understand the circumstance under which the approximation is valid. Whatever the relationship should be, it is important to seek it out because only by doing so does the model become a contributor to the overall understanding of nuclear structure theory. If a microscopic expression of a model can be found, one can proceed to determine the representations of its spectrum generating algebras in the Hilbert space of the shell model. The models then correspond to shell-model coupling schemes. Thus, they become submodels of the shell model and acquire relationships with other models corresponding to different coupling schemes.

We start, in Chapter 1, with an overview of the variety of experimental observations that are made on nuclei and the model ways in which they are interpreted. By compiling nuclear data and examining them from many perspectives, it appears that the data often suggest their own interpretation. What emerges from Chapter 1 is a view of the nucleus from many vantage points and a rationale for considering a corresponding set of phenomenological models. Later chapters present the models in some depth. This volume focuses on the four foundational models on which most developments in nuclear physics are based: the collective model, the shell model, the pair-coupling models, and the mean-field (Hartree-Fock) models. A subsequent volume will introduce more-microscopic models and emphasise the relationships between the models and their realisation as submodels of an all-embracing shell model.

It is assumed that the reader has a basic knowledge of quantum mechanics and angular momentum theory, as presented in any of the standard texts, and a familiarity with introductory nuclear physics, e.g., at the level of the book *"Introductory Nuclear Physics"* by K.S. Krane.[3] Some familiarity with elementary concepts in symmetry (group theory) will also be useful. No previous knowledge of nuclear structure is required. This book is intended to serve a wide range of uses. At one extreme, it provides a pedagogical introduction to nuclear physics. And, at the other, it provides a resource book for researchers at the frontiers of the subject.

---

[3]Krane K.S. (1988), *Introductory Nuclear Physics* (Wiley, New York).

Some aspects of this book were tackled at a much more elementary level in the book "Nuclear Collective Motion" published in 1970.[4] That book did not include the more recently developed algebraic perspectives and is superseded by the present two-volume text. However, the simple pragmatic description of the basic models given in 1970 is as relevant today as when it was published. Thus, for example, a first course in nuclear structure, could profitably be based on the model descriptions given in this 1970 book (to be republished by World Scientific) together with the overview given in Chapter 1 of this volume of our current, more complete, understanding of nuclear structure in terms of them.

The last 40 years have seen huge advances both in experimental and theoretical technology and, consequently, an explosion in the range of nuclei investigated. Thus, a much deeper understanding has developed of the circumstances under which the different nuclear models are applicable. The perspective for deriving the microscopic foundation of these models has also evolved. It is now time to assess how much of what we have learnt is built on firm foundations and where gaps in our understanding remain. It is by exploring failures of models that physics advances.

D.J. Rowe

J.L. Wood

2009

---

[4]Rowe D.J. (1970), *Nuclear Collective Motion: Models and Theory* (Methuen, London).

Acknowledgements

This book has evolved over a long period of time during which it has changed its shape as a result of numerous interactions and discussions with many more persons than we can name. Conversely, the concerns that arose in the logical development of the subject have much influenced both of our research programs. In particular, we are most indebted to our close collaborators, Ed Zganjar, Kris Heyde, Ken Krane, George Rosensteel, Juliana Carvalho, Joe Repka, and Chairul Bahri, as well as our several research associates and graduate students who have directly (and sometimes unwittingly) contributed to much of the material presented. Our perspectives have also been strongly influenced by interactions with Ted Hecht and Jerry Draayer, whose research programs have impacted directly on ours, by the emphasis on algebraic and dynamical symmetry methods of Brian Flowers, Phil Elliott, Larry Biedenharn, and Franco Iachello, and by our teachers Denys Wilkinson, Aage Bohr, Ben Mottelson, David Brink, Tony Lane, and Bruce French.

We are much indebted to Carol Nash, for numerous discussions and suggestions on the philosophical manner in which the subject should be treated and for preparing most of the difficult figures, and to Trevor Welsh, who spared no effort in proof reading the manuscript and in preparing the publishable version of the LaTeX file, including the index and bibliography. We also acknowledge, with appreciation, a critical reading and suggestions for improvement of the sections on effective interactions by Ryoji Okamoto.

Finally, we have a special debt of gratitude to Una Rowe for her continued support and for willingly and most generously looking after our material needs during the days and weeks (for one of us years) when we were totally engrossed in working on the book.

The data that we use to illustrate what is known experimentally is as up-to-date as we could make it. An approximate cut-off for the data used is June 2008. The evaluated data from Nuclear Data Sheets, http://www.nndc.bnl.gov/nds/, is used as widely as possible. In some cases the Evaluated Nuclear Structure Data File (ENSDF), http://www.nndc.bnl.gov/ensdf/, is the data source. We reference the usages generically (without specifying the volume and page number for Nuclear Data Sheets or indicating source location in ENSDF. We also cross-checked unevaluated data against the Unevaluated Data List (XUNDL), http://www.nndc.bnl.gov/ensdf/. We wish to acknowledge the extraordinary contribution of the Nuclear Data Group in maintaining this unique resource for nuclear structure physics.

Permissions: The following figures are copyrighted material and are reproduced with permission of the specified copyright holder:

Figures 1.3a, 1.22, 1.23, 1.54, 1.63, 1.68, 1.71, 1.75, 1.77, 1.78, 1.84, 2.4, 2.5, and 5.12 are reproduced with permission of Elsevier.

Figures 1.1, 1.3b, 1.12, 1.18, 1.24, 1.25, 1.47, 1.55, 1.82, 1.85, 5.1, 5.5, 5.6, 5.8, 5.9, 6.9, 6.10, 7.1, 7.2, 7.3, and 7.4 are reproduced with permission of The American Physical Society.

Figure 1.6 is reproduced with permission of World Scientific Publishing Co.

Figure 1.52 and 1.66 are reproduced with permission of the The Royal Danish Academy of Sciences and Letters.

Figure 1.7 is reproduced with permission of Dr. B. Frois.

Figure 1.49 is reproduced with permission of Dr. C.W. de Jager.

Figure 1.50a is reproduced with permission of Dr. L.-E. Svensson.

Figure 1.81 is reproduced with permission of Dr. T. Lauritsen.

Figure 1.83 is reproduced with permission of Dr. J. Sharpey-Shafer.

The specific sources of these figures are given in the respective figure captions.

Notations and conventions

We have tried to maintain a reasonably high level of mathematical rigour without being unduly pedantic. The primary objective is to use notations that are clear, informative, and unambiguous.

The use of the symbol := to distinguish a definition from a simple equality, is useful and we have adopted this notation. This enables us to use the symbol $\equiv$ to denote an equivalence relation or occasionally an expression, such as $A \equiv B$ to signify that A is identically equal to B (which usage should be clear from the context).

In general, we find it useful to distinguish a function f of a variable x and its value, $f(x)$, for any given x. However, when the meaning is clear, we sometimes follow the common practice of referring to $f(x)$ as a function when we wish to exhibit its argument explicitly. If $\hat{X}$ is a differential operator and $f' := \hat{X}f$, we also sometimes write $\hat{X}f(x)$ with the understanding that the operator acts on the function and not on its value at x; thus, $\hat{X}f(x) \equiv f'(x) \equiv [\hat{X}f](x)$. However, there are occasions in which more care is needed. Suppose, for example, that the variable g is an element of a group and $\hat{X}(g)$ is an operator for which $\hat{X}(g)f(x) = f(gx)$, where gx is a transformation of x. Then, it is easy to make the mistake of supposing that $\hat{X}(g_1)\hat{X}(g_2)f(x) = f(g_1g_2x)$ which is incorrect. In this case, putting in the brackets to signify the sequence of operations, i.e., $[\hat{X}(g_1)\hat{X}(g_2)f](x)$, leads to the correct answer. Proceeding a step at a time and defining $f' := \hat{X}(g_2)f$ and $f'' := \hat{X}(g_1)f'$, it is seen that $f''(x) = f'(g_1x)$ and $f'(x) = f(g_2x)$ together imply that $f''(x) = f'(g_1x) = f(g_2g_1x)$. Thus, in situations in which special care is needed, we insert the brackets to ensure that the given expression is correctly interpreted.

As in the last paragraph, it is useful to signify that a symbol denotes an operator by placing a caret over it. Thus, a quantum mechanical Hamiltonian is usually denoted by $\hat{H}$. This practice can be useful in situations in which a given symbol is naturally interpreted in more than one way and it is required to make a distinction between the alternatives. For example, $\hat{\mathbf{L}}^2 \equiv \hat{\mathbf{L}} \cdot \hat{\mathbf{L}}$ denotes a squared angular-momentum operator having eigenvalues given by $L(L+1)$ with integer values of L. As another example, consider a wave function ψ as a complex-valued function of a coordinate x. One can then regard $\hat{x}$ as a multiplicative operator on a Hilbert space of such wave functions defined by $[\hat{x}\psi](x) := x\psi(x)$. This notation is convenient because it means that the modified wave function $\psi' := \hat{x}\psi$ has values at any point, so that $[\hat{x}\psi](y) = y\psi(y)$ and not $x\psi(y)$. However, it is hard to be consistent in such a usage without being excessively pedantic especially when dealing with coordinates which can be interpreted in both ways. For example, when a potential energy function U, defined as a function of a coordinate x, appears as a term in a Hamiltonian, it could be expressed either as $\hat{U}$ or as $U(\hat{x})$ with the understanding that $[\hat{U}\psi](x) = [U(\hat{x})\psi](x) = U(x)\psi(x)$. The notation $U(\hat{x})$ is more informative as it displays the functional dependence of U on the x coordinate. However, it may also be distracting. Sometimes a given symbol, such as a nucleon creation operator $a^{\dagger}_{jm}$, may be explicitly defined as an operator and it would not be helpful to denote it by $\hat{a}^{\dagger}_{jm}$. Thus, we maintain a degree of flexibility in the usage of the $\hat{\ }$ notation; the primary criterion is always to be unambiguous and clear.

Another convention that we employ is upper indices to denote the contragredient components of a spherical tensor. Thus, for example, when the symbol $a^{\dagger}_{jm}$ is used to denote the creation operator for a nucleon with angular momentum quantum numbers jm, the corresponding annihilation operator is denoted by the symbol a^{jm} with the understanding that a^{jm} and $a^{\dagger}_{jm}$ are Hermitian adjoints of one another. The use of superscripts here draws attention to the fact that, if the operator $a^{\dagger}_{jm}$ is regarded as a component of a tensor, its Hermitian adjoint is the corresponding component of a contragredient tensor. Thus, whereas applying the operator $a^{\dagger}_{jm}$ to a state of angular momentum zero produces a state with angular momentum quantum numbers jm, application of the annihilation operator a^{jm} to an angular-momentum zero state produces a state with angular momentum quantum numbers $j, -m$. In this text we use lower indices to label the components of a standard (covariant) tensor and upper indices to label the components of a contravariant tensor. In this way, it is always clear what kind of a tensor we are dealing with and the potential for confusing the two kinds is avoided. Indices can be raised and lowered by use of a metric tensor g_j, which expresses a rotationally invariant (scalar), such as the number operator $\hat{n}_j := \sum_m a^{\dagger}_{jm} a^{jm}$ in the form

$$\hat{n}_j := \sum_{mm'} g^{mm'}_j a^{\dagger}_{jm} a_{jm'} = \sum_m (-1)^{j+m} a^{\dagger}_{jm} a_{j,-m}$$

so that

$$a^{jm} \equiv (-1)^{j+m} a_{j,-m}.$$

The reader is referred to Appendix A for a more detailed discussion of tensors and their manipulation.

Contents

Contents

Chapter 1

Elements of nuclear structure

1.1 Introduction

A nucleus is the core of an atom. By atomic standards, its dimensions are minuscule. Atoms have radii of the order of 10^{-9}m whereas nuclear radii are more like 10^{-14}m. Compared to an atom, a nucleus is like a grain of sand in a football stadium. Nevertheless, its mass is almost the entire mass of the atom. Approximately 99.97% of the mass of an atom resides in its nucleus. This means that, by mass, approximately 99.97% of the material world is nuclear matter.

Nuclei consist of nucleons of which there are two types: positively charged protons and uncharged neutrons. Both nucleon types have essentially the same mass and are approximately 2000 times as massive as an electron. They are held together in a nucleus by the so-called *strong interaction*. This interaction is much stronger but of much shorter range than the Coulomb interaction that binds the atomic electrons to their nuclei. At a separation distance of 1.0 fm, the strong attraction between two nucleons is some 30 times as strong as the Coulomb repulsion between two protons. However, at a distance of 20 fm, the strong interaction is smaller than the Coulomb interaction by a factor of 2000.

In spite of the predominance of nuclear matter in our world, it is easy to be oblivious to the existence of nuclei. This is because nuclei are hidden beneath protective clouds of atomic electrons which effectively keep them apart. In addition, nuclei are prevented from coming into contact with one another by the electrostatic repulsions that result from their positive charges. Thus, one has little direct experience of nuclei outside of the nuclear physics laboratory.

Because nuclei are so isolated, it is not surprising that most physical properties of the everyday world can be explained in terms of atoms and the electronic bonds they make with one another to form molecules and solids. Nevertheless, to understand the existence of atoms, one must first understand the existence of nuclei. Indeed, since an atom is stable only if its nucleus is stable, the question of which atoms can exist is one of nuclear physics. So also is the issue of how many chemical elements are accessible for use or study. At an even more fundamental level, the question of how atoms come into being at all is one of nuclear physics.

Nuclei are created in stars where the temperatures are literally astronomical. Such temperatures are necessary because nuclear reactions occur on an energy scale of hundreds of keV and room temperature corresponds only to an energy of 1/40 eV. Thus, for a nucleus, room temperature might just as well be at absolute zero. A nucleus at room temperature is very cold. It only starts to become "warm" at temperatures of one billion Kelvin. If such temperatures were not approached in stars, the rich variety of nuclei and the atoms they inhabit would not have materialised.

The nucleus was discovered by Geiger and Marsden[1,2] using a beam of alpha particles emitted by radioactive radium daughter nuclei. Geiger and Marsden directed an alpha particle beam at thin metal foils and to their surprise, observed some of the alpha particles to be scattered to large backward angles. From these experiments Rutherford[3] was able to infer the existence of the nucleus and an upper limit on its size. Figure 1.1 shows differential cross sections, as a function of bom-

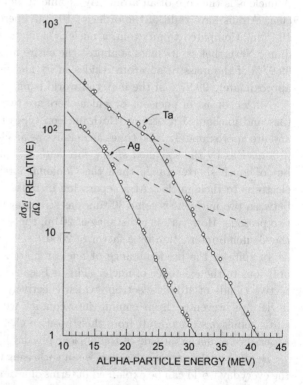

Figure 1.1: Differential cross sections for the elastic scattering of alpha particles from silver ($Z = 47$) and tantalum ($Z = 73$) as a function of bombarding energy observed at an angle of 60°. The dashed curves are the theoretical (Rutherford) cross sections for scattering alpha particles by a point charge distribution. These curves agree well with the data at low energies. The sudden departures from these curves that occur when the alpha particles have sufficient energy to come into contact with the nucleus provide a measure of the corresponding nuclear size. (The figure is from Eisberg R.M. and Porter C.E. (1961), *Rev. Mod. Phys.* **33**, 190.)

barding energy, for the scattering of alpha particles from silver and tantalum. At low energies, the results are similar to those obtained by Geiger and Marsden and are in excellent agreement with the so-called *Rutherford cross section*. Rutherford's

---

[1] Geiger H. and Marsden E. (1909), *Proc. Roy. Soc. London* **A82**, 495.

[2] Geiger H. (1910), *Proc. Roy. Soc. London* **A83**, 492.

[3] Rutherford E. (1911), *Phil. Mag.* **21**, 669.

cross section is what one expects if the alpha particles don't come into contact with the nucleus, i.e., if the scattering results purely from the long-range Coulomb interaction. Thus, the energy at which the cross section suddenly departs from the Rutherford curve (not seen by Geiger and Marsden but seen at the higher energies shown in Figure 1.1) provides a measure of the size of the corresponding nucleus.

Twentieth century science and technology have caused nuclear physics to impact on everyday life in a variety of ways. Some examples are: uses of nuclear radiation in such diverse areas as materials characterisation and cancer therapy; the uses of radioactive isotopes as tracers and diagnostic tools in medicine; nuclear magnetic resonance and positron emission tomography for medical imaging; uses of nuclear accelerators as spectrometers for carbon dating and the detection of rare trace elements (and to produce sources of nuclear radiation); nuclear fission in nuclear power reactors and nuclear weapons; and nuclear fusion in weapons and in the development of fusion reactors for our future energy needs.

To date, 117 chemical elements have been identified. Each element is characterised by an atom with a specified number, Z, of electrons and a nucleus with an equal number of protons. However, an element may have more than one isotopic form. An isotope is characterised by a nucleus with proton number Z and neutron number N. Approximately 2800 isotopes have been identified. They are shown in Figure 1.2. A total of 7000 are believed to be sufficiently long-lived for their identification to be possible with current techniques. However, the less stable nuclei have not yet been produced in sufficient numbers for experimental investigation.

Figure 1.2 also shows the approximate limits to the region of observable nuclei. Nuclei for which the binding energy of the last neutron, B_n, is negative are excluded. Nuclei for which the binding energy of the last proton, B_p, is insufficient to delay tunnelling through the Coulomb barrier by more than $\sim 10\,\text{ns}$ are also excluded. Figure 1.3 shows examples of (N, Z) combinations for which these limits have been reached.

In this introductory chapter, we present a summary of the important facets of nuclear structure illustrated liberally with experimental data. We also include elementary quantum mechanical descriptions of the two most important models of nuclear structure: the shell model (which explains the magic numbers) and the collective model.

The chapter is relatively self-contained and can be read on its own. Its purpose is to lay the phenomenological foundations of nuclear physics. We shall attempt to show that interpretations of nuclear phenomena often emerge naturally from a systematic examination of the data. To this end, we present the best and most up-to-date data that we could find rather than the data which constituted the original evidence. Details of the models and their implications for understanding the nuclear many-body problem are developed in subsequent chapters.

A phenomenological approach to nuclear structure physics is appropriate because there is no *a priori* theory which could have predicted the properties of nuclei

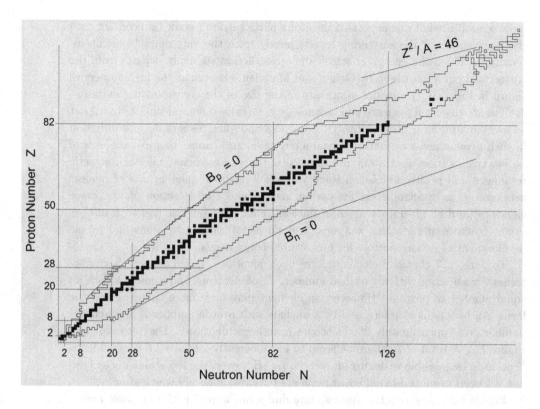

Figure 1.2: Chart of the nuclides (ca. 2005). The black squares denote the nuclei found in nature. The present extent of nuclei, for which at least one characteristic has been measured, are enclosed by the "stepped" border. The line marked $B_n = 0$ approximates the neutron drip line. The line marked $B_p = 0$ approximates the proton drip line. Outside of these borders, neutrons and protons are lost so rapidly from nuclei that the nuclei cannot be observed. The line marked $Z^2/A = 46$ indicates an approximate limit to nuclear stability with respect to spontaneous fission. The magic numbers 2, 8, 20, 28, 50, 82, and 126 (cf. Section 1.3) also are indicated.

before they were observed. There have been many experimental surprises. In this respect, nuclear physics is in "good company" with condensed matter physics. There are a number of parallels. For example, many nuclei and condensed matter systems exhibit superfluidity. In fact, it is sensible to regard nuclei as examples of condensed matter.

The future of nuclear physics, as also of condensed matter physics, depends in part on producing new condensates. For nuclei this means new combinations of N and Z. A consideration of Figure 1.2 reveals that we have studied only about 30% of the possibilities. To study the remaining 70% of the possibilities, which are very unstable, is one of the future challenges of nuclear physics.

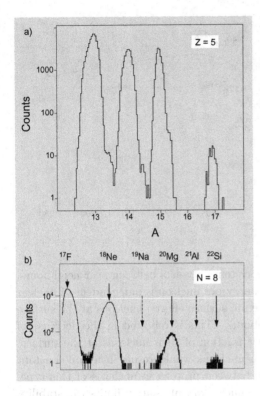

Figure 1.3: (a) Mass spectrum of fragments with $Z = 5$ (boron isotopes), following the bombardment of a tantalum target with a 1760 MeV beam of ^{40}Ar ions (Langevin M. *et al.* (1985), *Phys. Lett.* **B150**, 71). Evidently, $^{16}_{5}$B either does not exist or is too short lived to reach the detector system. (b) Mass spectrum of fragments with $N = 8$ in a similar nuclear reaction to (a) (Saint-Laurent M.G. *et al.* (1987), *Phys. Rev. Lett.* **59**, 33). Both ^{19}Na and ^{21}Al are absent, again implying either that they do not exist, or that they decay (e.g., by proton tunnelling through the Coulomb barrier) too quickly to be observed.

Exercises

1.1 Using $F_{\text{Coulomb}} = \frac{q_1 q_2}{4\pi\epsilon_0 r^2}$ where $\epsilon_0 = 8.85 \times 10^{-12}$ C^{2}m^{-2}N^{-1}, and $F_{\text{Yukawa}} = -\frac{d}{dr}\left(\frac{V_0 e^{-\lambda r}}{\lambda r}\right)$ where $V_0 = -40$ MeV and $\lambda = 0.7$ fm^{-1}, calculate the magnitudes of F_{Coulomb} and F_{Yukawa} at 1.0 and 20 fm. (The Yukawa force is an approximation to the strong interaction between two nucleons (cf. Section 5.7)).

1.2 Use Figure 1.1 to estimate the radii (in fm) of silver and tantulum nuclei. Assume $R(^4\text{He}) = 1.7$ fm and a range of 1.4 fm for the strong interaction.

1.2 Nuclear structure from gross properties

A first look at the gross properties of nuclei (e.g., binding energies, radii, and densities) suggests that they have many properties in common with liquid drops. They have binding energies and volumes that increase more or less linearly with nucleon number and they have densities that approach a saturation value at the centres of all but the lightest nuclei (cf. Figure 1.7).

The binding energy per nucleon, B/A, as a function of A ($= Z + N$) is shown in Figure 1.4. From $A \approx 20$ to $A \approx 200$ it has the approximately constant value of 8.3 MeV/nucleon. This near constancy (to within $\pm 5\%$) indicates that the range of the nucleon-nucleon force is short compared to the size of the nucleus. A short-range

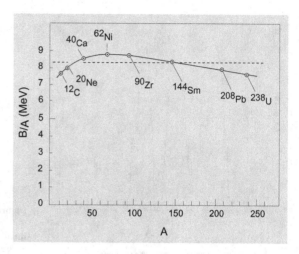

Figure 1.4: A plot of binding energy per nucleon, B/A, versus mass number A. The quantities B/A are experimental. Over the mass range $20 \leq A \leq 200$, the average value of B/A is ~ 8.3 MeV/nucleon (the horizontal dashed line). The most tightly bound nucleus is ^{62}Ni which has a binding energy of 8.795 MeV/nucleon. The solid curve is a smooth line through the data. The data points, for the nuclei shown, are chosen arbitrarily. (The data are from Audi G., Wapstra A.H. and Thibault C. (2003), *Nucl. Phys.* **A729**, 337.)

force between particles in a many-body system only acts between near neighbours. Thus, for a large nucleus, the binding energy per nucleon is independent of nuclear volume. However, for finite nuclei, there are surface effects; nucleons at the surface do not have a full contingent of neighbours. This is reflected in the decrease of B/A for light nuclei which have a larger fraction of their nucleons at the surface. In contrast, a long-range force between nucleons would result in a non-constant value of B/A. The Coulomb repulsion between protons in nuclei is of this type. It causes the decrease of B/A for heavy nuclei and ultimately limits the stability of very heavy nuclei. The practical consequence of this is spontaneous fission, as would occur for a droplet of water if an increasing electrostatic charge were applied. The $Z^2/A = 46$ line, beyond which spontaneous fission half-lives are estimated to be less than 1 μs, is shown in Figure 1.2. Another consequence of the Coulomb force in nuclei is the curvature of the stability line, evident in Figure 1.2, towards neutron excess in heavy nuclei.

Nuclear charge distributions are probed, for example, in electron scattering experiments. Nuclear mass radii are less easily measured. However, it is generally assumed that, with the exception of a few neutron-rich nuclei such as $^{11}_{3}$Li, the nuclear mass and charge distributions are proportional to one another.

Root-mean-square charge radii are shown as a function of $A^{1/3}$, for selected nuclei, in Figure 1.5. For heavy nuclei, the data approximate the straight line with

$$\left[\tfrac{5}{3} \langle r^2 \rangle_{\text{expt}} \right]^{\frac{1}{2}} \approx 1.1\, A^{1/3} + 0.65 \text{ fm.} \tag{1.1}$$

The factor $\tfrac{5}{3}$ is included in this expression because, for constant-density matter with a sharply-defined surface of radius $r = R_0 A^{1/3}$ (cf. Figure 1.6), the mean-square radius would be given by

$$\langle r^2 \rangle = \tfrac{3}{5} R_0^2 A^{2/3}. \tag{1.2}$$

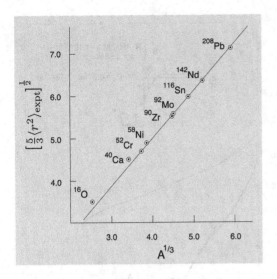

Figure 1.5: A plot of the root-mean-square charge radius, $\left[\frac{5}{3}\langle r^2\rangle_{\text{expt}}\right]^{\frac{1}{2}}$, versus $A^{1/3}$. Data are for selected nuclei. The solid line is a best (straight-line) fit for nuclei with $A \geq 90$. The deviation of the lighter nuclei from this line is due to surface diffuseness effects (cf. Figure 1.7). The units are femtometres (1 fm = 10^{-15}m). (The data are from Angeli I. (2004), *At. Data Nucl. Data Tables* **87**, 185.)

Thus, the data indicate that the charge density of nuclear matter is essentially constant in heavy nuclei. Variations from Equation (1.2) in light nuclei can be attributed to surface effects (cf. Figure 1.7).

Taken together, Figures 1.4 and 1.5 suggest that nuclear matter resembles a liquid and that nuclei can be modelled as liquid drops. Indeed, a liquid-drop model of the nucleus is quite successful at describing many gross properties of nuclei. However, whereas classical liquid drops have sharply defined surfaces, the nuclear density distribution has a diffuse surface. This is revealed by electron scattering cross sections as depicted, for ^{197}Au, in Figure 1.6. The cross section shown is fitted accurately by the expression for the nuclear charge density

$$\rho(r) = \rho_0 \left\{ 1 + \exp\left(\frac{r - R_B}{a}\right) \right\}^{-1}, \tag{1.3}$$

where ρ_0 is the central density, R_B is the radius at half density and a is a surface diffuseness parameter. Figure 1.7 shows the variation in nuclear charge density as a function of mass number.

The decreasing saturation density with increasing Z is a Coulomb repulsion effect.

Exercises

1.3 Derive the factor $\frac{3}{5}$ in Equation (1.2).

1.4 For a long-range force in a finite A-body system, show that the binding energy $B \propto A(A-1)$.

1.5 Using Equation (1.3) with $R_B = 6.38$ fm, $a = 0.53$ fm, determine the value of r for which $\rho = 0.1\rho_0$.

Figure 1.6: Differential cross section for elastic scattering of 153 MeV electrons by ^{197}Au nuclei. The diffraction pattern seen for the differential cross section as a function of angle reveals that the nucleus has a diffuse surface (B) rather than a sharp surface (A). A consequence of this (as shown in the inset) is that the nuclear radius at the half density is slightly smaller than it would be for a nuclear density with a sharp surface that gives the same $\langle r^2 \rangle^{1/2}$. The pattern that would be seen for scattering from a point charge is also shown (cf. Figure 1.1). (The figure is from Bohr A. and Mottelson B.R. (1969), *Nuclear Structure*, Vol. 1 (Benjamin, New York), (republished by World Scientific, Singapore), p. 159. The data are from: Hahn B., Ravenhall D.G. and Hofstadter R. (1956), *Phys. Rev.* **101**, 1131; Yennie D.R., Ravenhall D.G. and Wilson R.N. (1954), *Phys. Rev.* **95**, 500; and Herman R. and Hofstadter R. (1960), *High Energy Electron Scattering Tables*, Stanford Univ. Press, Stanford, California.)

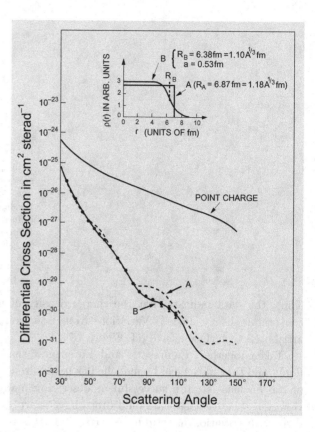

1.3 Nuclear shell structure

One of the big surprises in the development of nuclear physics was that nuclei, like atoms, exhibit properties characteristic of a system with a shell structure. The origin of shell structure in atoms is well understood. Atomic electrons move in the central (spherically symmetric) Coulomb field of a small, but relatively massive, and highly-charged nucleus. The field of the nucleus, as seen by an electron, is partially screened by other atomic electrons. But, apart from the screening effect, the interactions between electrons are relatively weak. Thus, to a first approximation, an atom can be regarded as a system of independent electrons moving in a partially-screened Coulomb field. Corrections can be made to include the correlation effects induced by the small, but well understood, residual electronic interactions and, with a sufficiently large computer, remarkably accurate results for atomic properties can be computed. The result is that atomic electrons occupy rather well-defined single-particle states. Furthermore, since electrons are spin-half fermions subject to the Pauli exclusion principle, at most two electrons (one spin up and one spin down) can occupy the same spatial state. Thus, in the ground state of an atom, the electrons fill the lowest energy states available. Shell structure arises because the energies of

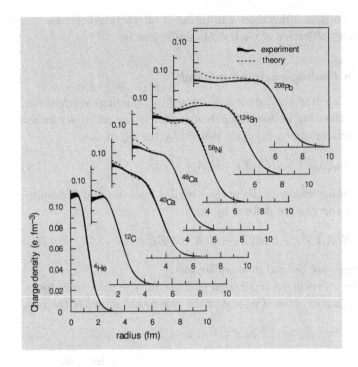

Figure 1.7: Ground-state charge densities for a range of nuclei obtained from elastic electron scattering cross sections. The experimental uncertainties are indicated by the thickness of the lines. The dashed curves are the results of independent-particle model calculations. (The figure shown here, with permission of Bernard Frois, is similar to one appearing in Frois B. and Papanicolas C. (1987), *Ann. Rev. Nucl. Part. Sci.* **37**, 133.)

states are grouped together into "shells" with energy gaps between them.

Shell structure in nuclei is more difficult to understand. A nucleus is composed of neutrons and protons which, like electrons, are spin-half fermions. But, apart from this common feature, atoms and nuclei are very different. Whereas atomic electrons move in the central field of the nucleus, the central field in which nucleons move is self generated. Atomic electrons are far apart and interact weakly by long-range forces. Nucleons inside a nucleus are close to one another and interact strongly by short-range forces. Consequently, the observation of shell structure in nuclei was not expected. Indeed, the interpretation of experimental data in shell-model terms was strongly resisted for many years. The success of the shell model was subsequently attributed to the Pauli exclusion principle which, to a large extent, inhibits the strong scattering of nucleons inside the nucleus by preventing nucleons from scattering into orbitals that are already occupied.

Nuclear shell effects are dominant features of nuclear structure. They lead directly to the nuclear shell model. The shell model is of fundamental importance because it provides the most detailed theoretical framework available for the description of nuclei in terms of interacting nucleons. One of the major aims in developing models and theories of nuclear structure is, therefore, to relate successful models to the shell model. If a model can be expressed in shell-model terms, one is justified in calling the model a *microscopic model*. Moreover, if a model description of some phenomena is successful and the model can be expressed in shell-model terms, one can claim to have explained the phenomena in terms of interacting neutrons and protons. This is a primary objective of nuclear physics research.

1.3.1 *Differences in binding energies and radii*

Important indicators of nuclear shell structure come from binding-energy differences. Binding-energy differences between neighbouring nuclei are called *separation energies*. One-neutron separation energies are defined by the expression

$$S_n(A, Z) = B(A, Z) - B(A - 1, Z), \tag{1.4}$$

where $B(A, Z)$ is the binding energy of the nucleus with A nucleons and Z protons. Two-neutron separation energies are defined by

$$S_{2n}(A, Z) = B(A, Z) - B(A - 2, Z). \tag{1.5}$$

Proton separation energies are defined in a similar way.

Figure 1.8 shows the one-neutron separation energies for the calcium isotopes. This figure has two striking features. The first is the saw-tooth nature of the plot.

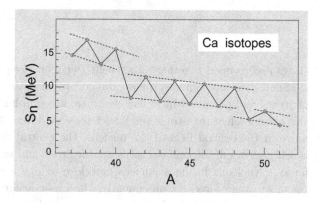

Figure 1.8: One-neutron separation energies, S_n, for the calcium isotopes. Note the odd-even staggering between neighbouring nuclei and the strong discontinuities that occur between $A = 40$ and 41 and between $A = 48$ and 49 (cf. Figure 1.9). (The data are from Audi G., Wapstra A.H. and Thibault C. (2003), *Nucl. Phys.* **A729**, 337.)

The second is the discontinuities that occur between $A = 40$ and 41 and between $A = 48$ and 49. The saw-tooth behaviour shows that it requires more energy to remove a neutron when the neutron number is initially even than when it is odd. This odd-even staggering effect is a manifestation of the predisposition of nucleons to form strongly-coupled pairs (cf. comments below and in Section 1.4). Thus, to remove a nucleon from an even nucleus, one has first to break apart the pair to which it belongs and this demands additional energy. The discontinuities at $A \approx 40$ and 48 are indicators of nuclear shell closures. They show up even more clearly in two-neutron separation energies.

Figure 1.9 shows the two-neutron separation energies for the ($Z = 20$) calcium isotopes. The odd-even staggering is now smoothed out but the discontinuities at $A = 40$ ($N = 20$) and 48 ($N = 28$) remain. One sees that it requires significantly

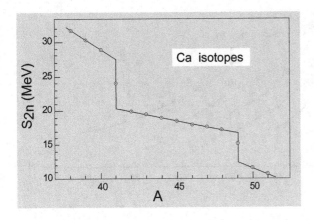

Figure 1.9: Two-neutron separation energies, S_{2n}, for the calcium isotopes. The odd-even staggering is smoothed away, leaving a clear indication of discontinuities at $A = 41$ and 49. (The data are from Audi G., Wapstra A.H. and Thibault C. (2003), *Nucl. Phys.* **A729**, 337.)

more energy to remove a pair of neutrons when $N \leq 20$ than when $N \geq 22$.[4] Thus, we say that the discontinuities occur at $N = 20$ and $N = 28$. Such discontinuities are observed both for protons and neutrons at $N, Z = 2, 8, 20, 28, 50$, and 82, and at $N = 126$. Examples of these discontinuities for $18 \leq N \leq 156$ are shown in Figure 1.10.

Differences in radii between neighbouring isotopes also change dramatically at $N = 2, 8, 20, 28, 50, 82$, and 126. This is shown for $24 \leq N \leq 144$ in Figure 1.11. One sees that the differences in radii increase from values close to local minima to values close to local maxima at the specified values of N. The numbers 2, 8, 20, 28, 50, 82 and 126 are called *magic numbers*. The occurrence of magic numbers suggests a shell structure in nuclei similar to that seen in atoms.

Atoms exhibit changes in binding energies (ionization potentials) and radii (covalent and ionic radii) due to changes in electronic shell filling. When atomic shell filling in atoms passes through the numbers 2, 10, 18, 36, 54, or 86, there are sudden decreases in ionization potentials and sudden increases in covalent and ionic radii. These changes reflect the exclusion of electrons from the "smaller" more strongly-bound configurations that are filled first in atoms. The implication of the data shown in Figures 1.10 and 1.11 is that nuclei also possess shell structure. In atoms, energy shells reflect the dominance of the independent-particle component of the Hamiltonian. A suggestion that the nuclear Hamiltonian contains a dominant independent-particle component, also comes from the observation that magic numbers have the same values for protons and neutrons, regardless of mass number, i.e., magic numbers for protons do not depend on the number of neutrons and vice versa. It is also suggested by the observation of *single-particle* states in the near neighbours of doubly closed-shell nuclei. These are states which, to a first approximation, are simple (uncorrelated) products of core states and single-particle states.

---

[4]The separation energy of a pair at $N = 21$ is approximately the average of the $N = 20$ and $N = 22$ values. This corresponds to the fact that the first neutron is removed from the $N = 21$ nucleus and the second from the $N = 20$ nucleus; the calcium isotopes have $Z = 20$.

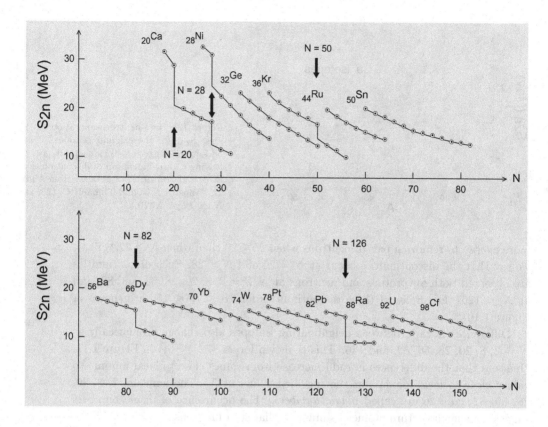

Figure 1.10: Two–neutron separation energies for selected isotopes with even N from $Z = 20$ (calcium) to $Z = 98$ (californium). The energy gaps at $N = 20, 28, 50, 82$ and 126 are clearly visible. Further, there is no evidence for large energy gaps at any other values of N between 20 and 156. (The data are from Audi G., Wapstra A.H. and Thibault C. (2003), *Nucl. Phys.* **A729**, 337.)

1.3.2 Nucleon transfer reactions and spectroscopy

A simple way to identify single-particle states in a nucleus is by means of single-nucleon transfer reactions. A transfer reaction is an interaction between a projectile nucleus and a target nucleus which results in a transfer of nucleons between the two. In a *pickup* reaction, nucleons are removed from the target and added to the projectile whereas in a *stripping* reaction the converse occurs. Transfer reactions are important sources of nuclear structure information because they measure the extent to which a final state of a nucleus differs from an initial state of a neighbouring nucleus by either the addition or removal of one or more nucleons.

Single-nucleon pickup and stripping reactions are particularly important because they enable one to infer the occupation probabilities of single-nucleon states. In an independent-particle model, single-nucleon states are either occupied or empty. However, due to correlations brought about by residual interactions, single-nucleon

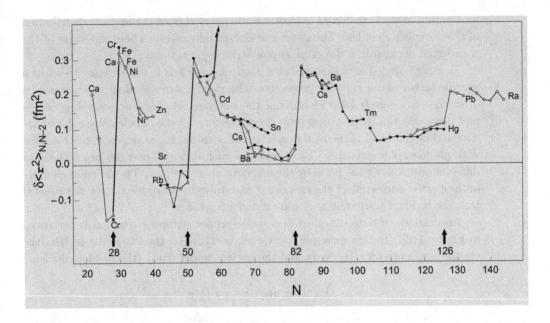

Figure 1.11: Differences in mean-square charge radii for isotopes differing in N by 2. Data are for selected isotopes and even N. Solid and open circles are used only to help distinguish isotopic sequences. Discontinuities are clearly evident at $N = 28$, 50, 82, and 126. The very large (off-scale) shifts for Rb and Sr at $N = 60$ are due to a sudden onset of deformation (see Figure 1.41). (The data are from: Otten E.W. (1989), in *Treatise on Heavy-Ion Science, Nuclei Far from Stability*, Vol. 8, edited by D.A. Bromley (Plenum Press, New York), p. 517; Nadjakov E.G., Marinova K.P. and Gangrsky Y.P. (1994), *At. Data Nucl. Data Tables* **56**, 133; Shera E.B. *et al.* (1976), *Phys. Rev.* **C14**, 731 – Fe, Ni, Zn; Mårtensson-Pendrill A.-M. *et al.* (1992), *Phys. Rev.* **A45**, 4675 – Ca; Wendt K. *et al.* (1988), *Z. Phys.* **A329**, 407 – Ba; and Alkhazov G.D. *et al.* (1988), *Nucl. Phys.* **A477**, 37 – Tm.)

states are, in general, fractionally occupied. Thus, to the extent that a single-nucleon state is occupied, it can give up a nucleon in a pickup reaction leaving behind a residual nucleus in a *hole state*. Conversely, to the extent that a single-nucleon state is empty, it can accept a nucleon in a stripping reaction thereby creating a residual nucleus in a *particle state*.

Transfer reactions are simplest to interpret when either the initial or the final state of the target nucleus has spin (i.e., total angular momentum) zero and when the conditions are such that the transition from the initial to the final state occurs, to a good approximation, in a single step. This happens when the interaction between the projectile and target nucleus is weak and can be treated in first order perturbation theory, i.e., in the Born approximation. One then describes the reaction as a *direct reaction*.

When either the state of the target nucleus or the state of the final nucleus has spin zero, the spin and parity of the transferred nucleon in a direct single-nucleon transfer reaction is simply the difference in the spin and parity of the initial and final nuclear states. Moreover, the orbital component of the total angular momentum of the transferred nucleon can be inferred from the angular distribution of the outgoing ejectiles. Thus, one can obtain information about the state of the transferred nucleon in a direct pickup or stripping reaction.

To study single-particle states in a particular nucleus it is necessary to select a specific target and a suitable projectile. The events corresponding to the reaction of interest are selected by identifying the outgoing ejectile nucleus using charge and mass selection with a magnetic spectrometer. Ejectile nuclei, so-analysed are detected in the focal plane of the spectrometer using, for example, a plate coated with photographic emulsion. Ejectile nuclei with different energies are focused at different positions on the plate by the magnetic spectrometer. The different energies of the ejectile nuclei reflect the energies of the states in the residual nucleus provided that the incident projectile nuclei are monoenergetic.

An example of data obtained in a single-proton stripping reaction is presented in Figure 1.12. In this example the target is $^{208}_{82}\text{Pb}_{126}$, the projectile is ^3_2He, the ejectile is a deuteron ($^2_1\text{H}_1$ or d), and the nucleus under study (the residual nucleus)

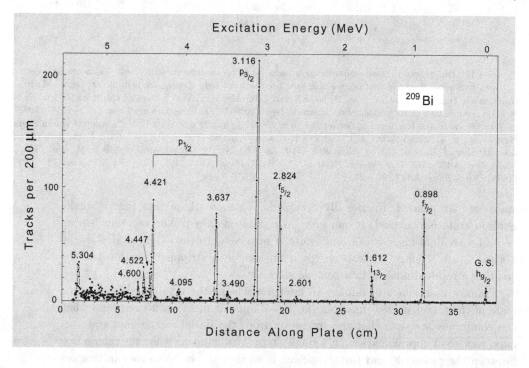

Figure 1.12: The spectrum of deuterons observed in the single-proton stripping reaction $^{208}\text{Pb}(^3\text{He,d})^{209}\text{Bi}$ at an angle of 110° with respect to the ^3He beam and at a bombarding energy of 18 MeV. The strong peaks correspond to particle states in ^{209}Bi (cf. Figure 1.13). These are labelled by energy (in MeV) and the single-particle quantum numbers l and j. A significant "fragmentation" of the $p_{1/2}$ independent-particle configuration is evident. The ^{209}Bi ground state is on the right. Weak peaks in the spectrum are non-independent-particle states. (The figure is from Ellegaard C., Patnaik B. and Barnes P.D. (1970), *Phys. Rev.* **C2**, 2450.)

is $^{209}_{83}\text{Bi}_{126}$. The number of deuteron tracks per $200\mu\text{m}$ in the photographic emulsion on the plate in the focal plane of the magnetic spectrometer is shown as a function of distance along the plate in centimetres. The distance along the plate is converted into energy by suitable calibration procedures. The nucleus $^{209}_{83}\text{Bi}_{126}$ is considered to have a closed neutron shell and a closed proton shell plus one proton. Thus, it is expected to have a low-energy spectrum corresponding to the allowed single-particle states of a proton outside of the $Z = 82$ closed proton shell. With this perspective in mind, the strong peaks in Figure 1.12 are associated with single-proton states and the weak peaks are associated with states which have a small component of a single-particle state mixed with a large component of other states; i.e., states in which the last proton is coupled to excited states of ^{208}Pb (cf. Figures 1.13 and 1.52).

Low-energy states in odd-mass nuclei adjacent to ^{208}Pb are shown in Figure 1.13. These states are single-particle states in ^{209}Bi and ^{209}Pb, and single-hole states in ^{207}Tl and ^{207}Pb. The nucleus ^{209}Bi is discussed above. The neutron particle states in ^{209}Pb can be identified using a single-neutron stripping reaction such as $^{208}\text{Pb}(\text{d, p})^{209}\text{Pb}$. Pickup reactions[5] such as $^{208}\text{Pb}(\text{d, t})^{207}\text{Pb}$ and $^{208}\text{Pb}(\text{d},$ $^{3}\text{He})^{207}\text{Tl}$ identify hole states. Some information is given also in Figure 1.13 on particle states in ^{207}Pb and hole states in ^{209}Bi and ^{209}Pb. For example, proton hole states in ^{209}Bi can be identified using the $^{210}\text{Po}(\text{t, }\alpha)^{209}\text{Bi}$ reaction (cf. Figure 1.23).

It is observed that only the ground state of ^{208}Pb and the several single-particle and single-hole states in the $A = 209$ and $A = 207$ nuclei are reasonably described as independent-particle states. In particular, it is found that the excited states of ^{208}Pb are not obtained simply by promoting a nucleon from an occupied to an unoccupied state, i.e., they are not simple one-particle-one-hole states. They are observed rather to be coherent linear combinations of one-particle-one-hole states. We shall have more to say about these states in Volume 2.

1.3.3 The shell model

The shell model is the basic microscopic model of nuclear structure theory. It is formulated on the premise that the nuclear Hamiltonian can be expressed as a sum of an independent-particle Hamiltonian and a residual interaction, $\hat{V}$, i.e.,

$$\hat{H} = \hat{H}_0 + \hat{V}. \tag{1.6}$$

The Hamiltonian $\hat{H}_0$ is a sum of single-particle Hamiltonians,

$$\hat{H}_0 = \sum_{i=1}^{A} \hat{h}_i, \tag{1.7}$$

---

[5]We use the standard abbreviations: $\text{d} \equiv {}^{2}_{1}\text{H}_1$, $\text{t} \equiv {}^{3}_{1}\text{H}_2$, $\alpha \equiv {}^{4}_{2}\text{He}_2$.

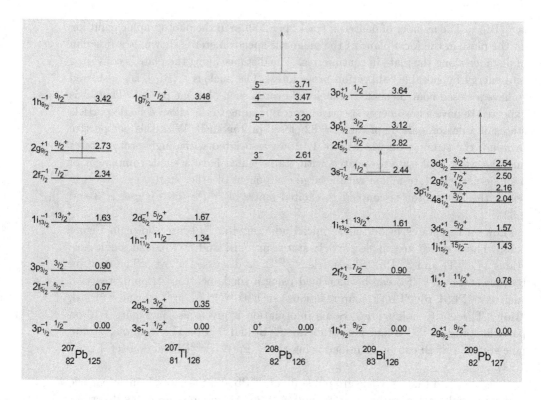

Figure 1.13: Low-energy states in odd-mass nuclei adjacent to ^{208}Pb. Excitation energies are in MeV. Levels are labelled by their spin and parity J^π and, for the odd-mass nuclei, the single-particle quantum numbers n, l, j are given, together with their particle (+1) or hole (−1) character. Parity is given by $(-1)^l$, where $l = 0, 1, 2, 3, 4, 5, 6, 7$ are labelled by s, p, d, f, g, h, i, j. The vertical arrows indicate the presence of higher energy excited states whose energies are known but are omitted from the figure. (The data are from *Nuclear Data Sheets*.)

where $\hat{h}_i$ governs the motion of the i'th nucleon. It corresponds to a system of nucleons moving independently in a spherically symmetric field, commonly referred to as the *single-particle potential*. The single-particle potential is chosen to represent the average interaction of a nucleon with the other nucleons.

In principle the nuclear Hamiltonian can always be expressed as a sum of two terms as in Equation (1.6). But, such an expression is only useful if $\hat{H}_0$ exhibits a shell structure and if the residual interaction is not so strong that it destroys this structure.

A simple single-particle Hamiltonian, $\hat{h}_i$, is of the form

$$\hat{h}_i = \frac{\hat{p}_i^2}{2M} + U(r_i), \qquad (1.8)$$

where $\hat{\mathbf{p}}_i$ $(= -i\hbar\nabla_i)$ is the momentum of the i'th nucleon, so that $\hat{p}_i^2 = \hat{\mathbf{p}}_i \cdot \hat{\mathbf{p}}_i = -\hbar^2\nabla^2$, M is the mass of a nucleon, $\mathbf{r}_i$ is its position vector $(r_i = |\mathbf{r}_i|)$, and U is a potential for a central single-particle field. Because all neutrons are identical and all protons are identical, the Hamiltonian $\hat{h}_i$ can only depend on the nucleon to the extent of distinguishing between neutrons and protons. Thus, for simplicity we shall drop the subscript on $\hat{h}_i$ and write

$$\hat{h} = \frac{\hat{p}^2}{2M} + U(r). \tag{1.9}$$

However, it is to be understood that the potential U may be different for neutrons and protons. Indeed, a repulsive Coulomb potential should be added for protons.

Because nuclear forces are of short range compared to the dimensions of a nucleus, a natural choice of single-particle potential is one of the form

$$U(r) = -V_0 \rho(r), \tag{1.10}$$

where ρ is the nuclear density (cf. Figures 1.6 and 1.7). To the extent that the density distribution of a nucleus resembles that of a liquid drop, we might consider the finite square-well potential,

$$U(r) = \begin{cases} -V_0 & \text{if } r < R, \\ 0 & \text{if } r \geq R. \end{cases} \tag{1.11}$$

A more realistic single-particle potential, proportional to the density distribution given by Equation (1.3), is the Woods-Saxon potential[6]

$$U(r) = -V_0 \left\{ 1 + \exp\left(\frac{r-R}{a}\right) \right\}^{-1}. \tag{1.12}$$

The disadvantage of these potentials is that single-particle wave functions have to be computed numerically.

There are two potentials which provide single-nucleon wave functions in analytic form: the spherical harmonic oscillator

$$U(r) = \tfrac{1}{2} M \omega^2 r^2 + \text{const.} \tag{1.13}$$

and the infinite square-well potential,

$$U(r) = \begin{cases} -V_0 & \text{if } r < R, \\ \infty & \text{if } r \geq R. \end{cases} \tag{1.14}$$

These two potentials and the Woods-Saxon potential are illustrated in Figure 1.14. The most commonly used potential is that of the spherical harmonic oscillator.

It can be seen from Figure 1.14 that the harmonic oscillator potential has some resemblance to the Woods-Saxon potential in the interior regions of the nucleus but it goes to infinity with increasing radius. The large-radius behaviour of the harmonic oscillator is less of a problem than it would appear to be at first sight because nuclear wave functions, for bound nucleons, fall off rapidly and approach

---

[6]Woods R.D. and Saxon D.S. (1954), *Phys. Rev.* **95**, 577.

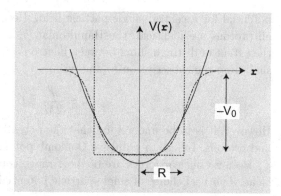

Figure 1.14: Single-particle shell-model potentials: an infinite square well potential (dashed line); a Woods-Saxon potential (dot-dashed line); and a harmonic oscillator potential (solid line).

zero at large radii. They simply fall off more rapidly in the oscillator potential than they would in a more realistic potential. Thus, a harmonic oscillator potential is not as realistic as a Woods-Saxon potential. But, with a little adjustment, it gives similar results. The reason for its popularity is that its wave functions and energy levels have analytical expressions. We shall also find, in subsequent chapters, that the harmonic oscillator shell model has many symmetries and provides a useful basis for the expression of collective models with a microscopic interpretation.

The spatial wave functions for a single particle in a spherically symmetric potential can be expressed in spherical polar coordinates in the form

$$\psi_{nlm}(r, \theta, \varphi) = R_{nl}(r) Y_{lm}(\theta, \varphi), \qquad (1.15)$$

where n is a radial quantum number, Y_{lm} is a spherical harmonic, and l, m are orbital angular momentum quantum numbers. These quantum numbers take values in the ranges $n = 1, 2, 3, \ldots$, $l = 0, 1, 2, \ldots$ and $m = -l, \ldots, +l$, respectively. Intrinsic spin degrees of freedom are suppressed for the moment but will be included shortly. Because a spherically symmetric potential is rotationally invariant, the single-particle energies, ε_{nl}, depend on n and l but not on m. The parity of a single-particle wave function ψ_{nlm} is $(-1)^l$; this follows from the symmetry of the spherical harmonics, $Y_{lm}(\theta, \varphi)$, under inversion.

For a spherical harmonic oscillator, cf. Section 5.8, the energy levels are given by

$$\varepsilon_{nl} = (N + {}^3\!/_2)\hbar\omega, \qquad (1.16)$$

where N (not to be confused with the neutron number) is given by

$$N = 2(n - 1) + l. \qquad (1.17)$$

These energies are shown in Figure 1.15. For each value of l, the quantum number m takes $2l + 1$ values. With two spin states for a nucleon (spin up and spin down), one finds that the multiplicity of states at each level is equal to $(N + 1)(N + 2)$. Because of the Pauli exclusion principle, no two identical nucleons in a nucleus can

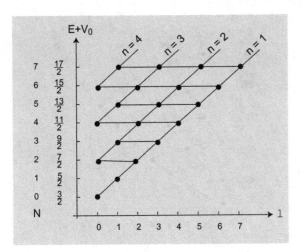

Figure 1.15: The energy eigenvalues of a spherical harmonic oscillator (cf. Equations (1.16) and (1.17)). Note the degeneracy in l. The energies are given in units of $\hbar\omega$.

have the same single-particle wave function. Thus, with harmonic oscillator wave functions, at most $(N+1)(N+2)$ neutrons and $(N+1)(N+2)$ protons can occupy the single-particle states of the N'th level. This number is called the *occupancy* of the level. If one adds the single-particle occupancies, starting from the bottom, to find the numbers of neutrons and protons required to fill succeeding harmonic oscillator shells, one finds closed shells when the neutron and proton numbers take the values 2, 8, 20, 40, 70, 112, 168, ... The first few of these numbers correspond to experimentally observed magic numbers.

The frequency, ω, of the harmonic oscillator potential, appropriate for a shell model description of an A-nucleon nucleus, is naturally chosen such that the mean-square radius of the nucleus, in its harmonic-oscillator ground state will accord with the measured value as shown, for a few nuclei, in Figure 1.5. By such means it is determined (cf. Exercise 5.32) that an appropriate choice is given by the formula

$$\hbar\omega \approx 41 A^{-1/3} \text{ MeV.} \qquad (1.18)$$

The energy levels for a particle in a finite square-well potential are easily determined; however, to obtain meaningful results, it is necessary to allow the radius of the potential to become progressively larger as the single-particle states are filled and the nucleus becomes bigger. The relationship between nuclear radius and nucleon number is given by Equations (1.1) and (1.2). For the spherical harmonic oscillator, the changing dimensions of the potential simply change the $\hbar\omega$ unit of energy and a corresponding unit of length; i.e., the energies and wave functions can all be expressed in dimensionless units which are simultaneously valid for all nuclei. This cannot be done for a finite potential. But, it can be done for an infinite square-well potential. The energy levels for a particle in an infinite isotropic square well of radius R are shown in Figure 1.16. The notable feature of the results is that the single-particle states of constant $N = 2(n-1) + l$ are no longer degenerate, as

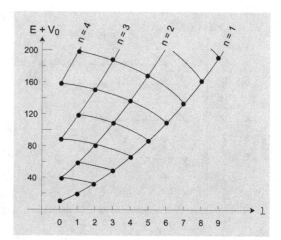

Figure 1.16: The energy eigenvalues of a particle in an infinite isotropic square well. The energies are given in units of $\hbar^2/2MR^2$. Compare the pattern as a function of angular momentum with that of Figure 1.15.

they are in the harmonic oscillator, but decrease in energy as l increases. Again there are gaps in the energy level spectrum. Moreover, the successive filling of the single-particle levels leads to shell closures at N, $Z = 2, 8, 20, 34, 58, 92, 138, 186$, ... As for the spherical harmonic oscillator potential, the first few of these numbers correspond to the experimental magic numbers.

To obtain the full sequence of magic numbers, it is necessary to include a strong *spin-orbit interaction* in the single-particle Hamiltonian. This important break-through in the development of the shell model was discovered independently by Goeppert-Mayer[7] and by Haxel, Jensen, and Suess.[8] Further, as noted by Nilsson,[9] one can simultaneously retain the primary advantage of the spherical harmonic oscillator, of giving analytical results, while simulating the effects of a more realistic, more square-shaped potential. All one has to do is add a term Dl^2 to the oscillator potential. The l degeneracy (cf. Figure 1.15) is then broken and, with $D < 0$, the energy of a state of angular momentum l is lowered by an amount proportional to $l(l+1)$. With a suitable value of D, this produces a pattern of single-particle energies similar to that shown for the infinite square-well potential in Figure 1.16, i.e., the l^2 term effectively interpolates between the oscillator and the infinite square-well. Physically, the recipe is sensible; higher l values have larger average orbital radii because of the centrifugal potential. Thus, a "flatter" potential, such as the square-well potential, results in lower energies for higher l states. With some adjustment of the value of D and the inclusion of a spin-orbit interaction, one can get analytical expressions for single-particle energies which reproduce the observed magic numbers.

Following Goeppert-Mayer, Jensen *et al.*, and Nilsson, we consider a single-

---

[7]Goeppert-Mayer M. (1949), *Phys. Rev.* **75**, 1969.

[8]Haxel O., Jensen J.H.D. and Suess H.E. (1949), *Phys. Rev.* **75**, 1766.

[9]Nilsson S.G. (1955), *Mat. Fys. Medd. Dan. Vid. Selsk.* **29** (16).

particle Hamiltonian

$$\hat{h} := \frac{\hat{p}^2}{2M} + \tfrac{1}{2}M\omega^2 r^2 + D\hat{\mathbf{l}}^2 + \xi\hat{\mathbf{l}}\cdot\hat{\mathbf{s}}, \qquad (1.19)$$

where $\hat{\mathbf{l}}$ is the orbital angular momentum and $\hat{\mathbf{s}}$ is the intrinsic spin of the nucleon. Empirically, $\xi < 0$ and, as already noted, $D < 0$. With the inclusion of intrinsic spin, the total angular momentum of a nucleon of orbital angular momentum l takes one or other of the values $j = l \pm \tfrac{1}{2}$. With the identity

$$\hat{\mathbf{l}}\cdot\hat{\mathbf{s}} = \tfrac{1}{2}(\hat{\mathbf{j}}^2 - \hat{\mathbf{l}}^2 - \hat{\mathbf{s}}^2), \qquad (1.20)$$

we see that the $\hat{\mathbf{l}}\cdot\hat{\mathbf{s}}$ operator has eigenvalues given by

$$\hat{\mathbf{l}}\cdot\hat{\mathbf{s}}\,\psi_{nljm} = \begin{cases} \tfrac{1}{2}l\,\psi_{nljm} & \text{for } j = l + \tfrac{1}{2}, \\ -\tfrac{1}{2}(l+1)\,\psi_{nljm} & \text{for } j = l - \tfrac{1}{2}. \end{cases} \qquad (1.21)$$

The single-particle energy level sequence for a Hamiltonian of the type given by Equation (1.19) is shown in Figure 1.17. Each level now has an occupancy equal to $(2j + 1)$. The figure shows these occupancies and the cumulative occupancies at the energy gaps, corresponding to the nucleon numbers which close a shell. It can be seen that these numbers are now the experimentally observed magic numbers.

Support for the independent-particle structure underlying nuclei in the ^{208}Pb region is provided by a comparison of electron scattering from ^{205}Tl and ^{206}Pb. These nuclei differ by a proton in the $3s_{1/2}$ shell-model orbit. The differences between elastic electron scattering from ^{205}Tl and ^{206}Pb reflect the density distribution of a proton in the $3s_{1/2}$ orbital. The density distribution deduced in this way is shown in Figure 1.18. The number of radial nodes agrees with the shell model radial quantum number.

The detailed properties of states in odd-mass nuclei adjacent to doubly-closed shells show that the independent-particle nature of these states is only approximate; there are observable correlation effects induced by the residual interaction, V, of Equation (1.6). For example, the ground state of $^{17}_8$O$_9$ (cf. Figure 1.19) has a non-zero electric quadrupole moment. This would be impossible if the ground state of ^{17}O were a pure closed-shell-plus-single-neutron configuration because the ^{16}O core has zero quadrupole moment and neutrons have no electrical charge. Thus, the non-zero quadrupole moment implies the presence of correlations between the extra-core neutron and the core. The nature of these correlations is easily understood in terms of *core polarization*. The effect is illustrated schematically in Figure 1.20 which shows the density distribution of the ^{16}O core pulled out of its spherical shape by the attractive forces of the extra-core neutron. The effect of the extra-core neutron on the core resembles the tidal motions of the Earth's oceans caused by the Moon's gravitational attraction.

A predominantly single-particle state, perturbed by small admixtures of non-single-particle configurations, is sometimes described as a *dressed* single-particle

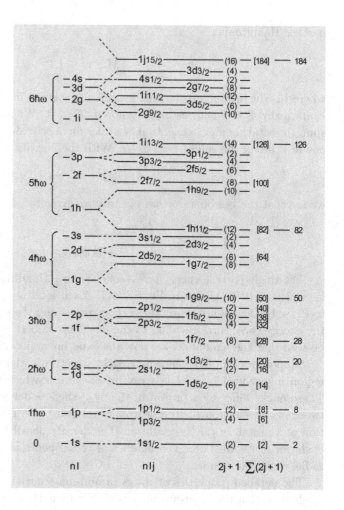

Figure 1.17: The energy-level sequence for a single particle moving in a spherical (isotropic three-dimensional) harmonic oscillator potential with well-flattening and spin-orbit coupling terms. The levels are labelled by the radial (n), orbital angular momentum (l), and total spin ($j = l \pm \frac{1}{2}$), quantum numbers; states with $l = 0, 1, 2, 3, \ldots$ are labelled $s, p, d, f, \ldots$. The maximum particle occupancies, $2j + 1$, of the subshells are given in parenthesis and the cumulative particle numbers are given in square brackets. The $\hat{l}^2$ term removes the angular momentum degeneracy of each major shell. The $\hat{l} \cdot \hat{s}$ term produces the shell gaps at the empirically established numbers 2, 8, 20, 28, 50, 82, 126 (184 has not yet been reached). There are separate such diagrams for protons and for neutrons. (The figure is based on a figure in Mayer M.G. and Jensen J.H.D. (1955), *Elementary Theory of Nuclear Shell Structure* (Wiley, New York), p. 58.)

state. This useful concept leads to the idea of *renormalised* or *effective* single-particle parameters. For example, both the ground-state electric quadrupole moment and the electric quadrupole transition rate between the ground state and first excited state in ^{17}O (cf. Figure 1.19) can be described within the framework of the independent-particle model if the odd neutron is given a nonzero effective charge. Using a single-particle model for the extra-core neutron and giving it an effective charge (equal to 43% of the proton charge) provides a consistent description of the electric quadrupole properties of ^{17}O.

The mixing of single-particle states with more complicated shell-model configurations is a relatively simple renormalisation effect in closed-shell-plus-one nuclei. This is because the more complicated configurations, to which the single-particle states couple, have considerably higher unperturbed energies relative to the independent-particle Hamiltonian H_0.

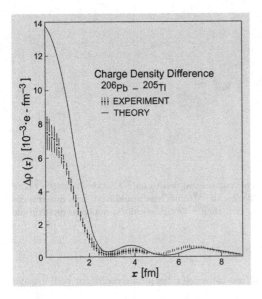

Figure 1.18: The charge density difference between ^{206}Pb and ^{205}Tl determined by elastic electron scattering. The nuclei ^{206}Pb and ^{205}Tl differ, according to the single-particle model, by a proton in the $3s_{1/2}$ orbital (cf. Figure 1.13). This is reflected in the figure by the two minima in $\Delta\rho(r)$ at 2.8 fm and 5.2 fm corresponding to the two radial nodes expected for an $n = 3$ orbital. (The figure is taken from Celenza L.S., Harindranath A. and Shakin C.M. (1985), *Phys. Rev.* **C32**, 2173.)

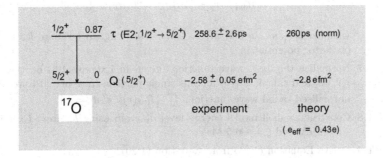

Figure 1.19: Some properties of the ground state and first excited state of ^{17}O. The ground state has spin-parity $5/2^+$ and an electric quadrupole moment of -2.58 e fm^2. The first excited state at 0.87 MeV has spin-parity $1/2^+$ and a mean lifetime of 258.6 ps. This mean lifetime reflects the rate of the electric quadrupole transition to the ground state. (The data are taken from Tilley D.R., Weller H.R. and Cheves C.M. (1993), *Nucl. Phys.* **A564**, 1.)

For nuclei having several valence-shell nucleons, there are generally many configurations with similar unperturbed energies. The residual interaction, $\hat{V}$ (cf. Equation (1.6)), then results in a large mixing of states and far more dramatic effects than simply the renormalisation of independent-particle model properties. Indeed, large correlations of various kinds emerge corresponding to different collective phenomena; such correlations are the subject of this book. Particularly prevalent in singly-closed shell nuclei are the so-called *pairing correlations*.

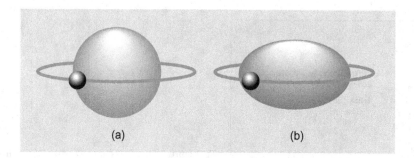

(a)   (b)

Figure 1.20: Schematic illustration of the ^{17}O nucleus as a neutron coupled to an ^{16}O core. If the ^{16}O core were spherical, as in (a), then, since the neutron has no charge, the ^{17}O nucleus would have zero quadrupole moment. However, if the neutron polarizes the core, as in (b), then ^{17}O can acquire a non-zero quadrupole moment corresponding to an oblate shape.

Exercises

1.6 Use Figure 1.18 and the wave function for a $3s_{1/2}$ harmonic oscillator radial wave function (cf. Equation (1.15)),

$$R_{3s_{1/2}}(r) \sim [4(\alpha r)^2 - 20(\alpha r) + 15]e^{-\alpha^2 r^2/2}, \qquad (1.22)$$

where $\alpha = \sqrt{m\omega/\hbar}$ and $m_p = 1.67 \times 10^{-27}$ kg, to estimate $\hbar\omega$ for a harmonic oscillator potential.

1.7 Normalise the $3s_{1/2}$ wave function (given in Exercise 1.6) on the interval $0 \leq r < \infty$, and calculate $\langle r^2 \rangle$ for this single-particle state in ^{206}Pb [recall that, for a normalised radial wave function, $\int_0^\infty \left[R_{nl}(r)\right]^2 r^2\, dr = 1$.]

1.8 Construct a shell model energy level diagram using $\hbar\omega$ from Exercise 1.6, $D = -0.5$ MeV and $\xi = -0.5$ MeV.

1.9 Derive Equation (1.21) from Equation (1.20).

1.4 Pairing in nuclei

Pairing in nuclei is the coupling of nucleons in pairs to states of zero angular momentum. The phenomenon is most evident in nuclei with two neutrons or two protons outside of doubly-closed shells. However, it occurs to some extent in all nuclei and is responsible for the fact that all even-even nuclei have zero angular momentum ($J = 0$) ground states.

The tendency of spin-$1/2$ particles to form $J = 0$ coupled pairs is well-known in a variety of many-fermion systems. For example, it is believed to be responsible for superconductivity[10,11,12,13] in macroscopic condensed systems. In this case, the

[10]Cooper L.N. (1956), *Phys. Rev.* **104**, 1189.
[11]Bardeen J., Cooper L.N. and Schrieffer J.R. (1957), *Phys. Rev.* **108**, 1175.
[12]Bogolyubov N.N. (1958), *Nuovo Cimento* **7**, 794.
[13]Bogolyubov N.N. (1959), *Sov. Phys.–Uspekhi* **2**, 236.

fermions are electrons and the pairs are commonly called *Cooper pairs*. The effects of pairing in a superconductor are dramatic because they cause a many-fermion system to exhibit properties, like persistent current flows, more usually associated with a (superfluid) many-boson system.

Recall that fermions have half-odd integer spins and are constrained by the Pauli principle; i.e., no two fermions can occupy the same state. Bosons, on the other hand, have integer spins and are not constrained by the Pauli principle. However, a pair of fermions has integer spin. Furthermore, if the total number of fermions in a many-fermion system is small relative to the maximum number that can be accommodated in the space available, the effects of the Pauli principle may be insignificant. It is possible then for pairs of fermions to behave as bosons or, more precisely, as *quasi-bosons*.

In a superconductor, the pair-coupled electrons exhibit superfluid (viscous-free) flow, like a boson quantum fluid. Because electrons have a charge, the superfluid flow carries a current and is referred to as a superconducting (resistance-free) current. Pairing and superfluidity are likewise believed to be important for understanding the nature of collective flows in nuclei. In this section, we present experimental evidence to show that pair coupling occurs and leads to directly observable phenomena in the neighbourhood of doubly-closed shell nuclei.

The nucleus $^{210}_{84}\text{Po}_{126}$ has two protons outside of the doubly-closed shell at $Z = 82$ and $N = 126$. If these two protons moved as independent particles, i.e., without interacting, ^{210}Po would possess a degenerate multiplet of states corresponding to the various total spins resulting from the coupling of the two single-particle spins. The low-energy spectrum of ^{210}Po is shown in Figure 1.21. Evidently, the

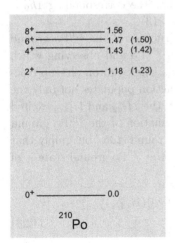

8^+	1.56
6^+	1.47 (1.50)
4^+	1.43 (1.42)
2^+	1.18 (1.23)
0^+	0.0

210
Po

Figure 1.21: The low-lying levels of ^{210}Po. Spin-parities and energies (in MeV) are given. The configuration of the pair of protons outside of the doubly-closed shell core is $(1h_{9/2})^2$. If there were no interaction between the protons constituting the pair, i.e., if they behaved as independent particles, the various $(1h_{9/2})^2$ spin couplings, which reflect the various orbital alignments, would be degenerate in energy. Calculated energies (see text) are given in parentheses. (The data are taken from *Nuclear Data Sheets*.)

multiplet of states corresponding to[14] $(\pi 1h_{9/2})^2_J$, $J = 0, 2, 4, 6, 8$ is not degenerate. The $J = 0$ coupling of pairs of like nucleons outside of doubly-closed shells is

---

[14]The symbol π signifies that the single particles are protons; for neutrons the symbol ν is used.

always observed to be strongly favoured energetically. It requires an energy of more than 1 MeV to break a $J = 0$ coupled proton pair. This favouring has already been noted as causing the odd-even staggering in one-nucleon separation energies (cf., for example, Figure 1.8). It can be attributed to a short-range attractive force between nucleons which favours the coupling of nucleons to pair states with maximally overlapping (closest proximity) wave functions. For identical nucleons, which are forbidden by the Pauli principle from having the same single-particle wave functions, this is achieved when they have opposite spin directions and, hence, total angular momentum zero. This interpretation is supported by a calculation (cf. also Figure 6.1) of the ^{210}Po spectrum for a $(1h_{9/2})^2$ configuration with a zero-range (delta function) interaction. With the strength of such an interaction adjusted such that the computed energy difference $E_8 - E_0$ agrees with the experimental value of 1.56 MeV, one obtains the energy levels shown (in parentheses) in Figure 1.21. Thus, the experimental energy levels of ^{210}Po indicate that the $J = 0$ proton pair is bound by some 1.5 MeV relative to a non-interacting pair and that, when recoupled to $J = 2$, the energy is reduced to approximately 25% of this value.

The single-proton transfer reaction ^{209}Bi$(\alpha, t)^{210}$Po reveals a number of two-proton multiplets in ^{210}Po. These are shown in Figure 1.22. All the multiplets have one proton in the $1h_{9/2}$ orbital because this is the ground-state configuration of the target. The other multiplets are seen to be non-degenerate; but the splitting of the $(1h_{9/2})^2$ multiplet, with the lowering of the $J = 0$ state, is by far the greatest (note that the peak, corresponding to the $J = 0$ ground state, is off scale to the right in Figure 1.22(a). However, the shell model continues to play a controlling influence as revealed by the fact that the centroids of the $(1h_{9/2})^2$, $1h_{9/2}2f_{7/2}$, and $1h_{9/2}1i_{13/2}$ multiplets are spaced with about the same intervals as the corresponding $1h_{9/2}$, $2f_{7/2}$, and $1i_{13/2}$ states in ^{209}Bi (cf. Figures 1.12 and 1.13).

Further inferences concerning the pair-coupled structure of the ^{210}Po ground state are obtained by removing a single proton from ^{210}Po and observing which single-particle states in ^{209}Bi remain. This is done by a single-proton pick-up reaction. As shown in Figure 1.23, the ^{210}Po$(t,\alpha)^{209}$Bi reaction populates not only the $1h_{9/2}$ ground state in ^{209}Bi, but also, to a lesser extent, the $2f_{7/2}$ and $1i_{13/2}$ excited states. Thus, the data are consistent with an approximation of the ^{210}Po ground state as a pair of $J = 0$ protons coupled to the ^{208}Pb ground state, but imply that the proton pair is in a mixed-configuration state; i.e., the ^{210}Po ground state is of the form

$$|^{210}\text{Po}; \text{g.s.}\rangle \approx \left[c_1 |(1h_{9/2})^2_{J=0}\rangle + c_2 |(2f_{7/2})^2_{J=0}\rangle + c_3 |(1i_{13/2})^2_{J=0}\rangle + \ldots \right]$$
$$\otimes |^{208}\text{Pb}; \text{g.s.}\rangle. \tag{1.23}$$

The two protons are said to form a *correlated pair*. Thus, the state of Equation (1.23) is an *independent-pair* state as opposed to an independent-particle state.

The mixed-configuration structure of the proton pair described by Equation (1.23) results from the off-diagonal elements of the short-range interaction; e.g.,

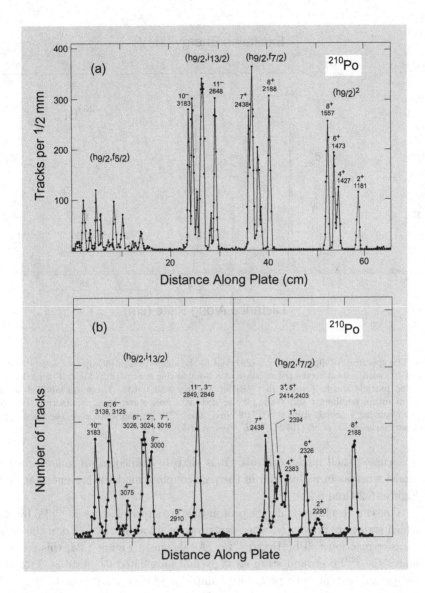

Figure 1.22: (a) The spectrum of tritons observed in the single-proton stripping reaction ^{209}Bi$(\alpha, t)^{210}$Po at an angle of $10°$ with respect to the α beam at a bombarding energy of 45 MeV. The peaks correspond to two-proton states in ^{210}Po where one of the protons is in the $1h_{9/2}$ target configuration. The states are labelled by their energies and the groups are labelled by their two-proton configurations. The ^{210}Po ground state is off scale to the right (cf. Figure 1.21). (b) The $1h_{9/2} - 2f_{7/2}$ and $1h_{9/2} - 1i_{13/2}$ multiplets with higher resolution. (The figure is adapted from Tickle R. and Bardwick J. (1971), *Phys. Lett.* **B36**, 32. The spins, parities, and excitation energies in keV are taken from *Nuclear Data Sheets*.)

$\langle (2f_{7/2})^2; J = 0 | V | (1h_{9/2})^2; J = 0 \rangle$. A consequence of this interaction is a *smearing* of the Fermi surface. The Fermi surface is the boundary line between occupied and

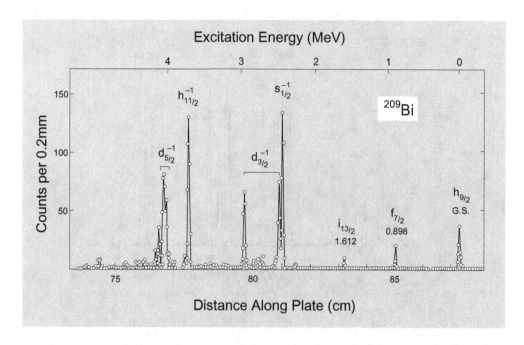

Figure 1.23: The spectrum of alpha particles observed in the single-proton pickup reaction ^{210}Po$(t, \alpha)^{209}$Bi at an angle of $40°$ with respect to the triton beam at a bombarding energy of 20 MeV. The peaks correspond to single-proton particle and hole states in ^{209}Bi (cf. Figures 1.12, 1.13). These are labelled by the single particle/hole quantum numbers l and j and, if the state is a hole state, by a superscript -1. Hole states correspond to removing a proton from the ^{208}Pb core. The ^{209}Bi ground state is on the right. (The figure is adapted from Barnes P.D. *et al.* (1972), *Nucl. Phys.* **A195**, 146.)

unoccupied shell model orbitals. It is sharply defined in an independent-particle model whereas in reality and in the pair-coupling model it is somewhat diffuse (cf. Figures 6.20 and 6.23).

One can also investigate the proton-pair structure of states in ^{210}Po by observing which states are populated when two protons are added to ^{208}Pb in the two-proton transfer reaction ^{208}Pb$(^{3}$He, n$)^{210}$Po. As shown in Figure 1.24, this reaction populates the ^{210}Po ground state strongly and an excited 0^{+} state at 2.62 MeV to a comparable extent. The probability amplitude for populating a state in ^{210}Po in a two-nucleon transfer reaction is, to a first approximation, a measure of the overlap between the state of the proton pair in ^{210}Po with that of the transferred pair. The transferred pair of protons is initially in a $J^{\pi} = 0^{+}$ configuration in ^{3}He. This explains why 0^{+} states, especially low–lying pair–correlated 0^{+} states, are preferentially populated. A similar behaviour is observed for proton-pair removal and for neutron-pair addition and removal to and from nuclei.

The two-neutron transfer reaction ^{210}Pb$(p, t)^{208}$Pb populates a number of excited states of ^{208}Pb with strengths comparable to or greater than that of the ground state. This is shown in Figure 1.25. The most strongly populated state is a

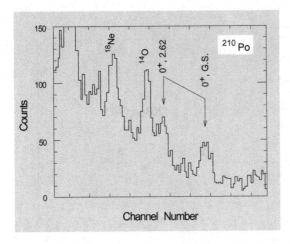

Figure 1.24: The spectrum of neutrons observed (by time-of-flight) in the two proton stripping reaction ^{208}Pb(^{3}He,n)^{210}Po at an angle of $0°$ with respect to the ^{3}He beam at a bombarding energy of 33.3 MeV. The ground state and an excited 0^+ state at 2.62 MeV in ^{210}Po are indicated. Other peaks are due to ^{12}C, ^{16}O and other target impurities. (The figure is taken from Anderson R.E. (1979), *Phys. Rev.* **C19**, 2138.)

$J^\pi = 0^+$ state at 4863 keV. This is a state corresponding to the removal of a pair of neutrons from below the closed shell at $N = 126$, leaving the pair above $N = 126$ undisturbed. The ground state of ^{208}Pb is populated by the removal of the neutron pair above the $N = 126$ closed shell leaving the bare closed shell. These two states are the only states with $J^\pi = 0^+$ that are strongly populated. The population of these states by a pair transfer reaction gives information on the structure of the removed pair in both cases. The occurrence of two such states can be attributed to the energy gap at $N = 126$; correlated pairs involving single-particle states from two different major shells underlie this structure.

The low-lying states of the doubly-odd nucleus $^{210}_{83}$Bi$_{127}$ are expected, on the basis of the shell model, to be formed primarily by adding a proton in the $1h_{9/2}$ orbital and a neutron in the $2g_{9/2}$ orbital to the ^{208}Pb ground state. Such a $(\pi 1h_{9/2}\nu 2g_{9/2})$ configuration can couple to the spin-parity values $J^\pi = 0^-$, 1^-, 2^-, 3^-, 4^-, 5^-, 6^-, 7^-, 8^-, 9^-. States with these J^π values are observed in $^{210}_{83}$Bi$_{127}$ and are shown in Figure 1.26. They are not degenerate in energy, as they would be in an independent-particle model. Nor would one expect them to be in view of the short-range interaction between the neutron and the proton.

It is interesting to compare the energy levels of the (predominantly) $(\pi 1h_{9/2}$ $\nu 2g_{9/2})$ states in the odd-odd ^{210}Bi nucleus with those of the (predominantly) $(\nu 2g_{9/2})^2$ and $(\pi 1h_{9/2})^2$ states in the neighbouring even-even nuclei ^{210}Pb and ^{210}Po, respectively. Comparisons are made, for the even-J energies, in Figure 1.27. It can be seen that the energy-level splittings differ in two key respects. One is the presence, as noted above, of odd values of J in ^{210}Bi; there are none in ^{210}Pb and ^{210}Po. The other is the much smaller favouring of the $J = 0$ coupling in ^{210}Bi compared to that in ^{210}Pb and ^{210}Po.

The absence/presence of odd-J states is seen in other doubly-even and doubly-odd nuclei. Compare, for example, the spectra for the nuclei $^{42}_{20}$Ca$_{22}$, $^{42}_{21}$Sc$_{21}$, $^{42}_{22}$Ti$_{20}$, shown in Figure 1.28. According to the shell model, one would expect the low-

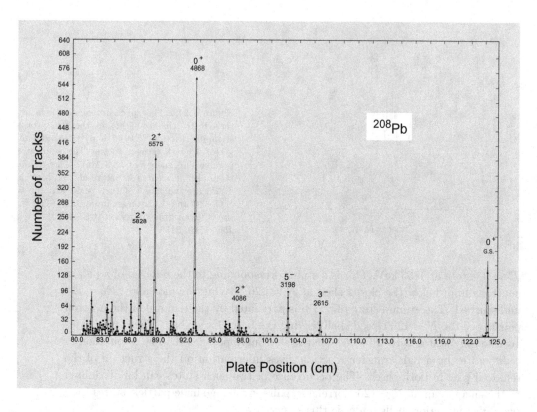

Figure 1.25: The spectrum of tritons in the two-neutron pickup reaction ^{210}Pb(p, t)^{208}Pb at a bombarding energy of 20 MeV and at an angle of 17° with respect to the proton beam. The peaks correspond to states in ^{208}Pb. They are labelled by energy (in keV), spin and parity. The ^{208}Pb ground state is on the right. The strong peak at 4868 keV is discussed in the text. (The figure is adapted from Igo G., Barnes P.D. and Flynn E.R. (1970), *Phys. Rev. Lett.* **24**, 470; energies below 5 MeV are taken from *Nuclear Data Sheets*.)

energy states of these nuclei to be generated primarily by coupling two nucleons in the $1f_{7/2}$ orbital to the ^{40}Ca ground state. Such a $(1f_{7/2})^2$ pair can couple to angular momentum $J = 0, 1, 2, 3, 4, 5, 6, 7$ and all such angular momentum states are observed in the low-energy spectrum of the doubly-odd $^{42}_{21}$Sc$_{21}$ nucleus. However, all the odd-J states are missing in the doubly-even nuclei. This has a natural explanation in terms of exchange symmetry. Whereas identical nucleons (i.e., two neutrons or two protons) in a common orbital are required to be in antisymmetrical states, a neutron and a proton can also form symmetrical pair states. Thus, it is possible for a neutron and a proton to maximise the overlaps of their wave functions in states of both maximum and minimum angular momentum. Indeed, one sees, in both Figures 1.26 and 1.28, a tendency for the exchange symmetric (odd-J) states to lie lower in energy than their even-J (antisymmetric) neighbours. Thus, there is a resultant staggering of energies with increasing J in doubly-odd nuclei in contrast to the monotonic behaviour in doubly-even nuclei. Results of a simple calculation

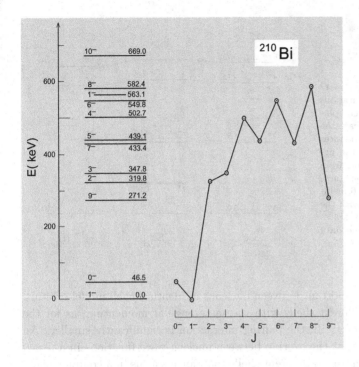

Figure 1.26: The low-lying levels of ^{210}Bi. Spin-parities and energies (in keV) are given on the left. The states shown correspond to the $\pi 1h_{9/2}\nu 2g_{9/2}$ configuration, together with the state $(\pi 1h_{9/2}\nu 1i_{11/2})_{J=10}$ and a 1^- state at 563 keV. On the right, the same states are shown in a plot of energy versus spin. (The data are taken from *Nuclear Data Sheets*.)

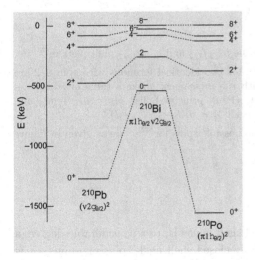

Figure 1.27: Comparison of the even-J energy levels of the odd-odd nucleus ^{210}Bi with those of the even-even nuclei ^{210}Pb and ^{210}Po. The energy scales for the three nuclei have been adjusted so that the $J = 8$ states of all nuclei are set at 0 keV. These nuclei are described in the shell model as having two nucleons outside of a ^{208}Pb core. (The data are taken from *Nuclear Data Sheets*.)

with a short-range (delta-function) interaction are given in Chapter 5 (cf. Figure 6.1). More general effects of exchange symmetry and the associated concepts of isospin are discussed in Section 5.5.

We conclude that $J = 0$ pair coupling is not favoured as much in odd-odd nuclei as it is in even-even nuclei. Moreover, experimental data indicate that, in general,

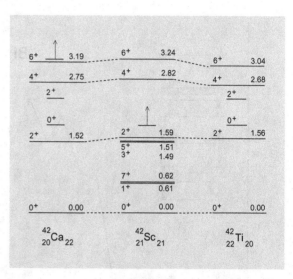

Figure 1.28: Low-energy states in the $A = 42$ isobars ^{42}Ca, ^{42}Sc, and ^{42}Ti. Excitations are in MeV. Levels are labelled by their spin-parity, J^π. The vertical arrows indicate the energies above which there are excited states known but which are omitted from the figure. The states shown for ^{42}Sc result from the various spin couplings of the configuration $\pi 1f_{7/2}\nu 1f_{7/2}$. The $J = 0, 2, 4, 6$ members of this multiplet are connected with the corresponding $(\pi 1f_{7/2})^2$ and $(\nu 1f_{7/2})^2$ states in ^{42}Ti and ^{42}Ca, respectively. (The data are taken from Endt P.M. (1990), *Nucl. Phys.* **A521**, 1.)

$J = 0$ pairing is most important between nucleons in identical shell model orbitals. For nucleons in two different shells with the same angular momentum, as for the $(\pi 1h_{9/2}\nu 2g_{9/2})$ states of ^{210}Bi, the pairing interaction is significantly smaller. An explanation for this is that the overlap between nucleon wave functions in a $J = 0$ state is smaller for nucleons in different shells than for nucleons in a common shell.

Exercises

1.10 Plot E_x vs. J (cf. Figure 1.26) for the $h_{9/2}f_{7/2}$ and $h_{9/2}i_{13/2}$ multiplets in ^{210}Po.

1.11 Show that the possible spin values for two identical fermions with spin 9/2 are $0, 2, 4, 6, 8$. [This can be done with the m scheme: make a table of m_j values (m_1, m_2), ensuring that $m_1 > m_2$ (Pauli principle), list $M := m_1 + m_2$, and identify the possible values of J (recall $M = J, J - 1, \ldots, -J$).]

1.12 Using the m scheme, show that the possible spins for ^{42}Sc are as given in Figure 1.28.

1.5 Singly-closed shell nuclei

The low-energy structure of singly-closed shell nuclei is, to a considerable degree, a direct extension of the structure of doubly-closed shell nuclei plus or minus a few nucleons. For example, the low-lying states in the doubly-even Sn ($Z = 50$) isotopes, shown in Figure 1.29, resemble those of ^{130}Sn. This resemblance is attributed to the pairing force which energetically favours the formation of multiple $J = 0$ pairs.

The even-mass tin isotopes are uniformly characterised by a $J^\pi = 0^+$ ground state and a large energy gap between the ground and first-excited states.[15]

[15]The interpretation of the energy gap in nuclei as an indication of the superfluid nature of

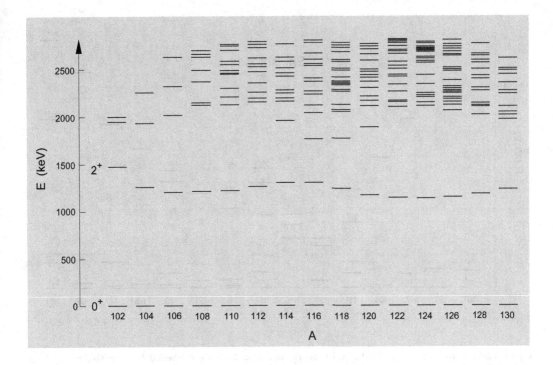

Figure 1.29: Low-energy states in the even-mass tin ($Z = 50$) isotopes. The 0^+ ground states and 2^+ first excited states are discussed in the text. (The data are taken from *Nuclear Data Sheets* and Juutinen S. *et al.* (1997), *Nucl. Phys.* **A617**, 74 – ^{106}Sn; Górska M. *et al.* (1998), *Phys. Rev.* **C58**, 108 – ^{104}Sn.)

In fact, the ground states of all doubly-even singly-closed shell nuclei have $J^\pi = 0^+$ and a significant energy gap separating them from the first excited state. This is consistent with the tendency of like nucleons to form $J = 0$ coupled pairs. Moreover, it suggests that the pair coupling of like nucleons persists with excitation energy; i.e., pairs tend to be broken one at a time and low-energy excitations are predominantly single-broken-pair configurations.

The second notable feature of Figure 1.29 is that the only excited state of an Sn isotope below ≈ 2 MeV is a single 2^+ state at a nearly constant energy of about 1 MeV. The near constant energy of the first excited 2^+ state is a widespread feature of singly-closed shell nuclei. The 2^+ state can be interpreted as a single broken pair of neutrons recoupled to $J^\pi = 2^+$. As seen for ^{210}Po (cf. Figure 1.21), a $J^\pi = 2^+$ pair is favoured energetically, although not as much as a $J^\pi = 0^+$ pair.

It is also instructive to interpret the first excited 2^+ states in singly-closed shell nuclei, such as the tin isotopes, as *one-phonon* quadrupole vibrational states (cf. Section 1.7.1). Such states may be compared with the superfluid persistent-flow states of a quantum fluid and with the persistent-current states of a superconductor. It

nuclear matter was quickly recognised by Bohr A., Mottelson B.R. and Pines D. (1958), *Phys. Rev.* **110**, 936, following the paper of Bardeen *et al.*, *op. cit.* Footnote 11 on Page 24.

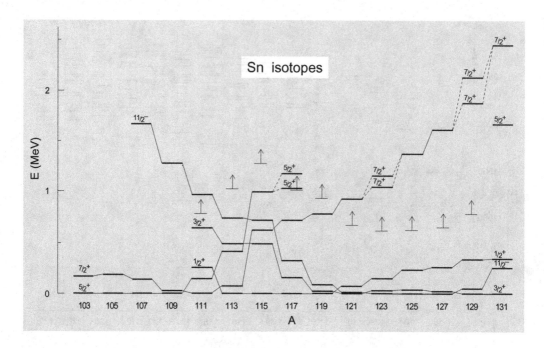

Figure 1.30: Low-energy states in the odd-mass tin isotopes. Levels are labelled by their spin-parity. The vertical arrows indicate the energies above which there are excited states known but which are omitted from the figure. The lowest three states are a selection from the spin-parities $5/2^+$, $7/2^+$, $1/2^+$, $3/2^+$, $11/2^-$, corresponding to the single-particle configurations $2d_{5/2}$, $1g_{7/2}$, $3s_{1/2}$, $2d_{3/2}$, $1h_{11/2}$, respectively. Information on states in ^{103,105,107,109}Sn is very limited. The identification of $2d_{5/2}$ in ^{117}Sn and $1g_{9/2}$ in ^{123,129}Sn is ambiguous. (The data are taken from *Nuclear Data Sheets* and Fahlander C. *et al.* (2001), *Phys. Rev.* **C63**, 021307(R) $- ^{103}$Sn.)

has already been remarked that $J = 0$ coupled pairs of nucleons, such as the Cooper pairs of a superconductor, can be regarded as quasi bosons. Because, for practical purposes, a nucleus is a zero-temperature system, the ground state of a pair-coupled nucleus may be compared with that of a zero-temperature quantum fluid or with the fully-paired ground state of a zero-temperature superconductor. It requires considerable energy to dissociate a Cooper pair and thereby create an excited state of a superconductor. However, the whole electron gas in a superconductor can be set in motion without breaking any pairs. Moreover, once in motion, this highly collective state (all electrons moving in unison) cannot easily dissipate its translational energy. In particular, it cannot give up translational energy by breaking a Cooper pair because the energy required to do so is too large. Thus, the current flow, associated with the collective translational motion of electrons in a superconductor, continues without resistance. In a similar way, one can understand the occurrence of collective vibrational 2^+ states with energies less than that required to break a two-nucleon pair. It should be noted, however, that the alternative descriptions of the 2^+ states as collective vibrational states and recoupled broken-pair states are not incompatible with one another. The collective vibrational interpretation simply implies a strongly-correlated structure for the recoupled $J^\pi = 2^+$ pair.

The systematic features of the low-lying states of the odd-mass tin isotopes, shown in Figure 1.30, can be understood qualitatively as resulting from the successive filling of the single-particle shell-model states $2d_{5/2}$, $1g_{7/2}$, $3s_{1/2}$, $2d_{3/2}$, $1h_{11/2}$, in the $50 \leq N \leq 82$ shell (cf. Figure 1.17); the ground state and first few excited states always match a selection from these single-particle states. The appearance of states with the same spins as single-particle shell-model states in odd nuclei with many (interacting) nucleons outside of a closed shell can be attributed (again) to the tendency of like nucleons to form $J = 0$ pairs, thereby leaving the odd unpaired nucleon in a specific shell-model state. Other excited states (indicated by vertical arrows) result from the "breaking" of $J = 0$ pairs and recoupling them to $J \neq 0$.

Transfer reactions measure the occupancies of single-particle orbitals. Consider, for example, adding a nucleon in a $d_{5/2}$ single-particle state to an even nucleus in its $J^\pi = 0^+$ ground state to form an odd-mass nucleus in a $J^\pi = 5/2^+$ state. Addition of a neutron to a nucleus is achieved in a (d,p) transfer reaction. The probability for the transfer to occur depends on the extent to which the $d_{5/2}$ neutron orbital in the target nucleus is vacant and, therefore, able to accommodate the additional neutron. Figure 1.31 shows the fractional occupation probabilities, v_j^2, for various shell-model orbitals as a function of neutron number in the Sn isotopes. It is

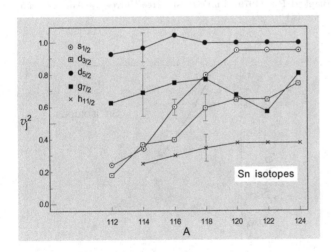

Figure 1.31: Fractional occupation probabilities, v_j^2, of single-particle orbitals in $^{112-124}$Sn. The uncertainties in v_j^2 shown are typical for each sub-shell (other uncertainties are omitted to avoid cluttering the figure). (The data are taken from Fleming D.G. (1982), *Can. J. Phys.* **60**, 428.)

evident from Figure 1.31 that the occupancies of the single-particle orbitals in the tin isotopes change rather smoothly, with no significant discontinuities, as a function of changing neutron number.

Two-neutron transfer reactions on doubly-even tin target nuclei show the persistent concentration of the transfer strength in the ground-state-to-ground-state transitions. This is illustrated in Figure 1.32. These data indicate that all of the ground states are simply related to their neighbours by the addition or removal of a spin-zero coupled neutron pair.

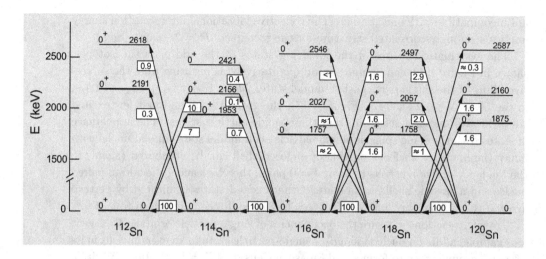

Figure 1.32: Two-neutron transfer reaction population of low-lying 0^+ states in the even-mass tin isotopes. The population intensity for each case, $^A\text{Sn}(\text{p,t})^{A-2}\text{Sn}$ or $^A\text{Sn}(\text{t,p})^{A+2}\text{Sn}$, is normalised to a ground-state-to-ground-state intensity of 100. The excited states depicted by heavy bars are strongly populated in the $(^3\text{He,n})$ reaction (cf. Figure 1.77). The cross sections are integrated over solid angle. Bombardment energies are in the range 10-20 MeV. (The data are taken from Bjerregaard J.H. *et al.* (1968), *Nucl. Phys.* **A110**, 1; Bjerregaard J.H. *et al.* (1969), *Nucl. Phys.* **A131**, 481; Fleming D.G. (1970), *Nucl. Phys.* **A157**, 1; Guazzoni P. (2004), *Phys. Rev.* **C69**, 024619; and Blankert P.J. (1979), Ph.D. thesis, Free University, Amersterdam. The energies are taken from *Nuclear Data Sheets.*)

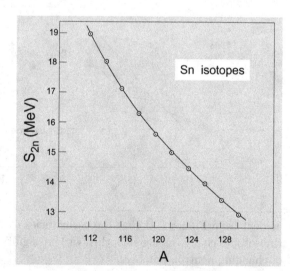

Figure 1.33: Two–neutron separation energies (in MeV) for the doubly-even tin isotopes with $112 \leq A \leq 130$. (The data are taken from Audi G., Wapstra A.H. and Thibault C. (2003), *Nucl. Phys.* **A729**, 337.)

Two-neutron separation energies, S_{2n}, for the tin isotopes are shown in Figure 1.33. They exhibit a smooth variation with neutron number. These data indicate that the substructure of shells is smoothed away by pairing correlations. A sharply defined set of subshells would show a series of steps in S_{2n} similar to those seen at major closed shells (cf. Figures 1.9 and 1.10), but on a smaller (energy) scale.

The changes in mean-square charge radii – the isotope shifts – for the tin isotopes are shown in Figure 1.34. The variations with neutron number are again smooth.

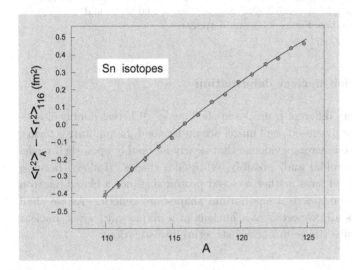

Figure 1.34: Differences in mean-square charge radii (isotope shifts) for the tin isotopes (referred to ^{116}Sn). (The data are taken from Anselment M. *et al.* (1986), *Phys. Rev.* **C34**, 1052.)

A sharply defined set of subshells would show a series of sharp increases followed by slow declines in $\delta\langle r^2\rangle$ similar to that shown in figure 1.11, but again on a smaller scale.

The dominance of pair coupling results in a great simplification of the low-lying states in singly-closed-shell nuclei. The doubly-even Sn isotopes closely resemble one another. The odd Sn isotopes show more differences, but the differences are easily understood in terms of increasing occupancies of single-particle states. Changes are smooth with no sudden "jumps" in the various properties exhibited in Figures 1.29-1.34. The smoothness is a consequence of the slow and continuous evolution of the structure of the pairs with increasing mass number. The pairs are not formed from a single j^2 configuration but are coherent linear combinations of many configurations of the form $\sum_i c_i |j_i^2; J = 0\rangle$, as noted in Section 1.3. Thus, the short-ranged forces, responsible for the pair correlations, effectively smooth out the occupancies of the shell model orbitals and the discontinuities that occur in the independent-particle model as the orbitals are sequentially filled.

Exercises

1.13 If the shell model configurations for the neutrons in the Sn isotopes were in the order $d_{5/2}$, $g_{7/2}$, $s_{1/2}$, $d_{3/2}$, $h_{11/2}$ with sequential filling, draw how Figure 1.31 would appear.

1.14 For the pair occupancies implied for ^{116}Sn from Exercise 1.13, sketch the expected strengths for neutron-pair pickup in the ^{116}Sn(p,t)^{114}Sn reaction. (Weight the possible ways of picking up the pairs by the number of pairs, $n_j = \frac{1}{2}(2j+1)$, in a j orbital.) Compare your answer with Figure 1.32 for ^{116}Sn(p,t)^{114}Sn.

1.15 Using the shell model configuration order in Exercise 1.13, and assigning binding energies in the shell model potential of $B(h_{11/2}) = -6.7$ MeV, $B(d_{3/2}) = -7.9$ MeV, $B(s_{1/2}) = -8.5$ MeV, $B(g_{7/2}) = -10.0$ MeV and $B(d_{5/2}) = -11.0$ MeV, draw how Figure 1.33 would appear.

1.6 Open-shell nuclei: nuclear deformation

Open-shell nuclei are very different from closed-shell nuclei. Whereas doubly closed-shell nuclei and most singly-closed shell nuclei are understood, in qualitative terms, as being spherical; there is strong evidence that nearly all doubly open-shell nuclei are deformed with spheroidal and, possibly, ellipsoidal shapes. Indeed, because the vast majority of nuclei have neither a closed proton shell nor a closed neutron shell, it appears that non-spherical equilibrium shapes are generic. As we shall see, deformation affects all properties of a nucleus in a major way. Thus, nuclear deformation is a pervasive theme in nuclear structure investigations.

1.6.1 *Even open-shell nuclei*

Changes in the structure of nuclei as their neutron or proton numbers move away from closed-shell values are dramatic. The systematics of the first $J^\pi = 2^+$ excited-state energies, $E(2_1^+)$, for the $N = 64, 66, 68$ isotones are shown in Figure 1.35. Taking the singly-closed-proton shell nucleus, ^{116}Sn, as reference, the addition or

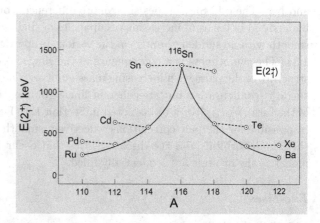

Figure 1.35: Energies of the first-excited 2^+ states in the even-mass isotones with $N = 64, 66, 68$. The $N = 66$ isotones are joined by a solid line. Isotopes are joined by dashed lines. (The data are taken from *Nuclear Data Sheets*.)

subtraction of a neutron pair, as already noted in the preceding section, has little effect on $E(2_1^+)$. However, when a proton pair is added or subtracted a large drop in $E(2_1^+)$ occurs. The systematics of $E(2_1^+)$ are very distinctive across open-shell regions. This is shown in Figure 1.36 for the $Z \geq 50$ and $N \leq 82$ region. A smooth

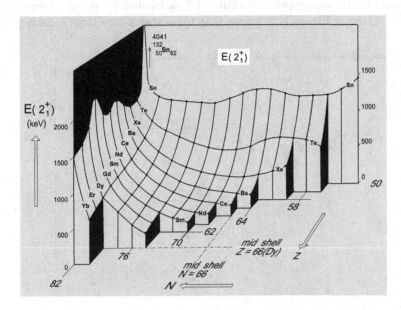

Figure 1.36: Energies of the first-excited 2^+ states in doubly-even nuclei in the region $Z \geq 50$, $N \leq 82$. (The data are taken from *Nuclear Data Sheets* and the *Evaluated Nuclear Structure Data File* (*ENSDF*).)

decrease in $E(2_1^+)$ is observed towards a minimum value near the centre of the open shell ($Z = 66$, $N = 66$). We shall see that the value of the $E(2_1^+)$ energy has a simple and direct correlation with the quadrupole deformation of a nucleus.

The electric quadrupole moments of doubly-even nuclei are necessarily zero in their ground states because doubly-even nuclei have $J^\pi = 0^+$ ground states. To have a non-zero quadrupole moment, a nucleus must have $J \geq 1$, otherwise the quantum fluctuations in the orientation of the nucleus (which increase as J decreases) average the quadrupole moment to zero, even when the nucleus is intrinsically deformed. Thus, we consider the electric quadrupole moments of first excited $J^\pi = 2^+$ states.

The quadrupole moment, Q, of a state of angular momentum J is defined as the expectation value of the $\hat{Q}_0$ component of the quadrupole tensor $\hat{Q}$ when the angular momentum of the state is maximally aligned along the axis of quantization, i.e.,[16]

$$Q := \langle JM = J | \hat{Q}_0 | JM = J \rangle, \tag{1.24}$$

where

$$\hat{Q}_0 := \sum_{i=1}^{Z} (2\hat{z}_i^2 - \hat{x}_i^2 - \hat{y}_i^2) \tag{1.25}$$

---

[16]Note that, as discussed in the "Notations and Conventions" paragraph following the Preface, we use the symbol := to denote a definition as opposed to a simple equality.

and the sum is over the protons; neutrons do not contribute to the charge quadrupole moment which is the moment measured.

A plot of the electric quadrupole moments of first excited 2^+ states of even open-shell nuclei, given in Figure 1.37, shows that these states have large negative

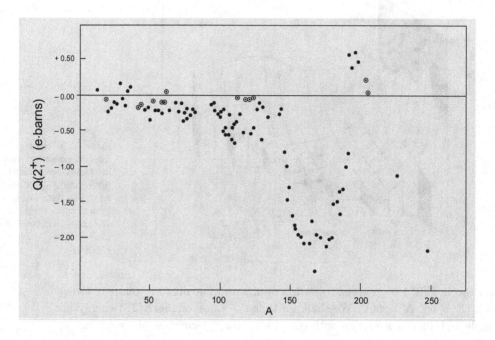

Figure 1.37: Electric quadrupole moments of the first excited 2^+ state, $Q(2_1^+)$ in doubly-even nuclei plotted versus mass number, A. Singly-closed shell nuclei are represented by open circles; doubly-open shell nuclei are represented by solid circles. (The data are taken from Raghavan P. (1989), *At. Data Nucl. Data Tables* **42**, 189 and Stone N.J. (2005), *At. Data Nucl. Data Tables* **90**, 75 (we omit cases for which there is serious disagreement between reported values); and from: Czosnyka T. *et al.* (1986), *Nucl. Phys.* **A458**, 123 – ^{248}Cm; Wollersheim H.J. *et al.* (1993), *Nucl. Phys.* **A556**, 261 – ^{226}Ra; Wu C.Y. *et al.* (1991), *Nucl. Phys.* **A533**, 359 – ^{182,184}W; Fahlander C. *et al.* (1992), *Nucl. Phys.* **A541**, 157 – ^{172}Yb; Kotliński B. *et al.* (1990), *Nucl. Phys.* **A517**, 365 – ^{168}Er; Svensson L.E. *et al.* (1995), *Nucl. Phys.* **A584**, 547 – ^{106,108}Pd; Kavka A.E. *et al.* (1995), *Nucl. Phys.* **A593**, 177 – ^{76,80,82}Se; and Svensson L.E. (1989), Ph.D. thesis, Univ. of Uppsala – ^{110}Pd.)

quadrupole moments for doubly-open shell nuclei and small quadrupole moments when N or Z is magic.

When a charged object with a large quadrupole deformation rotates, it emits electric quadrupole (E2) radiation. Thus, a broad overview of nuclear deformation in doubly-even nuclei is provided by the electric quadrupole transition rates for gamma-ray decay from first excited $J^\pi = 2^+$ states to the $J^\pi = 0^+$ ground states. These transition rates are given in terms of *reduced transition rates*, the so-called B(E2) values defined in the Appendix, Equations (B.3) - (B.5). The reduced transition rate $B(\text{E2}; 2_1^+ \rightarrow 0_1^+)$ is shown for nuclei in the range $142 < A < 152$ in

Figure 1.38. The very large values for E2 transition rates in the mid-shell regions are consistent with these nuclei possessing large deformations.

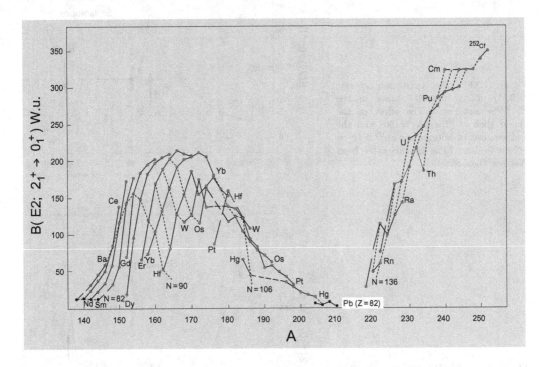

Figure 1.38: Electric quadrupole transition rates, $B(E2)$, between the first excited 2^+ state and the ground state in doubly-even nuclei with $142 \le A \le 252$. Isotopes are joined by solid lines and isotones by dashed lines. The $B(E2)$ values are in Weisskopf units, W.u.; see Appendix, Equation (B.12). (The data are taken from Raman S., Nestor Jr. C.W. and Tikkanen P. (2001), *At. Data Nucl. Data Tables* **78**, 1).

The above-mentioned results reveal the 2^+_1 excitation energy, $E(2^+_1)$, and $B(E2; 2^+_1 \rightarrow 0^+_1)$ reduced transition rate as sensitive indicators of nuclear deformation. To a first approximation, they show that singly-closed shell nuclei are essentially spherical, whereas doubly-open shell nuclei are deformed. However, the situation is not always so straightforward. Consider, for example, the neutron-rich strontium isotopes. These isotopes form a sequence that passes through the region where the implied deformation changes are more rapid than in any other mass region. The $E(2^+_1)$ values for the $^{76-100}$Sr isotopes are shown in Figure 1.39. The left-hand side of the figure shows the decrease in 2^+_1 excitation energy that, as the neutron number departs from the closed-shell value of $N = 50$, we have come to expect and associate with the onset of deformation. However, there is a distinct plateau on the right-hand side. The onset of deformation for $N > 50$ appears to be delayed until $N = 60$. This interpretation is supported by the $B(E2; 2^+_1 \rightarrow 0^+_1)$ systematics, shown in Figure 1.40. For N in the range 48 to 58, the reduced $E2$

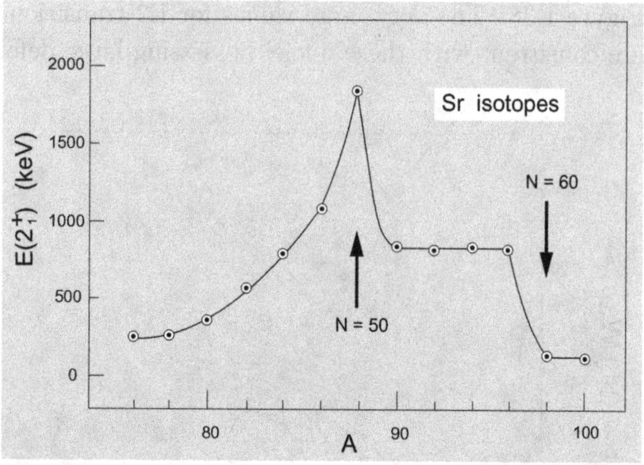

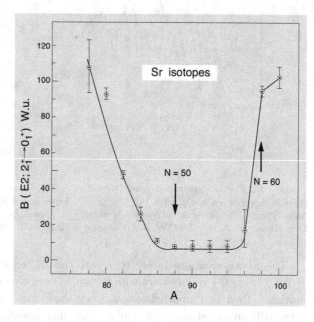

Figure 1.39: The excitation energy, $E(2_1^+)$, of the first 2^+ state in the $^{76-100}$Sr isotopes. The effect of the closed shell at ^{88}Sr ($N = 50$) and the sudden onset of deformation at ^{98}Sr ($N = 60$) are evident. (The data are taken from *Nuclear Data Sheets*.)

Figure 1.40: Electric quadrupole transition rates, $B(E2)$, between the first excited 2^+ state and ground state in $^{78-100}$Sr. The effect of the closed shell at $N = 50$ and the sudden onset of deformation at $N = 60$ again are evident (cf. Figure 1.39). (The data are taken from Raman S., Nestor Jr. C.W. and Tikkanen P. (2001), *At. Data Nucl. Data Tables* **78**, 1.)

transition rate is small which is consistent with these isotopes being spherical. But at $N = 60$ it increases dramatically implying a sudden onset of deformation.

The sudden onset of deformation at $N = 60$ in the strontium isotopes is evident also in isotope shift data. This can be seen in Figure 1.41 which shows a sudden increase in $\langle r^2 \rangle$ between $A = 97$ and $A = 98$ ($N = 59$ and $N = 60$). The quantity shown, $\langle r^2 \rangle = \langle \frac{1}{A} \sum_i r_i^2 \rangle$, is proportional to the nuclear monopole moment

$$M_0 = \langle \sum_i r_i^2 \rangle = \langle \sum_i \left(x_i^2 + y_i^2 + z_i^2 \right) \rangle. \tag{1.26}$$

The increase of this quantity between $N = 50$ and $N = 59$ can be attributed to the

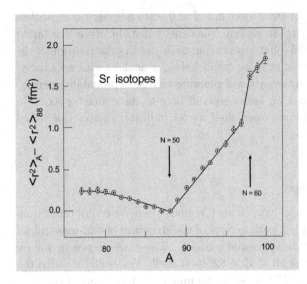

Figure 1.41: Isotope shifts, $\delta\langle r^2\rangle$, in fm^2 for the strontium isotopes versus mass number relative to $^{88}_{38}$Sr$_{50}$. Uncertainties are shown only where they are greater than the circles. The effect of the closed shell at $N = 50$ and the sudden onset of deformation at $N = 60$ are clearly evident (cf. Figure 1.11). (The data are taken from Buchinger F. *et al.* (1990), *Phys. Rev.* **C41**, 2883; Lievens P. *et al.* (1992), *Phys. Rev.* **C46**, 797 and Lievens P. *et al.* (1991), *Phys. Lett.* **B256**, 141.)

normal increase of the size of the nucleus with nucleon number. The extra increase at $N \approx 60$ can be understood by contrasting it with the continuous deformation of a sphere into an ellipsoid at constant density.

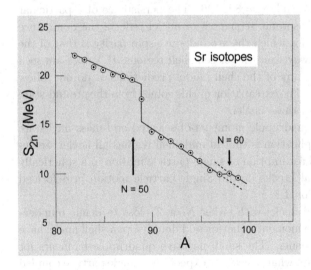

Figure 1.42: Two-neutron separation energies in the strontium isotopes versus mass number for $82 \leq A \leq 99$. Discontinuities corresponding to the closed shell at $N = 50$ and the increase in binding resulting from the sudden onset of deformation at $N = 60$ are evident. (The data are taken from Audi G., Wapstra A.H. and Thibault C. (2003), *Nucl. Phys.* **A729**, 337.)

The sudden onset of deformation at $N = 60$ in the strontium isotopes also has an effect on two-neutron separation energies, S_{2n}. This is shown in Figure 1.42. The increase in the energy required to remove a pair of neutrons, relative to the systematic trend, at $N = 60$ implies a gain in ground-state binding energy. The strontium nuclei with $N \geq 60$ are understood to gain binding energy by deforming.

The negative quadrupole moments observed for the $J^\pi = 2^+$ states imply that these states have oblate (disc-shaped) density distributions. However, as already noted, the density distribution of any object in a $J = 0$ state is spherical regardless of the intrinsic shape of the object. In general, the density distribution of a rotating object will not be zero, but it will differ from the intrinsic density distribution of the object. In particular, as discussed in more detail below, a nucleus with an axially-symmetric prolate (cigar-shaped) density will generate an oblate probability density distribution when rotating about an axis perpendicular to its symmetry axis. In fact, as the properties of odd-mass open-shell nuclei indicate, doubly open-shell nuclei are predominantly prolate.

1.6.2 *Odd open-shell nuclei*

The systematics of odd-mass nuclei in open-shell regions are more complicated than for even nuclei. One of the simplest perspectives of the structure of odd-mass nuclei is provided by their ground-state spins and parities. These are shown in Figure 1.43 for odd-proton nuclei of the $50 < Z < 82$ open shell. For nuclei close to the $Z = 50$ and 82 closed shells, the J^π values can be understood in terms of the shell model. The shell model orbitals which occur between $Z = 50$ and 82 are $2d_{5/2}$, $1g_{7/2}$, $1h_{11/2}$, $2d_{3/2}$, and $3s_{1/2}$ (cf. Figure 1.17). Thus, one expects ground states with $J^\pi = 5/2^+$, $7/2^+$, $11/2^-$, $3/2^+$, and $1/2^+$. This agrees with nuclei near to the closed shells at $Z = 50$, 82 and $N = 82$, 126. The persistence of a particular spin-parity along isotopic and isotonic sequences of nuclei reflects the dominance of pair coupling at low energy for which the ground-state spin-parity is that of the unpaired nucleon. However, away from the closed-shell regions J^π values are seen which either are different from any of the shell model predictions or do not follow the shell model spin sequences. The explanation of this comes from the ground-state electric quadrupole moments of these nuclei.

The ground-state electric quadrupole moments of selected odd-mass nuclei are shown in Figure 1.44. The implication is that in mid-shell regions all nuclei are deformed. Thus, the shell model assumption of single-particle motion in a spherically symmetrical (central) potential breaks down. Single-particle motion in deformed potentials is discussed in Section 1.7.

Figure 1.44 shows that odd nuclei with either N or Z close to magic numbers have small negative quadrupole moments whereas odd doubly-open-shell nuclei have large positive quadrupole moments. The small negative quadrupole moments for near singly-closed shell nuclei are what one would expect for a single particle coupled to a spherical even core. These nuclei have small oblate deformed shapes consistent with a nucleon circuiting a spherical core and polarizing the latter to some extent as illustrated in Figure 1.20. The large positive quadrupole moments of odd nuclei away from closed shells imply that the latter nuclei have large prolate deformations. This is in contrast to the large negative quadrupole (cf. Figure 1.37) observed for the 2^+ states of neighbouring doubly-even nuclei. A simple explanation of this observation is given by the nuclear rotor model.

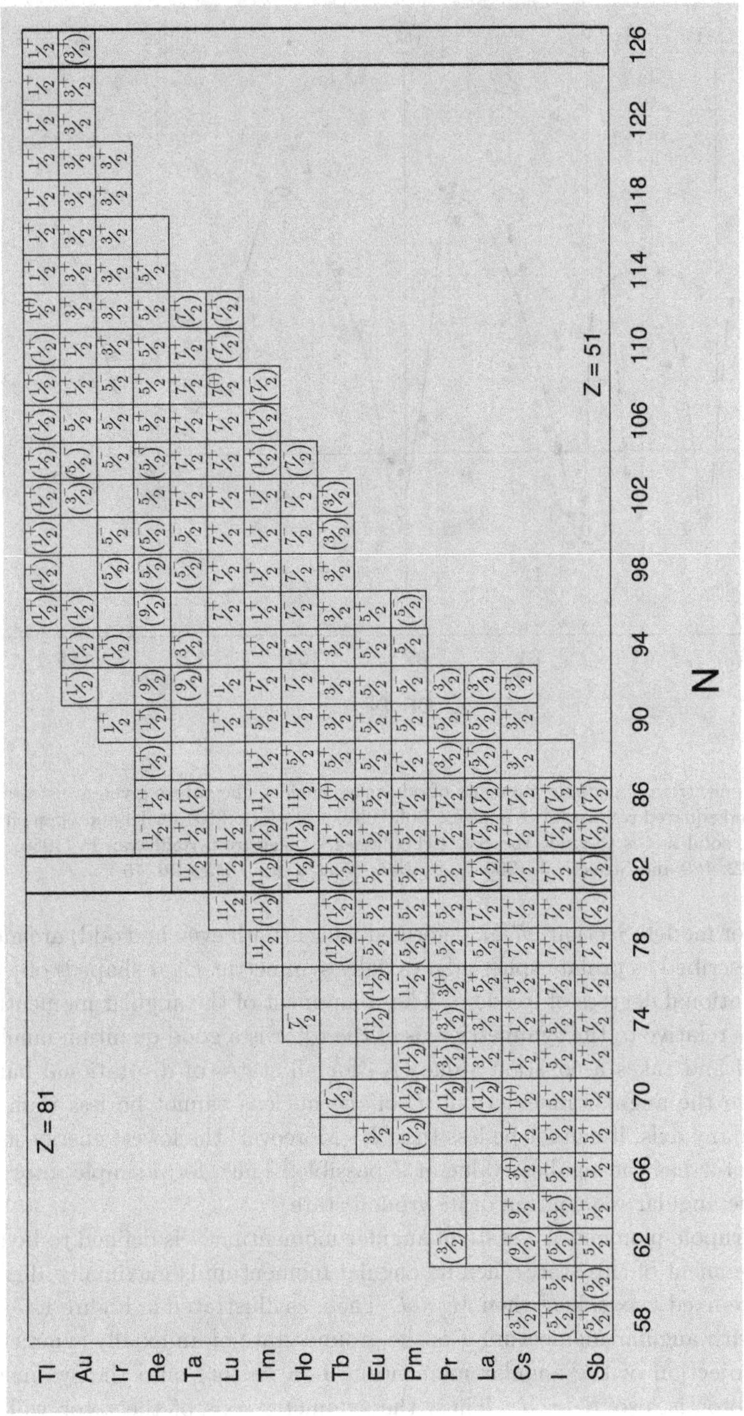

Figure 1.43: Ground-state spins and parities for odd-Z, even-N nuclei with $51 \leq Z \leq 81$. Values in parentheses are less reliable. (The data are taken from *Nuclear Data Sheets*.)

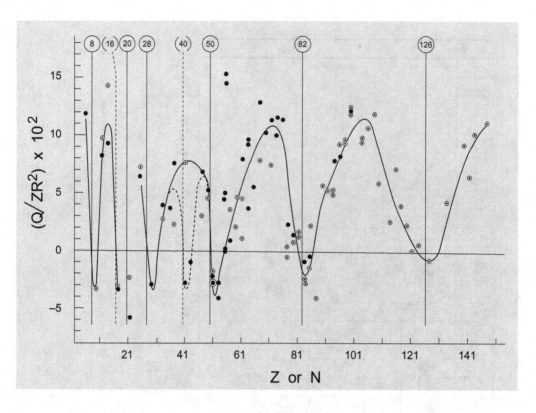

Figure 1.44: Ground-state electric quadrupole moments of odd-mass nuclei. The values given are scaled by ZR^2, i.e., by the charge and squared radius of the nucleus. Solid circles are for odd-Z nuclei and open circles are for odd-N nuclei. The solid line is to guide the eye. (The data are taken from Raghavan P. (1989), *At. Data Nucl. Data Tables* **42**, 189 and Stone N.J. (2005), *At. Data Nucl. Data Tables* **90**, 75.)

In the rotor model (Section 1.7.2), deformed nuclei (both even and odd) are most commonly described as prolate spheroidal (axially symmetric, cigar shaped) objects with only rotational degrees of freedom. The component of the angular momentum of the nucleus relative to the symmetry axis of the rotor is a good quantum number in this model and takes a constant value (K) for all states of a rotational band. Then, because the angular momentum, J, of the nucleus cannot be less than its projection on any axis, it cannot be less than K. Moreover, the lowest energy state of a simple rotor has the smallest value of J possible. Thus, for a simple rotor, K is equal to the angular momentum of its ground state.

The quadrupole moment of a state of angular momentum J is defined to be the quadrupole moment of this state when its angular momentum is maximally aligned with the space-fixed z axis, i.e., when $M = J$. Then, as illustrated in Figure 1.45(a), if the state with angular momentum J is the ground state of an axially symmetric rotor, the projection of the angular momentum $\mathbf{J}$ (a vector) onto the symmetry axis of the rotor is also $K = J$. Thus, the symmetry axis of the rotor will be

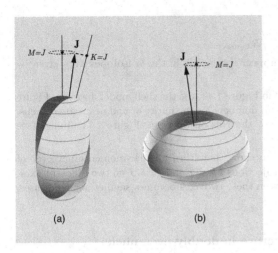

Figure 1.45: The maximum component of the angular momentum **J** of a rotor relative to the space-fixed z axis is $M = J$. For an axially-symmetric rotor. the component of J relative to its intrinsic symmetry axis, is K. For an odd-mass nucleus in its ground state, shown in (a), $K = J$ and the intrinsic symmetry axis is maximally aligned with the space-fixed z axis, subject to the limitations of quantum mechanics. For a doubly-even nucleus, shown in (b), $K = 0$ and the angular momentum of an excited $J > K = 0$ state is perpendicular to the intrinsic symmetry axis. The latter is as close to being perpendicular to the space-fixed z axis as quantum mechanics allows. Thus, the space-fixed quadrupole moment of a prolate rotor is positive for an odd-mass nucleus and negative (i.e., oblate) for a doubly-even nucleus.

maximally aligned with the space-fixed z axis to the extent allowed by quantum mechanics. Note that both M and K are constants of the motion (good quantum numbers), but the vector **J** is not. Therefore, loosely speaking, the orientation of the deformed nucleus, defined by its symmetry axis, precesses about the space-fixed z axis. This precession generates a modified density distribution, relative to the space-fixed axes, albeit one that still has a positive quadrupole moment for large enough values of J. The alignment of the symmetry axis with the space-fixed axis increases as $J = K$ increases. Thus, the larger the value of K for a simple prolate rotor, the closer the measured quadrupole moment of the ground state becomes to the positive intrinsic quadrupole moment (cf. Exercise 1.19). Conversely, the measured quadrupole moment becomes small for small values of $J = K$ as the orientation of the intrinsic symmetry axis becomes less well defined. In fact, it vanishes for $J \leq 1/2$. In particular, the quadrupole moment of a double-even nucleus in a $J = K = 0$ ground state is zero. The measured quadrupole moments for the rotational states of a $K = 0$ rotor are then those of its excited states for which $J > K = 0$, e.g., for its first excited $J = 2$ state. However, for an excited state of a $K = 0$ prolate rotor the quadrupole moment, observed relative to a space-fixed axis, is that of an oblate density distribution as illustrated in Figure 1.45(b). This follows because, if the projection of the angular momentum, **J**, onto the intrinsic symmetry axis is $K = 0$, then **J** is perpendicular to the symmetry axis. Consequently, in a state with $M = J$, the intrinsic symmetry axis of the rotor precesses in a plane approximately perpendicular to the space-fixed z axis.

Thus, we conclude that most nuclei away from closed shells are intrinsically deformed with prolate quadrupole shapes.[17]

---

[17]This interpretation was first made by Rainwater J. (1950), *Phys. Rev.* **79**, 432.

Exercises

1.16 From Figure 1.44, calculate $R_Z/R_\perp)_{\text{max}}$.

1.17 Draw how Figure 1.41 would appear if the radii of the Sr isotopes were given by Equation (1.2).

1.18 For the spin-parities presented in Figure 1.43, use the shell model diagram, Figure 1.17, to identify the n, l quantum numbers for the nuclei at and near closed shells. For nuclei far from closed shells, the Nilsson diagram, Figure 1.71 applies (see Exercise 1.32).

1.19 Use a geometric construction to show that if an angular-momentum vector of length $|\mathbf{J}| = \sqrt{J(J+1)}$ has projections $M = J$ and $K = J$ on two different axes, then the maximum angle between these two axes becomes smaller as J increases.

1.7 Low-energy collective structure in doubly-even nuclei

Collective behaviour in nuclei was discovered shortly after shell structure. The observation of shell structure (cf. Section 1.2) implies that nuclei have properties in common with systems of independent particles. Collective behaviour, on the other hand, occurs when many nucleons behave cooperatively and is more characteristic of strongly interacting systems. Thus, it appears that nuclei exhibit properties that range from one extreme to the other.

It will be shown in this section that the evidence for collective behaviour is even more compelling than for independent-particle behaviour. Consider, for example, the lifetimes of first excited 2^+ states of doubly-even nuclei. By expressing the transition rate for E2 gamma-ray decay of such a state to the ground state in single-particle (Weisskopf) units, as used in Figure 1.38, it is immediately apparent that many nucleons in the nucleus must participate in a highly-correlated manner to produce the huge transition rates observed.

The collective model was formulated by (Aage) Bohr and Mottelson[18] to describe such phenomena. It was inspired by the earlier liquid-drop model of (Niels) Bohr and Kalckar,[19,20] which had enjoyed considerable success in describing nuclear fission,[21] and by Rainwater's proposal that many nuclei have deformed intrinsic shapes.[22]

The collective model portrays the nucleus as a droplet of nuclear matter with only shape degrees of freedom. The model nucleus can vibrate and it can take on deformed equilibrium shapes and rotate. Thus, two simple limits of the model are: the harmonic vibrator model of a spherical nucleus and the rigid rotor model

---

[18]The early work of Bohr and Mottelson, that laid the foundations for describing collective structure in nuclei, is documented in their Nobel lectures and that of Rainwater which were published in *Reviews of Modern Physics*: Bohr A. (1976), *Rev. Mod. Phys.* **48**, 365; Mottelson B.R. (1976), *Rev. Mod. Phys.* **48**, 375; Rainwater J. (1976), *Rev. Mod. Phys.* **48**, 385.

[19]Bohr N. (1936), *Nature* **137**, 344.

[20]Bohr N. and Kalckar F. (1937), *Mat. Fys. Medd. Dan. Vid. Selsk* **14** (10).

[21]Bohr N. and Wheeler J.A. (1939), *Phys. Rev.* **56**, 426.

[22]Cf. Section 1.5, Footnote 17 on p. 47, and Footnote 18.

of a deformed nucleus. More generally, the model has the potential to describe anharmonic vibrations and coupled rotational-vibrational motions.

The collective model has been enormously successful to the extent that no-one would doubt its essential validity. Thus, one is faced with the challenge of explaining the coexistence of collective and independent-particle-like behaviour in nuclei. As will be shown later in this book, there is no fundamental incompatibility between the two. The shell model becomes very general with the inclusion of residual interactions between nucleons and, in principle, capable of explaining virtually all of nuclear structure physics; although, sometimes, the explanations can be complicated. Moreover, we shall find (in later chapters) that it is possible to formulate a basic collective model in such a way that it becomes a submodel of the shell model in which only a small number of collective degrees of freedom are active. Our objective here, however, is simply to examine the data, within the framework of the collective model, in an attempt to understand how the shell structure of a nucleus, whether it be a doubly-closed shell, a singly-closed shell, or a doubly-open shell nucleus, influences the kinds of collective states exhibited. To a first approximation, we find that doubly- and singly-closed shell nuclei are spherical and the best candidates for vibrators are located close to closed shells; the most deformed and best rotational nuclei are found in the middle of doubly-open shells.

1.7.1 The harmonic vibrator model

The harmonic vibrator model represents the nucleus as a droplet of nuclear matter with low-energy shape-vibrational degrees of freedom. The simplest and lowest frequency normal mode of an incompressible liquid drop is quadrupole vibration about a spherical equilibrium shape. The Hamiltonian for quantized harmonic quadrupole vibrations is

$$\hat{H}_{\mathrm{HV}} = \sum_{\mu} \hbar w_2 (O^{\dagger}_{2\mu} O^{2\mu} + \tfrac{1}{2}), \quad \mu = 0, \pm 1, \pm 2, \tag{1.27}$$

where $O^{\dagger}_{2\mu}$ and $O^{2\mu}$, respectively, create and annihilate oscillator quanta (phonons) with spin 2 and component of spin μ along the laboratory-fixed z-axis;[23] ω_2 is the angular frequency of the vibrations. These operators have positive parity. They satisfy the usual harmonic oscillator commutation relations

$$[O^{2\mu}, O^{\dagger}_{2\nu}] = \delta_{\mu\nu}. \tag{1.28}$$

The ground state of $\hat{H}_{\mathrm{vib}}$ is the phonon vacuum state, a state which satisfies the

---

[23]We adopt the convention that the $\{O^{\dagger}_{2\mu}\}$ operators transform under rotations as spherical harmonics $\{Y_{2\mu}\}$; in contrast, an annihilation operator $O^{2\mu} := (-1)^{\mu} O_{2,-\mu}$ is the Hermitian adjoint of the creation operator $O^{\dagger}_{2\mu}$ and transforms as the complex conjugate $Y^{*}_{2\mu} = (-1)^{\mu} Y_{2,-\mu}$; cf. Appendix A.6.

equation

$$O^{2\mu}|0\rangle = 0. \tag{1.29}$$

Excited states of n quanta are obtained by operating on the vacuum state with n phonon creation operators. The corresponding energy eigenvalues are

$$E_n = (n + \tfrac{5}{2})\hbar\omega, \quad n = 0, 1, 2, 3, \ldots, \tag{1.30}$$

where the contribution $\frac{5}{2}\hbar\omega$ is the zero-point vibrational energy. In the absence of intrinsic spin, the ground state of the vibrator model is a state of angular momentum[24] $I = 0$, as observed for the ground-state of any doubly-even nucleus. For an $n = 1$ excited state, the angular momentum is $I = 2$, the spin of a single phonon. For $n \geqslant 2$, the angular momentum takes a range of possible values which can be computed by counting the available magnetic substates (M states) as follows. Consider, for example, two identical (spin-2) phonons and recall that identical phonons (which are bosons) can have identical quantum numbers. Labelling the μ quantum numbers for the two phonons by m_1 and m_2 and letting them range over the values $0, \pm 1, \pm 2$, with the restriction $m_1 \geqslant m_2$ to avoid double counting, we determine the allowed combinations of ($M = m_1 + m_2$). Whence, for this example, we infer that $I = 0, 2, 4$.

Since the quadrupole-phonon creation operators $\{O_{2\mu}^\dagger\}$ have positive parity, it follows that all quadrupole-vibrational states of a nucleus with a positive-parity ground state are of positive parity.

If $\mathfrak{M}(E2; \mu)$ denotes the electric quadrupole operator, then an excited quadrupole state $|2\mu\rangle$ is said to be a *collective excitation* if the transition matrix element connecting it to the ground state, $\langle 2\mu|\mathfrak{M}(E2; \mu)|0\rangle$, is large compared with what it would be if the state $|2\mu\rangle$ were generated by simply changing the state of a single nucleon in the ground state.

Rates for E2 gamma-ray transitions between states of the harmonic vibrator model can be predicted if it is assumed that the effective E2 operator of the model is of the form

$$\mathfrak{M}(E2; \mu) = \alpha(O_{2\mu}^\dagger + O_{2\mu}), \tag{1.31}$$

where α is a constant to be fitted to the experimental data. One of the consequences of this assumption is that transitions can only occur between states of phonon number differing by one; i.e., there is a $\Delta n = \pm 1$ selection rule. In particular, the cross-over transitions from the two-phonon 0, 2, 4, triplet to the ground state are forbidden, and quadrupole moments (which are diagonal, i.e., $\Delta n = 0$) are zero. These selection rules are relaxed if one assumes a more general form for the E2 operator; for example, an added term $\beta[O_2^\dagger \times O_2]_\mu^2$ (where the notation indicates a

---

[24]We now use I in place of J to denote total spin/angular momentum. This practice, which is conventional in collective-model contexts, is introduced for the purpose of allowing a distinction to be made between collective and other (e.g., intrinsic) angular momenta.

coupling of the spherical tensors $O_2^\dagger$ and O_2 to a resultant rank-2 spherical tensor) would give rise to non-zero quadrupole moments at the expense of a new model parameter, β.

Measured gamma-ray transition rates are usually expressed in terms of reduced transition rates, the so-called $B(E\lambda)$ values (cf. Appendix, Equations (B.3), (B.4)). The reduced transition rate for an electric λ-pole transition from an initial state $|I_i\rangle$ to a final state $|I_f\rangle$ is defined by the expression

$$B(E\lambda; I_i \to I_f) := \frac{1}{2I_i+1} \sum_{M_i} \sum_{M_f \mu} |\langle I_f M_f | \mathfrak{M}(E2; \mu) | I_i M_i \rangle|^2. \qquad (1.32)$$

It is the squared transition matrix element, averaged over initial states and summed over final states. Thus, with the expression for $\mathfrak{M}(E2; \mu)$ given by Equation (1.31), one obtains a model prediction for the $B(E2)$ value for decay of the first excited 2^+ state to the 0^+ ground state, viz.

$$B(E2; 2_1 \to 0_1) = |\alpha|^2. \qquad (1.33)$$

Similarly, the reduced transition rates for decay of the two-phonon states are given by

$$B(E2; n{=}2 \to n{=}1) = 2B(E2; n{=}1 \to n{=}0). \qquad (1.34)$$

If the surface radius of a vibrating liquid drop is expanded as a sum of spherical harmonics

$$R_{\theta,\varphi}(t) = \sum_{\lambda\mu} C_{\lambda\mu}(t) Y_{\lambda\mu}(\theta, \varphi), \qquad (1.35)$$

then, in the terminology of electrostatics, one describes a vibrational mode in which only the expansion coefficients of a particular λ are non zero as a 2^λ-pole mode. Thus, a $\lambda = 0$ vibration is a *monopole* mode, $\lambda = 1$ is *dipole*, $\lambda = 2$ is *quadrupole*, and so on.

The vibrator model can describe vibrational modes of any multipolarity. However, monopole "breathing mode" vibrations are not expected to occur at low frequencies because nuclear matter is observed to be essentially incompressible. The next multipole mode, the dipole mode, cannot occur because it would correspond to translational motion of the centre of mass and would not be associated with an excitation of the nucleus. Vibrational modes of multipolarity higher than dipole are possible for a liquid drop and some candidates for corresponding vibrational states are observed in nuclei. Indeed, in addition to quadrupole vibrations, octupole vibrational spectra are observed and will be discussed shortly. There is also evidence for hexadecapole and 2^5-pole vibrations.

Note that an excited state can always be regarded as a vibrational excitation. Moreover, it can be regarded as a 2^λ-pole vibrational excitation if it decays by an Eλ-pole transition to the ground state. Thus, for a nucleus with a 0^+ ground state, a one-phonon 2^λ-pole vibrational state should have the spin and parity of the electric multipole operator $\mathfrak{M}(E\lambda; \mu)$; $I = \lambda$ and $\pi = (-1)^\lambda$. However, such a state is only "collective" if its $B(E\lambda)$ transition rate to the ground state is large compared to what it would be if the excited state were generated by rearranging the state of a single nucleon in the ground state. Note also that it is only meaningful to describe a state as "a harmonic vibration" if multi-phonon vibrational states are also observed at corresponding energies.

Unlike a simple charged liquid drop, nuclei also have vibrational modes involving their spin and isospin degrees of freedom. For example, because nuclei are made up of nucleons of two different types, they have dipole vibrational modes in which their protons and neutrons move in antiphase. Such vibrations are described as *isovector modes* as distinct from the *isoscalar modes* in which the neutrons and protons move in phase. Isovector dipole excitations are seen at high energies and are discussed in Section 1.10. An in-depth discussion of the range of possible vibrational models in nuclei will be given in Volume 2.

1.7.2 *The rotor model*

The quantum mechanical rotor has the Hamiltonian

$$\hat{H}_{\rm rot} := \frac{\hbar^2}{2}\left[\frac{\hat{R}_1^2}{\mathfrak{I}_1} + \frac{\hat{R}_2^2}{\mathfrak{I}_2} + \frac{\hat{R}_3^2}{\mathfrak{I}_3}\right] = \frac{\hbar^2}{2}\sum_{i=1}^{3}\frac{\hat{R}_i^2}{\mathfrak{I}_i}, \qquad (1.36)$$

where the $\mathfrak{I}_i$ are moments of inertia relative to a set of body-fixed axes and the R_i are components of rotational angular momentum $\mathbf{R}$ relative to these axes.[25]

A more complete model would include both rotational and intrinsic degrees of freedom and would have a Hamiltonian

$$\hat{H} = \hat{H}_{\rm rot} + \hat{H}_{\rm intr}. \qquad (1.37)$$

However, in spite of the separation of $\hat{H}$ into two parts, there are implicit coupling interactions between the rotational and intrinsic degrees of freedom contained in $\hat{H}_{\rm rot}$. These appear when it is recognised that the nucleus can have intrinsic angular momentum $\mathbf{J}$ in addition to the rotational angular momentum $\mathbf{R}$. In this situation, the total angular momentum is the sum

$$\hat{\mathbf{I}} = \hat{\mathbf{R}} + \hat{\mathbf{J}}. \qquad (1.38)$$

Thus, it is appropriate to write $\hat{R}_i = \hat{I}_i - \hat{J}_i$ and expand $\hat{H}_{\rm rot}$ as

$$\hat{H}_{\rm rot} = \sum_i \frac{\hbar^2}{2\mathfrak{I}_i}(\hat{I}_i - \hat{J}_i)^2 = \sum_i \frac{\hbar^2}{2\mathfrak{I}_i}\hat{I}_i^2 - \sum_i \frac{\hbar^2}{\mathfrak{I}_i}\hat{I}_i\hat{J}_i + \sum_i \frac{\hbar^2}{2\mathfrak{I}_i}\hat{J}_i^2. \qquad (1.39)$$

---

[25] The appearance of $\hbar^2$ in the Hamiltonian is a result of a convention, in nuclear physics, of defining angular momenta in units of $\hbar$. Thus, when we speak of a system as having angular momentum (spin), $\hat{\mathbf{I}}$, $\hat{\mathbf{R}}$ or $\hat{\mathbf{J}}$, we really mean that it has angular momentum (spin) $\hbar\hat{\mathbf{I}}$, $\hbar\hat{\mathbf{R}}$ or $\hbar\hat{\mathbf{J}}$.

The last term in Equation (1.39) acts only on the intrinsic degrees of freedom; it does not produce rotational-intrinsic coupling and it can be incorporated into $\hat{H}_{\text{intr}}$. We therefore write

$$\hat{H} = \hat{H}'_{\text{rot}} - \sum_i \frac{\hbar^2}{\mathfrak{S}_i} \hat{I}_i \hat{J}_i + \hat{H}_{\text{intr}}, \qquad (1.40)$$

where

$$\hat{H}'_{\text{rot}} = \sum_i \frac{\hbar^2}{2\mathfrak{S}_i} \hat{I}_i^2. \qquad (1.41)$$

One now sees two coupling interactions. The term containing $\hat{I}_i \hat{J}_i$ is the so-called *Coriolis interaction*, some manifestations of which are presented in Section 1.8. A *centrifugal coupling* also appears if the moments of inertia are regarded as functions of the intrinsic variables and not simply assigned rigidly-defined values.

In the simplest version of the rotor model, the moments of inertia are treated as constants and the Coriolis coupling is ignored. It is not possible to justify these approximations by any *a priori* theoretical argument. One can only attempt to understand why they should be valid in situations where well-defined rotational bands are seen and the rotor model works. In such situations, it appears that the nuclei have relatively rigid intrinsic structures that are not easily perturbed by the Coriolis and centrifugal interactions. More importantly, when the rotor model applies, rotational energy differences are small compared to the complementary intrinsic excitation energies; in classical terms, the rotational motions (in low angular momentum states) are *adiabatic*, i.e., slow, compared to the intrinsic motions to which they would couple. Thus, the neglect of Coriolis and centrifugal coupling interactions is known as an *adiabatic approximation*.

In general, the rotor Hamiltonian, $\hat{H}'_{\text{rot}}$, does not have analytical solutions and its energy levels are determined by diagonalisation. An exception is when the rotor has two equal moments of inertia, e.g.,

$$\mathfrak{S}_1 = \mathfrak{S}_2 = \mathfrak{S} \neq \mathfrak{S}_3. \qquad (1.42)$$

Such a rotor is described as a *symmetric top* or as being *axially symmetric*. The Hamiltonian for a symmetric top can be written

$$\hat{H}'_{\text{rot}} = \frac{\hbar^2}{2} \left[\frac{\hat{\mathbf{I}}^2}{\mathfrak{S}} + \left(\frac{1}{\mathfrak{S}_3} - \frac{1}{\mathfrak{S}} \right) \hat{I}_3^2 \right], \qquad (1.43)$$

and its energy eigenvalues are given by

$$E'_{KI} = \frac{\hbar^2}{2} \left[\frac{I(I+1)}{\mathfrak{S}} + \left(\frac{1}{\mathfrak{S}_3} - \frac{1}{\mathfrak{S}} \right) K^2 \right]. \qquad (1.44)$$

The corresponding eigenvectors of $\hat{H}'_{\text{rot}}$, $|KIM\rangle$, satisfy the equations

$$\hat{\mathbf{I}}^2|KIM\rangle = I(I+1)|KIM\rangle, \tag{1.45}$$

$$\hat{I}_z|KIM\rangle = M|KIM\rangle, \tag{1.46}$$

$$\hat{I}_3|KIM\rangle = K|KIM\rangle, \tag{1.47}$$

where $\hat{I}_z$ and $\hat{I}_3$ are the components of $\hat{\mathbf{I}}$ along the laboratory-fixed z-axis and the the body-fixed 3-axis, respectively.[26] The situation is illustrated in Figure 1.46.

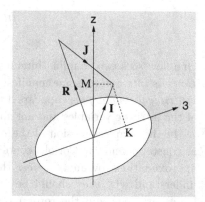

Figure 1.46: Relationships between the total angular momentum, **I**, the intrinsic angular momentum, **J**, the rotational angular momentum, **R**, and the components of **I** along the laboratory-fixed z axis, M, and the symmetry 3 axis in the body-fixed frame, K.

Equation (1.44) defines a band of states for each value of K. For even-even nuclei, the ground-state spin is zero. Hence $K = 0$ for this band. If the (intrinsic) energy of the ground state is E_0, then the energies of the ground-state rotational band are given in the rotor model by

$$E_I = E_0 + \frac{\hbar^2}{2\Im}I(I+1). \tag{1.48}$$

Excited bands can have other K values and moments of inertia that are not necessarily equal to those of the ground-state band. Thus, in general,

$$E_{\alpha K I} = E_{\alpha K} + \frac{\hbar^2}{2\Im_\alpha}I(I+1). \tag{1.49}$$

The moments of inertia $\Im_\alpha$ and the band-head energies $E_{\alpha K}$ are parameters of the model to be fitted to experimental data.

A distinctive characteristic of $K = 0$ bands, in doubly-even nuclei, is the spin sequence $I = 0, 2, 4, 6, 8, \ldots$; i.e., the states of odd I are missing. A similar feature is observed for rigid molecules which are both axially symmetric and have a plane of reflection symmetry perpendicular to the symmetry axis (homonuclear diatomic

---

[26]It might appear that it is not possible for M and K, which denote components of the angular momentum relative to different axes, to be simultaneously good quantum numbers. As will be shown when the rotor model is developed in more detail in Volume 2, M and K are, in fact, eigenvalues of commuting operators. This is possible because the intrinsic symmetry axis of the rotor is not a fixed axis; it rotates with the rotor.

molecules and symmetric top molecules will be discussed in Volume 2). Thus, within the framework of the rotor model, one interprets the even-I spin sequence as implying a reflection symmetry of the nuclear rotor in a plane perpendicular to its symmetry axis. An implication of such a symmetry is that the states $|KIM\rangle$ and $|-K, IM\rangle$ appear in linear combination

$$|KIM\rangle + \varepsilon(-1)^{I+K}|-K, IM\rangle, \tag{1.50}$$

where $\varepsilon = \pm 1$ according as the intrinsic wave function is symmetric or antisymmetric under rotation through an angle π about an axis perpendicular to the symmetry axis. One sees that, for a symmetric ($\varepsilon = 1$) combination, the states of odd I vanish when $K = 0$; only even values of I survive. For $\varepsilon = -1$, a $K = 0$ band has an odd-I only spin sequence; such bands are also seen.

In addition to their characteristic $I(I+1)$ spectra, some distinguishing features of a rotational nucleus are the huge values of the quadrupole moments of its $I \geqslant 1$ states. These are given in the rotor model by the product of an intrinsic quadrupole moment, $\bar{Q}_0(\alpha K)$, characteristic of the rotor band, and a geometric factor which depends on the angular momentum of the particular state;

$$Q(\alpha KI) = \frac{3K^2 - I(I+1)}{(I+1)(2I+3)} e\bar{Q}_0(\alpha K), \tag{1.51}$$

where α distinguishes bands with the same K. (Note, that as I increases for a given K, $Q(\alpha KI)$ will change sign for $K > 1/2$.) The intrinsic quadrupole moment of a rotational state can also be determined from the E2 transition rates between the states of a rotor band, which have B(E2) values given by

$$B(E2; \alpha KI_i \to \alpha KI_f) = \frac{5}{16\pi}(I_iK, 20|I_fK)^2 e^2 |\bar{Q}_0(\alpha K)|^2, \tag{1.52}$$

where $(I_iK, 20|I_fK)$ is a Clebsch-Gordan coefficient.

1.7.3 Low-energy vibrational states in doubly-even nuclei

Figure 1.47 shows states at low energy in the singly-closed shell nucleus ^{118}Sn that are strongly excited in inelastic electron scattering. From the strength of excitation, it is deduced that the first excited state at 1.23 MeV is a collective quadrupole excitation and the excited state at 2.33 MeV is a collective octupole excitation. The $I^\pi = 4^+$ states at 2.28, 2.49, and 2.73 MeV indicate some hexadecapole collectivity, but it is fragmented.

To infer the characters of collective excitations, such as seen in Figure 1.47, it is necessary to consider them within the context of a larger pattern of states. Figure 1.48(a) shows the states in ^{114,116,118}Sn that have strong electric quadrupole transitions to the first 2^+ states.

The pattern is approximately that of a harmonic vibrator in all three nuclei. Recall (Section 1.7.1) that, for harmonic vibrations, one expects a degenerate two-

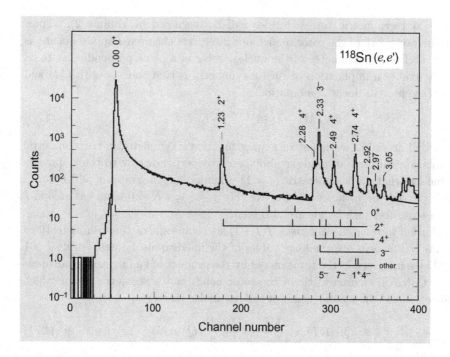

Figure 1.47: Inelastic electron scattering from ^{118}Sn. The data are for an incident energy of 252 MeV and a scattering angle of 68° with respect to the incident beam. The peaks corresponding to excited states reveal those states which are connected to the ground state by the largest electromagnetic matrix elements. The $I^\pi = 2^+$ state at 1.23 MeV and the $I^\pi = 3^-$ state at 2.33 MeV are deduced to be excited by strong electric quadrupole and electric octupole processes, respectively. The location of other excited states in ^{118}Sn is indicated. The figure is discussed further in the text. (The figure is taken from Peterson R.J. *et al.* (1991), *Phys. Rev.* **C44**, 136; spins, parities, and excitation energies (in MeV) are taken from *Nuclear Data Sheets*.)

phonon triplet with $I^\pi = 0^+, 2^+, 4^+$ at twice the energy of the first excited 2^+ state with $B(E2; n = 2 \to n = 1) = 2 \times B(E2; n = 1 \to n = 0)$. Evidently, the two-phonon E2 strength is fragmented. This is explained by the fact that the two-phonon quadrupole vibrational excitations in the tin isotopes couple to neutron one-broken pair states of which there are many above 2 MeV (cf. Figure 1.29). But, in spite of this coupling, the lowest-energy degree of freedom in the even-mass tin isotopes is seen to be essentially quadrupole vibrational.

The cadmium ($Z = 48$) and palladium ($Z = 46$) isotopes in the neighbourhood of ^{114,116,118}Sn, specifically ^{110,112,114}Cd and ^{106,108,110}Pd, are shown in Figures 1.48(b) and (c). These isotopes show less fragmentation of the two-phonon quadrupole strength than do the ($Z = 50$ closed-shell) Sn isotopes. The reason is simple; the energy of a quadrupole phonon becomes lower when moving away from $Z = 50$; thus, the two-phonon states lie lower relative to other degrees of freedom and, consequently, there is less coupling.

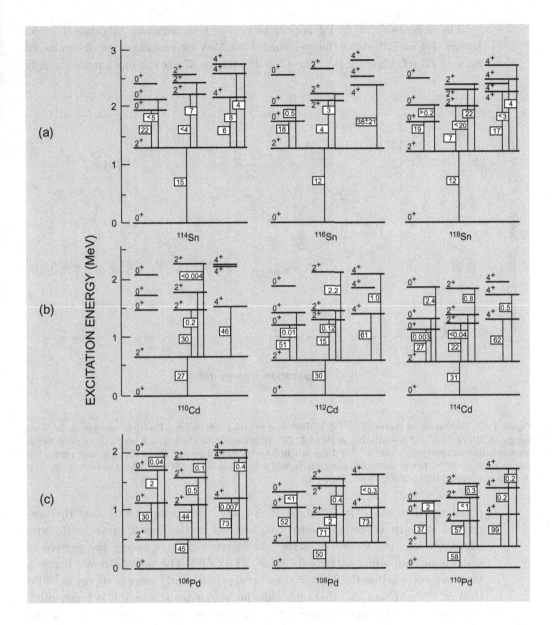

Figure 1.48: Examples of nuclei that have excited states with strong electric quadrupole transitions to the first 2^+ states, including ^{118}Sn (cf. Figure 1.47) and ^{110}Pd (cf. Figures 1.49 and 1.50). The numbers in the boxes are $B(\text{E2})$ values in W.u. The uncertainty in the $B(\text{E2}; 4_1^+ \rightarrow 2_1^+)$ value in ^{116}Sn is shown because it is unusually large. (The data are taken from *Nuclear Data Sheets*; Svensson L.E. *et al.* (1995), *Nucl. Phys.* **A584**, 547; Svensson L.E. (1989), Ph.D. thesis, Univ. of Uppsala – ^{106,108,110}Pd; Corminbouef F. *et al.* (2000), *Phys. Rev.* **C63**, 014305 – ^{110}Cd; Garrett P.E. *et al.* (2007), *Phys. Rev.* **C75**, 054310 – ^{112}Cd; Jonsson N.-G. *et al.* (1981), *Nucl. Phys.* **A371**, 333 – ^{114}Sn.)

The isotopes ^{106,108,110}Pd appear to be the best harmonic vibrational nuclei known. Figure 1.49 shows the spectrum of 346 MeV electrons inelastically scattered from ^{110}Pd (cf. Figure 1.47). The $I^\pi = 2^+$ state at 374 keV is much more strongly

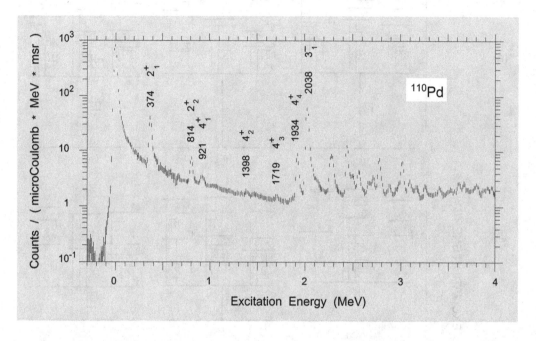

Figure 1.49: Excitation of states of ^{110}Pd by inelastic electron scattering. The data are for an incident energy of 345.72 MeV and a scattering angle of 32.72° with respect to the incident beam. Scattering peaks corresponding to low-lying states in ^{110}Pd are labelled by excitation energy (in keV), spin, and parity (cf. Figure 1.48(c)). (The figure was made available to us by C.W. de Jager, NIKHEF-K, Amsterdam; cf. also Wesseling J. *et al.* (1991), *Nucl. Phys.* **A535**, 285.)

populated than the $I^\pi = 2^+$ state at 814 keV. This demonstrates that the one-phonon strength is predominantly concentrated in the first 2^+ state with little admixture into the second 2^+ state. At higher excitation energy the pattern of states populated is qualitatively the same as for ^{118}Sn. The main difference between the two spectra is that the 2^+_1 excitation occurs significantly lower in energy in ^{110}Pd than in ^{118}Sn. (Note also that although the hexadecapole strength is fragmented, it is concentrated in states above 1900 keV and very little strength is seen in the 4^+ states at 921 and 1398 keV.)

Multi-phonon structure in quadrupole vibrational nuclei is revealed by the spectra of gamma rays emitted following heavy-ion Coulomb excitation. This is shown for ^{108,110}Pd in Figure 1.50. Heavy-ion Coulomb excitation induces multi-step excitation through states connected sequentially to the ground state by large electromagnetic matrix elements. In addition to the two-phonon quadrupole vibrational triplet, the three-phonon quintuplet with $I^\pi = 0^+, 2^+, 3^+, 4^+, 6^+$ is seen in Figure

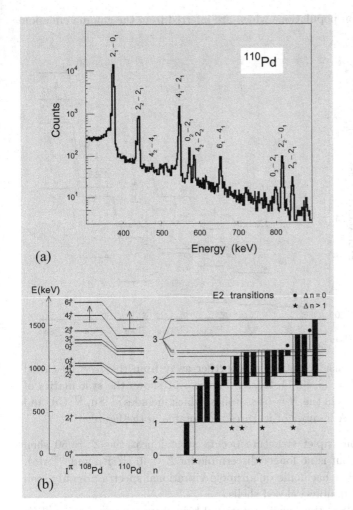

(a)

(b)

Figure 1.50: (a) Spectra of gamma rays emitted following the Coulomb excitation of ^{110}Pd with a beam of ^{58}Ni ions. The transitions are labelled by the spins of the initial and final states; a subscript, e.g., 2_1, indicates the first $I = 2$ state, etc. (cf. Figure 1.48c). (The figure is taken from Svensson L.E. (1989), Ph.D. thesis, Univ. of Uppsala). (b) The low-lying states in ^{108,110}Pd and the E2 reduced transition strengths deduced from Coulomb excitation with beams of ^{16}O, ^{58}Ni, and ^{208}Pb. Thick vertical bars indicate strong E2 transitions, thin vertical lines indicate weak E2 transitions. Levels are grouped into "phonon" multiplets, labelled by n. In a harmonic vibrator model, the transitions labelled by stars would be forbidden. Transitions labelled by solid circles are forbidden in the simplest harmonic vibrator models but can be accommodated using a modified E2 operator (this will be discussed in Volume 2). The horizontal bars with vertical arrows indicate excitation energies in ^{108,110}Pd above which levels have been omitted. (The data are from Svensson L.E. et al. (1995), Nucl. Phys. **A584**, 547; Svensson L.E. (1989), Ph.D. thesis, Univ. of Uppsala; and Nuclear Data Sheets.)

1.50. The pattern of equi-spaced multiplets of degenerate multi-phonon states expected for a harmonic quadrupole vibrator deteriorates progressively with increasing excitation energy in these nuclei. However, multi-phonon ($\Delta n > 1$) transitions are strongly hindered, indicating that even at the three-phonon excitation level, the states retain some characteristics of a harmonic quadrupole vibrator.

A subtle point may be noted in Figure 1.50: the $2_2 \rightarrow 0_1$, $\Delta n = 2$ transition is clearly seen in the gamma-ray spectrum at 814 keV (as also is the $2_3 \rightarrow 2_1$, $\Delta n = 2$ transition at 840 keV). This is because the gamma-ray spectrum reflects the spontaneously radiating de-excitation paths, not the Coulomb-field induced excitation paths, and spontaneous emission of electric quadrupole radiation is enhanced by a phase space factor $(\Delta E)^5$ (cf. Appendix, Equation (B.4)). For this reason a number of strongly collective low-energy transitions are not seen in the gamma-ray spectrum; their ΔE values are too small. (The strengths of such transitions are deduced from Coulomb excitation populations which are inferred from the gamma emission strengths.)

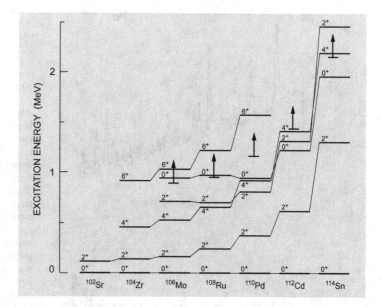

Figure 1.51: Systematics of the first few excited states in the $N = 64$ isotones. The vertical arrows indicate the energies above which there are excited states known but which are omitted from the figure. (The data are from *Nuclear Data Sheets.*)

It is instructive to consider nuclei even further away from the $Z = 50$ closed shell in the sequence $Z = 48, 46, 44, 42, \ldots$. Figure 1.51 shows the systematics of the first few excited states in the $N = 64$ isotones, which includes ^{114}Sn, ^{112}Cd, and ^{110}Pd (cf. Figure 1.48). A number of features are notable in the figure:

(i) The two-phonon triplet structure occurs at and near the $Z = 50$ shell closure (^{114}Sn) but is no longer discernible for $Z < 46$ (^{108}Ru and ^{106}Mo). This suggests that harmonic quadrupole vibrational spectra are only seen near (singly- or doubly-) closed shells.

(ii) Far from the ^{114}Sn closed shell, rotational behaviour appears, i.e., in ^{104}Zr and probably in ^{102}Sr. This is anticipated from the large quadrupole moments and electric quadrupole transition rates observed far from closed shells (cf. Figures 1.44 and 1.38).

(iii) The trend in $E(2_1^+)$, $E(4_1^+)$, and $E(6_1^+)$ is smooth (cf. Figures 1.36) and suggests that the change from quadrupole vibrations to rotations (at least for the $N = 64$ isotones) is smooth.

Table 1.1 presents the best known examples of low-energy collective quadrupole vibrational behaviour. Evidently none of the nuclei tabulated exhibit all the criteria of a good harmonic quadrupole vibrator.

Other multipole vibrational modes, besides quadrupole, are also seen at low energy (this has been noted already in Figures 1.47 and 1.49). Figure 1.52 shows low-lying states in ^{208}Pb populated by inelastic deuteron scattering. The first excited

Table 1.1: Some nuclei which exhibit characteristics of near-harmonic quadrupole vibrational behaviour. The boxes highlight the best cases of $R_4 := (E(4_1^+) - E(0_1^+))/(E(2_1^+) - E(0_1^+)) \approx 2.00$, near-degenerate two-phonon triplets, and very small $B(E2; 2_2^+ \to 0_1^+)/B(E2; 2_2^+ \to 2_1^+)$ ratios (denoted by $B_{2_2 0_1}/B_{2_2 2_1}$). (The notation 1.2^2, for example, is used for 1.2 ± 0.2.) (The data are taken from *Nuclear Data Sheets*.)

Isotope	R_4	$E(0_2^+)$	$E(2_2^+)$	$E(4_1^+)$	$\dfrac{B_{2_2 0_1}}{B_{2_2 2_1}}$	$\dfrac{B_{4_1 2_1}}{B_{2_1 0_1}}$	$\dfrac{B_{2_2 2_1}}{B_{2_1 0_1}}$	$\dfrac{B_{0_2 2_1}}{B_{2_1 0_1}}$
^{62}Ni	1.99	2048	2302	2336	0.045	–	–	–
^{80}Kr	2.33	1321	1256	1436	0.012	1.2^2	0.7^1	–
^{82}Sr	2.32	1311	1176	1328	≈ 0.006	2.3^5	–	–
^{98}Ru	2.14	1322	1414	1398	0.022	0.4^1	1.4^5	–
^{104}Pd	2.38	1334	1342	1324	0.055	1.4^2	0.6^1	0.4^1
^{106}Pd	2.40	1134	1128	1229	0.027	1.6^1	1.0^1	0.8^1
^{108}Pd	2.42	1053	931	1048	0.011	1.5^2	1.4^1	1.1^1
^{110}Pd	2.46	947	814	921	0.014	1.7^2	1.0^2	0.6^1
^{118}Te	1.99	957	1151	1206	>0.006	–	–	–
^{120}Te	2.07	1103	1201	1162	0.026	–	–	–
^{122}Te	2.09	1357	1257	1181	0.011	–	–	–

state, at 2.62 MeV, has $I^\pi = 3^-$ and the second excited state, at 3.20 MeV, has $I^\pi = 5^-$. These states are interpreted as collective octupole and collective 2^5-pole states, respectively. (It is instructive to compare and contrast the results of Figure 1.52 with those of Figure 1.25.)

Systematic study of low-energy excitations in nuclei using inelastic scattering of electrons and light ions reveals that octupole vibrations are the best-defined multipole mode after quadrupole vibrations. Figure 1.53 illustrates the systematics of the excitation energies of the lowest-lying collective 3^- states in selected regions of the mass surface (the nuclei excluded from Figure 1.53 are those with strongly-deformed ground states with $N > 90$, $Z < 78$ (Pt) and $Z > 92$ (U). Figure 1.54 illustrates the systematics of octupole transition strength to 3_1^- states for all known cases.

A number of features are notable in Figures 1.53 and 1.54

(i) Octupole vibrations lie much higher in energy than quadrupole vibrations (except in some doubly-closed-shell nuclei, such as ^{208}Pb). (The two regions of low-lying 3_1^- states evident in Figure 1.53 are discussed in the following section because they are regions of large ground-state deformation.)

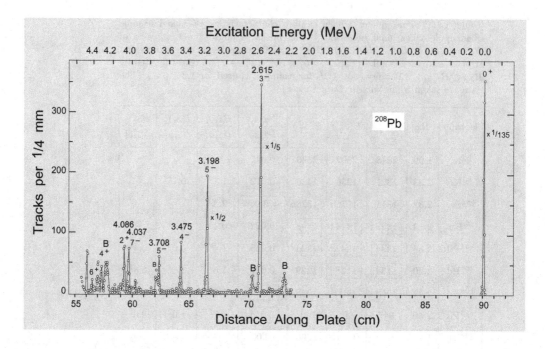

Figure 1.52: Inelastic deuteron scattering from ^{208}Pb. The data are for an incident energy of 13.1 MeV and a scattering angle of 150° with respect to the incident beam. Deuteron lines are labelled by the spins and parities of the corresponding levels. Lines marked with a B are "background" events resulting from target contaminants. Note the scale reduction factors. (The figure is from Ungrin J. *et al.* (1971), *Mat. Fys. Medd. Dan. Vid. Selsk.* **38** (8); the energies and spin-parities are from *Nuclear Data Sheets*.)

(ii) Electric octupole transitions are relatively much weaker, when expressed in Weisskopf (single-particle) units, than electric quadrupole transitions.

(iii) The regions of smallest $E(3_1^-)$ are not in mid open-shell regions.

(iv) The nuclei with the largest $B(E3)$ values are not in mid open-shell regions. Indeed, they are close to closed shells.

Octupole vibrations differ from quadrupole vibrations because they have different underlying microscopic structures and negative parity. In particular, to produce a negative-parity excitation, it is necessary to promote a particle from one shell to an adjacent shell of opposite parity; i.e., create a so-called *particle-hole* excitation. Collective octupole vibrations are interpreted as coherent combinations of such particle-hole excitations. This provides a partial explanation of features (iii) and (iv). From Figures 1.47, 1.49, and 1.52, we can arrive at a number of inferences regarding low-energy vibrational modes in nuclei:

(i) The important vibrational modes are those with $I^\pi = 2^+$, 3^-, 4^+ and 5^-.

(ii) Vibrational states increase in excitation energy and decrease in strength with increasing multipole order.

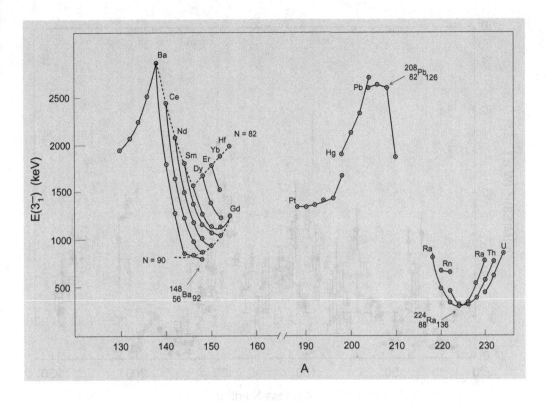

Figure 1.53: Systematics of the energies of the first excited states with spin-parity 3^- for $56 \leq Z \leq 64$, $78 \leq Z \leq 82$, and $88 \leq Z \leq 92$. Isotopes are joined by solid lines and isotones by dashed lines. (The data are taken from Kibédi T. and Spear R.H. (2002), *At. Data Nucl. Data Tables* **80**, 35.)

Multi-phonon excitations are revisited in Section 2.3.1. The available information suggests that (with the exception of some two-phonon quadrupole states) multi-phonon states generally lie in regions of relatively large energy-level density where coupling to other degrees of freedom and fragmentation of the collective strength is almost inevitable. Multi-phonon excitations will be discussed further in Volume 2.

With only sparse data available on multi-phonon excitations, it is appropriate to question the usefulness of the vibrator model. From a conceptual point of view it is enormously valuable. Its most important contribution to the understanding of nuclei is to explain the character of the low-energy 2^+ and 3^- excited states of many doubly-even nuclei. These states appear systematically with excitation energies that vary smoothly with mass number, and have very large E2 and E3 transition rates to the ground state. The latter is a clear indication that they are "collective excitations". Moreover, the absence of large (diagonal) quadrupole moments for the $I \neq 0$ states of nuclei near single closed-shells (cf. Figure 1.37), implies that these nuclei are not significantly deformed. The vibrator model then

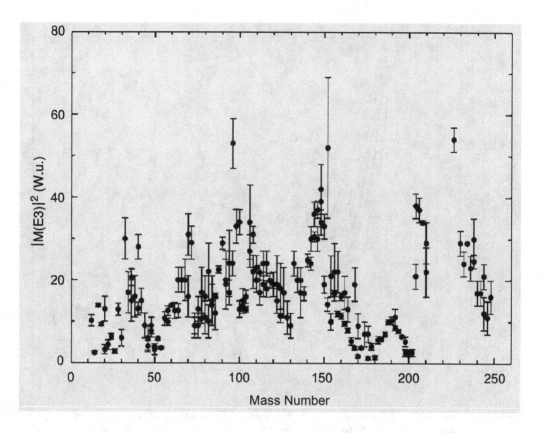

Figure 1.54: Systematics of electric octupole transition strength $|M(E3)|^2$ between the ground state and first excited 3^- state in doubly-even nuclei. The $|M(E3)|^2$ values are in Weisskopf units, W.u. (see Appendix, Equation (B.11)). (The figure is taken from Kibédi T. and Spear R.H. (2002), *At. Data Nucl. Data Tables* **80**, 35.)

provides an interpretation of their character. The vibrator model is related to the rotor model which ultimately helps understand how, with the progression to the middle of doubly-open shells, nuclei become less rigidly spherical, they vibrate more slowly and eventually become unstable against assuming a deformed equilibrium shape. Thus, the model paves the way for the eventual understanding of low-energy collective states of all nuclei in terms of the correlated many-nucleon shell model.

The fact that the vibrator model enables one to interpret the angular distributions of, e.g., inelastically scattered electrons and protons that leave the nucleus in vibrational states is particularly important. Such experiments not only support the essential validity of the vibrational concept, they also measure the so-called *transition densities* and reveal the ways in which the nuclear density vibrates. Examples of such transition densities will be given in Volume 2.

The absence of many well-defined multi-phonon vibrational states implies that vibrational motions are not very harmonic and/or that, at the relatively high energies of these states, there are many other states of the nucleus, involving other degrees of freedom, that mix with the multi-phonon vibrational states and cause the latter to become fragmented. This is an experimental fact that must be taken into account in the development of a more detailed model.

1.7.4 *Rotational states in doubly-even nuclei*

Figure 1.55 shows spectra of gamma rays emitted by ^{242}Pu and ^{244}Pu following the excitation of these nuclei by Coulomb interaction with a beam of ^{208}Pb ions. The spectra are characteristic of rotational bands. The excitation energies of states of the band, inferred from the gamma rays, increase approximately with $I(I+1)$, where I is the angular momentum. The quadratic dependence on I means that, in the decay of a state of angular momentum I to one of angular momentum $I - 2$ the energy difference ΔE, and hence the energy of the emitted gamma ray, is proportional to I; in other words, the cascading gamma rays occur, to a first approximation, at equally spaced energy intervals. Thus, the more or less equally spaced gamma ray peaks, seen in the upper part of the figure, are immediate indicators of rotational spectra. The fit to the rotor model is not perfect because as one sees, the gamma-ray energies tend to become closer to one another with increasing energy (and even to decrease in value in ^{244}Pu at around $I \approx 22$: a feature called *backbending* which will be discussed in Volume 2). Thus, a more accurate description of rotational energy levels is given by an expression

$$E_I = E_0 + A_I I(I+1), \tag{1.53}$$

where A_I decreases slowly with I. However, this does not invalidate the rotor model interpretation of the states. It simply implies that the moment of inertia for the rotor band increases slowly with angular momentum as it might, for example, if there were centrifugal stretching; we return to this question below.

It is observed in Figure 1.51 that the spectra of the $N = 64$ isotones reveal a continuous evolution from a vibrational spectrum for ^{114}Sn (at $Z = 50$) to a rotational spectrum for ^{104}Zr (at $Z = 40$). Such smooth transitions can be seen in many isotopic and isotonic sequences. A convenient way to characterise the extent to which a nucleus is rotational is by the ratio $R_4 = (E_4 - E_0)/(E_2 - E_0) \equiv E(4_1^+)/E(2_1^+)$. Figure 1.56 shows values of R_4 across the open shell region defined by $50 < Z < 82$, $82 < N < 126$. For a number of nuclei in this region, the value of R_4 is seen to approach the rigid-rotor value of 3.333. Only for a few nuclei does it approach the harmonic vibrator ratio of 2.0.

Other measures of the rigidity of a rotor are given by the ratios of energy differences $r_4 = (E_4 - E_2)/(E_2 - E_0)$ and $r_6 = (E_6 - E_4)/(E_2 - E_0)$. For a rigid rotor these ratios have the values $2\frac{1}{3}$ and $3\frac{2}{3}$ whereas for a rotor that is subject to centrifugal stretching, for example, the ratios take smaller values. Figure 1.57

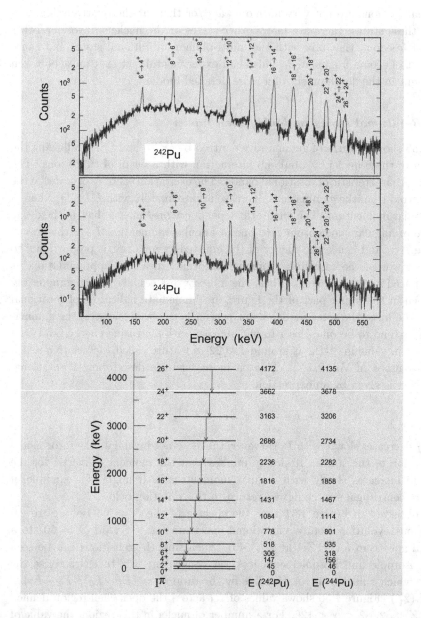

Figure 1.55: Gamma-ray spectra from the Coulomb excitation of ^{242,244}Pu with a beam of ^{208}Pb ions. The energy levels of the decay schemes corresponding to the observed E2 transitions are also shown and are good examples of rotational bands. The absence of gamma-ray peaks corresponding to the transitions $2^+ \rightarrow 0^+$ and $4^+ \rightarrow 2^+$ is a consequence of experimental conditions and need not concern us. The irregularities above spin 20 are discussed in Volume 2. (The data and spectra are taken from Spreng W. *et al.* (1983), *Phys. Rev. Lett.* **51**, 1522.)

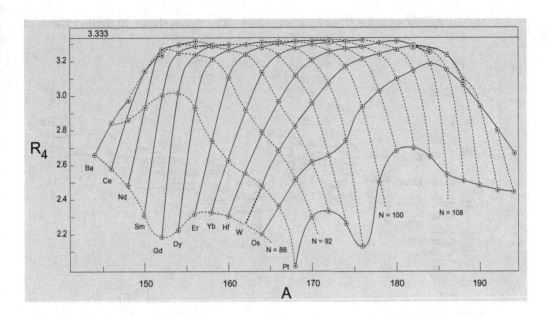

Figure 1.56: The ratio $R_4 = E(4_1^+)/E(2_1^+)$ for selected doubly-even nuclei with $144 \leq A \leq 194$. Isotopes are joined by solid lines and isotones by dashed lines. The rotational limit, $R_4 = 3.333$ is indicated. (The plot is based on data taken from *Nuclear Data Sheets*.)

shows some of the most rotational nuclei known on a plot of r_4 versus r_6. A typical

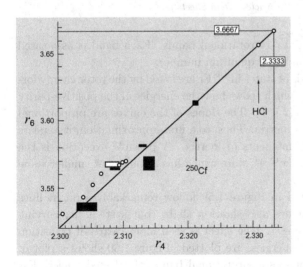

Figure 1.57: Some of the most rotational nuclei known, shown on a plot of r_4 versus r_6 (see text). The "boxes" indicate uncertainties in the experimental quantities. The circles imply that the uncertainties are too small to be shown. The hydrogen chloride molecule is shown for comparison. The diagonal line corresponds to the rotational energy relationship $E_I = E_0 + AI(I+1) + BI^2(I+1)^2$, which will be discussed in Volume 2. The rigid rotor limits of $r_4 = 2.333$ and $r_6 = 3.667$ are shown. (The plot is based on data taken from *Nuclear Data Sheets* and corresponds to the nuclei: ^{250}Cf, ^{244,246}Cm, ^{240}Pu, ^{164,166}Dy, ^{172,174,176}Yb, ^{168}Er, ^{180}Hf, and ^{236,238}U.)

molecule is shown for comparison. One sees that nuclei are not as rigid as molecules.

The rigid-rotor model is successful at describing excited rotational bands, as well as ground-state bands, of deformed nuclei. An extensive system of rotational bands can be seen for ^{168}Er in Figure 1.58. The figure shows 70 excited states of

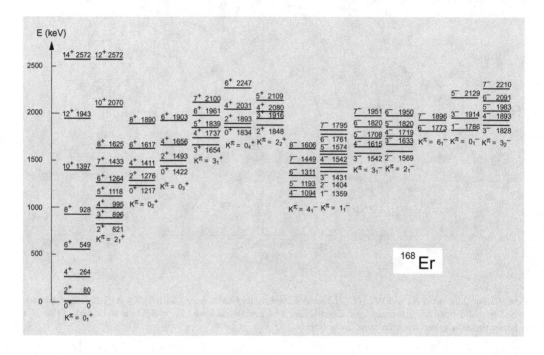

Figure 1.58: The low-lying states of ^{168}Er arranged into rotational bands. The positive-parity bands are shown also in Figure 1.59. (The data are taken from *Nuclear Data Sheets*.)

^{168}Er classified unambiguously into 14 rotational bands. Each band is associated with a different intrinsic state and a K quantum number.

The extent to which the bands of states in ^{168}Er are fitted by the rotor energy formula can be seen in Figure 1.59, which shows how the energies of the positive-parity states of Figure 1.58 vary with $I(I+1)$. The slopes of the curves are proportional to the inverses of the moments of inertia. Thus, to a first approximation, it appears that all the bands have similar moments of inertia. A possible exception is the $K^\pi = 0^+$ band built on the 1217 keV 0^+ state which has a somewhat smaller value for $\hbar^2/2\Im$, i.e., a shallower slope.

Although the energies plotted in Figure 1.59 follow remarkably straight lines as functions of $I(I+1)$, the figure also shows a slight, but systematic, curving to shallower slopes with increasing I. The effect is seen more clearly if transition energies, rather than excitation energies, are plotted. Figure 1.60 shows a plot of $(4I-2)/\Delta E_{I,I-2}$ vs. I for ground-state rotational bands of selected nuclei. For rigid rotors these bands would appear as horizontal lines. The non-rigidity or non-adiabaticity of these nuclei is dramatically illustrated. Whatever is changing with increasing angular momentum is changing smoothly.

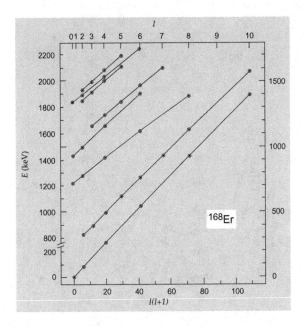

Figure 1.59: Plots of the excitation energies of the low-lying positive-parity states of ^{168}Er (cf. Figure 1.58) versus $I(I+1)$, cf. Equation (1.53) to reveal rotational behaviour. Note the left and right energy scales, with the left-hand scale applying to all but the ground-state band.

It is tempting to suppose that nuclei, being liquid-drop like, should stretch under increasing angular momentum. Indeed, such behaviour would explain the upward curvature of the plots in Figure 1.60. Remarkably, shape changes are not indicated by quadrupole moment data and electric quadrupole transition data. Such data are presented in Figure 1.61 for the same nuclei as shown in Figure 1.60. Within experimental errors (which are quite large because the measurements are not easy) the data indicate that the quadrupole moments and E2 transition rates are consistent with a rotor of constant intrinsic quadrupole moment, i.e., the parameter $\bar{Q}_0$ (cf. Equations (1.51) and (1.52)), for each nucleus is a constant. Another possibility is that the nuclear fluid flows are those of a superfluid and that the degree of superfluidity decreases with increasing angular momentum due to the Coriolis interaction.[27] Whatever is changing, the small energy differences between the states of a rotational band are much more sensitive to minor changes in the intrinsic structure of the states than are their quadrupole moments.

Some insight into the internal dynamics of a rotating nucleus can be gained by considering empirical moments of inertia compared to classical rigid-body estimates. The classical rigid-body moment of inertia for a nucleus of mass A is $\frac{2}{3}MA^2\langle r^2\rangle$, where M is the mass of a nucleon and $\langle r^2\rangle$ is the mean-square radius of the deformed nucleus. Using a typical value of $\langle r^2\rangle$ for a strongly deformed nucleus, one estimates

---

[27]The similarity between the force on a charged particle in a magnetic field and the Coriolis force on a particle in a rotating frame of reference was noticed many years ago and led Mottelson B.R. and Valatin J.G. (1960), *Phys. Rev. Lett.* **5**, 511, to predict that a breakup of nucleon pairs should occur in rotational nuclei at high rotational angular momenta similar to the destruction of superconductivity in the Meissner effect.

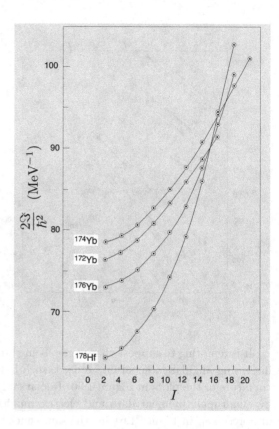

Figure 1.60: The quantity $2\Im/\hbar^2 = (4I - 2)/\Delta E_{I,I-2}$ plotted against I, where $E_I = (\hbar^2/2\Im)I(I+1)$, $\Delta E_{I,I-2} = E_I - E_{I-2}$, shown for ^{172,174,176}Yb and ^{178}Hf to illustrate variations of moments of inertia, $\Im$, as a function of spin. (The data are taken from *Nuclear Data Sheets*.)

that $\hbar^2/2\Im \approx 150$ MeV^{-1}. The empirical values shown in Figure 1.60 are smaller by a factor of two. This could be interpreted to mean that not all of the nucleus participates in the rotational motion. Alternatively, it could mean that rotational flows are more like those of a quantum superfluid than a rigid body. There are many possibilities. Moreover, the situation is sufficiently complex that, to date, one has little indication from experiment as to the nature of rotational flows in nuclei. The only indication from the behaviour of moments of inertia as a function of the angular momentum is that rotational flows appear to become more rigid-body-like with increasing angular momentum. These issues will be discussed in Volume 2.

Apart from moments of inertia and intrinsic quadrupole moments, the properties of intrinsic states of rotational bands are not given by the simple rotor model. However, some inferences about the symmetries of intrinsic states can be gained from rotor model analyses. The most pervasive symmetry, namely axial symmetry, is inferred from the goodness of the K quantum number. The goodness of the K quantum number is assessed by looking at K-selection rules. According to the rotor model, radiative electromagnetic transitions cannot occur between rotational states of a symmetric top if the change, ΔK, in the K quantum numbers exceeds the multipolarity of the electromagnetic transition (cf. Section 2.3.3). This "K forbid-

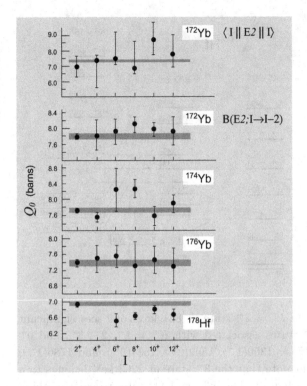

Figure 1.61: Plots of intrinsic quadrupole moments $\bar{Q}_0$ ($= \bar{Q}_0(\alpha K)$) versus spin using diagonal (Equation (1.51)) and off-diagonal (Equation (1.52)) matrix elements for the ground-state rotational bands in selected rare-earth nuclei (cf. Figure 1.60). The shaded bands are the $\bar{Q}_0$ values for a rigid rotor, fitted to the observed $B(E2; 2^+ \rightarrow 0^+)$ values for the respective nuclei. (The plotted values are deduced from matrix element data taken from Fahlander C. *et al.* (1992), *Nucl. Phys.* **A541**, 157, and $B(E2)$ data are taken from *Nuclear Data Sheets*.)

denness" arises from the fact that a λ-pole operator cannot change any component of the angular momentum of the rotor (in this case the component along the symmetry axis) by more than λ units. In particular, if K is a good quantum number, an Eλ transition matrix element is zero for any $\lambda < \Delta K$. Figure 1.62 shows a selection of excited states in ^{178}Hf. The states shown are classified into bands which, according to the (symmetric top) rotor model, have $K^\pi = 0^+$, 8^-, and 16^+. Now, the K-selection rule forbids Eλ transitions between states of the bands shown in Figure 1.62 for any $\lambda < 8$. Thus, the goodness of the K quantum number is reflected in the extreme retardation of the interband electromagnetic transitions. Indeed, these transitions are among the very slowest known for their respective multipolarities.

Reflection symmetry of the intrinsic states of even-even nuclei is inferred directly from the observation of an even, 0^+, 2^+, $4,^+$, ... sequence of states in the $K = 0$ ground state band, as discussed briefly in Section 1.6.2 (and in more depth in Volume 2). Together with axial symmetry, one concludes that nuclear rotor bands are generally consistent with intrinsic states having spheroidal symmetry; the precise meaning of this observation will be given in Volume 2.

Although we are primarily concerned here with rotational states, it is of interest to note that a lot can be learnt about the structure of a deformed nucleus from its intrinsic excitations. Figure 1.63 shows the spectrum of deuterons inelastically scattered from ^{168}Er. Besides members of the ground state, $K^\pi = 0^+$, band and

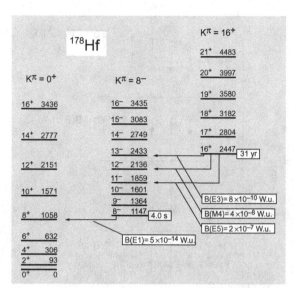

Figure 1.62: Selected rotational bands in ^{178}Hf which show K-forbidden interband gamma-ray transitions. The half lives of the $K^\pi = 8^-$ and 16^+ band heads are shown. (The data are taken from *Nuclear Data Sheets* and Smith M.B. *et al.* (2003), *Phys. Rev.* **C68**, 031302(R).)

the $K^\pi = 2^+$ band built on the $I^\pi = 2^+$ state at 821 keV, one sees significant population of $I^\pi = 3^-$ and 4^+ states associated with $K^\pi = 1^-$, 2^-, 3^+, and 0^- bands built on states with $E(I^\pi) = 1359(1^-)$, $1569(2^-)$, $1654(3^+)$, and $1786(1^-)$, respectively (cf. Figure 1.58, but note that the 3^- state at 2263 keV belongs to a $K^\pi = 3^-$ band not shown in this figure). These 3^- and 4^+ states reveal something about the "fate" of octupole and hexadecapole vibrations in a deformed nucleus. In particular, the octupole collective strength appears in four 3^- states associated with bands having $K^\pi = 0^-, 1^-, 2^-$, and 3^-. These K values are the possible alignments of a vibrational phonon of spin $J = 3$ in the deformed field of the nucleus. Note, however, cf. Figures 1.63 and 1.58, that not all 3^- states carry octupole collectivity; specifically, the 3^- states at 1542 and 1828 keV are not significantly populated by inelastic deuteron scattering. The E2 transitions between the ground-state and $K^\pi = 2^+$ band built on the 2^+ state at 821 keV also reveal that the ground-state band rotational quadrupole strength does not "exhaust" all of the quadrupole collectivity at low energy.

In spite of the consistency of most rotational data in even-even nuclei with the hypothesis of spheroidal (axial and reflection) symmetry, it is well to look out for other possibilities. There are, for example, hints that some nuclei might have intrinsic states with octupole shapes. One would expect to find such nuclei in regions where the first-excited odd-parity band falls particularly low in energy. Likely candidates are clearly seen in Figure 1.53 which presents the systematics of 3_1^- state excitation energies. Two examples are shown in Figure 1.64. The nuclei ^{220}Ra and ^{222}Th have particularly low-lying negative parity states. Moreover, above spin 3, their states form a monotonic spin-parity sequence: $5^-, 6^+, 7^-, 8^+, \ldots$. The pattern of excitation shown in Figure 1.64 has the characteristics of a rotating symmetric

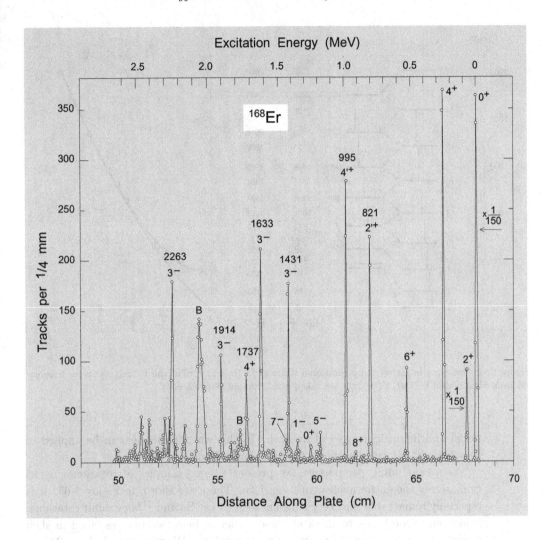

Figure 1.63: Inelastic deuteron scattering from ^{168}Er. The data are for an incident energy of 12.098 MeV and a scattering angle of 125° with respect to the incident beam. Deuteron lines are labelled by the spins and parities of the corresponding levels (cf. Figure 1.58). Lines marked with a B are "background" events resulting from target contaminants. Note the scale reduction factor for the ground state and first excited 2$^+$ state. See the text for a discussion of the ^{168}Er levels observed. (The figure is from Tjøm P.O. and Elbek B. (1968), *Nucl. Phys.* **A107**, 385. The spin-parities and energies of states are from *Nuclear Data Sheets*.)

top without reflection symmetry. Nuclei with such rotational spectra can be usefully interpreted as possessing a significant octupole component to their deformations. The implications of such intrinsic deformations will be discussed in Volume 2.

The rotor model is a powerful tool for organising and interpreting the spectroscopy of strongly-deformed nuclei. The rigidity of the simple model can be "softened" and the model generalised in many ways to include coupling to vibra-

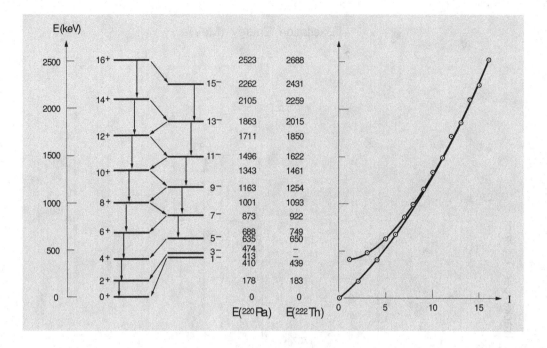

Figure 1.64: Low-lying negative-parity rotational states in ^{220}Ra and ^{222}Th (the 1^- and 3^- states have not yet been identified in ^{222}Th). (The data are taken from *Nuclear Data Sheets*.)

tional and intrinsic degrees of freedom. In this way, the model can be applied to less-strongly-deformed and less-rigidly-deformed nuclei.

As a concluding perspective, we present a few examples of rotational nuclei from across the entire nuclear mass surface. These are shown in Figure 1.65. It is especially remarkable that very light nuclei such as ^{8}Be and ^{24}Mg exhibit rotational behaviour. One tends to think of these nuclei as being simply described in shell model terms because of their small particle numbers. We will see in Volume 2 that, by studying nuclei such as these, a bridge can be built between the rotational model and microscopic models with single-particle (shell model) degrees of freedom.

Exercises

1.20 Show that the $n = 3$ states of the harmonic quadrupole-vibrator model have spin and parity $I^\pi = 0^+, 2^+, 3^+, 4^+, 6^+$.

1.21 Show that if R is the distance from the origin of a point lying on a sphere of radius R_0 centred about a point with (x, y, z) coordinates $(0, 0, \alpha R_0)$, then

$$R = R_0 \left[1 + \alpha \cos\theta + 0(\alpha^2) \right], \tag{1.54}$$

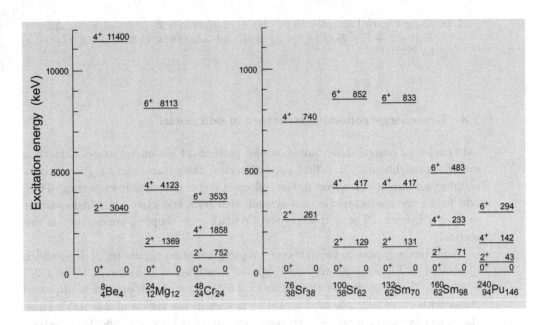

Figure 1.65: Examples of rotational nuclei from across the entire mass surface with proton and neutron numbers corresponding to mid-open-shell regions (cf. Figure 1.2). (The data are from *Nuclear Data Sheets*, from Ajzenberg-Selove F. (1988), *Nucl. Phys.* **A490**, 1, and from Endt P.M. (1990), *Nucl. Phys.* **A521**, 1.)

where $\cos\theta = z/R_0$ and that, to leading order in α, the points at distance

$$R(\theta, \varphi) = R_0 \left[1 + \alpha Y_{10}(\theta, \varphi) \right] \tag{1.55}$$

from the coordinate origin in spherical coordinates, form a sphere centred about the point in $(0, 0, \sqrt{\frac{3}{4\pi}} \alpha R_0)$ in (x, y, z) coordinates, i.e., a Y_{10} shape component is equivalent to a translation of the centre of mass.

1.22 Show that if

$$E_I = E_0 + AI(I+1) + BI^2(I+1)^2, \tag{1.56}$$

then $r_6 = \frac{22}{5} r_4 - \frac{33}{5}$.

1.23 Calculate $\Im_{\text{rigid}}$ for the nuclei in Figures 1.60 and 1.65. (Convert the units of your answers so that direct comparisons can be made.)

1.24 Derive Equation (1.43) from Equations (1.41) and (1.42).

1.25 For $\bar{Q}_0(\alpha K) = \bar{Q}_0$ for all K and α, evaluate Equation (1.51) for $Q(\alpha K I)$ with $K = 0$, $I = 0, 2, 4$; $K = 2$, $I = 2, 3, 4$; $K = 4$, $I = 4$.

1.26 For Equation (1.52) with $K = 0$, and $B(\text{E2}; \alpha 0 I_i \to \alpha 0 I_f) = B(\text{E2}; I \to I - 2) = B_{I,I-2}$, show that $B_{42}/B_{20} = 10/7$, $B_{64}/B_{20} = 225/143$ and $B_{86}/B_{20} = 28/17$.

1.27 From Equations (1.51) and (1.52), derive the relationship

$$Q(2_1^+) = -\frac{2}{7} \sqrt{16\pi B(\text{E2}; 2_1^+ \to 0_1^+)} \tag{1.57}$$

for the ground state ($K = 0$) rotational band of a doubly-even nucleus.

1.28 Using Figure 1.55, plot E_I vs. $I(I+1)$, $E_\gamma(I)$ $(= E_I - E_{I-2})$ vs. I, ΔE_γ $(= E_\gamma[I+2 \to I] - E_\gamma[I \to I-2])$ vs. I, and $\Delta E_\gamma(I)$ $(= E_\gamma(I+2) - E_\gamma(I))$ vs. I, for ^{242,244}Pu.

1.8 Low-energy collective structure in odd nuclei

Odd nuclei, in general, have more complex patterns of low-energy states than their even-even neighbours. To a first approximation, their states can be generated by coupling an unpaired nucleon to the collective states of a doubly-even core. Thus, odd nuclei are characteristic of a system with both collective and single-particle degrees of freedom. The simplest types of particle-core coupling are outlined in this section.

A particularly simple particle-core coupling situation occurs in $^{209}_{83}$Bi$_{126}$ and is illustrated in Figure 1.66. A closely-spaced multiplet of states with $I^\pi = 3/2^+$, $5/2^+$, ..., $15/2^+$ is seen in the region of 2.6 MeV. A closely-spaced multiplet of states with $I^\pi = 3/2^+$, $5/2^+$, ..., $15/2^+$ is seen in the region of 2.6 MeV. These states can be understood as comprising an unpaired proton in the $1h_{9/2}$ orbital coupled to the 3^- collective excitation at 2.615 MeV of the ^{208}Pb core (cf. Figure 1.52). A second closely-spaced multiplet in the region of 3.1 MeV results from the coupling of the $1h_{9/2}$ proton to the 5^- excitation at 3.198 MeV of the ^{208}Pb core (the ^{209}Bi collective states revealed in Figure 1.66 can be compared with other ^{209}Bi states seen in Figures 1.12 and 1.23, respectively). An example of the coupling of an unpaired proton to a 2^+ core excitation, to form a closely-spaced multiplet, occurs in $^{115}_{49}$In$_{66}$ and is shown in Figure 1.67. The nuclei $^{209}_{83}$Bi$_{126}$ and $^{115}_{49}$In$_{66}$ are adjacent to doubly- and singly-closed shells, respectively. The pattern of closely-spaced multiplets clustered around the energy of the core excitation is characteristic of *weak coupling* and is seen in odd-mass nuclei in the neighbourhood of closed shells.

The nucleus $^{175}_{71}$Lu$_{104}$ provides an example of particle-core coupling far from closed shells. A gamma-ray spectrum from the Coulomb excitation of ^{175}Lu with ^{40}Ca projectiles is shown in Figure 1.68. The complete picture of low-lying excited states in ^{175}Lu is shown in Figure 1.69. The first notable feature is that both ^{175}Lu and its even core (^{174}Yb) exhibit rotational bands with similar moments of inertia: $2\Im/\hbar^2 \approx 80$ MeV^{-1} for ^{175}Lu and ≈ 80 MeV^{-1} for ^{174}Yb (cf. Figure 1.60), when compared for low spin. This suggests that the collective (rotational) dynamics are essentially the same in the two nuclei. However, in contrast to the situation in ^{209}Bi and ^{115}In, where each collective excitation of the core becomes a cluster of states in the odd nucleus, the levels of ^{175}Lu have no such multiplet structure. The interpretation of the structure of deformed odd-mass nuclei is that the unpaired nucleon is *strongly coupled* to the deformed core and gets "carried around" by the rotating core as opposed to being a "spectator" as it is in ^{209}Bi and ^{115}In.

The strong-coupling model is appropriate when the rotational motions of the

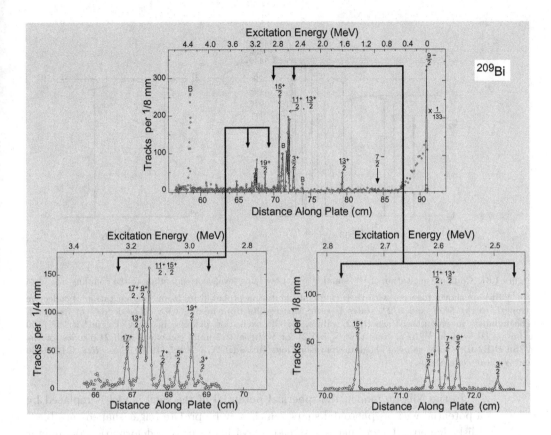

Figure 1.66: Inelastic deuteron scattering from ^{209}Bi. The data are for an incident energy of 13.1 MeV and scattering angles of 150° (upper spectrum) and 120° (lower spectra) with respect to the incident beam. Deuteron lines are labelled by the spins and parities of the corresponding levels or by B for background. The lower part of the figure shows expanded detail of the multiplets at ~ 2.6 and ~ 3.1 MeV excitation energy in ^{209}Bi. The details are discussed in the text. (The figures are taken from Ungrin J. *et al.* (1971), *Mat. Fys. Medd. Dan. Vid. Selsk.* **38** (8). The spin-parity assignments are taken from *Nuclear Data Sheets.*)

core are slow (adiabatic) relative to higher frequency motions of the unpaired nucleon. In this situation, the deformed core provides a rotating frame of reference in which the extra nucleon moves. Thus, when the rotational motion is relatively slow, the extra nucleon "sees" a deformed field as opposed to the spherical field of the spherical shell model. The rotational motions of the coupled system are then again those of a simple rotor.

The strong-coupling interpretation is supported by the observation of various bands in ^{175}Lu which differ in the I^π values of their band heads. The different bands are associated with different states of the unpaired proton in the deformed field. Some details of the motion of a nucleon in a (Nilsson model[28]) deformed field follow.

---

[28] Nilsson, *op. cit.* Footnote 9 on Page 20.

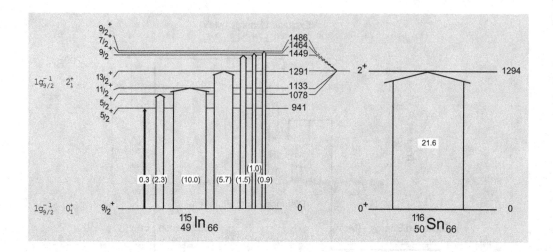

Figure 1.67: Coulomb excitation of ^{115}In and ^{116}Sn. Low-lying levels corresponding to the coupling $\pi 1g_{9/2}^{-1} \otimes$ $2_1^+ (^{116}$Sn$)$ are shown together with their Coulomb excitation probabilities (some fragmentation of collective strength to the $5/2^+$ and $9/2^+$ states is seen; this results from mixing with other degrees of freedom). The numbers in parentheses are $B(E2)$ values, for the excitation process, in units of e^2fm$^4 \times 10^2$ (100 e^2fm$^4 \equiv 60$ W.u. in ^{115}In.). Note that $\sum B(E2)$ for ^{115}In is 21.7 units compared with 21.6 units for the ^{116}Sn $B(E2; 0_1^+ \rightarrow 2_1^+)$ value. (The data are taken from Tuttle III W.K. *et al.* (1976), *Phys. Rev.* **C13**, 1036 and *Nuclear Data Sheets*.)

In the Nilsson model, the spherical potential of the shell model is replaced by a potential with a spheroidal shape which may be prolate (cigar shaped) or oblate (disk shaped). The former is assigned a positive value of a deformation parameter ϵ and the latter a negative value. The deformation parameter ϵ is defined below in Equation (1.60). Quadrupole moments indicate that deformed nuclei with prolate shapes ($\epsilon > 0$) are most common and, for such nuclei, it is appropriate to use Nilsson potentials with corresponding prolate deformations.

With a non-central interaction between the odd particle and the core, the angular momentum of the particle ceases to be a good quantum number. The total angular momentum of the nucleus is conserved but it becomes partitioned between the single particle and the rotational core. On the other hand, the magnitude $|\Omega|$ of the projection[29] Ω of the particle's angular momentum along the symmetry axis of the spheroidal Nilsson potential is conserved. Thus, single-particle motion in a spheroidal potential is characterised by the quantum number $|\Omega|$.

For a particle in a spherical potential moving in an orbital with total single-particle spin j, the states of z-component $m = \pm^1/_2, \pm^3/_2, \pm^5/_2, \ldots, \pm j$ all have the

---

[29]By convention, the projection of an odd-particle angular momentum along the symmetry axis of a spheroidal rotor is assigned the quantum number Ω and the projection of the total angular momentum of the rotor plus particle, is assigned the quantum number K, cf. Figure 1.46. For the ground-state band of an even nucleus, K is invariable zero. For the neighbouring odd-mass nuclei, K is then usually equal to Ω.

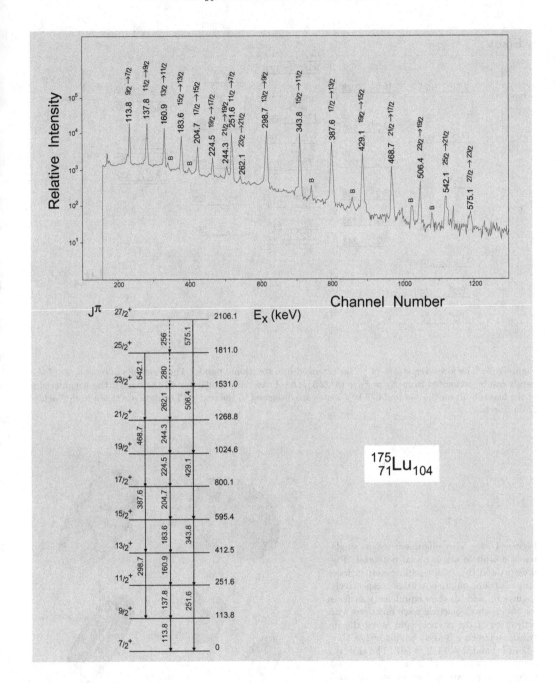

Figure 1.68: Coulomb excitation of ^{175}Lu with ^{40}Ca ions. Lines marked with a B are due to a target impurity. The upper part of the figure shows the gamma-ray spectrum observed and the lower part shows the energy levels of the associated decay scheme which form a good example of a rotational band. All energies are given in keV. (The data are taken from Skensved P. *et al.* (1981), *Nucl. Phys.* **A366**, 125.)

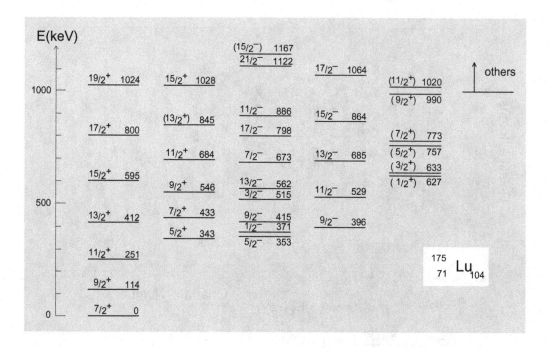

Figure 1.69: The low-lying states of ^{175}Lu arranged into rotational bands. The moments of inertia, $\Im$, of the bands can be estimated from $E_I = E_0 + (\hbar^2/2\Im)I(I+1)$ and are similar for all the bands (the irregularities of the bands built on the 354 and 627 keV states are discussed in the text). (The data are taken from *Nuclear Data Sheets.*)

Figure 1.70: The alignment of a single-particle orbit in a spheroidal potential. Figures (a) and (b) show equipotential surfaces for prolate and oblate potentials, respectively. Figures (c) and (d) show equidensity surfaces for $1h_{11/2}$ single-particle wave functions with projection of the particle spin along the intrinsic symmetry 3-axis, having value $\Omega = 1/2$ in (c) and $\Omega = 11/2$ in (d). The $\Omega = 1/2$ state and the $\Omega = 11/2$ state have lowest energy in a prolate and an oblate potential, respectively.

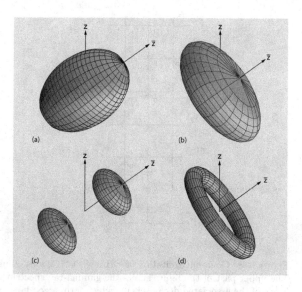

same energy. However, if the potential is allowed to become spheroidal, the states of $|\Omega| = {}^1\!/_2, \, {}^3\!/_2, \, {}^5\!/_2, \, \ldots$, separate in energy. For a prolate spheroidal potential, the $|\Omega| = {}^1\!/_2$ orbital lies lowest and the energy increases with increasing $|\Omega|$. This is due to the fact that the density distribution of a single-particle state of low $|\Omega|$ fits more readily inside an equipotential surface of a prolate potential than in an equipotential surface of an oblate potential, and vice-versa. This is shown in Figure 1.70 for a $1h_{11/2}$ single-particle state. For more than one available shell-model j orbital, mixing of configurations with the same Ω^π value from different j orbitals occurs.

The Nilsson model is essentially a deformed version of the single-particle shell model. It replaces the spherical harmonic oscillator potential of Equation (1.13) with a spheroidal potential. Thus, for deformed nuclei, the spherically symmetric single-particle Hamiltonian of Equation (1.19) is replaced by an axially symmetric Nilsson model Hamiltonian

$$\hat{h} := \frac{\hat{p}^2}{2M} + \tfrac{1}{2}M\left[\omega_\perp^2(\bar{x}^2 + \bar{y}^2) + \omega_z^2\bar{z}^2\right] + D\hat{\mathbf{l}}^2 + \xi\hat{\mathbf{l}}\cdot\hat{\mathbf{s}}, \qquad (1.58)$$

where the $\bar{x}$, $\bar{y}$ and $\bar{z}$ coordinates are defined relative to the intrinsic axes of the rotor.

An equipotential surface for this Hamiltonian is a spheroid with semi axes $R_x = R_y = R_\perp$, R_z related by the equation

$$\omega_\perp^2 R_x^2 = \omega_\perp^2 R_y^2 = \omega_z^2 R_x^2. \qquad (1.59)$$

For consistency, $\omega_\perp$ and ω_z, should be chosen such that the shape of this equipotential surface approximates the shape of a corresponding equidensity surface of the deformed core. Thus, based on the constancy of the nuclear matter density in the interior of a nucleus, it is appropriate to express the frequencies of the Nilsson model Hamiltonian in terms of a volume-conserving scale transformation

$$\omega_\perp = \omega_0 e^{\epsilon/3}, \quad \omega_z = \omega_0 e^{-2\epsilon/3}, \qquad (1.60)$$

for which $\omega_x\omega_y\omega_z = \text{const.} := \omega_0^3$ (cf. Exercises (1.34) and (1.35)). When $\epsilon > 0$, the equipotential surface is then a prolate spheroid and when $\epsilon < 0$ it is oblate.

The total Hamiltonian for a nucleon coupled to the states of a $K = 0$ ground-state rotational band of an even-nucleus core is given by

$$\hat{H} := \frac{\hbar^2\hat{\mathbf{R}}^2}{2\Im} + \hat{h}, \qquad (1.61)$$

where $\hat{\mathbf{R}}$ is the angular momentum of the rotor (cf. Equation (1.43)). If $\hat{\mathbf{j}}$ denotes the angular momentum of the odd nucleon and $\hat{\mathbf{I}} := \hat{\mathbf{R}} + \hat{\mathbf{j}}$ is the total angular momentum of the odd-mass nucleus, then

$$\hat{H} = \frac{\hbar^2\hat{\mathbf{I}}^2}{2\Im} + \hat{h} + \frac{\hbar^2\hat{\mathbf{j}}^2}{2\Im} - \frac{\hbar^2}{\Im}\hat{\mathbf{I}}\cdot\hat{\mathbf{j}}. \qquad (1.62)$$

The last two terms in this expansion, known as *rotation-particle coupling* inter-actions, are due to the centrifugal and Coriolis forces, respectively, which appear because the particle is in a rotating reference frame. They are usually neglected, in a first approximation. The energies of the odd-mass nucleus are then given in the Nilsson model by

$$E_{I\alpha} = \frac{\hbar^2 I(I+1)}{2\Im} + \varepsilon_\alpha, \qquad (1.63)$$

where $\{\varepsilon_\alpha\}$ are the eigenvalues of $\hat{h}$. However, as we shall show in Volume 2, the rotation-particle coupling interactions can be included in perturbation theory and often have significant observable effects.

The Nilsson model Hamiltonian, $\hat{h}$, is analytically solvable in two limits: the small deformation *spherical limit* in which $\epsilon \to 0$ and the large-deformation *asymptotic limit* in which $\epsilon \to \infty$.

For zero deformation, the Nilsson single-particle energies are defined, as in the spherical shell model, by the values of N, l, and j, where $N = n_x + n_y + n_z$ is the number of oscillator quanta. For small deformation, the non-spherical component of the potential can also be included, to leading order in ϵ, in perturbation theory. The Nilsson model Hamiltonian, (1.58), then has leading-order expansion in the deformation parameter, ϵ, given by

$$\hat{h} = \hat{h}_0 + D\hat{\mathbf{l}}^2 + \xi \hat{\mathbf{l}} \cdot \hat{\mathbf{s}} + \epsilon \hat{V}_{\text{def}}, \qquad (1.64)$$

with

$$\hat{V}_{\text{def}} := -\frac{4}{3}\sqrt{\frac{\pi}{5}} M\omega_0^2 r^2 Y_{20}(\theta, \varphi). \qquad (1.65)$$

Thus, if $\{\varepsilon_{Nljm}\}$ denotes the single-particle energies of the spherical shell model then, to first order in ϵ, the corresponding Nilsson model energies are given by

$$\begin{aligned}\varepsilon_{Nlj\Omega}(\epsilon) &= \varepsilon_{Nljm} + \epsilon\langle NLj\Omega|\hat{V}_{\text{def}}|Nlj\Omega\rangle \\ &= \varepsilon_{Nljm} + \frac{2}{3}\epsilon(N + 3/2)\hbar\omega_0 \frac{3\Omega^2 - j(j+1)}{4j(j+1)},\end{aligned} \qquad (1.66)$$

where $\Omega = m$ is the component of the single-particle angular momentum along the $\bar{z}$-axis of the potential. Thus, for fixed values of N, l, and j, and a prolate ($\epsilon > 0$) deformation, the Nilsson single-particle energies increase linearly with $\epsilon\Omega^2$. A two-fold, $\pm\Omega$, degeneracy results from the reflection symmetry of the potential with respect to the $\bar{x}$-$\bar{y}$ plane.

For large deformations, the component

$$\hat{h}_\epsilon := \frac{\hat{p}^2}{2M} + \frac{1}{2}M\left[\omega_\perp^2(\bar{x}^2 + \bar{y}^2) + \omega_z^2\bar{z}^2\right] \qquad (1.67)$$

of the Hamiltonian $\hat{h}$ can be solved analytically in cylindrical coordinates and the term $D\hat{\mathbf{l}}^2 + \xi \hat{\mathbf{l}} \cdot \hat{\mathbf{s}}$ included as a first-order perturbation. In the asymptotic limit, when $\hbar\omega_z - \hbar\omega_\perp \gg \langle D\hat{\mathbf{l}}^2 + \xi\,\hat{\mathbf{l}}\cdot\hat{\mathbf{s}}\rangle$, the mixing of different eigenstates of $\hat{h}_\epsilon$ is small and the single-particle energies are defined by the good (asymptotic) quantum numbers N, n_z, Λ (the eigenvalue of $\hat{l}_{\bar{z}}$), and Ω (the eigenvalue of $\hat{j}_{\bar{z}}$).

A Nilsson model energy level diagram for the $50 < Z < 82$ deformed region is shown in Figure 1.71. As expected, each energy level is two-fold degenerate which

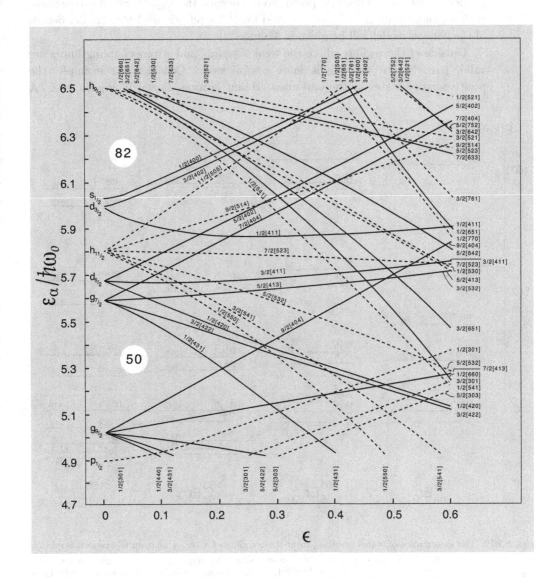

Figure 1.71: A Nilsson diagram for protons in nuclei with $50 \le Z \le 82$. Energies, in units of $\hbar\omega_0$, are plotted against the deformation parameter, ϵ. Energy levels are labelled by their spherical shell model quantum numbers, l and j, at $\epsilon = 0$, and by the asymptotic quantum numbers $\Omega[Nn_z\Lambda]$ for $\epsilon \ne 0$. (The figure is adapted from Lamm I.-L. (1969), *Nucl. Phys.* **A125**, 504.)

means that Nilsson levels are filled pairwise. Dashed lines are used in the figure to distinguish negative-parity levels from positive-parity levels which are depicted by solid lines.

The Nilsson states expected for deformation $\epsilon = 0.3$ and $Z = 71$ are seen to agree qualitatively with the various bands observed at low energy in ^{175}Lu. In particular, for $\epsilon = 0.3$ the 71st proton would occupy the $^7/_2[404]$ which is consistent with a ground-state $K = ^7/_2$ as observed for ^{175}Lu (cf. Figure 1.69). (Fuller details of the Nilsson model will be given in Volume 2.)

Outside of the regions where the weak-coupling and strong-coupling limits are valid, particle-core coupling is more complicated. Consider, for example, the positive-parity states in the odd-mass erbium isotopes shown in Figure 1.72. A

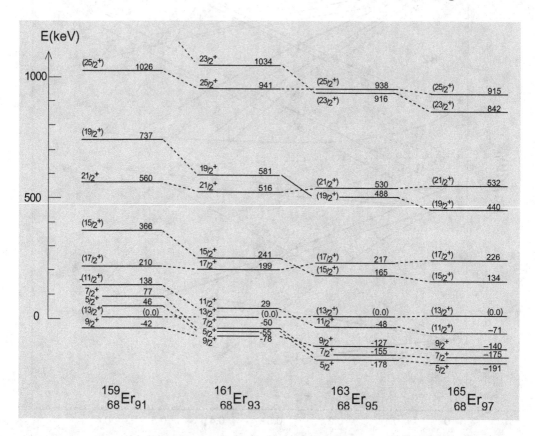

Figure 1.72: The systematics of states resulting from the coupling of a $1i_{13/2}$ neutron to even-mass erbium cores. The energies are shown relative to the $13/2^+$ states. Note the emergence of a "strongly-coupled" spin sequence, i.e., $5/2^+, 7/2^+, 9/2^+, 11/2^+, \ldots$, in ^{163,165}Er. The change in the spin sequence in ^{159,161}Er is discussed in the text. (The data are taken from *Nuclear Data Sheets*.)

strongly-coupled pattern is emerging for the heaviest Er isotopes, but the lighter

Er isotopes show a loss of the strongly-coupled pattern. The reduction of strong coupling is termed *decoupling*.

Decoupling is a consequence of the Coriolis interaction, given in Equation 1.39. Its expectation value reduces to $-(\hbar^2/\Im)\langle \vec{I} \cdot \vec{j} \rangle$ for a spheroidal (axially symmetric) nucleus. The Coriolis interaction cannot be neglected if $\Im$ is small or if $\langle \vec{I} \cdot \vec{j} \rangle$ is large. The systematic behaviour shown in Figure 1.72 is attributed to a decreasing deformation (decreasing $\Im$) with decreasing mass and a consequent increase in the influence of the Coriolis interaction. The Coriolis interaction evidently favours the alignment of $\vec{j}$ with $\vec{I}$. This "rotation alignment" effect of the Coriolis interaction opposes the "deformation alignment" of the spheroidal potential (cf. Figures 1.70 and 1.71) which favours the alignment of the probability density distribution of the particle with that of the slowly rotating rotor core. (Fuller details of Coriolis decoupling will be given in Volume 2.)

Coriolis decoupling provides an explanation for the irregularities of the rotational bands built on the 371 and 627 keV states of ^{175}Lu (cf. Figure 1.69). These two bands have $|\Omega| = 1/2$ (equal to the lowest spin in each band). In fact, the Coriolis interaction is invariably important for $|\Omega| = 1/2$ bands. This is because it makes diagonal contributions to the energies of $|\Omega| = 1/2$ states; a fact that becomes evident when $\langle \vec{I} \cdot \vec{j} \rangle$ is expressed in the form

$$\langle \vec{I} \cdot \vec{j} \rangle = \langle \frac{1}{2}(\hat{I}_+\hat{j}_- + \hat{I}_-\hat{j}_+) + \hat{I}_z\hat{j}_z \rangle. \tag{1.68}$$

One finds that the operator $\hat{I}_+\hat{j}_- + \hat{I}_-\hat{j}_+$ has non-zero matrix elements between the $\Omega = \pm 1/2$ components of a $|\Omega| = 1/2$ rotational state. The modified rotational energy formula that results is

$$E_I = E_0 + A[I(I+1) + (-1)^{I+1/2}(I + 1/2)\, a\, \delta_{K,1/2}], \tag{1.69}$$

where a, the so-called *decoupling parameter*, is characteristic of the intrinsic state of the nucleus. (This will be discussed in more detail in Volume 2.)

Exercises

1.29 Obtain values of a that appear in Equation (1.69) by fitting this equation to the bands in ^{175}Lu built on the 353 and 627 keV states shown in Figure 1.69.

1.30 Obtain $\Im$ for the bands shown in Figure 1.69 and compare with ^{174}Yb (Figure 1.60).

1.31 Plot E_I vs. $I(I+1)$ for the band shown in Figure 1.68.

1.32 For the nuclei with $N = 91$ in Figure 1.43, use the Nilsson model diagram, Figure 1.71, to identify N, n_Z, Λ quantum numbers.

1.33 Identify the Nilsson configurations, $[N, n_Z, \Lambda]$, associated with the bands shown in Figure 1.69.

1.34 Show that if the density function ρ_0 is spherically symmetric, i.e., if $\rho_0(x,y,z) = \rho_0(r)$, then the volume enclosed by an equidensity surface of ρ_ϵ, defined by

$$\rho_\epsilon(x,y,z) := \rho_0(e^{\epsilon/3}x, e^{\epsilon/3}y, e^{-2\epsilon/3}z), \tag{1.70}$$

is independent of ϵ.

1.35 Show that if the spherical harmonic oscillator potential

$$U_0(x,y,z) := \tfrac{1}{2}M\omega^2 r^2 \tag{1.71}$$

is appropriate for a nucleus with a spherically symmetric density ρ_0, then the deformed harmonic oscillator potential that is appropriate for the density ρ_ϵ, defined in Exercise (1.34), is given by

$$U_\epsilon(x,y,z) = U_0(e^{\epsilon/3}x, e^{\epsilon/3}y, e^{-2\epsilon/3}z). \tag{1.72}$$

1.36 Expand the potential

$$U(x,y,z) := \tfrac{1}{2}M\left[\omega_\perp^2(x^2+y^2) + \omega_z^2 z^2\right] \tag{1.73}$$

to leading order in ϵ and use Equation (1.59) and the identity $r^2 Y_{20}(\theta,\varphi) = \sqrt{\frac{5}{16\pi}}(3z^2 - r^2)$ to derive the expression for $\hat{V}_{\text{def}}$ given by Equation (1.65).

1.9   Shape coexistence in nuclei

The most deformed states observed in nuclei are not among the low-lying states considered so far. They are found at higher excitation energies. Excited states occasionally have smaller deformation than the ground state; but often they have greater deformation and sometimes much greater deformation. The existence of different deformations in a single nucleus is referred to as *shape coexistence*. (Such variations in deformations in a single nucleus are also sometimes called *shape isomerism*.[30])

The first proposed[31] examples of shape coexistence were in the nuclei ^{16}O and ^{40}Ca. The relevant data for these nuclei are shown in Figures 1.73 and 1.74. What is remarkable is that these nuclei have doubly-closed shells and, in their ground states, they are among the best examples of spherical nuclei. Nevertheless, as indicated in Figures 1.73 and 1.74, their first excited states are associated with bands that are basically rotational. In both ^{16}O and ^{40}Ca there is evidence of second bands, starting at 9.585 MeV in ^{16}O and at 5.212 MeV in ^{40}Ca, and possibly other bands. The second band in ^{40}Ca has even greater deformation than the first band.

---

[30]The word "isomerism" comes from chemistry where it is used to describe the same (iso) building blocks (atoms) assembled in different ways. The historical use in nuclear physics refers to long-lived excited states. In the current context, shape isomerism refers to states of a single nucleus with different shapes.

[31]Morinaga H. (1956), *Phys. Rev.* **101**, 254.

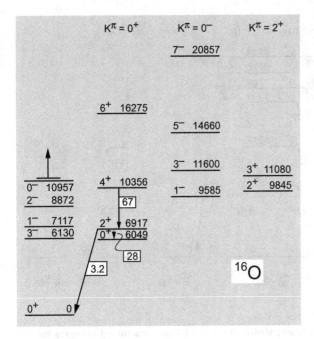

Figure 1.73: The excited states of ^{16}O showing all known states up to 11.1 MeV and selected other states. The first excited state at 6.049 MeV is a 0^+ proton-pair-neutron-pair excitation across the 8-nucleon shell gaps. Associated with it is a band of states (shown directly above it) with spin-parities 2^+, 4^+, 6^+. Their inter-relationship is supported by gamma-ray decay data for which $B(E2)$ values are given in Weisskopf units. The close spacing of the 0^+, 2^+, 4^+, 6^+ band members and the large $B(E2)$ values indicate large deformation. There is also evidence for a $K^\pi = 0^-$ and a $K^\pi = 2^+$ band. (The states on the left are believed to be predominantly non-deformed one-particle-one-hole excitations.) (The data are taken from Tilley D.R., Weller H.R. and Cheves C.M. (1993), *Nucl. Phys.* **A564**, 1.)

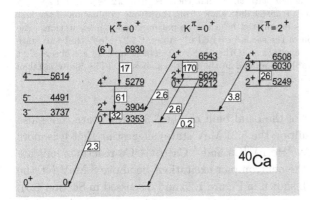

Figure 1.74: The excited states of ^{40}Ca showing all known states up to 5.9 MeV and selected other states. The first excited state at 3.35 MeV is a 0^+ proton-pair-neutron-pair excitation across the 20-nucleon shell gaps. Associated with it is a band of states (shown directly above it) with spin-parities 2^+, 4^+, 6^+ (cf. Figure 1.73 for ^{16}O). There is evidence for an even more deformed band built on a 0^+ state at 5.21 MeV. There is also evidence for a $K^\pi = 2^+$ band built on the 2^+ state at 5.2 MeV. (The data are taken from *Nuclear Data Sheets*.)

This is reflected in its very close energy spacing and a *very* large $B(E2)$ value between the 4^+ and 2^+ states.

The structures of the deformed excited states in ^{40}Ca are revealed by the multi-nucleon transfer reactions summarised in Figure 1.75. The first excited 0^+ state in ^{40}Ca can be interpreted as resulting from the excitation of both a proton pair and a neutron pair across the closed shells at $N, Z = 20$. This is supported by the strong population of this state in the ^{36}Ar$(^6$Li, d$)^{40}$Ca reaction. Even more dramatic is the evidence from the ^{32}S$(^{12}$C, $\alpha)^{40}$Ca reaction which suggests that the 5.21 MeV state results from the excitation of two proton pairs and two neutron pairs across the $N, Z = 20$ shells. The energy spacing of the band built on the 5.21

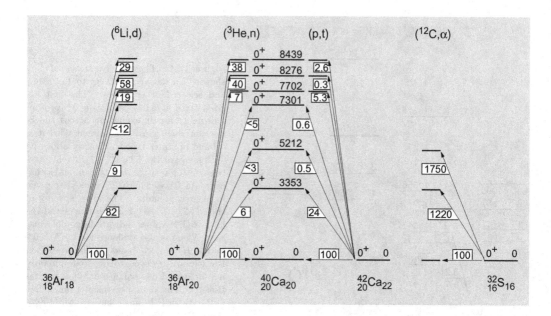

Figure 1.75: Multi-nucleon transfer reaction data which reveal the particle-hole structure of low-lying excited 0^+ states in ^{40}Ca. The (^{3}He,n) and (p,t) transfer data show that there is little $\pi 2p - 2h$ or $\nu 2p - 2h$ configuration strength at low energy. The (^{6}Li,d) transfer data show, from the strong population of the first excited 0^+ state at 3.35 MeV, that this state is dominated by a proton-pair-neutron-pair excitation. The (^{12}C,α) data show, from the strong population of the second excited 0^+ state at 5.21 MeV, that this state is dominated by a two-proton-two-neutron-pair excitation. (Other excited 0^+ states in ^{40}Ca are observed at 7815, 8019, and 8484 keV, but are not observably populated in the reactions shown.) (The figure is taken from Wood J.L. *et al.* (1992), *Phys. Repts.* **215**, 101.)

MeV state is smaller than that of the band built on the 3.35 MeV state (note also the $B(E2)$ value in the band built on the 5.21 MeV state) suggesting that it is more deformed. In contrast, the ^{38}Ar(^{3}He, n)^{40}Ca and ^{42}Ca(p, t)^{40}Ca reactions indicate that both the proton pair and the neutron pair excitations lie above 8 MeV (cf. the neutron pair excitation in ^{208}Pb shown in Figure 1.25 and discussed in Section 1.3). The implication is that there is a strong attractive interaction between the excited proton pairs and neutron pairs to explain the low energies of the 3.35 and 5.21 MeV states.

The above details are consistent with the observation that it is only when both proton and neutron shells are open that nuclear ground states exhibit deformation[32] (cf. Section 1.6.1). In ^{16}O and ^{40}Ca, the excitation of both protons and neutrons across closed shells[33] results in "open-shell-like" structures. The structures underlying the shape coexistence in ^{16}O and ^{40}Ca support this view.

Shape coexistence at closed shells is now established for a number of regions.

---

[32]de Shalit A. and Goldhaber M. (1953), *Phys. Rev.* **92**, 1211.
[33]Brown G.E. and Green A.M. (1966), *Nucl. Phys.* **75**, 401.

For example, the systematics of deformed bands in the doubly-even tin isotopes are shown in Figure 1.76 (cf. also Figure 1.29). Transfer reactions again reveal the

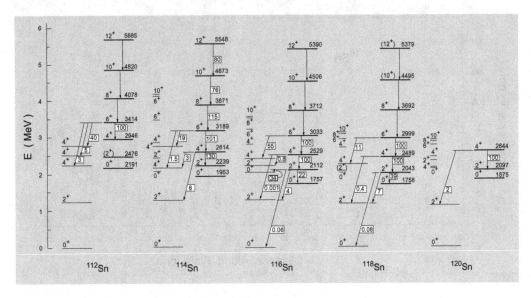

Figure 1.76: Deformed bands in ^{112,114,116,118,120}Sn. The B(E2) values (in Weisskopf units) which support the bands are shown. The B(E2) values given as 100 are relative. (Compare with Figure 1.48(a).) (The data are taken from *Nuclear Data Sheets*.)

role of $J = 0$ pairs of nucleons excited across closed shells in the formation of these states: Figure 1.77 shows the population of states in ^{116}Sn by the (^{3}He, n) reaction; Figure 1.78 shows the states populated by the ^{115}In(^{3}He, d)^{116}Sn reaction. In the latter case, the strongly-populated states are proton one-particle-one-hole states

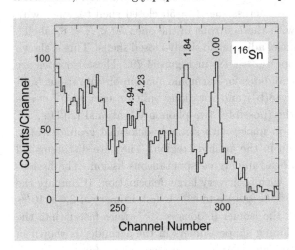

Figure 1.77: The spectrum of neutrons observed (by time-of-flight) in the two-proton stripping reaction ^{114}Cd(^{3}He,n)^{116}Sn at an angle of 0° with respect to the ^{3}He beam at a bombarding energy of 25.4 MeV. The peak labelled 1.84 MeV corresponds to the excited 0^+ state in ^{116}Sn at 1757 keV and indicates that the state is a proton pair excitation. (The figure is taken from Fielding H.W. *et al.* (1977), *Nucl. Phys.* **A281**, 389.)

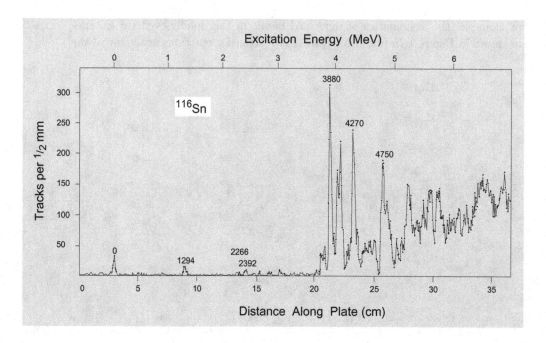

Figure 1.78: The spectrum of deuterons observed in the single-proton stripping reaction ^{115}In(^{3}He, d)^{116}Sn at an angle of $20°$ with respect to the ^{3}He beam at a bombarding energy of 25.3 MeV. The spectrum dramatically illustrates the effect of the proton shell gap at $Z = 50$ on the proton degrees of freedom in ^{116}Sn. (The figure is taken from Shoup R. *et al.* (1969), *Nucl. Phys.* **A135**, 689.)

and all lie above 3.8 MeV. A comparison of these energies with the $\pi(2p{-}2h)$ state at 1.757 MeV indicates an interaction energy between these protons and the neutrons in ^{116}Sn of ~ 6 MeV. It is clear that these deformed bands in the even-Sn isotopes result from a proton *pair* excitation across the $Z = 50$ closed shell together with the interaction of these protons with the neutrons in the open $50 < N < 82$ shell.

Shape coexistence occurs in nuclei adjacent to singly-closed shells. This is shown for the even-mass cadmium isotopes ($Z = 48$) in Figure 1.79. These isotopes are seen to exhibit a coexistence of candidates for harmonic quadrupole vibrations built on the ground state (cf. Figure 1.48(b)) and rotations built on excited 0^+ states that are nearly degenerate with the (possibly) two-phonon vibrational triplets.

Shape coexistence also occurs in nuclei with strongly-deformed ground states. A number of such cases are known in the actinide region. They are distinguished by the ease with which the band head undergoes spontaneous fission. The fission-ability of these states is consistent with their very large deformation. (Compare the spontaneous fission half lives for ^{238g}U (g.s.) and ^{238m}U (s.f. isomer) of 2.6×10^{23}s and 4.5×10^{-6}s, respectively, i.e., the isomer fissions $\sim 10^{29}$ times faster than the ground state.) A view of the fissioning shape isomers in the actinides is shown in Figure 1.80. States with such large deformation are termed *superdeformed*.

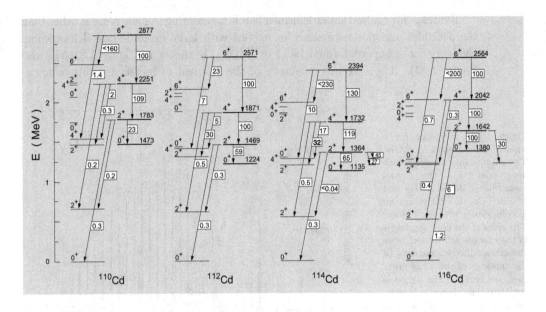

Figure 1.79: Deformed bands in ^{110,112,114,116}Cd. The $B(E2)$ values are given in Weisskopf units. For levels for which the strongest $B(E2)$ is equal to 100, the other $B(E2)$'s are given relative to 100. (Compare with Figure 1.48(b).) (The data are taken from *Nuclear Data Sheets*; and from Corminbouef F. *et al.* (2000), *Phys. Rev.* **C63**, 014305 – ^{110}Cd; Kadi M. *et al.* (2003), *Phys. Rev.* **C68**, 031306(R) – ^{116}Cd; Lehmann H. *et al.* (1996), *Phys. Lett.* **B387**, 259 – ^{112}Cd; Juutinen S. *et al.* (1996), *Phys. Lett.* **B386**, 80 – ^{114,116}Cd.)

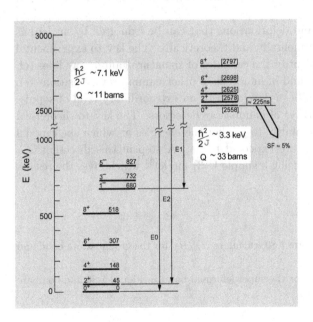

Figure 1.80: A composite view of the fissioning shape isomers in the actinides. The data up to and including the 2558 keV state are for ^{238}U. The data for the band built on the 2558 keV fission isomer are typical for rotational bands built on fission isomers (cf. the rotational parameter of 3.3 keV). The state at 2558 keV decays by very hindered $E0$, $E1$, and $E2$ transitions to low-lying less-deformed states and by spontaneous fission (s.f.). The rotational parameters, $\hbar^2/2\Im$, are for a fit of the rotational energy formula, Equation (1.49). The quadrupole moments, Q_0, are determined from mean lifetimes in these bands. Note the much larger quadrupole moments and the implied much larger moment of inertia for the band built on the fission isomer. (The data are taken from *Nuclear Data Sheets* and the $\hbar^2/2\Im$ and Q_0 values are averages of data reported in Metag V., Habs D. and Specht H.J. (1980), *Phys. Repts.* **65**, 1.)

Evidence for excited superdeformed bands is emerging in mass regions other than the actinides and not necessarily in regions with large ground-state deformation. Evidence for a superdeformed band in $^{152}_{66}\text{Dy}_{86}$ is shown in Figure 1.81 (see also Figure 1.83). The extreme constancy of the moment of inertia for this band, as

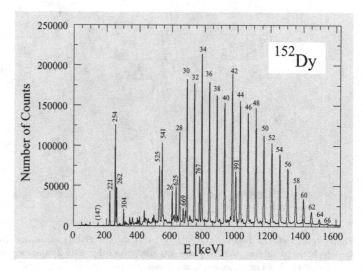

Figure 1.81: Evidence for a superdeformed band in ^{152}Dy. The gamma rays shown, which are seen in the reactions $^{76}\text{Ge}(^{80}\text{Se}, 4n)^{152}\text{Dy}$ and $^{108}\text{Pd}(^{48}\text{Ca}, 4n)^{152}\text{Dy}$, are all in coincidence with each other. The spins of the decaying levels in the cascade are shown. Note the extraordinary constancy of the spacings of the γ-ray energies (see text). (The figure is similar to one shown in Lauritsen T. *et al.* (2002), *Phys. Rev. Lett.* **88**, 042501 and was made available to us by T. Lauritsen.)

reflected by a constant value of $\Delta E_{I,I-2}/(4I-2)$ (cf. Equation (1.49)), and constant $\Delta^2 E_{I,I-2}$ (i.e., differences in $\Delta E_{I,I-2}$) suggests a very rigid nuclear deformation (cf. Figure 1.60).

The full range of shapes and deformations that can be exhibited by nuclei has yet to be explored both experimentally and theoretically. The key to experimental exploration is the ability to identify the cascades of signature gamma-ray lines (cf. Figure 1.81) against a background of hundreds of other gamma-ray transitions (cf. Figure 1.83). This ability has been developed to an extraordinary level of sophistication by the use of large arrays of gamma-ray detectors. The key to theoretical exploration is likely to be the ability to predict the energies at which shell model states of different deformation are expected to lie. A step in this direction is a calculation of the energy levels using a simple U(3) model[34] which gives the results for ^{16}O shown in Figure 1.82.

Exercises

1.37 From information in Figure 1.80, calculate $R_Z/R_\perp$ for the ground-state band and the superdeformed band.

1.38 Calculate $\mathfrak{I}_{\text{rigid}}$ from Q for the superdeformed band in Figure 1.80 and compare with $\hbar^2/2\mathfrak{I}$.

---

[34]Rowe D.J., Thiamova G. and Wood J.L. (2006), *Phys. Rev. Lett.* **97**, 202501.

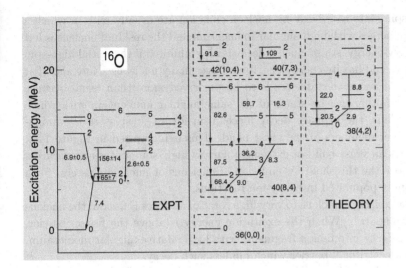

Figure 1.82: Positive-parity energy levels of an algebraic model for ^{16}O in comparison with those observed below 15 MeV (cf. Figure 1.73 and note that, for ^{16}O, $1\,e^2\text{fm}^4 = 2.32\,\text{W.u.}$). Energy levels associated with an irreducible U(3) representation $N(\lambda\mu)$ are shown in groups. Reduced E2 transition rates are indicated (in $e^2\text{fm}^4$ units) beside the arrows. (The figure is from Rowe *et al.*, *op. cit.* Footnote 34 on Page 92.)

1.10 Nuclear structure at high excitation energy

Whereas nuclei generally exhibit discrete levels with narrow (gamma-ray) decay widths at low excitation energies, they are characterised at high excitation energies by large level densities and overlapping states with broad particle-emission and other radioactive decay widths.

The densities of levels in nuclei increase with increasing excitation energy (cf., for example, Figures 1.25, 1.29, 1.49, 1.78). The rate of increase tends to be greater in heavy than in light nuclei, and greater in open-shell than in closed-shell nuclei. It is greater in odd nuclei than in even nuclei and greater still in odd-odd nuclei. In all nuclei, the level density eventually increases to the extent that it becomes impossible to resolve individual levels. Moreover, at energies above the threshold for emission of nucleons or clusters of nucleons, levels acquire emission widths (recall the relationship between lifetime τ and decay width $\Gamma = \hbar/\tau$). The "continuum" is said to have been reached when the levels start to overlap.

A study of level densities is of considerable interest from the perspective of statistical models of the nucleus. However, what is remarkable is the appearance of very non-statistical structures that stand out in the continuum because they are preferentially excited in various nuclear reactions. These structures come in the form of broad resonances and narrow isolated states with collective properties.

An example of a narrow isolated state is a high-spin state. The states of lowest energy for different values of the angular momentum are known as *yrast* states.[35] Such states have narrow decay widths for several reasons. Although they may lie at excitation energies above the threshold for emission of nucleons or nucleon clusters, their decay by particle emission is inhibited by the centrifugal barrier; the states into

---

[35]The term *yrast* is taken from the Swedish word for dizziest.

which a high-spin state can decay by nucleon emission are usually ones in which a single nucleon emerges with high angular momentum and the residual nucleus is left in a low-spin, low-energy, state. Such a decay is also inhibited if the initial high-spin state is collective and has its energy distributed over many nucleons. Note also that the collectivity of an yrast state, tends to be preserved more than usual because of the absence of other nearby states of the same angular momentum with which it could mix. Thus, the primary decay mechanism for an yrast state is gamma-ray decay to other members of the yrast band. This is illustrated in Figure 1.83. The figure shows an yrast band of extremely narrow high–spin states at excitation energies far above the threshold for emission of nucleons or nucleon clusters. Such states are strongly populated in heavy-ion reactions.

The limit to exciting a nucleus to very high rotational states is fission; the nucleus eventually "flies apart". When the excitation energy is above the fission barrier, the nucleus can decay into fission fragments with high relative angular momentum. Again there is a centrifugal barrier which hinders such decay.

Another example of a level with narrow decay width observed at high excitation energy is a so-called *isobaric analog state* (IAS) or *isobaric analog resonance* (IAR). Such a state (or resonance) can be excited by a transfer reaction in which a proton replaces a neutron in the nucleus. An illustration of this is given in Figure 1.84. The figure shows a very narrow $J^\pi = 0^+$ state at an energy which, when allowance is made for the extra Coulomb energy of ^{120}Sb, has the same energy as the ground state of the target nucleus ^{120}Sn. The implication is that the narrow state populated in ^{120}Sb is the isobaric analog of the ground state of ^{120}Sn.

Similar to the high-spin states, isobaric analogs of neighbouring ground states have particularly stable structures and different symmetries (in this case isospin) to those of neighbouring states (cf. Figure 1.28). Thus, one understands why isobaric analogs of ground states have such narrow widths. More extensive data on isobaric analog states are presented in Section 5.5.

The most important high-energy collective structure is undoubtedly the so-called *giant dipole resonance* or GDR. This structure shows up dramatically in every nucleus as a huge broad resonance in high-energy photon absorption experiments. An example is shown in Figure 1.85. The GDR and its interpretation has been most influential in the development of the theory of nuclear collective structure and in showing how the collective model relates, at a microscopic level, to the shell model.

The predominantly dipole character of the resonance is inferred from angular distribution measurements both on gamma rays emitted following excitation of the resonance and from (γ, n) reactions. It is found that the integrated gamma-ray absorption cross section for exciting the resonance comes close to exhausting the maximum possible strength for dipole excitation of a nucleus by photon absorption (as expressed in the so-called *dipole sum rule*). This gives an immediate interpretation of the structure underlying the resonance as corresponding to a one-phonon dipole vibrational state in a highly-collective mode in which the neutrons and pro-

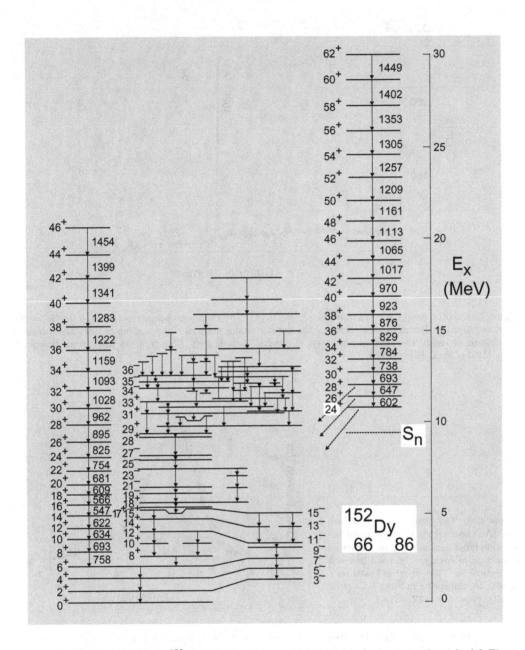

Figure 1.83: High-spin states in ^{152}Dy including the superdeformed band, shown on the right (cf. Figure 1.81). Transition energies are in keV. The threshold for neutron emission (the neutron separation energy, S_n) is indicated. (The figure is similar to one shown in Bentley M.A. *et al.* (1991), *J. Phys. G: Nucl. Part. Phys.* **17**, 481 and was made available to us by J. Sharpey-Shafer. Spins in the superdeformed band have been changed to conform with Lauritsen T. *et al.* (2002), *Phys. Rev. Lett.* **88**, 042501.)

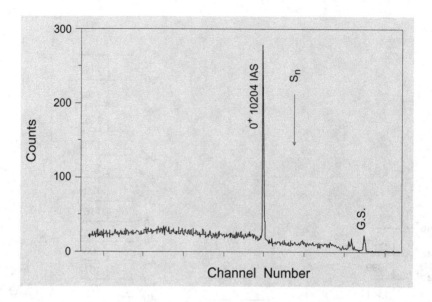

Figure 1.84: Triton energy spectra from the ^{120}Sn(^{3}He, t)^{120}Sb charge-exchange reaction with E(^{3}He) = 200 MeV and $\theta_t = 0°$. The excitation energy of the isobaric analog state in ^{120}Sb is given in keV. The threshold for neutron emission (the neutron separation energy, S_n) is indicated. (The figure is taken from Jänecke J. *et al.* (1991), *Nucl. Phys.* **A526**, 1.)

Figure 1.85: The giant dipole resonance in ^{197}Au observed via the total photoabsorption cross section for gamma rays as a function of energy. The solid line is a theoretical fit. (The figure is based on a similar one appearing in Fultz S.C. *et al.* (1962), *Phys. Rev.* **127**, 1273.)

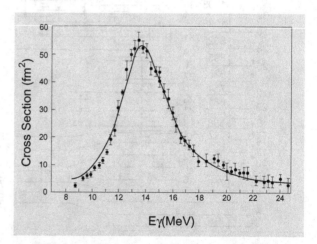

tons oscillate in antiphase. Such a mode occurs at high energy because the neutrons and protons resist being separated. However, the broad width of the GDR implies that the states of a collective dipole vibrator are not eigenstates of the nucleus. Thus, the one-phonon dipole states are mixed with many other states spread over the width of the resonance. More details of the GDR and its interpretation will be given in Volume 2.

Chapter 2

The Bohr collective model

From the survey of nuclear data presented in Chapter 1, a picture emerges of nuclei made up of many nucleons which, to a first approximation, sometimes behave independently (near doubly-closed shells), sometimes interact strongly to form correlated two-nucleon pairs (notably in singly-closed shell nuclei), and at other times behave collectively in a fluid-like manner to generate vibrational and rotational states of the whole nucleus. These perspectives are overly simplistic. Nevertheless, together they provide the foundational models for a qualitative, and often quantitative, understanding of most nuclear structure data. Moreover, they provide the language used by nuclear physicists. The current chapter outlines the collective model of the nucleus as introduced by Aage Bohr.[1]

The Bohr collective model was introduced as a hydrodynamic collective model based on an earlier liquid drop model.[2,3] However, it turns out that many of the liquid drop model concepts, such as a sharply-defined surface, are overly restrictive and unnecessary. This aspect of the model is therefore avoided in the presentation given here. In this chapter, we also restrict consideration to the quadrupole degrees of freedom. The configuration space of the model is then a real five-dimensional vector space, $\mathbb{R}^5$, whose elements are are interpreted as quadrupole tensors for a nucleus and a model Hamiltonian is naturally expressed in the form

$$\hat{H} = -\frac{\hbar^2}{2B}\nabla^2 + \hat{V}, \tag{2.1}$$

where ∇^2 is the Laplacian and $\hat{V}$ is a rotationally-invariant potential energy function of the $\mathbb{R}^5$ coordinates; B is a mass parameter.

In looking ahead to understanding how the Bohr model fits into an overall microscopically-based theory of nuclear structure, it is worth noting that, already in Bohr's original paper and in an earlier paper by Rainwater,[4] a mechanism was identified for combining independent-particle and collective degrees of freedom in a

---

[1]Bohr A. (1952), *Mat. Fys. Medd. Dan. Vid. Selsk.* **26** (14).
[2]Bohr, *op. cit.* Footnote 19 on Page 48.
[3]Bohr and Kalckar, *op. cit.* Footnote 20 on Page 48.
[4]Rainwater, *op. cit.* Footnote 17 on Page 47.

single unified model. Since its inception, the Bohr model has also been developed in many other ways and is now known in its most general form as the Bohr-Mottelson model.[5],[6] In particular contexts, it is also known as the *rotation-vibration model*[7] and as the Bohr-Mottelson-Frankfurt model.[8] Sometimes it is called the *geometric model* to distinguish it from other models. Although sometimes convenient, this distinction is misleading because the Bohr model also has rich algebraic and analytical structures which complements its geometry. Identification of the algebraic structure of a collective model is particularly useful for understanding what extensions of the model are possible, such as the inclusion of vorticity degrees of freedom, and how to embed the model in a microscopic many-nucleon theory. Development of the Bohr model as an algebraic collective model is given in Chapter 4. The unified model, other vibrational and rotational models of a many-nucleon system that evolved from the basic Bohr model, and their relationships to the microscopic shell-model theory of nuclei will be developed in Volume 2.

2.1   Coordinates and observables

The Bohr model was originally formulated in terms of surface deformation parameters $\{\alpha_{\lambda\mu}\}$ in an expansion $R(\theta, \varphi) = R_0\big(1 + \sum_{\lambda\mu} \alpha_{\lambda\mu} Y_{\lambda\mu}^*(\theta, \varphi)\big)$ in terms of spherical harmonics, $\{Y_{\lambda\mu}\}$, of a radial function for the surface of a liquid drop description of the nucleus. However, because nuclear surfaces are diffuse and ill-defined, it is more meaningful to characterise nuclear shapes in terms of multipole moments of the nuclear density distributions. Multipole moments of the nuclear density distribution are much more useful as collective coordinates than surface shape parameters, firstly because they relate directly to observable moments of the nuclear charge distribution (when the neutrons and protons move in phase) and secondly because multipole moments have a microscopic expression in terms of the many-nucleon coordinates of a microscopic theory. We restrict consideration here to quadrupole deformation degrees of freedom.

2.1.1   *Quadrupole moments as collective coordinates*

A distribution of A point nucleons in ordinary three-dimensional space with Cartesian coordinates $\{x_{ni}, n = 1, \ldots, A, i = 1, 2, 3\}$ (or a distribution of nuclear matter

---

[5]Bohr A. and Mottelson B.R. (1953), *Mat. Fys. Medd. Dan. Vid. Selsk.* **27** (16).

[6]Bohr A. and Mottelson B.R. (1975), *Nuclear Structure*, Vol. 2 (Benjamin, Reading, Mass.), (republished by World Scientific, Singapore).

[7]The rotation-vibration model, introduced by Faessler and Greiner, is reviewed in Chapter 6 of Eisenberg and Greiner's book, and numerous references are given therein.

[8]Eisenberg J.M. and Greiner W. (1987), *Nuclear Models* (North Holland, Amsterdam), third edn.

of density ρ) is characterised by the Cartesian quadrupole moments

$$Q_{ij} := \sum_n x_{ni}x_{nj} \quad (\text{or} \int x_i x_j \, \rho(\mathbf{r}) \, dv), \quad i,j = 1,2,3. \tag{2.2}$$

However, because of the need to construct a model with angular-momentum properties, it is more useful to define a spherical-tensor basis of quadrupole moments by

$$q_m := \sum_{n=1}^{A} r_n^2 Y_{2m}(\theta_n, \varphi_n) \quad (\text{or} \int r^2 Y_{2m}(\theta, \varphi) \, \rho(\mathbf{r}) \, dv), \quad m = 0, \pm 1, \pm 2. \tag{2.3}$$

The standard expressions for the spherical harmonics then give the relationships

$$q_0 = \sqrt{\frac{5}{16\pi}} (2Q_{33} - Q_{11} - Q_{22}),$$

$$q_{\pm 1} = \mp\sqrt{\frac{15}{8\pi}} (Q_{31} \pm iQ_{32}), \tag{2.4}$$

$$q_{\pm 2} = \sqrt{\frac{15}{32\pi}} (Q_{11} - Q_{22} \pm 2iQ_{12}).$$

The spherical-tensor quadrupole moments, $\{q_m\}$, define a point q in a real five-dimensional Euclidean space $\mathbb{R}^5$, which is spanned by a subset of the Cartesian quadrupole moments. They also define an ellipsoidal shape of a model nucleus. The complete set of Cartesian quadrupole moments would define a point in a real six dimensional space and the monopole moment, $\mathfrak{m} := \sum_i Q_{ii}$, as well as the quadrupole moments of a model nucleus.

2.1.2 Spherical polar coordinates

Spherical polar coordinates for the Euclidean space $\mathbb{R}^5$ include a radial coordinate, equal to the distance of a point in the space from the origin, and a set of angle coordinates which specify its location on a unit four-sphere,[9] centred at the origin.

A radial coordinate β for $\mathbb{R}^5$ is defined, in parallel with the definition, $r^2 = x^2 + y^2 + z^2$, of a radial coordinate r for the three-dimensional space $\mathbb{R}^3$, by $\beta^2 := \sum_m |q_m|^2$. Angle coordinates for a four-sphere in $\mathbb{R}^5$ are defined as follows.

Recall that Cartesian quadrupole moments transform under SO(3) rotations according to the equation

$$Q_{ij} \rightarrow (\Omega Q \Omega^{-1})_{ij} = \sum_{kl} \Omega_{ik} Q_{kl} \Omega_{jl}, \quad \Omega \in SO(3), \tag{2.5}$$

where $\Omega \in SO(3)$ is here regarded as a 3×3 real orthogonal matrix with $\det \Omega = 1$. In contrast, spherical-tensor quadrupole moments (cf. Appendix A.3) transform

---

[9]A sphere in $\mathbb{R}^5$ is the set of points in $\mathbb{R}^5$ that lie at a fixed distance from the origin. Thus, it is a four-dimensional manifold and is referred to as a four-sphere, S_4.

according to an $L = 2$ irreducible representation (irrep) of SO(3)

$$q_m \to \sum_\mu q_\mu \mathscr{D}^2_{\mu m}(\Omega), \quad \Omega \in \text{SO}(3),$$
$$q_m \to q_m e^{im\theta}, \quad \theta \in \text{SO}(2),$$

(2.6)

where $\mathscr{D}^2(\Omega)$ is the $L = 2$ matrix representing the rotation $\Omega \in \text{SO}(3)$ and m is an SO(2) quantum number for rotations about the 3-axis. Every Q (being real and symmetric) can be brought to diagonal form $\bar{Q}$ by some rotation. It follows that spherical-tensor quadrupole moments are equivalently expressed by

$$q_m = \sum_\mu \bar{q}_\mu \mathscr{D}^2_{\mu m}(\Omega),$$

(2.7)

with

$$\bar{q}_0 = \sqrt{\frac{5}{16\pi}} \, (2\bar{Q}_{33} - \bar{Q}_{11} - \bar{Q}_{22}), \quad \bar{q}_{\pm 1} = 0, \quad \bar{q}_{\pm 2} = \sqrt{\frac{15}{32\pi}} \, (\bar{Q}_{11} - \bar{Q}_{22}), \quad (2.8)$$

for some $\Omega \in \text{SO}(3)$. We shall refer to $\bar{q}_0$ and $\bar{q}_{+2} = \bar{q}_{-2}$ as *intrinsic quadrupole moments*. The radial coordinate β, being invariant under rotations, is given in terms of the intrinsic quadrupole moments by

$$\beta^2 = \sum_m |q_m|^2 = \bar{q}_0^2 + 2\bar{q}_2^2.$$

(2.9)

This suggests defining an angle coordinate γ for $\mathbb{R}^5$, to complement the radial coordinate β and the three Euler angle coordinates for the SO(3) rotation Ω, by setting

$$\bar{q}_0 := \beta \cos\gamma, \quad \bar{q}_2 := \bar{q}_{-2} := \frac{1}{\sqrt{2}} \beta \sin\gamma.$$

(2.10)

Thus, the quadrupole moments are given in terms of spherical polar coordinates by the identity

$$q_m = \beta \left[\cos\gamma \, \mathscr{D}^2_{0m}(\Omega) + \frac{1}{\sqrt{2}} \sin\gamma \left(\mathscr{D}^2_{2m}(\Omega) + \mathscr{D}^2_{-2m}(\Omega) \right) \right].$$

(2.11)

Equivalently, they are expressed as

$$q_m(\beta, \gamma, \Omega) = \beta \mathcal{Q}_m(\gamma, \Omega),$$

(2.12)

where

$$\mathcal{Q}_m(\gamma, \Omega) = \cos\gamma \, \mathscr{D}^2_{0m}(\Omega) + \frac{1}{\sqrt{2}} \sin\gamma \left(\mathscr{D}^2_{2m}(\Omega) + \mathscr{D}^2_{-2m}(\Omega) \right)$$

(2.13)

is proportional to an elementary spherical harmonic for the four-sphere.

The spherical polar coordinates (β, γ, Ω) for $\mathbb{R}^5$ correspond to a sequence of transformations that parallel those for the familiar spherical polar coordinates of three-dimensional space, $\mathbb{R}^3$. In three dimensions, one can start from the point with coordinates $(x = 0, y = 0, z = 1)$ and transform it radially to a point $(0, 0, r)$.

This point, regarded as the north pole of a two-sphere, S_2, is then rotated about the y-axis to a point with coordinates $(r\sin\theta, 0, r\cos\theta)$, and again rotated about the z-axis to a point $(r\sin\theta\cos\varphi, r\sin\theta\sin\varphi, r\cos\theta)$. In a similar way, the above spherical polar coordinates for $\mathbb{R}^5$ correspond to a radial transformation from the point $(q_0 = 1, q_{\pm 1} = q_{\pm 2} = 0)$ to the point $(q_0 = \beta, q_{\pm 1} = q_{\pm 2} = 0)$ followed by a rotation in the two-dimensional plane with coordinates $(X = q_0, Y = \frac{1}{\sqrt{2}}(q_2 + q_{-2}))$ which sends $X = \beta \to \beta\cos\gamma$, $Y = 0 \to \beta\sin\gamma$, and

$$q_0 \to \bar{q}_0 = \beta\cos\gamma, \quad q_{\pm 1} \to \bar{q}_{\pm 1} = 0, \quad q_{\pm 2} \to \bar{q}_{\pm 2} = \frac{1}{\sqrt{2}}\beta\sin\gamma. \tag{2.14}$$

Finally, the point $\{\bar{q}_m\}$ with these *intrinsic coordinates* is rotated to a generic point by the SO(3) rotation of Equation (2.11); the latter is achieved by a sequence of three SO(2) $\subset$ SO(3) rotations through Euler angles in the standard way.

Recall that, for ordinary three-dimensional space, the spherical polar coordinates (r, θ, φ) have the restricted domains $0 \le r$, $0 \le \theta \le \pi$, $0 \le \varphi \le 2\pi$. This is to avoid assigning different coordinate values to a single point. Similar restrictions apply to the collective model coordinates. The β coordinate is naturally restricted to non-negative values. The restriction on γ is found by observing that the intrinsic quadrupole moments defined in terms of β and γ by Equations (2.8) and (2.10), are given by

$$\bar{Q}_{11} = \tfrac{1}{3}\left(Q^{(0)} + \sqrt{\tfrac{16\pi}{5}}\,\beta\cos(\gamma - \tfrac{2\pi}{3})\right),$$

$$\bar{Q}_{22} = \tfrac{1}{3}\left(Q^{(0)} + \sqrt{\tfrac{16\pi}{5}}\,\beta\cos(\gamma - \tfrac{4\pi}{3})\right), \tag{2.15}$$

$$\bar{Q}_{33} = \tfrac{1}{3}\left(Q^{(0)} + \sqrt{\tfrac{16\pi}{5}}\,\beta\cos\gamma\right),$$

where $Q^{(0)} = \sum_i \bar{Q}_{ii}$. It follows from these identities that the transformations $\gamma \to -\gamma$ and $\gamma \to \gamma - \frac{2\pi k}{3}$, for k an integer, simply permute the intrinsic quadrupole moments $\{\bar{Q}_{ii}\}$ among themselves and therefore leave the shape of the nucleus invariant. Thus, the above transformations of γ have the same effect as a change of the rotational coordinates Ω. A set of (β, γ) that all correspond to the same quadrupole shape is shown in Figure 2.1(a). It follows from this figure that γ should be restricted to the range $0 \le \gamma \le \pi/3$.

It is similarly observed that the matrix of intrinsic quadrupole moments is invariant under any rotation belonging to the $D_2 \subset$ SO(3) subgroup.[10] However, in this case, rather than restricting the ranges of the (γ, Ω) coordinates, it is generally simpler to impose symmetry constraints on the wave functions such that they take the same values for different but equivalent sets of coordinates. The mechanism for doing this will be discussed at some length in the treatment of rotor models in Volume 2 and is briefly introduced in Section 2.3.3.

---

[10]The subgroup D_2, known as the dihedral group, is the group of rotations through multiples of π about the intrinsic axes.

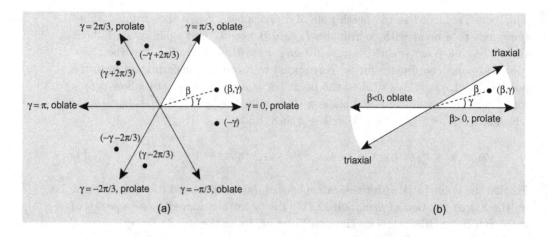

Figure 2.1: Sets of (β, γ) coordinates representing precisely the same nuclear shape, albeit sometimes oriented differently. The unshaded area in (a) indicates the basic domain of (β, γ) with $\beta \geq 0$. The unshaded area in (b) indicates the basic domain when β is allowed to take both positive and negative values. (Based on Figure 6B-1 of Bohr and Mottelson, *op. cit.* Footnote 6 on Page 98.)

For some purposes, it is also convenient to allow β to take negative values and to restrict γ to the range $0 \leq \gamma \leq \pi/6$. The domain of the (β, γ) coordinates is then as indicated in Figure 2.1(b) with the understanding that the extended range of β implies the extra equivalence relationship

$$(-\beta, \gamma) \equiv (\beta, \pi/3 - \gamma). \tag{2.16}$$

A geometrical interpretation of the (β, γ) coordinates is obtained by considering the deformed density distribution of nuclear matter, ρ, generated by a scale transformation of a spherical density, ρ_0, i.e., by setting

$$\rho(x, y, z) = \rho_0(x/a, y/b, z/c). \tag{2.17}$$

The (x, y, z) coordinates for an equi-density surface of ρ then satisfy the equation of an ellipsoid

$$\frac{x^2}{a^2} + \frac{y^2}{b^2} + \frac{z^2}{c^2} = R^2. \tag{2.18}$$

Moreover, one obtains

$$\bar{Q}_{11} = \int x^2 \, \rho(x, y, z) \, dv = \frac{1}{3} a^2 \langle r^2 \rangle,$$

$$\bar{Q}_{22} = \int y^2 \, \rho(x, y, z) \, dv = \frac{1}{3} b^2 \langle r^2 \rangle, \tag{2.19}$$

$$\bar{Q}_{33} = \int z^2 \, \rho(x, y, z) \, dv = \frac{1}{3} c^2 \langle r^2 \rangle,$$

where $\langle r^2 \rangle$ is the mean-square radius of the spherical matter distribution ρ_0. Thus, comparing these expressions with those of Equation (2.15) in terms of β and γ, it is seen that: when $\beta = 0$, the ellipsoid becomes spherical; when $\beta > 0$ and $\gamma = 0$, it is a prolate spheroid; when $\beta > 0$ and $\gamma = \pi/3$, it is an oblate spheroid; and, when $\beta > 0$ and $0 < \gamma < \pi/3$, it has three unequal axes (cf. Exercise 2.2) and is *ellipsoidal.*

2.1.3 *Canonical momenta and momentum operators*

If an object moves along a path in the Euclidean space $\mathbb{R}^5$ with $\{q_m\}$ evolving as a function of time then, at any point on the path, the object has momentum coordinates $\{\pi_m\}$ given classically by $\pi_m = B\dot{q}_m$, where B is the mass parameter of Equation (2.1) and $\dot{q}_m = dq_m/dt$.

In quantum mechanics, the quadrupole momenta are interpreted as infinitesimal generators of deformation. In terms of the spherical $\{q_m\}$ coordinates they are given by the standard Dirac prescription

$$\hat{\pi}_m := -i\hbar \frac{\partial}{\partial q^m}, \tag{2.20}$$

where[11] $q^m := q_m^* = (-1)^m q_{-m}$. Thus, one can regard quantization of the collective model as a map from the classical variables (q_m, π_m) to quantal observables (operators) $(\hat{q}_m, \hat{\pi}_m)$ which act on wave functions of the Hilbert space $\mathcal{L}^2(\mathbb{R}^5)$ according to the identities[12]

$$\hat{q}_m \Psi(q) = q_m \Psi(q), \quad \hat{\pi}_m \Psi(q) = -i\hbar \frac{\partial \Psi}{\partial q^m}(q). \tag{2.21}$$

2.1.4 *Other observables of the collective model*

The basic observables of the collective model, as expressed above, are the mass quadrupole moments $\{\hat{q}_m\}$ and their canonical momenta $\{\hat{\pi}_m\}$. Other observables, such as the Hamiltonian and angular momentum operators are constructed as functions of these basic observables.

---

[11] A minor complication, cf. Section A.6, arises because of the use of complex spherical coordinates for a real space. It is handled in this book by the use of upper and lower indices such that lower indices denote the standard components of a (covariant) spherical tensor whereas upper indices denote components of a contravariant tensor. This notation lets us take advantage of the powerful Einstein convention for contracting pairs of tensors to scalars by summing over upper and lower indices. For example, the scalar product $q \cdot q$ is given by $q \cdot q = \sum_m q^m q_m = \sum_m |q_m|^2$.

[12] In writing an expression, such as $\hat{X}\Psi(q)$, it is to be understood that the operator $\hat{X}$ acts on the function Ψ and not on its value, $\Psi(q)$, at q. Moreover, when there are products of operators, the operators act in sequence from right to left. Thus, $\hat{X}\Psi(q) \equiv [\hat{X}\Psi](q)$ and $\hat{Y}\hat{X}\Psi(q) \equiv [\hat{Y}[\hat{X}\Psi]](q)$; cf. Exercise 2.6.

With the understanding that the collective model is only intended to be applied to a subset of nuclear states in which all the many complementary (non-collective) degrees of freedom are dormant, it is also possible to infer appropriate forms for other observables. For example, with the assumption that the neutrons and protons are uniformly distributed over the nucleus in its ground state and move essentially in unison, when in states described by collective model dynamics, it is appropriate to define electric quadrupole operators as being proportional to the mass quadrupole operators and given by

$$\hat{\mathfrak{M}}(\text{E2}; m) := \frac{Ze}{A}\,\hat{q}_m, \tag{2.22}$$

where Ze is the nuclear charge and A is the total nucleon number.

Magnetic dipole moments are $(L = 1)$ vector operators with positive parity and transform under rotations in the same way as angular momenta. A magnetic dipole operator $\vec{\mu}$ is therefore assumed to be given in the collective model, in units of the nuclear magneton $e\hbar/(2Mc)$, by

$$\vec{\mu} = g\,\hat{\mathbf{I}}\ (= g\hat{\mathbf{L}}) \tag{2.23}$$

where g is a *gyromagnetic ratio*, which can be adjusted to fit magnetic moment data, and $\hat{\mathbf{I}}$ is the total angular momentum. Note that spin degrees of freedom are not included in the basic hydrodynamic model but could be included, along with vorticity and other degrees of freedom, in an extended model. Thus, the total angular momentum $\hat{\mathbf{I}}$ of the basic model is equated with the orbital angular momentum $\hat{\mathbf{L}}$ and, for the collective dynamics of a uniformly charged nucleus, one might guess that $g \approx Z/A$. However, without specification of the actual current flows in the collective model, the parameter g, like the mass parameter B in the collective model kinetic energy, is regarded as an adjustable parameter.

Exercises

2.1 Show that

$$[q \otimes q]_0 = \frac{1}{\sqrt{5}}\beta^2, \quad [q \otimes q \otimes q]_0 = -\sqrt{\frac{2}{35}}\,\beta^3 \cos 3\gamma. \tag{2.24}$$

2.2 Evaluate the semi-major axes for the ellipsoid defined by Equation (2.18) for $\gamma = 0$, $\gamma = \pi/3$, and $0 < \gamma < \pi/3$.

2.3 By expanding

$$\frac{\partial}{\partial \beta} = \sum_m \frac{\partial q^m}{\partial \beta} \frac{\partial}{\partial q^m}, \tag{2.25}$$

show that the infinitesimal generator of a beta scale transformation, $\hat{P} = -\mathrm{i}\,\beta \partial/\partial\beta$, is given in terms of the $\{q_m\}$ coordinates by

$$\hat{P} = \frac{1}{\hbar} \sum_m \hat{q}^m \hat{\pi}_m. \tag{2.26}$$

2.4 Show that the infinitesimal generator, $\hat{\sigma} = -i\partial/\partial\gamma$, of a gamma rotation can be expressed as

$$\hat{\sigma} = \frac{\beta}{\hbar}\sum_m\left[-\sin\gamma\,\mathscr{D}^2_{0m}(\Omega) + \frac{1}{\sqrt{2}}\cos\gamma\left(\mathscr{D}^2_{2m}(\Omega) + \mathscr{D}^2_{-2m}(\Omega)\right)\right]\hat{\pi}^m. \qquad (2.27)$$

2.5 The angular momentum operators, $\{\hat{L}_m\}$, are components of an $L = 1$ tensor and therefore proportional to

$$\left[\hat{q}\otimes\hat{\pi}\right]_{1m} = \sum_{m_1 m_2}(2m_1\,2m_2|1m)\,\hat{q}_{m_2}\hat{\pi}_{m_1}, \qquad (2.28)$$

where $(2m_1\,2m_2|1m)$ is a Clebsch-Gordan coupling coefficient. Use the identity

$$[\hat{L}_0, \hat{q}_m] = m\hat{q}_m \qquad (2.29)$$

to infer that the correctly normalised expression is given by

$$\hat{L}_k = -\sqrt{10}\,\frac{i}{\hbar}\left[\hat{q}\otimes\hat{\pi}\right]_{1k}. \qquad (2.30)$$

2.6 Let G denote a group of transformations of a variable x. Let $\{\hat{T}(g); g\in G\}$ denote a corresponding group of transformations of a set of functions of x defined by

$$\Psi_g(x) := \hat{T}(g)\Psi(x) = \Psi(g^{-1}x), \quad g\in G. \qquad (2.31)$$

Show that

$$\hat{T}(g_1)\hat{T}(g_2)\Psi(x) = \Psi(g_2^{-1}g_1^{-1}x) = \Psi((g_1 g_2)^{-1}x) = \hat{T}(g_1 g_2)\psi(x). \qquad (2.32)$$

(This example exhibits the importance of paying attention to the order of the operations in the expression $\hat{T}(g_1)\hat{T}(g_2)\Psi(x)$ to avoid making the mistake of supposing that $\hat{T}(g_1)\hat{T}(g_2)\Psi(x) = \Psi(g_1^{-1}g_2^{-1}x)$.)

2.2 The volume element and Laplacian

Expressions for the volume element and Laplacian are needed for construction of the collective model Hilbert space and to determine the action of the model kinetic energy on its wave functions. An assumption of the Bohr model (see Section 4.8) is that its configuration space, $\mathbb{R}^5$, has the geometry of a Euclidean space. In terms of Cartesian coordinates, $\{x_i\}$, for $\mathbb{R}^5$, the volume element is assumed to be $dv = d^5x \equiv \prod_i dx_i$ and the Laplacian is assumed to be $\nabla^2 = \sum_i \partial^2/\partial x_i^2$. In terms of (curvilinear) spherical polar coordinates, the volume element is then

$$dv = \beta^4 d\beta\,\sin 3\gamma\,d\gamma\,d\Omega, \qquad (2.33)$$

where $d\Omega$ is the SO(3) volume element, Equation (A.107), and the Laplacian is

$$\nabla^2 = \frac{1}{\beta^4}\frac{\partial}{\partial\beta}\beta^4\frac{\partial}{\partial\beta} - \frac{\hat{\Lambda}^2}{\beta^2}, \qquad (2.34)$$

where $\hat{\Lambda}^2$, the SO(5) Casimir invariant, is expressed in terms of intrinsic SO(3) angular momenta $\{\bar{L}_k\}$ by

$$\hat{\Lambda}^2 = -\frac{1}{\sin 3\gamma}\frac{\partial}{\partial\gamma}\sin 3\gamma\frac{\partial}{\partial\gamma} + \sum_{k=1}^{3}\frac{\bar{L}_k^2}{4\sin^2(\gamma - 2\pi k/3)}. \qquad (2.35)$$

The remainder of this section, which the reader may choose to bypass, gives a detailed derivation[13] of these expressions by use of the *metric tensor*.

2.2.1 *The metric tensor*

The geometrical properties of an arbitrary Riemannian manifold are characterised by a metric tensor from which distances along paths, volume elements, curvature, and other geometric properties can be determined. In particular, an element of length ds^2 on a Riemannian manifold is defined in terms of a set of local coordinates, $\{\xi^\sigma\}$, by the expression

$$ds^2 = \sum_{\sigma\tau} g_{\sigma\tau}\, d\xi^\sigma d\xi^\tau, \qquad (2.36)$$

where $g_{\sigma\tau}$ is a component of the metric tensor.

When the manifold of interest is a Euclidean space or a submanifold of a Euclidean space, as are the collective model space, $\mathbb{R}^5$, and its four-sphere submanifold, S_4, there are many simplifications because the underlying Euclidean space is flat. An element of length is then defined in terms of Cartesian coordinates, $\{x_i\}$, by

$$ds^2 := \sum_i dx_i^2. \qquad (2.37)$$

Thus, the metric tensor at any point of a Euclidean space has components, relative to a system of Cartesian coordinates, given by $g_{ij} = \delta_{ij}$. However, an element of length in a Euclidean space can also be expressed relative to arbitrary curvilinear coordinates, $\{\xi^\sigma\}$, in the general form of Equation (2.36), with components of the metric tensor given by

$$g_{\sigma\tau} = \sum_i \frac{\partial x_i}{\partial\xi^\sigma}\frac{\partial x_i}{\partial\xi^\tau}. \qquad (2.38)$$

Moreover, because

$$\sum_i \frac{\partial x_i}{\partial\xi^\sigma}\frac{\partial\xi^\tau}{\partial x_i} = \delta_{\sigma,\tau}, \qquad (2.39)$$

the inverse, $\bar{g}$, of the metric tensor has matrix elements

$$\bar{g}^{\sigma\tau} = \sum_i \frac{\partial\xi^\sigma}{\partial x_i}\frac{\partial\xi^\tau}{\partial x_i}. \qquad (2.40)$$

---

[13]The original derivation was outlined by Bohr, *op. cit.* Footnote 1 on Page 97.

The volume element $dv := d^n x = \prod_i dx_i$ of an n-dimensional Euclidean space is expressed in terms of curvilinear coordinates by

$$dv = J\, d^5\xi \equiv J \prod_\sigma d\xi^\sigma, \tag{2.41}$$

where J, the Jacobian of the $\{dx_i\} \to \{d\xi^\sigma\}$ transformation, is the determinant of the matrix $\{\partial x_i / \partial \xi^\sigma\}$. Thus, because the metric matrix of Equation (2.38) is the product of the matrix $\{\partial x_i / \partial \xi^\sigma\}$ and its transpose, it follows that

$$J = \sqrt{\det(g)}\,. \tag{2.42}$$

2.2.2 The Laplacian for $\mathbb{R}^5$ in terms of collective coordinates

When expressed in terms of curvilinear coordinates, the Laplacian, ∇^2, for a Euclidean space takes the form

$$\nabla^2 = \sum_{\kappa\lambda} \frac{1}{J} \frac{\partial}{\partial \xi^\kappa} J \bar{g}^{\kappa\lambda} \frac{\partial}{\partial \xi^\lambda}, \tag{2.43}$$

of the so-called *Pauli operator* or, more generally, the *Laplace-Beltrami operator*, as shown in Box 2.I. We now specialise to the space $\mathbb{R}^5$ and derive Equation (2.34).

First observe that the 'length' β of a vector $q \in \mathbb{R}^5$ is expressed in terms of the $\{q_m\}$ quadrupole moments by

$$\beta^2 = q \cdot q = \sum_m |q_m|^2 = \sum_m q^m q_m. \tag{2.44}$$

Thus, an element of length is given by

$$ds^2 = \sum_m dq^m\, dq_m = \sum_{mn} g_{mn}\, dq^m\, dq^n, \tag{2.45}$$

with

$$g_{mn} = (-1)^m \delta_{m,-n}, \tag{2.46}$$

and the Laplacian is

$$\nabla^2 = \sum_m \frac{\partial^2}{\partial q_m \partial q^m}. \tag{2.47}$$

We next express ∇^2 in terms of the derivative operators

$$\frac{\partial}{\partial \beta}, \quad \frac{\partial}{\partial \gamma}, \quad \frac{\partial}{\partial \theta^k}, \quad k = 1, 2, 3, \tag{2.48}$$

where $\{\theta_k\}$ are local orientation angle coordinates.

2.I   THE LAPLACE-BELTRAMI OPERATOR

The Laplacian ∇^2 on a real Euclidean space is defined in terms of Cartesian coordinates $\{x_i\}$ by $\nabla^2 = \sum_i \partial^2/\partial x_i^2$. It is also expressed in general curvilinear coordinates by the so-called Pauli operator[a] (known in a more general context as the Laplace-Beltrami operator[b])

$$\nabla^2 = \sum_{\kappa\lambda} \frac{1}{J} \frac{\partial}{\partial \xi^\kappa} J \bar{g}^{\kappa\lambda} \frac{\partial}{\partial \xi^\lambda}, \tag{2.I.i}$$

where $J := \sqrt{\det(g)}$ is the Jacobian of the $\{x_i\} \to \{\xi^\lambda\}$ transformation. Equation (2.I.i) has the advantage that it is defined on any Riemannian manifold.

For a Euclidean space, the Laplace-Beltrami operator is derived as follows. First expand $\partial/\partial x_i$ in arbitrary coordinates

$$\frac{\partial}{\partial x_i} = \sum_\lambda \frac{\partial \xi^\lambda}{\partial x_i} \frac{\partial}{\partial \xi^\lambda}. \tag{2.I.ii}$$

Then, with the matrix elements of the metric and its inverse expanded according to the equations

$$g_{\kappa\lambda} = \sum_i \frac{\partial x_i}{\partial \xi^\kappa} \frac{\partial x_i}{\partial \xi^\lambda}, \quad \bar{g}^{\kappa\lambda} = \sum_i \frac{\partial \xi^\kappa}{\partial x_i} \frac{\partial \xi^\lambda}{\partial x_i}, \tag{2.I.iii}$$

Equation (2.I.ii) is written

$$\frac{\partial}{\partial x_i} = \sum_{\kappa\lambda} \frac{\partial x_i}{\partial \xi^\kappa} \bar{g}^{\kappa\lambda} \frac{\partial}{\partial \xi^\lambda}. \tag{2.I.iv}$$

Noting that $\partial/\partial x_i$ is skew Hermitian, matrix elements of the Laplacian are expressed as

$$\langle \psi | \nabla^2 | \varphi \rangle = -\sum_i \int \left(\frac{\partial \psi}{\partial x_i} \right)^* \frac{\partial \varphi}{\partial x_i} \, d^5 x \tag{2.I.v}$$

and re-expressed in terms of curvilinear coordinates by

$$\langle \psi | \nabla^2 | \varphi \rangle = -\sum_{i\kappa\lambda\kappa'\lambda'} \int d^5 x \left(\frac{\partial \psi}{\partial \xi^\lambda} \right)^* \frac{\partial x_i}{\partial \xi^\kappa} \bar{g}^{\kappa\lambda} \frac{\partial x_i}{\partial \xi^{\kappa'}} \bar{g}^{\kappa'\lambda'} \frac{\partial \varphi}{\partial \xi^{\lambda'}}$$

$$= -\sum_{\kappa\lambda} \int d^5 x \left(\frac{\partial \psi}{\partial \xi^\kappa} \right)^* \bar{g}^{\kappa\lambda} \frac{\partial \varphi}{\partial \xi^\lambda}$$

$$= -\sum_{\kappa\lambda} \int d^5 \xi \left(\frac{\partial \psi}{\partial \xi^\kappa} \right)^* J \bar{g}^{\kappa\lambda} \frac{\partial \varphi}{\partial \xi^\lambda}. \tag{2.I.vi}$$

The operator $\partial/\partial \xi^\kappa$ is skew-Hermitian relative to the $d^5 \xi$ volume element. Thus, we obtain

$$\langle \psi | \nabla^2 | \varphi \rangle = \sum_{\kappa\lambda} \int d^5 \xi \, \psi^* \frac{\partial}{\partial \xi^\kappa} J \bar{g}^{\kappa\lambda} \frac{\partial \varphi}{\partial \xi^\lambda} = \sum_{\kappa\lambda} \int d^5 x \, \psi^* \frac{1}{J} \frac{\partial}{\partial \xi^\kappa} J \bar{g}^{\kappa\lambda} \frac{\partial \varphi}{\partial \xi^\lambda} \tag{2.I.vii}$$

and, hence, the desired expression of the Laplace-Beltrami operator.

---

[a]Pauli W. (1933), *Handbuch der Physik Bd.*, Vol. XXIV/1 (Springer-Verlag, Berlin).
[b]Hicks N.J. (1971), *Notes on Differential Geometry* (Van Nostrand, London).

Before proceeding, we first express the angle derivatives in terms of infinitesimal generators of rotations. There are two ways to do this. If the orientation of a rotor is characterised by the rotation $\Omega \in SO(3)$ that takes the body-fixed axes into the space-fixed axes, then, under a rotation $\delta\Omega$ of the space-fixed axes, Ω is mapped to $\Omega \cdot \delta\Omega$. Conversely, for a rotation relative to the body-fixed axes, Ω is mapped to $\delta\Omega \cdot \Omega$. The former is a so-called *right rotation* and the latter a *left rotation*. The corresponding infinitesimal generators of right and left rotations are interpreted in the collective model as components of angular-momentum relative to the space-fixed and body-fixed axes respectively.

Consider the Wigner $\mathscr{D}$ function (discussed in Section A.8.2)

$$\mathscr{D}^L_{KM}(\Omega) := \langle LK|\hat{R}(\Omega)|LM\rangle. \tag{2.49}$$

We can parameterise $\delta\Omega$ in terms of local $\{\theta_k\}$ coordinates and define

$$\mathscr{D}^L_{KM}(\Omega \cdot \delta\Omega) := \langle LK|\hat{R}(\Omega)e^{i\sum_k \theta_k \hat{L}_k}|LM\rangle, \tag{2.50}$$

$$\mathscr{D}^L_{KM}(\delta\Omega \cdot \Omega) := \langle LK|e^{i\sum_k \theta_k \hat{L}_k}\hat{R}(\Omega)|LM\rangle. \tag{2.51}$$

These expressions provide alternative definitions of the local angle coordinates $\{\theta_k\}$. Because moments of inertia relative to the intrinsic (body-fixed) axes are invariant under rotations, the latter are more convenient. We then define

$$\frac{\partial}{\partial\theta^k}\mathscr{D}^L_{KM}(\Omega) := \frac{\partial}{\partial\theta^k}\langle LK|e^{i\theta_k \hat{L}_k}\hat{R}(\Omega)|LM\rangle\bigg|_{\theta_k=0} = i\bar{L}_k\mathscr{D}^L_{KM}(\Omega), \tag{2.52}$$

where $\bar{L}_k$ is an intrinsic angular-momentum operator for which

$$\bar{L}_k\mathscr{D}^L_{KM}(\Omega) = \sum_N \langle LK|\hat{L}_k|LN\rangle\langle LN|\hat{R}(\Omega)|LM\rangle$$

$$= \sum_N \langle LK|\hat{L}_k|LN\rangle\mathscr{D}^L_{NM}(\Omega). \tag{2.53}$$

Thus, from the so(3) matrix elements of $\hat{L}_0 := \hat{L}_3$ and $\hat{L}_\pm := \hat{L}_1 \pm i\hat{L}_2$,

$$\langle LK|\hat{L}_0|LN\rangle = N\delta_{K,N},$$
$$\langle LK|\hat{L}_\pm|LN\rangle = \sqrt{(L \mp N)(L \pm N + 1)}\,\delta_{K,N\pm1}, \tag{2.54}$$

we obtain, for the intrinsic angular-momentum operators

$$\bar{L}_0\mathscr{D}^L_{KM}(\Omega) = K\mathscr{D}^L_{KM}(\Omega),$$
$$\bar{L}_\pm\mathscr{D}^L_{KM}(\Omega) = \sqrt{(L \pm K)(L \mp K + 1)}\,\mathscr{D}^L_{K\mp1,M}(\Omega). \tag{2.55}$$

These expressions differ from the expressions

$$\hat{L}_0\mathscr{D}^L_{KM}(\Omega) = M\mathscr{D}^L_{KM}(\Omega),$$
$$\hat{L}_\pm\mathscr{D}^L_{KM}(\Omega) = \sqrt{(L \mp M)(L \pm M + 1)}\,\mathscr{D}^L_{K\mp1,M}(\Omega), \tag{2.56}$$

for the standard angular-momentum operators, relative to space-fixed axes.

Applying the differential operators of (2.48) to the quadrupole moments,

$$q_m(\beta,\gamma,\Omega) = \beta\cos\gamma\,\mathscr{D}^2_{0m}(\Omega) + \sqrt{\tfrac{1}{2}}\sin\gamma\left(\mathscr{D}^2_{2m}(\Omega) + \mathscr{D}^2_{-2m}(\Omega)\right), \qquad (2.57)$$

gives

$$\frac{\partial q_m}{\partial\beta} = \cos\gamma\,\mathscr{D}^2_{0m}(\Omega) + \sqrt{\tfrac{1}{2}}\sin\gamma\left(\mathscr{D}^2_{2m}(\Omega) + \mathscr{D}^2_{-2m}(\Omega)\right), \qquad (2.58)$$

$$\frac{\partial q_m}{\partial\gamma} = \beta\left[-\sin\gamma\,\mathscr{D}^2_{0m}(\Omega) + \sqrt{\tfrac{1}{2}}\cos\gamma\left(\mathscr{D}^2_{2m}(\Omega) + \mathscr{D}^2_{-2m}(\Omega)\right)\right], \qquad (2.59)$$

$$\frac{\partial q_m}{\partial\theta^k} = \mathrm{i}\beta\left[\cos\gamma\,(\bar{L}_k\mathscr{D}^2_{0m})(\Omega) + \sqrt{\tfrac{1}{2}}\sin\gamma\,(\bar{L}_k\,[\mathscr{D}^2_{2m} + \mathscr{D}^2_{-2m}])(\Omega)\right]. \qquad (2.60)$$

It follows, from Equation (2.55), that

$$\bar{L}_1\mathscr{D}^2_{0m} = \sqrt{\tfrac{3}{2}}\left[\mathscr{D}^2_{1m} + \mathscr{D}^2_{-1,m}\right], \qquad (2.61)$$

$$\bar{L}_1\left[\mathscr{D}^2_{2m} + \mathscr{D}^2_{-2,m}\right] = \left[\mathscr{D}^2_{1m} + \mathscr{D}^2_{-1,m}\right] \qquad (2.62)$$

and, hence, that

$$\frac{\partial q_m}{\partial\theta^1} = -\mathrm{i}\beta\left(\sqrt{\tfrac{3}{2}}\cos\gamma + \sqrt{\tfrac{1}{2}}\sin\gamma\right)\left[\mathscr{D}^2_{1m}(\Omega) + \mathscr{D}^2_{-1,m}(\Omega)\right]. \qquad (2.63)$$

Similarly, we obtain

$$\frac{\partial q_m}{\partial\theta^2} = \beta\left(\sqrt{\tfrac{3}{2}}\cos\gamma - \sqrt{\tfrac{1}{2}}\sin\gamma\right)\left[\mathscr{D}^2_{1m}(\Omega) - \mathscr{D}^2_{-1,m}(\Omega)\right], \qquad (2.64)$$

$$\frac{\partial q_m}{\partial\theta^3} = \mathrm{i}\beta\sqrt{2}\sin\gamma\left[\mathscr{D}^2_{2m}(\Omega) - \mathscr{D}^2_{-2,m}(\Omega)\right]. \qquad (2.65)$$

Finally, by means of the identity

$$\sum_m(-1)^m\mathscr{D}^2_{\mu,m}(\Omega)\mathscr{D}^2_{\mu',-m}(\Omega) = \sqrt{5}\,(2\mu'\,2\mu|0,0) = (-1)^\mu\delta_{\mu',-\mu}, \qquad (2.66)$$

we obtain the non-zero elements of the metric matrix

$$g_{\beta,\beta} = \sum_m\frac{\partial q^m}{\partial\beta}\frac{\partial q_m}{\partial\beta} = 1, \quad g_{\gamma,\gamma} = \beta^2, \qquad (2.67)$$

and, with g_{θ^k,θ^l} denoted simply by $g_{k,l}$,

$$g_{1,1} = 2\beta^2\left(\sqrt{\tfrac{3}{2}}\cos\gamma + \sqrt{\tfrac{1}{2}}\sin\gamma\right)^2 = 4\beta^2\sin^2\left(\gamma - \tfrac{2\pi}{3}\right), \qquad (2.68)$$

$$g_{2,2} = 2\beta^2\left(\sqrt{\tfrac{3}{2}}\cos\gamma - \sqrt{\tfrac{1}{2}}\sin\gamma\right)^2 = 4\beta^2\sin^2\left(\gamma - \tfrac{4\pi}{3}\right), \qquad (2.69)$$

$$g_{3,3} = 4\beta^2\sin^2\gamma. \qquad (2.70)$$

Thus, for the above coordinates the metric matrix is diagonal and has determinant

$$J^2 = 64\beta^8\sin^2\gamma\,\sin^2(\gamma - 2\pi/3)\,\sin^2(\gamma - 4\pi/3) = 4\beta^8\sin^2 3\gamma. \qquad (2.71)$$

Insertion of these expressions into Equation (2.43) finally gives Equation (2.34). The volume element (2.33) is likewise obtained from the definition (2.41) to within a convenient proportionality constant, where $d\Omega$ is the standard SO(3) volume element.[14]

Exercises

2.7 Show that if the metric on a Euclidean space with curvilinear coordinates $\{\xi^\sigma\}$ and Cartesian coordinates $\{x_i\}$ is given by

$$g_{\sigma\tau}(x) = \sum_i \frac{\partial x_i}{\partial \xi^\sigma} \frac{\partial x_i}{\partial \xi^\tau}, \qquad (2.72)$$

then the inverse of the metric is given by

$$\bar{g}^{\sigma\tau}(x) = \sum_i \frac{\partial \xi^\sigma}{\partial x_i} \frac{\partial \xi^\tau}{\partial x_i}. \qquad (2.73)$$

2.8 Given a system of (complex) spherical coordinates $\{q_m, m = 0, \pm1, \pm2\}$ for the real Euclidean space $\mathbb{R}^5$ which satisfy the relationship

$$q_m^* = (-1)^m q_{-m}, \qquad (2.74)$$

define real Cartesian coordinates by

$$x_0 = q_0, \qquad \begin{aligned} x_m &= \tfrac{1}{\sqrt{2}}(q_m + q_m^*), \\ x_{-m} &= \tfrac{-i}{\sqrt{2}}(q_m - q_m^*), \end{aligned} \qquad \text{for } m > 0. \qquad (2.75)$$

Then, with the assumption that $g_{m,n} = \delta_{m,n}$, in terms of the $\{x_m\}$ coordinates, show that in terms of the complex $\{q_m\}$ coordinates, $g_{m,n} = (-1)^m \delta_{m,-n}$.

2.3 Solvable submodels of the Bohr collective model

There are three standard submodels of the Bohr collective model which are analytically solvable: one for spherical nuclei, and two for deformed nuclei. However, the most useful version of the collective model for deformed nuclei is based on an adiabatic approximation and is not, strictly speaking, a submodel. It is described in Section 2.4. A full algebraic treatment of the Bohr model is given in Chapter 4.

---

[14]If an SO(3) rotation Ω is parameterised in terms of Euler angles, $\Omega = \Omega(\psi, \theta, \varphi)$, in the usual way, then, to within a convenient norm, the SO(3)-invariant volume element is $d\Omega = \sin\theta \, d\psi \, d\theta \, d\varphi$.

2.3.1 *The harmonic spherical vibrator submodel*

(i) *Energy levels*

The Hamiltonian for the harmonic spherical vibrator submodel of the Bohr model,

$$\hat{H}_{\mathrm{HV}} = -\frac{\hbar^2}{2B}\nabla^2 + \tfrac{1}{2}C\beta^2, \tag{2.76}$$

which can also be expressed as

$$\hat{H}_{\mathrm{HV}} = \frac{1}{2B}\sum_m \left(\hat{\pi}_m\hat{\pi}^m + B^2\omega^2\hat{q}_m\hat{q}^m\right), \tag{2.77}$$

is that of a five-dimensional harmonic oscillator with $\omega^2 = C/B$. The Schrödinger equation for this Hamiltonian is solved in the standard way by writing it in the form

$$\hat{H}_{\mathrm{HV}} = (\hat{N} + \tfrac{5}{2}\hat{I})\hbar\omega, \quad \hat{N} = \sum_m d_m^\dagger d^m; \tag{2.78}$$

where $\hat{I}$ is the identity operator,

$$d_m^\dagger := \frac{1}{\sqrt{2}}\left(a\hat{q}_m - \frac{\mathrm{i}}{a\hbar}\hat{\pi}_m\right), \quad d^m := \frac{1}{\sqrt{2}}\left(a\hat{q}^m + \frac{\mathrm{i}}{a\hbar}\hat{\pi}^m\right) = (-1)^m d_{-m}, \tag{2.79}$$

are raising and lowering operators for the five-dimensional harmonic oscillator, and $a := \sqrt{\omega B/\hbar}$ is an inverse width parameter. The $\{d_m^\dagger, d^m\}$ operators, referred to as *d*-phonon operators,[15] satisfy the boson commutation relations

$$[d^m, d^{m'}] = [d_m^\dagger, d_{m'}^\dagger] = 0, \quad [d^m, d_{m'}^\dagger] = \delta_{m'}^m, \tag{2.80}$$

$$[\hat{N}, d_m^\dagger] = d_m^\dagger, \quad [\hat{N}, d^m] = -d^m, \tag{2.81}$$

of a Heisenberg-Weyl algebra. Thus, if $|0\rangle$ is the *d*-phonon vacuum state, i.e., the state $|0\rangle$ annihilated by all the lowering operators

$$d_m|0\rangle = 0, \quad m = 0, \pm1, \pm2, \tag{2.82}$$

then the states created from this state by homogeneous polynomials of degree N in the *d*-phonon creation operators $\{d_m^\dagger\}$ are eigenstates of $\hat{H}_{\mathrm{HV}}$ with energies

$$E_N = (N + \tfrac{5}{2})\hbar\omega, \quad N = 0, 1, 2, \ldots \tag{2.83}$$

---

[15]Whereas the harmonic oscillator quanta of electromagnetic radiation are called *photons*, those of vibrational systems are often called *phonons*. Both are massless bosons.

(ii) *Basis states and eigenstates*

A simple orthonormal basis of eigenstates for the harmonic spherical vibrator is given by the uncoupled states

$$|N_{-2}\ldots N_2\rangle = \prod_{m=-2}^{2} \frac{1}{\sqrt{N_m!}}(d_m^\dagger)^{N_m}|0\rangle. \tag{2.84}$$

These states have energies given by Equation (2.83) with $N = \sum_m N_m$. However, for practical purposes, one needs states labelled by both energy and angular-momentum quantum numbers, L, M. Thus, we suppose that $\{|N\alpha LM\rangle\}$ is a set of eigenstates of $\hat{H}_{\mathrm{HV}}$, where an additional multiplicity label α is included to distinguish states of the same N, L and M.

For values of $N \leq 2$, for which at most one SO(3) multiplet of a given L occurs, the multiplicity label, α, can be omitted and the states $\{|NLM\rangle\}$ are given by

$$|12M\rangle = d_M^\dagger|0\rangle, \tag{2.85}$$

$$|2LM\rangle = \tfrac{1}{\sqrt{2}}[d^\dagger \otimes d^\dagger]_{LM}|0\rangle, \quad L = 0,2,4, \tag{2.86}$$

where $|0\rangle$ is the d-phonon vacuum state, and the norm of a coupled state is determined by expanding it in the uncoupled, but orthonormal basis (2.84), e.g.,

$$[d^\dagger \otimes d^\dagger]_{LM}|0\rangle := \sum_{m_1 m_2}(2m_1\,2m_2|LM)\,d_{m_2}^\dagger d_{m_1}^\dagger|0\rangle. \tag{2.87}$$

For higher values of N, the states of maximum L and M are given by

$$|N,L=M=2N\rangle = \frac{1}{\sqrt{n!}}(d_2^\dagger)^N|0\rangle. \tag{2.88}$$

A list of the number of occurrences of the L values for a given N is obtained by a so-called "peeling-off" procedure as follows. For a given value of N, start by listing all the values of N_m for which $\sum_m N_m = N$ and $\sum_m N_m m = M$, for $M \geq 0$. Each entry in the list, shown in Table 2.1 for $N = 3$, corresponds to an N-phonon state $|N_{-2}\ldots N_2\rangle$. Then, if $\mathcal{N}(M)$ is the number of states obtained in this way for

Table 2.1: The possible values of n_m for given values of $M \geq 0$ and $n = 3$.

M	6	5	4		3			2				1				0				
n_2	3	2	2	1	2	1		2	1	1		1	1			1	1			
n_1		1		2		1	3		1		2	1		2	1			2	1	
n_0						1				2	1		1		2	1			1	3
n_{-1}					1				1				1	1			2		1	
n_{-2}								1				1				1		1		
$\mathcal{N}(M)$	1	1	2		3			4				4				5				

each M, the multiplicity of distinct states of a given value of L is $\mathcal{N}(L) - \mathcal{N}(L+1)$.

Table 2.1 implies that, for $N = 3$, L can take the values 6, 4, 3, 2, and 0, each with multiplicity one. Thus, the $N = 3$ states are uniquely defined to within normalisation factors by coupling an extra $d^\dagger$ phonon-creation operator to the already known $N = 2$ states as follows. First observe that each of the $N = 3$, $L = 0$ and 6 states can be reached from the $N = 2$ states in only one way, i.e.,

$$|300\rangle \propto [d^\dagger \otimes |22\rangle]_0, \quad |36M\rangle \propto [d^\dagger \otimes |24\rangle]_{6M}. \tag{2.89}$$

Thus, these coupled states can be expanded in terms of the orthonormal $\{|N_{-2} \ldots N_2\rangle\}$ basis and their norms determined. An $N = 3$, $L = 4$ state is similarly obtained, to within a norm factor, by coupling a $d^\dagger$ phonon-creation operator to an $N = 2$ state of either $L = 2$ or 4. However, because there is only one $N = 3$, $L = 4$ state, the two states constructed in this way must be proportional and when normalised, both will give an identical $L = 4$ state.

This simple brute-force method for constructing a basis works well for states of small phonon number. Moreover, the expansion of the $\{|NLM\rangle\}$ states obtained in the orthonormal $\{|N_{-2} \ldots N_2\rangle\}$ basis has the useful property that it makes the calculation of matrix elements of the type $\langle N + 1, L'M'|d_m^\dagger|NLM\rangle$ very straightforward. Thus, low-energy harmonic vibrational states and their properties can be determined in this way. Figure 2.2 shows the harmonic vibrational states up to excitation energy $3\hbar\omega$.

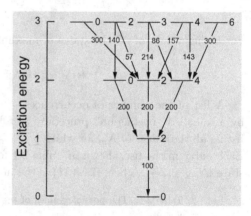

Figure 2.2: Some energy levels and SO(3)-reduced E2 transition rates for the harmonic spherical-vibrator model. The excitation energies are shown relative to the excitation energy of the one- phonon state and SO(3)-reduced E2 transition rates are shown in units such that the reduced transition rate for decay of the one-phonon state is 100. Note, that the summed SO(3)-reduced transition rates for decay of any N-phonon harmonic vibrational states is $100N$. Thus, for example, the reduced E2 transition rate for the $2_3 \rightarrow 4_1$ transition (not shown) is 103 to within round-off errors. These results are derived in Chapter 4.

It is shown in Chapter 4, that algebraic methods provide analytical ways of computing harmonic spherical-vibrator states and matrix elements of the quadrupole transitions which enable them to be used effectively as basis states for the diagonalisation of more general collective-model Hamiltonians.

(iii) *E2 matrix elements and transitions*

The operator, $\widehat{\mathfrak{M}}(E2; m)$, for nuclear E2 transitions is the electric quadrupole moment which is presumed, in the Bohr model, to be proportional to the mass quadrupole moment and given by Equation (eq:2.E2op). In the harmonic spherical-vibrator model, it is expressed in terms of the d-phonon creation and annihilation operators of Equation (2.79) by

$$\hat{\mathfrak{M}}(\text{E2}; m) = \frac{Ze}{A}\hat{q}_m = \frac{Ze}{\sqrt{2}Aa}(d_m^\dagger + d_m). \tag{2.90}$$

A reduced E2 transition rate between two multiplets of SO(3) states is defined by summing the squared matrix elements of the $\hat{\mathfrak{M}}(\text{E2}; m)$ operators over the states of the final level and averaging over the initial states. Thus, an SO(3)-reduced E2 transition rate is given (cf. Exercises 2.12 – 2.14) by

$$B(\text{E2}; \alpha_i L_i \to \alpha_f L_f) = \sum_{M_i m M_f} (2L_i + 1)^{-1}|\langle\alpha_f L_f M_f|\hat{\mathfrak{M}}(\text{E2}; m)|\alpha_i L_i M_i\rangle|^2$$

$$= \frac{|\langle\alpha_f L_f\|\hat{\mathfrak{M}}(\text{E2})\|\alpha_i L_i\rangle|^2}{2L_i + 1}, \tag{2.91}$$

where reduced matrix elements are defined for an arbitrary spherical tensor by the Wigner-Eckart theorem (Section A.3.3) as follows.

The action of any spherical tensor operator $\hat{T}_L$ of angular momentum L on a state $|N_i\alpha_i L_i M_i\rangle$ of an orthonormal basis is given by

$$\hat{T}_{LM}|N_i\alpha_i L_i M_i\rangle = \sum_{\alpha_f N_f L_f M_f} |N_f\alpha_f L_f M_f\rangle\langle N_f\alpha_f L_f M_f|\hat{T}_{LM}|N_i\alpha_i L_i M_i\rangle. \tag{2.92}$$

Thus, by use of the Wigner-Eckart theorem,

$$\langle N_f\alpha_f L_f M_f|\hat{T}_{LM}|N_i\alpha_i L_i M_i\rangle = (L_i M_i\, LM|L_f M_f)\frac{\langle N_f\alpha_f L_f\|\hat{T}_L\|N_i\alpha_i L_i\rangle}{\sqrt{2L_f + 1}}, \tag{2.93}$$

we obtain the useful expressions

$$\left[\hat{T}_L \otimes |N_i\alpha_i L_i\rangle\right]_{L_f M_f} = \sum_{\alpha_f N_f}|N_f\alpha_f L_f M_f\rangle\frac{\langle N_f\alpha_f L_f\|\hat{T}_L\|N_i\alpha_i L_i\rangle}{\sqrt{2L_f + 1}},$$

$$\langle N_f\alpha_f L_f\|\hat{T}_L\|N_i\alpha_i L_i\rangle = \sqrt{2L_f + 1}\,\langle N_f\alpha_f L_f M_f|\left[\hat{T}_L \otimes |N_i\alpha_i L_i\rangle\right]_{L_f M_f}, \tag{2.94}$$

where

$$\left[\hat{T}_L \otimes |N_i\alpha_i L_i\rangle\right]_{L_f M_f} := \sum_{MM_i}\hat{T}_{LM}|N_i\alpha_i L_i M_i\rangle\,(L_i M_i\, LM|L_f M_f). \tag{2.95}$$

For example, Equations (2.85) and (2.86) give

$$[d^\dagger \times |0\rangle]_{2M} = d_M^\dagger|0\rangle = |12M\rangle \tag{2.96}$$

and

$$[d^\dagger \times |12\rangle]_{LM} = [d^\dagger \otimes d^\dagger]_{LM}|0\rangle = \sqrt{2}\,|2LM\rangle. \tag{2.97}$$

Hence, by Equation (2.94), we obtain the reduced matrix elements

$$\langle 12\|a\hat{q}\|0\rangle = \sqrt{\tfrac{5}{2}}, \quad \langle 2L\|a\hat{q}\|12\rangle = \sqrt{2L+1}, \tag{2.98}$$

and the reduced E2 transition rates (see Exercises 2.12 and 2.13)

$$B(E2; 12 \to 0) = \left(\frac{Ze}{A}\right)^2 \frac{\hbar}{2B\omega}, \tag{2.99}$$

$$B(E2; 2L \to 12) = \left(\frac{Ze}{A}\right)^2 \frac{\hbar}{B\omega}. \tag{2.100}$$

Reduced E2 transition rates between states up to $n = 3$ are shown in Figure 2.2.

The static electric quadrupole moment $Q(L_i)$ of a state $|L_iM_i\rangle$ of angular momentum L_i is defined by

$$Q(L_i) := \sqrt{\frac{16\pi}{5}}\, \langle L_iM = L_i|\hat{\mathfrak{M}}(E2; 0)|L_iM = L_i\rangle, \tag{2.101}$$

where the factor $\sqrt{16\pi/5}$ is the difference between the Cartesian quadrupole moment $2Q_{33} - Q_{11} - Q_{22}$ and q_0, as given by Equation (2.4). Thus, with the definition Equation (2.90) of the Bohr model electric quadrupole operator, it is seen that all static electric quadrupole moments are predicted to vanish in the harmonic vibrational limit of the model.

Efficient algebraic methods for computing matrix elements and E2 transition rates are given in Chapter 4.

(iv) *Comparisons of the harmonic spherical vibrator model with experiment*

The highly collective nature of many low-energy excitations, particularly in doubly-even nuclei, as considered in Section 1.7.3, has been established experimentally. The Bohr model provides a macroscopic perspective of such phenomena. In particular, the simply-solvable harmonic spherical-vibrator limit of the model makes detailed predictions that can be confronted with the available data.

The occurrence of harmonic spherical-vibrational behaviour in nuclei has been considered in Section 1.7.3. Taking Table 1.1 as a guide, the isotopes ^{106,108,110}Pd are among the best candidates for spherical nuclear vibrators for which substantial $B(E2)$ data are available. The relevant $B(E2)$ data for these nuclei are summarised in Table 2.2.

The reduced E2 transition rate for decay of the lowest $L = 2$ state to the ground state can be seen in Figure 1.48 to lie between 45 and 55 W.u. for the ^{106,108,110}Pd isotopes; this indicates a high degree of collectivity. However, the results shown in Table 2.2 for $B(E2)$ ratios, particularly those involving the $L = 0$, 3 candidates for three-phonon states, do not provide convincing support for the harmonic spherical-vibrator interpretation of these nuclei. Table 2.2 also contains information on diagonal quadrupole moments which indicate that the ^{106,108,110}Pd isotopes

Table 2.2: Electric quadrupole properties of the ^{106,108,110}Pd isotopes compared with those of a harmonic spherical vibrator. The quantities tabulated are $B(E2)$ ratios, e.g., $b_{4_1 2_1} \equiv B(E2; 4_1 \to 2_1)/B(E2; 2_1 \to 0_1)$ and quadrupole moments, $Q(L_i)$ as defined in the text, in units of e·b, where $|L_i\rangle$ denotes the i'th lowest state of angular momentum L. The data are taken from Svensson L.E. (1989), Ph.D. thesis, Univ. of Uppsala and Svensson L.E. et al. (1995), Nucl. Phys. **A584**, 547.

	th.	^{106}Pd expt.	^{106}Pd $\frac{\text{expt.}}{\text{th.}}$	^{108}Pd expt.	^{108}Pd $\frac{\text{expt.}}{\text{th.}}$	^{110}Pd expt.	^{110}Pd $\frac{\text{expt.}}{\text{th.}}$
$b_{4_1 2_1}$	2	1.72	0.86	1.55	0.78	1.64	0.82
$b_{2_2 2_1}$	2	0.93	0.47	1.02	0.51	0.88	0.44
$b_{0_2 2_1}$	2	1.03	0.52	1.08	0.54	0.52	0.26
$b_{2_2 0_1}$	0	0.02		0.01		0.01	
$b_{6_1 4_1}$	3	2.16	0.72	2.05	0.68	1.97	0.66
$b_{4_2 4_1}$	1.43	0.56	0.39	0.61	0.43	0.58	0.41
$b_{4_2 2_2}$	1.57	0.85	0.54	1.11	0.71	0.62	0.39
$b_{3_1 4_1}$	0.86	0.14	0.16	–	–	0.23	0.27
$b_{3_1 2_2}$	2.14	0.41	0.19	0.52	0.24	0.46	0.21
$b_{2_3 2_2}$	0.57	0.25	0.44	0.11	0.19	0.26	0.46
$b_{2_3 0_2}$	1.4	0.93	0.66	1.54	1.10	1.41	1.01
$b_{2_3 4_1}$	1.03	0.13	0.12	1.12	1.09	1.06	1.03
$b_{0_3 2_2}$	3	0.34	0.11	0.12	0.04	0.95	0.36
$b_{4_2 2_1}$	0	2×10^{-4}		4×10^{-3}		3×10^{-3}	
$b_{3_1 2_1}$	0	0.01		0.01		7×10^{-3}	
$b_{2_3 0_1}$	0	3×10^{-3}		2×10^{-3}		6×10^{-3}	
$b_{2_3 2_1}$	0	0.01		0.03		0.02	
$b_{0_3 2_1}$	0	0.06		0.01		0.04	
$Q(2_1)$	0	−0.72		−0.83		−0.87	
$Q(2_2)$	0	+0.52		+0.73		+0.70	
$Q(4_1)$	0	−1.02		−0.82		−1.6	

possess significant static deformations. We take these results as an indication that, while collective quadrupole vibrations are potentially an important component of the dynamics of spherical and near spherical nuclei, there is little evidence to suggest that the vibrations are harmonic. In fact, it appears likely that collective vibrational states are not only anharmonic but that multi-phonon states also mix strongly with other, non-collective, states of nuclei.

2.3.2 The Wilets-Jean submodel

A complication in solving the Schrödinger equation for a Bohr model Hamiltonian

$$\hat{H} = -\frac{\hbar^2}{2B}\nabla^2 + V(\beta, \gamma) \tag{2.102}$$

with an arbitrary potential energy $V(\beta, \gamma)$, arises from the coupling between the several degrees of freedom. Because of rotational invariance, the potential $\hat{V}$ is independent of Ω (the orientation angles). A major simplification, observed by

Wilets and Jean,[16] results when the potential is also independent of γ. Thus, Wilets-Jean models are invariant under all SO(5) transformations. They are said to be *gamma soft*.

Because a Wilets-Jean Hamiltonian is SO(5) invariant, its eigenstates occur in multiplets that span irreducible representations (irreps) of the SO(5) group. Energy eigenstates are then labelled by SO(5) quantum numbers and their wave functions are products of orbital SO(5) wave functions (SO(5) spherical harmonics) and radial (beta) functions, i.e.,

$$\Psi_{\nu v\alpha LM}(\beta,\gamma,\Omega) = R_\nu(\beta)\mathcal{Y}_{v\alpha LM}(\gamma,\Omega). \tag{2.103}$$

We know from Equation (2.34) that $\nabla^2 = \hat{\Delta} - \hat{\Lambda}^2/\beta^2$, where

$$\hat{\Delta} = \frac{1}{\beta^4}\frac{\partial}{\partial\beta}\beta^4\frac{\partial}{\partial\beta} \tag{2.104}$$

and $\hat{\Lambda}^2$ is the SO(5) Casimir operator. As discussed further in Chapter 4 (cf. Equation (4.242)), $\hat{\Lambda}^2$ has eigenfunctions given by SO(5) spherical harmonics and satisfies the equation

$$\hat{\Lambda}^2\mathcal{Y}_{v\alpha LM}(\gamma,\Omega) = v(v+3)\mathcal{Y}_{v\alpha LM}(\gamma,\Omega), \quad v = 0,1,2,\ldots, \tag{2.105}$$

where v is an SO(5) angular-momentum quantum number (often referred to as SO(5) seniority). Thus, the energies and beta wave functions of a Wilets-Jean model are solutions of the eigenvalue equation

$$\left[-\frac{\hbar^2}{2B}\left(\hat{\Delta} - \frac{v(v+3)}{\beta^2}\right) + V(\beta)\right]R_\nu(\beta) = E_{\nu v}R_\nu(\beta). \tag{2.106}$$

A rigid-beta Wilets-Jean model assumes, in addition, that the β coordinate is frozen at some non-zero value, β_0. The beta degree of freedom is then suppressed and the Hamiltonian reduces to

$$\hat{H}_{\text{WJ}} = \frac{\hbar^2}{2B\beta_0^2}\hat{\Lambda}^2. \tag{2.107}$$

The energies of the model are then given by

$$E_v = \frac{\hbar^2}{2B\beta_0^2}v(v+3) \tag{2.108}$$

and the energy-level spectrum is as shown in Figure 2.3.

The spectrum of angular-momentum states in a given energy level of SO(5) seniority v is easily inferred from the harmonic spherical vibrator spectrum of angular-momentum states. As shown in Section 4.1.3, a harmonic spherical vibrator energy level of N oscillator quanta, comprises states that span SO(5) irreps of seniority $v = N, n - N, \ldots, 1$ or 0. Thus, for example, the $N = 0$ ground state, which has angular-momentum $L = 0$, also has seniority $v = 0$. Similarly, the $N = 1$ level,

[16]Wilets L. and Jean M. (1956), *Phys. Rev.* **102**, 788.

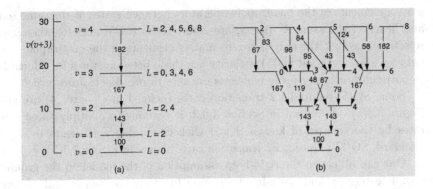

Figure 2.3: Energy levels of the rigid-beta Wilets-Jean model in units of $\hbar^2/(2B\beta_0^2)$. SO(5)-reduced E2 transition rates are shown in (a) in units such that the $v = 1$ to $v = 0$ reduced transition rate is 100. Conventional SO(3)-reduced E2 transition rates are shown in (b). Note that the summed SO(3)-reduced E2 transition rates for decay of any state of a given seniority add up to the SO(5)-reduced E2 transition rate for decay of a state of that seniority. Thus, for example, the summed SO(3)-reduced E2 transition rates for decay of the $v = 4$ state of angular momentum $L = 2$ add up to 182 which means that the $2_3 \rightarrow 4_2$ E2 transition rate (not shown) has value 32 and the $4_3 \rightarrow 6_1$ is 2 (to within round-off errors). These results are derived in Chapter 4.

which has angular-momentum $L = 2$, also has seniority $v = 1$. The $N = 2$ level, consists of states of angular momentum $L = 0$, 2 and 4 and seniority $v = 0$ and 2. Knowing that the $v = 0$ state has angular momentum $L = 0$, we then infer that the $v = 2$ states have $L = 2$ and 4. Continuing in this way, we derive the angular-momentum states shown for the Wilets-Jean model in Figure 2.3.

With the beta coordinate frozen at $\beta = \beta_0$, the expression (2.12) for a quadrupole moment operator becomes

$$\hat{q}_m = \beta_0 \hat{\mathcal{Q}}_m. \tag{2.109}$$

Thus, the calculation of reduced matrix elements of $\hat{q}$ and E2 transition rates require only the evaluation of SO(5) integrals in the expression (cf. Equation (2.94))

$$\frac{\langle v'\alpha'L'\|\hat{q}\|v\alpha L\rangle}{\sqrt{2L'+1}} = \beta_0 \int \mathcal{Y}^*_{v'\alpha'L'M'}(\gamma,\Omega)\,[\mathcal{Q}(\gamma,\Omega)\otimes\mathcal{Y}_{v\alpha L}(\gamma,\Omega)]_{L'M'}\,\sin 3\gamma\,d\gamma\,d\Omega. \tag{2.110}$$

It is shown how to calculate such matrix elements and E2 transition rates algebraically in Chapter 4 (Section 4.6.3). The algebraic calculations make use of the fact that the quadrupole moments, $\{\hat{q}_m\}$, are components of a $v = 1$, SO(5) tensor. As a result, all matrix elements of the quadrupole operators between states of SO(5) seniority differing by more than $\Delta v = 1$ are zero. In addition, with respect to inversion in the five-dimensional space of the quadrupole moments, in which $\hat{q}_m \rightarrow -\hat{q}_m$, it is determined that states of SO(5) seniority v have parity $(-1)^v$.[17] As a result, all

---

[17]Note that inversion in the five-dimensional space of the quadrupole moments, which we refer to as $\mathbb{R}^5$ inversion, differs from the more familiar $\mathbb{R}^3$ inversion in which the coordinates of a

matrix elements of the quadrupole operators between states of the same seniority, and hence of the same $\mathbb{R}^5$ parity, are zero. These two results together imply the selection rule that the only non-zero matrix elements of the quadrupole operators between states of good SO(5) seniority are those between states for which $\Delta v = 1$. In particular, *all quadrupole moments are zero in the Wilets-Jean model.*

Some SO(5)-reduced E2 transition rates (defined in Chapter 4) are shown in Figure 2.3(a). As shown in Section 4.6.3, it remains to multiply these transition rates by the squares of known SO(5) Clebsch-Gordan coefficients to obtain the standard, SO(3)-reduced, E2 transition rates shown in Figure 2.3(b).

One can object to the rigid-beta assumption of the model on the grounds that complete rigidity and delta-function wave functions are unphysical. However, a physical situation in which the model might be applied would be one in which the β wave functions were considered to be sharply peaked about a mean equilibrium value and for which excitations of the β degree of freedom occurred at high energies. Thus, the model can be viewed as a limit of a more realistic physical model in which, because of energy differences, the beta degree of freedom played a negligible role in the dynamics of the low-energy collective states. Similar comments apply to the following rigid-rotor model.

Although there is no reason to expect that a Wilets-Jean model should be realised in actual nuclei, Figure 1.51 and specifically the energy-level spectrum of ^{108}Ru, with ratios of energy levels given experimentally by $E_{4_1}/E_{2_1} = 2.75$ and $E_{2_2}/E_{2_1} = 2.93$, compared with a common value of 2.50 in the Wilets-Jean model, suggests that one should keep an eye open for its possible applications. Figure 1.51 further suggests that the Wilets-Jean model might be applicable in regions between spherical quadrupole vibrational nuclei and nuclei that exhibit characteristic rotor model behaviour. This possibility will be considered in more detail in Volume 2. The Wilets-Jean model is also a useful starting point for a collective model with low-energy gamma-vibrational states that are adiabatically decoupled from higher-energy beta vibrational excitations, as shown in Chapter 4.

2.3.3 *The rigid-rotor submodel*

Further simplifications result when both the β and γ coordinates are frozen. The only remaining degrees of freedom are then rotations. Thus, with the SO(5) Casimir operator given by Equation (2.35) and with β and γ taking the fixed values β_0 and γ_0, respectively, the collective model Hamiltonian reduces to that of a rotor,

$$\hat{H}_{\text{RRot}} = \sum_{k=1}^{3} \frac{\hbar^2 \bar{L}_k^2}{2\Im_k}, \tag{2.111}$$

---

three-component vector, such as a position vector, $\mathbf{r} \equiv (x, y, z)$, is mapped to its negative. With respect to $\mathbb{R}^3$ inversion, the five-component quadrupole tensor, which is quadratic in the nucleon coordinates, has positive $\mathbb{R}^3$ parity in contrast to its negative parity with respect to $\mathbb{R}^5$ inversion.

with moments of inertia

$$\Im_k = 4B\beta_0^2 \sin^2(\gamma_0 - 2\pi k/3). \tag{2.112}$$

These moments of inertia are shown in Figure 2.4.

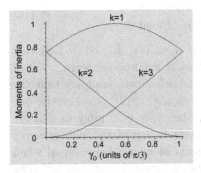

Figure 2.4: Moments of inertia, in units of $4B\beta_0^2$, as functions of γ_0 in the rotor limit of the Bohr collective model. Note that $\Im_2 = \Im_3$ when $\gamma_0 = \pi/6$ (cf. Equation (2.114)) for any value of β_0. (Based on a figure of Meyer-ter-Vehn J. (1975), *Nucl. Phys.* **A249**, 111.)

It is important to note that, although the mass parameter, B, was assigned the irrotational-flow value that it would have for a quantum fluid, in Bohr's original formulation of his model,[18] an assumption of irrotational flow is not an essential component of the model and is not made in the current treatment. Thus, B can be treated as a parameter and its value adjusted so that Equation (2.100) fits observed E2 transition rates for vibrational nuclei. In applications of the model to deformed nuclei, its value might be adjusted so that Equation (2.112) reproduces, as far as possible, observed moments of inertia. It turns out that the values obtained experimentally in such fits are far from irrotational-flow values which suggests that nuclear collective flows are not irrotational, i.e., they involve vorticity degrees of freedom This possibility will be discussed in Volume 2. Note also that the γ-dependence of the moments of inertia of the Bohr model is determined by the expansion of the SO(5) Casimir invariant, Equation (2.35), and is not adjustable without a generalisation of the model. In particular, the ratios of the moments of inertia as functions of γ, arise from the SO(5) structure of the Bohr model and do not depend on any liquid-drop model assumptions.[19]

---

[18]Bohr, *op. cit.* Footnote 1 on Page 97.

[19]A recent analysis by Wood J.L. *et al.* (1992), *Phys. Repts.* **215**, 101 suggests that in real nuclei, the ratios of the moments of inertia also deviate considerably from those of an SO(5) model.

(i) *Symmetric and asymmetric tops*

Figure 2.4 shows that two moments of inertia become equal when γ_0 takes one of the values 0, $\pi/6$, or $\pi/3$; at these values

$$\gamma_0 = 0: \quad \Im_1 = \Im_2, \quad \Im_3 = 0$$
$$\bar{Q}_{11} = \bar{Q}_{22} < \bar{Q}_{33} \quad \text{(prolate)}, \tag{2.113}$$

$$\gamma_0 = \pi/6: \quad \Im_1 > \Im_2 = \Im_3 \neq 0$$
$$\bar{Q}_{22} < \bar{Q}_{11} < \bar{Q}_{33} \quad \text{(triaxial)}, \tag{2.114}$$

$$\gamma_0 = \pi/3: \quad \Im_1 = \Im_3, \quad \Im_2 = 0$$
$$\bar{Q}_{11} = \bar{Q}_{33} > \bar{Q}_{22} \quad \text{(oblate)}. \tag{2.115}$$

At $\gamma_0 = 0$ and $\pi/3$, two of the intrinsic quadrupole moments, defined by Equation (2.15), are equal implying that the model nucleus is axially symmetric. When $\gamma_0 = 0$, one quadrupole moment is larger than the other two, implying that the mass distribution is that of a *prolate spheroid* (cigar shaped). Conversely, one quadrupole moment is smaller than the other two, when $\gamma_0 = \pi/3$, implying that the mass distribution defined by the quadrupole moments is that of an *oblate spheroid* (discus shaped). However, in spite of two moments of inertia being equal when $\gamma_0 = \pi/6$, the three intrinsic quadrupole moments all have different values.[20] The shape of the mass distribution defined by the quadrupole moments is then *ellipsoidal* and is said to be triaxial.

A rotor with two equal moments of inertia is said to be a *symmetric top*. For other values of γ, the three moments of inertia are all distinct and the rotor is an *asymmetric top*.

(ii) *Rotor model wave functions*

Claim 2.1   *An orthonormal basis of wave functions $\{\Phi_{KLM}; L \geq 0\}$ for the rotor submodel of the Bohr model, with volume element $d\Omega$, is given by*

$$\Phi_{KLM}(\Omega) = \sqrt{\frac{2L+1}{16\pi^2(1+\delta_{K0})}} \left[\mathscr{D}^L_{KM}(\Omega) + (-1)^L \mathscr{D}^L_{-K,M}(\Omega) \right], \tag{2.116}$$

with $K = 0, 2, 4, \ldots$, when none of the three moments of inertia is zero. When there is one vanishing moment of inertia, K is restricted to $K = 0$.

Proof: According to the Peter-Weyl theorem,[21] an orthonormal basis for the Hilbert space of rotational states for a completely asymmetric body is given by the normalised Wigner $\mathscr{D}$ functions

$$\phi_{KLM}(\Omega) = \sqrt{\frac{2L+1}{8\pi^2}} \, \mathscr{D}^L_{KM}(\Omega), \tag{2.117}$$

---

[20]Meyer-ter-Vehn J. (1975), *Nucl. Phys.* **A249**, 111.
[21]See Theorem A.1, Page 602.

with L running over all non-negative integers and K, M running over the integer values between $\pm L$ for each L.[22] However, even the asymmetric top submodel of the Bohr model is not completely asymmetric; it has an intrinsic D_2 symmetry, where D_2 is the four-element subgroup consisting of an identity element and SO(3) rotations through angle π about any of the intrinsic axes. This follows, as noted in Section 2.1.2, from the observation that Cartesian quadrupole moments of the Bohr model are diagonal in the intrinsic frame, i.e.,

$$\bar{Q}_{kl} = \delta_{kl}\bar{Q}_{kk}. \tag{2.118}$$

Hence, they are invariant under rotations through an angle π about any of the intrinsic axes. It follows, for fixed values of (β, γ, Ω), that each element of the set of quadrupole coordinates

$$\{q_m(\beta, \gamma, \omega\Omega); \omega \in D_2\} \tag{2.119}$$

has precisely the same value. Moreover, any function of these quadrupole moments must also be invariant under such rotations. In particular, rotor model wave functions in the Bohr model must be restricted to linear combinations of the Wigner $\mathscr{D}$ functions that satisfy the symmetry constraint

$$\Phi_{KLM}(\omega\Omega) = \Phi_{KLM}(\Omega), \quad \forall \omega \in D_2. \tag{2.120}$$

To ensure that they do, let $\omega_k = e^{-i\pi L_k}$ denote a rotation through angle π about the k axis. From the definition of a Wigner rotation matrix as an SO(3) representation matrix, it follows that

$$\mathscr{D}^L_{KM}(\omega_k\Omega) = \sum_N \mathscr{D}^L_{KN}(\omega_k)\,\mathscr{D}^L_{NM}(\Omega). \tag{2.121}$$

The identity

$$\mathscr{D}^L_{KN}(e^{-i\pi L_3}) = e^{-iK\pi}\delta_{K,N} = (-1)^K\delta_{K,N} \tag{2.122}$$

then restricts K to even values and

$$\mathscr{D}^L_{KN}(e^{-i\pi L_2}) = (-1)^{L+K}\delta_{K,-N} \tag{2.123}$$

requires the inclusion of the second term on the right side of Equation (2.116). Because $\omega_1 = \omega_2\omega_3$, there are no further symmetry conditions. Thus, it can be seen that the wave functions of the Claim are a complete basis of wave functions that satisfy the symmetry constraint (2.120).

When $\gamma_0 = 0$ and $\pi/3$, one moment of inertia vanishes and two moments of inertia are equal, as shown by Figure 2.4 and Equations (2.113) – (2.115). The rotor submodel of the Bohr model is then axially symmetric. Moreover, the quadrupole

---

[22]Some properties of the Wigner $\mathscr{D}$ functions are reviewed in Section A.8. Their use as bases for rotor model wave functions will be discussed in more depth in Volume 2.

moment coordinates in the intrinsic frame are then invariant under the larger symmetry group D_∞ which includes rotations through all angles about the symmetry axis plus rotations through angle π about perpendicular axes. Thus, if we take the symmetry axis to be the 3 axis (which can always be done by relabelling the coordinate axes if necessary), the set of rotor-model wave functions must be further restricted by requiring them to satisfy the constraint

$$\Phi_{KLM}(e^{-i\theta \hat{L}_3}\Omega) = (-1)^{-i\theta K}\Phi_{KLM}(\Omega) = \Phi_{KLM}(\Omega), \quad \forall \theta \le 2\pi. \tag{2.124}$$

This restricts K to the value $K = 0$ and completes the proof of the Claim. □

It is of interest to observe that because the Hilbert space of the Bohr rotor is a subspace of that spanned by the complete set of states of Equation (2.116), there must exist more general rotational models which do not share the same invariance properties under D_2 or D_∞. Further generalisations of the rotor model to include intrinsic degrees of freedom will be discussed in Volume 2.

(iii) *Energy levels*

The energy-level spectrum of a symmetric top is particularly easy to derive. Suppose, for example, that

$$\Im_1 = \Im_2 \ne \Im_3. \tag{2.125}$$

The rigid-rotor Hamiltonian can then be expressed as

$$\hat{H}_{\text{RRot}} = \frac{\hbar^2 \hat{\mathbf{L}}^2}{2\Im_1} + \left(\frac{\hbar^2}{2\Im_3} - \frac{\hbar^2}{2\Im_1}\right)\bar{L}_3^2, \tag{2.126}$$

where it is noted that the square of the angular momentum, $\hat{\mathbf{L}}^2$, is a scalar and is independent of the reference frame so that

$$\hat{\mathbf{L}}^2 = \sum_k \hat{L}_k^2 = \bar{\mathbf{L}}^2 = \sum_k \bar{L}_k^2. \tag{2.127}$$

Thus, if neither $\Im_1$ nor $\Im_3$ is zero, the rotor model wave functions $\{\Phi_{KLM}\}$ are eigenfunctions of the Hamiltonian $\hat{H}_{\text{RRot}}$ with energies given by

$$E_{KL} = \frac{\hbar^2}{2\Im_1}L(L+1) + \left(\frac{\hbar^2}{2\Im_3} - \frac{\hbar^2}{2\Im_1}\right)K^2 \tag{2.128}$$

and with the ranges of L and K values allowed by Claim 2.1. There is one situation in which this occurs in the Bohr model; namely in the Meyer-ter-Vehn limit for which $\gamma = \pi/6 = 30°$. We then have the situation (well-known in molecular physics) of a rotor that is not axially symmetric but is nevertheless an oblate symmetric top with respect to its moments of inertia. The energy spectra for the three symmetric-top limits are shown on the right side of Figure 2.5.

If one moment of inertia vanishes, which (with a relabelling of the axes if necessary) we can take to be $\Im_3$, then, according to Claim 2.1, the basis wave functions

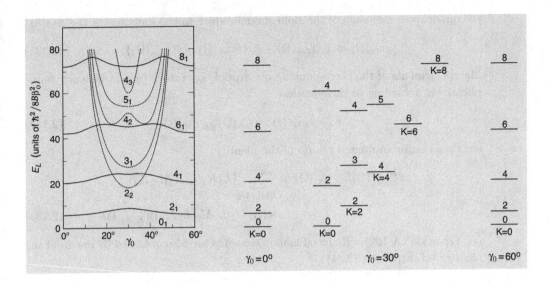

Figure 2.5: The left side of the figure shows the energy spectrum of the rotor limit of the Bohr model, computed by Meyer-ter-Vehn, *op. cit.* Footnote 20 on Page 122, as a function of γ_0. The energy levels for the three symmetric top solutions, together with their K quantum number assignments, are shown on the right side of the figure. Note that, for $\gamma_0 = 30°$, the excitation energy of a state with a given value of L decreases as its K quantum number increases. This is a result that is characteristic of an oblate symmetric top for which the distinct moment of inertia is larger than the other two as observed for some molecules.

reduce to the $K = 0$ subset $\{\Phi_{LM} \equiv \Phi_{0LM}\}$.[23] The energy levels are then given simply by

$$E_L = \frac{\hbar^2}{2\Im_1} L(L+1). \tag{2.129}$$

A general asymmetric top Hamiltonian is not already diagonal in the $\{\Phi_{KLM}\}$ basis. However, its matrix elements are readily computed and the Hamiltonian can be diagonalised as will be shown, in a more general rotor model context, in Volume 2. The energy-level spectra obtained for the moments of inertia given by Equation (2.112) are shown in Figure 2.5.

(iv) *Quadrupole moments and E2 transition rates for the rotor limit*

When the intrinsic quadrupole moments have the frozen values

$$\bar{q}_0 = \beta_0 \cos\gamma_0, \quad \bar{q}_2 = \beta_0 \frac{1}{\sqrt{2}} \sin\gamma_0, \tag{2.130}$$

---

[23]One can interpret this result by considering that, as $\Im_3 \to 0$, the $K \neq 0$ excitation energies increase and go to infinity and become unobservable in the $\Im_3 = 0$ limit.

the quadrupole moments of the Bohr model reduce to the functions of $\Omega \in SO(3)$

$$q_m(\Omega) = \bar{q}_0 \mathscr{D}^2_{0m}(\Omega) + \bar{q}_2 \left(\mathscr{D}^2_{2m}(\Omega) + \mathscr{D}^2_{-2m}(\Omega) \right). \tag{2.131}$$

Matrix elements of the corresponding quadrupole operators $\{\hat{q}_m\}$, which act multiplicatively according to the equation

$$\hat{q}_m \phi_{KLM}(\Omega) := q_m(\Omega)\phi_{KLM}(\Omega) \tag{2.132}$$

are then readily evaluated by use of the identity

$$\mathscr{D}^{L_2}_{K_2 M_2}(\Omega)\mathscr{D}^{L_1}_{K_1 M_1}(\Omega) = \sum_{L_3 K_3 M_3} (L_1 K_1 \, L_2 K_2 | L_3 K_3)$$
$$\times (L_1 M_1 \, L_2 M_2 | L_3 M_3)\mathscr{D}^{L_3}_{K_3 M_3}(\Omega) \tag{2.133}$$

(cf. Equation (A.102)). Reduced matrix elements are also obtained by means of the identity (cf. Equation (2.94))

$$[\hat{q} \otimes \Phi_{KL}]_{L'M} = \sum_{K'} \Phi_{K'L'M} \frac{\langle K'L' \| \hat{q} \| KL \rangle}{\sqrt{2L'+1}}. \tag{2.134}$$

We can similarly define multiplicative $\hat{\mathscr{D}}$-function operators by

$$\hat{\mathscr{D}}^{L_2}_{K_2 M_2} \mathscr{D}^{L_1}_{K_1 M_1}(\Omega) := \mathscr{D}^{L_2}_{K_2 M_2}(\Omega)\mathscr{D}^{L_1}_{K_1 M_1}(\Omega). \tag{2.135}$$

Then, from Equation (2.133) and the expression (2.116) for the wave functions, it is determined that

$$\left[\hat{\mathscr{D}}^2_0 \otimes \Phi_{KL} \right]_{L'M'} = \sqrt{\frac{2L+1}{2L'+1}} \, (LK\,20|L'K)\Phi_{KL'M'}, \tag{2.136}$$

and

$$\left[\left(\hat{\mathscr{D}}^2_2 + \hat{\mathscr{D}}^2_{-2} \right) \otimes \Phi_{KL} \right]_{L'M'}$$
$$= \sqrt{\frac{(2L+1)(1+\delta_{K,0})}{2L'+1}} \, (LK\,22|L',K+2)\Phi_{K+2,L'M'}$$
$$+ \sqrt{\frac{(2L+1)(1+\delta_{K,2})}{2L'+1}} \, (LK\,2,-2|L',K-2)\Phi_{K-2,L'M'}. \tag{2.137}$$

Hence, we derive the reduced matrix elements

$$\langle KL' \| \hat{q} \| KL \rangle = (2L+1)^{1/2}(LK\,20|L'K)\bar{q}_0, \tag{2.138}$$

$$\langle K+2, L' \| \hat{q} \| KL \rangle = [(2L+1)(1+\delta_{K0})]^{1/2}(LK\,22|L',K+2)\bar{q}_2, \tag{2.139}$$

$$\langle K-2, L' \| \hat{q} \| KL \rangle = [(2L+1)(1+\delta_{K2})]^{1/2}(LK\,2,-2|L',K-2)\bar{q}_2. \tag{2.140}$$

When K is a good quantum number, as it is for the symmetric top submodel of the rigid-rotor model, static electric quadrupole moments are given, in accordance with the definition (2.101), by

$$Q(KL) = \sqrt{\frac{16\pi}{5}} \frac{Ze}{A} \langle KL, M = L | \hat{q}_0 | KL, M = L \rangle$$

$$= \sqrt{\frac{16\pi}{5}} (LK\,20|LK)(LL\,20|LL) \frac{Ze}{A} \bar{q}_0$$

$$= \sqrt{\frac{16\pi}{5}} \frac{3K^2 - L(L+1)}{(L+1)(2L+3)} \frac{Ze}{A} \bar{q}_0 \qquad (2.141)$$

and the non-zero reduced E2 transition rates (cf. Equation (2.91)) are given by

$$B(\text{E2}; KL \rightarrow KL') = (LK\,20|L'K)^2 \left(\frac{Ze\bar{q}_0}{A}\right)^2, \qquad (2.142)$$

$$B(\text{E2}; KL \rightarrow K{+}2, L') = (1 + \delta_{K0})(LK\,22|L', K{+}2)^2 \left(\frac{Ze\bar{q}_2}{A}\right)^2, \qquad (2.143)$$

$$B(\text{E2}; KL \rightarrow K{-}2, L') = (1 + \delta_{K2})(LK\,2, -2|L', K{-}2)^2 \left(\frac{Ze\bar{q}_2}{A}\right)^2. \qquad (2.144)$$

These expressions show that E2 transitions between rotor bands with K quantum numbers differing by any value other than 0 or 2 are forbidden in the model. This will be shown in Volume 2 to be a special case of a general selection rule for symmetric top rotors which forbids Eλ or Mλ transitions between states belonging to rotational bands with $\Delta K > \lambda$, as discussed in Section 1.7.4.

For the axially symmetric cases, in which $\gamma_0 = 0$ or $\pi/3$, the rigid-rotor limit of the Bohr model admits only $K = 0$ rotational bands. Thus, in the strict rigid-rotor limit of the Bohr model, with only quadrupole degrees of freedom, only the symmetric top model, given by the accidental coincidence of two moments of inertia for a non-axially symmetric $\gamma_0 = \pi/6$ intrinsic shape, exhibits rotational bands of states with good, non-zero, K quantum numbers, cf. Figure 2.5. However, sequences of K bands, with even values of K, occur for axially symmetric rotors when gamma vibrational excitations are included (as discussed in the following section). They also occur, in the Bohr model, when deformations of higher multipole than quadrupole are included. Moreover, when the Bohr model is extended to include intrinsic degrees of freedom, as in the more general Bohr-Mottelson model,[24] a plethora of rotational bands with a variety of K quantum numbers become possible, as observed (see, for example, Figure 1.58). Collective models with intrinsic degrees of freedom will be considered in Volume 2.

(v) *Significance of the rigid-rotor model*

In spite of its simplicity, the solvable rigid-rotor submodel of the Bohr model was a major development in nuclear physics. It is important conceptually because it provides a qualitative description of the essential rotational dynamics observed in

---

[24]Bohr and Mottelson, *op. cit.* Footnote 6 on Page 98.

a wide range of nuclei which prepares the way for more realistic descriptions of nuclear rotational structure.

A significant achievement of the Bohr rotor model is its prediction that $K = 0$ bands comprise only states of even angular momentum, as observed for the ground-state rotational bands of doubly-even nuclei. This prediction emerges as a consequence of the D_2 intrinsic symmetry of the model. Such symmetries play an essential role in the formulation of more sophisticated rotor models but would not have been so readily understood if one did not have the Bohr model as a prototype.

A valuable use of the model is to investigate the intrinsic shapes of rotational nuclei by observations of E2 moments and transition rates. What emerges, is an interpretation of the widespread evidence for K as a good quantum number, particularly in doubly-even nuclei, in terms of symmetric top behaviour. Conversely, the model provides means to identify asymmetric top behaviour, when it occurs. Such investigations indicate that prolate shapes of nuclei are considerably more prevalent than oblate shapes (cf. Figure 1.37); an observation that needs to be explained. The intrinsic shapes of rotational nuclei will be explored in Volume 2.

As reviewed in Sections 1.7.4, the model is remarkably successful in predicting ratios of rotational energy levels, in terms of moments of inertia (treated as parameters), and ratios of E2 transition rates between rotational states. Thus, the data leave little doubt that the observed energy-level spectra of many (most) doubly-open shell nuclei are dominated by rotational bands. However, the particular rotational sequences of the rigid-rotor limit of the Bohr model (cf. Figure 2.5) are rarely observed as predicted. In particular, rotational nuclei, such as ^{168}Er (cf. Figure 1.58), exhibit many more values of the K quantum number than the rigid-rotor limit of the Bohr model allows. Moreover, rotational bands are also seen widely in odd-mass nuclei, with half-odd integer angular momenta. Another concern for the rotor model is the relationship between the inertia tensor and the quadrupole tensor. As noted at the beginning of this section and in Footnote 19 (on Page 121), there is experimental evidence to suggest that observed moments of inertia do not relate to the deformation parameters of the quadrupole tensor in the manner prescribed by the Bohr model expression (2.112). Evidently, and not surprisingly, there is much more going on inside a deformed nucleus than can be represented by the simple rotational dynamics of the Bohr model.

Exercises

2.9 By use of the symmetry properties of Clebsch-Gordan coefficients

$$(j_1 m_1 j_2 m_2 | jm) = (-1)^{j_1 + j_2 - j} (j_2 m_2 j_1 m_1 | jm),$$
$$= (-1)^{j_1 + j_2 - j} (j_1, -m_1 j_2, -m_2 | j, -m), \tag{2.145}$$

show that

$$[d^\dagger \otimes d^\dagger]_{LM} = 0 \quad \text{for } L = 1, 3. \tag{2.146}$$

2.10 Show that the $n = 3$ states of maximum L are given by

$$|36M\rangle = \frac{1}{\sqrt{6}}[d^\dagger \otimes d^\dagger \otimes d^\dagger]_{6M}|0\rangle. \qquad (2.147)$$

(Hint: Prove this identity first for $M = 6$.)

2.11 Use the symmetry property of Clebsch-Gordan coefficients,

$$(j_1m_1\, j_2m_2|jm) = (-1)^{j_3-j_1-m_2}\sqrt{\frac{2j_3+1}{2j_1+1}}\,(j_3m_3\, j_2, -m_2|j_1, m_1), \qquad (2.148)$$

to show that if a spherical tensor $\hat{T}_L$ satisfies the Hermiticity relationship

$$\left(\hat{T}_{LM}\right)^\dagger = (-1)^M \hat{T}_{L,-M} \qquad (2.149)$$

then the reduced matrix elements, defined by Equation (2.93), satisfy

$$\langle n_2\alpha_2 L_2\|\hat{T}_L\|n_1\alpha_1 L_1\rangle = (-1)^{L_1-L_2}\langle n_1\alpha_1 L_1\|\hat{T}_L\|n_2\alpha_2 L_2\rangle^*. \qquad (2.150)$$

2.12 Show that a reduced transition probability, given by Equation (2.91), is also given by

$$B(E2; \alpha_i L_i \to \alpha_f L_f) = \frac{|\langle\alpha_i L_i\|\hat{\mathfrak{M}}(E2)\|\alpha_f L_f\rangle|^2}{2L_i+1}. \qquad (2.151)$$

2.13 Show that

$$B(E2; \alpha_1 L_1 \to \alpha_2 L_2) = \frac{2L_2+1}{2L_1+1}B(E2; \alpha_2 L_2 \to \alpha_1 L_1). \qquad (2.152)$$

2.14 Show that the $B(E2)$ transition rates between the states of maximum L, for a given number of oscillator quanta, are given in the harmonic spherical vibrator limit of the Bohr model by

$$B(E2; n, L=2n \to n-1, L'=2n-2) = \left(\frac{Ze}{A}\right)^2\frac{n\hbar}{2B\omega}. \qquad (2.153)$$

2.15 Show that the rigid-rotor submodel of the Bohr model does not have an $L = 1$ state.

2.4 The adiabatic Bohr model

When the potential energy of the Bohr model has a steep-sided minimum at $\beta = \beta_0$ and $\gamma = \gamma_0$ then, for a sufficiently large value of β_0, the model becomes solvable because of an *adiabatic decoupling* of the vibrational and rotational degrees of freedom. The decoupling arises when the rotational motions of low-angular momentum states are slow (adiabatic) in comparison to the relatively high-frequencies of the (β, γ)-vibrational states to which they would otherwise couple.

The concept of adiabatically-decoupled beta and gamma vibrations and rotations is intuitively natural and provides a much more realistic approximation to the

Bohr model than its rigid-beta and rigid-rotor limits. Moreover, it is an approximation for which corrections can be calculated. It has been applied widely in simple physical interpretations of the rotational and vibrational states of deformed nuclei, and has evolved into the *standard* model of nuclear collective structure known widely as the *Bohr-Mottelson model*.[25] The essential differences between the Bohr-Mottelson model and the Bohr model are mentioned in the following, concluding, section of this chapter.[26]

2.4.1 *The adiabatic Hamiltonian*

Consider, for example, a Bohr model Hamiltonian

$$\hat{H} = -\frac{\hbar^2}{2B}\nabla^2 + V(\beta,\gamma). \qquad (2.154)$$

If V has a deep steep-sided minimum at (β_0, γ_0), then the dependence of the wave function on β and γ will be sharply peaked about $\beta = \beta_0$ and $\gamma = \gamma_0$. A reasonable approximation is then to make a Taylor expansion of V about its minimum

$$V(\beta,\gamma) = V(\beta_0,0) + \tfrac{1}{2}B\omega_\beta^2(\beta - \beta_0)^2 + \tfrac{1}{2}B\beta_0^2\omega_\gamma^2(\gamma - \gamma_0)^2 + \dots \qquad (2.155)$$

and retain only the leading terms shown.

More significantly, the expression

$$\nabla^2 = \frac{1}{\beta^4}\frac{\partial}{\partial\beta}\beta^4\frac{\partial}{\partial\beta} + \frac{1}{\beta^2\sin 3\gamma}\frac{\partial}{\partial\gamma}\sin 3\gamma\frac{\partial}{\partial\gamma} - \sum_{k=1}^{3}\frac{\bar{L}_k^2}{4\beta^2\sin^2(\gamma - 2\pi k/3)}, \qquad (2.156)$$

given by Equations (2.34) and (2.35), also simplifies in the adiabatic approximation. There are three distinct situations for $\beta_0 \neq 0$: (i) the generic situation, in which the equilibrium deformation is that of an asymmetric top; (ii) the special $\gamma_0 = \pi/6$ situation in which the equilibrium shape is triaxial but the moments of inertia are those of a symmetric top; and (iii) the two axially symmetric equilibrium situations, for which $\gamma_0 = 0$ or $\pi/3$. In each case, the adiabatic model simply extends the rigid-rotor model for rotations about the equilibrium deformation by the addition of harmonic beta and gamma vibrations. When $\beta_0 = 0$, the model has a spherical equilibrium shape and reduces to the analytically solvable harmonic spherical vibrator model.

---

[25]Bohr and Mottelson's model includes their collective model and the unified model (cf. introduction to this chapter) both of which will be developed in Volume 2.

[26]An adiabatic version of the Bohr model was also developed within the framework of a so-called *rotation-vibration model*, by Faessler and Greiner; cf. Footnote 7 on Page 98.

2.4.2 E2 moments and transitions in the adiabatic limit

The space-fixed quadrupole moments of the Bohr model, in its adiabatic limit, continue to be expressed in the form

$$q_m(\Omega) = \bar{q}_0 \, \mathscr{D}^2_{0m}(\Omega) + \bar{q}_2 \left(\mathscr{D}^2_{2m}(\Omega) + \mathscr{D}^2_{-2m}(\Omega) \right); \tag{2.157}$$

but the intrinsic quadrupole moments can be approximated to leading order by

$$\bar{q}_0 = \beta \cos\gamma \approx [\beta_0 + (\beta - \beta_0)] \cos\gamma_0 - \beta_0(\gamma - \gamma_0) \sin\gamma_0, \tag{2.158}$$

$$\bar{q}_2 = \frac{1}{\sqrt{2}} \beta \sin\gamma \approx \frac{1}{\sqrt{2}} [\beta_0 \sin\gamma_0 + (\beta - \beta_0) \sin\gamma_0 + \beta_0(\gamma - \gamma_0) \cos\gamma_0]. \tag{2.159}$$

With the extra β and γ harmonic vibrational degrees of freedom, the quadrupole matrix elements of Equations (2.138)–(2.140) are given by the expressions:

$$\langle n'_\beta n'_\gamma KL' \| \hat{q} \| n_\beta n_\gamma KL \rangle = (2L+1)^{1/2}(LK\,20|L'K)\langle n'_\beta n'_\gamma | \bar{q}_0 | n_\beta n_\gamma \rangle, \tag{2.160}$$

$$\langle n'_\beta n'_\gamma K+2, L' \| \hat{q} \| n_\beta n_\gamma KL \rangle = [(2L+1)(1+\delta_{K,0})]^{1/2}(LK\,22|L', K+2)$$
$$\times \langle n'_\beta n'_\gamma | \bar{q}_2 | n_\beta n_\gamma \rangle, \tag{2.161}$$

$$\langle n'_\beta n'_\gamma K-2, L' \| \hat{q} \| KL \rangle = [(2L+1)(1+\delta_{K,2})]^{1/2}(LK\,2,-2|L', K-2)$$
$$\times \langle n'_\beta n'_\gamma | \bar{q}_2 | n_\beta n_\gamma \rangle. \tag{2.162}$$

Thus, when K is a good quantum number, the reduced E2 transition rates of Equations (2.142) - (2.144) extend to

$$B(\text{E2}; n_\beta n_\gamma KL \to n'_\beta n'_\gamma KL') = (LK\,20|L'K)^2 \left(\frac{Ze}{A}\right)^2$$
$$\times |\langle n'_\beta n'_\gamma | \bar{q}_0 | n_\beta n_\gamma \rangle|^2, \tag{2.163}$$

$$B(\text{E2}; n_\beta n_\gamma KL \to n'_\beta n'_\gamma K+2, L') = (1+\delta_{K,0})(LK\,22|L', K+2)^2 \left(\frac{Ze}{A}\right)^2$$
$$\times |\langle n'_\beta n'_\gamma | \bar{q}_2 | n_\beta n_\gamma \rangle|^2, \tag{2.164}$$

$$B(\text{E2}; n_\beta n_\gamma KL \to n'_\beta n'_\gamma K-2, L') = (1+\delta_{K,2})(LK\,2,-2|L', K-2)^2 \left(\frac{Ze}{A}\right)^2$$
$$\times |\langle n'_\beta n'_\gamma | \bar{q}_2 | n_\beta n_\gamma \rangle|^2. \tag{2.165}$$

2.4.3 Beta and gamma vibrations of a triaxial rotor

If neither β_0 nor γ_0 is zero, ∇^2 is approximated by

$$\nabla^2 \approx \frac{\partial^2}{\partial\beta^2} + \frac{1}{\beta_0^2} \frac{\partial^2}{\partial\gamma^2} - \sum_{k=1}^{3} \frac{\bar{L}_k^2}{4\beta_0^2 \sin^2(\gamma_0 - 2\pi k/3)}. \tag{2.166}$$

The Hamiltonian then separates into three uncoupled terms,

$$\hat{H} \approx \hat{H}_{\text{rotor}} + \hat{H}_\beta + \hat{H}_\gamma, \tag{2.167}$$

with

$$\hat{H}_{\text{rotor}} = \sum_{k=1}^{3} \frac{\hbar^2 \bar{L}_k^2}{8B\beta_0^2 \sin^2(\gamma_0 - 2\pi k/3)}, \tag{2.168}$$

$$\hat{H}_\beta = -\frac{\hbar^2}{2B}\frac{\partial^2}{\partial\beta^2} + \frac{1}{2}B\omega_\beta^2(\beta - \beta_0)^2, \tag{2.169}$$

$$\hat{H}_\gamma = -\frac{\hbar^2}{2B\beta_0^2}\frac{\partial^2}{\partial\gamma^2} + \frac{1}{2}B\beta_0^2\omega_\gamma^2(\gamma - \gamma_0)^2. \tag{2.170}$$

Thus, the beta and gamma vibrations decouple from the rotations in this adiabatic limit and the energy level spectrum is given by

$$E(\alpha L n_\beta n_\gamma) = E_{\alpha L}^{\text{rotor}} + (n_\beta + \tfrac{1}{2})\hbar\omega_\beta + (n_\gamma + \tfrac{1}{2})\hbar\omega_\gamma, \tag{2.171}$$

with $n_\beta = 0, 1, 2, \ldots$ and $n_\gamma = 0, 1, 2, \ldots$. Note that the beta and gamma vibrations of a *triaxial rotor* carry no angular momentum; they are like radial vibrations in their respective spaces. Thus, in the adiabatic limit, the rigid-rotor spectrum, given by $E_{\alpha L}^{\text{rotor}}$, gets repeated for each integer value of n_β and n_γ.

The Meyer-ter-Vehn, $\gamma_0 = \pi/6$, limit

The triaxial rotor with $\gamma_0 = \pi/6$ is a special case because two of its moments of inertia are equal. Thus, it is a symmetric top and its spectral properties are determined analytically. In particular, the rotor component of its Hamiltonian (cf. Section 2.3.3) is given by

$$\hat{H}_{\text{rotor}} = \frac{\hbar^2}{8B\beta_0^2}(4\hat{\mathbf{L}}^2 - 3\bar{L}_1^2) \tag{2.172}$$

and the energy spectrum of $\hat{H}_{\text{rotor}} + \hat{H}_\beta + \hat{H}_\gamma$ is

$$E(KL n_\beta n_\gamma) = \frac{\hbar^2}{8B\beta_0^2}(4L(L+1) - 3K^2) + (n_\beta + \tfrac{1}{2})\hbar\omega_\beta + (n_\gamma + \tfrac{1}{2})\hbar\omega_\gamma, \tag{2.173}$$

where K is an eigenvalue of $\bar{L}_1$. This energy spectrum is shown in Figure 2.6.

In the calculation of quadrupole matrix elements, it must be remembered that the symmetry axis in the Meyer-ter-Vehn limit is the 1-axis whereas, in the expression of symmetric-top rotor bands, the K quantum numbers were defined relative to the 3-axis as symmetry axis. Thus, in order to use the standard expressions in the Meyer-ter-Vehn limit, it is necessary to first relabel the axes such that the 1-axis becomes the 3-axis. The corresponding change in the value of γ_0 is inferred from Equation (2.15) which shows that if $\bar{Q}_{11}$ is set equal to $\bar{Q}_{33}'$ then $\cos(\gamma_0 - 2\pi/3)$ should be set equal to $\cos\gamma_0'$. Therefore, in relabelling the axis in this way, $\gamma_0 = \pi/6$ has to be replaced by $\gamma_0' = -\pi/2$. Consequently, $\cos\gamma_0 \to \cos\gamma_0' = 0$

132

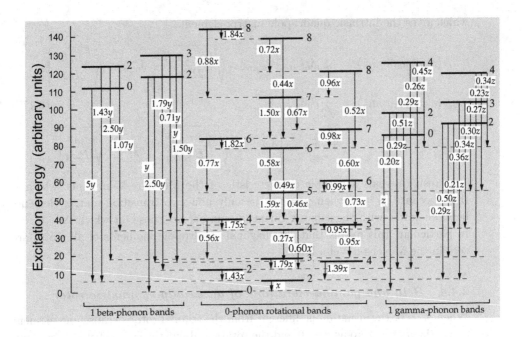

Figure 2.6: Typical spectrum of the adiabatic Bohr model in the triaxial symmetric-top limit in which $\gamma_0 = \pi/6$. The figure shows the beginnings of a sequence of zero-vibrational-phonon $K = 0, 2, 4, \ldots$ rotational bands and parallel excited one-beta-phonon and one-gamma-phonon bands. The dominant reduced E2 transition rates are shown in terms of adjustable parameters x, y, and z. A notable characteristic of the $\gamma_0 = \pi/6$ limit is that the only non-zero $\Delta K = 0$, E2 transitions are those between states with $\Delta n_\gamma = 1$.

and $\sin \gamma_0 \to \sin \gamma_0^{(1)} = 1$. Hence, in the Meyer-ter-Vehn limit,

$$\bar{q}_0 = -\beta_0(\gamma - \gamma_0'), \quad \bar{q}_2 = \frac{1}{\sqrt{2}}[\beta_0 + (\beta - \beta_0)]. \tag{2.174}$$

The reduced matrix elements of Equations (2.160) – (2.162) are then determined from the observation that, for harmonic oscillator β and γ wave functions,

$$\langle \psi_{n_\beta - 1} | (\beta - \beta_0) | \psi_{n_\beta} \rangle = \sqrt{\frac{n_\beta \hbar}{2B\omega_\beta}}, \tag{2.175}$$

$$\langle \varphi_{n_\gamma - 1} | (\gamma - \gamma_0) | \varphi_{n_\gamma} \rangle = \sqrt{\frac{n_\gamma \hbar}{2B\omega_\gamma}}, \tag{2.176}$$

which gives the intrinsic quadrupole matrix elements

$$\langle n'_\beta n'_\gamma | \bar{q}_0 | n_\beta n_\gamma \rangle = -\beta_0 \delta_{n'_\beta, n_\beta} \left[\delta_{n'_\gamma, n_\gamma+1} \sqrt{\frac{(n_\gamma+1)\hbar}{2B\omega_\gamma}} + \delta_{n'_\gamma, n_\gamma-1} \sqrt{\frac{n_\gamma \hbar}{2B\omega_\gamma}} \right],$$

$$\langle n'_\beta n'_\gamma | \bar{q}_2 | n_\beta n_\gamma \rangle = \frac{1}{\sqrt{2}} \beta_0 \delta_{n'_\beta, n_\beta} \delta_{n'_\gamma, n_\gamma} \qquad (2.177)$$

$$+ \frac{1}{2} \delta_{n'_\gamma, n_\gamma} \left[\delta_{n'_\beta, n_\beta+1} \sqrt{\frac{(n_\beta+1)\hbar}{B\omega_\beta}} + \delta_{n'_\beta, n_\beta-1} \sqrt{\frac{n_\beta \hbar}{B\omega_\beta}} \right].$$

This result reveals a notable characteristic of the Meyer-re-Vehn limit, which is that, because $\bar{q}_0 = 0$ when $\gamma = \gamma'_0$, the only non-zero transition matrix elements between states with the same K quantum number are those for which $\Delta n_\gamma = \pm 1$. In particular, this means that the diagonal quadrupole moments of all states are zero in the Meyer-ter-Vehn limit.

2.4.4 *Beta and gamma vibrations of an axially symmetric rotor*

When $\gamma_0 = 0$, the situation is different from the triaxial rotor because one moment of inertia goes to zero as $\gamma \to 0$ and the rotation about the symmetry axis becomes non-adiabatically coupled to the γ degree of freedom. In this situation, the leading-order expansion of ∇^2, for small deviations of β and γ from their minima, is given by

$$\nabla^2 \approx \frac{\partial^2}{\partial \beta^2} + \frac{1}{\beta_0^2 \gamma} \frac{\partial}{\partial \gamma} \gamma \frac{\partial}{\partial \gamma} - \frac{\bar{L}_3^2}{4\beta_0^2 \gamma^2} - \frac{\hat{\mathbf{L}}^2 - \bar{L}_3^2}{3\beta_0^2}. \qquad (2.178)$$

The Hamiltonian again separates into three uncoupled terms,

$$\hat{H} \approx \hat{H}_{\text{rotor}} + \hat{H}_\beta + \hat{H}_\gamma, \qquad (2.179)$$

but now with

$$\hat{H}_{\text{rotor}} = \text{const.} + \frac{\hbar^2}{6B\beta_0^2} [\hat{\mathbf{L}}^2 - \bar{L}_3^2], \qquad (2.180)$$

$$\hat{H}_\beta = -\frac{\hbar^2}{2B} \frac{\partial^2}{\partial \beta^2} + \frac{1}{2} B\omega_\beta^2 (\beta - \beta_0)^2, \qquad (2.181)$$

$$\hat{H}_\gamma = -\frac{\hbar^2}{2B\beta_0^2} \left[\frac{1}{\gamma} \frac{\partial}{\partial \gamma} \gamma \frac{\partial}{\partial \gamma} - \frac{\bar{L}_3^2}{4\gamma^2} \right] + \frac{1}{2} B\beta_0^2 \omega_\gamma^2 \gamma^2. \qquad (2.182)$$

The Hamiltonian $\hat{H}_{\text{rotor}}$ is that of an axially symmetric rotor with energy-level spectrum

$$E_{KL}^{\text{rotor}} = \frac{\hbar^2}{6B\beta_0^2} [L(L+1) - K^2] \qquad (2.183)$$

and $\hat{H}_\beta$ is the Hamiltonian for a harmonic vibrator with energy levels

$$E_{n_\beta}^{\beta-\text{vib}} = (n_\beta + \tfrac{1}{2})\hbar\omega_\beta. \tag{2.184}$$

By defining the SO(2) angular momentum $\hat{S}_0 := \bar{L}_3/2$, it is also seen that the γ component of ∇^2 is the Laplacian for a two-dimensional Euclidean space,[27]

$$\nabla_\gamma^2 = \frac{1}{\gamma}\frac{\partial}{\partial\gamma}\gamma\frac{\partial}{\partial\gamma} - \frac{\hat{S}_0^2}{\gamma^2}. \tag{2.185}$$

Thus, the Hamiltonian

$$\hat{H}_\gamma = -\frac{\hbar^2}{2B\beta_0^2}\nabla_\gamma^2 + \tfrac{1}{2}B\beta_0^2\omega_\gamma^2\gamma^2 \tag{2.186}$$

is that of an isotropic two-dimensional harmonic oscillator and has energy levels given by

$$E_{n_\gamma}^{\gamma-\text{vib}} = (n_\gamma + 1)\hbar\omega_\gamma. \tag{2.187}$$

The above shows that, in the adiabatic limit, γ is the radial coordinate for an isotropic two-dimensional harmonic oscillator. For such a harmonic oscillator, the eigenvalues of $\hat{S}_0$ are given, for each value of n_γ, by

$$|S_0| = n_\gamma, \quad n_\gamma - 2, \quad n_\gamma - 4, \quad \ldots, \quad 1 \text{ or } 0. \tag{2.188}$$

The corresponding eigenvalues of $|\bar{L}_3|$ are then

$$K = 2n_\gamma, \quad 2n_\gamma - 4, \quad 2n_\gamma - 8, \quad \ldots, \quad 2 \text{ or } 0. \tag{2.189}$$

Thus, the eigenfunctions of the adiabatic Bohr model, with $\gamma_0 = 0$, are given by

$$\Psi_{n_\beta n_\gamma KLM}(\beta, \gamma, \Omega) = \psi_{n_\beta}(\beta)\,\varphi_{n_\gamma K}(\gamma)\,\Phi_{KLM}(\Omega), \tag{2.190}$$

where ψ_{n_β} is a harmonic oscillator beta-vibrational wave function, $\varphi_{n_\gamma K}$ is a radial wave function for the two-dimensional harmonic oscillator with n_γ harmonic oscillator quanta and $\hat{S}_0$ angular momentum $K/2$, and Φ_{KLM} is a rotor wave function given by Equation (2.116). The corresponding energy levels are given by

$$E_{n_\beta n_\gamma KL} = \frac{\hbar^2}{6B\beta_0^2}\left[L(L+1) - K^2\right] + (n_\beta + \tfrac{1}{2})\hbar\omega_\beta + (n_\gamma + 1)\hbar\omega_\gamma, \tag{2.191}$$

with $n_\beta = 0, 1, 2, \ldots$ and $n_\gamma = 0, 1, 2, \ldots$.

When $\gamma_0 = 0$, the intrinsic quadrupole moments are given to leading order by

$$\bar{q}_0 = \beta\cos\gamma = (\beta - \beta_0) + \beta_0, \quad \bar{q}_2 = \frac{1}{\sqrt{2}}\beta\sin\gamma = \frac{1}{\sqrt{2}}\beta_0\gamma. \tag{2.192}$$

---

[27]Dumitrescu T.S. and Hamamoto I. (1982), *Nucl. Phys.* **A383**, 205.

The space-fixed quadrupole moments are then

$$q_m(\Omega) = \beta_0 \, \mathscr{D}^2_{0m}(\Omega) + (\beta - \beta_0) \, \mathcal{D}^2_{0m}(\Omega)$$
$$+ \frac{1}{\sqrt{2}} \beta_0 \gamma \left(\mathscr{D}^2_{2m}(\Omega) + \mathscr{D}^2_{-2m}(\Omega) \right). \tag{2.193}$$

The first term contributes only to $\Delta K = \Delta n_\beta = \Delta n_\gamma = 0$ matrix elements. The second part contributes only to matrix elements between bands for which $\Delta n_\beta = 1$ and $\Delta K = \Delta n_\gamma = 0$, and the third part contributes only to matrix elements between bands for which $\Delta n_\beta = 0$, $\Delta n_\gamma = 1$ and $\Delta K = \pm 2$. Thus, in addition to rotational states, the repeated action of the quadrupole moment operators on the $K = 0$ ground-state band generates a sequence of beta vibrational bands, a sequence of gamma vibrational bands, and combinations of the two. The beta vibrational bands all have $K = 0$. The one-phonon gamma band has $K = 2$. A two-phonon gamma band can have $K = 0$ or 4. A three-phonon gamma band can have $K = 2$ or 6, and so on. The first few bands are shown in Figure 2.7.

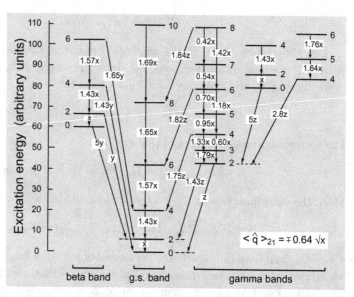

Figure 2.7: Typical spectrum of the adiabatic Bohr model in the axially symmetric limit showing a ground-state band, a one-phonon beta vibrational band, and one- and two-phonon gamma vibrational bands. Some reduced E2 transition rates are shown in terms of adjustable parameters x, y, and z. The quadrupole moment of the first excited $L = 2$ state, $<\hat{q}>_{21}$ takes positive or negative values when γ_0 is, respectively, zero or $\pi/3$.

The reduced E2 transition rates for the $\gamma_0 = 0$ limit are obtained from Equations (2.142)–(2.144) with the additional observation that the intrinsic matrix elements of $(\beta - \beta_0)$ and γ are given by

$$\left| \langle \psi_{n_\beta - 1} | (\beta - \beta_0) | \psi_{n_\beta} \rangle \right|^2 = \frac{n_\beta \hbar}{2B\omega_\beta}, \tag{2.194}$$

$$\left| \langle \varphi_{n_\gamma - 1, K+2} | \frac{1}{\sqrt{2}} \beta_0 \gamma | \varphi_{n_\gamma K} \rangle \right|^2 = \frac{n_\gamma \hbar}{4B\omega_\gamma} \left(n_\gamma - \tfrac{1}{2} K \right), \tag{2.195}$$

$$\left| \langle \varphi_{n_\gamma - 1, K-2} | \frac{1}{\sqrt{2}} \beta_0 \gamma | \varphi_{n_\gamma K} \rangle \right|^2 = \frac{n_\gamma \hbar}{4B\omega_\gamma} \left(n_\gamma + \tfrac{1}{2} K \right). \tag{2.196}$$

(Cf. Exercise 2.17 for a guide to evaluation of the radial gamma matrix elements.) The result is

$$B(E2; n_\beta n_\gamma KL \to n_\beta n_\gamma KL') = (LK\,20|L'K)^2 \left(\frac{Ze}{A}\right)^2 \beta_0^2, \qquad (2.197)$$

$$B(E2; n_\beta n_\gamma KL \to n_\beta - 1, n_\gamma KL') = (LK\,20|L'K)^2 \left(\frac{Ze}{A}\right)^2 \frac{n_\beta \hbar}{2B\omega_\beta}, \qquad (2.198)$$

$$B(E2; n_\beta n_\gamma KL \to n_\beta, n_\gamma - 1, K+2, L') = (1+\delta_{K0})(LK\,22|L', K+2)^2$$
$$\times \left(\frac{Ze}{A}\right)^2 \frac{\hbar}{4B\omega_\gamma}(n_\gamma - \tfrac{1}{2}K), \qquad (2.199)$$

$$B(E2; n_\beta n_\gamma KL \to n_\beta, n_\gamma - 1, K-2, L') = (1+\delta_{K2})(LK\,2, -2|L', K-2)^2$$
$$\times \left(\frac{Ze}{A}\right)^2 \frac{\hbar}{4B\omega_\gamma}(n_\gamma + \tfrac{1}{2}K). \qquad (2.200)$$

Static electric quadrupole moments are given, in accordance with the definition (2.101), as in the axially-symmetric rigid-rotor limit by

$$Q(KL) = \sqrt{\frac{16\pi}{5}} \frac{3K^2 \mp L(L+1)}{(L+1)(2L+3)} \frac{Ze}{A} \beta_0, \qquad (2.201)$$

according as $\gamma_0 = 0$ or $\pi/3$.

2.4.5 *Comparison of the adiabatic Bohr model predictions with experiment*

The adiabatic Bohr model for an axially symmetric rotor-vibrator predicts an energy level spectrum in terms of one mass parameter, B, and two vibrational frequencies, ω_β and ω_γ. It also predicts the ratios of many E2 transition rates and quadrupole moments. With the other parameter β_0, it further predicts the absolute values of the quadrupole moments and E2 transition rates. Thus, it is possible to confront the many predictions of the model with data. For the strongly-deformed rare-earth nuclei, many rotational bands have been observed and there are extensive $B(E2; 2_i^+ \to 0_1^+)$ data for E2 transitions from excited 2^+ states to 0^+ ground states.

An obvious limitation of the model is that it only predicts bands with even integer values of the K quantum number whereas many other K bands are observed in rotational nuclei. Clearly a nucleus has many more degrees of freedom than those of the Bohr model. Nevertheless, it is instructive to assess the extent to which the model describes a subset of observed bands.

The ground state of any even-even nucleus always has angular momentum and parity $L^\pi = 0^+$. Thus, the $K = 0$ ground-state band of a well-deformed nucleus can be identified with that of the adiabatic Bohr model and it is possible to compare properties of states belonging to the lowest excited $K = 2$ band with those of a one-phonon gamma band and states belonging to the lowest excited $K = 0$ band with those of a one-phonon beta band. Table 2.3 shows the following measured quantities for a set of rare-earth isotopes: the excitation energy E_{2_1} of the first excited 2^+ state of the ground-state band; the ratio of the excitation energies $4_1^+/2_1^+$ for the ground-state band; the excitation energies, 2_β^+, 2_γ^+, of the 2^+ states of the ostensible beta and gamma bands; the reduced E2 transition rate for decay of the 2_1^+ to the ground state

$$x = B(\text{E2}; 2_1^+ \rightarrow 0_1^+); \tag{2.202}$$

and the reduced E2 transition rates

$$y = B(\text{E2}; 2_\beta^+ \rightarrow 0_1^+), \tag{2.203}$$
$$z = B(\text{E2}; 2_\gamma^+ \rightarrow 0_1^+), \tag{2.204}$$

for decay of the 2^+ states of the "beta" and "gamma" bands to the ground state.

The model predicts that $E_{4_1}^+/E_{2_1}^+$ should be equal to 10/3 and that

$$x = \frac{1}{5}\left(\frac{Ze}{A}\right)^2 \beta_0^2, \tag{2.205}$$

$$y = \frac{1}{5}\left(\frac{Ze}{A}\right)^2 \frac{E_{2_1}}{2\hbar\omega_\beta}\beta_0^2, \tag{2.206}$$

$$z = \frac{1}{5}\left(\frac{Ze}{A}\right)^2 \frac{E_{2_1}}{\hbar\omega_\gamma}\beta_0^2. \tag{2.207}$$

Hence, it predicts that

$$\frac{y}{x} = \frac{E_{2_1}}{2\hbar\omega_\beta}, \quad \frac{z}{x} = \frac{E_{2_1}}{\hbar\omega_\gamma}. \tag{2.208}$$

Table 2.3 shows a few results for the $K = 0$ ground-state band and first-excited $K = 2$ and $K = 0$ bands. The observed properties of the ground-state band are generally in good agreement with the predictions of the model. For example, the $E_{4_1}^+/E_{2_1}^+$ energy ratio is generally only a little less than the extreme-rigid-rotor value of 3.33 and the ratios of E2 transition rates (not shown in the table) are also essentially as predicted (cf. comments in Section 1.7.4). The first-excited $K = 2$ band, when interpreted as a one-phonon gamma excitation, is also in qualitative agreement with experiment. However, the $B(\text{E2}; 2_2 \rightarrow 0_1^+)$ transition rate observed for the $K = 2$ to ground-band transition is typically $\sim 25\%$ of that predicted by the model for a one-phonon gamma band at the observed energy. This result is possibly due to a fragmentation of the gamma collectivity due to mixing with other non-collective bands.[28] The $B(\text{E2}; 2_\beta^+ \rightarrow 0_1^+)$ rate for the excited $K = 0$ to ground band

---

[28]If the pure gamma-vibrational states of the model were mixed with other states, not in the model, then the centroid of the gamma transition strength would lie at a higher energy than would

Table 2.3: Collective properties of even rare-earth isotopes as defined in the text. The model predicts that $E_{4_1}^+/E_{2_1}^+$ should be equal to 10/3, that y/x should equal $E_{2_1}/(2\hbar\omega_\beta)$, and z/x should equal $E_{2_1}/(\hbar\omega_\gamma)$. (The data are taken from Garrett P.E. (2001), *J. Phys. G: Nucl. Part. Phys.* **27**, R1 and *Nuclear Data Sheets*.)

Isotope	E_{2_1} keV	E_{4_1}/E_{2_1}	x W.u.	E_{2_β} keV	y W.u.	E_{2_γ} keV	z W.u.	y/x	$E_{2_1}/2\hbar\omega_\beta$	z/x	$E_{2_1}/\hbar\omega_\gamma$
^{154}Sm	82.0	3.25	177	1178	0.94^{23}	1440	3.2^{5}	0.005	0.037	0.018	0.057
^{156}Gd	89.0	3.24	185	1129	0.63^{6}	1154	4.69^{17}	0.003	0.043	0.025	0.078
^{158}Gd	79.5	3.29	197	1260	0.31^{4}	1187	3.4^{4}	0.002	0.033	0.018	0.066
^{160}Gd	75.3	3.30	202	1377		988	3.80^{22}		0.028	0.019	0.075
^{158}Dy	78.9	3.21	183	1086	2.1^{5}	946	5.9^{12}	0.011	0.039	0.032	0.084
^{160}Dy	86.8	3.27	198	1350	0.65^{8}	966	4.5^{3}	0.004	0.034	0.023	0.090
^{162}Dy	80.7	3.29	203	1453		888	4.59^{31}		0.030	0.023	0.090
^{164}Dy	73.4	3.30	209	1716		762	4.0^{4}			0.019	0.096
^{162}Er	102.0	3.23	190	1171	1.6^{10}	901	6.2^{3}	0.008	0.048	0.032	0.114
^{164}Er	91.4	3.28	203	1315	0.23^{12}	860	5.2^{6}	0.001	0.037	0.024	0.105
^{166}Er	80.6	3.29	214	1528		786	5.5^{4}			0.026	0.102
^{168}Er	79.8	3.31	209	1276	0.06^{1}	821	4.80^{17}	0.0003	0.033	0.023	0.096
^{170}Er	78.6	3.31	207	960	0.28^{3}	934	3.68^{11}	0.001	0.045	0.018	0.084
^{168}Yb	87.7	3.27	201	1233	1.8^{2}	984	4.60^{10}	0.009	0.039	0.022	0.090
^{170}Yb	84.3	3.29	206	1139	1.08^{21}	1146	2.7^{6}	0.005	0.040	0.013	0.075
^{172}Yb	78.7	3.31	211	1118	0.24^{1}	1466	1.33^{11}	0.001	0.038	0.0063	0.054
^{174}Yb	76.5	3.31	205	1561	0.54^{23}	1634	2.5^{5}		0.026	0.0087	0.048
^{176}Yb	82.1	3.31	180	1200		1261	1.8^{2}			0.0093	0.066
^{174}Hf	91.0	3.27	168	900	2.1^{6}	1227	4.8^{22}	0.014	0.057	0.032	0.075
^{176}Hf	88.3	3.28	179	1227	1.0^{2}	1341	3.9^{6}	0.006	0.039	0.022	0.066
^{178}Hf	93.2	3.29	161	1277	0.061^{25}	1175	3.9^{5}	0.005	0.039	0.025	0.078
^{180}Hf	93.3	3.31	154	1183		1200	3.8^{6}		0.044	0.025	0.078
^{182}W	100.1	3.29	136	1257	0.91^{8}	1221	3.40^{9}	0.007	0.044	0.025	0.081
^{184}W	111.2	3.27	121	1121	0.21^{3}	903	4.41^{22}	0.002	0.049	0.037	0.111
^{186}W	122.3	3.23	110	1015		738	4.63^{23}		0.067	0.042	0.165

transition is typically less than 10% of that predicted by the model for a one-phonon beta band at the observed excitation energy. A simple interpretation of this result is that the ground-state band is much more beta-rigid than indicated by interpreting the first-excited $K = 0$ band to be a pure beta band; this interpretation would suggest that the predicted beta vibrational strength should be reduced and most of it should lie at higher energies,[29] which would be consistent with the high rigidity of the ground-state band against centrifugal stretching that is commonly observed. There are undoubtedly many non-collective excited states, at the higher energies which can be expected to mix with the beta vibrational states. Unfortunately, it is difficult to investigate this possibility experimentally because the spectroscopic characterisation of low-spin states above ~ 1300 keV is very demanding and there is a severe lack of data.

---

otherwise be the case and would mean that the ground-state band was more gamma-rigid than supposed.

[29] Cf. Garrett P.E. *et al.* (1977), *Phys. Lett.* **B400**, 250.

As shown in Figure 2.7, the model also predicts two-phonon gamma vibrational bands: one with $K = 0$ and $B(E2; 0_{\gamma\gamma} \rightarrow 2_\gamma) = 5z$, and one with $K = 4$ and $B(E2; 4_{\gamma\gamma} \rightarrow 2_\gamma) = 2.8z$. The best characterised candidates for such two-phonon gamma bands are in ^{166}Er, details of which are shown in Figure 2.8. To our knowledge, there are no other known candidates for $0^+_{\gamma\gamma}$ excitations. Similar to spherical nuclei, a natural interpretation is that the gamma vibrations are anharmonic and that the multi-phonon excitations mix strongly with non-collective excitations.

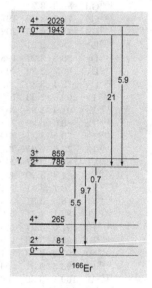

Figure 2.8: Selected energy levels and $B(E2)$ transition rates for the candidate one-phonon and two-phonon gamma vibrational states in ^{166}Er. (The data are from Garrett P.E. *et al.* (1997), *Phys. Rev. Lett.* **78**, 4545 and Fahlander C. *et al.* (1996), *Phys. Lett.* **B388**, 475.)

Exercises

2.16 Show that $\hat{H}_\gamma$, as given by Equation (2.170), can be expressed in terms of a harmonic oscillator unit a in the form

$$\hat{H}_\gamma = \frac{1}{2}\left[-\frac{1}{a^2}\frac{\partial^2}{\partial\gamma^2} + a^2(\gamma - \gamma_0)^2\right]. \tag{2.209}$$

2.17 Use the expressions, obtained from Claim 4.3,

$$\langle \lambda+1, \nu-1|(a\gamma)|\lambda\nu\rangle = \sqrt{\nu}, \tag{2.210}$$
$$\langle \lambda-1, \nu|(a\gamma)|\lambda\nu\rangle = \langle \lambda\nu|(a\gamma)|\lambda-1,\nu\rangle = \sqrt{\lambda+\nu-1}, \tag{2.211}$$

for the radial matrix elements of a two-dimensional harmonic oscillator, with

$$\nu = \tfrac{1}{2}(n_\gamma - \tfrac{1}{2}K), \quad \lambda = \tfrac{1}{2}K + 1, \tag{2.212}$$

to derive the gamma matrix elements of Equations (2.195) and (2.196).

2.18 Use Equations (2.197)–(2.200) to derive expressions for the parameters x, y, z of Figure 2.7. Derive a relationship between y/z and the ratio of ω_y to ω_z.

2.5 Conclusions and implications

In spite of its shortcomings, indeed partly because of its shortcomings, the Bohr model contributes a great deal of insight into the way a system of many nucleons can behave collectively and exhibit properties in which only a few macroscopic degrees of freedom participate. It suggests that the many-nucleon dynamics of a nucleus separates, approximately, into collective and complementary subdynamics such that in some low-energy states only the collective component of the dynamics is active. It also suggests that a more complete description of nuclear states could be given by a Hamiltonian of the form

$$\hat{H} = \hat{H}_{\text{coll}} + \hat{H}_{\text{intr}} + V_{\text{coup}}, \tag{2.213}$$

where $\hat{H}_{\text{coll}}$ is a collective model Hamiltonian, $\hat{H}_{\text{intr}}$ is a complementary intrinsic Hamiltonian which controls all the many-nucleon dynamics suppressed in the collective model, and V_{coup} is a coupling interaction. With such a perspective, it might be supposed that low-energy states of nuclei, which can be described by a collective Hamiltonian, are states whose intrinsic components are in their ground states and for which the coupling interaction has small effect.

The adiabatic limit of the Bohr model, considered in the previous section, shows how such a decoupling comes about. One can expect there to be an adiabatic decoupling of the collective and intrinsic degrees of freedom when the system is in its lowest-energy intrinsic states and so long as the collective excitation energies remain small in comparison to intrinsic excitations of the same angular momentum. It would appear that beta and gamma vibrational excitations do not meet the latter criterion.

The above perspective will be adopted in the more general Bohr-Mottelson collective model developed in Volume 2. This more general model admits the presence of many intrinsic states and assigns phenomenological parameters to them (excitation energies, deformation parameters, and moments of inertia) to be determined by fits to experimental data.

A basic objective for a fundamental understanding of nuclear structure is to understand how collective dynamics are embedded in the more complete many-nucleon dynamics described by the shell model. As the Bohr model was originally formulated, in terms of surface deformation parameters of a liquid drop model, this objective posed a major challenge. However, by expressing the model in terms of quadrupole moments as collective coordinates the challenge is already partially met. This is because quadrupole moments are directly expressible in terms of nucleon coordinates.

The above challenge will be taken up in Volume 2. There are two natural approaches: the first is to make a separation of variables from many-nucleon coordinates to collective and intrinsic coordinates; the second is the algebraic approach. The algebraic approach seeks to discover how the dynamical symmetries of the collective model are realised within the framework of the microscopic shell model and proves to be more successful. In this volume, we develop first the collective model and then the shell model in algebraic terms following a chapter on the fundamental role of algebraic descriptions of quantum mechanical systems. It will subsequently be shown in Volume 2, that the algebraic Bohr model has a microscopic counterpart[30] which is fully expressible in terms of many-nucleon coordinates and includes intrinsic vortex spin degrees of freedom whose inclusion remedies many of the deficiencies of the Bohr model.

---

[30]The microscopic collective model is the so-called *symplectic model* of Rosensteel G. and Rowe D.J. (1977), *Phys. Rev. Lett.* **38**, 10 and Rosensteel G. and Rowe D.J. (1980), *Ann. Phys. (NY)* **126**, 343.

Chapter 3

Symmetry, dynamical groups and spectrum generating algebras

A unifying theme that connects the many models described in this book is obtained by relating their dynamical symmetries and algebraic structures. The importance of identifying the algebraic structures of a model cannot be overemphasised. At a practical level, it enables the well-developed mathematics of group theory to be applied to the solution of model problems. When different models are expressed in algebraic terms, it frequently becomes possible to relate them through their so-called *spectrum generating algebras* and/or their *dynamical groups* (defined below). Examples of such relationships can be found throughout our treatment of the fundamentals of nuclear models. In particular, it will be shown in Volume 2 that the dynamical group of the collective model can be extended to a larger dynamical group of a microscopic collective model which, thereby, gives it a microscopic foundation in the nuclear shell model.

In this chapter, we outline some of the basic group-theoretical concepts and techniques that will be used in subsequent chapters.

3.1 Symmetry groups

A group G of transformations of a Hilbert space $\mathbb{H}$ with Hamiltonian $\hat{H}$ is called a *symmetry group* for $\hat{H}$, or simply a symmetry of $\hat{H}$, if it transforms eigenstates of $\hat{H}$ into other eigenstates of the same eigenvalue. More precisely, if $\hat{T}$ is a unitary representation of G on $\mathbb{H}$, then G is a symmetry group for $\hat{H}$ if and only if the transformation $\hat{H} \to \hat{T}(g)\hat{H}\hat{T}(g^{-1})$ of $\hat{H}$ by any element $g \in G$ leaves $\hat{H}$ invariant; i.e., if and only if $\hat{T}(g)\hat{H}\hat{T}(g^{-1}) = \hat{H}$ for all $g \in G$.

If a Hamiltonian is invariant under SU(2) rotations, for example, then its equal-energy eigenstates span representations of SU(2) and can be labelled by angular momentum quantum numbers. Thus, if $\{|\alpha J M\rangle\}$ are orthonormal basis states for a system, where J and M are angular momentum quantum numbers and α distinguishes different SU(2) irreps of the same J, then the Hamiltonian is block-diagonal in this basis in as much as it satisfies the equation

$$\langle \alpha J M | \hat{H} | \beta J' M' \rangle = \delta_{JJ'}\delta_{MM'}\mathcal{H}^J_{\alpha\beta}, \tag{3.1}$$

where $\mathcal{H}^J$ is the M-independent matrix with elements

$$\mathcal{H}^J_{\alpha\beta} = \langle \alpha JM|\hat{H}|\beta JM\rangle \qquad (3.2)$$

(see Exercise 3.1). Thus, to determine a complete set of eigenstates of a Hamiltonian, it only remains to diagonalise the $\mathcal{H}^J$ Hamiltonians with respect to the multiplicity indices which distinguish states with a common set of *good, symmetry-defined, quantum numbers*, which in this example are J and M.

Because of the above properties, symmetry groups are widely employed in constructing simply-solvable models and in computing the properties of many systems. However, it turns out that *dynamical symmetries* are much more common and even more useful in practice.

3.2 Dynamical groups

Several variations on the definition of a dynamical group for a model can be found in the literature.[1] Fundamentally a dynamical group for a model is a Lie group that has an irreducible unitary representation as a group of transformations of the model Hilbert space. A model can have more than one dynamical group. To be useful, the dynamical group should be such that its representation theory and that of its Lie algebra are of assistance in determining the spectral properties of the model.

A useful way to think of a dynamical group is in terms of the dynamical motions embraced by a model. For example, for the Bohr model, a possible dynamical group is the set of all translations of the position and momentum coordinates for the model. This is the Heisenberg-Weyl group, HW(5), whose infinitesimal generators include the quadrupole moment coordinates and their canonical momenta. A more useful dynamical group for the Bohr model is one that also includes rotational and vibrational transformations as well as the Heisenberg-Weyl translations.

One can also think of a dynamical group in more abstract terms. For example, for a model with an n-dimensional Hilbert space, a possible dynamical group is the group U(n) of all unitary transformations of this space. However, in general, there are smaller and, hence, simpler dynamical groups that are more useful in practice. An ideal dynamical group is one that is simple enough that its irreps are known, or can be calculated easily, and rich enough to contain the dynamics of the model of interest.

Numerous examples of such dynamical groups are given throughout this book. It is also useful to consider groups whose representations on the model space are direct sums of irreps. Such groups are typically useful when one wishes to focus on the subdynamics of a model. We shall speak of them as *dynamical groups for subsets of states* or *dynamical groups for submodels*, as best fits the situation.

---

[1] Cf., for example, the compendium Bohm A., Ne'eman Y. and Barut A.O., eds. (1988), *Dynamical Groups and Spectrum Generating Algebras*, Vol. 1 & 2 (World Scientific, Singapore).

The identification of a dynamical group for a model serves several important purposes. In the first place, it provides a mechanism for quantizing a classical version of the model. This is basically because quantizing a model is equivalent to determining the relevant unitary irreps of a dynamical group for the model.[2] Moreover, constructing a basis for an irrep of a dynamical group for the model provides a basis for the corresponding model Hilbert space and facilitates the calculation of matrix elements of physically relevant model observables.

Exercise

3.1 Let G denote a dynamical group for a model with a Hilbert space that carries a unitary irrep of G labelled by λ. Let $\{|\lambda\rho\kappa\nu\rangle\}$ denote an orthonormal basis for this Hilbert space, where ν indexes a basis for an irrep, labelled by κ, of a subgroup $G_1 \subset G$ and ρ indexes multiple copies of the irrep κ of G_1 in the irrep λ of G. Show that the Hamiltonian can be expressed in such a basis by a matrix with entries of the form

$$\langle\lambda\rho\kappa\nu|\hat{H}|\lambda\rho'\kappa'\nu'\rangle = \delta_{\kappa,\kappa'}\delta_{\nu,\nu'}\mathcal{H}^{\kappa}_{\rho\rho'}. \tag{3.3}$$

3.3 Dynamical subgroup chains

Particularly useful basis states for a model (without spin degrees of freedom)[3] are ones labelled by quantum numbers associated with a chain of subgroups, e.g.,

$$\begin{array}{ccccccc} G & \supset & G_1 & \supset & \mathrm{SO}(3) & \supset & \mathrm{SO}(2), \\ \lambda & & \rho \ \kappa & & \alpha \ L & & M \end{array} \tag{3.4}$$

where λ, κ, L, and M, respectively, label irreps of the groups G, G_1, SO(3) and SO(2), in the chain. (The subgroups SO(3) and SO(2) are included in such subgroup chains when one wishes to define basis states with good angular momentum quantum numbers.) Irrep labels provide quantum numbers for labelling basis states. However, there are generally *missing labels* in the sense that the set of quantum numbers given by the representation labels of the groups in the chain may be insufficient to identify a state uniquely. *Multiplicity labels* must then be added to the set for a complete classification of basis states. For example, if multiple equivalent irreps of the group G_1, labelled by κ, occur within the space of a single irrep of G, labelled by λ, these irreps are distinguished by a multiplicity index, ρ.

Basis states, as defined above, are said to reduce the given subgroup chain. The representation theory for the groups in the chain then provides powerful tools for deducing the energy spectrum and other properties of the system. A chain of subgroups of a dynamical group, G, for a model is also of interest because it implies the existence of a corresponding chain of submodels each of which has a dynamical

---

[2]Bartlett S.D., Rowe D.J. and Repka J. (2002), *J. Phys. A: Math. Gen.* **35**, 5599, 5625.

[3]If the model includes spin degrees of freedom, the subgroup SO(3) in the chain (3.4) should be replaced by SU(2).

group given by an associated subgroup in the chain. For example, the Hilbert space spanned by all states $\{|\lambda\rho\kappa\alpha LM\rangle\}$, defined by Equation (3.4) with fixed $(\lambda\rho\kappa)$ quantum numbers, carries an irrep κ of the subgroup $G_1 \subset G$ and is the Hilbert space for a submodel with G_1 as its dynamical group.

Frequently there are alternative choices of subgroup chains that can be used to construct basis states for a model. As will be illustrated in the context of the Bohr model, the alternatives are naturally associated with submodel choices. Such choices have Hamiltonians that are diagonal in bases that reduce associated choices of subgroup chain. Simple examples are Hamiltonians that are sums of the Casimir invariants and other invariants of the groups in a chain (see Section 3.8). Examples of such Hamiltonians are naturally expressed in terms of the Lie algebras of the subgroups in the chain and are discussed below.

When a subgroup chain of a dynamical group contains a symmetry group of the Hamiltonian, then the corresponding irrep label is a good quantum number, i.e., a quantum number that is conserved by the Hamiltonian and can be used to label eigenstates. For example, if a model Hamiltonian has a dynamical group G, with subgroups given by Equation (3.4) and is rotationally invariant so that SO(3) and SO(2) are symmetry groups, then the labels L and M are good quantum numbers for the model. However, as the above discussion of dynamical groups and their subgroups shows, a subgroup in a chain need not be a symmetry group of the Hamiltonian to provide good quantum numbers. In particular, the labels associated with the dynamical groups of submodels may, in situations of interest, also provide good quantum numbers. The following discussion of spectrum generating algebras shows explicitly how this can happen. In general, when a complete set of eigenstates of a model Hamiltonian, $\hat{H}$, reduce some subgroup chain of a dynamical group, the Hamiltonian is said to have a *dynamical symmetry* defined by the subgroup chain.[4,5]

3.4   Lie algebras of physical observables

Most Lie algebras that arise in the quantum mechanics of physical systems have structures that can be derived from the Heisenberg-Weyl algebra, whose elements include the position and momentum coordinates, (x, y, z, p_x, p_y, p_z), of a particle in a Euclidean space. These observables are represented in quantum mechanics as

---

[4]Dynamical symmetries achieved widespread recognition in their applications to the *Interacting Boson Model*; cf. Iachello F. and Arima A. (1987), *The Interacting Boson Model* (Cambridge University Press, Cambridge), and references therein. However, it is to be noted that the concept of a dynamical group, in the context of the Interacting Boson Model, is used in a more restrictive sense than that defined here.

[5]When a subset of eigenstates of the model Hamiltonian, $\hat{H}$, reduce some subgroup chain of a dynamical group, the Hamiltonian is said to have *a partial dynamical symmetry* as defined by Alhassid Y. and Leviatan A. (1992), *J. Phys. A: Math. Gen.* **25**, L1265, and by Leviatan A. (1996), *Phys. Rev. Lett.* **77**, 818.

operators, $(\hat{x}, \hat{y}, \hat{z}, \hat{p}_x, \hat{p}_y, \hat{p}_z)$, which satisfy the commutation relations

$$\begin{align}
[\hat{x}, \hat{p}_x] = [\hat{y}, \hat{p}_y] = [\hat{z}, \hat{p}_z] &= i\hbar\hat{I}, \\
[\hat{x}, \hat{y}] = [\hat{x}, \hat{p}_y] = [\hat{p}_x, \hat{p}_y] = \cdots = [\hat{I}, \hat{x}] = [\hat{I}, \hat{p}_x] &= \cdots = 0.
\end{align}$$
(3.5)

Corresponding commutation relations follow for any algebra whose elements are quadratic functions of the many-particle position and momentum coordinates; e.g., the commutation relations for the angular-momentum components of a particle in a three-dimensional Euclidean space (in units of $\hbar$),

$$\hbar\hat{L}_x = \hat{y}\hat{p}_z - \hat{z}\hat{p}_y, \quad \hbar\hat{L}_y = \hat{z}\hat{p}_x - \hat{x}\hat{p}_z, \quad \hbar\hat{L}_z = \hat{x}\hat{p}_y - \hat{y}\hat{p}_x, \tag{3.6}$$

are determined (cf. Exercise 3.5). to be

$$[\hat{L}_x, \hat{L}_y] = i\hat{L}_z, \quad [\hat{L}_y, \hat{L}_z] = i\hat{L}_x, \quad [\hat{L}_z, \hat{L}_x] = i\hat{L}_y. \tag{3.7}$$

For any Lie algebra **g** derived in this way, it is found that, if $\hat{X}$ and $\hat{Y}$ are elements of **g**, their commutator can be expressed in the form

$$[\hat{X}, \hat{Y}] = i\hat{Z}, \quad \text{with } \hat{Z} \in \mathbf{g}. \tag{3.8}$$

This result is consistent with the axiom of quantum mechanics that physical observables should be represented by Hermitian operators; i.e., because the commutator, $[\hat{X}, \hat{Y}] = \hat{X}\hat{Y} - \hat{Y}\hat{X}$, of two Hermitian operators is an anti-Hermitian operator, it must be multiplied by $\pm i$ to turn it into a Hermitian operator. Operators representing physical observables are required to be Hermitian in quantum mechanics, so that their eigenvalues take only real values. Consequently, a Lie algebra of physical observables is a *real* Lie algebra in the sense that it is a set of real linear combinations of a basis of physical observables; it must be real because complex linear combinations of Hermitian operators include operators that are not Hermitian.

Often, in the study of Lie algebras, the factor "i", which is included in Equation (3.8) for convenience in applications in quantum mechanics, is considered to be undesirable and removed, notably in mathematical texts. This is easily done. For example, a Lie algebra of Hermitian operators, having a basis with commutation relations expressed in terms of a set of real structure constants, $\{C_{\mu\nu}^{\lambda}\}$, by the equation

$$[\hat{X}_\mu, \hat{X}_\nu] = i\sum_\lambda C_{\mu\nu}^{\lambda}\hat{X}_\lambda, \tag{3.9}$$

is simply replaced by an isomorphic real Lie algebra of anti-Hermitian operators with a basis $\{X_\nu = i\mathcal{X}_\nu\}$ that satisfies the commutation relations

$$[\hat{\mathcal{X}}_\mu, \hat{\mathcal{X}}_\nu] = \sum_\lambda C_{\mu\nu}^{\lambda}\hat{\mathcal{X}}_\lambda \tag{3.10}$$

with exactly the same structure constants.

3.5 The Lie algebra of a Lie group

The elements of a Lie algebra can be regarded as infinitesimal generators of a corresponding Lie group. For example, in its fundamental representation, the rotation group, SO(3), is a group of real 3×3 orthogonal matrices with unit determinant and group product given by matrix multiplication. A rotation through angle ϕ about the z axis is represented by the matrix

$$
R_z(\phi) = \begin{pmatrix} \cos \phi & -\sin \phi & 0 \\ \sin \phi & \cos \phi & 0 \\ 0 & 0 & 1 \end{pmatrix}. \tag{3.11}
$$

An infinitesimal generator of such a rotation is then defined by the matrix

$$
L_z = \mathrm{i} \frac{\partial}{\partial \phi} R_z(\phi) \Big|_{\phi=0} = \begin{pmatrix} 0 & -\mathrm{i} & 0 \\ \mathrm{i} & 0 & 0 \\ 0 & 0 & 0 \end{pmatrix}. \tag{3.12}
$$

The matrices representing the infinitesimal generators for rotations about the x, y, and z axes span the so(3) Lie algebra.

Conversely, when the elements of a Lie algebra, with basis $\{X_\nu\}$, are represented as Hermitian operators, $\{\hat{X}_\nu\}$, corresponding Lie group elements are represented by unitary transformations of the form

$$
\hat{R}(\xi) := \exp \Big[-\mathrm{i} \sum_\nu \xi^\nu \hat{X}_\nu \Big], \tag{3.13}
$$

where $\{\xi^\nu\}$ is a set of real parameters. These Lie group representations are said to be generated by exponentiation of a Lie algebra representation. Thus, when we speak of a unitary representation of a Lie algebra, we mean that its elements are represented by Hermitian operators (or matrices) and that corresponding group elements are represented by unitary transformations (or matrices). For example, the time evolution operator $\exp(-\frac{\mathrm{i}}{\hbar} \hat{H} t)$ is unitary when $\hat{H}$ is Hermitian and, as a consequence, it preserves the norms of quantum states, i.e.,

$$
\langle \psi(t) | \psi(t) \rangle = \langle \psi(0) | e^{\frac{\mathrm{i}}{\hbar} \hat{H} t} e^{-\frac{\mathrm{i}}{\hbar} \hat{H} t} | \psi(0) \rangle = \langle \psi(0) | \psi(0) \rangle. \tag{3.14}
$$

There are also complex Lie algebras, whose elements are complex linear combinations of a basis. However, it is meaningless to speak of a unitary representation of a complex Lie algebra because, if a Lie algebra element, X, is represented by a Hermitian operator, $\hat{X}$, then iX is represented by the skew-Hermitian operator i$\hat{X}$ (see Exercises 3.2 and 3.4). Nevertheless, it is frequently useful to construct complex linear combinations of the physical observables to form so-called *raising and lowering operators* in the construction of irreps. Such operators occur in Hermitian conjugate pairs; they do not belong to a real Lie algebra but belong to its so-called *complex extension*.

Exercises

3.2 Show that the Lie algebra of the U(1) group $\{e^{i\theta}; \theta \in \mathbb{R}\}$ is set of real numbers $0 \leq \theta \leq 4\pi$.

3.3 Starting with the expression (3.11) derive the matrices for $R_x(\phi)$, $R_y(\phi)$, L_x, and L_y, and show that the matrices L_x, L_y, and L_z satisfy Equation (3.7).

3.4 Show that the group GL(1,$\mathbb{C}$) of non-zero complex numbers has no unitary representations.

3.5 Use the so-called *Leibnitz rule* for linear operators, $[X, YZ] = [X, Y]Z + Y[X, Z]$, to derive the angular-momentum commutation relations of Equation (3.7) from those of the Heisenberg-Weyl commutation relations. Also use it to derive the *Jacobi identity*

$$[X, [Y, Z]] = [[X, Y], Z] + [Y, [X, Z]]. \tag{3.15}$$

3.6 Use the Leibnitz rule, given in Exercise 3.5, to derive the commutation relations (3.7) of the angular momentum operators from their definition (3.6).

3.6 Spectrum generating algebras

As for a dynamical group, there are various definitions in the literature of a spectrum generating algebra (SGA) for a model.[6] The essential requirement of an SGA, **g**, is that it have a unitary representation in which its elements are Hermitian operators on the model Hilbert space, $\mathbb{H}$, such that the corresponding representation theory of **g** can assist in determining the spectral properties of the model. It is often convenient to require that the representation of an SGA is irreducible on the Hilbert space of the model. The Lie algebra of a dynamical group is often the natural choice of SGA for a model.

In practice, an SGA for a model is most useful when the Hamiltonian and other observables of the model are expressible as polynomials in its elements. In such a situation, the matrix elements of model observables can be determined by the representation theory of the SGA.[7] We shall also speak of a Lie algebra **g** as an *SGA for a Hamiltonian*, as opposed to an SGA for a model, if the Hamiltonian is a function of the elements of **g** but some other observables of the model are not. Such an SGA facilitates the computation of energy levels of a model, although a larger SGA that includes other observables may be needed to compute transition matrix elements and further properties of the model.

Chains of subalgebras, corresponding to subgroup chains of a dynamical group, are also useful. Consider, for example, a model with a dynamical symmetry given by the subgroup chain (3.4) and let **g**, $\mathbf{g}_1$, so(3) and so(2), respectively, denote the

---

[6]The properties of spectrum generating algebras were studied in depth by Dothan Y. (1970), *Phys. Rev.* **D2**, 2944. Cf. also Bohm *et al.*, *op. cit.* Footnote 1 on Page 144.

[7]We shall also encounter situations in which the Hamiltonian is a non-polynomial function of some elements of an SGA.

Lie algebras of the groups G, G_1, SO(3) and SO(2) in the chain.[8] A simple example of a Hamiltonian with such a dynamical symmetry is

$$\hat{H} = \hat{\Lambda} + \hat{\Lambda}_1 + k\hat{\mathbf{L}}^2, \tag{3.16}$$

where $\hat{\Lambda}$, $\hat{\Lambda}_1$, and $\hat{\mathbf{L}}^2$ are, respectively, Casimir invariants or sums of Casimir invariants of the Lie algebras $\mathbf{g}$, $\mathbf{g}_1$ and so(3), It can be seen that the sequence of Hamiltonians $\hat{H} = \Lambda + \hat{H}_1$, $\hat{H}_1 = \Lambda_1 + \hat{H}_2$, and $\hat{H}_2 = k\hat{\mathbf{L}}^2$, are Hamiltonians for a sequence of submodels with respective dynamical groups G, G_1, and SO(3). The group SO(3) is also a symmetry group for each of these Hamiltonians.

Exercise

3.7 Give a dynamical group and a symmetry group for a Hamiltonian that is a linear combination of Casimir invariants for the groups SU(3) and its subgroups, SO(3) $\subset$ U(3) and SO(2) $\subset$ SO(3).

3.7 Irreps of Lie algebras with raising and lowering operators

A standard technique for deriving irreps of many Lie algebras is to use raising and lowering operators. The prototype for this technique arises from the solutions of the simple harmonic oscillator Schrödinger equation. It should be recognised, however, that not all Lie algebra irreps can be constructed in this way. Other techniques are introduced in the text as needed.

3.7.1 *The Heisenberg-Weyl algebra, hw(1)*

The simple harmonic oscillator (SHO) is governed by a Hamiltonian

$$\hat{H} = \frac{1}{2M}\hat{p}^2 + \frac{1}{2}M\omega^2\hat{x}^2, \tag{3.17}$$

where $\hat{x}$ and $\hat{p}$ are mutually canonical position and momentum observables that, in quantum mechanics, satisfy the commutation relations,

$$[\hat{x},\hat{p}] = \mathrm{i}\hbar\hat{I}, \quad [\hat{I},\hat{x}] = [\hat{I},\hat{p}] = 0, \tag{3.18}$$

of a Heisenberg-Weyl algebra, hw(1). These commutation relations are obtained, for example, by interpreting: $\hat{x}$ as the multiplicative operator whose action on a wave function is given by $[\hat{x}\psi](x) = x\psi(x)$; $\hat{p}$ as the differential operator $\hat{p} = -\mathrm{i}\hbar d/dx$; and $\hat{I}$ as the identity operator. The SHO Hamiltonian is then a second order differential

---

[8]When it is desirable to distinguish between a Lie group and its Lie algebra, we follow the standard convention of using upper case letters for the group and lower case for its Lie algebra.

operator and the corresponding Schrödinger equation is

$$\hat{H}|\psi\rangle \equiv \left[-\frac{\hbar^2}{2M}\frac{d^2}{dx^2} + \frac{1}{2}M\omega^2\hat{x}^2\right]\psi = E\psi. \tag{3.19}$$

The standard approach to solving Equation (3.19) is by the *factorisation method*, as follows. When applied to the SHO, this method introduces operators

$$c^\dagger = \sqrt{\frac{M\omega}{2\hbar}}\left(\hat{x} - \frac{i}{M\omega}\hat{p}\right), \quad c = \sqrt{\frac{M\omega}{2\hbar}}\left(\hat{x} + \frac{i}{M\omega}\hat{p}\right), \tag{3.20}$$

in terms of which the Hamiltonian assumes the simple form

$$\hat{H} = \hbar\omega(c^\dagger c + \tfrac{1}{2}), \tag{3.21}$$

and for which

$$[c, c^\dagger] = \hat{I}, \quad [\hat{I}, c^\dagger] = [\hat{I}, c] = 0, \tag{3.22}$$

$$[\hat{H}, c^\dagger] = \hbar\omega\, c^\dagger, \quad [\hat{H}, c] = -\hbar\omega\, c. \tag{3.23}$$

It follows that, if $|n\rangle$ is an eigenstate of $\hat{H}$ of energy eigenvalue E_n, then

$$\hat{H}c^\dagger|n\rangle = (E_n + \hbar\omega)c^\dagger|n\rangle, \quad \hat{H}c|n\rangle = (E_n - \hbar\omega)c|n\rangle. \tag{3.24}$$

Thus, $c^\dagger$ and c, respectively, increase and decrease the energy of a state by $\hbar\omega$. We therefore describe $c^\dagger$ and c as raising and lowering operators, respectively, for the hw(1) Lie algebra. They are also known as creation and annihilation operators of SHO quanta.

The SHO Hamiltonian, (3.17), being a sum of the squares of Hermitian operators, is positive definite and has no negative-energy eigenvalues. Thus, it has a lowest-energy (ground) state, $|0\rangle$, that is annihilated by the lowering operator, i.e.,

$$c|0\rangle = 0. \tag{3.25}$$

Hence, the ground state has energy given by the so-called *zero-point energy*, $E_0 = \frac{1}{2}\hbar\omega$. The corresponding ground-state wave function is obtained by solution of Equation (3.25) (see Exercise 3.8). However, wave functions are not often needed because most results of interest can be obtained without them by purely algebraic methods. In particular, matrix elements of the raising and lowering operators, and hence the $\hat{x}$ and $\hat{p}$ operators and polynomials in these operators, are obtained as follows.

Let $|n\rangle$ denote a normalised eigenstate of $\hat{H}$ of energy

$$E_n = E_0 + n\hbar\omega = (n + \tfrac{1}{2})\hbar\omega \tag{3.26}$$

and suppose that

$$c^\dagger|n\rangle = C_n|n+1\rangle, \tag{3.27}$$

where C_n is a normalisation factor. According to the definition, (3.20), if $\hat{x}$ and $\hat{p}$ are Hermitian operators, $c^\dagger$ and c are Hermitian adjoints of one another (as recognised in the notation) and their non-zero matrix elements are related by

$$\langle n|c|n+1\rangle = \langle n+1|c^\dagger|n\rangle^* = C_n^*. \tag{3.28}$$

Thus, from the equation $\langle n|[c,c^\dagger]|n\rangle = 1$, we obtain the identity

$$\langle n|c|n+1\rangle\langle n+1|c^\dagger|n\rangle - \langle n|c^\dagger|n-1\rangle\langle n-1|c|n\rangle = 1, \tag{3.29}$$

and the recursion relation

$$|C_n|^2 = |C_{n-1}|^2 + 1. \tag{3.30}$$

From the identity $\langle 1|1\rangle = \langle 0|cc^\dagger|0\rangle = |C_0|^2 = 1$, it follows that

$$|C_n|^2 = n+1, \quad c^\dagger|n\rangle = \sqrt{n+1}\,|n+1\rangle, \quad c|n\rangle = \sqrt{n}\,|n-1\rangle. \tag{3.31}$$

3.7.2 *The angular momentum algebras, so(3) and su(2)*

The angular momentum operators of Equation (3.6), with commutation relations given by Equation (3.7), span an so(3) Lie algebra. Irreps of this Lie algebra are constructed by first choosing one angular momentum operator which should be diagonalised (we choose $\hat{L}_z$) and defining raising and lowering operators relative to this choice by

$$\hat{L}_0 := \hat{L}_z, \quad \hat{L}_\pm := \hat{L}_x \pm \mathrm{i}\hat{L}_y. \tag{3.32}$$

These operators satisfy commutation relations

$$[\hat{L}_0, \hat{L}_\pm] = \pm\hat{L}_\pm, \quad [\hat{L}_+, \hat{L}_-] = 2\hat{L}_0. \tag{3.33}$$

The element $\hat{L}_0$ spans a one-dimensional so(2) algebra which is a Cartan subalgebra of so(3).[9]

Let $\{|m\rangle\}$ denote an orthonormal basis of eigenstates of $\hat{L}_0$ for an so(3) irrep for which

$$\hat{L}_0|m\rangle = m|m\rangle. \tag{3.34}$$

It then follows, from the equation $[\hat{L}_0, \hat{L}_\pm] = \pm\hat{L}_\pm$, that

$$\hat{L}_0\hat{L}_\pm|m\rangle = (m \pm 1)\hat{L}_\pm|m\rangle. \tag{3.35}$$

---

[9]For practical purposes, a Cartan subalgebra is a maximal Abelian subalgebra. A more precise definition can be found, for example, in Wybourne B.G. (1974), *Classical Groups for Physicists* (Wiley, New York).

Suppose that

$$\hat{L}_+|m\rangle = C_m|m+1\rangle, \tag{3.36}$$

where C_m is a normalisation factor. According to the definition, (3.32), if $\hat{L}_x$ and $\hat{L}_y$ are Hermitian operators, $\hat{L}_\pm$ are Hermitian adjoints of one another and their non-zero matrix elements are related by

$$\langle m+1|\hat{L}_+|m\rangle = \langle m|\hat{L}_-|m+1\rangle^* = C_m. \tag{3.37}$$

Thus, from the equation $\langle m|[\hat{L}_+, \hat{L}_-]|m\rangle = \langle m|2\hat{L}_0|m\rangle = 2m$, we obtain the identity

$$\langle m|\hat{L}_+|m-1\rangle\langle m-1|\hat{L}_-|m\rangle - \langle m|\hat{L}_-|m+1\rangle\langle m+1|\hat{L}_+|m\rangle = 2m, \tag{3.38}$$

and the recursion relation

$$|C_m|^2 = |C_{m-1}|^2 - 2m. \tag{3.39}$$

This equation implies that the sequence of $|C_m|^2$ coefficients, with m increasing in unit integer steps, must terminate with $|C_m|^2 = 0$ for some positive value of $m = j$ for which

$$\hat{L}_+|j-1\rangle \neq 0, \quad \hat{L}_+|j\rangle = 0. \tag{3.40}$$

Otherwise $|C_m|^2$ will become negative. Solving the recursion relation downwards, we then obtain

$$|C_j|^2 = 0, \quad |C_{j-1}|^2 = 2j, \quad |C_{j-2}|^2 = 2(2j-1), \quad |C_{j-3}|^2 = 3(2j-2), \ldots, \tag{3.41}$$

and the general expression

$$|C_m|^2 = (j-m)(j+m+1). \tag{3.42}$$

The coefficient $|C_m|^2$ is now seen to vanish when $m = j$, as intended, and also when $m + 1 = -j$. However, because the lowering operator can only decrease the value of m in integer steps, it follows that the sequence can only terminate at a minimum as well as at a maximum value of m, if $2j$ is an integer. We then have so-called *highest-weight* and *lowest-weight* states $|\pm j\rangle$ which satisfy the equations

$$\hat{L}_+|j\rangle = \hat{L}_-|-j\rangle = 0. \tag{3.43}$$

Thus, if we relabel the basis states for an so(3) irrep with highest weight $m = j$ by $\{|jm\rangle\}$, we obtain the general expressions

$$\begin{aligned}
\hat{L}_0|jm\rangle &= m|jm\rangle, \\
\hat{L}_+|jm\rangle &= \sqrt{(j-m)(j+m+1)}\,|j,m+1\rangle, \\
\hat{L}_-|jm\rangle &= \sqrt{(j+m)(j-m+1)}\,|j,m-1\rangle.
\end{aligned} \tag{3.44}$$

The above results show that the Lie algebra so(3) has two classes of irreps: those for which j is an integer, and those for which it is a half-odd integer. However, for the SO(3) group there are extra considerations. In particular, as Equation (3.12) shows, an SO(3) rotation through an angle 2π, about any axis, is equivalent to no rotation at all. Under a rotation through angle ϕ about the z axis, a state $|jm\rangle$ is multipled by a phase factor according to the equation

$$\hat{R}_z(\phi)|jm\rangle = e^{-i\phi\hat{L}_z}|jm\rangle = e^{-im\phi}|jm\rangle. \tag{3.45}$$

This gives $\hat{R}_z(2\pi)|jm\rangle = (-1)^{2m}|jm\rangle$. Thus, for an SO(3) irrep, m and hence j are restricted to integer values and only for such values of j are the elements of the so(3) Lie algebra the infinitesimal generators of SO(3) rotations. In contrast, for the $2j$-odd representations of so(3), for which every $2m$ is odd, the elements of so(3) are infinitesimal generators of so-called *spinor* representations which are true representations of the group SU(2) appropriate for the description of rotations of particles with half-odd integer intrinsic spins.[10] For example, in the two-dimensional irrep, with basis states $|\frac{1}{2}, \pm\frac{1}{2}\rangle$, the operators $\hat{L}_0$ and $\hat{L}_\pm$ have matrix elements given by the 2×2 matrices

$$s_0 = \begin{pmatrix} \frac{1}{2} & 0 \\ 0 & -\frac{1}{2} \end{pmatrix}, \quad s_+ = \begin{pmatrix} 0 & 1 \\ 0 & 0 \end{pmatrix}, \quad s_- = \begin{pmatrix} 0 & 0 \\ 1 & 0 \end{pmatrix}, \tag{3.46}$$

which are proportional to Pauli spin matrices.

The fact that so(3) has irreps with $2j$ taking all integer values is a reflection of the fact that the so(3) and su(2) Lie algebras are isomorphic to one another. However, in accordance with convention, we usually denote the infinitesimal generators of SU(2) irreps by $\{\hat{J}_i\}$, and reserve the symbols $\{\hat{L}_i\}$ for the infinitesimal generators of SO(3) irreps. Thus, for su(2), we usually write the commutation relations of Equation (3.33) as

$$[\hat{J}_x, \hat{J}_y] = i\hat{J}_z, \quad [\hat{J}_y, \hat{J}_z] = i\hat{J}_x, \quad [\hat{J}_z, \hat{J}_x] = i\hat{J}_y. \tag{3.47}$$

3.7.3 *The su(1,1) Lie algebra*

The commutation relations for the su(1,1) $\simeq$ so(2,1) algebra are given by

$$[\hat{S}_x, \hat{S}_y] = -i\hat{S}_z, \quad [\hat{S}_y, \hat{S}_z] = i\hat{S}_x, \quad [\hat{S}_z, \hat{S}_x] = i\hat{S}_y. \tag{3.48}$$

They are similar to those of su(2), given by Equation (3.47), except for one change of sign. If we similarly define $\hat{S}_0$ and raising and lowering operators, $\hat{S}_\pm$, by

$$\hat{S}_0 := \hat{S}_z, \quad \hat{S}_\pm := \hat{S}_x \pm i\hat{S}_y, \tag{3.49}$$

---

[10]It was shown by Uhlenbeck G.E. and Goudsmit S. (1925), *Naturwissenschaften* **47**, 953, that the splitting of excited states of the hydrogen atom into doublets would follow if the electron has an intrinsic spin of $\frac{1}{2}\hbar$. A fascinating historical review, by Goudsmit, of the background to their discovery can be found at the website http://www.lorentz.leidenuniv.nl/history/spin/goudsmit.html.

we obtain equations similar to those of Equation (3.33),

$$[\hat{S}_0, \hat{S}_\pm] = \pm \hat{S}_\pm, \quad [\hat{S}_+, \hat{S}_-] = -2\hat{S}_0, \tag{3.50}$$

except for a change of sign in the second equation.

Unitary irreps of the su(1,1) algebra can now be derived from these commutation relations in essentially the same way as they were derived for su(2) $\simeq$ so(3). However, in contrast to su(2) $\simeq$ so(3), we find that there are many classes of su(1,1) irreps. Some have highest weights, some have lowest weights, some have both highest and lowest weights, and some have neither. In applications to nuclear physics, we are primarily interested in unitary irreps with lowest weights, and non-unitary irreps with both highest and lowest weights. From the definitions, (3.49), a representation of su(1,1) will be unitary if $\hat{S}_0$ and $\hat{S}_\pm$ satisfy the Hermitian adjoint relationships $\hat{S}_0^\dagger = \hat{S}_0$ and $\hat{S}_\pm^\dagger = \hat{S}_\mp$, and non-unitary otherwise.

(i) *Unitary irreps with lowest-weight states*

Let $\{|\lambda\nu\rangle; \nu = 0, 1, 2, \dots\}$ denote an orthonormal basis for a unitary irrep of su(1,1) consisting of eigenstates of $\hat{S}_0$, and suppose that $|\lambda 0\rangle$ is a lowest-weight state with eigenvalue $\lambda/2$, so that

$$2\hat{S}_0|\lambda 0\rangle = \lambda|\lambda 0\rangle. \tag{3.51}$$

Such an irrep is said to have lowest weight λ. Define the basis states by

$$\hat{S}_+|\lambda\nu\rangle = C_{\lambda\nu}|\lambda, \nu + 1\rangle, \tag{3.52}$$

where $C_{\lambda\nu}$ is a normalisation factor. The equation $[\hat{S}_0, \hat{S}_\pm]|\lambda\nu\rangle = \pm\hat{S}_\pm|\lambda\nu\rangle$ then implies that

$$2\hat{S}_0|\lambda\nu\rangle = (\lambda + 2\nu)|\lambda\nu\rangle. \tag{3.53}$$

According to the definition, (3.49), if $\hat{S}_x$ and $\hat{S}_y$ are Hermitian operators, $\hat{S}_\pm$ are Hermitian adjoints of one another and their non-zero matrix elements are related by

$$\langle\lambda\nu|\hat{S}_-|\lambda\mu\rangle = \langle\lambda\mu|\hat{S}_+|\lambda\nu\rangle^* = \delta_{\mu,\nu+1}C_{\lambda\nu}^*. \tag{3.54}$$

The equation $[\hat{S}_+, \hat{S}_-] = -2\hat{S}_0$ then also implies that

$$\langle\lambda\nu|\hat{S}_-|\lambda\nu+1\rangle\langle\lambda\nu+1|\hat{S}_+|\lambda\nu\rangle - \langle\lambda\nu|\hat{S}_+|\lambda\nu-1\rangle\langle\lambda\nu-1|\hat{S}_-|\lambda\nu\rangle = \lambda + 2\nu, \tag{3.55}$$

and we obtain the recursion relation

$$|C_{\lambda\nu}|^2 = |C_{\lambda,\nu-1}|^2 + \lambda + 2\nu, \tag{3.56}$$

with $|C_{\lambda 0}|^2 = \lambda \geq 0$. The solution of this equation leads to a non-trivial infinite-dimensional unitary irrep for all $\lambda > 0$ and non-negative integer values of ν, with

$$\hat{S}_0|\lambda\nu\rangle = \tfrac{1}{2}(\lambda + 2\nu)|\lambda\nu\rangle,$$
$$\hat{S}_+|\lambda\nu\rangle = \sqrt{(\lambda + \nu)(\nu + 1)}\,|\lambda, \nu + 1\rangle, \qquad (3.57)$$
$$\hat{S}_-|\lambda\nu\rangle = \sqrt{(\lambda + \nu - 1)\nu}\,|\lambda, \nu - 1\rangle.$$

An example of an su(1,1) Lie algebra is obtained by setting

$$\hat{S}_+ = \tfrac{1}{2}c^\dagger c^\dagger, \quad \hat{S}_- = \tfrac{1}{2}cc, \quad \hat{S}_0 = \tfrac{1}{4}(c^\dagger c + cc^\dagger) = \tfrac{1}{2}(c^\dagger c + \tfrac{1}{2}\hat{I}), \qquad (3.58)$$

where $c^\dagger$ and c are, respectively, creation and annihilation operator of SHO quanta as defined by Equation (3.20). One easily ascertains, by use of the basic hw(1) commutation relations of Equation (3.22) (see Exercise 3.9), that the operators of Equation (3.58) satisfy the su(1,1) commutation relations (3.50). The operators of Equation (3.58) are said to define a *realisation* of su(1,1). It is then seen, for example, that the SHO ground state, $|0\rangle$, is annihilated by the $\hat{S}_-$ operator and is an eigenstate of $2\hat{S}_0$ with eigenvalue $\lambda = 1/2$. Thus, the SHO ground state is the lowest-weight state for an su(1,1) irrep of lowest weight $\lambda = 1/2$. Such harmonic-oscillator representations are discussed further in Sections 4.2.2 and 5.8.1.

(ii) *Non-unitary tensor irreps*

We next consider an orthonormal basis $\{|\lambda\nu\rangle; \nu = 0, 1, 2, \dots\}$ for an su(1,1) irrep, consisting of eigenstates of $\hat{S}_0$, as before, and again suppose that $|\lambda 0\rangle$ is a lowest-weight state with eigenvalue given by $2\hat{S}_0|\lambda 0\rangle = \lambda|\lambda 0\rangle$. However, instead of requiring $(\hat{S}_-)^\dagger = \hat{S}_+$, as for a unitary representation, we now consider an irrep for which $(\hat{S}_-)^\dagger = -\hat{S}_+$. For such an irrep, we have the non-zero matrix elements

$$\langle\lambda\nu|\hat{S}_-|\lambda, \nu + 1\rangle^* = -\langle\lambda, \nu + 1|\hat{S}_+|\lambda\nu\rangle = -C_{\lambda\nu}. \qquad (3.59)$$

The equation $[\hat{S}_-, \hat{S}_+]|\lambda\nu\rangle = 2\hat{S}_0|\lambda\nu\rangle$ now yields the recursion relation

$$|C_{\lambda\nu}|^2 = |C_{\lambda, \nu-1}|^2 - \lambda - 2\nu, \qquad (3.60)$$

with $|C_{\lambda 0}|^2 = -\lambda \geq 0$, which has solution

$$|C_{\lambda\nu}|^2 = -(\lambda + \nu)(\nu + 1). \qquad (3.61)$$

However, because $|C_{\lambda\nu}|^2$ cannot take negative values, the sequence must terminate with $|C_{\lambda\nu}|^2$ vanishing when $\nu = -\lambda$. Thus, for negative integer values of λ and non-negative integer values of ν we obtain finite-dimensional irreps of su(1,1) with

$$\hat{S}_0|\lambda\nu\rangle = \tfrac{1}{2}(\lambda + 2\nu)|\lambda\nu\rangle,$$
$$\hat{S}_+|\lambda\nu\rangle = \pm\sqrt{-(\lambda + \nu)(\nu + 1)}\,|\lambda, \nu + 1\rangle, \qquad (3.62)$$
$$\hat{S}_-|\lambda\nu\rangle = \mp\sqrt{-(\lambda + \nu - 1)\nu}\,|\lambda, \nu - 1\rangle.$$

Consider, for example, the $\lambda = -1$ irrep. This irrep is two-dimensional, i.e., $\nu = 0$ or 1, and from Equation (3.62), we derive the matrices:

$$2S_0 = \begin{pmatrix} 1 & 0 \\ 0 & -1 \end{pmatrix}, \quad S_+ = \begin{pmatrix} 0 & \pm 1 \\ 0 & 0 \end{pmatrix}, \quad S_- = \begin{pmatrix} 0 & 0 \\ \mp 1 & 0 \end{pmatrix}. \tag{3.63}$$

It is interesting to note that, with the su(1,1) operators realised as in Equation (3.58), the SHO raising and lowering operators satisfy the commutation relations

$$\begin{aligned} [\hat{S}_-, c^\dagger] &= c, & [2\hat{S}_0, c^\dagger] &= c^\dagger, & [\hat{S}_+, c^\dagger] &= 0, \\ [\hat{S}_-, c] &= 0, & [2\hat{S}_0, c] &= -c, & [\hat{S}_+, c] &= -c^\dagger. \end{aligned} \tag{3.64}$$

These expression show that the SHO raising and lowering operators transform under su(1,1) as basis vectors for an su(1,1) irrep of lowest weight $\lambda = -1$ (and highest weight $+1$). They are said to be components of an su(1,1) tensor.

Exercises

3.8 Show that, if $\sqrt{2}\,c = a\hat{x} + \frac{i}{a\hbar}\hat{p}$, $\hat{x}\psi(x) = x\psi(x)$, and $\hat{p} = -i\hbar d/dx$, then the equation $c\psi = 0$ has solution $\psi(x) = e^{-(ax)^2/2}$.

3.9 Use the Leibnitz rule, given Exercise 3.5, to show that the operators defined by Equation (3.58) satisfy the su(1,1) commutation relations of Equation (3.50).

3.8 Casimir operators

When a Lie algebra, **g**, is an SGA, an important role is played by subsets of elements and combinations of elements of **g** that commute with every element of **g**. Such elements and combinations of elements are described as *invariants* of the Lie algebra. A *first-order invariant* of **g** is, by definition, a non-zero element of **g**. For example, every u(d) has a u(1) subalgebra of first-order invariants given by multiples of the d-dimensional identity matrix. However, not every Lie algebra has a first-order invariant. In fact, a particularly important class of Lie algebras, namely the set of so-called *semi-simple Lie algebras*, have no first-order invariants. Examples of semi-simple Lie algebras encountered in this text include su(d), su(1,1), so(d), usp($2d$), and sp($d, \mathbb{R}$). Counterexamples include u(d) and $[\mathbb{R}^5]$so(5); these have Abelian subalgebras of invariant operators, u(1) and $\mathbb{R}^5$, respectively.

A *Casimir invariant* was originally defined[11] as an operator of second order in the elements of a semisimple Lie algebra that commutes with every element of that algebra. However, it has become common practice to refer to any combination of elements of any Lie algebra (or its complex extension) that commutes with all elements of the algebra as a Casimir invariant.[12] Important properties of these

---

[11] Casimir H. (1931), *Proc. Roy. Acad. Amsterdam* **34**, 844.

[12] Note that, as defined in this way, a Casimir invariant is only specified to within an arbitrary normalisation factor.

invariants are: (i) that they commute with one another and with all Casimir invariants of the subalgebras; and (ii) that all states of an irrep are eigenstates of a Casimir invariant with a common eigenvalue.

The Casimir invariant of $\mathrm{su}(2) \simeq \mathrm{so}(3)$ is

$$\hat{\mathcal{K}}_{\mathrm{su}(2)} := \hat{\mathbf{J}} \cdot \hat{\mathbf{J}} = \hat{J}_x^2 + \hat{J}_y^2 + \hat{J}_z^2 = \hat{J}_0(\hat{J}_0 + 1) + \hat{J}_-\hat{J}_+. \tag{3.65}$$

With the labelling of irreps of $\mathrm{su}(2)$ by j, as in Section 3.7.2, the operator $\hat{\mathcal{K}}_{\mathrm{su}(2)}$ has eigenvalues in this representation (see Exercise 3.10) given by

$$\mathcal{K}_{\mathrm{su}(2)} = j(j+1). \tag{3.66}$$

The Casimir invariant of $\mathrm{su}(1,1)$ is

$$\hat{\mathcal{K}}_{\mathrm{su}(1,1)} := \hat{S}_0^2 - \hat{S}_x^2 - \hat{S}_y^2 = \hat{S}_0(\hat{S}_0 - 1) - \hat{S}_+\hat{S}_-. \tag{3.67}$$

From the matrix elements given by Equation (3.62), it is determined (see Exercise 3.10) to have the eigenvalues

$$\mathcal{K}_{\mathrm{su}(1,1)} = \tfrac{1}{4}\lambda(\lambda - 2) \tag{3.68}$$

on the states of a unitary irrep with lowest weight λ.

Exercise

3.10 By application of the expression (3.65) for the $\mathrm{su}(2)$ Casimir invariant to an $\mathrm{su}(2)$ highest-weight state, derive Equation (3.66). Similarly apply the expression (3.67) for the $\mathrm{su}(1,1)$ Casimir invariant to an $\mathrm{su}(1,1)$ lowest-weight state to derive Equation (3.68).

3.9 Irreps of Lie groups

Because the elements of a Lie algebra are infinitesimal generators of a group, an irrep of a Lie algebra is defined by an irrep of the corresponding group. Such irreps are derived by differentiation as illustrated by Equations (3.11) and (3.12). Conversely, an irrep of a Lie algebra defines a corresponding irrep of a group by integration. This amounts simply to exponentiating the Lie algebra irrep. For example, the representation of the element L_z of $\mathrm{so}(3)$ given by the matrix (3.12) exponentiates to the SO(3) group elements

$$R_z(\phi) = \exp\left(-\mathrm{i}\phi L_z\right) = \begin{pmatrix} \cos\phi & -\sin\phi & 0 \\ \sin\phi & \cos\phi & 0 \\ 0 & 0 & 1 \end{pmatrix}, \tag{3.69}$$

as given by Equation (3.11). The representation of any SO(3) element is similarly obtained by exponentiating the representation of some $\mathrm{so}(3)$ element in this way. However, as we have observed for $\mathrm{so}(3)$, not all irreps of a Lie algebra integrate to true irreps of its Lie group; sometimes they integrate to multivalued (e.g., spinor) irreps which are true irreps of a so-called *covering group* for which the parameter space of the group is larger. For example, we saw in Section 3.7.2 that, even though so(3) is isomorphic to su(2), the half-odd integer irreps of so(3) integrate to spinor irreps of SO(3). The latter are two-valued irreps of SO(3) and correspond to true irreps of SU(2). This is a manifestation of the fact that SU(2) is the covering group of SO(3); in fact, it is a two-fold cover because every element of SO(3) corresponds to two elements of SU(2).

Different techniques for constructing the irreps of Lie groups and other groups will be addressed as they are needed throughout the text.

3.10 Compact and non-compact groups

Essential differences between the representation theory of su(1,1) and that of su(2) result from the fact that, whereas su(2) is the Lie algebra of a compact Lie group, su(1,1) is the Lie algebra of a non-compact Lie group.

The group SU(2) has a realisation as transformations of the two-component boson operators for a two-dimensional harmonic oscillator,

$$\begin{pmatrix} c_1^\dagger \\ c_2^\dagger \end{pmatrix} \rightarrow \begin{pmatrix} b_1^\dagger \\ b_2^\dagger \end{pmatrix} = \begin{pmatrix} \alpha & \beta \\ -\beta^* & \alpha^* \end{pmatrix} \begin{pmatrix} c_1^\dagger \\ c_2^\dagger \end{pmatrix}, \tag{3.70}$$

which conserve the commutation relations

$$[b_i, b_j^\dagger] = [c_i, c_j^\dagger] = \delta_{i,j}, \quad [b_i, b_j] = [c_i, c_j] = [b_i^\dagger, b_j^\dagger] = [c_i^\dagger, c_j^\dagger] = 0. \tag{3.71}$$

To satisfy these commutation relations, it is sufficient to require that $|\alpha|^2 + |\beta|^2 = 1$. Thus, SU(2) is defined as the set of complex 2×2 matrices

$$\mathrm{SU}(2) := \left\{ \begin{pmatrix} \alpha & \beta \\ -\beta^* & \alpha^* \end{pmatrix} ; \; |\alpha|^2 + |\beta|^2 = 1 \right\}, \tag{3.72}$$

with group product given by matrix multiplication. It follows that an element of SU(2) is labelled by two complex parameters, $\alpha := x + iy$ and $\beta := z + iw$, or, equivalently, by four real parameters, x, y, z, w, subject to the constraint that

$$x^2 + y^2 + z^2 + w^2 = 1. \tag{3.73}$$

Hence, the *group manifold* (loosely referred to as the parameter space) for SU(2) is the surface of a three-sphere.

Similarly, SU(1,1) has a realisation as transformations of SHO boson operators,

$$\begin{pmatrix} c^\dagger \\ c \end{pmatrix} \rightarrow \begin{pmatrix} b^\dagger \\ b \end{pmatrix} = \begin{pmatrix} \alpha & \beta \\ \beta^* & \alpha^* \end{pmatrix} \begin{pmatrix} c^\dagger \\ c \end{pmatrix}, \tag{3.74}$$

which conserve the commutation relations

$$[b, b^\dagger] = [c, c^\dagger] = 1, \quad [b, b] = [c, c] = [b^\dagger, b^\dagger] = [c^\dagger, c^\dagger] = 0. \tag{3.75}$$

To satisfy these commutation relations, it is now required that $|\alpha|^2 - |\beta|^2 = 1$. Thus, SU(1,1) is defined as the group of complex 2×2 matrices

$$\text{SU}(1,1) := \left\{ \begin{pmatrix} \alpha & \beta \\ \beta^* & \alpha^* \end{pmatrix} ; \ |\alpha|^2 - |\beta|^2 = 1 \right\}. \tag{3.76}$$

Thus, the group manifold for SU(1,1) is the subspace of a four-dimensional Euclidean space defined by the equation

$$x^2 + y^2 - z^2 - w^2 = 1. \tag{3.77}$$

Loosely speaking, a Lie group is said to be compact if the set of all group elements has a finite volume. Thus, SU(2) is compact and SU(1,1) is non-compact. More precisely, the volume of a group manifold requires consideration of a so-called *invariant measure* of the volume of a subset of group elements (see Footnote 15 on Page 602). The concept of an invariant measure is easily visualised. For example, if all points inside an element of area on the surface of a two-sphere (given in terms of spherical polar coordinates by $r^2 dr \sin\theta \, d\theta \, d\phi$) are rotated they comprise a new element on the surface of the sphere with exactly the same area. Thus, the standard Euclidean measure of area on the sphere is said to be invariant under rotations.

An important consequence of the definition is that the unitary irreps of a compact Lie group are all finite dimensional. However, the converse is not necessarily true. A unitary irrep of a non-compact group may be finite but is most commonly infinite-dimensional. And, if infinite-dimensional, it may have upper or lower bounds to its weights, but need not have either. For example, the above two-dimensional defining irreps of SU(2) and SU(1,1) given, respectively, by Equations (3.72) and (3.76), are seen to be unitary for SU(2) and non-unitary for SU(1,1) (see Exercise 3.11); all SU(2) irreps are unitary and finite-dimensional whereas the unitary SU(1,1) irreps considered have lowest weights but no highest weights and are of infinite dimension.

Exercise

3.11 A matrix is said to be unitary if it has an inverse given by the complex conjugate of its transpose. Show that the SU(2) matrices of Equation (3.72) are unitary and that the SU(1,1) matrices of Equation (3.76) are, generally, not unitary.

3.11 Unitary representations as quantizations

In the Hamiltonian formulation of classical mechanics, the physical observables of a model are regarded as functions on a phase space that obey Poisson bracket relationships.[13] These Poisson bracket relationships define a classical realisation of the model's algebra of observables. The quantization of a classical model is then seen as the construction of a unitary representation of its algebra of observables in which the classical observables are replaced by Hermitian operators on a Hilbert space. In general, a given Lie algebra has as many quantizations as it has unitary representations. The many discrete quantizations of the angular momentum algebra mean, for example, that the square magnitude of the angular momentum of a model cannot have continuously variable values, as it can in classical mechanics; it can only take a value, $j(j+1)$, for which there exists a unitary representation, i.e., a value for which $2j$ is a non-negative integer.

From this perspective, it is of interest to note that, in a famous theorem of Stone and von Neumann,[14] it was shown that every Heisenberg-Weyl algebra has a single irreducible unitary representation to within unitary equivalence. In consideration of the fact that virtually every other Lie algebra encountered in physics has numerous unitary irreps, and therefore many quantizations, it would appear that this uniqueness of the quantization of the Heisenberg-Weyl algebras was most fortuitous for the development of quantum mechanics.

It is also notable that the irrep of the hw(1) Lie algebra, given in Section 3.7.1, is unique in spite of the fact that its lowest-weight state is the ground state of a Hamiltonian with adjustable parameters (M and ω). The important observation is that the Hilbert space is the same and an orthonormal basis $\{|n\rangle\}$ can be defined for it with common matrix elements, given by

$$\langle m|c^\dagger|n\rangle = \langle n|c|m\rangle = \delta_{m,n+1}\sqrt{n+1}, \tag{3.78}$$

for any finite (non-zero) values of M and ω.

---

[13]The relationship between classical and quantal representations of a Lie algebra is discussed in Chapter 7.

[14]Discussed, for example, by Mackey G.W. (1976), *The Theory of Unitary Group Representations* (University of Chicago Press, Chicago).

Chapter 4

The algebraic collective model

The Bohr model, in common with many models in quantum mechanics, has a well-defined algebraic structure. This means that it is expressed in terms of observables, e.g., coordinates and momenta, that have well-defined commutation relations and belong to a Lie algebra of observables. In fact, as this chapter shows, the Bohr model has a rich algebraic structure and becomes more powerful and conceptually more useful when expressed in algebraic terms. Such an expression does not change the physical content of the model. However, it opens up new avenues for application of the model in the interpretation of nuclear phenomena. For convenience, we shall refer to this formulation as the *algebraic collective model* (or ACM).

4.1 The harmonic oscillator basis for the Bohr model

The Bohr collective model has traditionally been expressed[1,2] in a harmonic spherical vibrator basis of states labelled by the quantum numbers of the subgroups in the chain

$$
\begin{array}{ccccc}
\mathrm{U}(5) \supset \mathrm{SO}(5) \supset \mathrm{SO}(3) \supset \mathrm{SO}(2) \\
N & v & \alpha & L & M
\end{array}, \tag{4.1}
$$

where the U(5) irrep label N denotes the number of harmonic oscillator quanta (phonons) in a state, the SO(5) label v is the 5-dimensional analogue of angular momentum, L and M are standard SO(3) $\supset$ SO(2) angular momentum quantum numbers, and α is a multiplicity label to distinguish multiple states of a common L for a given value of v.

The Lie algebras of the groups in this chain are simply expressible in terms of the position and momentum observables $\{\hat{q}_m, \hat{\pi}_m; m = 0, \pm 1, \pm 2\}$ of an underlying Heisenberg-Weyl algebra.

The basis defined by the subgroup chain, (4.1), is useful for the description of spherical and near spherical nuclei. However, as discussed in the following section,

---

[1] Chacón E., Moshinsky M. and Sharp R.T. (1976), *J. Math. Phys.* **17**, 668.
[2] Chacón E. and Moshinsky M. (1977), *J. Math. Phys.* **18**, 870.

it is just one of a continuous set of bases, defined by subgroups of SU(1,1) × SO(5), that are appropriate for the description of nuclei ranging from spherical to well deformed.

4.1.1   *The Heisenberg-Weyl algebra, hw(5)*

The Bohr model is initially expressed in terms of position and momentum observables $\{\hat{q}_m, \hat{\pi}^m; m = 0, \pm1, \pm2\}$ that satisfy the commutation relations

$$[\hat{q}^m, \hat{q}_n] = [\hat{\pi}^m, \hat{\pi}_n] = 0, \quad [\hat{q}^m, \hat{\pi}_n] = [\hat{q}_n, \hat{\pi}^m] = i\hbar\delta^m_n \hat{I}, \qquad (4.2)$$

of a Heisenberg-Weyl Lie algebra, hw(5), where $\hat{q}^m = \hat{q}^\dagger_m = (-1)^m\hat{q}_{-m}$, $\hat{\pi}^m = \hat{\pi}^\dagger_m = (-1)^m\hat{\pi}_{-m}$. As discussed in Chapter 2, this Lie algebra is realised by interpreting $\hat{q}_m$, $\hat{\pi}^m$, and $\hat{I}$ as operators on collective model wave functions such that $\hat{q}_m\Psi(q) = q_m\Psi(q)$, $\hat{\pi}_m\Psi(q) = -i\hbar\frac{\partial\Psi}{\partial q^m}(q)$, and $\hat{I}\Psi(q) = \Psi(q)$. All other observables of the model and their algebraic structures are expressed in terms of these basic observables. Thus, the Hilbert space of the Bohr model is the Hilbert space for an irrep of hw(5) and hw(5) is an SGA for the model.

(i) *Harmonic oscillator phonon (boson) operators*

For many purposes, it is convenient to regard the elements of the hw(5) Lie algebra as Hermitian linear combinations of the operators $\{d^m, d^\dagger_m, \hat{I}\}$, where d^m and $d^\dagger_m$ are the d-phonon operators of the harmonic spherical vibrator limit of the model,[3]

$$d^\dagger_m := \frac{1}{\sqrt{2}}\left(a\hat{q}_m - \frac{i}{a\hbar}\hat{\pi}_m\right), \quad d^m := \frac{1}{\sqrt{2}}\left(a\hat{q}^m + \frac{i}{a\hbar}\hat{\pi}^m\right), \qquad (4.3)$$

and where a is a dimensional unit such that $d^\dagger_m$ and d^m are dimensionless (see Section 2.3.1). These operators obey the boson commutation relations

$$[d^\dagger_m, d^\dagger_n] = [d^m, d^n] = 0, \quad [d^m, d^\dagger_n] = \delta^m_n \hat{I}, \qquad (4.4)$$

and are the creation and annihilation operators of harmonic spherical-vibrator quanta as described in Section 2.3.1.

In general, an algebraic model has many realisations in as much as its SGA has many representations. However, as remarked in Section 3.11, a Heisenberg-Weyl algebra, such as hw(5), has only a single irrep to within unitary equivalence. This

---

[3]The Heisenberg-Weyl algebra, hw(5), is a real Lie algebra which comprises only real linear combinations of the Hermitian operators $\{d^\dagger_m + d^m, i(d^\dagger_m - d^m), \hat{I}\}$. Thus, the operators $d^\dagger_m$ and d^m are not contained in the real hw(5) Lie algebra; they are elements of the complex extension of hw(5). However, to avoid being overly pedantic, we may not always be careful to distinguish between a real Lie algebra and its complex extension when the distinction is self-evident. We may also take liberties in referring to elements of the complex extension of a Lie algebra as *infinitesimal generators* of the corresponding group when, strictly speaking, only the elements of the real Lie algebra should be so described.

can be seen from the fact that the lowest-weight state of an irrep of a Heisenberg-Weyl algebra is always a phonon vacuum state. Thus, from an algebraic perspective, *there is only one representation of the Bohr model*, although it may be expressed in many ways. The Bohr model nevertheless has a rich algebraic structure. For example, its Hilbert space carries an infinite number of U(5) irreps labelled by the phonon number N, with N taking all non-negative integer values.

(ii) *Covariant and contravariant tensors*

The use of upper and lower indices in Equations (4.2) and (4.3) is to distinguish between the components of covariant and contravariant tensors. The raising operators $\{d_m^\dagger\}$ are defined as (covariant) components of a spherical tensor of angular momentum $L = 2$, i.e., they transform under SO(3) rotations in the standard way according to the equation

$$\hat{R}(\Omega)d_m^\dagger \hat{R}(\Omega^{-1}) = \sum_n d_n^\dagger \mathscr{D}_{nm}^2(\Omega), \quad \Omega \in \mathrm{SO}(3), \tag{4.5}$$

where $\mathscr{D}^2$ is an $L = 2$ rotation matrix. However, the Hermitian adjoints of these operators, $\{(d_m^\dagger)^\dagger\}$, transform differently; they transform as contravariant components of a spherical tensor (see Exercise 4.1). Thus, according to the Einstein convention, we label them with upper indices such that

$$d^m := (d_m^\dagger)^\dagger, \quad (d^m)^\dagger = d_m^\dagger. \tag{4.6}$$

Indices can be lowered by observing that the SO(3) scalar,

$$[d^\dagger \otimes d]_0 = \sum_{mn}(2n\,2m|00)d_m^\dagger d_n = \sum_m \frac{(-1)^m}{\sqrt{5}}d_m^\dagger d_{-m}, \tag{4.7}$$

formed by coupling $d^\dagger$ and d to angular momentum zero, should be proportional to the number operator $\hat{N} = \sum_m d_m^\dagger d^m$. Thus, it is appropriate to define

$$d_{-m} := (-1)^m d^m \tag{4.8}$$

so that

$$\hat{N} = \sqrt{5}\,[d^\dagger \otimes d]_0. \tag{4.9}$$

The relationships between covariant and contravariant tensors are described in more generality in Appendix A.6.

4.1.2 *The unitary group U(5) and its u(5) Lie algebra*

The group U(5) is fundamentally a group of invertible 5×5 unitary matrices[4] with group product given by matrix multiplication. A unitary matrix, g, can be

---

[4]To be invertible the matrices must have non-vanishing determinant.

expressed as an exponential $g = e^{iX}$, where X is a Hermitian matrix. Thus, the Lie algebra u(5) is the set of Hermitian 5×5 matrices.

Let C_{mn} denote the 5×5 matrix with a single non-zero entry of unity in row m and column n, i.e., the matrix with elements

$$\left(C_{mn}\right)_{ij} = \delta_{mi}\delta_{nj}. \tag{4.10}$$

The matrices $\{C_{mn}\}$ then satisfy the commutation relations

$$[C_{mn}, C_{pq}] = \delta_{np}C_{mq} - \delta_{mq}C_{pn}, \tag{4.11}$$

and are a basis for the complex extension of u(5). The u(5) Lie algebra is then spanned by real linear combinations of the Hermitian matrices $\{C_{mn} + C_{nm}\}$ and $\{i(C_{mn} - C_{nm})\}$.

The group U(5) has a realisation in the Bohr model as unitary transformations of the five-dimensional complex vector space spanned by the d-phonon raising operators of Equation (4.3). In this realisation, its Lie algebra, u(5) is spanned by Hermitian linear combinations of the operators

$$\hat{C}_{mn} := d_m^\dagger d^n\,, \quad m, n = 0, \pm 1, \pm 2. \tag{4.12}$$

These operators satisfy the commutation relations

$$[\hat{C}_{mn}, d_n^\dagger] = d_m^\dagger, \tag{4.13}$$

and hence

$$[\hat{C}_{mn}, \hat{C}_{pq}] = \delta_{np}\hat{C}_{mq} - \delta_{mq}\hat{C}_{pn}. \tag{4.14}$$

The d-phonon number operator

$$\hat{N} = d^\dagger \cdot d := \sum_m d_m^\dagger d^m = \sum_m (-1)^m d_m^\dagger d_{-m}, \tag{4.15}$$

is then seen to be the element of u(5) that commutes with all u(5) elements. Thus, U(5) is a symmetry group of the harmonic spherical-vibrator Hamiltonian

$$\hat{H}_{\mathrm{HV}} = (\hat{N} + {}^5\!/_2\hat{I})\hbar\omega. \tag{4.16}$$

A Cartan subalgebra for u(5) is spanned by the commuting operators $\{\hat{N}_m := \hat{C}_{mm} = d_m^\dagger d^m\}$; raising operators are given by $\{\hat{C}_{mn}; m > n\}$, and lowering operators by $\{\hat{C}_{mn}; m < n\}$.[5] Thus, the state

$$|N, \mathrm{hwt}\rangle = \frac{1}{\sqrt{N!}}(d_2^\dagger)^N|0\rangle, \tag{4.17}$$

---

[5]Sometimes it is convenient to choose the reverse ordering and describe $\{\hat{C}_{mn}; m < n\}$ as raising operators and $\{\hat{C}_{mn}; m > n\}$ as lowering operators. The choice is, in principle, arbitrary.

is a highest-weight state for a U(5) irrep in the Bohr model which contains all the states of phonon number, N. This is shown by observing that the N-phonon space is spanned by states of the form $\prod (d_m^\dagger)^{n_m}|0\rangle$ with $\sum_m n_m = N$, where $|0\rangle$ is the phonon vacuum state, and by the observation that all such states can be reached by a sequence of U(5) lowering operators from the highest-weight state $|N, \text{hwt}\rangle$. Such an irrep is labelled by the symbol $\{N\}$. It follows that the Hilbert space of the Bohr model is a direct sum of subspaces which carry the U(5) irreps $\{N\}$ with $N = 0, 1, 2, \ldots$.

4.1.3 *The orthogonal group O(5) and its so(5) Lie algebra*

For present purposes, the group O(5) is defined as the subgroup of U(5) transformations of the $\{d^\dagger\}$ phonon operators that leave the scalar product

$$d^\dagger \cdot d^\dagger := \sum_m (-1)^m d_m^\dagger d_{-m}^\dagger \qquad (4.18)$$

invariant. It consists of combinations of SO(5) "rotations" and discrete O(1) parity inversions in the five-dimensional configuration space, $\mathbb{R}^5$, of the Bohr model. However, because the discrete subgroup O(1) has no infinitesimal generator, the Lie algebra of O(5) is identical to the Lie algebra, so(5), of its SO(5) subgroup and is spanned by the generalised angular-momentum operators

$$\hat{\mathcal{L}}_{mn} := \mathrm{i}\big(\hat{C}_{mn} - \hat{C}_{nm}\big), \qquad (4.19)$$

where the $\hat{C}$ operators are defined by Equation (4.12).

(i) *Unitary irreps*

The SO(5) irreps that occur in the Bohr model are uniquely characterised by their generalised (five-dimensional) angular momenta, known as *seniority* (for historic reasons) and denoted by v; the corresponding irrep is labelled by $[v]$. In fact, because the states of an SO(5) irrep $[v]$ in the Bohr model also belong to an O(5) irrep $[v]$ with an $\mathbb{R}^5$ (also called O(5)) parity $(-1)^v$, it is sufficient to focus on the irreps of the SO(5) subgroup of O(5). These irreps are conveniently constructed by a building-up process starting from the d-phonon vacuum state $|0\rangle$ which spans the trivial one-dimensional identity irrep $[0]$.

From the vacuum state, one can construct an infinite sequence of $[v = 0]$ irreps by adding pairs of SO(5) scalar-coupled phonons to form the states, given to within normalisation factors, by

$$|N = 2\nu, v = 0\rangle \propto (d^\dagger \cdot d^\dagger)^\nu |0\rangle. \qquad (4.20)$$

The 1-phonon states $\{d_m^\dagger |0\rangle\}$ span the fundamental 5-dimensional $[v = 1]$ irrep of SO(5), for which the highest-weight state is the state $d_2^\dagger |0\rangle$. Highest-weight states for an infinite sequence of $[v = 1]$ irreps are then given by the states

$$|N = 1 + 2\nu, 1, \text{hwt}\rangle \;\propto\; (d^\dagger \cdot d^\dagger)^\nu d_2^\dagger |0\rangle, \quad \nu = 0, 1, 2, \ldots. \tag{4.21}$$

Now, each of the N-phonon U(5) highest-weight states, given by Equation (4.17), is created by a product of N phonon creation operators of SO(5) highest weight; it is therefore also a highest-weight state for an SO(5) irrep $[v = N]$. One can then construct an infinite sequence of $[v]$ irreps by adding pairs of SO(5) scalar-coupled phonons to such U(5) $\supset$ SO(5) highest-weight states to give SO(5) irreps, all of seniority v, with highest-weight states

$$|N = v + 2\nu, v, \text{hwt}\rangle \;\propto\; (d^\dagger \cdot d^\dagger)^\nu |v, \text{hwt}\rangle, \quad \nu = 0, 1, 2, \ldots. \tag{4.22}$$

where $|v, \text{hwt}\rangle$ is given by Equation (4.17) with $N = v$.

This construction gives all the SO(5) irreps that occur in the space of the Bohr model, as can be seen from the pattern of states of the five-dimensional harmonic oscillator shown in Figure 4.1; this figure may be compared with that of Figure 2.2. Each SO(5) irrep, $[v]$, is labelled in Figure 4.1 by (ν, v), where ν and v are

Figure 4.1: Energy levels of the harmonic spherical vibrator model, with excitation energies $N\hbar\omega$, $N = 0, 1, 2, \ldots$, grouped into U(5) irreps (horizontal shaded areas) and SO(5) irreps (horizontal lines). Each SO(5) irrep is labelled by (ν, v) as defined in the text. Equivalent SO(5) irreps, having the same value of v, are put into vertical columns. As shown in Section 4.2.2, the states of all SO(5) irreps in a single column span an SU(1,1)$\times$SO(5) irrep. The integers above each level indicate the SO(3) angular momenta $\{L\}$ of states in the corresponding SO(5) irrep. The figure also shows SO(5)-reduced E2 transition rates between states in units of the reduced transition rate for decay of the one-phonon state as derived by the methods discussed in Section 4.6.1.

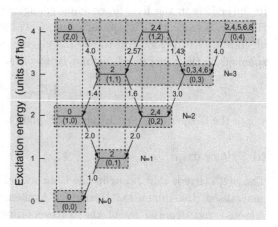

defined by Equation (4.22). A parallel, and more familiar, figure for the three-dimensional harmonic oscillator states is shown in Figure 5.13, where states are labelled by U(3) $\supset$ SO(3) quantum numbers, (ν, l), as discussed in Section 5.8.2, and for which the SO(3) irreps are put into columns of equivalent irreps, of common angular momentum, l.

(ii) *U(5) $\supset$ SO(5) branching rules*

Equation (4.22) and Figure 4.1 show that the set of states of $N\hbar\omega$ excitation energy, with $N = v + 2\nu$, break up into subsets of states: (i) states having no zero-coupled d-phonon pairs and, hence, seniority $v = N$; (ii) states having one zero-coupled d-phonon pair and seniority $v = N - 2$; (iii) states having two zero-coupled pairs and seniority $v = N - 4$; and so on. In other words, for a given value of N, the seniority quantum number v takes the values

$$v = N, \quad N - 2, \quad N - 4, \quad \ldots, \quad 1 \text{ or } 0. \tag{4.23}$$

In algebraic terminology, the Hilbert space of a U(5) irrep $\{N\}$ carries a reducible representation of SO(5) that can be decomposed into a direct sum of SO(5) irreps, $[v]$, according to the branching rule

$$\text{U}(5) \downarrow \text{SO}(5) \;;\; \{N\} \downarrow [N] \oplus [N - 2] \oplus \cdots \oplus [1] \text{ or } [0]. \tag{4.24}$$

Such U(d) $\downarrow$ O(d) branching rules for arbitrary d are considered in more depth in Section 5.12.7.

(iii) *Casimir invariant of SO(5)*

The SO(5) angular momentum operators take the form

$$\hat{\mathcal{L}}_{ij} = -\mathrm{i}(d_i^\dagger d_j - d_j^\dagger d_i), \tag{4.25}$$

when expressed in a Cartesian basis[6] of raising and lowering operators, defined in terms of an adjustable unit a (so that $a\hat{q}_j$ is dimensionless)[7] by

$$\hat{q}_j = \frac{1}{\sqrt{2}a}(d_j^\dagger + d_j), \quad \hat{\pi}_j = \frac{\mathrm{i}a\hbar}{\sqrt{2}}(d_j^\dagger - d_j), \tag{4.26}$$

Inserting these operators into the expression $\hat{\Lambda}^2 = \sum_{i<j} \hat{\mathcal{L}}_{ij}^2$ for the SO(5) Casimir operator and reordering the d-phonon operators, gives the identity

$$\hat{\Lambda}^2 := \hat{N}(\hat{N} + 3) - (d^\dagger \cdot d^\dagger)(d \cdot d). \tag{4.27}$$

Application of $\hat{\Lambda}^2$, in this form, to the states of a lowest-energy $(\nu = 0, v)$ multiplet of SO(5) states of seniority v (cf. Figure 4.1) shows that they have the common eigenvalue

$$\Lambda^2 = v(v + 3). \tag{4.28}$$

This expression of the eigenvalues of $\hat{\Lambda}^2 = \sum_{i<j} \hat{\mathcal{L}}_{ij}^2$ in terms of v, confirms that v is an SO(5) angular momentum quantum number. Moreover, because all states of a column (of states) in Figure 4.1 belong to equivalent SO(5) irreps, they all share a common eigenvalue of $\hat{\Lambda}^2$.

---

[6]A vector in the Euclidean space $\mathbb{R}^5$ is said to be expressed in a Cartesian basis if its components are Cartesian coordinates, e.g., $\{x_i, i = 1, \ldots, 5\}$. Alternatively, it would be expressed in a spherical tensor basis, if its components were spherical polar coordinates, e.g., $\{q_m, m = 0, \pm 1, \pm 2\}$ as defined in Section 2.1.2. Tensors with Cartesian components are special because for them there is no distinction between upper and lower indices. Thus, for example, if $\hat{q}$ is a Euclidean vector with Cartesian components $\{\hat{x}_i\}$, then $\hat{x}^i = \hat{x}_i$ and $\hat{q} \cdot \hat{q} = \sum_{m=-2}^{+2} q_m q^m = \sum_{i=1}^{5} \hat{x}_i^2$.

[7]In the harmonic spherical vibrator limit, a natural unit is given by $a = \sqrt{B\omega/\hbar}$; cf. Equation (2.79).

4.1.4 *The rotation group SO(3) and its so(3) Lie algebra*

Whereas the five components $\{\hat{q}_m\}$ transform as a basis for a $v = 1$ irrep of SO(5), they transform under SO(3) as a basis for an $L = 2$ irrep. i.e.

$$\hat{R}(\Omega)\hat{q}_m\hat{R}(\Omega^{-1}) = \sum_k \hat{q}_k \mathscr{D}^2_{km}(\Omega), \quad \Omega \in \text{SO}(3). \tag{4.29}$$

The angular-momentum operators $\{\hat{L}_i\}$, which are the infinitesimal generators of SO(3) and span the so(3) Lie algebra, can be expressed as a subset of generalised so(5) angular-momentum operators. In fact, they are proportional to components of the $L = 1$ coupled tensor product $[d^\dagger \otimes d]_1$. What is important is that the angular-momentum operators have commutation relations with the components, $\{\hat{X}_{LM}\}$ of any SO(3) tensor operator $\hat{X}_L$ according to the standard commutation relations

$$[\hat{L}_0, \hat{X}_{LM}] = M\hat{X}_{LM}, \quad [\hat{L}_\pm, \hat{X}_{LM}] = \sqrt{(L \mp M)(L \pm M + 1)}\,\hat{X}_{J,M\pm1}. \tag{4.30}$$

The range of SO(3) angular momentum states in a given SO(5) irrep can be inferred by a peeling-off procedure as described in Section 2.3.2. A general expression has been derived[8] by consideration of SO(5) $\downarrow$ SO(4) $\downarrow$ SO(3) branching rules. The result is that, on restriction of SO(5) to its SO(3) subgroup, an SO(5) irrep $[v]$ becomes a direct sum of SO(3) irreps of angular momenta L given by

$$\begin{aligned} L &= 2K, \quad 2K-2, \quad 2K-3, \quad \ldots, \quad K, \\ K &= v, \quad v-3, \quad v-6, \quad \ldots, \quad K_{\min}, \end{aligned} \tag{4.31}$$

where $K_{\min} = 0$, 1, or 2; e.g., for the $v = 6$ irrep of SO(5), K takes the values 6, 3, and 0, and the irrep contains the SO(3) irreps of angular momentum $L = 12, 10, 9, 8, 7, 6^2, 4, 3, 0$. Note that the angular momentum 6 occurs with multiplicity 2. Equation (4.31) is confirmed by the use of plethysms (cf. Section 5.11.10) and by the generating function methods for SO(5) spherical harmonics outlined in Section 4.9.

Exercises

4.1 Show that a lowering operator, d^m, of the Heisenberg-Weyl algebra, hw(5), transforms under an SO(3) rotation, Ω, according to the equation

$$\hat{R}(\Omega)d^m\hat{R}(\Omega^{-1}) = \sum_n d^n \mathscr{D}^{2*}_{nm}(\Omega). \tag{4.32}$$

4.2 Show that the quadrupole moments of a nucleus can be represented (in a Cartesian basis) as elements of a real, traceless, symmetric 3×3 matrices and that they transform under rotations according to an $L = 2$ irrep of SO(3). Hence show that the quadrupole moments can be combined with the infinitesimal generators of SO(3) to form a semi-direct sum Lie algebra as defined in Box 4.I on Page 181. (This Lie algebra is an SGA for the rigid-rotor model.)

---

[8]Williams S.A. and Pursey D.L. (1968), *J. Math. Phys.* **9**, 1230.

4.3 Derive the values of L for the SO(3) irreps contained in the SO(5) irreps with $v = 1, \ldots 5$ and compare your results with those shown in Figure 4.15.

4.2 Algebraic and geometric structure of the ACM

The algebraic expression of the Bohr model, which we refer to as the ACM,[9,10,11] makes explicit use of the factorisation of collective model wave functions into products of beta (radial) wave functions and SO(5) angular wave functions, where the beta wave functions carry irreps of an SU(1,1) Lie group and the angular wave functions carry irreps of SO(5). Thus, basis states for the ACM are classified by the quantum numbers of the subgroup chain

$$
\begin{array}{cccccc}
\text{SU}(1,1) \times \text{SO}(5) \supset \text{U}(1) \times \text{SO}(3) \supset \text{SO}(2) \\
\lambda \quad\quad v \quad\quad \alpha \quad\quad \nu \quad\quad L \quad\quad\quad M
\end{array}, \tag{4.33}
$$

where U(1) is a subgroup of SU(1,1) that labels a basis for an SU(1,1) irrep.

It is shown in this section that, for particular values of the SU(1,1) quantum number λ, related to the seniority v of the SO(5) wave function by $\lambda = v + 5/2$, the ACM basis wave functions are identical to those of the harmonic spherical vibrator basis as defined above. This basis is appropriate for the description of spherical nuclei. However, as we show, the group SU(1,1) also has a continuous (modified oscillator) series of irreps that are more appropriate for the description of deformed nuclei.[12] Succeeding sections show that the matrix elements needed for collective model calculations can be calculated algebraically in any of these $\text{SU}(1,1) \times \text{SO}(5)$ bases.

4.2.1 *Factorisation of the Bohr model Hilbert space*

It was observed in Chapter 2, that the configuration space $\mathbb{R}^5$ of the Bohr model can be considered as a tensor product of a radial line, $\mathbb{R}_+$, and a four sphere, S_4. It was also shown that the volume element, dv, for $\mathbb{R}^5$ is a product of volume elements for $\mathbb{R}_+$ and S_4 given in Equation (2.33) by

$$
dv = (\beta^4 d\beta) \times (\sin 3\gamma \, d\gamma \, d\Omega). \tag{4.34}
$$

---

[9] Rowe D.J. (2004), *Nucl. Phys.* **A735**, 372.

[10] Rowe D.J., Welsh T.A. and Caprio M.A. (2009), *Phys. Rev.* **C79**, 054304.

[11] An algebraic expression of the Bohr model has also been pursued by De Baerdemacker S., Heyde K. and Hellemans V. (2007), *J. Phys. A: Math. Gen.* **40**, 2733.

[12] Although, for linguistic convenience, we speak of irreps of SU(1,1), we are really considering irreps of the su(1,1) Lie algebra which define multivalued irreps of the SU(1,1) group in the same way that the irreps of the so(3) Lie algebra (isomorphic to su(2)) define double-valued (spinor) irreps of SO(3).

It follows that the Hilbert space for the Bohr model, $\mathbb{H} := \mathcal{L}^2(\mathbb{R}^5)$, is the tensor product space

$$\mathbb{H} = \mathcal{L}^2(\mathbb{R}_+) \otimes \mathcal{L}^2(S_4), \qquad (4.35)$$

where $\mathcal{L}^2(\mathbb{R}_+)$ is a Hilbert space of square-integrable beta functions with volume element $\beta^4 d\beta$ and $\mathcal{L}^2(S_4)$ is a Hilbert space of square-integrable functions on S_4 with volume element $\sin 3\gamma \, d\gamma \, d\Omega$. Thus, it is appropriate to define basis wave functions for $\mathbb{H}$ of the product form

$$\Psi_{\lambda\nu;v\alpha LM}(\beta, \gamma, \Omega) = \frac{1}{\beta^2} \mathcal{R}_\nu^\lambda(\beta) \, \mathcal{Y}_{v\alpha LM}(\gamma, \Omega), \qquad (4.36)$$

where the set of beta wave functions $\{\mathcal{R}_\nu^\lambda\}$, with a given value of λ, is an orthonormal basis for an SU(1,1) irrep with inner product

$$\int_0^\infty \mathcal{R}_\mu^\lambda(\beta) \mathcal{R}_\nu^\lambda(\beta) \, d\beta = \delta_{\mu\nu} \qquad (4.37)$$

and the functions $\{\mathcal{Y}_{v\alpha LM}\}$ are SO(5) spherical harmonics. The SO(5) spherical harmonics are an orthonormal basis for the Hilbert space $\mathcal{L}^2(S_4)$ and are labelled by the quantum numbers of the groups in the chain

$$\begin{array}{ccc} \text{SO(5)} \supset \text{SO(3)} \supset \text{SO(2)} \\ v \quad \alpha \quad L \qquad M \end{array}, \qquad (4.38)$$

where α is a multiplicity index. Thus, for each value of the SO(5) angular momentum (seniority) quantum number, v, they provide an orthonormal basis for the corresponding SO(5) irrep. They satisfy the inner product relationship

$$\int_{S_4} \mathcal{Y}_{v\alpha LM}^*(\gamma, \Omega) \, \mathcal{Y}_{v'\beta L'M'}(\gamma, \Omega) \sin 3\gamma \, d\gamma \, d\Omega = \delta_{vv'}\delta_{\alpha\beta}\delta_{LL'}\delta_{MM'}. \qquad (4.39)$$

Their properties are discussed in more detail in Sections 4.5.2 and 4.9.

A significant consequence of the above factorisation is that matrix elements of most observables of interest become products of radial (beta) matrix elements and so(5) matrix elements. For example, if $|\lambda\nu; v\alpha LM\rangle \equiv |\lambda\nu\rangle \otimes |v\alpha LM\rangle$ denotes the product state with wave function given by Equation (4.36) then the matrix elements of the quadrupole moments, $\{\hat{q}_m = \hat{\beta}\hat{Q}_m\}$, factor according to the equation

$$\langle \lambda'\nu'; v'\alpha'L'M'|\hat{q}_m|\lambda\nu; v\alpha LM\rangle = \langle \lambda'\nu'|\hat{\beta}|\lambda\nu\rangle\langle v'\alpha'L'M'|\hat{Q}_m|v\alpha LM\rangle. \qquad (4.40)$$

Consistent with the above factorisation of $\mathbb{H}$, Equation (2.34) shows that the Laplacian, ∇^2, is given by

$$\nabla^2 = \hat{\Delta} - \frac{1}{\beta^2}\hat{\Lambda}^2, \qquad (4.41)$$

where

$$\hat{\Delta} = \frac{1}{\beta^4} \frac{\partial}{\partial \beta} \beta^4 \frac{\partial}{\partial \beta}, \tag{4.42}$$

is the Laplacian for $\mathcal{L}^2(\mathbb{R}_+)$, and $\hat{\Lambda}^2$, defined in Section 4.1.3(iii) (see also Equation (2.35)), is both an SO(5) Casimir operator and the Laplacian for $\mathcal{L}^2(S_4)$.

4.2.2 The su(1,1) Lie algebra and its harmonic series irrep

The non-compact su(1,1) Lie algebra (see Section 3.10), with commutation relations

$$[\hat{S}_0, \hat{S}_\pm] = \pm \hat{S}_\pm, \quad [\hat{S}_-, \hat{S}_+] = 2\hat{S}_0, \tag{4.43}$$

has a realisation in terms of the angular-momentum-zero and O(5)-invariant scalar products of the collective model phonon operators,

$$\hat{S}_+ := \tfrac{1}{2} d^\dagger \cdot d^\dagger, \quad \hat{S}_- := \tfrac{1}{2} d \cdot d,$$
$$\hat{S}_0 := \tfrac{1}{4}(d^\dagger \cdot d + d \cdot d^\dagger) = \tfrac{1}{2}\left(d^\dagger \cdot d + \tfrac{5}{2}\hat{I}\right) = \tfrac{1}{2}\left(\hat{N} + \tfrac{5}{2}\hat{I}\right). \tag{4.44}$$

(i) *Harmonic series irreps*

The su(1,1) algebra, as realised by Equation (4.44), has irreps on the collective model Hilbert space known as *harmonic series* irreps.[13] The harmonic series of irreps are a subset of the unitary irreps defined by Equation (3.57);

$$\hat{S}_0|\lambda\nu\rangle = \tfrac{1}{2}(\lambda + 2\nu)|\lambda\nu\rangle,$$
$$\hat{S}_+|\lambda\nu\rangle = \sqrt{(\lambda + \nu)(\nu + 1)}\,|\lambda, \nu + 1\rangle, \tag{4.45}$$
$$\hat{S}_-|\lambda\nu\rangle = \sqrt{(\lambda + \nu - 1)\nu}\,|\lambda, \nu - 1\rangle.$$

with λ taking values for which $\lambda - 1/2$ is a non-negative integer. These irreps have lowest-weight states but no highest-weight states and are infinite-dimensional.

Equation (4.44) shows that $\hat{S}_0$ is proportional to the harmonic spherical vibrator Hamiltonian, $\hat{H}_{\text{HV}} = (\hat{N} + 5/2\hat{I})\hbar\omega = 2\hbar\omega\hat{S}_0$. It is also seen that the su(1,1) raising operator, $\hat{S}_+$, is an operator that shifts the states of the SO(5) irrep (ν, v), as depicted in Figure 4.1, to the equivalent SO(5) irrep $(\nu + 1, v)$. Likewise, the su(1,1) lowering operator, $\hat{S}_-$, shifts the states of the SO(5) irrep (ν, v) to the equivalent SO(5) irrep $(\nu - 1, v)$. Thus, for $v = 0, 1, 2, \ldots$, all the states of an SO(5) irrep, $(0, v)$, are lowest-weight states for harmonic series su(1,1) irreps (λ) with $\lambda = v + 5/2$.

(ii) *A duality relationship between the irreps of SU(1,1) and O(5)*

The above results reveal a deep *duality relationship* between the harmonic series irreps of SU(1,1) and irreps of O(5). They show that all the states of the Bohr model that have SO(5) seniority v, i.e., all states that belong to the column of equivalent

---

[13]These harmonic series irreps of su(1,1) irreps, considered here, integrate to spinor irreps of the SU(1,1) group.

SO(5) irreps labelled by (ν, v) in Figure 4.1, belong to a single irrep of the direct product group SU(1,1) × SO(5). This irrep of SU(1,1) × SO(5) is characterised by a state $|v, \text{hwt}\rangle = (d_2^\dagger)^v |0\rangle / \sqrt{v!}$, which, as shown in Section 4.1.3, is an SO(5) highest-weight state and is now observed to be simultaneously an SU(1,1) lowest-weight state. What is more, as shown in Section 4.1.3, the state $|v, \text{hwt}\rangle$ is also a highest-weight state for a U(5) irrep. Thus, we complete the interpretation of Figure 4.1 in which the horizontal lines, corresponding to the states of SO(5) irreps, combine to gives bases for U(5) irreps, shown as horizontal shaded areas, and bases for SU(1,1) × SO(5) irreps, shown as columns.

Because of the relationship, $\lambda = v + \frac{5}{2}$, it is clear that there is also a relationship between the eigenvalues of the SU(1,1) and SO(5) Casimir operators. The relationship is exposed by inserting the expressions given by Equation (4.44) for the SU(1,1) operators into Equation (3.67) to obtain

$$4\hat{\mathcal{K}}_{\mathrm{SU}(1,1)} = \hat{N}(\hat{N} + 3) - (d^\dagger \cdot d^\dagger)(d \cdot d) + \tfrac{5}{4}\hat{I} \tag{4.46}$$

and, from Equation (4.27), the relationship

$$\hat{\Lambda}^2 = 4\hat{\mathcal{K}}_{\mathrm{SU}(1,1)} - \tfrac{5}{4}\hat{I}. \tag{4.47}$$

This one-to-one correspondence between the irreps of the two groups, SU(1,1) and SO(5), on the Hilbert space of the model is characteristic of a duality relationship. Such duality relationships occur in many places in nuclear physics and are reviewed in Sections 5.10–5.12.

(iii) *Beta wave functions for the harmonic series of irreps*

Substituting the expansions of the d-phonon operators, given by Equation (4.3), into the definitions of the su(1,1) operators given by Equation (4.44) with the momentum operators $\hat{\pi}_m = -i\hbar \partial/\partial q^m$, leads to the expressions

$$\hat{S}_0 = \frac{1}{4}\left[-\frac{\nabla^2}{a^2} + (a\beta)^2 \right], \tag{4.48}$$

$$\hat{S}_\pm = \frac{1}{4}\left[\frac{\nabla^2}{a^2} + (a\beta)^2 \mp (2q \cdot \nabla + 5) \right], \tag{4.49}$$

which, with the identity, $\nabla^2 = \hat{\Delta} - \hat{\Lambda}^2/\beta^2$, from Equation (4.41), become

$$\hat{S}_0 = \frac{1}{4}\left[-\frac{\hat{\Delta}}{a^2} + \frac{\hat{\Lambda}^2}{(a\beta)^2} + (a\beta)^2 \right], \tag{4.50}$$

$$\hat{S}_\pm = \frac{1}{4}\left[\frac{\hat{\Delta}}{a^2} - \frac{\hat{\Lambda}^2}{(a\beta)^2} + (a\beta)^2 \mp (2q \cdot \nabla + 5) \right]. \tag{4.51}$$

Here a is variously interpreted as a *scale* parameter and as a *stiffness* parameter which may be used to fix the widths of the basis wave functions.

These expressions show how the su(1,1) operators, $\hat{S}_0$ and $\hat{S}_{\pm}$, act on products of beta and SO(5) wave functions. From the identities

$$\hat{\Delta}\frac{1}{\beta^2} = \frac{1}{\beta^4}\frac{d}{d\beta}\beta^4\frac{d}{d\beta}\frac{1}{\beta^2} = \frac{1}{\beta^2}\left(\frac{d^2}{d\beta^2} - \frac{2}{\beta^2}\right) \tag{4.52}$$

and $\hat{\Lambda}^2 \mathcal{Y}_{v\alpha LM} = v(v+3)\mathcal{Y}_{v\alpha LM}$, from Equations (4.27) and (4.28), it follows that, when acting on the product wave functions of Equation (4.36), with $\lambda = v + \tfrac{5}{2}$, the su(1,1) operator $\hat{S}_0$, gives

$$\hat{S}_0 \frac{1}{\beta^2}\mathcal{R}_v^{v+5/2}\,\mathcal{Y}_{v\alpha LM} = \frac{1}{\beta^2}\,\mathcal{Y}_{v\alpha LM}\,\hat{S}_0^{(v+5/2)}\mathcal{R}_v^{v+5/2}, \tag{4.53}$$

where

$$\begin{aligned}
\hat{S}_0^{(v+5/2)} &:= \frac{1}{4}\left[-\frac{1}{a^2}\left(\frac{d^2}{d\beta^2} - \frac{2}{\beta^2}\right) + \frac{v(v+3)}{(a\beta)^2} + (a\beta)^2\right] \\
&= \frac{1}{4}\left[-\frac{1}{a^2}\frac{d^2}{d\beta^2} + \frac{(v+1)(v+2)}{(a\beta)^2} + (a\beta)^2\right].
\end{aligned} \tag{4.54}$$

Unlike the operator $\hat{S}_0$, the operator $\hat{S}_0^{(v+5/2)}$ acts only on the beta wave functions. A similar replacement is obtained for $\hat{S}_{\pm}$ with

$$\hat{S}_{\pm}^{(v+5/2)} := \frac{1}{4}\left[\frac{1}{a^2}\frac{d^2}{d\beta^2} - \frac{(v+1)(v+2)}{(a\beta)^2} + (a\beta)^2 \mp \left(2\beta\frac{d}{d\beta} + 1\right)\right]. \tag{4.55}$$

Equivalently, with the substitution $v = \lambda - \tfrac{5}{2}$, we obtain

$$\begin{aligned}
\hat{S}_0^{(\lambda)} &= \frac{1}{4}\left[-\frac{1}{a^2}\frac{d^2}{d\beta^2} + \frac{(\lambda - 3/2)(\lambda - 1/2)}{(a\beta)^2} + (a\beta)^2\right], \\
\hat{S}_{\pm}^{(\lambda)} &= \frac{1}{4}\left[\frac{1}{a^2}\frac{d^2}{d\beta^2} - \frac{(\lambda - 3/2)(\lambda - 1/2)}{(a\beta)^2} + (a\beta)^2 \mp \left(2\beta\frac{d}{d\beta} + 1\right)\right].
\end{aligned} \tag{4.56}$$

These operators continue to satisfy the su(1,1) commutation relations, given by Equation (3.50), for all values of λ.

It follows, from these expressions and Equation (3.57), that the beta wave functions of the harmonic spherical vibrator satisfy the linear differential equations

$$\begin{aligned}
\hat{S}_0^{(\lambda)}\mathcal{R}_v^{\lambda} &= \tfrac{1}{2}(\lambda + 2\nu)\mathcal{R}_v^{\lambda}, \\
\hat{S}_+^{(\lambda)}\mathcal{R}_v^{\lambda} &= \sqrt{(\lambda + \nu)(\nu + 1)}\,\mathcal{R}_{\nu+1}^{\lambda}, \\
\hat{S}_-^{(\lambda)}\mathcal{R}_{\nu+1}^{\lambda} &= \sqrt{(\lambda + \nu)(\nu + 1)}\,\mathcal{R}_v^{\lambda},
\end{aligned} \tag{4.57}$$

whenever $\lambda = v + \tfrac{5}{2}$, for any integer $v \geq 0$. The radial wave functions, for any harmonic oscillator, appearing in the equations have the known solutions given by

$$\mathcal{R}_v^{\lambda}(\beta) = (-1)^{\nu}\sqrt{\frac{2\nu!\,a}{\Gamma(\nu + \lambda)}}\,(a\beta)^{\lambda - 1/2}\,e^{-a^2\beta^2/2}\,L_{\nu}^{(\lambda - 1)}(a^2\beta^2), \tag{4.58}$$

where $L_\nu^{(\lambda-1)}$ is a generalised Laguerre polynomial.[14] In fact, as shown below, Equations (4.57), with radial wave functions defined by (4.58), are valid for any positive real λ. Moreover, for any such λ, the set of beta wave functions $\{\mathcal{R}_\nu^\lambda, \nu = 0, 1, 2, \dots\}$ is a complete orthonormal basis for the Hilbert space $\mathcal{L}^2(\mathbb{R}_+)$.

(iv) *The harmonic-oscillator phonon operators as su(1,1) tensors*

With the su(1,1) operators of Equation (4.44), the collective model phonon operators satisfy the commutation relations

$$[\hat{S}_-, d_m^\dagger] = d_m, \quad [2\hat{S}_0, d_m^\dagger] = d_m^\dagger, \quad [\hat{S}_+, d_m^\dagger] = 0,$$
$$[\hat{S}_-, d_m] = 0, \quad [2\hat{S}_0, d_m] = -d_m, \quad [\hat{S}_+, d_m] = -d_m^\dagger. \tag{4.59}$$

which show that each member of a pair of d-phonon operators, $(d_m^\dagger, d_m)$, is a component of an SU(1,1) tensor operator. If these equations are regarded as transformations in which

$$\hat{S}_0 : (d_m^\dagger, d_m) \rightarrow \tfrac{1}{2}(d_m^\dagger, -d_m) = (d_m^\dagger, d_m) \begin{pmatrix} \frac{1}{2} & 0 \\ 0 & -\frac{1}{2} \end{pmatrix}, \tag{4.60}$$

$$\hat{S}_+ : (d_m^\dagger, d_m) \rightarrow (0, -d_m^\dagger) = (d_m^\dagger, d_m) \begin{pmatrix} 0 & -1 \\ 0 & 0 \end{pmatrix}, \tag{4.61}$$

$$S_- : (d_m^\dagger, d_m) \rightarrow (d_m, 0) = (d_m^\dagger, d_m) \begin{pmatrix} 0 & 0 \\ 1 & 0 \end{pmatrix}, \tag{4.62}$$

it is seen that the operators $d_m^\dagger$ and d_m transform as basis vectors for a two-dimensional irrep of su(1,1) in which $\hat{S}_0$ and $\hat{S}_\pm$ are represented by the matrices

$$S_0 = \begin{pmatrix} 1/2 & 0 \\ 0 & -1/2 \end{pmatrix}, \quad S_+ = \begin{pmatrix} 0 & -1 \\ 0 & 0 \end{pmatrix}, \quad S_- = \begin{pmatrix} 0 & 0 \\ 1 & 0 \end{pmatrix}. \tag{4.63}$$

Thus, this tensor irrep has highest and lowest weights given, respectively, by ± 1.

4.2.3 *The modified oscillator series of SU(1,1) irreps*

In applications of the Bohr model to deformed nuclei, it will be shown in the following that the calculations are much more rapidly convergent in a basis for a suitably chosen modified oscillator su(1,1) irrep. The use of modified oscillator irreps in determining the radial wave functions for central force problems has a long history.[15,16] They are used here to define beta wave functions for SU(1,1) irreps, obtained by solution of Equations (4.57) for arbitrary values of $\lambda > 0$.

---

[14] Abramowitz M. and Stegun I.A. (1968), *Handbook of Mathematical Functions* (Dover Publications, New York), fifth printing.

[15] Wybourne B.G. (1974), *Classical Groups for Physicists* (Wiley, New York).

[16] Reviews have been given by Čížek J. and Paldus J. (1977), *Int. J. Quantum Chem.* **12**, 875, and by Cooke T.H. and Wood J.L. (2002), *Amer. J. Phys.* **70**, 945.

A natural starting point is to consider the operation of $\hat{S}_0^{(\lambda)}$ on $\mathcal{R}_\nu^\lambda$, as defined by Equation (4.58) for an arbitrary real value of λ, but without an a priori assumption that $L_\nu^{(\lambda-1)}$ is a generalised Laguerre polynomial. One finds that, for $\mathcal{R}_\nu^\lambda$ to be a non-vanishing eigenfunction of $\hat{S}_0^{(\lambda)}$, λ must be positive and the equation that results for $L_\nu^{(\lambda-1)}$ is, in fact, the defining equation of a generalised Laguerre polynomial.[17] This shows that the beta wave functions defined by Equation (4.58) satisfy Equations (4.57) and the set $\{\mathcal{R}_\nu^\lambda ; \nu = 0, 1, 2, \dots\}$ is an orthonormal basis for an su(1,1) irrep for any real $\lambda > 0$. Thus, the beta wave functions of the harmonic series of su(1,1) irreps are special cases of the more general, modified oscillator series with continuously variable λ.

As shown in the following, explicit expressions of the beta wave functions are not needed in ACM calculations because the matrix elements of relevant observables are determined by algebraic methods. Nevertheless, it is insightful to see them. A few beta wave functions are shown in Figure 4.2 for $\lambda = 5/2$ (appropriate for a harmonic spherical vibrator ground state) and for $\lambda = 37$ (as appropriate for a nucleus with mean deformation $\beta_0 = 6$).

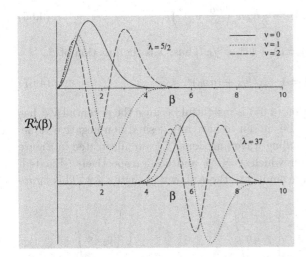

Figure 4.2: Radial (beta) wave functions for $\lambda = 5/2$ and 37 with scale parameter $a = 1$.

It is important to recognise that the wave functions, $\{\mathcal{R}_\nu^\lambda\}$, depend on the scale parameter, a, as well as on λ. Increasing the value of λ increases the value of β about which the wave functions are localised and changing the value of a scales the whole figure. Thus, by choosing appropriate values of λ and a, it is possible to choose beta wave functions appropriate to any desired mean deformation and width. With optimally chosen values of λ and a, they then provide basis wave functions for collective models of deformed nuclei in terms of which calculations converge an order of magnitude faster than in a harmonic spherical vibrator basis.[18]

---

[17] Abramowitz and Stegun, *op. cit.* Footnote 14 on Page 176.
[18] Rowe D.J. and Turner P.S. (2005), *Nucl. Phys.* **A753**, 94.

(i) *Physical interpretation of the beta wave functions*

Whereas the beta wave functions, $\{\mathcal{R}_\nu^{\nu+5/2}\}$ of the harmonic series representations are the radial wave functions for the eigenstates of a harmonic spherical-oscillator Hamiltonian, those of the modified oscillator series are radial wave functions for Hamiltonians obtained by artificially changing the magnitude of the SO(5) centrifugal potential in the harmonic spherical-oscillator Hamiltonian as follows.

The beta wave function $\mathcal{R}_\nu^\lambda$ is an eigenfunction of the Hamiltonian

$$\hat{H}^{(\lambda)} = 2\hbar\omega\hat{S}_0^{(\lambda)} = \frac{1}{2}\hbar\omega\left[-\frac{1}{a^2}\left(\frac{d^2}{d\beta^2} - \frac{2}{\beta^2}\right) + \frac{(\lambda - 5/2)(\lambda + 1/2)}{(a\beta)^2} + (a\beta)^2\right]. \quad (4.64)$$

A physical interpretation of this Hamiltonian is obtained by expressing it the form

$$\hat{H}^{(\lambda)} = -\frac{\hbar\omega}{2a^2}\left(\frac{d^2}{d\beta^2} - \frac{2}{\beta^2}\right) + V_{\beta_0}(\beta), \quad (4.65)$$

where

$$V_{\beta_0}(\beta) := \frac{1}{2}\hbar\omega a^2\left(\frac{\beta_0^4}{\beta^2} + \beta^2\right), \quad (4.66)$$

and $(a\beta_0)^4 := (\lambda - 5/2)(\lambda + 1/2) = (\lambda - 1)^2 - 9/4$, Thus, λ is the function of β_0

$$\lambda = 1 + \sqrt{9/4 + (a\beta_0)^4}. \quad (4.67)$$

Expressing $\hat{H}^{(\lambda)}$ as in Equation (4.65) is insightful because the potential V_{β_0} has a minimum at $\beta = \beta_0$ which can be chosen to have any desired non-negative value. Such potentials and the corresponding wave functions, $\mathcal{R}_0^\lambda$, are illustrated in Figure 4.3 for $\beta_0 = 0$, 8 and $a = 1$, for which $\lambda = 5/2$ and 65.1, respectively. Excited, wave functions, $\mathcal{R}_\nu^\lambda$, for $\nu > 0$, are similar to those shown in Figure 4.2. This figure

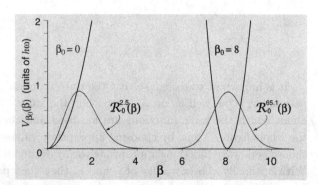

Figure 4.3: The potential, $V_{\beta_0}(\beta)$, of Equation (4.66) shown for two values of β_0. Also shown are the corresponding $\mathcal{R}_0^\lambda(\beta)$ wave functions as functions of β for a unit value of the scale parameter a.

shows the advantages of choosing bases of beta wave functions $\{\mathcal{R}_\nu^\lambda; \nu = 0, 1, 2, \dots\}$ with values of λ and a chosen such that β_0 and a correspond, respectively, to the required mean deformation and fluctuations about this mean.

(ii) *The Davidson model*

The potential V_{β_0} of Equation (4.66) is a five-dimensional analogue of a three-dimensional potential proposed by Davidson for use in molecular physics.[19] A five-dimensional version of the Davidson model, with Hamiltonian

$$\hat{H}_{\mathrm{DM}} := -\hbar\omega \frac{\nabla^2}{2a^2} + V_{\beta_0}(\beta), \qquad (4.68)$$

has been suggested for use in nuclear physics.[20] The radial Hamiltonian for eigenstates of $\hat{H}_{\mathrm{DM}}$ of seniority v is then given by

$$\hat{H}_{\mathrm{DM}}^{(v)} = \frac{1}{2}\hbar\omega \left[-\frac{1}{a^2}\left(\frac{d^2}{d\beta^2} - \frac{2}{\beta^2}\right) + \frac{v(v+3)}{(a\beta)^2} + \frac{a^2\beta_0^4}{\beta^2} + (a\beta)^2 \right], \qquad (4.69)$$

which can be re-expressed as

$$\hat{H}_{\mathrm{DM}}^{(v)} = -\frac{\hbar\omega}{2a^2}\left(\frac{d^2}{d\beta^2} - \frac{2}{\beta^2}\right) + V_{\beta_v}(\beta), \qquad (4.70)$$

where now β_0 is replaced by β_v defined by

$$(a\beta_v)^4 := v(v+3) + (a\beta_0)^4. \qquad (4.71)$$

Thus, the Davidson model has a complete set of eigenfunctions of the form

$$\Psi_{\lambda_v v;v\alpha LM}(\beta,\gamma,\Omega) = \frac{1}{\beta^2}\mathcal{R}_\nu^{\lambda_v}(\beta)\,\mathcal{Y}_{v\alpha LM}(\gamma,\Omega), \qquad (4.72)$$

with

$$\lambda_v := 1 + \sqrt{9/4 + (a\beta_v)^4} = 1 + \sqrt{(v+3/2)^2 + (a\beta_0)^4}. \qquad (4.73)$$

Note that the radial wave functions of the Davidson model belong to su(1,1) irreps with seniority-dependent values of $\lambda = \lambda_v$. This is a consequence of the SO(5) centrifugal term $v(v+3)/(a\beta)^2$ in the kinetic energy component of the Hamiltonian. While this might seem to be desirable, the use of many su(1,1) irreps, related in this way, makes it difficult to calculate beta matrix elements of observables of interest between states of different v and, hence, different λ_v, except by brute-force numerical methods. As we now show, the calculation of matrix elements is simplified if, rather than the Davidson relationship, (4.73), λ_v is chosen such that

$$\lambda_{v+1} = \lambda_v \pm 1, \qquad (4.74)$$

in parallel with the harmonic spherical vibrator limit, for which $\lambda_v = v + 5/2$.

---

[19]Davidson P.M. (1932), *Proc. Roy. Soc. London* **135**, 459.

[20]Such a Davidson model was introduced into nuclear physics by Rohoziński S.G., Srebrny J. and Horbaczewska K. (1974), *Z. Phys.* **268**, 401, and by Elliott J.P., Evans J.A. and Park P. (1986), *Phys. Lett.* **B169**, 309. Its su(1,1) structure was developed, in the nuclear collective model context, by Rowe D.J. and Bahri C. (1998), *J. Phys. A: Math. Gen.* **31**, 4947.

Exercises

4.4 Show that, for the three-dimensional isotropic harmonic oscillator,

$$r\nabla^2 \frac{1}{r} = \frac{d^2}{dr^2} - \frac{\hat{l}^2}{r^2},\tag{4.75}$$

where $\hat{l}^2$ is the square of the three-dimensional angular momentum operator, and that, for the Bohr model,

$$\beta^2\nabla^2 \frac{1}{\beta^2} = \frac{d^2}{d\beta^2} - \frac{1}{\beta^2}(\hat{\Lambda} + 2).\tag{4.76}$$

4.5 Show that, if a wave function for the three-dimensional isotropic harmonic oscillator is expressed in the form

$$\Psi_{\nu lm}(r,\theta,\varphi) = \frac{1}{r}u_{\nu l}(r)\,Y_{lm}(\theta,\varphi),\tag{4.77}$$

where Y_{lm} is a standard SO(3) spherical harmonic, then the radial wave function u_{nl} is a solution of the equation

$$\left[\frac{\hbar^2}{2m}\left(-\frac{d^2}{dr^2} + \frac{l(l+1)}{r^2}\right) + \frac{1}{2}m\omega^2 r^2\right]u_{\nu l}(r) = (l + 2\nu + \tfrac{3}{2})\hbar\omega u_{nl}(r).\tag{4.78}$$

Hence, by defining $a = \sqrt{m\omega/\hbar}$, show that $u_{\nu l}(r) = \mathcal{R}_\nu^{l+3/2}(r)$.

4.3 Dynamical groups and symmetries of solvable submodels

All the observables of the Bohr model are ultimately expressible in terms of the basic position and momentum observables, $\{\hat{q}_m, \hat{\pi}_m\}$, that satisfy the Heisenberg commutation relations given by Equation (4.1). Thus, the Heisenberg-Weyl Lie algebra is an SGA for the model. However, neither the Heisenberg-Weyl algebra nor the group it generates have sufficiently rich structures to describe the variety of dynamics and solvable submodels of the Bohr model. Much richer algebraic structures are generated by combinations of the Heisenberg-Weyl algebra with the Lie algebras of U(5) or of SU(1,1)×SO(5), as defined in Sections 4.1.2 and 4.2. In this way, one constructs two natural dynamical groups for the ACM: the semi-direct product, [HW(5)]U(5), of the Heisenberg-Weyl and U(5) groups; and the semi-direct product, [HW(5)](SU(1,1)×SO(5)), of the Heisenberg-Weyl and SU(1,1)×SO(5) groups (see Box 4.I for a definition of semi-direct product groups). The latter dynamical group proves to be considerably more versatile.

This section shows that the standard solvable submodels of the Bohr model, described in Chapter 2, are naturally associated with dynamical subgroup chains

4.I Direct product and semi-direct product groups

The direct product, $G_1 \times G_2$, of two groups is the group whose elements are pairs (g_1, g_2), with $g_1 \in G_1$ and $g_2 \in G_2$, and whose product is given by $(g_1, g_2) \circ (g_1', g_2') = (g_1 g_1', g_2 g_2')$. Thus, if U and V are representations of G_1 and G_2 on a common Hilbert space, a representation of $G_1 \times G_2$ is given by the product of the representations U and V on this space, i.e., by $(g_1, g_1) \to U(g_1)V(g_2)$. Note that, to preserve the direct product structure, it is required that the operators $V(g_2)$ and $U(g_1')$ commute with one another so that $U(g_1)V(g_2)U(g_1')V(g_2') = U(g_1 g_1')V(g_2 g_2')$.

Many pairs of groups can also be combined to form semi-direct products, denoted variously by $[G_1]G_2$ and $G_1 \rtimes G_2$. For example, the Euclidean group, $\mathrm{E}^3 = [T^3]\mathrm{SO}(3)$, is a semi-direct product of translation and rotation groups. Suppose a translation $\mathbf{a} \in T^3$ maps a vector $\mathbf{r} \in \mathbb{R}^3$ to the vector $\mathbf{a} + \mathbf{r}$ and a rotation $\Omega \in \mathrm{SO}(3)$ maps $\mathbf{r} \in \mathbb{R}^3$ to $\Omega \mathbf{r}$. A combined translation-rotation element, $(\mathbf{a}, \Omega)$, of the group E^3 then maps a vector $\mathbf{r} \in \mathbb{R}^3$ to $\mathbf{a} + \Omega \mathbf{r}$. The composition of group elements for E^3 can now be inferred by consideration of the sequence of maps

$$(\mathbf{a}, \Omega) \circ (\mathbf{b}, \Omega') : \mathbf{r} \to \mathbf{a} + \Omega \mathbf{b} + \Omega \Omega' \mathbf{r}. \tag{4.I.i}$$

Thus, the group product for E^3 is given by $(\mathbf{a}, \Omega) \circ (\mathbf{b}, \Omega') = (\mathbf{a} + \Omega \mathbf{b}, \Omega \Omega')$. Other examples of semi-direct product groups are the space groups of crystal lattices, the Poincaré group, and the dynamical group of a rigid-rotor model (considered later in this chapter).

In general, a semi-direct product, $[G_1]G_2$, can be defined for two groups G_1 and G_2 whenever G_2 has a representation as a group of transformations, T, of the elements of G_1. The semi-direct product $[G_1]G_2$ then has a group product defined by the equation

$$(g_1, g_2) \circ (g_1', g_2') := (g_1 T(g_2)g_1', g_2 g_2'). \tag{4.I.ii}$$

Suppose, for example, that G_1 is the Heisenberg-Weyl group, HW(5), and G_2 is the unitary group U(5) and suppose that G_1 and G_2 have representations W and U, respectively, on a model Hilbert space. Then, the combination of group elements (g_1, g_2), with $g_1 \in \mathrm{HW}(5)$ and $g_2 \in \mathrm{U}(5)$ is naturally represented by the product of operators $W(g_1)U(g_2)$ on the model Hilbert space. A product of combinations should then be represented by the sequence of transformations

$$(g_1, g_2) \circ (g_1', g_2') \to W(g_1)U(g_2)W(g_1')U(g_2') = W(g_1)\big(U(g_2)W(g_1')U(g_2^{-1})\big)U(g_2)U(g_2') \tag{4.I.iii}$$

which, with $U(g_2)W(g_1')U(g_2^{-1})$ expressed as

$$U(g_2)W(g_1')U(g_2^{-1}) = W\big(T(g_2)g_1'\big), \tag{4.I.iv}$$

defines the group product in accord with Equation (4.I.ii) and shows that the combinations (g_1, g_2) are elements of the semi-direct product group, [HW(5)]U(5).

The Lie algebras of direct and semi-direct product Lie groups are known, respectively, as direct and semi-direct sum Lie algebras. A direct sum, $\mathbf{g} = \mathbf{g}_1 \oplus \mathbf{g}_2$, of two Lie algebras has the property that the elements of $\mathbf{g}_1$ commute with those of $\mathbf{g}_2$. On the other hand, a semi-direct sum, denoted variously by $[\mathbf{g}_1]\mathbf{g}_2$ and $\mathbf{g}_1 \otimes_s \mathbf{g}_2$, has the property that

$$[X_1, X_1'] \in \mathbf{g}_1, \quad [X_2, X_2'] \in \mathbf{g}_2, \quad [X_1, X_2] \in \mathbf{g}_1, \tag{4.I.v}$$

for all $X_1, X_1' \in \mathbf{g}_1$ and all $X_2, X_2' \in \mathbf{g}_2$. The subalgebra $\mathbf{g}_1 \subset \mathbf{g}$ is said to be an *ideal* of $\mathbf{g}$.

corresponding to different paths through the set of groups

$$
\begin{array}{ccccc}
[\mathrm{HW}(5)]\mathrm{U}(5) & \supset & [\mathbb{R}^5]\mathrm{SO}(5) & \supset & [\mathbb{R}^5]\mathrm{SO}(3) \\
\cup & & \cup & & \cup \\
\mathrm{U}(5) & \supset & \mathrm{SO}(5) & \supset & \mathrm{SO}(3) \quad \supset \quad \mathrm{SO}(2)
\end{array}
\tag{4.79}
$$

starting with $[\mathrm{HW}(5)]\mathrm{U}(5)$ and ending with $\mathrm{SO}(2)$, where $\mathbb{R}^5$ is the Abelian group with real Lie algebra spanned by the real linear combinations, $\{(q_m+q_m^*), \mathrm{i}(q_m-q_m^*)\}$ of the quadrupole moments $\{q_m\}$ (cf. Exercise 4.2).[21] We now indicate how these submodels, as well as the adiabatic Bohr model, are described and extended in the ACM in terms of the dynamical subgroup chain

$$
\begin{array}{cccccc}
[\mathrm{HW}(5)]\big(\mathrm{SU}(1,1)\times\mathrm{SO}(5)\big) \supset \mathrm{SU}(1,1)\times\mathrm{SO}(5) \supset \mathrm{U}(1)\times\mathrm{SO}(3) \supset \mathrm{SO}(2) \\
\qquad\qquad\quad \lambda \qquad\qquad\quad v \qquad\quad \alpha \quad v \qquad\quad L \qquad\qquad M
\end{array}
\tag{4.80}
$$

4.3.1 *The harmonic spherical vibrator submodel*

Because the Hamiltonian of the harmonic spherical vibrator model, as shown by Equation (4.16), is expressed in terms of the d-phonon number operator, $\hat{N}$ by $\hat{H}_{\mathrm{HV}} = (\hat{N} + 5/2)\hbar\omega$, its eigenstates have well-defined d-phonon number, N, and, as shown in Section 2.3.1, have energies given by $E_N = (N + 5/2)\hbar\omega$.

As noted in Section 4.1.2, $\hat{N}$ is an element of a Cartan subalgebra for the group $\mathrm{U}(5)$, which implies that $\mathrm{U}(5)$ is a symmetry group of the Hamiltonian $\hat{H}_{\mathrm{HV}}$. Moreover, the irreps of $\mathrm{U}(5)$ in the space of the five-dimensional harmonic oscillator are uniquely labelled by their phonon number, N. The rotation groups $\mathrm{SO}(3) \supset \mathrm{SO}(2)$ are subgroups of $\mathrm{U}(5)$ and their irreps provide angular momentum quantum numbers, L and M, for the basis states of these $\mathrm{U}(5)$ irreps. In addition, the group $\mathrm{SO}(5)$, intermediate between $\mathrm{U}(5)$ and $\mathrm{SO}(3)$, provides the seniority quantum number v, defined by Equation (4.28). It follows that the basis states $\{|Nv\alpha LM\rangle\}$ for a unitary irrep of the dynamical group $[\mathrm{HW}(5)]\mathrm{U}(5)$, labelled by the quantum numbers of the subgroups in the chain given by Equation (4.33), i.e.,

$$
\begin{array}{cccc}
\mathrm{U}(5) \supset \mathrm{SO}(5) \supset \mathrm{SO}(3) \supset \mathrm{SO}(2) \\
\quad N \qquad\quad v \qquad\quad \alpha \quad L \qquad\quad M
\end{array}
\tag{4.81}
$$

where α is a multiplicity index, are eigenstates of $\hat{H}_{\mathrm{HV}}$.

Because of the factorisation of the harmonic spherical-vibrator-model wave functions into products of radial, $\mathrm{SU}(1,1)$, and angular, $\mathrm{SO}(5)$, wave functions, and because of the duality relationship between the $\mathrm{SO}(5)$ irreps and the harmonic series of $\mathrm{SU}(1,1)$ irreps, discussed in Section 4.2.2, the harmonic spherical vibrator model

---

[21] The quadrupole moments $\{q_m\}$, being complex in a spherical basis, span the complex extension of the $\mathbb{R}^5$ Lie algebra.

Hamiltonian also has a dynamical symmetry given by the subgroup chain

$$
\begin{array}{ccccccc}
\mathrm{SU}(1,1) & \times & \mathrm{SO}(5) & \supset & \mathrm{U}(1) \times \mathrm{SO}(3) & \supset & \mathrm{SO}(2) \\
\lambda = v + {}^5\!/_2 & & v & & \alpha \quad \nu \qquad L & & M
\end{array}, \qquad (4.82)
$$

with $\nu = \frac{1}{2}(N - v)$; cf. Figure 4.1.

Thus, we have two powerful subgroup chains and corresponding Lie algebra chains which facilitate Bohr model calculations for more general Hamiltonians in a harmonic spherical vibrator basis. As we show in the following, the algebraic structures of the second chain are particularly powerful in enabling the calculation of matrix elements of operators of relevance to the model, such as quadrupole transition matrix elements. For example, the E2 transition rates shown in Figure 2.2 are determined by such means (see Section 4.6.1).

4.3.2 *Wilets-Jean submodels*

Eigenstates of the rigid-beta Wilets-Jean Hamiltonian, $\hat{H}_{\mathrm{WJ}}$, of Equation (2.107) have the quantum numbers of the subgroups in the chain

$$
\begin{array}{ccccccccc}
[\mathrm{HW}(5)]\mathrm{U}(5) & \supset & [\mathbb{R}^5]\mathrm{SO}(5) & \supset & \mathrm{SO}(5) & \supset & \mathrm{SO}(3) & \supset & \mathrm{SO}(2) \, . \\
\beta_0 & & & & v & & \alpha \quad L & & M
\end{array} \qquad (4.83)
$$

The group $[\mathbb{R}^5]\mathrm{SO}(5)$ in this chain is a semi-direct product group with infinitesimal generators given by the quadrupole moments $\{q_m\}$ (actually their real linear combinations, as discussed above) and the infinitesimal generators of $\mathrm{SO}(5)$ rotations of the five-dimensional space spanned by the quadrupole moments. Note that $\beta^2 = \sum_m |q_m|^2$ commutes with all the quadrupole moments and is invariant under $\mathrm{SO}(5)$ transformations. Thus, an irrep of the group $[\mathbb{R}^5]\mathrm{SO}(5)$ is characterised by a fixed value, β_0, of β. Then, with a frozen value of $\beta = \beta_0$, a basis of wave functions for an $[\mathbb{R}^5]\mathrm{SO}(5)$ irrep is given by a complete set of $\mathrm{SO}(5)$ spherical harmonics $\{\mathcal{Y}_{v\alpha LM}\}$ (as constructed by the algorithm given in Section 4.9) and the quadrupole moments $\{q_m = \beta \mathcal{Q}_m(\gamma, \Omega)\}$ are represented by operators $\{\hat{q}_m\}$, whose actions on these wave functions are defined by

$$
[\hat{q}_m \mathcal{Y}_{v\alpha LM}](\gamma, \Omega) := \beta_0 \mathcal{Q}_m(\gamma, \Omega) \mathcal{Y}_{v\alpha LM}(\gamma, \Omega), \qquad (4.84)
$$

in accordance with Equation (2.12). Note that the angular part, $\mathcal{Q}_m$, of a quadrupole moment q_m is proportional to the elementary, $v = 1$, $\mathrm{SO}(5)$ spherical harmonic $\mathcal{Y}_{112m}$ (see Section 4.5.2) and that a product of spherical harmonics is a linear combination of other spherical harmonics.

Because the beta coordinate is frozen at a fixed value, β_0, in the Wilets-Jean model, there is no need for beta wave functions in this model. Equivalently, one can regard the states of the Wilets-Jean model as having a common beta wave function that is sharply-peaked about β_0 with a small but non-zero width. This avoids the problem that such a wave function is not normalisable in the zero-width limit when it becomes proportional to a delta function. The rigid-beta limit is unphysical; it can be approached but never quite reached. Thus, rigid-beta Wilets-Jean model states, classified by the subgroup chain (4.83), do not provide a basis for the Bohr model,[22] but, soft-beta Wilets-Jean models, that can approach as closely as one wishes to the rigid-beta limit, are readily constructed by use of the dynamical symmetry chain (4.80). The states, $\{|\lambda\nu; v\alpha LM\rangle\}$, defined by the irreps of this subgroup chain diagonalise a Wilets-Jean Hamiltonian of the form

$$\hat{H}_{\mathrm{WJ}} = A\hat{S}_+^{(\lambda)}\hat{S}_-^{(\lambda)} + B\hat{S}_0^{(\lambda)} + C\hat{\Lambda}^2, \qquad (4.85)$$

for arbitrary values of the coupling constants, A, B, and C. This Hamiltonian has an SU(1,1) dynamical group and an SO(5) symmetry group. Moreover, for some relatively large value of λ and/or a large value of the scale parameter, a in Equation (4.56), the low-energy SU(1,1) wave functions are sharply peaked about a value of β_0 given by Equation (4.67) but remain normalisable and provide a basis for the Bohr model for finite values of these parameters (cf. Figure 4.3). An example of a sequence of Bohr-model Hamiltonians whose low-energy states approach those of the rigid-beta Wilets-Jean model is given in Section 4.7.2.

4.3.3 *Rotor submodels*

The rigid-rotor submodel of the Bohr model is defined by the dynamical subgroup chain

$$[\mathrm{HW}(5)]\mathrm{U}(5) \supset [\mathbb{R}^5]\mathrm{SO}(5) \supset [\mathbb{R}^5]\mathrm{SO}(3) \supset \mathrm{SO}(3) \supset \mathrm{SO}(2). \atop \beta_0 \qquad (\beta_0,\gamma_0) \quad K \quad L \qquad M \qquad (4.86)$$

As for the rigid-beta Wilets-Jean model, an irrep of the group $[\mathbb{R}^5]\mathrm{SO}(5)$ is defined by a fixed value, β_0^2, of the SO(5) invariant $|q_m|^2$. An irrep of the $[\mathbb{R}^5]\mathrm{SO}(3)$ subgroup is similarly defined by fixed values of the two SO(3) invariants $\beta^2 = |q_m|^2$ and $\beta^3\cos 3\gamma \propto (q \otimes q \otimes q)_0$. Thus, a rigid-rotor irrep is characterised by constant values, (β_0,γ_0), of β and γ such that the quadrupole moments of the model are represented, as described in Section 2.3.3, by the map

$$q_m \to \hat{q}_m = \beta_0\cos\gamma_0\,\mathscr{D}_{0m}^2(\Omega) + \frac{1}{\sqrt{2}}\beta_0\sin\gamma_0\left(\mathscr{D}_{2m}^2(\Omega) + \mathscr{D}_{-2m}^2(\Omega)\right). \qquad (4.87)$$

Basis wave functions and matrix elements for the rigid-rotor submodel of the Bohr model are then as given in Section 2.3.3.

The basis states of a rigid-rotor model, like those of the rigid-beta Wilets-Jean model, correspond to an unphysical limit of the Bohr model that can be approached but never quite reached. Examples of Bohr-model Hamiltonians which give results that approach those of the rigid-rotor submodel are given in Section 4.7.5.

---

[22]Basis wave functions for the Bohr model must be square-integrable and cannot be delta functions.

4.3.4 The general ACM

The above solvable submodels provide approximate descriptions of a subset of nuclear collective states. However, harmonic vibrational spectra are rarely, if ever, seen experimentally and, while rotational states are prevalent in nuclei, they are not as rigid as the rigid-beta Wilets-Jean and rigid-rotor models represent them to be. More realistic applications of the Bohr model are given by Hamiltonians that relax the rigidity constraints and admit rotational motions combined with beta and gamma vibrations. Determining the properties of such Hamiltonians is greatly assisted by the use of algebraic methods.

Within the framework of the dynamical symmetries corresponding to different paths through the system of subgroups of the dynamical group [HW(5)]U(5) to its SO(2) subgroup, shown in Equation (4.79), only the chain

$$[HW(5)]U(5) \supset U(5) \supset SO(5) \supset SO(3) \supset SO(2) \atop N \qquad v \qquad \alpha \quad L \qquad M \, , \tag{4.88}$$

which diagonalises a harmonic spherical-vibrator Hamiltonian, provides a basis of states for a general collective model calculation. This is because the rigid-beta and rigid-gamma wave functions of the other subgroup chains contain unphysical delta functions. Thus, the Frankfurt version of the collective model[23,24,25,26,27] makes use of an algorithm of Chacón et al.[28,29] for constructing basis states $\{|nv\alpha LM\rangle\}$, labelled by the quantum numbers of this subgroup chain, and for computing matrix elements of a general collective model Hamiltonian in this basis.

On the other hand, the ACM makes use of a subgroup chain involving the SU(1,1)×SO(5) group to generate basis wave functions as products of beta wave functions and SO(5) spherical harmonics. For spherical nuclei, the appropriate SU(1,1)×SO(5) basis, which also reduces the subgroup chain of Equation (4.88), is given by the harmonic series of SU(1,1) irreps; cf. Section 4.2.2. However, for deformed nuclei, the modified oscillator series of SU(1,1) irreps, defined in Section 4.2.3, give much more rapidly convergent results.

For all SU(1,1)×SO(5) irreps used in the ACM, the matrix elements of collective model observables, which are polynomials in the basic position and momentum variables $\{q_m, \pi_m\}$, can be computed algebraically.[30] Thus, Hamiltonians of interest are scalar (i.e., SO(3)-invariant) and time-reversal invariant polynomials of modest

---

[23]Gneuss G. and Greiner W. (1971), *Nucl. Phys.* **A171**, 449.

[24]Hess P.O. *et al.* (1980), *Z. Phys.* **A296**, 147.

[25]Hess P.O., Maruhn J.A. and Greiner W. (1981), *J. Phys. G: Nucl. Phys.* **7**, 737.

[26]Eisenberg and Greiner, *op. cit.* Footnote 8 on Page 98.

[27]Troltenier D., Maruhn J.A. and Hess P.O. (1991), in *Computational Nuclear Physics 1*, edited by K. Langanke, J.A. Maruhn and S.E. Koonin (Springer, Berlin), p. 105.

[28]Chacón *et al.*, *op. cit.* Footnote 1 on Page 163.

[29]Chacón and Moshinsky, *op. cit.* Footnote 2 on Page 163.

[30]We remind the reader that when, for convenience, we speak of SU(1,1) irreps, we include multivalued irreps which are, strictly speaking, only true irreps of a covering group as noted in Footnote 12 on Page 171.

degree in these variables. It is known that any scalar polynomial in the quadrupole moments is a polynomial in the two elementary SO(3) scalars (cf. Exercise 4.8): $(q \otimes q)_0 \propto \beta^2$ and $(q \otimes q \otimes q)_0 \propto \beta^3 \cos 3\gamma$.[31] In fact, we can be more general and include polynomials containing terms in $1/\beta^2$. Thus, a typical Hamiltonian of interest might, for example, be of the form

$$\hat{H} = -\frac{\hbar^2}{2B}\nabla^2 + V(\beta, \gamma), \qquad (4.89)$$

with

$$V(\beta, \gamma) = c_1\beta^2 + c_2\beta^4 + c_3\frac{1}{\beta^2} + c_4\beta \cos 3\gamma + c_5[\cos 3\gamma]^2. \qquad (4.90)$$

The following sections of this chapter show how to perform ACM calculations with Hamiltonians of this type.

4.3.5 *The adiabatic limit*

An important characteristic of the ACM is that it enables rigid-beta and rigid-gamma limits to be approached in a continuous way with increasingly narrow, but nevertheless square-integrable beta and gamma wave functions. The Wilets-Jean and rotor submodels of the Bohr model are then seen as special cases of the more general, and more useful adiabatic model described in Section 2.4. Thus, it is of interest to explore the expression of the adiabatic model within the framework of the ACM.

The adiabatic limit of the Bohr model is derived in Chapter 2 by consideration of a Hamiltonian of the form given by Equation (4.89). When the potential energy $V(\beta, \gamma)$ has a steep-sided minimum about some values, β_0 and γ_0, of β and γ, which define the equilibrium shape of the nucleus modelled, then a leading-order Taylor expansion of $V(\beta, \gamma)$ about the point (β_0, γ_0) immediately gives a solvable adiabatic model with rotational plus uncoupled harmonic β and γ vibrational modes.

For example, with suitably chosen values of the constants c_1, c_2, and c_3, the potential energy function $V(\beta, \gamma)$, given by Equation (4.90), will give a potential with a minimum at any desired value β_0. With c_4 negative and $c_5 = 0$, $V(\beta, \gamma)$ has a minimum at $\gamma_0 = 0$; and, with $c_4 = 0$ and c_5 positive, it has a minimum at the Meyer-ter-Vehn value (cf. Section 2.4), $\gamma_0 = \pi/6$. Thus, by uniformly increasing these values of the c_i constants, the potential develops steeper sides about its minimum and the approach to the adiabatic limit can be explored. Examples of such calculations are given in Section 4.7.

---

[31]Such polynomial potentials were considered by Eisenberg and Greiner, *op. cit.* Footnote 8 on Page 98.

Exercises

4.6 Show that the $\{q_m\}$ quadrupole moments span the complex extension of $\mathbb{R}^5$ (see Footnote 3 on Page 164 for a definition of the complex extension of a Lie algebra).

4.7 Give the eigenfunctions and corresponding eigenvalues for the Wilets-Jean Hamiltonian of Equation (4.85).

4.8 Use the observation that any real symmetric 3×3 matrix M is uniquely defined, to within a rotation, by its three eigenvalues, to show that these eigenvalues are defined by the three rotational invariants: $\text{Tr}(M)$, $\text{Tr}(M^2)$, and $\text{Tr}(M^3)$. Hence, show that a complete set of rotationally-invariant polynomials in the elements of a real symmetric matrix M of zero trace, is given by polynomials in $\text{Tr}(M^2)$ and $\text{Tr}(M^3)$.

4.9 Derive a potential energy function, $V(\beta, \gamma)$, of the form given by Equation (4.90) that has a minimum at some specified values, β_0 and γ_0, of β and γ.

4.4 Beta matrix elements for the ACM

We consider a basis of states, $\{|\lambda\nu; v\alpha LM\rangle\}$, for the Bohr model which reduce the subgroup chain of Equation (4.33) and have product wave functions $\Psi_{\lambda\nu;v\alpha LM}(\beta, \gamma, \Omega) = \frac{1}{\beta^2} \mathcal{R}_\nu^\lambda(\beta) \, \mathcal{Y}_{v\alpha LM}(\gamma, \Omega)$, as given by Equation (4.36).

As shown in this and the following sections, matrix elements of operators of interest are simply evaluated in the ACM from products of beta matrix elements, SO(5)-reduced matrix elements, and SO(5) Clebsch-Gordan coefficients. In this section, we give analytical expressions for matrix elements of β, β^2, $1/\beta$, $1/\beta^2$, $d/d\beta$, and $d^2/d\beta^2$. Although they are not actually needed, because these matrix elements are derived algebraically, the SU(1,1) beta wave functions, $\{\mathcal{R}_\nu^\lambda\}$, are given explicitly by Equation (4.58).

4.4.1 *Summary of beta matrix elements*

The matrix elements of an operator $\hat{X}$ that acts only on the beta wave functions are given by the expression[32,33]

$$\langle\lambda\mu; v\alpha LM|\hat{X}|\lambda'\nu; v'\alpha'L'M'\rangle = F_{\lambda\mu;\lambda'\nu}\left(\beta^2\hat{X}\frac{1}{\beta^2}\right)\delta_{vv'}\delta_{\alpha\alpha'}\delta_{LL'}\delta_{MM'}, \quad (4.91)$$

where

$$F_{\lambda\mu;\lambda'\nu}(\hat{Z}) := \int_0^\infty \mathcal{R}_\mu^\lambda(\beta)\,\hat{Z}\mathcal{R}_\nu^{\lambda'}(\beta)\,d\beta. \quad (4.92)$$

so that

$$F_{\lambda\mu;\lambda'\nu}\left(\beta^2\hat{X}\frac{1}{\beta^2}\right) = \int_0^\infty \left[\frac{1}{\beta^2}\mathcal{R}_\mu^\lambda(\beta)\right]\hat{X}\left[\frac{1}{\beta^2}\mathcal{R}_\nu^{\lambda'}(\beta)\right]\beta^4\,d\beta. \quad (4.93)$$

---

[32] Rowe D.J. (2005), *J. Phys. A: Math. Gen.* **38**, 10181, and references therein.
[33] Rowe *et al.*, *op. cit.* Footnote 10 on Page 171.

Note that the β^4 term in the volume element, $dv = \beta^4 \sin 3\gamma \, d\beta \, d\gamma \, d\Omega$, is factored out in Equation (4.92) by the inclusion of a factor $1/\beta^2$ in the expansion of the wave function $\Psi_{\lambda\nu;v\alpha LM}$, as in Equation (4.36).

Claim 4.1

$$F_{\lambda\mu;\lambda\nu}(\hat{S}_0^{(\lambda)}) = \tfrac{1}{2}(\lambda + 2\nu)\,\delta_{\mu,\nu}, \tag{4.94}$$

$$F_{\lambda\mu;\lambda\nu}(\hat{S}_+^{(\lambda)}) = \sqrt{(\lambda+\nu)(\nu+1)}\,\delta_{\mu,\nu+1}, \tag{4.95}$$

$$F_{\lambda\mu;\lambda\nu}(\hat{S}_-^{(\lambda)}) = \sqrt{(\lambda+\nu-1)\nu}\,\delta_{\mu,\nu-1}, \tag{4.96}$$

$$F_{\lambda\mu;\lambda\nu}(a^2\beta^2) = \delta_{\mu,\nu+1}\sqrt{(\lambda+\nu)(\nu+1)} + \delta_{\mu,\nu-1}\sqrt{(\lambda+\nu-1)\nu}$$
$$+\,\delta_{\mu,\nu}(\lambda + 2\nu), \tag{4.97}$$

$$F_{\lambda\mu;\lambda\nu}\left(\frac{1}{a^2\beta^2}\right) = \frac{(-1)^{\mu-\nu}}{\lambda-1}\sqrt{\frac{\mu!\,\Gamma(\lambda+\nu)}{\nu!\,\Gamma(\lambda+\mu)}}, \qquad \text{for } \mu \geq \nu, \ \lambda > 1, \tag{4.98}$$

$$F_{\lambda\mu;\lambda\nu}\left(\frac{1}{a^2}\frac{d^2}{d\beta^2}\right) = \delta_{\mu,\nu+1}\sqrt{(\lambda+\nu)(\nu+1)} + \delta_{\mu,\nu-1}\sqrt{(\lambda+\nu-1)\nu}$$
$$-\,\delta_{\mu,\nu}(\lambda+2\nu) + (\lambda - \tfrac{3}{2})(\lambda - \tfrac{1}{2})F_{\lambda\mu;\lambda\nu}\left(\frac{1}{a^2\beta^2}\right), \tag{4.99}$$

$$F_{\lambda+1,\mu;\lambda\nu}(a\beta) = \delta_{\mu,\nu}\sqrt{\lambda+\nu} + \delta_{\mu,\nu-1}\sqrt{\nu}, \tag{4.100}$$

$$F_{\lambda-1,\mu;\lambda\nu}\left(\frac{1}{a\beta}\right) = \begin{cases} 0 & \text{if } \mu > \nu, \\[2mm] (-1)^{\mu-\nu}\sqrt{\dfrac{\nu!\,\Gamma(\lambda+\mu-1)}{\mu!\,\Gamma(\lambda+\nu)}} & \text{if } \mu \leq \nu, \end{cases} \tag{4.101}$$

$$F_{\lambda-1,\mu;\lambda\nu}\left(\frac{1}{a}\frac{d}{d\beta}\right) = \delta_{\mu,\nu}\sqrt{\lambda+\nu-1} - \delta_{\mu,\nu+1}\sqrt{\nu+1}$$
$$-(\lambda - \tfrac{3}{2})F_{\lambda-1,\mu;\lambda\nu}\left(\frac{1}{a\beta}\right). \tag{4.102}$$

Other matrix elements are obtained by symmetry. For example, interchange of μ and ν in Equations (4.100), (4.98), and (4.101) leads to the equations

$$F_{\lambda-1,\mu;\lambda\nu}(a\beta) = \delta_{\mu,\nu}\sqrt{\lambda+\nu-1} + \delta_{\mu,\nu+1}\sqrt{\nu+1}, \tag{4.103}$$

$$F_{\lambda\mu;\lambda\nu}\left(\frac{1}{a\beta}\right) = \frac{(-1)^{\mu-\nu}}{\lambda-1}\sqrt{\frac{\nu!\,\Gamma(\lambda+\mu)}{\mu!\,\Gamma(\lambda+\nu)}}, \qquad \text{for } \mu \leq \nu, \ \lambda > 1, \tag{4.104}$$

$$F_{\lambda+1,\mu;\lambda\nu}\left(\frac{1}{a\beta}\right) = \begin{cases} 0 & \text{if } \mu < \nu, \\[2mm] (-1)^{\mu-\nu}\sqrt{\dfrac{\mu!\,\Gamma(\lambda+\nu)}{\nu!\,\Gamma(\lambda+\mu+1)}} & \text{if } \mu \geq \nu. \end{cases} \tag{4.105}$$

Interchanging μ and ν in Equation (4.102) and remembering that $d/d\beta$ is skew-Hermitian relative to the volume element $d\beta$ leads to the equation

$$F_{\lambda+1,\mu;\lambda\nu}\left(\frac{1}{a}\frac{d}{d\beta}\right) = -\delta_{\mu,\nu}\sqrt{\lambda+\nu} + \delta_{\mu,\nu-1}\sqrt{\nu} + (\lambda - 1/2)F_{\lambda+1,\mu;\lambda\nu}\left(\frac{1}{a\beta}\right). \quad (4.106)$$

Matrix elements for operators that have polynomial expressions in β, $1/\beta$, and $d/d\beta$, are evaluated by applying the above expressions sequentially. Note also that it is possible to consider the above matrix elements in the approach to the rigid-beta limit. This is the limit in which a and $\lambda \to \infty$ but $\sqrt{\lambda}/a$ remains finite. As this limit is approached

$$F_{\lambda+1,\mu;\lambda\nu}(a\beta) \to \delta_{\mu,\nu}\sqrt{\lambda} + \delta_{\mu,\nu-1}\sqrt{\nu}. \quad (4.107)$$

Thus, with β_0 defined by $a\beta_0 = \sqrt{\lambda}$, we find that

$$F_{\lambda+1,\mu;\lambda\nu}(a\beta) \to \delta_{\mu,\nu}a\beta_0 + \delta_{\mu,\nu-1}\sqrt{\nu}, \quad (4.108)$$

consistent with the adiabatic limit of the Bohr model.

The following sections provide derivations of the beta matrix elements given by Claim 4.1.

4.4.2 Beta matrix elements from the su(1,1) irreps

Beta matrix elements can be calculated by algebraic means when the basis wave functions span irreps of an su(1,1) Lie algebra. Matrix elements of the su(1,1) operators, $\hat{S}_0^{(\lambda)}$ and $\hat{S}_\pm^{(\lambda)}$, given by Equations (4.94)–(4.96), are obtained immediately from Equations (4.57). In this section, we derive the actions of β^2, $1/\beta^2$ and $d^2/d\beta^2$ on basis wave functions belonging to any modified oscillator series irrep of su(1,1) as defined in Section 4.2.3.

Claim 4.2   *For $\lambda > 1$,*

$$a^2\beta^2\mathcal{R}_\nu^\lambda(\beta) = \sqrt{(\lambda+\nu)(\nu+1)}\,\mathcal{R}_{\nu+1}^\lambda(\beta) + \sqrt{(\lambda+\nu-1)\nu}\,\mathcal{R}_{\nu-1}^\lambda(\beta)$$
$$+ (\lambda + 2\nu)\,\mathcal{R}_\nu^\lambda(\beta), \quad (4.109)$$

$$\frac{1}{a^2\beta^2}\mathcal{R}_\nu^\lambda(\beta) = \sum_{\mu<\nu}\frac{(-1)^{\mu-\nu}}{\lambda-1}\sqrt{\frac{\nu!\,\Gamma(\lambda+\mu)}{\mu!\,\Gamma(\lambda+\nu)}}\,\mathcal{R}_\mu^\lambda(\beta)$$

$$+ \sum_{\mu\geq\nu}\frac{(-1)^{\mu-\nu}}{\lambda-1}\sqrt{\frac{\mu!\,\Gamma(\lambda+\nu)}{\nu!\,\Gamma(\lambda+\mu)}}\,\mathcal{R}_\mu^\lambda(\beta), \quad (4.110)$$

$$\frac{1}{a^2}\frac{d^2}{d\beta^2}\mathcal{R}_\nu^\lambda(\beta) = \sqrt{(\lambda+\nu)(\nu+1)}\,\mathcal{R}_{\nu+1}^\lambda(\beta) + \sqrt{(\lambda+\nu-1)\nu}\,\mathcal{R}_{\nu-1}^\lambda(\beta)$$

$$- (\lambda + 2\nu)\mathcal{R}_\nu^\lambda(\beta) + (\lambda - 3/2)(\lambda - 1/2)\frac{1}{(a\beta)^2}\mathcal{R}_\nu^\lambda(\beta). \quad (4.111)$$

Proof: The first of these equations follows from Equation (4.57) and noting, via (4.56), that $(a\beta)^2$ is an element of the su(1,1) Lie algebra with expansion

$$(a\beta)^2 = \hat{S}_+^{(\lambda)} + \hat{S}_-^{(\lambda)} + 2\hat{S}_0^{(\lambda)}. \tag{4.112}$$

Equation (4.110) is obtained by defining

$$f_{\mu\nu}^\lambda := \int_0^\infty \mathcal{R}_\mu^\lambda(\beta) \frac{1}{a^2\beta^2} \mathcal{R}_\nu^\lambda(\beta)\, d\beta = f_{\nu\mu}^\lambda, \tag{4.113}$$

which, by Equation (4.109), satisfies the recursion relation

$$(\lambda + 2\nu) f_{\mu\nu}^\lambda + \sqrt{(\lambda+\nu-1)\nu}\, f_{\mu,\nu-1}^\lambda + \sqrt{(\lambda+\nu)(\nu+1)}\, f_{\mu,\nu+1}^\lambda = \delta_{\mu\nu}. \tag{4.114}$$

Starting with the value of $f_{00}^\lambda = 1/(\lambda-1)$, which can be obtained by evaluating the integral expression (4.113) for f_{00}^λ, or by use of Equation (4.136) (see Exercise 4.10), it is determined (see Exercise 4.11) that

$$f_{\mu\nu}^\lambda = \frac{(-1)^{\mu-\nu}}{\lambda-1} \sqrt{\frac{\nu!\,\Gamma(\lambda+\mu)}{\mu!\,\Gamma(\lambda+\nu)}}, \quad \text{for } \mu \le \nu, \tag{4.115}$$

which is equivalent to Equation (4.110).

Equation (4.111) follows from the identity

$$\frac{1}{a^2}\frac{d^2}{d\beta^2} - \frac{(\lambda - 3/2)(\lambda - 1/2)}{(a\beta)^2} = \hat{S}_+^{(\lambda)} + \hat{S}_-^{(\lambda)} - 2\hat{S}_0^{(\lambda)}, \tag{4.116}$$

obtained from Equationn (4.56). This completes the proof of Claim 4.2. □

The action of any polynomial in β^2, $1/\beta^2$, or $d^2/d\beta^2$ on the beta wave functions $\{\mathcal{R}_\nu^\lambda\}$ is obtained by repeated use of the results of Claim 4.2. This provides an efficient way of computing the beta matrix elements of many Hamiltonians.

4.4.3 *Beta matrix elements from the factorisation method*

Matrix elements of β, $1/\beta$, and $d/d\beta$ are obtained[34] by the so-called *factorisation* method. This method was introduced by Schrödinger[35,36,37] and employed extensively by Infeld and Hull.[38] It is reviewed in several texts on quantum mechanics.[39]

Let $A(X)$ and $A^\dagger(X)$, where X is a real number, denote the operators

$$A(X) := \frac{1}{a}\frac{d}{d\beta} + \frac{X}{a\beta} + a\beta, \quad A^\dagger(X) := -\frac{1}{a}\frac{d}{d\beta} + \frac{X}{a\beta} + a\beta, \tag{4.117}$$

[34]Rowe, *op. cit.* Footnote 32 on Page 187.
[35]Schrödinger E. (1940-41), *Proc. Roy. Irish Acad.* **A46**, 9.
[36]Schrödinger E. (1940-41), *Proc. Roy. Irish Acad.* **A46**, 183.
[37]Schrödinger E. (1941), *Proc. Roy. Irish Acad.* **A47**, 53.
[38]Infeld L. and Hull T.E. (1951), *Rev. Mod. Phys.* **23**, 21.
[39]For example, Hecht K.T. (2000), *Quantum Mechanics* (Springer, New York).

where it is noted that, with respect to the inner product (4.37), β and $1/\beta$ are Hermitian and $d/d\beta$ is skew-Hermitian. It then follows that

$$A(X)A^\dagger(X) = -\frac{1}{a^2}\frac{d^2}{d\beta^2} + \frac{X(X-1)}{(a\beta)^2} + (a\beta)^2 + 2X + 1, \qquad (4.118)$$

$$A(X)^\dagger A(X) = -\frac{1}{a^2}\frac{d^2}{d\beta^2} + \frac{X(X+1)}{(a\beta)^2} + (a\beta)^2 + 2X - 1, \qquad (4.119)$$

and that

$$A(\lambda - \tfrac{1}{2})A^\dagger(\lambda - \tfrac{1}{2}) = 4\hat{\mathcal{S}}_0^{(\lambda)} + 2\lambda, \qquad (4.120)$$

$$A^\dagger(\lambda - \tfrac{1}{2})A(\lambda - \tfrac{1}{2}) = 4\hat{\mathcal{S}}_0^{(\lambda+1)} + 2\lambda - 2. \qquad (4.121)$$

It follows that

$$A(\lambda - \tfrac{1}{2})A^\dagger(\lambda - \tfrac{1}{2})\mathcal{R}_\nu^\lambda = 4(\lambda + \nu)\mathcal{R}_\nu^\lambda, \qquad (4.122)$$

$$A^\dagger(\lambda - \tfrac{1}{2})A(\lambda - \tfrac{1}{2})\mathcal{R}_\nu^{\lambda+1} = 4(\lambda + \nu)\mathcal{R}_\nu^{\lambda+1}, \qquad (4.123)$$

and hence that

$$A^\dagger(\lambda - \tfrac{1}{2})A(\lambda - \tfrac{1}{2})\left[A^\dagger(\lambda - \tfrac{1}{2})\mathcal{R}_\nu^\lambda\right] = 4(\lambda + \nu)\left[A^\dagger(\lambda - \tfrac{1}{2})\mathcal{R}_\nu^\lambda\right], \qquad (4.124)$$

$$A(\lambda - \tfrac{1}{2})A^\dagger(\lambda - \tfrac{1}{2})\left[A(\lambda - \tfrac{1}{2})\mathcal{R}_\nu^{\lambda+1}\right] = 4(\lambda + \nu)\left[A(\lambda - \tfrac{1}{2})\mathcal{R}_\nu^{\lambda+1}\right]. \qquad (4.125)$$

By comparing these equations with the eigenvalue equations (inferred from Equation (4.57))

$$(4\hat{\mathcal{S}}_0^{(\lambda+1)} + 2\lambda - 2)\mathcal{R}_\nu^{\lambda+1} = 4(\lambda + \nu)\mathcal{R}_\nu^{\lambda+1}, \qquad (4.126)$$

$$(4\hat{\mathcal{S}}_0^{(\lambda)} + 2\lambda)\mathcal{R}_\nu^\lambda = 4(\lambda + \nu)\mathcal{R}_\nu^\lambda. \qquad (4.127)$$

we conclude that, to within phase factors (which we choose to be positive to accord with Equation (4.58)),

$$A^\dagger(\lambda - \tfrac{1}{2})\mathcal{R}_\nu^\lambda = 2\sqrt{\lambda + \nu}\,\mathcal{R}_\nu^{\lambda+1}, \qquad (4.128)$$

$$A(\lambda - \tfrac{1}{2})\mathcal{R}_\nu^{\lambda+1} = 2\sqrt{\lambda + \nu}\,\mathcal{R}_\nu^\lambda. \qquad (4.129)$$

From the properties of the operators $A(-\lambda + \tfrac{3}{2})$ and $A^\dagger(-\lambda + \tfrac{3}{2})$ (see Exercise 4.12), one similarly derives the identities

$$A^\dagger(-\lambda + \tfrac{3}{2})\mathcal{R}_\nu^\lambda = 2\sqrt{\nu + 1}\,\mathcal{R}_{\nu+1}^{\lambda-1}, \quad \text{for } \lambda > 1, \qquad (4.130)$$

$$A(-\lambda + \tfrac{3}{2})\mathcal{R}_{\nu+1}^{\lambda-1} = 2\sqrt{\nu + 1}\,\mathcal{R}_\nu^\lambda, \quad \text{for } \lambda > 1. \qquad (4.131)$$

From Equations (4.128)–(4.131), we now obtain the expressions:

Claim 4.3 *For $\lambda > 1$*

$$(a\beta)\,\mathcal{R}_\nu^\lambda(\beta) = \sqrt{\lambda+\nu}\,\mathcal{R}_\nu^{\lambda+1}(\beta) + \sqrt{\nu}\,\mathcal{R}_{\nu-1}^{\lambda+1}(\beta), \tag{4.132}$$

$$\frac{1}{a\beta}\mathcal{R}_\nu^\lambda(\beta) = \sum_{\mu=0}^{\nu}(-1)^{\mu-\nu}\sqrt{\frac{\nu!\,\Gamma(\lambda+\mu-1)}{\mu!\,\Gamma(\lambda+\nu)}}\,\mathcal{R}_\mu^{\lambda-1}(\beta), \tag{4.133}$$

$$\frac{1}{a}\frac{d}{d\beta}\mathcal{R}_\nu^\lambda(\beta) = \sqrt{\lambda+\nu-1}\,\mathcal{R}_\nu^{\lambda-1}(\beta) - \sqrt{\nu+1}\,\mathcal{R}_{\nu+1}^{\lambda-1}(\beta)$$

$$-(\lambda-\tfrac{3}{2})\frac{1}{a\beta}\,\mathcal{R}_\nu^\lambda(\beta). \tag{4.134}$$

Proof: The first of these equations follows from the definition, (4.117), and Equations (4.128), which give (4.131) with

$$2a\beta = A^\dagger(\lambda-\tfrac{1}{2}) + A(-\lambda+\tfrac{1}{2}). \tag{4.135}$$

Equation (4.133) is obtained by expressing Equation 4.132) as a recursion relation

$$\frac{1}{a\beta}\mathcal{R}_\nu^\lambda(\beta) = \frac{1}{\sqrt{\lambda+\nu-1}}\mathcal{R}_\nu^{\lambda-1}(\beta) - \sqrt{\frac{\nu}{\lambda+\nu-1}}\frac{1}{a\beta}\mathcal{R}_{\nu-1}^\lambda(\beta), \tag{4.136}$$

which is readily solved to give the desired result. The last equation of Claim 4.3 is obtained from the identity

$$\frac{2}{a}\left(\frac{d}{d\beta}+\frac{\lambda-\tfrac{3}{2}}{\beta}\right) = A(\lambda-\tfrac{3}{2}) - A^\dagger(-\lambda+\tfrac{3}{2}). \tag{4.137}$$

This completes the proof of Claim 4.3, □

The matrix elements given by Equations (4.100)–(4.102) are obtained immediately from these actions.

Exercises

4.10 For $\lambda > 1$, derive the identities

$$\frac{1}{a\beta}\mathcal{R}_0^\lambda(\beta) = \frac{1}{\sqrt{\lambda-1}}\mathcal{R}_0^{\lambda-1}(\beta), \tag{4.138}$$

$$\frac{1}{a\beta}\mathcal{R}_1^\lambda(\beta) = \frac{1}{\sqrt{\lambda}}\mathcal{R}_1^{\lambda-1}(\beta) - \frac{1}{\sqrt{\lambda(\lambda-1)}}\mathcal{R}_0^{\lambda-1}(\beta) \tag{4.139}$$

from Equation (4.136), and show that, for $\lambda > 1$,

$$f_{00}^\lambda = \int_0^\infty \frac{1}{(a\beta)^2}\left|\mathcal{R}_0^\lambda(\beta)\right|^2 d\beta = \frac{1}{\lambda-1}, \quad f_{11}^\lambda = -\sqrt{\lambda}f_{10}^\lambda = \frac{1}{\lambda-1}. \tag{4.140}$$

4.11 Assuming that $f_{0\nu}^\lambda$ satisfies Equation (4.115), for some value of ν, as Equation (4.140) shows that it does for $\nu = 0$, use the recursion relation (4.114) to show that it satisfies this equation for all ν. Then, use the identity $f_{\nu,0} = f_{0,\nu}$ and Equation (4.114), to show that $f_{\mu,\nu} = f_{\nu,\mu}$ satisfies Equation (4.115) for all $\mu \le \nu$.

4.12 Starting from the $X = -\lambda + {}^3\!/_2$ case of the definitions (4.117), derive Equations (4.130) and (4.131).

4.5 SO(5) matrix elements

We now consider matrix elements of SO(5) tensor operators between angular states classified by SO(5) quantum numbers. An enormous gain in efficiency is achieved by expressing these matrix elements in terms of reduced matrix elements using the Wigner-Eckart theorem in a natural extension of the way the matrix elements of spherical SO(3) tensors are customarily expressed. According to the Wigner-Eckart theorem, matrix elements of the components, $\{\hat{T}_{LM}\}$, of an SO(3) tensor $\hat{T}_L$ between states with good angular momentum quantum numbers, factor according to the equation

$$\langle \alpha L_f M_f | \hat{T}_{LM} | \beta L_i M_i \rangle = (L_i M_i \, LM | L_f M_f) \frac{\langle \alpha L_f \| \hat{T}_L \| \beta L_i \rangle}{\sqrt{2L_f + 1}}, \qquad (4.141)$$

where $(L_i M_i \, LM | L_f M_f)$ is a standard SO(3) Clebsch-Gordan coefficient and $\langle \alpha L_f \| \hat{T}_L \| \beta L_i \rangle$ is an SO(3)-reduced matrix element[40] (see Appendix A.3.3 for a review of the Wigner Eckart theorem for SO(3) and SU(2)).

4.5.1 *The Wigner-Eckart theorem for SO(5)*

Suppose that $\{\hat{T}^v_{\alpha LM}\}$ are the components of an SO(5) tensor operator $\hat{T}^v$ of seniority v that act on the space $\mathcal{L}^2(S_4)$. Then, the matrix elements of these components between states that have good SO(5) quantum numbers, factor according to the equation

$$\langle v_f \alpha_f L_f M_f | \hat{T}^v_{\alpha LM} | v_i \alpha_i L_i M_i \rangle = (v_i \alpha_i L_i M_i, v \alpha L M | v_f \alpha_f L_f M_f)$$
$$\times \langle v_f \| | \hat{T}^v \| | v_i \rangle, \qquad (4.142)$$

where $(v_i \alpha_i L_i M_i, v \alpha L M | v_f \alpha_f L_f M_f)$ is an SO(5) Clebsch-Gordan coefficient and $\langle v_f \| | \hat{T}^v \| | v_i \rangle$ is an SO(5)-reduced matrix element.

Further simplification is achieved by expressing the SO(5) Clebsch-Gordan coefficients in terms of SO(3)-reduced SO(5) Clebsch-Gordan coefficients,[41] denoted

---

[40]Note that the factor $\sqrt{2L_f + 1}$ is conventionally included in the denominator for SO(3) so that the reduced matrix elements have desirable symmetry properties. However, its presence is not an essential part of the Wigner-Eckart theorem as it can be absorbed into adjusted values of the reduced matrix elements.

[41]An SO(3)-reduced SO(5) Clebsch-Gordan coefficient $(v_i \alpha_i L_i, v \alpha L \| v_f \alpha_f L_f)$ is sometimes referred to as an *isoscalar factor*. They could be described more informatively as SO(5)/SO(3) (pronounced SO(5) mod SO(3)) Clebsch-Gordan coefficients.

by $(v_i\alpha_i L_i, v\alpha L\|v_f\alpha_f L_f)$ and defined by the identity

$$(v_i\alpha_i L_i M_i, v\alpha LM|v_f\alpha_f L_f M_f)$$
$$= (L_i M_i\, LM|L_f M_f)(v_i\alpha_i L_i, v\alpha L\|v_f\alpha_f L_f). \qquad (4.143)$$

The above results then give the SO(3)-reduced matrix elements of the operators $\{\hat{T}^v_{\alpha LM}\}$, as components of SO(3) tensors $\{\hat{T}^v_{\alpha L}\}$, by means of the expression

$$\frac{\langle v_f\alpha_f L_f\|\hat{T}^v_{\alpha L}\|v_i\alpha_i L_i\rangle}{\sqrt{2L_f+1}} = (v_i\alpha_i L_i, v\alpha L\|v_f\alpha_f L_f)\,\langle v_f\|\|\hat{T}^v\|\|v_i\rangle. \qquad (4.144)$$

4.5.2 *SO(5) spherical harmonics as basis wave functions*

In a product basis of states for the ACM that span irreps of the dynamical group SU(1,1) × SO(5), the wave functions for the angular SO(5) states are generalisations to 5-dimensional space of the standard SO(3) spherical harmonics $\{Y_{LM}\}$. Thus, we denote the SO(5) spherical harmonics by $\{\mathcal{Y}_{v\alpha LM}\}$ (they have been denoted elsewhere by $\{\Psi_{v\alpha LM}\}^{42,43,44}$).

The SO(5) spherical harmonics $\{\mathcal{Y}_{v\alpha LM}\}$ form an orthonormal basis for the SO(5) irrep $[v]$ of seniority v. Moreover, the SO(5) spherical harmonics of all seniorities satisfy the orthonormality relationship given by Equation (4.39) and are an orthonormal basis for the Hilbert space $\mathcal{L}^2(S_4)$.

The fundamental, $v=1$, SO(5) spherical harmonics are proportional to the components, $\{\mathcal{Q}_m, m=0,\pm1,\pm2\}$, of the unit quadrupole moments, defined such that $\mathcal{Q}\cdot\mathcal{Q}=1$. Their normalisation is determined simply from the observation that

$$\int_{S_4}\mathcal{Q}\cdot\mathcal{Q}\,\sin 3\gamma\,d\gamma\,d\Omega = \int_{S_4}\sin 3\gamma\,d\gamma\,d\Omega = \frac{16\pi^2}{3}. \qquad (4.145)$$

Thus, with the multiplicity index assigned the value $\alpha=1$ for the multiplicity-free $v=1$ irrep, we obtain the $v=1$, SO(5) spherical harmonics

$$\mathcal{Y}_{112m}(\gamma,\Omega) = \frac{\sqrt{15}}{4\pi}\,\mathcal{Q}_m(\gamma,\Omega). \qquad (4.146)$$

As a unit vector in the five-dimensional vector space $\mathbb{R}^5$, $\mathcal{Q}$ transforms to its negative under inversion through the $\mathbb{R}^5$ origin. It follows that a polynomial of even degree in the components of $\mathcal{Q}$ has positive O(5) parity and a polynomial of odd degree has negative parity.[45] Thus, the SO(5) spherical harmonics of a given seniority v carry irreps of both SO(5) and O(5) with O(5) parity $(-1)^v$. These

[42]Rowe D.J., Turner P.S. and Repka J. (2004), *J. Math. Phys.* **45**, 2761.
[43]Caprio M.A., Rowe D.J. and Welsh T.A. (2009), *Comp. Phys. Comm.* **180**, 1150–1163.
[44]Rowe *et al.*, *op. cit.* Footnote 10 on Page 171.
[45]Note that the degree of an arbitrary polynomial in the components $\mathcal{Q}_m$ is not well-defined. This is because the identity, $\mathcal{Q}\cdot\mathcal{Q}=\sum_m|\mathcal{Q}_m|^2=1$, means that an arbitrary polynomial is unchanged when multiplied by $\mathcal{Q}\cdot\mathcal{Q}$. However, because a polynomial that is even (odd) remains even (odd), when multiplied by $\mathcal{Q}\cdot\mathcal{Q}$, it has a well-defined O(5) parity.

properties parallel those of the SO(3) spherical harmonics with angular momentum l and O(3) parity $(-1)^l$.

Following the algorithm of Rowe, Turner, and Repka,[46] outlined in Section 4.9, it is straightforward to construct an ordered basis for the Hilbert space $\mathcal{L}^2(S_4)$ by taking multiple coupled-tensor products of the $v = 1$, $\mathcal{Q}$ tensors. This basis can then be orthogonalised by a Gram-Schmidt transformation and normalised to give SO(5) spherical harmonics. Computer codes have been developed for this purpose and used to compute SO(5) Clebsch-Gordan coefficients by the method outlined in the following section.

The SO(5) spherical harmonic $\mathcal{Y}_{112m}$ is of fundamental importance in the ACM because it appears in the expansion of the quadrupole operator

$$q_m = \frac{4\pi}{\sqrt{15}}\beta\mathcal{Y}_{112m} = \beta\mathcal{Q}_m. \tag{4.147}$$

Also important are the expressions of polynomials in $\cos 3\gamma$ in terms of SO(5) spherical harmonics. This is because any SO(3) scalar polynomial in the collective-model quadrupole moments is expressible as a polynomial in the two basic SO(3) scalars: $q \cdot q = \beta^2$ and $[q \otimes q \otimes q]_0 \propto \beta^3 \cos 3\gamma$. Thus, the only SO(5) spherical harmonics that appear in the expression of an SO(3)-invariant collective model Hamiltonian are polynomials in $\cos 3\gamma$. It follows from the identity

$$\int_0^{\pi/3} P_n(\cos 3\gamma) P_m(\cos 3\gamma) \sin 3\gamma \, d\gamma = \frac{1}{3}\int_{-1}^{+1} P_n(z) P_m(z) \, dz = \frac{2}{3(2n+1)} \tag{4.148}$$

for Legendre polynomials, $P_n(z)$, that

$$\mathcal{Y}_{3n,100}(\gamma,\Omega) = \frac{1}{4\pi}\sqrt{3(2n+1)}\, P_n(\cos 3\gamma). \tag{4.149}$$

In particular, $\mathcal{Y}_{3100}(\gamma,\Omega) = \frac{3}{4\pi}\cos 3\gamma$. Apart from their essential role in the computation of SO(5) Clebsch-Gordan coefficients, there is, in fact, little need for explicit expressions for the SO(5) spherical harmonics, in applications of the ACM, because SO(5)-reduced matrix elements are most easily derived by algebraic methods.

4.5.3 *SO(5) Clebsch-Gordan coefficients and their symmetries*

The SO(5) Clebsch-Gordan coefficients needed for applications of the collective model, and their symmetry properties in an SO(3)-coupled basis, are derived easily from the integral expression,

$$\int_{S_4} \mathcal{Y}^*_{v_3\alpha_3 L_3 M_3}(\gamma,\Omega)\, [\mathcal{Y}_{v_2\alpha_2 L_2}(\gamma,\Omega) \otimes \mathcal{Y}_{v_1\alpha_1 L_1}(\gamma,\Omega)]_{L_3 M_3}\sin 3\gamma\, d\gamma\, d\Omega$$

$$= (v_1\alpha_1 L_1, v_2\alpha_2 L_2 \| v_3\alpha_3 L_3)\langle v_3\||\hat{\mathcal{Y}}_{v_2}\||v_1\rangle, \tag{4.150}$$

---

[46] Rowe *et al.*, *op. cit.* Footnote 42 on Page 194.

where $\hat{\mathcal{Y}}_v$ is a tensor operator whose components act multiplicatively on wave functions in $\mathcal{L}^2(S_4)$ according to the definition

$$\hat{\mathcal{Y}}_{v\alpha LM}\Psi(\gamma,\Omega) := \mathcal{Y}_{v\alpha LM}(\gamma,\Omega)\Psi(\gamma,\Omega). \tag{4.151}$$

First observe that the absolute values of the Clebsch-Gordan coefficients are fixed, in the standard way, by requiring that

$$\sum_{\alpha_1 L_1 \alpha_2 L_2} (v_1\alpha_1 L_1, v_2\alpha_2 L_2 \| v_3\alpha_3 L_3)^2 = 1, \tag{4.152}$$

$$\sum_{v_3\alpha_3 L_3} (v_1\alpha_1 L_1, v_2\alpha_2 L_2 \| v_3\alpha_3 L_3)^2 = 1. \tag{4.153}$$

Their phases are also defined by choosing the reduced matrix elements, $\langle v_3 \|\| \hat{\mathcal{Y}}_{v_2} \|\| v_1 \rangle$, to be real and positive. Thus, the required Clebsch-Gordan coefficients are computed by evaluating the integrals of Equation (4.150) and multiplying by normalisation factors such that Equations (4.152) and (4.153) are satisfied. The normalisation factors then define the $\langle v_3 \|\| \hat{\mathcal{Y}}_{v_2} \|\| v_1 \rangle$ reduced matrix elements of interest. Such $SO(5) \supset SO(3)$ Clebsch-Gordan coefficients computed in this way are readily available.[47] A few coefficients are given in Section 4.9.

From the symmetry relationship

$$(L_1 M_1\, L_2 M_2 | L_3 M_3) = (-1)^{L_1 + L_2 - L_3}(L_2 M_2\, L_1 M_1 | L_3 M_3) \tag{4.154}$$

for $SO(3)$ Clebsch-Gordan coefficients, it follows from Equation (4.150) that $SO(5)$ Clebsch-Gordan coefficients have a similar symmetry, i.e.,

$$(v_2\alpha_2 L_2, v_1\alpha_1 L_1 \| v_3\alpha_3 L_3) = (-1)^{L_1 + L_2 - L_3}(v_1\alpha_1 L_1, v_2\alpha_2 L_2 \| v_3\alpha_3 L_3). \tag{4.155}$$

A second symmetry relationship is obtained by using the identity

$$\mathcal{Y}^*_{v\alpha LM} = (-1)^{L+M}\mathcal{Y}_{v\alpha L,-M}, \tag{4.156}$$

(derived from the corresponding symmetry of the rotor-model wave functions given by Claim 2.1 on Page 122) to show that Equation (4.150) can be re-expressed as

$$\int_{S_4} \left[\mathcal{Y}_{v_3\alpha_3 L_3}(\gamma,\Omega) \otimes \mathcal{Y}_{v_2\alpha_2 L_2}(\gamma,\Omega) \otimes \mathcal{Y}_{v_1\alpha_1 L_1}(\gamma,\Omega)\right]_0 \sin 3\gamma\, d\gamma\, d\Omega$$

$$= \sqrt{2L_3 + 1}\, (v_1\alpha_1 L_1, v_2\alpha_2 L_2 \| v_3\alpha_3 L_3) \langle v_3 \|\| \hat{\mathcal{Y}}_{v_2} \|\| v_1 \rangle, \tag{4.157}$$

---

[47]Some $SO(5) \supset SO(3)$ Clebsch-Gordan coefficients were given by Rowe *et al.*, *op. cit.* Footnote 42 on Page 194. Machine readable tables of coefficients can be obtained from Caprio *et al.*, *op. cit.* Footnote 43 on Page 194, and also from Welsh T.A. (2008), (accessible from http://www.physics.utoronto.ca/~rowe/group.html).

where the integrand on the left-hand side is simply an angular-momentum $L = 0$, SO(3)-coupled, product. The symmetry relationship

$$\left[\mathcal{Y}_{v_3\alpha_3 L_3} \otimes \mathcal{Y}_{v_2\alpha_2 L_2} \otimes \mathcal{Y}_{v_1\alpha_1 L_1}\right]_0$$
$$= (-1)^{L_1+L_2-L_3}\left[\mathcal{Y}_{v_1\alpha_1 L_1} \otimes \mathcal{Y}_{v_2\alpha_2 L_2} \otimes \mathcal{Y}_{v_3\alpha_3 L_3}\right]_0 \qquad (4.158)$$

then shows that

$$\sqrt{2L_1 + 1}\,(v_3\alpha_3 L_3, v_2\alpha_2 L_2 \| v_1\alpha_1 L_1)\,\langle v_1 \|\| \hat{\mathcal{Y}}_{v_2} \|\| v_3\rangle$$
$$= (-1)^{L_1+L_2-L_3}\sqrt{2L_3 + 1}\,(v_1\alpha_1 L_1, v_2\alpha_2 L_2 \| v_3\alpha_3 L_3)\,\langle v_3 \|\| \hat{\mathcal{Y}}_{v_2} \|\| v_1\rangle. \qquad (4.159)$$

From the sum rule

$$\sum_{\alpha_1 L_1 M_1, \alpha_2 L_2 M_2, \alpha_3 L_3 M_3} (v_1\alpha_1 L_1 M_1, v_2\alpha_2 L_2 M_2 | v_3\alpha_3 L_3 M_3)^2 = \dim(v_3), \qquad (4.160)$$

where

$$\dim(v) = \tfrac{1}{6}(v + 1)(v + 2)(2v + 3) \qquad (4.161)$$

is the dimension of the SO(5) irrep[48] of seniority v, and from Equation (4.152), we also obtain the identity

$$\sum_{\alpha_1 L_1, \alpha_2 L_2, \alpha_3 L_3} (2L_3 + 1)(v_1\alpha_1 L_1, v_2\alpha_2 L_2 \| v_3\alpha_3 L_3)^2 = \dim(v_3). \qquad (4.162)$$

Thus, by squaring and summing both sides of Equation (4.159), we obtain

$$\frac{\langle v_1 \|\| \hat{\mathcal{Y}}_{v_2} \|\| v_3\rangle}{\langle v_3 \|\| \hat{\mathcal{Y}}_{v_2} \|\| v_1\rangle} = \sqrt{\frac{\dim(v_3)}{\dim(v_1)}} \qquad (4.163)$$

and the symmetry relation for SO(5) $\supset$ SO(3) Clebsch-Gordan coefficients

$$(v_3\alpha_3 L_3, v_2\alpha_2 L_2 \| v_1\alpha_1 L_1) = (-1)^{L_1+L_2-L_3}\sqrt{\frac{\dim(v_1)\,(2L_3 + 1)}{\dim(v_3)\,(2L_1 + 1)}}$$
$$\times (v_1\alpha_1 L_1, v_2\alpha_2 L_2 \| v_3\alpha_3 L_3). \qquad (4.164)$$

4.5.4 SO(5)-reduced matrix elements

The most important SO(5)-reduced matrix elements are those of the quadrupole tensor, $\mathcal{Q}$, required for the evaluation of E2 matrix elements, and those of the SO(3) scalar, $\cos 3\gamma = (4\pi/3)\mathcal{Y}_{3100}(\gamma, \Omega)$, in terms of which collective model potential energies are expressed.

---

[48]The dimension of an SO(n) irrep is given in Section 5.12.12.

Claim 4.4 *SO(5)-reduced matrix elements of the SO(5) components $\{\hat{\mathcal{Q}}_m\}$ of the collective quadrupole moment operators $\{\hat{q}_m = \hat{\beta}\hat{\mathcal{Q}}_m\}$ are given by*

$$\langle v'|||\hat{\mathcal{Q}}|||v\rangle = \sqrt{\frac{v+1}{2v+5}}\,\delta_{v',v+1} + \sqrt{\frac{v+2}{2v+1}}\,\delta_{v',v-1}. \tag{4.165}$$

Proof: First observe that, because $\hat{\mathcal{Q}}$ is an SO(5), $v = 1$ tensor of negative O(5) parity, its product with an SO(5) spherical harmonic of seniority v can only produce a linear combination of SO(5) spherical harmonics of seniority $v \pm 1$. Thus, to prove the claim, it is required only to determine the values of the $\langle v \pm 1|||\hat{\mathcal{Q}}|||v\rangle$ matrix elements. This is done by considering a subset of matrix elements of the $d_2^\dagger$ raising operator of the harmonic spherical vibrator model. Let $|0v, 2v\rangle$ denote, in abbreviated notation, a harmonic spherical-vibrator state

$$|0v, 2v\rangle \equiv |n = 0, v, L = 2v, M = 2v\rangle = \frac{1}{\sqrt{v!}}(d_2^\dagger)^v|0\rangle, \tag{4.166}$$

where $|0\rangle$ is the zero-phonon vacuum state. For such states, one immediately determines the matrix elements

$$\langle 0v', 2v'|d_2^\dagger|0v, 2v\rangle = \delta_{v',v+1}\sqrt{v+1}. \tag{4.167}$$

Because the stretched SO(5) Clebsch-Gordan coupling coefficient for this matrix element has value $(v, 2v, 1, 2, \|v+1, 2v+2) = 1$, the reduced matrix element is

$$\langle 0v'|||d^\dagger|||0v\rangle = \delta_{v',v+1}\sqrt{v+1}. \tag{4.168}$$

Now recall that

$$\hat{q}_m = \frac{1}{\sqrt{2}\,a}(d_m^\dagger + d_m). \tag{4.169}$$

It follows that $\hat{q}$ has a subset of SO(5)-reduced matrix elements given by

$$\langle 0, v+1|||a\hat{q}|||0v\rangle = \sqrt{\frac{v+1}{2}}. \tag{4.170}$$

Replacing $\hat{q}_m$ with $\hat{\beta}\hat{\mathcal{Q}}_m$, we also note that

$$\langle 0, v+1|||a\hat{q}|||0v\rangle = F_{\lambda+1,0;\lambda,0}(a\beta)\,\langle v+1|||\hat{\mathcal{Q}}|||v\rangle, \tag{4.171}$$

where, for the harmonic spherical vibrator, $\lambda = v + 5/2$. Thus, with $F_{\lambda+1,0;\lambda,0}(a\beta)$ given by Equation (4.103), we obtain

$$\langle v+1|||\hat{\mathcal{Q}}|||v\rangle = \sqrt{\frac{v+1}{2v+5}}. \tag{4.172}$$

From Equation (4.163), we also obtain

$$\frac{\langle v|||\hat{\mathcal{Q}}|||v+1\rangle}{\langle v+1|||\hat{\mathcal{Q}}|||v\rangle} = \sqrt{\frac{\dim(v+1)}{\dim(v)}} = \sqrt{\frac{(v+3)(2v+5)}{(v+1)(2v+3)}}. \tag{4.173}$$

It follows that

$$\langle v|||\hat{\mathcal{Q}}|||v+1\rangle = \sqrt{\frac{v+3}{2v+3}}, \tag{4.174}$$

and the Claim is proved. □

Empirical formula: *The non-zero SO(5)-reduced matrix elements of a general SO(5) spherical harmonic are given by*

$$\langle v_3||||\hat{\mathcal{Y}}_{v_2}|||v_1\rangle = \frac{1}{4\pi} \frac{(\frac{\sigma}{2}+1)!}{(\frac{\sigma}{2}-v_1)!(\frac{\sigma}{2}-v_2)!(\frac{\sigma}{2}-v_3)!} \sqrt{\frac{(2v_1+3)(2v_2+3)}{(v_3+2)(v_3+1)}}$$

$$\times \sqrt{\frac{(\sigma+4)(\sigma-2v_1+1)!(\sigma-2v_2+1)!(\sigma-2v_3+1)!}{(\sigma+3)!}}, \tag{4.175}$$

where $\sigma = v_1 + v_2 + v_3$.

Equation (4.175) is invaluable for turning Equation (4.150) into an explicit expression for the SO(5) Clebsch-Gordan coefficients as an integral over products of SO(5) spherical harmonics.[49]

From Equation (4.175) we obtain, for example, the useful matrix elements

$$\langle v+1|||\hat{\mathcal{Y}}_3|||v\rangle = \frac{9}{4\pi} \sqrt{\frac{5v(v+1)(v+4)}{2(2v+1)(2v+5)(2v+7)}}, \tag{4.176}$$

$$\langle v+3|||\hat{\mathcal{Y}}_3|||v\rangle = \frac{3}{4\pi} \sqrt{\frac{35(v+1)(v+2)(v+3)}{2(2v+5)(2v+7)(2v+9)}}. \tag{4.177}$$

Exercises

4.13 By evaluating the integral $\int_{S_4} [\cos 3\gamma]^2 \sin 3\gamma \, d\gamma \, d\Omega$, show that $\mathcal{Y}_{3100}(\gamma, \Omega) = \frac{3}{4\pi} \cos 3\gamma$.

---

[49]The formula (4.175) was inferred, by T.A. Welsh (private communication and reported in Caprio *et al.*, *op. cit.* Footnote 43 on Page 194) from the pattern of numerical values of the integrals in Equation (4.150). To prove its validity in any particular case, one has only to ascertain that the Clebsch-Gordan coefficients derived with its use satisfy the sum rules of Equations (4.152) and (4.153). In this way, it has been verified by extensive numerical calculations to hold in all cases in which we have determined SO(5) Clebsch-Gordan coefficients. However, it remains to be proved in complete generality.

4.14 Use the identity $\mathcal{Y}^*_{v\alpha LM} = (-1)^{L+M}\mathcal{Y}_{v\alpha L,-M}$ to show that

$$\int_{S_4} \left[\mathcal{Y}_{v_3\alpha_3 L_3}(\gamma,\Omega) \otimes \mathcal{Y}_{v_2\alpha_2 L_2}(\gamma,\Omega) \otimes \mathcal{Y}_{v_1\alpha_1 L_1}(\gamma,\Omega)\right]_0 \sin 3\gamma \, d\gamma \, d\Omega$$

$$= \sqrt{2L_3+1}\,(v_1\alpha_1 L_1, v_2\alpha_2 L_2\|v_3\alpha_3 L_3)\,\langle v_3\|\|\hat{\mathcal{Y}}_{v_2}\|\|v_1\rangle. \quad (4.178)$$

4.6 Combined matrix elements

Other operators of interest in the collective model are expressed in terms of products of operators having one factor that acts on the SU(1,1) beta wave functions and another factor that acts on the SO(5) wave functions. In this section, we give analytical expressions for the SO(5)-reduced matrix elements of the basic collective model observables, $\{\hat{q}_m, \hat{\pi}_m\}$, of the d-phonon operators, and of ∇^2.

Claim 4.5 *SO(5)-reduced matrix elements of the basic observables, $\{\hat{q}_m, \hat{\pi}_m\}$ and ∇^2, of the collective model are given, for any of the modified oscillator series of beta wave functions, by*

$$\langle \lambda'\mu; v'\|\|\hat{q}\|\|\lambda\mu; v\rangle = F_{\lambda'\mu;\lambda\nu}(\beta)\left[\delta_{v',v+1}\sqrt{\frac{v+1}{2v+5}} + \delta_{v',v-1}\sqrt{\frac{v+2}{2v+1}}\right], \quad (4.179)$$

$$\langle \lambda'\mu; v+1\|\|\hat{\pi}\|\|\lambda\mu; v\rangle = -i\hbar F_{\lambda'\mu;\lambda\nu}\left(\frac{d}{d\beta} - \frac{v+2}{\beta}\right)\sqrt{\frac{v+1}{2v+5}}, \quad (4.180)$$

$$\langle \lambda'\mu; v-1\|\|\hat{\pi}\|\|\lambda\mu; v\rangle = -i\hbar F_{\lambda'\mu;\lambda\nu}\left(\frac{d}{d\beta} + \frac{v+1}{\beta}\right)\sqrt{\frac{v+2}{2v+1}}, \quad (4.181)$$

$$\langle \lambda\mu; v\|\|\frac{\nabla^2}{a^2}\|\|\lambda\nu; v\rangle = F_{\lambda\mu;\lambda\nu}\left(\frac{d^2}{d\beta^2} - \frac{v(v+3)+2}{\beta^2}\right)$$

$$= -\delta_{\mu,\nu}(\lambda+2\nu) + \delta_{\mu,\nu+1}\sqrt{(\lambda+\nu)(\nu+1)} + \delta_{\mu,\nu-1}\sqrt{(\lambda+\nu-1)\nu}$$

$$+ \left[(\lambda-\tfrac{1}{2})(\lambda-\tfrac{3}{2}) - (v+1)(v+2)\right]F_{\lambda\mu;\lambda\nu}\left(1/(a\beta)^2\right), \quad (4.182)$$

where the $F_{\lambda'\mu;\lambda\nu}$ matrix elements are given, for $\lambda' = \lambda \pm 1$, by Claim 4.1.

This claim shows that matrix elements of any collective model that is a polynomial in the basic $\{\hat{q}_m, \hat{\pi}_m\}$ observables can be calculated analytically, albeit with summations over intermediate states.

Equation (4.179) follows immediately from Equations (4.40) and (4.165) which imply that the SO(5)-reduced matrix elements of $\hat{q} = \hat{\beta}\hat{\mathcal{Q}}$ are given by

$$\langle \lambda'\nu', v'\|\|\hat{q}\|\|\lambda\nu, v\rangle = \langle \lambda'\nu'|\hat{\beta}|\lambda\nu\rangle\left[\delta_{v',v+1}\sqrt{\frac{v+1}{2v+5}} + \delta_{v',v-1}\sqrt{\frac{v+2}{2v+1}}\right]. \quad (4.183)$$

Matrix elements of the Laplacian operator are obtained from the identities

$$\nabla^2 = \hat{\Delta} - \frac{1}{\beta^2}\hat{\Lambda}^2, \tag{4.184}$$

$$\hat{\Delta} = \frac{1}{\beta^4}\frac{\partial}{\partial\beta}\beta^4\frac{\partial}{\partial\beta}, \qquad \beta^2\hat{\Delta}\frac{1}{\beta^2} = \frac{d^2}{d\beta^2} - \frac{2}{\beta^2}, \tag{4.185}$$

which lead to the first line of Equation (4.182). Then, from the expression (4.99) for $F_{\lambda\mu;\lambda\nu}\left(\frac{1}{a^2}\frac{d^2}{d\beta^2}\right)$, we obtain the second line. However, Equations (4.180) and (4.181) are much less obvious. We start by showing that they are true in the harmonic vibrational basis and then generalise to arbitrary modified oscillator series bases.

4.6.1 *Matrix elements in a harmonic spherical vibrator basis*

In the harmonic spherical vibrator basis, λ is related to v by $\lambda = v + 5/2$. Beta matrix elements are then given by Equations (4.100) and (4.103) and we immediately obtain the non-zero matrix elements

$$\langle\nu; v+1|||a\hat{q}|||\nu; v\rangle = \sqrt{\frac{(\nu+v+5/2)(v+1)}{2v+5}}, \tag{4.186}$$

$$\langle\nu+1; v-1|||a\hat{q}|||\nu; v\rangle = \sqrt{\frac{(\nu+1)(v+2)}{2v+1}}, \tag{4.187}$$

$$\langle\nu; v-1|||a\hat{q}|||\nu; v\rangle = \sqrt{\frac{(\nu+v+3/2)(v+2)}{2v+1}}, \tag{4.188}$$

$$\langle\nu-1; v+1|||a\hat{q}|||\nu; v\rangle = \sqrt{\frac{\nu(v+1)}{2v+5}}. \tag{4.189}$$

By replacing $a\hat{q}_m$ with its expression $\frac{1}{\sqrt{2}}(d_m^\dagger + d_m)$ in terms of the d-boson operators of the harmonic spherical vibrator model, given by Equation (4.3), and noting that only one of the $d_m^\dagger$ and d_m operators contributes to a given matrix element (cf. Figure 4.4), it is determined that the non-zero SO(5)-reduced matrix elements of

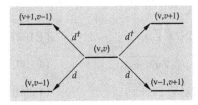

Figure 4.4: States of the harmonic spherical vibrator model that can be reached by d-phonon raising and lowering operators from a state of a harmonic spherical vibrator irrep (ν, v), as labelled in Figure 4.1. Thus, if we denote the set of states in the irrep (ν, v) by $|\nu, v\rangle := \{|\nu, v\alpha LM\rangle\}$, the figure illustrates the fact that $d^\dagger \otimes |\nu, v\rangle = |\nu+1, v-1\rangle \oplus |\nu, v+1\rangle$ and $d \otimes |\nu, v\rangle = |\nu, v-1\rangle \oplus |\nu-1, v+1\rangle$.

the d-phonon operators are given, in the harmonic spherical vibrator basis, by

$$\langle \mu; v + 1 |||d^\dagger|||\nu; v \rangle = \delta_{\mu,\nu}\sqrt{\frac{(2\nu + 2v + 5)(v + 1)}{2v + 5}}, \tag{4.190}$$

$$\langle \mu; v - 1 |||d^\dagger|||\nu; v \rangle = \delta_{\mu,\nu+1}\sqrt{\frac{2(\nu + 1)(v + 2)}{2v + 1}}, \tag{4.191}$$

$$\langle \mu; v - 1 |||d|||\nu; v \rangle = \delta_{\mu,\nu}\sqrt{\frac{(2\nu + 2v + 3)(v + 2)}{2v + 1}}, \tag{4.192}$$

$$\langle \mu; v + 1 |||d|||\nu; v \rangle = \delta_{\mu,\nu-1}\sqrt{\frac{2\nu(v + 1)}{2v + 5}}. \tag{4.193}$$

From these matrix elements one can also derive other matrix elements. In particular, for the momentum operators, $\hat{\pi}_m = \frac{ia\hbar}{\sqrt{2}}(d_m^\dagger - d_m)$, one derives

$$\langle \nu; v + 1 |||\hat{\pi}|||\nu; v \rangle = ia\hbar\sqrt{\frac{(\nu + v + {}^5\!/_2)(v + 1)}{2v + 5}}, \tag{4.194}$$

$$\langle \nu + 1; v - 1 |||\hat{\pi}|||\nu; v \rangle = ia\hbar\sqrt{\frac{(\nu + 1)(v + 2)}{2v + 1}}, \tag{4.195}$$

$$\langle \nu; v - 1 |||\hat{\pi}|||\nu; v \rangle = -ia\hbar\sqrt{\frac{(\nu + v + {}^3\!/_2)(v + 2)}{2v + 1}}, \tag{4.196}$$

$$\langle \nu - 1; v + 1 |||\hat{\pi}|||\nu; v \rangle = -ia\hbar\sqrt{\frac{\nu(v + 1)}{2v + 5}}. \tag{4.197}$$

4.6.2 *Matrix elements in general modified oscillator series irreps*

The above matrix elements in a harmonic spherical vibrator basis are special cases of more general expressions which are also products of beta matrix elements and reduced matrix elements of $\hat{Q}$. To see this, first use the definition (4.92) for $F_{\lambda\mu;\lambda\nu}$ to re-express Equations (4.128)–(4.131)

$$F_{\lambda+1,\mu;\lambda\nu}\big(A^\dagger(\lambda - {}^1\!/_2)\big) = \delta_{\mu,\nu}2\sqrt{\lambda + \nu}, \tag{4.198}$$

$$F_{\lambda-1,\mu;\lambda\nu}\big(A^\dagger(-\lambda + {}^3\!/_2)\big) = \delta_{\mu,\nu+1}2\sqrt{\nu + 1}, \tag{4.199}$$

$$F_{\lambda-1,\mu;\lambda\nu}\big(A(\lambda - {}^3\!/_2)\big) = \delta_{\mu,\nu}2\sqrt{\lambda - 1 + \nu}, \tag{4.200}$$

$$F_{\lambda+1,\mu;\lambda\nu}\big(A(-\lambda + {}^1\!/_2)\big) = \delta_{\mu,\nu-1}2\sqrt{\nu}. \tag{4.201}$$

Then, with $\lambda_v = v + {}^5\!/_2$, the d-phonon matrix elements in a harmonic spherical vibrator basis, given by Equations (4.190)–(4.193), can be re-expressed in the form

$$\langle \mu; v' |||d^\dagger |||v; v\rangle = \delta_{v',v+1}\frac{1}{\sqrt{2}}F_{\lambda_{v'}\mu;\lambda_v\nu}\big(A^\dagger(v+2)\big)\,\langle v+1|||\hat{\mathcal{Q}}|||v\rangle$$

$$+\delta_{v',v-1}\frac{1}{\sqrt{2}}F_{\lambda_{v'}\mu;\lambda_v\nu}\big(A^\dagger(-v-1)\big)\,\langle v-1|||\hat{\mathcal{Q}}|||v\rangle, \quad (4.202)$$

$$\langle \mu; v' |||d|||v; v\rangle = \delta_{v',v-1}\frac{1}{\sqrt{2}}F_{\lambda_{v'}\mu;\lambda_v\nu}\big(A(v+1)\big)\,\langle v-1|||\hat{\mathcal{Q}}|||v\rangle$$

$$+\delta_{v',v+1}\frac{1}{\sqrt{2}}F_{\lambda_{v'}\mu;\lambda_v\nu}\big(A(-v-2)\big)\,\langle v+1|||\hat{\mathcal{Q}}|||v\rangle. \quad (4.203)$$

These equations can now be generalised.

Claim 4.6 *SO(5)-reduced matrix elements of the d-phonon operators, are given, for arbitrary modified-oscillator beta wave functions, by*

$$\langle \lambda'\mu; v' |||d^\dagger |||\lambda\nu; v\rangle = \delta_{v',v+1}\frac{1}{\sqrt{2}}F_{\lambda'\mu;\lambda\nu}\big(A^\dagger(v+2)\big)\,\langle v+1|||\hat{\mathcal{Q}}|||v\rangle$$

$$+\delta_{v',v-1}\frac{1}{\sqrt{2}}F_{\lambda'\mu;\lambda\nu}\big(A^\dagger(-v-1)\big)\,\langle v-1|||\hat{\mathcal{Q}}|||v\rangle, \quad (4.204)$$

$$\langle \lambda'\mu; v' |||d|||\lambda\nu; v\rangle = \delta_{v',v-1}\frac{1}{\sqrt{2}}F_{\lambda'\mu;\lambda\nu}\big(A(v+1)\big)\,\langle v-1|||\hat{\mathcal{Q}}|||v\rangle$$

$$+\delta_{v',v+1}\frac{1}{\sqrt{2}}F_{\lambda'\mu;\lambda\nu}\big(A(-v-2)\big)\,\langle v+1|||\hat{\mathcal{Q}}|||v\rangle, \quad (4.205)$$

with $A(X)$ and $A^\dagger(X)$ as defined in Section 4.4.3.

Proof: Observe that Equations (4.202) and (4.203) hold for all values of μ, ν. It follows that parallel equations hold for any beta basis functions. In particular, they hold for the basis wave functions of the modified oscillator series SU(1,1) irreps. $\square$

Combining these results, by means of the relationships $\hat{q}_m = \frac{1}{\sqrt{2}\,a}(d_m^\dagger + d_m)$ and $\hat{\pi}_m = \frac{\mathrm{i}a\hbar}{\sqrt{2}}(d_m^\dagger - d_m)$, and the explicit expressions for $A^\dagger(X)$ and $A(X)$ given by Equation (4.117), completes the proof of Claim 4.5. $\square$

4.6.3 *Reduced E2 transition rates between states of good SO(5) seniority*

Reduced transition rates are defined, for the E2 operator $\hat{\mathfrak{M}}(\mathrm{E2}; m) = (Ze/A)\hat{q}_m$, by summing its squared transition matrix elements over final states and averaging over initial states. Such reduced E2 transition rates are customarily considered for transitions between energy-degenerate multiplets of states that carry SO(3), or SU(2), irreps. However, they are also useful for transitions between multiplets of states that carry irreps of larger symmetry groups. Thus, SO(5)-reduced E2 transition rates are defined for transitions between SO(5) multiplets of states by

$$\bar{B}(E2; \lambda_i \mu v_i \to \lambda_f \nu v_f) = \sum_{\substack{\alpha_i L_i M_i \\ m \alpha_f L_f M_f}} \frac{|\langle \lambda_i \mu; v_i \alpha_i L_i M_i | \hat{\mathfrak{M}}(E2; m) | \lambda_f \nu; v_f \alpha_f L_f M_f \rangle|^2}{\dim(v_i)}$$

$$= \left(\frac{Ze}{A} \right)^2 |\langle \lambda_i \mu; v_i || |\hat{q}| || \lambda_f \nu; v_f \rangle|^2. \qquad (4.206)$$

Note that in obtaining this expression we have made use of the identity

$$\sum_{M_i m M_f} |\langle \lambda_i \mu; v_i \alpha_i L_i M_i | \hat{\mathfrak{M}}(E2; m) | \lambda_f \nu; v_f \alpha_f L_f M_f \rangle|^2$$

$$= \sum_{M_i m M_f} |\langle \lambda_f \nu; v_f \alpha_f L_f M_f | \hat{\mathfrak{M}}(E2; m) | \lambda_i \mu; v_i \alpha_i L_i M_i \rangle|^2. \qquad (4.207)$$

Standard, and more detailed, SO(3)-reduced transition rates are readily obtained from these SO(5)-reduced rates, by use of Equation (4.144), which gives the equation

$$B(E2; \lambda_i \mu v_i \alpha_i L_i \to \lambda_f \nu v_f \alpha_f L_f)$$
$$= (2L_i + 1)^{-1} |\langle \lambda_i \mu v_i \alpha_i L_i || \hat{\mathfrak{M}}(E2) || \lambda_f \nu v_f \alpha_f L_f \rangle|^2$$
$$= (v_f \alpha_f L_f, 112 || v_i \alpha_i L_i)^2 \bar{B}(E2; \lambda_i \mu v_i \to \lambda_f \nu v_f). \qquad (4.208)$$

A few such SO(3)-reduced E2 transition rates are shown in Figure 2.2.

For states labelled by SU(1,1) × SO(5) quantum numbers, with matrix elements given by Claim 4.5, the SO(5)-reduced E2 transition rates are given by

$$\bar{B}(E2; \lambda_i \mu v_i \to \lambda_f \nu v_f) = \left(\frac{Ze}{A} \right)^2 \left| F_{\lambda_i \mu; \lambda_f \nu}(\beta) \right|^2,$$

$$\times \left[\delta_{v_f, v_i - 1} \frac{v_i}{2v_i + 3} + \delta_{v_f, v_i + 1} \frac{v_i + 3}{2v_i + 3} \right]. \qquad (4.209)$$

Thus, in the harmonic spherical-vibrator limit, for which $\lambda_i = v_i + 5/2$, $\lambda_f = v_f + 5/2$,

$$\bar{B}(E2; \mu v \to \nu, v - 1)_{\text{HV}} = \left(\frac{Ze}{Aa} \right)^2 \frac{v}{2v + 3} \left[(v + \tfrac{3}{2} + \mu) \delta_{\nu, \mu} + (\mu + 1) \delta_{\nu, \mu + 1} \right], \qquad (4.210)$$

$$\bar{B}(E2; \mu v \to \nu, v + 1)_{\text{HV}} = \left(\frac{Ze}{Aa} \right)^2 \frac{v + 3}{2v + 3} \left[(v + \tfrac{5}{2} + \mu) \delta_{\mu, \nu} + \mu \delta_{\nu, \mu - 1} \right]. \qquad (4.211)$$

These equations give the SO(5)-reduced E2 transition rates shown in Figure 4.1.

In the rigid-beta Wilets-Jean model, for which $\langle \lambda \pm 1, \nu | \beta | \lambda \mu \rangle = \beta_0 \delta_{\mu, \nu}$ (cf.

Equation (4.108)), we obtain

$$\bar{B}(\text{E2}; \mu v \to \nu, v-1)_{\text{WJ}} = \delta_{\mu,\nu} \left(\frac{Ze}{A}\right)^2 \frac{v}{2v+3}\beta_0^2, \qquad (4.212)$$

$$\bar{B}(\text{E2}; \mu v \to \nu, v+1)_{\text{WJ}} = \delta_{\mu,\nu} \left(\frac{Ze}{A}\right)^2 \frac{v+3}{2v+3}\beta_0^2. \qquad (4.213)$$

Exercise

4.15 Use the symmetry relations of Section 4.5.3 to show that, whereas SO(3)-reduced E2 transition rates satisfy the equation

$$B(\text{E2}; v_i \alpha_i L_i \to v_f \alpha_f L_f) = \frac{2L_f+1}{2L_i+1} B(\text{E2}; \lambda_f \nu v_f \to \lambda_i \mu v_i), \qquad (4.214)$$

SO(5)-reduced E2 transition rates satisfy

$$\bar{B}(\text{E2}; \lambda_i \mu v_i \to \lambda_f \nu v_f) = \frac{\dim(v_f)}{\dim(v_i)} \bar{B}(\text{E2}; \lambda_f \nu v_f \to \lambda_i \mu v_i). \qquad (4.215)$$

4.7 Collective model calculations

With the availability of SO(5) Clebsch-Gordan coefficients and the analytical expressions, given above for beta matrix elements and SO(5) reduced matrix elements, ACM calculations are quick and easy for any Hamiltonian that is a polynomial of modest degree in the position and momentum observables, $\{q_m, \pi_m\}$, of the model. In addition, Hamiltonians that include terms in $1/\beta^2$, where $\beta^2 = q \cdot q$, can also be accommodated. For physical reasons, one would normally restrict consideration to Hamiltonians that are time-reversal and rotationally (i.e., SO(3)) invariant. For example, one might consider a Hamiltonian of the form

$$\hat{H} = -\frac{\hbar^2}{2B}\nabla^2 + V(\beta, \cos 3\gamma), \qquad (4.216)$$

where $V(\beta, \cos 3\gamma)$ is a polynomial in β^2, $1/\beta^2$ and $\beta \cos 3\gamma \propto \frac{1}{\beta^2}[\hat{q} \otimes \hat{q} \otimes \hat{q}]_0$.

Note that separate terms, such as β and $\cos 3\gamma$, are irrational functions of the $\{q_m\}$ coordinates, e.g., $\beta = \sqrt{q \cdot q}$. Thus, except for special cases, the matrix elements of such terms, if they appeared in the Hamiltonian, would have to be computed numerically. This is easily done.[50] However, for present purposes we restrict to rational Hamiltonians whose matrix elements can be computed analytically.

---

[50]Because β is positive definite, the infinite matrix $F(\beta)$, with elements $F_{\lambda\mu;\lambda\nu}(\beta)$, is the unique positive-definite square root of the matrix $F(\beta^2)$. Thus, accurate approximations to a finite submatrix of $F(\beta)$ can be determined by taking the unique positive-definite square root of a larger $F(\beta^2)$ matrix as proposed by Rowe *et al.*, *op. cit.* Footnote 10 on Page 171.

4.7.1 Choice of a basis

The first step towards determining the spectrum of a given collective model Hamiltonian is to choose a basis of SU(1,1)×SO(5) product states, $\{|\lambda\nu; v\alpha LM\rangle = |\lambda\nu\rangle \otimes |v\alpha LM\rangle\}$, with wave functions $\{\Psi_{\lambda\nu;v\alpha LM} = \beta^{-2}\mathcal{R}_\nu^\lambda \mathcal{Y}_{v\alpha LM}\}$. As Equation (4.37) implies, any set of beta wave functions $\{R_\nu^\lambda; \nu = 0, 1, 2, \dots\}$, with $\lambda > 0$, is an orthonormal basis for an SU(1,1) irrep. Thus, for spherical and near spherical nuclei, it is appropriate to use a harmonic spherical-vibrator basis in which the beta quantum number λ is related to the seniority v of the accompanying SO(5) spherical harmonic by $\lambda_v = v + 5/2$. For deformed nuclei, it is much more efficient to employ basis wave functions for larger, optimally-chosen, values of λ with the constraint that $\lambda_{v\pm 1} = \lambda_v \pm 1$; this constraint enables use of the analytical expressions for beta matrix elements given in Section 4.4.1, for which it is noted that matrix elements of the O(5)-parity changing quadrupole moment operators $\{\hat{q}_m\}$, which are linear in β (and, hence, not in the su(1,1) Lie algebra), are only determined analytically between states of SU(1,1) irreps with λ differing by 1.[51] A simple choice is

$$\lambda_v = \begin{cases} \lambda_0 & \text{if } v \text{ is even,} \\ \lambda_0 + 1 & \text{if } v \text{ is odd.} \end{cases} \tag{4.217}$$

Because the expressions for beta matrix elements are given in terms of a scale parameter, a, it is also necessary, in an application, to fix the value a.

Ideally, we want values of λ_0 and a for which the most accurate results are obtained for a given (small) number of basis states. However, it is also desirable to make a reasonable choice quickly and easily, in the knowledge that with enough basis states it shouldn't matter what values are chosen because the complete set of states, for any λ and a, spans the space. The best choices of λ_0 and a are presumably ones for which $\langle\hat{H}\rangle := \langle\lambda_0\nu = 0; v = 0|\hat{H}|\lambda_0\nu = 0; v = 0\rangle$ is a minimum. A simpler choice makes use of the physical interpretation of the beta wave functions, given in Section 4.2.3, to set $\lambda_0 = 1 + \sqrt{(a\beta_0)^4 + 9/4}$, where β_0 is the value of β for which the potential energy component, $V(\beta, \cos\gamma)$, of the Hamiltonian has its minimum. It then remains to determine the value of a for which the expectation $\langle\hat{H}\rangle$ is minimised for this value of λ_0. Alternatively, a good value of a is determined by considering the curvature of the potential energy, $V(\beta, \cos\gamma)$, at its minimum.

4.7.2 Calculations for an SO(5)-invariant Wilets-Jean model

As a first example, consider the SO(5)-invariant Hamiltonian[52]

$$\hat{H}(\alpha) = -\frac{\hbar^2}{2B_0}\nabla^2 + \frac{1}{2}B_0\omega^2\left[(1 - 2\alpha)\beta^2 + \alpha\frac{\beta^4}{b^2}\right], \tag{4.218}$$

---

[51] To relax this constraint, one can evaluate matrix elements of β between states of a common SU(1,1) irrep numerically, e.g., as indicated in Footnote 50.

[52] The properties of this Hamiltonian were explored by Turner P.S. and Rowe D.J. (2005), *Nucl. Phys.* **A756**, 333.

where b is a dimensional unit in which quadrupole moments (and hence β) are usefully expressed. This Hamiltonian is simplified by a suitable choice of units. If we first express it in the form

$$\hat{H}(\alpha) = \frac{1}{2}\hbar\omega\left(-\frac{\hbar}{\omega B_0}\nabla^2 + \frac{\omega B_0}{\hbar}\left[(1 - 2\alpha)\beta^2 + \alpha\frac{\beta^4}{b^2}\right]\right), \qquad (4.219)$$

then, by choosing $\hbar\omega$ as the unit of energy and expressing β in units of b (which corresponds to setting $\hbar\omega$ and b equal to unity), $\hat{H}(\alpha)$ simplifies to

$$\hat{H}(\alpha) = -\frac{\nabla^2}{2B} + \frac{1}{2}B\left[(1 - 2\alpha)\beta^2 + \alpha\beta^4\right], \qquad (4.220)$$

where $B := \omega B_0/\hbar$, the mass parameter in the new units (see Exercise 4.16), is equal to the square of the harmonic oscillator unit $a = \sqrt{\omega B_0/\hbar}$ as defined in Section 2.3.1.

(i) Parameter values

The potential energy function

$$V_\alpha(\beta) = \tfrac{1}{2}B\left[(1 - 2\alpha)\beta^2 + \alpha\beta^4\right] \qquad (4.221)$$

is shown for various values of α in Figure 4.5. For $\alpha = 0$, the potential, $V_0(\beta) =$

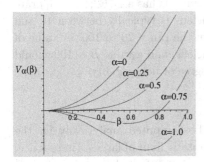

$V_\alpha(\beta)$

$\alpha=0$

$\alpha=0.25$

$\alpha=0.5$

$\alpha=0.75$

0.2 0.4 β 0.6 0.8 1.0

$\alpha=1.0$

Figure 4.5: The potential energy, V_α, shown as a function of β for different values of α.

$B\beta^2/2$, is that of a harmonic oscillator and $\hat{H}(0)$ is the Hamiltonian of the harmonic spherical vibrator (cf. Figure 2.2). As α increases, V_α goes through a critical point at $\alpha = 0.5$; its minimum is at $\beta_0 = 0$ for $\alpha \leq 0.5$ and at $\beta_0 = \sqrt{1 - 1/(2\alpha)}$ for $\alpha \geq 0.5$. In terms of the unit in which β is expressed, the equilibrium deformation β_0, at which $V_\alpha(\beta)$ is a minimum for a given α, approaches a maximum value of 1.0 as $\alpha \to \infty$. However, the magnitude of this unit can be selected as desired so that, in fact, β_0 can take any value. Note also that, while β_0 depends on α, the mean-square deviations of the deformation β about β_0, as defined by expectations values of $(\beta - \beta_0)^2$ for the model wave functions, depend on the depth of the potential which is proportional to $1/\sqrt{B}$. Thus, the fluctuations in β about β_0 become smaller as the magnitude of B are increased. Appropriate values of α and B for the description of a given nucleus can be determined by fits to the desired energy-level spectrum and E2 transition rates of the selected nucleus.

For the harmonic spherical vibrator limit, $\alpha = 0$, the model predicts E2 transition rates for decay of the first excited ($L = 2$) state to the ($L = 0$) ground state to be given, according to the expressions of Section 4.6.3, by

$$B(\text{E2}; 2_1 \to 0_1)_{\alpha=0} = \frac{1}{2} \left(\frac{Ze}{Aa} \right)^2 = \frac{1}{2B} \left(\frac{Ze}{A} \right)^2. \tag{4.222}$$

For $\alpha > 0.5$, the quadrupole vibrational fluctuations of the ground state about the equilibrium deformation β_0 are increasingly suppressed with increase in the value of B, and the E2 transition rate for decay of the first excited state to the ground state rapidly approaches the asymptotic value,

$$B(\text{E2}; 2_1 \to 0_1)_{\alpha \gg 0.5} \sim \frac{1}{5} \left(\frac{Ze}{A} \right)^2 \beta_0^2, \tag{4.223}$$

of the rigid-beta Wilets-Jean limit, given by Equation (4.212).

When $\alpha = 2$, the equilibrium deformation of the model is $\beta_0^2 = 0.75$. The value of $B(\text{E2}; 2_1 \to 0_1)_{\alpha=2}$ is close to its asymptotic limit and we obtain the ratio

$$\frac{B(\text{E2}; 2_1 \to 0_1)_{\alpha=2}}{B(\text{E2}; 2_1 \to 0_1)_{\alpha=0}} \approx \frac{3B}{10}. \tag{4.224}$$

Figure 1.41 indicates that a maximally deformed nucleus has a $B(\text{E2}; 2_1 \to 0_1)$ rate of ~ 300 W.u., whereas that for a spherical nucleus is typically between 10 and 50 W.u. This suggests considering values of B in the range 20 to 100. A value of $B \sim 25$ should then correspond to a beta-soft nucleus, and a value $B \sim 100$ should correspond to a nucleus with a relatively high degree of beta rigidity.

(ii) *Analytical limits*

The energy levels and E2 transition rates can be computed analytically for the Hamiltonian $\hat{H}(\alpha)$ when α is zero and in the asymptotic limit of large α (in practice, large means $\alpha \gtrsim 2$). When $\alpha = 0$ the results are those of the harmonic spherical vibrator limit and, when $\alpha \gtrsim 2$ and $B \gtrsim 75$, they are given by a near-rigid-beta Wilets-Jean ground-state band and high-lying beta-vibrational bands. In the large-α asymptotic limit, the vibrational fluctuations of β about β_0 become small; the beta-vibrational energy increases with increasing α and, for large α and/or B, a single rigid-beta Wilets-Jean ground-state band remains at finite energies.

Some energy levels and SO(5)-reduced E2 transition rates are shown for $\alpha = 0$ and 2 in Figure 4.6. Because the model Hamiltonian has an SO(5) symmetry, the energy levels are degenerate multiplets of the angular-momentum states that occur in an SO(5) irrep, as shown explicitly in Figure 2.3. These multiplets of eigenstates are conveniently labelled by $\{|n; v\rangle; n = 0, 1, \dots \}$, where n is here a beta-vibrational quantum number.

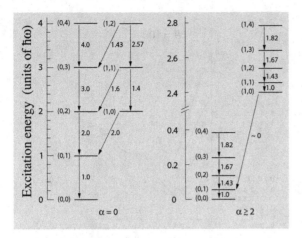

Figure 4.6: Analytical results for the lowest-energy bands of states for the Hamiltonian $\hat{H}(\alpha)$ for $\alpha = 0$ and 2. Energy levels are labelled by (n,v). The SO(5)-reduced E2 transition rates are indicated by numbered arrows. Note the expanded and broken energy scale for the $\alpha \geq 2$ energy levels.

(iii) *Calculation of matrix elements*

Although the energy eigenstates of the Hamiltonian $\hat{H}(\alpha)$, of Equation (4.220), have good SO(5) seniority, the SU(1,1) quantum number v is only a good quantum number in the harmonic spherical-vibrator limit. Thus, for $\alpha \gtrsim 0.5$, it is appropriate to expand the beta wave functions in terms of bases $\{|\lambda v\rangle\}$ for an optimal SU(1,1) irrep.

For the selected irrep, matrix elements of the Laplacian are given by Equation (4.182) and matrix elements of β^2, $\langle \lambda\mu; v|(a\beta)^2|\lambda v; v\rangle = F_{\lambda\mu;\lambda v}(a^2\beta^2)$, are given by Equation (4.97). Matrix elements of β^4 are then given by

$$\langle \lambda\mu; v|(a\beta)^4|\lambda v; v\rangle = \sum_{v'} F_{\lambda\mu;\lambda v'}(a^2\beta^2)F_{\lambda v';\lambda v}(a^2\beta^2). \qquad (4.225)$$

The parameter $a = \sqrt{B}$ in these expressions defines the scale of the beta wave functions, in accordance with Equation (4.58).

(iv) *Results*

Low-lying energy levels of $\hat{H}(\alpha)$, labelled by n and v, are shown for $B = 50$ and 100, as functions of α, in Figure 4.7. The E2 transition rates between the levels are shown as functions of α for $B = 100$ in Figure 4.8.

The results exhibit the manner in which the model progresses from one analytically solvable limit to the other. For $\alpha < 0.5$, the model behaves as a spherical vibrator with anharmonicities and beta vibrational fluctuations that become increasingly large as α approaches the critical value of $\alpha_{\text{crit}} = 0.5$. For values of $\alpha > 0.5$, a ground-state band of SO(5) (gamma-soft) rotational states emerges and beta-vibrational excited bands appear at increasingly higher energies. It is particularly notable that as B increases, the phase transition from the spherical vibrator to a (gamma-soft) rotor becomes increasingly sharp. For moderately large values

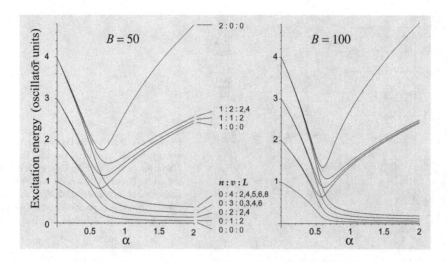

Figure 4.7: Low-lying excitation energies for the SO(5)-Hamiltonian $\hat{H}(\alpha)$ as functions of α for $B = 50$ and 100. The beta quantum number n, the SO(5) seniority quantum number v, and the angular momenta are shown for each energy level on the $B = 50$ figure. The energy levels for $B = 100$ are labelled in the same way. (From Turner and Rowe, *op. cit.* Footnote 52 on Page 206.)

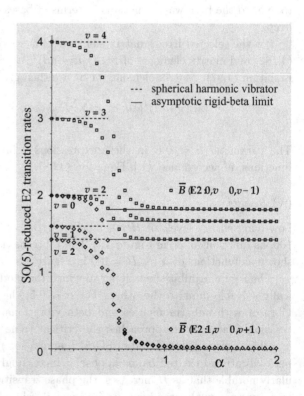

Figure 4.8: Some SO(5)-reduced E2 transition rates shown, in units of the $(n, v) = (0, 1)$ to (0,0) transition rate, as a function of α for the Hamiltonian, $\hat{H}(\alpha)$, with $B = 100$. The horizontal lines are the rates predicted by the harmonic spherical vibrator for $\alpha = 0$ and by their asymptotic $\alpha \to \infty$, rigid-beta Wilets-Jean, expressions (cf. Figure 4.6). Levels are labelled as in Figures 4.6 and 4.7. The significance and labelling of the figure, which exhibits the progression of the low-energy states of the Hamiltonian, $\hat{H}(\alpha)$, from those of a harmonic vibrator to those of a rigid-beta Wilets-Jean rotor, is most readily understood by reference to the limiting results shown in Figure 4.6. (From Turner and Rowe, *op. cit.* Footnote 52 on Page 206.)

of α and/or B, the beta vibrational fluctuations become small relative to the mean value β_0, the energies of the ($n > 0$) beta-vibrational states become high relative to the gamma-soft ($n = 0$) rotational states, and the E2 transition rates for their decay to states of the ground-state band become small and vanish as B and/or $\alpha \to \infty$.

The wave functions that result from the above calculations are also informative. The beta wave functions for the $v = 0$ and $v = 10$ states of the ground and first excited beta band are shown, for $\alpha = 1.0, 2.0,$ and 5.0, in Figure 4.9. It is remarkable

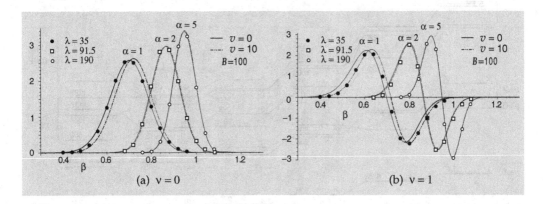

Figure 4.9: Comparison of (a) lowest ($\nu = 0$) and (b) first excited ($\nu = 1$) beta wave functions for $v = 0$ and 10 eigenstates of $\hat{H}(\alpha)$ for $B = 100$ and selected values of α. Also shown, as sequences of points, are the values of the beta wave functions $\mathcal{R}^\lambda_\nu(\beta)$ for the values of λ indicated and $a^4 = 2B^2(2\alpha - 1)$. (Based on a figure from Turner and Rowe, *op. cit.* Footnote 52 on Page 206.)

that those shown for $v = 0$ are all essentially identical to single basis wave functions, $\mathcal{R}^\lambda_0$ for the ground state band and $\mathcal{R}^\lambda_1$ for the beta band, with a suitable value of λ. Moreover, this result holds for states up to $v = 10$ with an accuracy that increases with increasing values of α. Thus, it is seen that the beta wave functions for states of a given band become essentially independent of their seniority for sufficiently large values of α. Evidently, the SO(5) centrifugal stretching effects are relatively small for $\alpha \gtrsim 2$ and $v \lesssim 10$. This result is a consequence of an adiabatic decoupling of the beta and SO(5) degrees of freedom that is expected to become increasingly pronounced with increasing deformation and decreasing fluctuations of the deformation about its mean. It shows that few SU(1, 1) × SO(5) basis states are needed, even for beta-soft nuclei, when the beta basis wave functions are those of optimally chosen SU(1,1) irreps, as defined in Section 4.7.1.[53] Indeed, for moderately large values of B and α, single SU(1, 1) × SO(5) basis wave functions already give good approximations for many low-energy states.[54]

---

[53]Comparison of the number of optimal basis states required versus the number required in the spherical vibrational basis has been given by Rowe and Turner, *op. cit.* Footnote 18 on Page 177.

[54]This has been shown for the Hamiltonian $\hat{H}(\alpha)$ by Rowe, *op. cit.* Footnote 32 on Page 187.

4.7.3 *A possible nuclear application*

In Section 2.3.2 (cf. also Figure 1.51), it was noted that ^{108}Ru is a potential manifestation of the Wilets-Jean model. Figure 4.10 shows the energy levels of the doubly-even Ru isotopes for $A = 100 - 110$. It would appear that ^{100}Ru is a candi-

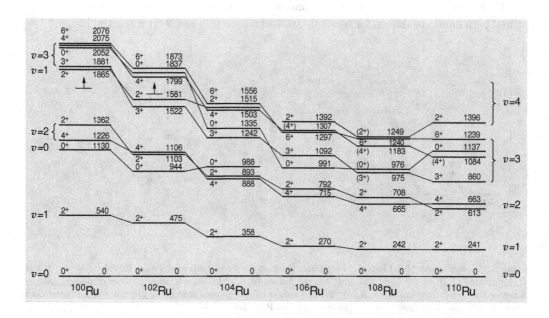

Figure 4.10: Low-lying excitations in the $^{100-110}$Ru isotopes. The vertical arrows indicate the energies (in ^{100,102}Ru) above which there are known excited states that are not included in the figure. Excitation energies are in keV. The inter-nuclear connections and tentative v assignments suggest how these isotopes may be a realisation of Figure 4.7 in the neighbourhood of $\alpha \sim 0.5$. (The data are taken from *Nuclear Data Sheets*.)

date for a near-harmonic spherical vibrator and that the energy-level spectra of the $A > 102$ isotopes correspond qualitatively to those of the above Wilets-Jean model with $\alpha \sim 0.5$.

Table 4.1 presents the $B(\mathrm{E}2)$ data for ^{100}Ru and ^{104}Ru, which are the only Ru isotopes for which extensive data are available. The corresponding predictions for the model Hamiltonian, with $B = 40$ and for a few values of α are shown for comparison. It is seen that, for some values of α, the model results are qualitatively successful at describing much of the data shown. For example, the model explains the small observed values of $B(\mathrm{E}2; 2_2 \to 0_1)$ by interpreting the $2_2 \to 0_1$ transition as a $\Delta v = 2$ transition which is identically zero in the collective model because the E2 operator has seniority $v = 1$. A vanishing $2_2 \to 0_1$ E2 transition rate is also explained by the fact that the E2 operator has negative O(5) parity, where the O(5) parity of a state of seniority v is given by $(-1)^v$; cf. Section 4.5. The parity selection

Table 4.1: Some E2 transition rates, $B(E2; J_i \to J_f)$, shown in units in which $B(E2; 2_1 \to 0_1) = 1$, for ^{100,104}Ru. The entries in the row for the $3_1 \to 2_2$ transitions, for example, are experimental and model transition rates from an initial lowest-energy state of $J = 3$ to a final second lowest-energy state of $J = 2$. For comparison, the corresponding values (see Figure 4.8) for the model Hamiltonian, $\hat{H}(\alpha)$ of Equation (4.220), are shown for $B = 40$ and a few values of α. Also shown is the diagonal E2 matrix element $\mathfrak{M}(E2)_{2_1}$ of the 2_1^+ state of ^{104}Ru (i.e., the lowest-energy state of angular momentum and parity $J^\pi = 2^+$). Data are taken from *Nuclear Data Sheets* and the references cited in the footnotes.

$J_i \to J_f$	^{100}Ru	^{104}Ru	$\alpha = 0$	$\alpha = 0.5$	$\alpha = 1.0$	$\alpha = \infty$	$\hat{H}(40, 0.5, 0.5, 0)$	$\hat{H}(40, 1.0, 0.5, 0)$
$4_1 \to 2_1$	1.45^{11}	1.40^4 ‡	2	1.83	1.45	1.43	1.84	1.50
$2_2 \to 2_1$	0.64^{11}	0.77^{10} ‡	2	1.83	1.45	1.43	1.70	0.56
$2_2 \to 0_1$	0.041^5	0.032^2 ‡	0	0	0	0	0.02	0.07
$0_2 \to 2_1$	0.98^{14}	0.42^3 ‡	2	1.42	0	0	1.45	0.06
$6_1 \to 4_1$		1.97^{16} ‡	3	2.56	1.72	1.67	2.58	1.77
$4_2 \to 4_1$	0.50^{34} †	0.39^2 ‡	1.43	1.22	0.82	0.80	1.13	0.45
$4_2 \to 2_2$	1.4^8 †	0.68^6 ‡	1.57	1.34	0.90	0.87	1.33	0.88
$3_1 \to 4_1$	1.38^{14} †	0.28^{14} ‡	0.86	0.73	0.49	0.48	0.69	0.31
$3_1 \to 2_2$	0.26^{17} †	1.28^{21} ‡	2.14	1.83	1.23	1.19	1.82	1.31
$2_3 \to 4_1$	0.35^8 †	0.16^3 ‡	1.03	0.66	0	0	0.66	0.02
$2_3 \to 2_2$	< 0.6 †	0.06^2 ‡	0.57	0.37	0	0	0.37	0
$2_3 \to 0_2$	1.06^{25} †	0.66^{13} ‡	1.40	1.13	0	0.67	1.12	0.80
$\mathfrak{M}(E2)_{2_1}$		-0.75^{10}	0	0	0	0	-0.39	-1.00

† Genilloud L. *et al.* (2001), *Nucl. Phys.* **A683**, 287.

‡ Stachel J. *et al.* (1982), *Nucl. Phys.* **A383**, 429; Stachel J. *et al.* (1984), *Nucl. Phys.* **A419**, 589; and Srebrny J. *et al.* (2006), *Nucl. Phys.* **A766**, 25.

rule also implies that the quadrupole moment (diagonal E2 matrix element) of any state of good O(5) parity is zero. Thus, diagonal E2 matrix elements shown in Table 4.1 for the 2_1^+ states of ^{104}Ru suggests that a more realistic Hamiltonian should include O(5)-parity mixing interactions. Such interactions are naturally included in a Hamiltonian with a gamma-dependent potential.

The results shown in the last two columns of Table 4.1 were obtained for the Hamiltonians $\hat{H}(40, 0.5, 0.5, 0)$ and $\hat{H}(40, 1.0, 0.5, 0)$ defined in Section 4.7.5. These are the same Hamiltonians as used for the $\alpha = 0.5$ and 1.0 columns but include a small $\beta \cos 3\gamma$, SO(5) symmetry breaking, interaction. With this small symmetry-breaking interaction, the $v = 0, 1$, and 2 energy levels for $\alpha = 0.5$ are close to those observed for ^{104}Ru. The observed E2 transition rates and the diagonal E2 matrix elements similarly lie between the symmetry broken results for $\alpha = 0.5$ and 1.0. However, for a simple $\beta \cos 3\gamma$ SO(5) symmetry breaking interaction, one finds that the highest angular-momentum members of an SO(5) multiplet fall lower in energy than the other members (see Figure 4.12) which is not seen in Figure 4.10 for the $L = 6$ states for any of the Ru isotopes. As we would expect, there is more going on in these isotopes than the simple collective model can be expected to describe.

4.7.4 *Calculations for a beta-rigid Hamiltonian*

For a fixed value of β, the kinetic energy of the Bohr model (see Equation (4.184)) is proportional to the SO(5) Casimir invariant, $\hat{\Lambda}^2$. Thus, we now consider a rigid-beta model with Hamiltonian

$$\hat{\mathcal{H}}(\chi) = \hat{\Lambda}^2 - \chi \cos 3\gamma, \tag{4.226}$$

and only rotational and gamma-vibrational degrees of freedom. The $\cos 3\gamma$ potential favours a prolate deformation when $\chi > 0$ and an oblate deformation when $\chi < 0$.

In accordance with Equation (4.28), the operator, $\hat{\Lambda}^2$, is diagonal in an SO(5)-coupled basis and has non-vanishing matrix elements

$$\langle v\alpha LM|\hat{\Lambda}^2|v\alpha LM\rangle = v(v+3). \tag{4.227}$$

Because $\cos 3\gamma$ is proportional to the Ω-independent SO(5) spherical harmonic $\mathcal{Y}_{3100}(\gamma,\Omega) = \frac{3}{4\pi}\cos 3\gamma$, its reduced matrix elements are given explicitly by Equations (4.176) and (4.177). Thus, from Equation (4.144), we have the M-independent matrix elements given in terms of SO(5) Clebsch-Gordan coefficients[55] by

$$\langle v'\alpha'LM|\cos 3\gamma|v\alpha LM\rangle = \frac{4\pi}{3}(v\alpha L, 310\|v'\alpha'L)\,\langle v'\|\|\hat{\mathcal{Y}}_3\|\|v\rangle. \tag{4.228}$$

with

$$\langle v+1\|\|\hat{\mathcal{Y}}_3\|\|v\rangle = \frac{9}{4\pi}\sqrt{\frac{5v(v+1)(v+4)}{2(2v+1)(2v+5)(2v+7)}}\,,$$

$$\langle v+3\|\|\hat{\mathcal{Y}}_3\|\|v\rangle = \frac{3}{4\pi}\sqrt{\frac{35(v+1)(v+2)(v+3)}{2(2v+5)(2v+7)(2v+9)}}\,. \tag{4.229}$$

Figure 4.11 shows the spectrum of the Hamiltonian $\hat{\mathcal{H}}(\chi)$ for $\chi = \pm 50$. A comparison of this spectrum with that of Figure 2.7 reveals that the results are already very close to those of the Adiabatic Collective Model of Section 2.4 (with no beta-vibrational excitations). Because β is assigned a frozen value in these calculations, the beta vibrations and centrifugal β-stretching effects are suppressed. Thus, the relatively small differences between the calculated results of Figure 4.11 and those of the adiabatic model are primarily due to SO(5) centrifugal coupling interactions between the gamma-vibrational and rotational degrees of freedom.[56] As expected, it is found that, with an increasing value of the parameter χ, the gamma-vibrational bands rise in energy and the centrifugal stretching effects are increasingly suppressed.

---

[55]Cf. Footnote 47 on page 196.

[56]The results shown in Figure 4.11 differ negligibly from similar results reported in Rowe, *op. cit.* Footnote 9 on Page 171. The earlier calculations were done in a truncated space and without the benefit of known reduced matrix elements and SO(5) Clebsch-Gordan coefficients. Their accuracy is confirmed by the above results which did not change, to the level of accuracy that would be visible in Figure 4.11, when carried out in spaces of seniorities $1 \le v \le 12$ and $1 \le v \le 24$.

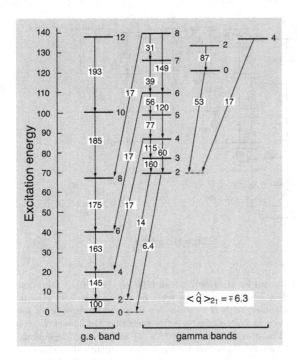

Figure 4.11: Low-energy spectrum of the Hamiltonian $\hat{\mathcal{H}}(\chi)$ of Equation (4.226) for $\chi = \pm 50$. Reduced E2 transition rates are shown in units for which $B(\text{E2}; 2_1 \to 0_1) = 100$. Energy levels are given in units such that the lowest $L = 2$ state has energy $E_{2_1} = 6$. Thus, if there were no centrifugal coupling between the rotational and γ-vibrational degrees of freedom, the lowest $L = 10$ state, for example, would have energy $E_{10_1} = 110$. The quadrupole moment of the first-excited $L = 2$ state is denoted by $\langle \hat{q} \rangle_{2_1}$. (Based on a figure from Rowe, *op. cit.* Footnote 9 on Page 171.)

It is interesting to note that the energy-level spectrum is independent of the sign of χ. However, the quadrupole moment, $\langle \hat{q} \rangle_{2_1}$, of the first-excited $L = 2$ state is negative when $\chi > 0$, as expected for a prolate rotor, and positive when $\chi < 0$, as expected for an oblate rotor.

4.7.5 Calculations for beta- and gamma-dependent potentials

We next consider Hamiltonians of the general form (see Exercise 4.16)[57]

$$\hat{H}(B, \alpha, \chi, \kappa) = -\frac{\nabla^2}{2B} + \frac{1}{2}B\big[(1 - 2\alpha)\beta^2 + \alpha\beta^4\big] - \chi\beta\cos 3\gamma + \kappa[\cos 3\gamma]^2. \quad (4.230)$$

When $\kappa = 0$, the potential-energy component of this Hamiltonian has a spheroidal minimum that is prolate if $\chi > 0$ and oblate if $\chi < 0$. For other values of κ and χ, the potential at its minimum is ellipsoidal. We consider three examples.

(i) *A model with $B = 40$, $\alpha = 1.0$, $\chi = \pm 0.5$, and $\kappa = 0$*

Reference to Figure 4.7 shows that a Hamiltonian with a mass parameter $B \lesssim 50$ and $\alpha = 1.0$ corresponds to a model with relatively-large beta fluctuations about a small equilibrium deformation. From experience gained with a number of model calculations, one also finds that $\chi = 0.5$ corresponds to a weak gamma potential. The spectrum that results is shown in Figure 4.12(c).

---

[57]Rowe *et al.*, *op. cit.* Footnote 10 on Page 171.

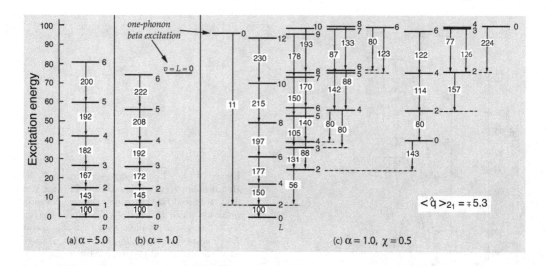

Figure 4.12: Energy levels and reduced E2 transition rates for the Hamiltonian $\hat{H}(B, \alpha, \chi, \kappa = 0)$ of Equation (4.230). Figure (a) is for the near beta-rigid Wilets-Jean Hamiltonian with $B = 40$, $\alpha = 5.0$, $\chi = 0$. Figure (b) is for a relatively soft Wilets-Jean Hamiltonian with $B = 40$, $\alpha = 1.0$, $\chi = 0$. Figure (c) is for $B = 40$, $\alpha = 1.0$, $\chi = \pm 0.5$. Energy levels are given in each figure in units such that the lowest $L = 2$ state has an excitation energy of $L(L + 1) = 6$. The energy levels shown in (a) and (b) are degenerate SO(5) multiplets and the E2 transition rates shown are SO(5)-reduced; thus they are identical in value to the standard $B(E2; L_i = 2v \to L_f = 2v - 1)$ transition rates between the states of maximum L of each multiplet. The transition rates shown in (c) are standard SO(3)-reduced $B(E2; L_i \to L_f)$ transition rates. All transition rates are given in units such that $B(E2; 2_1 \to 0_1) = 100$. (From Rowe et al., op. cit. Footnote 10 on Page 171.)

For comparison, we show in Figures 4.12(a) and 4.12(b) the $\chi = 0$ Wilets-Jean spectra for $\alpha = 5.0$ and 1.0, respectively. For $\chi = 0$, the SO(5) multiplets of states of a given seniority v are degenerate. As α is lowered from 5.0 to 1.0, the fluctuations of β about its mean increase and centrifugal stretching occurs, as can be seen in Figure 4.12(b). In addition, a one-phonon beta-vibrational excitation appears at a relatively low energy. When χ is increased to 0.5, the spectrum that results, Figure 4.12(c), retains many characteristics of the Wilets-Jean limit. It also shows the dominant effect of the small $\cos 3\gamma$ interaction to be the beginning of an alignment of the energy levels into rotational bands. This is most pronounced at lower energies where there are signs of an approach to a rotor-vibrator spectrum (cf. Figure 4.13), which is most clearly observed in the yrast band. However, with increasing energy, the energy levels rapidly revert to near degenerate SO(5) multiplets. Note also that the first excited $L = 2$ state acquires a quadrupole moment that is zero in the Wilets-Jean limit. An interpretation of these results is that the low-energy states tend to be trapped in the shallow well around $\gamma = 0$ (or $\gamma = \pi/3$ for $\chi = -0.5$), whereas higher-energy states are much less affected by this well. Similar results were observed by Caprio et al.[58]

[58]Caprio M.A., Cejnar P. and Iachello F. (2008), Ann. Phys. (NY) 323, 1106.

(ii) *A model with* $B = 20$, $\alpha = 1.5$, $\chi = \pm 2.0$, *and* $\kappa = 0$

The parameters of this example are chosen to obtain a ground-state rotational band for a moderately-deformed axially-symmetric model with beta- and gamma-vibrational bands in the low-energy domain. The resulting spectrum is shown in Figure 4.13(a).

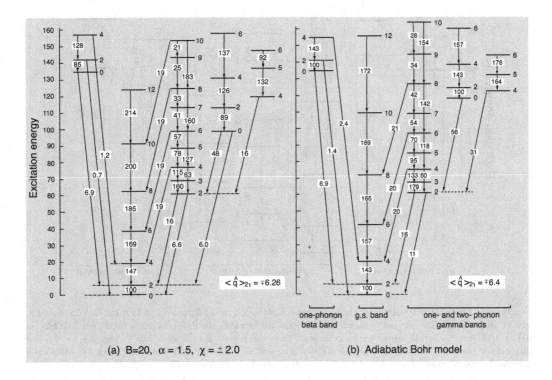

Figure 4.13: (a) Low-energy spectrum of the Hamiltonian $\hat{H}(B, \alpha, \chi, \kappa)$ of Equation (4.230) for $B = 20$, $\alpha = 1.5$, $\chi = \pm 2.0$, and $\kappa = 0$. The E2 transition decay rates (only the largest are shown) are in units for which $B(E2; 2_1 \to 0_1) = 100$. Energy levels are given in units such that the lowest $L = 2$ state has energy $E_{2_1} = 6$. Thus, if there were no centrifugal stretching, the lowest $L = 10$ state, for example, would have energy $E_{10_1} = 110$. (b) Corresponding results for the adiabatic Bohr model in the axially symmetric limit showing a ground-state band, a one-phonon beta-vibrational band, and one- and two-phonon gamma-vibrational bands. (From Rowe *et al.*, *op. cit.* Footnote 10 on Page 171.)

The primary differences between Figure 4.13(a) and that of the adiabatic model, Figure 4.13(b), may be attributed to small gamma anharmonicities, due to the difference between $\cos 3\gamma$ and a harmonic γ^2 potential, and to substantial centrifugal stretching effects. In particular, it is found that, whereas a small value of B is needed to obtain a beta band at moderately low excitation energies, the centrifugal perturbations to the rotational bands increase dramatically as B is decreased.

(iii) *A model with $B = 22$, $\alpha = 1.5$, $\chi = 0$, and $\kappa = 4.0$*

The parameters are now chosen to give a ground-state rotational band for a moderately-deformed triaxial Meyer-ter-Vehn model (cf. Section 2.4) with low energy beta- and gamma-vibrational bands. The spectrum that results is shown in Figure 4.14.

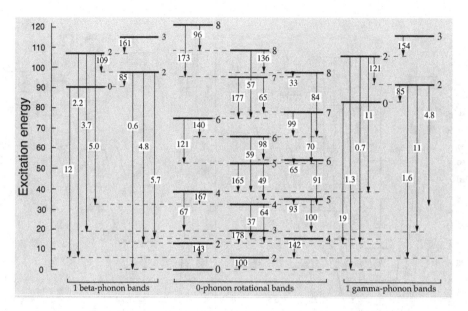

Figure 4.14: Low-energy spectrum of the Hamiltonian $\hat{H}(B, \alpha, \chi, \kappa)$ of Equation (4.230) for $B = 22$, $\alpha = 1.5$, $\chi = 0$, and $\kappa = 4.0$. The largest reduced E2 transition decay rates are shown in units for which $B(E2; 2_1 \to 0_1) = 100$. Energy levels are given in units such that the lowest $L = 2$ state has energy $E_{2_1} = 6$. The quadrupole moments of all states are identically zero and all E2 intraband transition rates are identically zero in this calculation. (From Rowe *et al.*, *op. cit.* Footnote 10 on Page 171.)

This figure may be compared with Figure 2.6 of the adiabatic model which it reproduces to a remarkable degree. But again, with the small mass parameter B, needed to get low-energy beta bands, there are significant centrifugal stretching corrections. Note also that the moments of inertia for the beta- and gamma-vibrational bands are smaller than those for the ground-state band.

It is also interesting to note that the ACM provides a simple explanation of the selection rule, of the adiabatic $\gamma_0 = \pi/6$ limit, which forbids all non-zero quadrupole moments and all $\Delta K = 0$, E2 transitions for which $\Delta n_\gamma \neq \pm 1$. This selection rule follows immediately from the observation that *any Hamiltonian, for which the potential energy is of even degree in $\cos 3\gamma$, conserves the $\mathbb{R}^5$ parity.* Therefore, because the quadrupole operator has odd $\mathbb{R}^5$ parity, its matrix elements between states of the same $\mathbb{R}^5$ parity are all identically zero. Thus, the calculations bring to light a remarkable property of the Meyer-ter-Vehn model which is that the one-phonon $L = 0$ beta excitation has the same $\mathbb{R}^5$ parity as the $L = 0$ ground state and that the one-phonon $L = 0$ gamma excitation has the opposite parity to both the ground and one-phonon beta states. A rationale for this result is obtained from the following considerations. In the first place it is to be expected that the $L = 0$ ground state should be a state of maximal (i.e., positive) $\mathbb{R}^5$ symmetry. The $L = 0$ beta and gamma one-phonon vibrational excitations must then be of either positive or negative $\mathbb{R}^5$ parity. It then comes as no surprise to learn from the calculations that the beta-vibrational $L = 0$ state, which is predominantly an SU(1,1) excitation for which the infinitesimal generators are of even $\mathbb{R}^5$ parity (e.g., $\beta^2 \propto [Q \otimes Q]_0$), is of positive $\mathbb{R}^5$ parity. Conversely, it is understandable that the gamma-vibrational $L = 0$ state proves to be of negative $\mathbb{R}^5$ parity because the simplest, angular-momentum zero combination of quadrupole moment operators that involve the γ coordinate, $\beta^3 \cos 3\gamma \propto [Q \otimes Q \otimes Q]_0$, is of negative $\mathbb{R}^5$ parity.

The above examples exhibit the large diversity of results that can be obtained with simple collective model Hamiltonians. They also illustrate the fact that, with the ready availability of SO(5) Clebsch-Gordan coefficients and analytical expressions for reduced matrix elements, such calculations can be done quickly and easily.

Exercises

4.16 Show that the Hamiltonian

$$\hat{H} = -\frac{\hbar^2}{2B_0}\nabla^2 + \frac{1}{2}B_0\omega^2\left[(1 - 2\alpha)\beta^2 + \alpha\frac{\beta^4}{b^2}\right] - \chi\frac{\beta}{b}\cos 3\gamma + \kappa\cos^2 3\gamma, \quad (4.231)$$

where b is a unit in which the quadrupole moments (and hence β) are expressed, can be re-expressed as

$$\hat{H}(B, \alpha, \chi, \kappa) = -\frac{\nabla^2}{2B} + \frac{1}{2}B\left[(1 - 2\alpha)\beta^2 + \alpha\beta^4\right] - \chi\beta\cos 3\gamma + \kappa\cos^2 3\gamma, \quad (4.232)$$

if energies are defined in units of $\hbar\omega$, β is expressed in units of b, and $B := B_0\omega/\hbar$.

4.17 Use Equation (4.208) and the SO(5) CG coefficients from Table 4.3 to derive the SO(3)-reduced E2 transition rates shown in Figures 2.2 and 2.3 from the SO(5)-reduced matrix elements given in Figure 4.6.

4.18 Use Equation (4.208) and the SO(5) CG coefficients from Table 4.3 to derive the SO(3)-reduced E2 transition rates for $B = 100$, corresponding to those quoted in Table 4.1 for $B = 40$, from Figure 4.8.

4.8 Successes and limitations of the Bohr model

Although the Bohr model contributes enormously to the understanding of collective dynamics in nuclei, one must remember that it is a model with only a few macroscopic degrees of freedom and an assumed geometric and algebraic structure that may not be the most appropriate. Nevertheless, interpretation of nuclear data in terms of the model makes it possible to assess the extent to which a few macroscopic quadrupole collective degrees of freedom of a nucleus actually do decouple, to some level of approximation, from other degrees of freedom. As discussed in Section 4.3.5, such an adiabatic decoupling would be expected if the intrinsic excitations of a nucleus occurred at much higher energies than the macroscopic collective excitations. The collective model would then have a chance of describing the low-energy states.

The appearance of remarkably robust rotational bands in a wide range of nuclei provides compelling evidence for the widespread existence of nuclei with deformed equilibrium shapes that are perturbed only weakly by coupling to other degrees of freedom. The degree of rigidity of observed bands is particularly notable in view of the results, reported above, which show that the rotational bands of the ACM exhibit rather large centrifugal stretching effects, as a result of coupling to the beta- and gamma-vibrations, when the beta- and/or gamma-vibrational bands fall into the low-energy domain. In fact, the magnitude of this coupling calls into question the consistency of interpreting low-energy excited bands as beta- or gamma-vibrational bands. A possible interpretation of this difference is that the beta- and gamma-vibrational excitations actually lie at higher energies, than supposed in the model, and mix with intrinsic excitations of the nucleus. Another interpretation is that an ostensible one-phonon gamma band of an axially symmetric nucleus is more realistically described as a member of a multiband sequence of a triaxial near-rigid rotor. There are many possibilities. However, the potential for detailed ACM calculations makes it possible to see if any such interpretations are feasible within the framework of the Bohr model.

The large variety of rotational bands and widespread evidence of quadrupole collectivity in most doubly-open shell nuclei, as reviewed in Section 1.7, show that, in fact, the collective motions described by the Bohr model are not restricted to the lowest-energy states in which the competing intrinsic degrees of freedom remain dormant. This suggests a more general model perspective in which the Hilbert space of a nucleus is viewed as a tensor product of a collective model Hilbert space and a complementary space of intrinsic states. The Bohr-Mottelson unified models of nuclear collective structure, to be discussed in Volume 2, can be viewed from such a perspective.[59]

A potential source of information on collective dynamics is provided by a study of moments of inertia. Moments of inertia are characteristic of the dynamical flows that occur in a rotating nucleus. In the Bohr model, the ratios of moments of inertia are fixed by the SO(5) Casimir operator and turn out to have the same ratios as those of an irrotational-flow quantum fluid. Their magnitudes are determined by a single mass parameter, B. This parameter is usually treated as adjustable on the grounds that the model is not committed to any particular collective flows. In the original formulation of his model, Bohr proposed that B should be assigned an irrotational-flow value. With the Bohr model shape coordinates replaced by

---

[59]Bohr and Mottelson, *op. cit.* Footnote 6 on Page 98.

quadrupole moments, it would appear that this choice is appropriate because, as shown explicitly in Section 8.2.6, the quadrupole moment operators are, in fact, infinitesimal generators of irrotational flows. However, observed rotational bands require much larger moments of inertia. Moreover, there is evidence to suggest that the ratios of moments of inertia that best fit nuclear data may also differ significantly from those of the Bohr model.[60]

From a liquid-drop model perspective, the moments of inertia observed for nuclear rotations suggest that a better description of nuclear dynamics might be given by including vorticity degrees of freedom as done in superfluid hydrodynamics.[61] Thus, in addition to moment-of-inertia considerations, it would be instructive to have direct experimental evidence of the effects of vorticity degrees of freedom in nuclei. An obvious place to look for such evidence is in transverse electron scattering form factors which depend on the nuclear current flows.[62,63] Unfortunately, measurement of the relevant transverse form factors appears to be very difficult.

Another concern is that the geometric and algebraic structures assumed for the Bohr model, may not be the most appropriate. A vital step in facilitating a study of these questions was achieved by replacing the surface shape parameters of Bohr's original model by quadrupole moments. Thus, if the Cartesian quadrupole moments for a nucleus with density $\rho(\mathbf{r})$ at a point $\mathbf{r} = (x_1, x_2, x_3) \in \mathbb{R}^3$, given by

$$Q_{ij} = \int_{\mathbb{R}^3} x_i x_j \rho(\mathbf{r}) \, dv, \tag{4.233}$$

are interpreted as the quadrupole moments of a discrete distribution of nucleons with position coordinates $\{\mathbf{r}_n = (x_{n1}, x_{n2}, x_{n3})\}$, then

$$Q_{ij} = \sum_n x_{ni} x_{nj}, \tag{4.234}$$

and the time derivatives of these quadrupole moments are

$$\dot{Q}_{ij} = \sum_n (x_{ni}\dot{x}_{nj} + \dot{x}_{ni}x_{nj}). \tag{4.235}$$

Thus, they correspond to moments of momentum given by

$$P_{ij} := M\dot{Q}_{ij} = \sum_n (x_{ni}\dot{p}_{nj} + \dot{p}_{ni}x_{nj})., \tag{4.236}$$

where $p_{ni} = M\dot{x}_{ni}$ is the momentum of the i'th nucleon. The quantization of the $\{Q_{ij}, P_{ij}\}$ observables is then given by mapping them to operators

$$\hat{Q}_{ij} = \sum_n \hat{x}_{ni}\hat{x}_{nj}, \quad \hat{P}_{ij} = -i\hbar \sum_n \left(\hat{x}_{ni}\frac{\partial}{\partial x_{nj}} + \frac{\partial}{\partial x_{ni}}\hat{x}_{nj} \right). \tag{4.237}$$

---

[60] Wood J.L. *et al.* (1992), *Phys. Repts.* **215**, 101.
[61] Putterman S.J. (1974), *Superfluid Hydrodynamics* (North-Holland, Amsterdam).
[62] Hotta A. *et al.* (1987), *Phys. Rev.* **C36**, 2212.
[63] Carvalho M.J. and Rowe D.J. (1997), *Nucl. Phys.* **A618**, 65.

Moreover the kinetic energy for a system of nucleons, given in the usual way by

$$T = \frac{1}{M} \sum_{n,i} p_{ni}^2, \tag{4.238}$$

has the standard quantization

$$\hat{T} = -\frac{\hbar^2}{2M} \sum_{n,i} \frac{\partial^2}{\partial x_{ni}^2}, \tag{4.239}$$

where M is the nucleon mass. These operators generate a Lie algebra spanned by the operators $\{\hat{Q}_{ij}, \hat{P}_{ij}, \hat{K}_{ij}\}$, where

$$\hat{K}_{ij} := \sum_n \frac{\partial^2}{\partial x_{ni}\partial x_{nj}}, \tag{4.240}$$

which is the Lie algebra, sp(3, $\mathbb{R}$), of the non-compact symplectic group. Thus, with these new observables, the collective model becomes the so-called sp(3, $\mathbb{R}$) symplectic model,[64,65,66] which was constructed as a microscopic collective submodel of the nuclear shell model. This model will be discussed in depth in Volume 2, where it will be shown to include vorticity as well as irrotational-flow degrees of freedom.

For present purposes, it is useful to note that any Bohr model Hamiltonian, of the form $\hat{H} = -\frac{\hbar^2}{2B}\nabla^2 + \hat{V}(\beta, \gamma)$, immediately defines a microscopic shell-model Hamiltonian in which $-\frac{\hbar^2}{2B}\nabla^2$ is replaced by the many-nucleon kinetic energy $\hat{T}$ and $\hat{V}(\beta, \gamma)$ is expressed in terms of the microscopic quadrupole moment operators. Moreover, following this substitution, rotational bands are obtained with moments of inertia close to those observed.[67,68]

4.9 The computation of SO(5) spherical harmonics and Clebsch-Gordan coefficients

The ACM relies on the availability of SO(5) Clebsch-Gordan coefficients in an SO(3) basis. These coefficients are readily computed from the overlaps of corresponding SO(5) spherical harmonics. We conclude this chapter with a brief outline of the algorithm[69] for determining these spherical harmonics and coefficients.

Standard SO(3) spherical harmonics $\{Y_{LM}\}$ are an orthonormal basis of square-integrable wave functions on the two-sphere, S_2. For each $L = 0, 1, 2, \ldots$, these functions span an irrep of SO(3) in an SO(2) basis. Similarly, SO(5) spherical harmonics $\{\mathcal{Y}_{v\alpha LM}\}$ are an orthonormal basis of square-integrable wave functions

[64]Rosensteel G. and Rowe D.J. (1977), *Phys. Rev. Lett.* **38**, 10.
[65]Rosensteel G. and Rowe D.J. (1980), *Ann. Phys. (NY)* **126**, 343.
[66]Rowe D.J. (1985), *Rep. Prog. Phys.* **48**, 1419.
[67]Park P. *et al.* (1984), *Nucl. Phys.* **A414**, 93.
[68]Bahri C. and Rowe D.J. (2000), *Nucl. Phys.* **A662**, 125.
[69]Rowe *et al.*, *op. cit.* Footnote 42 on Page 194.

on the four-sphere, S_4, and, for each $v = 0, 1, 2, \ldots$, they span an irrep of SO(5) in an SO(3) $\supset$ SO(2) basis. Thus, whereas the SO(3) spherical harmonics are eigenfunctions of $\hat{\mathbf{L}}^2$ and $\hat{L}_0$,

$$\hat{\mathbf{L}}^2 Y_{LM} = L(L+1)Y_{LM}, \quad \hat{L}_0 Y_{LM} = MY_{LM}, \tag{4.241}$$

the SO(5) spherical harmonics are eigenfunctions of the SO(5) Casimir invariant, $\hat{\Lambda}^2$, in addition to $\hat{\mathbf{L}}^2$ and $\hat{L}_0$,

$$\hat{\Lambda}^2 \mathcal{Y}_{v\alpha LM} = v(v+3)\mathcal{Y}_{v\alpha LM},$$
$$\hat{\mathbf{L}}^2 \mathcal{Y}_{v\alpha LM} = L(L+1)\mathcal{Y}_{v\alpha LM}, \quad \hat{L}_0 \mathcal{Y}_{v\alpha LM} = M\mathcal{Y}_{v\alpha LM}. \tag{4.242}$$

A solution of these equations, as differential equations, was tackled many years ago by Bès.[70] Such a direct approach is not easy because of the coupling between the gamma and SO(3) degrees of freedom. Bès succeeded in obtaining explicit expressions for the SO(5) spherical harmonics with $L \leq 6$. However, for larger values of L, the many coupled differential equations that result rapidly become intractable. This section outlines a simple algorithm for deriving these spherical harmonics.

4.9.1 An algorithm for computing SO(5) spherical harmonics

A first step is to display the angular-momentum states that occur in the SO(5) irreps of seniority $v = 0, 1, 2, \ldots$ in a systematic way. This is done by using the SO(5)$\downarrow$SO(3) branching rule of Equation (4.31) to determine the spectrum of states of angular momentum L, as shown for $v \leq 6$ in Figure 4.15. For each value of v,

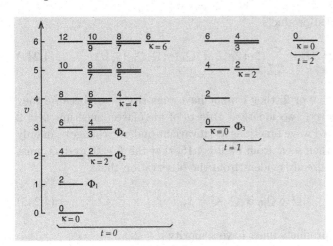

Figure 4.15: The spectrum of SO(3) (angular momentum L) irreps in $\mathcal{L}^2(S_4)$ put into columns in one-to-one correspondence with rotor-model K bands. Basis wave functions for the subspace of $\mathcal{L}^2(S_4)$ spanned by SO(3) highest-weight ($M = L$) wave functions are generated by the wave functions Φ_1, Φ_2, Φ_3, and Φ_4, for the $M = L$ states of the, respective, $L = 2_1$, 2_2, 0_2, and 3_1, irreps of SO(3) as shown. (Based on a figure from Rowe et al., op. cit. Footnote 42 on Page 194.)

as indicated on the vertical axis of this figure, the SO(3) irreps, labelled by angular momentum L, contained in the SO(5) irrep v, are in one-to-one correspondence with

---

[70]Bès D.R. (1959), *Nucl. Phys.* **10**, 373.

the horizontal lines shown. For example, the $v = 3$ irrep of SO(5), contains one copy of each of the SO(3) irreps with $L = 6, 4, 3$, and 0. By arranging the SO(3) irreps into vertical columns, as they would appear in corresponding rotor-model K bands, their pattern is recognised immediately.

In fact, the spectrum of SO(3) angular momenta, L, is seen to be the same as that of the rigid-beta Wilets-Jean model. Moreover, the spectrum of $t = 0$ states is that of a triaxial rigid rotor, and this spectrum is repeated for each value of t as it would for the t-phonon gamma-vibrational states. This is expected because, if β^2 and $\cos 3\gamma$ take fixed values, the spectrum of states should be that of a rigid rotor, whereas if β^2 remains fixed but $\cos 3\gamma$ is expanded,

$$\cos 3\gamma = \cos 3\gamma_0 - 3(\gamma - \gamma_0)\sin 3\gamma_0 + \ldots, \qquad (4.243)$$

it is seen that the rigid-rotor spectrum will be coupled to that of a gamma vibrator.

As observed in Section 4.5.2, SO(5) spherical harmonics of seniority v can be expressed as polynomials in the components of the unit quadrupole moments, $\{\mathcal{Q}_m, m = 0, \pm 1, \pm 2\}$ (modulo factors of $\mathcal{Q} \cdot \mathcal{Q} = 1$ as explained in Footnote 45, Page 194). By definition (see Sections 4.1.3 and 4.5.2), the $\{\mathcal{Q}_m\}$ quadrupole moments span the basic 5-dimensional, $v = 1$, irrep of SO(5). Moreover, they have negative parity under inversion in the 5-dimensional vector space, $\mathbb{R}^5$, spanned by the quadrupole moments. It follows that polynomials of even and odd degree have even and odd $\mathbb{R}^5$ parity, respectively. These observations and a knowledge of the inner product, which gives the orthonormality property of the SO(5) spherical harmonics in accordance with Equation (4.39), provide a simple algorithm for their construction.

Observe that the coupled-products

$$[\mathcal{Q} \otimes \mathcal{Q}]_{LM}(\gamma, \Omega) = \sum_{mn} (2m\, 2n | LM)\mathcal{Q}_n(\gamma, \Omega)\mathcal{Q}_m(\gamma, \Omega) \qquad (4.244)$$

can only have seniority $v = 0$ or 2; they cannot have seniority $v = 1$ because they are of even degree. Moreover, two identical objects of angular momentum $L = 2$ (i.e., objects whose combined wave functions are invariant under exchange) can only couple to $L = 0, 2, 4$. It is then seen, from Figure 4.15, that the $L = 0$ product must have seniority $v = 0$, as is already evident from the observation that

$$[\mathcal{Q} \otimes \mathcal{Q}]_0 \propto \mathcal{Q} \cdot \mathcal{Q} = 1, \qquad (4.245)$$

and that the $L = 2$ and 4 products must have seniority $v = 2$.

Similarly, the coupled products $[\mathcal{Q} \otimes \mathcal{Q} \otimes \mathcal{Q}]_{LM}$ can only have seniority $v = 1$ or 3 and again, from the SO(5)↓SO(3) branching rules exhibited in Figure 4.15, their seniorities are uniquely defined by their angular momenta. Thus, the SO(5) spherical harmonics for $v \leq 3$ are all known to within normalisation factors that can be determined from the inner product relationship

$$\langle v\alpha LM | v'\alpha' L'M' \rangle = \int_{S_4} \mathcal{Y}^*_{v\alpha LM}(\gamma, \Omega)\, \mathcal{Y}_{v'\alpha' L'M'}(\gamma, \Omega)\, \sin 3\gamma\, d\gamma\, d\Omega$$
$$= \delta_{vv'}\delta_{\alpha\alpha'}\delta_{LL'}\delta_{MM'}, \tag{4.246}$$

given in Section 4.2.1.

The above results imply that the SO(3) highest-weight, $M = L$, components of the $v \leq 3$ subset of spherical harmonics have simple expressions in terms of the four angular-momentum-coupled functions

$$\begin{aligned}
\Phi_1 &\propto \mathcal{Q}_{22}, & v &= 1,\, L=M=2, \\
\Phi_2 &\propto [\mathcal{Q} \otimes \mathcal{Q}]_{22}, & v &= 2,\, L=M=2, \\
\Phi_3 &\propto [\mathcal{Q} \otimes \mathcal{Q} \otimes \mathcal{Q}]_0, & v &= 3,\, L=M=0, \\
\Phi_4 &\propto [\mathcal{Q} \otimes \mathcal{Q} \otimes \mathcal{Q}]_{33}, & v &= 3,\, L=M=3,
\end{aligned} \tag{4.247}$$

which we refer to as *generating functions*. In terms of these functions, the SO(5) spherical harmonics of highest SO(3) weight, $\{\mathcal{Y}_{v\alpha L, M=L}\}$, for $v \leq 3$, are given uniquely by

$$\begin{aligned}
&\mathcal{Y}_{0100} \propto 1, & &\mathcal{Y}_{1122} \propto \Phi_1, & &\mathcal{Y}_{2122} \propto \Phi_2, & &\mathcal{Y}_{2144} \propto [\Phi_1]^2, \\
&\mathcal{Y}_{3100} \propto \Phi_3, & &\mathcal{Y}_{3133} \propto \Phi_4, & &\mathcal{Y}_{3144} \propto \Phi_1\Phi_2, & &\mathcal{Y}_{3166} \propto [\Phi_1]^3,
\end{aligned} \tag{4.248}$$

where we make use of the observation that the highest-weight component of a stretched coupled product of two tensors, defined as the simple product of highest-weight components of the tensors, e.g.,

$$[\mathcal{Q} \otimes \mathcal{Q}]_{44} = \mathcal{Q}_{22} \times \mathcal{Q}_{22}. \tag{4.249}$$

Some SO(5) spherical harmonics of $v > 3$ are also expressible as monomials in the generating functions. For example,

$$\mathcal{Y}_{v=n,\alpha=1,L=M=2n} \propto [\Phi_1]^n. \tag{4.250}$$

This is guaranteed because a polynomial of maximal degree n in the $v = 1$ quadrupole moments cannot have seniority $v > n$. On the other hand, there is no such polynomial with angular momentum $L = 2v$ of degree $n < v$. Thus, the functions $[\Phi_1]^n$ can only have seniority $v = n$. Such examples are special cases.

In the general construction of SO(5) spherical harmonics, it is necessary to take account of the fact that, because $\mathcal{Q} \cdot \mathcal{Q} = 1$, the degree of a polynomial in the components of $\mathcal{Q}$ is not, in general, well defined; i.e., any function is unchanged when multiplied by $\mathcal{Q} \cdot \mathcal{Q}$. This potential problem, discussed in Footnote 45 on Page 194, is circumvented by restricting consideration to polynomials that contain no such factors. This is achieved by use of the following completeness theorem.

Theorem 4.7 *The subspace of SO(3) highest-weight functions in the Hilbert space $\mathcal{L}^2(S_4)$ is spanned by the monomials,*

$$\Phi_{NtL} := \Phi_1^{n_1} \Phi_2^{n_2} \Phi_3^{n_3} \Phi_4^{n_4}, \qquad (4.251)$$

in the four generating functions defined by Equation (4.247), where n_1, n_2, n_3, take all non-negative integer values and $n_4 = 0$ or 1. A monomial, Φ_{NtL} is a polynomial of degree N in the unit quadrupole moments, $\{\mathcal{Q}_m\}$, and has angular momentum, L, given by

$$N := n_1 + 2n_2 + 3n_3 + 3n_4, \quad L := 2n_1 + 2n_2 + 3n_4; \qquad (4.252)$$

the label $t := n_3$ is a multiplicity index which denotes the number of zero-coupled triplets Φ_3 appearing in its definition.

A proof of this theorem follows easily from the observation that a complete set of monomials in the quadrupole moments $\{q_m = \beta \mathcal{Q}_m; m = 0, \pm 1, \pm 2\}$ spans the Hilbert space $\mathcal{L}^2(\mathbb{R}^5)$.

Using Theorem 4.7, the SO(5) spherical harmonics, with $M = L$, can be constructed sequentially by Gram-Schmidt orthonormalisation.[71] For example, for $v = 4$, it is clear that

$$\mathcal{Y}_{41LL} \propto \Phi_{40L}, \quad \text{for } L = 5, 6, \text{and } 8. \qquad (4.253)$$

This follows because there are no SO(3) irreps with these values of L in the $v = 2$ or 0 irreps of SO(5). However, there are SO(3) irreps with $L = 2$ and 4 in the $v = 2$ irrep of SO(5). Thus, the $v = 2$ and $v = 4$ spherical harmonics with $L = 2, 4$ are expressible in the form

$$\mathcal{Y}_{21LL} = c_1(L)\Phi_{20L}, \qquad (4.254)$$

$$\mathcal{Y}_{41LL} = c_{22}(L)\Phi_{40L} + c_{21}(L)\Phi_{20L}, \qquad (4.255)$$

with coefficients, $c_1(L)$, $c_{22}(L)$ and $c_{21}(L)$, to be determined such that the SO(5) spherical harmonics satisfy the orthonormality requirements

$$\langle 41LL|41LL \rangle = 1, \quad \langle 21LL|21LL \rangle = 1, \quad \langle 41LL|21LL \rangle = 0, \qquad (4.256)$$

where $|v\alpha LM\rangle$ denotes a state whose wave function is the SO(5) spherical harmonic $\mathcal{Y}_{v\alpha LM}$.

The Gram-Schmidt orthonormalisation procedure is based on a determination of the overlap integrals for the non-orthogonal basis wave functions. These are evaluated by making an expansion,

$$\Phi_{NtL}(\gamma, \Omega) = \sum_K f_K^{(NtL)}(\gamma) \mathscr{D}_{KL}^L(\Omega), \qquad (4.257)$$

---

[71]Arfken G. (1985), *Mathematical Methods for Physicists* (Academic Press, San Diego).

in terms of Wigner $\mathscr{D}$ functions, which are described in Appendix A.8. With the inner product relationship for the Wigner functions given by Equation (A.104), the required overlaps are then given by

$$\langle \Phi_{NtL} | \Phi_{N't'L'} \rangle = \delta_{LL'} \frac{8\pi^2}{2L+1} \int_0^{\pi/3} \sum_K f_K^{(NtL)}(\gamma) f_K^{(N't'L)}(\gamma) \sin 3\gamma \, d\gamma. \quad (4.258)$$

The latter integrals are readily carried out in exact arithmetic by noting that each $f_K^{(NtL)}$ is a polynomial in $\sin\gamma$ and $\cos\gamma$ and, thus, has a simple expansion in powers of $e^{\pm i\gamma}$. The SO(5) spherical harmonics are then easily determined by the above-outlined Gram-Schmidt orthonormalisation procedure with the help of a computer to keep track of the coefficients.

A simplification of the procedure arises because, as shown in Claim 2.1 of Section 2.3.3, any function of the quadrupole moments of the Bohr model has an intrinsic D_2 symmetry. Thus, as for the rotor-model wave functions of Equation (2.116), only even-integer values of K occur in Equation (4.257) and $f_{-K}^{(NtL)} = (-1)^L f_K^{(NtL)}$. This symmetry is explicitly incorporated into the equivalent expansion

$$\Phi_{NtL}(\gamma, \Omega) = \sum_{K \geq 0}^{\text{even}} F_K^{(NtL)}(\gamma) \xi_{KL}^{(L)}(\Omega), \quad (4.259)$$

where

$$\xi_{KM}^{(L)} := \frac{1}{\sqrt{1 + \delta_{K,0}}} \left[\mathscr{D}_{KM}^L + (-1)^L \mathscr{D}_{-K,M}^L \right], \quad (4.260)$$

is proportional to a rotor-model wave function. When expanded as in (4.259), the basis functions $\{\Phi_{NtL}\}$ for the SO(3) highest-weight subspace of $\mathcal{L}^2(S_4)$ have an immediate extension to a basis $\{\Phi_{NtLM}\}$ for all $\mathcal{L}^2(S_4)$, given by

$$\Phi_{NtLM}(\gamma, \Omega) = \sum_{K \geq 0}^{\text{even}} F_K^{(NtL)}(\gamma) \xi_{KM}^{(L)}(\Omega). \quad (4.261)$$

An expansion in the form (4.259) of the functions, $\{\Phi_{NtL}\}$, defined in terms of the generating functions by Theorem 4.7, starts with the generating functions of Equation (4.247) given with convenient normalisations by the expansions (cf. Exercise (4.20))

$$\Phi_1(\gamma, \Omega) = \cos\gamma \, \xi_{02}^{(2)}(\Omega) + \sin\gamma \, \xi_{22}^{(2)}(\Omega), \quad (4.262)$$

$$\Phi_2(\gamma, \Omega) = \cos 2\gamma \, \xi_{02}^{(2)}(\Omega) - \sin 2\gamma \, \xi_{22}^{(2)}(\Omega), \quad (4.263)$$

$$\Phi_3(\gamma, \Omega) = \cos 3\gamma \, \xi_0^{(0)}(\Omega), \quad (4.264)$$

$$\Phi_4(\gamma, \Omega) = \sin 3\gamma \, \xi_{23}^{(3)}(\Omega). \quad (4.265)$$

The functions $\{\Phi_{NtL}\}$ of Equation (4.251) are then built up step by step in the form (4.259) by use of the coupling formula for Wigner $\mathscr{D}$ functions

$$\mathscr{D}^{L_2}_{K_2L_2}(\Omega)\mathscr{D}^{L_1}_{K_1L_1}(\Omega) = (L_1K_1\,L_2K_2|LK)\mathscr{D}^{L}_{KL}(\Omega), \qquad (4.266)$$

where $L = L_1 + L_2$ and $K = K_1 + K_2$ (see Exercises 4.23—4.25).

Spherical harmonics for $v \leq 6$, computed with a Maple code[72] based on this algorithm, are given in Table 4.2. However, in using this table, it is important to be aware that, when there is a multiplicity of SO(5) spherical harmonics with common values of v, L, and M, the spherical harmonics are only defined to within unitary transformations that preserve these quantum numbers. Thus, for example, the two $L = 6, v = 6$ spherical harmonics, given in Table 4.2, are equally well replaced by unitary transformations of the ones given.

The expansions of SO(5) spherical harmonics in terms of the Φ_{NtLM} basis, as given in Table 4.2, are concise which is useful for storage purposes. However, to make use of them, it is also necessary to have have available the Φ_{NtLM} basis functions.

4.9.2   *Computation of SO(5) Clebsch-Gordan coefficients*

Once the SO(5) spherical harmonics, $\{\mathcal{Y}_{v\alpha LM}\}$, have been determined as (M-independent) linear combinations of the $\{\Phi_{NtLM}\}$ basis functions, it remains to evaluate the overlap integrals

$$\int_{S_4} \Phi^*_{N_3t_3L_3M_3}(\gamma,\Omega)\,[\Phi_{N_2t_2L_2}(\gamma,\Omega)\otimes\Phi_{N_1t_1L_1}(\gamma,\Omega)]_{L_3M_3}\,\sin 3\gamma\,\mathrm{d}\gamma\,\mathrm{d}\Omega \quad (4.267)$$

and, hence, the overlap integrals

$$\int_{S_4} \mathcal{Y}^*_{v_3\alpha_3L_3M_3}(\gamma,\Omega)\,[\mathcal{Y}_{v_2\alpha_2L_2}(\gamma,\Omega)\otimes\mathcal{Y}_{v_1\alpha_1L_1}(\gamma,\Omega)]_{L_3M_3}\,\sin 3\gamma\,\mathrm{d}\gamma\,\mathrm{d}\Omega \quad (4.268)$$

to determine the SO(5) Clebsch-Gordan coefficients according to the strategy outlined in Section 4.5.3.

Making use of the combination rule for Wigner $\mathscr{D}$ functions, obtained from Equation (A.102),

$$\left[\mathscr{D}^{L_2}_{K_2M_2}(\Omega)\otimes\mathscr{D}^{L_1}_{K_1M_1}(\Omega)\right]_{LM} = (L_1K_1\,L_2K_2|L,K_1+K_2)\mathscr{D}^{L}_{K_1+K_2,M}(\Omega), \quad (4.269)$$

it follows that Equation (4.267) evaluates to

$$\frac{8\pi^2}{(2L_3+1)}\sum_{K_1K_2K_3}(L_1K_1\,L_2K_2|L_3K_3)$$

$$\times \int_0^{\pi/3} f^{(N_3t_3L_3)}_{K_3}(\gamma)\,f^{(N_2t_2L_2)}_{K_2}(\gamma)\,f^{(N_1t_1L_1)}_{K_1}(\gamma)\,\sin 3\gamma\,d\gamma. \quad (4.270)$$

---

[72]Rowe *et al.*, *op. cit.* Footnote 42 on Page 194.

Table 4.2: SO(5) spherical harmonics for $v \leq 6$ and $M = L$.

v	α	L	$4\pi/(2L+1)^{1/2}\,\mathcal{Y}_{v\alpha LL}$
0	1	0	$\sqrt{3}$
1	1	2	$\sqrt{3/2}\,\Phi_{102}$
2	1	2	$\sqrt{3/2}\,\Phi_{202}$
		4	$\sqrt{35/12}\,\Phi_{204}$
3	1	0	$\sqrt{9/2}\,\Phi_{310}$
		3	$\sqrt{9/4}\,\Phi_{303}$
		4	$\sqrt{105/22}\,\Phi_{304}$
		6	$\sqrt{315/52}\,\Phi_{306}$
4	1	2	$\sqrt{27/4}\,[\Phi_{412} - \tfrac{1}{3}\Phi_{202}]$
		4	$\sqrt{189/52}\,[\Phi_{404} - \tfrac{4}{9}\Phi_{204}]$
		5	$\sqrt{15/2}\,\Phi_{405}$
		6	$\sqrt{693/52}\,\Phi_{406}$
		8	$\sqrt{3465/272}\,\Phi_{408}$
5	1	2	$\sqrt{27/4}\,[\Phi_{512} - \tfrac{1}{3}\Phi_{102}]$
		4	$\sqrt{77/4}\,[\Phi_{514} - \tfrac{6}{11}\Phi_{304}]$
		5	$\sqrt{15/2}\,\Phi_{505}$
		6	$\sqrt{1089/68}\,[\Phi_{506} - \tfrac{4}{11}\Phi_{306}]$
		7	$\sqrt{1001/48}\,\Phi_{507}$
		8	$\sqrt{45045/1292}\,\Phi_{508}$
		10	$\sqrt{429/16}\,\Phi_{50,10}$
6	1	0	$\sqrt{135/8}\,\left[\Phi_{620} - \tfrac{\sqrt{2}}{3}\right]$
		3	$\sqrt{45/4}\,\Phi_{613}$
		4	$\sqrt{975/34}\,[\Phi_{614} - \tfrac{3}{13}\Phi_{404} - \tfrac{3}{13}\Phi_{204}]$
		6	$\sqrt{225225/22364}\,[\Phi_{606} - \tfrac{12}{13}\Phi_{406}]$
	2	6	$\sqrt{307505/5168}\,[\Phi_{616} + \tfrac{568}{5591}\Phi_{606} - \tfrac{4395}{5591}\Phi_{406}]$
	1	7	$\sqrt{5005/136}\,\Phi_{607}$
		8	$\sqrt{139425/2584}\,[\Phi_{608} - \tfrac{4}{13}\Phi_{408}]$
		9	$\sqrt{4095/76}\,\Phi_{609}$
		10	$\sqrt{32175/368}\,\Phi_{6,0,10}$
		12	$\sqrt{9009/160}\,\Phi_{6,0,12}$

A few coefficients, determined in this way, are given in Table 4.3. Many more are available as indicated in Footnote 47 on Page 196.

Table 4.3: Some reduced SO(5) Clebsch-Gordan coefficients $(v_3\alpha_3 L_3, v_2\alpha_2 L_2 \| v_1\alpha_1 L_1)$. The first two columns of each of the five arrays give the values of (v_1, L_1) and (v_2, L_2). Subsequent columns are headed by the values of (v_3, L_3). For the examples shown, the multiplicity index α is redundant because, for $v < 6$, the multiplicity is one.

(v_1, L_1)	(v_2, L_2)	$(2,2)$	$(2,4)$
$1,2$	$1,2$	-1	1

(v_1, L_1)	(v_2, L_2)	$(3,0)$	$(3,3)$	$(3,4)$	$(3,6)$
$(2,2)$	$(1,2)$	1	$-\dfrac{\sqrt{35}}{7}$	$\dfrac{\sqrt{231}}{21}$	0
$(2,4)$	$(1,2)$	0	$-\dfrac{\sqrt{14}}{7}$	$-\dfrac{\sqrt{210}}{21}$	1

(v_1, L_1)	(v_2, L_2)	$(4,2)$	$(4,4)$	$(4,5)$	$(4,6)$	$(4,8)$
$(3,0)$	$(1,2)$	$\dfrac{\sqrt{330}}{30}$	0	0	0	0
$(3,3)$	$(1,2)$	$-\dfrac{\sqrt{66}}{12}$	$-\dfrac{\sqrt{4290}}{90}$	$\dfrac{\sqrt{210}}{20}$	0	0
$(3,4)$	$(1,2)$	$\dfrac{\sqrt{70}}{20}$	$-\dfrac{\sqrt{2002}}{66}$	$-\dfrac{\sqrt{462}}{44}$	$\dfrac{\sqrt{330}}{22}$	0
$(3,6)$	$(1,2)$	0	$-\dfrac{4\sqrt{165}}{495}$	$-\dfrac{\sqrt{715}}{55}$	$-\dfrac{\sqrt{154}}{22}$	1

(v_1, L_1)	(v_2, L_2)	$(5,2)$	$(5,4)$	$(5,5)$	$(5,6)$	$(5,7)$	$(5,8)$	$(5,10)$
$(4,2)$	$(1,2)$	$-\dfrac{\sqrt{910}}{35}$	$\dfrac{\sqrt{3003}}{77}$	0	0	0	0	0
$(4,4)$	$(1,2)$	$\dfrac{3\sqrt{35}}{35}$	$\dfrac{24\sqrt{21}}{385}$	$-\dfrac{3\sqrt{231}}{55}$	$\dfrac{3\sqrt{23205}}{715}$	0	0	0
$(4,5)$	$(1,2)$	0	$-\dfrac{\sqrt{1001}}{55}$	$-\dfrac{\sqrt{6}}{5}$	$-\dfrac{2\sqrt{6545}}{385}$	$\dfrac{2\sqrt{210}}{35}$	0	0
$(4,6)$	$(1,2)$	0	$\dfrac{7\sqrt{5}}{55}$	$-\dfrac{2\sqrt{55}}{55}$	$-\dfrac{6\sqrt{34}}{55}$	$-\dfrac{\sqrt{3}}{5}$	$\dfrac{\sqrt{19}}{5}$	0
$(4,8)$	$(1,2)$	0	0	0	$-\dfrac{16\sqrt{1001}}{5005}$	$-\dfrac{\sqrt{238}}{35}$	$-\dfrac{\sqrt{6}}{5}$	1

(v_1, L_1)	(v_2, L_2)	$(2,2)$	$(2,4)$	$(4,2)$	$(4,4)$	$(4,5)$	$(4,6)$	$(4,8)$
$(2,2)$	$(2,2)$	$\dfrac{\sqrt{735}}{147}$	$\dfrac{4\sqrt{735}}{441}$	$-\dfrac{2\sqrt{66}}{21}$	$\dfrac{\sqrt{2145}}{63}$	0	0	0
$(2,4)$	$(2,2)$	$\dfrac{4\sqrt{147}}{147}$	$-\dfrac{5\sqrt{3234}}{441}$	$\dfrac{\sqrt{330}}{42}$	$\dfrac{2\sqrt{78}}{63}$	$\dfrac{\sqrt{2}}{2}$	$\dfrac{\sqrt{70}}{14}$	0
$(2,2)$	$(2,4)$	$\dfrac{4\sqrt{147}}{147}$	$-\dfrac{5\sqrt{3234}}{441}$	$\dfrac{\sqrt{330}}{42}$	$\dfrac{2\sqrt{78}}{63}$	$-\dfrac{\sqrt{2}}{2}$	$\dfrac{\sqrt{70}}{14}$	0
$(2,4)$	$(2,4)$	$-\dfrac{\sqrt{16170}}{147}$	$\dfrac{\sqrt{21021}}{441}$	$\dfrac{2\sqrt{3}}{21}$	$\dfrac{20\sqrt{3}}{63}$	0	$-\dfrac{\sqrt{14}}{7}$	1

Exercises

4.19 Show that the expansion

$$\psi_{L_2 M_2} \times \psi_{L_1 M_1} = \sum_{LM} C_{LM} \Psi_{L_1 L_2 LM} \qquad (4.271)$$

has a non-vanishing component with $L = L_1 + L_2$ and $M = M_1 + M_2$. Hint: Show that the raising operator $\hat{L}_+ = L_{2+} + L_{1+}$ can be applied to the product wave function $(L_1 + L_2 - M_1 - M_2)$ times before it gives a vanishing result.

4.20 Starting from the expression

$$\mathcal{Q}_m(\gamma, \Omega) = \cos\gamma \, \mathscr{D}^2_{0m}(\Omega) + \frac{1}{\sqrt{2}} \sin\gamma \left(\mathscr{D}^2_{2m}(\Omega) + \mathscr{D}^2_{-2,m}(\Omega) \right), \qquad (4.272)$$

derive Equations (4.263)–(4.265).

4.21 Show that the function $\Phi = \left[\mathcal{Q} \times \mathcal{Q} \right]_0$ with values

$$\Phi(\gamma, \Omega) = \sum_M (2M\, 2, -M|00) \, \mathcal{Q}_{-M}(\gamma, \Omega) \mathcal{Q}_M(\gamma, \Omega), \qquad (4.273)$$

where $(2M\, 2, -M|00)$ is an SO(3) Clebsch-Gordan coefficient, is a constant.

4.22 Show that, if $\{\varphi_i\}$ is a non-orthogonal basis with inner products given by $\langle \bullet | \bullet \rangle$, then an orthonormal basis $\{\psi_i\}$ is given by

$$\begin{aligned}
\psi_1 &= N_1 \varphi_1, \\
\psi_2 &= N_2 \left[\varphi_2 - \psi_1 \langle \psi_1 | \varphi_2 \rangle \right], \\
\psi_3 &= N_3 \left[\varphi_3 - \psi_1 \langle \psi_1 | \varphi_3 \rangle - \psi_2 \langle \psi_2 | \varphi_3 \rangle \right], \\
\psi_4 &= N_4 \left[\varphi_4 - \psi_1 \langle \psi_1 | \varphi_4 \rangle - \psi_2 \langle \psi_2 | \varphi_4 \rangle - \psi_3 \langle \psi_3 | \varphi_4 \rangle \right], \\
&\text{etc.,}
\end{aligned} \qquad (4.274)$$

where N_i is defined such that $\langle \psi_i | \psi_i \rangle = 1$.

4.23 Use the definition of Φ_{NtL} given by Theorem 4.7 and the expressions for SO(5) spherical harmonics given in Table 4.2, to derive the identities

$$\begin{aligned}
&\mathcal{Y}_{1122} = \frac{1}{4\pi}\sqrt{\frac{15}{2}}\, \Phi_1, \quad \mathcal{Y}_{2122} = \frac{1}{8\pi}\sqrt{\frac{15}{2}}\, \Phi_2, \quad \mathcal{Y}_{2144} = \frac{1}{8\pi}\sqrt{105}\, \Phi_1^2, \\
&\mathcal{Y}_{3100} = \frac{3}{4\pi}\sqrt{\frac{1}{2}}\, \Phi_3, \quad \mathcal{Y}_{3133} = \frac{3}{8\pi}\sqrt{7}\, \Phi_4.
\end{aligned} \qquad (4.275)$$

4.24 Use Equation (4.266) to show that

$$\begin{aligned}
\xi^{(2)}_{02} \xi^{(2)}_{02} &= \frac{6}{\sqrt{35}}\, \xi^{(4)}_{04}, \quad \xi^{(2)}_{02} \xi^{(2)}_{22} = \sqrt{\frac{3}{7}}\, \xi^{(4)}_{24}, \\
\xi^{(2)}_{22} \xi^{(2)}_{22} &= \xi^{(4)}_{44} + \frac{1}{\sqrt{35}}\, \xi^{(4)}_{04},
\end{aligned} \qquad (4.276)$$

and that

$$\Phi_1^2 = \sqrt{\frac{1}{35}}\, (6\cos^2\gamma + \sin^2\gamma)\, \xi^4_{04} + \sqrt{\frac{3}{7}} \sin 2\gamma\, \xi^4_{24} + \sin^2\gamma\, \xi^4_{44}. \qquad (4.277)$$

4.25 Use the expressions for Φ_1 to Φ_4 given by Equations (4.262)–(4.265) to obtain

$$\mathcal{Y}_{112M} = \frac{1}{4\pi}\sqrt{\frac{15}{2}}\left[\cos\gamma\,\xi_{0M}^{(2)} + \sin\gamma\,\xi_{2M}^{(2)}\right],$$

$$\mathcal{Y}_{212M} = \frac{1}{8\pi}\sqrt{\frac{15}{2}}\left[\cos 2\gamma\,\xi_{0M}^{(2)} - \sin 2\gamma\,\xi_{2M}^{(2)}\right],$$

$$\mathcal{Y}_{214M} = \frac{1}{16\pi}\left[\sqrt{\frac{3}{2}}(5\cos 2\gamma + 7)\xi_{0M}^{(4)} + 3\sqrt{10}\sin 2\gamma\,\xi_{2M}^{(4)} + \sqrt{\frac{105}{2}}(1 - \cos 2\gamma)\xi_{4M}^{(4)}\right],$$

$$\mathcal{Y}_{310} = \frac{3}{8\pi}\sqrt{2}\cos 3\gamma\,\xi_0^{(0)}, \qquad \mathcal{Y}_{313M} = \frac{3}{8\pi}\sqrt{7}\sin 3\gamma\,\xi_{2M}^{(3)}.$$

$$(4.278)$$

Compare your results with those of Caprio *et al.*, *op. cit.* Footnote 43 on Page 194.

Chapter 5

The shell model

5.1 Introduction

There is strong experimental evidence for shell structure in nuclei. This is reviewed in Section 1.3 where it is shown that the observed shell structure is reproduced by an IPM (independent-particle model) Hamiltonian which includes a central potential and a spin-orbit interaction. Also, as reviewed in Chapter 1, such an IPM provides an explanation of many nuclear properties: notably the extra stability of closed-shell nuclei, properties of nucleon binding energies, and spins and parities of many odd-mass nuclei. Particularly important for the development of a theory of nuclear structure is that the IPM provides a *basis* for the shell model which, in turn, provides the framework for the many-nucleon quantum theory of nuclei.

A primary objective of this book is to explore the experimental and theoretical foundations of nuclear structure models with a particular focus on their relationships to the microscopic theory of the nucleus as a system of interacting neutrons and protons. With this objective in mind, the Bohr collective model is phrased in Chapter 2 in terms of quadrupole moments which have expressions in terms of nucleon coordinates, instead of surface shape parameters which do not. In Chapter 4, the collective model is developed in terms of its dynamical symmetries and algebraic structures in preparation for giving it a microscopic interpretation, i.e., viewing it as describing the collective subdynamics of a many-nucleon system. The chapters of Volume 2 will endeavour to identify these many-nucleon subdynamics and, in the process, discover the extent to which collective models can be identified with submodels of the shell model. The shell model is presented in this chapter with these goals in mind.

It is shown in this and subsequent chapters that the many facets of nuclear structure, such as the emergence of collective states, arise from the ways that nucleons correlate their dynamics. In particular, it is shown that most of the successful models of nuclear structure have microscopic counterparts associated with shell-model coupling schemes. A connection between models and coupling schemes is made by showing that, like collective models, shell-model coupling schemes are naturally defined by chains of dynamical symmetry groups. Thus, the presentation of the shell model in this chapter emphasises more than usual the role played by groups of transformations and their subgroups in the construction of basis states and coupling schemes.

The essential strategy of the shell model is to separate the many-nucleon Hamiltonian into two parts:

$$\hat{H} = \hat{H}_0 + \hat{V}_{\text{res}}, \tag{5.1}$$

where $\hat{H}_0$ is a Hamiltonian for a solvable IPM and $\hat{V}_{\text{res}} := \hat{H} - \hat{H}_0$ is a residual interaction. Such a separation could be made in many ways. For example, $\hat{H}_0$ could be the sum of the nucleon kinetic energies and $\hat{V}_{\text{res}}$ the sum of their interaction potentials. However, it is more useful if the eigenstates of $\hat{H}_0$ provide basis states, for a given nucleus, that are *stratified into shells* of increasing energy such that low-lying states of the nucleus can be expanded as a rapidly convergent sequence in a corresponding partially-ordered basis;[1] expansion of nuclear states on a basis of kinetic energy eigenstates would amount to multi-dimensional Fourier analysis and would converge extremely slowly. A much more useful IPM Hamiltonian, $\hat{H}_0$, is one for which the low-energy eigenstates already have the observed bulk properties of corresponding physical states; e.g., they should have (approximately) the observed nuclear density distributions.

Large multiplicities of states of equal energy occur for a rotationally-invariant Hamiltonian, $\hat{H}_0$, with a central potential. A single-particle orbital of spin (i.e., total angular momentum) j has a multiplicity of $2j+1$ equal-energy single-particle states. A *nuclear shell* is then a cluster of independent-particle states (also called orbitals) of similar energies. The multiplicity of states of n nucleons in a so-called $(j)^n$ *configuration* initially increases with n, reaches a maximum when the j-configuration is half filled, and then decreases due to the Pauli principle as n approaches its maximum value; a fully-occupied orbital is unique (i.e., has multiplicity one) and has total angular momentum zero. Thus, the mid-shell multiplicity is largest when the particles are distributed uniformly over the j-orbitals of the shell.

If an orbital of spin j_i has single-particle energy ε_i, then the unperturbed energy of all many-particle states of a $(j_1)^{n_1}(j_2)^{n_2} \ldots$ configuration, i.e., the energy relative to the Hamiltonian $\hat{H}_0$, is $\sum_i n_i \varepsilon_i$. Thus, an independent-particle model Hamiltonian gives a partially-ordered basis for the shell model. An important characteristic of this partial ordering arises from the clustering of single-particle energy levels into shells. For example, if j_1 and j_2 are the spins of two single-particle orbitals belonging to the same shell, so that $\varepsilon_1 \approx \varepsilon_2$, then the energies, $n_1\varepsilon_1 + n_2\varepsilon_2$, of states of all configurations $(j_1)^{n_1}(j_2)^{n_2}$ of the same $n_1 + n_2$ are similar.[2] However, gaps appear in the many-nucleon spectrum of $\hat{H}_0$ when there are corresponding gaps between single-nucleon energy levels.

---

[1] The eigenstates of $\hat{H}_0$ are only partially ordered by increasing energy; they cannot be fully ordered, in general, because of degeneracies among their energy eigenvalues.

[2] We refer to a distribution of nucleons over a specified set of single-particle states as a *shell-model configuration*.

The shell structure of an IPM Hamiltonian is particularly important because it provides a sequence of approximations to the many-nucleon problem. The zero'th order approximation is the IPM itself. A higher-level approximation is obtained by diagonalising a Hamiltonian of the form given by Equation (5.1) in a restricted Hilbert space spanned by the states of a relatively small number of lowest-energy IPM configurations. Increasing the number of configurations is then expected to give a better approximation. This is the standard shell-model strategy. When it is valid, better and better results can be obtained by sequentially increasing the size of the shell-model space. However, the limited capacity of computers severely constrains the maximum size of a shell-model space that can be handled. The quality of the results also depends critically on the choice of Hamiltonian.

One can think of the interaction $\hat{V}_{\text{res}}$ as responsible for correlating the motions of otherwise independent nucleons. As shown in Chapter 6, an important component of the residual interaction is one that induces correlations between pairs of nucleons. Hence a second approximation, after the IPM, is the so-called *independent-pair approximation*. More general cluster models can also be considered. However, it soon becomes apparent that, to obtain a microscopic description of collective model states, an enormous gain in efficiency can be achieved by selecting shell-model basis states and subspaces that are tailored to the particular collective dynamics of interest. In practice, this means choosing an appropriate many-nucleon coupling scheme. Moreover, as we later discover, the appropriate coupling scheme does not always fully respect the partial ordering of the IPM basis.

To determine if the shell model strategy is valid for a given Hamiltonian, one might attempt to determine if its eigenstates have convergent expansions in an IPM basis. This is a difficult mathematical problem that received considerable attention in the early days of the shell model.[3,4] The concern was that, because of the observed strong repulsion of nucleons at short ranges, it was unclear that there could be convergence of shell-model states with realistic nuclear Hamiltonians. Such considerations delayed the introduction of the shell model until the overwhelming experimental evidence of shell structure in nuclei emboldened nuclear physicists to take it seriously. From its brave acceptance began the modern paradigm of nuclear structure theory. A rationale for why the shell model is not destroyed by strong short-range repulsive forces is provided by a theory of effective interactions (cf. Section 5.7).

The material in this chapter is organised as follows. After an introduction to the techniques of fermion second quantization, an outline is given of isospin degrees of freedom and effective shell-model interactions in nuclear physics. Then follows an in-depth treatment of the rich algebraic structures, symmetries, and coupling schemes available in the harmonic-oscillator shell model. These symmetries and coupling schemes provide algebraic solutions to many of the technical problems that

---

[3]Brueckner K.A. (1955), *Phys. Rev.* **97**, 1353.
[4]Bethe H.A. and Goldstone J. (1957), *Proc. Roy. Soc. London* **A238**, 551.

arise in applications of the shell model; e.g., the calculation of matrix elements and separation of centre-of-mass degrees of freedom. But, most notably for our purposes, they provide natural mechanisms for identifying the relevant shell-model spaces and bases in which different kinds of nuclear dynamics are most naturally expressed. Thus, although other shell models can be constructed, based on more realistic IPM potentials, such as the Woods-Saxon (Section 1.3) or Hartree-Fock (Chapter 7) potentials, the harmonic-oscillator shell model has huge practical advantages and incorporates the macroscopic collective models much more directly.

5.2   The representation of antisymmetric states

The indistinguishability of identical particles and the associated invariance of the Hamiltonian of a many-particle system under the exchange of all coordinates (space and spin) of any two identical particles are fundamental symmetries of quantum mechanics. Moreover, because the exchange of identical particles cannot be detected experimentally, the wave functions for a many-particle system can at most change by a phase factor under such a permutation.

The group of permutations of the coordinates of a system of A particles is known as the *symmetric group*, S_A. This is a discrete group comprising $A!$ elements. Its representations are well known and, in particular, it is known to have just two one-dimensional representations in which exchange operations result only in the multiplication of a wave function by a phase factor: (i) the *fully symmetric* (identity) representation, in which the wave functions are invariant under permutations of the coordinates associated with different particles; and (ii) the *fully antisymmetric* representation, in which the wave functions change sign under interchange of the coordinates of any two particles. Particles whose wave functions are fully symmetric are known as *bosons* and particles whose wave functions are fully antisymmetric are known as *fermions*. A profound theorem, known as the *spin-statistics theorem*,[5],[6] is that bosons have integer intrinsic spins whereas fermions have half-odd integer intrinsic spins.

Totally antisymmetric states can be handled by expressing them in bases of Slater determinants. This is useful for some purposes. However, a more elegant and efficient way of keeping track of exchange symmetries is provided by the methods of second quantization. We introduce Slater determinants first and then develop the methods of second quantization under the heading of fermion creation and annihilation operators.

---

[5]Fierz M. (1939), *Helv. Phys. Acta* **12**, 3.
[6]Pauli W. (1940), *Phys. Rev.* **58**, 716.

5.2.1 *Slater determinants*

Let $\{\xi_i\}$ denote all the space and spin coordinates of the i'th particle. An A-particle wave function is then a function of the coordinates $(\xi_1, \ldots, \xi_A)$. However, to be an acceptable wave function for a many-fermion system, a simple product of single-particle wave functions

$$\Phi_\nu(\xi_1, \xi_2, \ldots, \xi_A) := \psi_{\nu_1}(\xi_1)\psi_{\nu_2}(\xi_2) \ldots \psi_{\nu_A}(\xi_A), \qquad (5.2)$$

must be replaced by the antisymmetric product

$$\Psi_\nu(\xi_1, \xi_2, \ldots, \xi_A) := \frac{1}{\sqrt{A!}} \sum_P (-1)^P [P\Phi_\nu](\xi_1, \xi_2, \ldots, \xi_A), \qquad (5.3)$$

where P is a permutation operator, that rearranges the order of the coordinates $(\xi_1, \xi_2, \ldots, \xi_A)$, the factor $(-1)^P$ is equal to ± 1 according as P is an even or an odd permutation, and the sum is over all $A!$ permutations of the A-particle coordinates. For example, the transposition P_{12} of particles 1 and 2 is odd. Thus, $(-1)^{P_{12}} = -1$ and

$$(-1)^{P_{12}}[P_{12}\Phi](\xi_1, \xi_2, \xi_3, \ldots, \xi_A) = -\Phi(\xi_2, \xi_1, \xi_3, \ldots, \xi_A). \qquad (5.4)$$

The above antisymmetrised wave function is a so-called *Slater determinant* of single-particle wave functions

$$\Psi_\nu(\xi_1, \ldots, \xi_A) = \frac{1}{\sqrt{A!}} \begin{vmatrix} \psi_{\nu_1}(\xi_1) & \psi_{\nu_1}(\xi_2) & \cdots & \psi_{\nu_1}(\xi_A) \\ \psi_{\nu_2}(\xi_1) & \psi_{\nu_2}(\xi_2) & \cdots & \psi_{\nu_2}(\xi_A) \\ \vdots & \vdots & & \vdots \\ \psi_{\nu_A}(\xi_1) & \psi_{\nu_A}(\xi_2) & \cdots & \psi_{\nu_A}(\xi_A) \end{vmatrix}. \qquad (5.5)$$

A Slater determinant can be constructed from any set of single-particle wave functions. However, if any two of the selected set are identical, the determinant vanishes. In fact, it vanishes unless the single-particle wave functions form a linearly-independent set. Thus, for most purposes, it is customary to focus on Slater determinants constructed from orthonormal subsets of single-particle wave functions.

Although useful for conceptual purposes, the expression of a Slater determinant as given by Equation (5.5), is extremely cumbersome. However, it can be seen that the A-particle state $|\Psi_\nu\rangle$ represented by the Slater determinant Ψ_ν is succinctly defined by specification of an ordered sequence of occupied single-particle states; viz.

$$|\Psi_\nu\rangle = |\nu_1\nu_2 \cdots \nu_A\rangle. \qquad (5.6)$$

Because the state (5.6) changes sign under an odd permutation, the ordering of the single-particle state labels is important, but only to within even permutations.

5.2.2 *Fermion creation and annihilation operators*

We now show that the antisymmetric states, with Slater-determinant wave functions, are a basis for a representation of a powerful algebra of fermion creation and annihilation operators.

The use of creation and annihilation operators for particles is sometimes known as *second quantization*. At the time this terminology was introduced, it was common to regard standard quantum mechanics, in which the classical dynamics of particles are replaced by wave mechanics, as *first quantization*. On the other hand, quantization of electromagnetic radiation,[7] in which electromagnetic waves are replaced by photons, was regarded as second quantization. This terminology persists although it is now understood that first and second quantization are really complementary aspects of a single quantization procedure.

Quantization of the electromagnetic field can be understood at an elementary level as extending the Hamiltonian for a finite number of harmonic oscillators

$$\hat{H} = \sum_k \hbar\omega_k \left(c_k^\dagger c^k + \tfrac{1}{2} \right),$$ (5.7)

with raising and lowering operators that satisfy the commutation relations

$$[c_i^\dagger, c_k^\dagger] = [c^i, c^k] = 0, \quad [c_i^\dagger, c^k] = \delta_i^k,$$ (5.8)

and the Hermitian adjoint relations $(c^k)^\dagger = c_k^\dagger$ (cf. Appendix A.6), to an infinite number characterised by continuously variable frequencies. The raising operators for states of the electromagnetic field, thought of as field operators, are then interpreted as creation operators for photons. They provide the link between the wave and particle theories of light.

A parallel step towards a fermionic field theory was taken by Jordan and Wigner[8] who introduced operators, $(a_k^\dagger, a^k)$, that would create and annihilate fermions. They recognised that the Pauli principle, which forbids two identical fermions from occupying the same state, would automatically be accommodated if the fermion operators were required to obey anticommutation relations rather than commutation relations.

In the language of second quantization, Slater determinants are replaced by fermion creation and annihilation operators. These operators act on a so-called *Fock space*. Such a space comprises the Hilbert spaces of all many-fermion states: the zero-particle (vacuum) state $|-\rangle$, one-fermion states $\{|\nu\rangle\}$, two-fermion states $\{|\mu\nu\rangle\}$, and so on. Single-fermion creation operators are in one-to-one correspondence with single-fermion states, which they create from the vacuum according to the identity $a_\nu^\dagger|-\rangle = |\nu\rangle$. Acting on a one-fermion state they generate a two-fermion

[7]Born M., Heisenberg W. and Jordan P. (1926), *Z. Phys.* **35**, 557, (reprinted in van der Waerden B.L., ed. (1968), *Sources of Quantum Mechanics* (Dover Publications, New York)).

[8]Jordan P. and Wigner E.P. (1928), *Z. Phys.* **47**, 631, (reprinted in Schwinger J., ed. (1958), *Selected Papers on Quantum Electrodynamics* (Dover Publications, New York)).

state, e.g., $a_\mu^\dagger |\nu\rangle = |\mu\nu\rangle$; on a general A-fermion state they generate an $(A+1)$-fermion state. Thus, a Slater determinant is a wave function for a state

$$|\nu_1\nu_2\cdots\nu_A\rangle = a_{\nu_1}^\dagger a_{\nu_2}^\dagger \cdots a_{\nu_A}^\dagger |-\rangle. \qquad (5.9)$$

The Hermitian adjoints of the creation operators, defined by $a^\nu := (a_\nu^\dagger)^\dagger$, satisfy the identity

$$\langle \nu_1\nu_2\cdots\nu_A| = \langle -|a^{\nu_A}\cdots a^{\nu_2}a^{\nu_1}. \qquad (5.10)$$

Thus, the equation

$$\langle \mu\nu|\mu\nu\rangle = \langle\nu|\nu\rangle = 1 \qquad (5.11)$$

implies that

$$\langle -|a^\nu a^\mu a_\mu^\dagger a_\nu^\dagger|-\rangle = \langle -|a^\nu a_\nu^\dagger|-\rangle = \langle -|-\rangle = 1, \qquad (5.12)$$

and that $a^\mu a_\mu^\dagger a_\nu^\dagger|-\rangle = a_\nu^\dagger|-\rangle$. More generally, one finds that

$$a^\mu|\alpha\beta\rangle = \delta_\alpha^\mu|\beta\rangle - \delta_\beta^\mu|\alpha\rangle. \qquad (5.13)$$

The $\{a^\mu\}$ operators map the A-fermion Hilbert space to the $(A-1)$-fermion space. They are fermion annihilation operators.

The above results are expressed succinctly in terms of the algebraic properties of the creation and annihilation operators. To satisfy the antisymmetry properties of a many-fermion state, the creation operators, and hence their Hermitian adjoints, are required to satisfy the anticommutation relations

$$\{a_\mu^\dagger, a_\nu^\dagger\} = \{a^\mu, a^\nu\} = 0, \qquad (5.14)$$

where an anticommutator of two operators is defined by

$$\{\hat{A}, \hat{B}\} := \hat{A}\hat{B} + \hat{B}\hat{A}. \qquad (5.15)$$

Equation (5.13) implies the additional anticommutation relations

$$\{a^\mu, a_\nu^\dagger\} = \delta_\nu^\mu. \qquad (5.16)$$

The anticommutation relations of Equations (5.14) and (5.16) are special cases of a so-called *Jordan algebra*.[9]

5.2.3 *Neutrons and protons*

Because neutrons and protons are non-identical fermions, wave functions need not be antisymmetric under neutron-proton exchange. Thus, one can construct wave functions for a nucleus that are simple products of many-neutron and many-proton

---

[9]Jordan P., von Neumann J. and Wigner E.P. (1934), *Annals of Math.* **35**, 29.

wave functions each of which is separately antisymmetric. However, for both physical and practical reasons, it is useful to regard neutrons and protons as different states of a single particle, known as a *nucleon*. The physical motivation is that the strong interactions between pairs of nucleons in the same relative states, are almost equal regardless of whether the two nucleons are a pair of neutrons, a pair of protons, or one of each (cf. Figure 1.28). The practical motivation for talking of two-state nucleons is that it gives simplicity and elegance to the formalism without cost (or the necessity of approximation) and helps enormously in keeping track of nuclear symmetries (cf. Sections 5.5.1 and 5.11).

In the nucleon picture, one simply adds an extra label τ to the set of quantum numbers that characterise a single-particle state, where τ takes one of two values: $\tau = \mathrm{n}$ (equivalently $\tau = 1/2$) for a neutron and $\tau = \mathrm{p}$ (equivalently $\tau = -1/2$) for a proton.[10] Thus, for example, a neutron in a single-particle state ν is assigned a wave function $\psi_{\mathrm{n}\nu} := \psi_{1/2,\nu}$ whereas a proton in the same single-particle state is assigned a wave function $\psi_{\mathrm{p}\nu} := \psi_{-1/2,\nu}$. The $\tau = \pm 1/2$ states of a nucleon parallel the two possible intrinsic spin states of a nucleon as being up or down (equivalently as having directional component of spin $m = \pm 1/2$). Thus, whereas a nucleon has intrinsic spin $s = 1/2$, it is also said to have *isospin* $t = 1/2$. The advantages of the isospin language are discussed in Section 5.5.

Regarding neutrons and protons as different states of a nucleon means that a generalised Pauli exclusion principle applies. It also means that nucleon operators, defined with an extended set of indices, satisfy the anticommutation relations

$$\{a^\dagger_{\tau_1\mu}, a^\dagger_{\tau_2\nu}\} = \{a^{\tau_1\mu}, a^{\tau_2\nu}\} = 0, \quad \{a^{\tau_1\mu}, a^\dagger_{\tau_2\nu}\} = \delta^{\tau_1}_{\tau_2}\delta^\mu_\nu. \tag{5.17}$$

In the following, we assume the τ index to be absorbed into the set ν of all quantum numbers needed to specify a single-particle state $|\nu\rangle$.

5.2.4 *Operators on Fock space*

Although fermion creation and annihilation operators are introduced as a technical device to simplify the manipulation of many-fermion states, they have a profound influence on the way one thinks of many-fermion systems. In particular, they generate a single Fock space for all nuclei rather than a separate Hilbert space for each nucleus. Thus, for example, although nucleon numbers are conserved quantities in non-relativistic quantum mechanics, it nevertheless becomes possible to express the "scattering" of a nucleon in a nucleus, e.g., through absorption of a gamma ray, in terms of an operator that annihilates the nucleon in an initial state and recreates it in a new final state. Such an operator is described as a *one-body* operator. An interaction between a pair of nucleons, which causes both to change their states, can

---

[10]The opposite, and seemingly more natural, convention is used in particle physics. The reason neutrons are assigned a positive component of isospin in nuclear physics is because most nuclei have more neutrons than protons and so the sum $T_0 = \sum \tau$ for most nuclei is positive.

likewise be represented by a *two-body* operator which annihilates the two nucleons and recreates them in new states.

(i) *One-body operators*

An arbitrary non-singular[11] one-fermion operator $\hat{T}$ is completely defined by its matrix elements,

$$T_{\mu\nu} := \langle\mu|\hat{T}|\nu\rangle, \tag{5.18}$$

in an orthonormal basis of single-particle states. These matrix elements characterise the action,

$$\hat{T}|\nu\rangle = \sum_{\mu}|\mu\rangle\langle\mu|\hat{T}|\nu\rangle, \tag{5.19}$$

of $\hat{T}$ on any one-fermion state. Now, within the Hilbert space of one-particle states, the operator $\hat{T}$ can be represented in the language of second quantization by

$$\hat{T} = \sum_{\mu\nu}T_{\mu\nu}a_{\mu}^{\dagger}a^{\nu}. \tag{5.20}$$

This follows from the observation that

$$\langle\mu|\hat{T}|\nu\rangle = \sum_{\mu'\nu'}T_{\mu'\nu'}\langle-|a^{\mu}a_{\mu'}^{\dagger}a^{\nu'}a_{\nu}^{\dagger}|-\rangle = T_{\mu\nu}. \tag{5.21}$$

The remarkable fact is that the expression of a one-fermion operator in the form (5.20) immediately extends the action of $\hat{T}$ in the appropriate way to the whole many-fermion Fock space. For example, if $|\nu_1\nu_2\ldots\rangle$ denotes a Slater determinant of an orthonormal set of single-particle states, then the expectation value of $\hat{T}$ in this state is given by

$$\langle\nu_1\nu_2\ldots|\hat{T}|\nu_1\nu_2\ldots\rangle = \sum_{i}T_{\nu_i\nu_i}. \tag{5.22}$$

Moreover, the matrix element of $\hat{T}$ between two Slater determinants of orthonormal single-particle states that differ by the state, $\mu_1 \neq \nu_1$, of one particle is given, for example, by

$$\langle\mu_1\nu_2\nu_3\ldots|\hat{T}|\nu_1\nu_2\nu_3\ldots\rangle = T_{\mu_1\nu_1}, \tag{5.23}$$

whereas a matrix element between such Slater determinants differing by two or more fermions in orthogonal states is zero. It can be checked that these results are identical to those obtained in the more cumbersome Slater determinant notation by the extension of $\hat{T}$ to an N-nucleon Hilbert space given by setting $\hat{T} = \sum_{i=1}^{N}\hat{T}_i$, where i is used here to index a nucleon.

---

[11]The qualification that the operator must be non-singular is to ensure that the matrix elements of the operator have well-defined (i.e., finite) values.

Examples of one-body operators are the kinetic energy and the quadrupole moments of a nucleus defined, respectively, as sums of single-nucleon kinetic energies and single-nucleon quadrupole moments. Another example is the nucleon number operator

$$\hat{n} := \sum_{\nu} a_{\nu}^{\dagger} a^{\nu}, \tag{5.24}$$

which satisfies the commutation relations

$$[\hat{n}, a_{\nu}^{\dagger}] = a_{\nu}^{\dagger}, \quad [\hat{n}, a_{\mu}^{\dagger} a_{\nu}^{\dagger}] = 2 a_{\mu}^{\dagger} a_{\nu}^{\dagger}. \tag{5.25}$$

(ii) *Two-body operators*

An arbitrary non-singular two-fermion operator (cf. Footnote 11 on Page 241), such as an interaction potential between two fermions, $\hat{V}$, is completely defined by its two-fermion matrix elements

$$V_{\mu\nu,\mu'\nu'} = \langle \mu\nu | \hat{V} | \mu'\nu' \rangle, \tag{5.26}$$

which, for a two-body interaction expressed in terms of coordinates of identical particles by $\hat{V}(\xi, \xi') = \hat{V}(\xi', \xi)$, have values given (for an orthonormal single-particle basis) by

$$\langle \mu\nu | \hat{V} | \mu'\nu' \rangle = \iint \psi_{\mu}^{*}(\xi) \, \psi_{\nu}^{*}(\xi') \, V(\xi, \xi') \, \psi_{\mu'}(\xi) \, \psi_{\nu'}(\xi') \, dv(\xi) \, dv(\xi')$$

$$- \iint \psi_{\mu}^{*}(\xi) \, \psi_{\nu}^{*}(\xi') \, V(\xi, \xi') \, \psi_{\mu'}(\xi') \, \psi_{\nu'}(\xi) \, dv(\xi) \, dv(\xi'), \quad (5.27)$$

where the integrals are over all coordinates for the two particles. The second term of this expression is the so-called *exchange* term. With the inclusion of this term, the two-fermion matrix elements satisfy the antisymmetry relationships

$$V_{\mu\nu,\mu'\nu'} = -V_{\nu\mu,\mu'\nu'} = -V_{\mu\nu,\nu'\mu'} = V_{\nu\mu,\nu'\mu'}. \tag{5.28}$$

In the language of second quantization, the interaction, $\hat{V}$, is represented by

$$\hat{V} = \tfrac{1}{4} \sum_{\mu\nu\mu'\nu'} V_{\mu\nu,\mu'\nu'} a_{\mu}^{\dagger} a_{\nu}^{\dagger} a^{\nu'} a^{\mu'}, \tag{5.29}$$

where the factor $1/4$ takes account of the fact that

$$\sum_{\mu\nu\mu'\nu'} V_{\mu\nu,\mu'\nu'} \langle \alpha\beta | a_{\mu}^{\dagger} a_{\nu}^{\dagger} a^{\nu'} a^{\mu'} | \gamma\delta \rangle = V_{\alpha\beta,\gamma\delta} - V_{\alpha\beta,\delta\gamma} - V_{\beta\alpha,\gamma\delta} + V_{\beta\alpha,\delta\gamma}$$

$$= 4 V_{\alpha\beta,\gamma\delta}. \tag{5.30}$$

Again the representation of a two-fermion operator in this formalism extends its action in the appropriate way to the whole many-fermion Fock space. This can be checked by ascertaining that all matrix elements are the same as those obtained with Slater determinants.

Exercises

5.1 In the so-called np formalism, neutrons and protons are considered to be distinct particles and their wave functions are not antisymmetrised. In the isospin formalism, neutrons and protons are regarded as different states of a nucleon, and consequently their wave functions are required to be antisymmetric with respect to exchange of the states (including the isospin states) of any pair of nucleons. Show in an example that the matrix elements of a one-body operator between many-nucleon states are the same regardless of whether the many-body wave functions are antisymmetrised or not with respect to neutron-proton exchange; the same is true of a two-body operator.

5.2 Show that, in any orthonormal single-fermion basis, the operator $\hat{n} = \sum_\nu a_\nu^\dagger a^\nu$ is the number operator, i.e., its eigenvalue is the number of fermions in any state on which it operates.

5.3 Show that $\{a_\nu^\dagger, a_\nu^\dagger\} = 0$ implies the Pauli principle (i.e., that no two fermions can occupy the same single-particle state) and that, if $\hat{n}_\nu := a_\nu^\dagger a^\nu$, then $\hat{n}_\nu^2 = \hat{n}_\nu$. Note that this identity implies that $\hat{n}_\nu$ is a projection operator and its eigenvalues are either zero or one.

5.4 Show that matrix elements of a two-body interaction satisfy the antisymmetry relations of Equation (5.28).

5.3 Upper and lower indices

For many purposes, it is useful to regard a set of creation operators $\{a_\nu^\dagger\}$ as the components of a spherical tensor (cf. Appendix A.6). For example, if single-particle states $\{|jm\rangle; m = -j, \ldots, +j\}$ are labelled by the angular-momentum quantum numbers, jm, for an irreducible representation of SU(2), then these states transform under an SU(2) rotation, $\Omega \in$ SU(2) (see Appendix A.8), according to the equation

$$|jm\rangle \to \hat{R}(\Omega)|jm\rangle = \sum_n |jn\rangle \, \mathcal{D}_{nm}^j(\Omega). \tag{5.31}$$

The corresponding creation operators, $\{a_{jm}^\dagger; m = -j, \ldots, +j\}$, which transform according to the equation

$$a_{jm}^\dagger \to \hat{R}(\Omega)a_{jm}^\dagger \hat{R}(\Omega^{-1}) = \sum_n a_{jn}^\dagger \mathcal{D}_{nm}^j(\Omega), \tag{5.32}$$

are then said to be the components of a spherical tensor $a_j^\dagger$.

By taking the Hermitian adjoints of both sides of Equation (5.32), we find that

$$a^{jm} \to \hat{R}(\Omega)a^{jm}\hat{R}(\Omega^{-1}) = \sum_n a^{jn}\mathcal{D}_{nm}^{j*}(\Omega), \tag{5.33}$$

which shows that the annihilation operators do not transform in quite the same way as the creation operators. In fact, because the sum $\sum_m a_{jm}^\dagger a^{jm}$ is invariant under rotations, it follows that the annihilation operators, $\{a^{jm}; m = -j, \ldots, j\}$, transform under rotations as components of a *contravariant* spherical tensor and consequently, following Einstein's convention, we label them with upper indices (cf. Appendix A.6 for a further discussion of covariant and contravariant tensors.)

Spherical tensors can be combined to form scalar products by vector coupling them to angular momentum zero, e.g.,

$$[a_j^\dagger \otimes a_j^\dagger]_0 := \sum_m (j, -m\, jm|00)\, a_{jm}^\dagger a_{j,-m}^\dagger = \frac{1}{\sqrt{2j+1}} \sum_m (-1)^{j+m}\, a_{jm}^\dagger a_{j,-m}^\dagger. \quad (5.34)$$

This suggests that the factor $(-1)^{j+m}$ can be regarded as a component,

$$g^{jm\,j'm'} := \delta_{j',j}\delta_{m',-m}(-1)^{j+m}, \quad (5.35)$$

of a metric tensor g in terms of which the scalar product of the covariant tensors $a_{j'}^\dagger$ and $a_j^\dagger$, for example, is given by the expression[12]

$$a_j^\dagger \cdot a_{j'}^\dagger := \sum_{mm'} a_{jm}^\dagger\, g^{jm\,j'm'}\, a_{j'm'}^\dagger = \delta_{j',j}\sqrt{2j+1}\,[a_j^\dagger \otimes a_j^\dagger]_0. \quad (5.36)$$

The metric tensor g can also be used to raise and lower indices. Thus, we can define a covariant tensor, a_j, with components $\{a_{jm}\}$ related to those of a^j by

$$a^{jm} = \sum_{m'} g^{jm\,jm'} a_{jm'} = (-1)^{j+m} a_{j,-m}, \quad a_{jm} = (-1)^{j-m} a^{j,-m}. \quad (5.37)$$

The nucleon number operator can then be expressed in the various forms

$$\hat{n} = \sum_{jm} a_{jm}^\dagger a^{jm} = \sum_j a_j^\dagger \cdot a_j = \sum_j \sqrt{2j+1}\,[a_j^\dagger \otimes a_j]_0, \quad (5.38)$$

and a coupled product, $[a_j^\dagger \otimes a_{j'}]_{JM}$, is expressed generally by

$$[a_j^\dagger \otimes a_{j'}]_{JM} := \sum_{mm'} (j'm'\, jm|JM)\, a_{jm}^\dagger a_{j'm'}. \quad (5.39)$$

We also obtain the useful Hermiticity relationships

$$\left(a_{jm}^\dagger\right)^\dagger = a^{jm} = (-1)^{j+m} a_{j,-m}, \quad \left(a_{jm}\right)^\dagger = (-1)^{j-m} a_{j,-m}^\dagger. \quad (5.40)$$

It should be noted, however, that the components of a tensor are not always expressed in a spherical, SU(2), basis. For example, in Chapter 2, it was found to be useful on occasion to consider Cartesian components $\{Q_{ij}; i,j = 1,\ldots 3\}$ of the nuclear quadrupole tensor. Recall that two vectors **x** and **y** with Cartesian

---

[12]It is important to be aware that Equation (5.35) is only uniquely defined to within an overall ± 1 phase factor. For example, one could choose to define $g^{jm\,j'm'} := \delta_{j',j}\delta_{m',-m}(-1)^{j-m}$. For practical purposes, it would be inconvenient to have a phase factor that was not real, such as $(-1)^m$ when $2m$ is an odd integer. However, for SO(3) tensors with integer angular momenta, it is usually convenient to set $g^{lm\,l'm'} := \delta_{l',l}\delta_{m',-m}(-1)^m$ in order to accord with the standard relationship, $Y_{lm}^* = (-1)^m Y_{l,-m}$, for spherical harmonics.

components $\{x_i\}$ and $\{y_i\}$ have scalar product given by $\mathbf{x} \cdot \mathbf{y} = \sum_i x_i y_i$. This expression reflects the fact that the metric tensor is given in a Cartesian basis by

$$g^{ij} = \delta_{ij}. \tag{5.41}$$

Thus, whereas the scalar $q \cdot q$ is given for the spherical quadrupole tensor by

$$q \cdot q = \sum_\nu (-1)^\nu q_\nu q_{-\nu} = \sqrt{5} \, [q \otimes q]_0, \tag{5.42}$$

it is given for the Cartesian quadrupole tensor by

$$Q \cdot Q = \sum_{ij} Q_{ij} Q_{ji}. \tag{5.43}$$

The important point is that, in a Cartesian basis, there is no need for a distinction between upper and lower indices because, for example, $Q^{ij} = Q_{ij}$.

An application of the above expressions is to derive a symmetry relationship for SU(2)-reduced matrix elements of the fermion operators. Starting from the identity

$$\langle \beta J_2 M_2 | a^\dagger_{jm} | \alpha J_1 M_1 \rangle = (-1)^{j+m} \langle \alpha J_1 M_1 | a_{j,-m} | \beta J_2 M_2 \rangle^*, \tag{5.44}$$

using the Wigner-Eckart theorem to give

$$\langle \beta J_2 M_2 | a^\dagger_{jm} | \alpha J_1 M_1 \rangle = (J_1 M_1 \, jm | J_2 M_2) \frac{\langle \beta J_2 \| a^\dagger_j \| \alpha J_1 \rangle}{\sqrt{2J_2 + 1}},$$

$$\langle \alpha J_1 M_1 | a_{j,-m} | \beta J_2 M_2 \rangle = (J_2 M_2 \, j, -m | J_1 M_1) \frac{\langle \alpha J_1 \| a_j \| \beta J_2 \rangle}{\sqrt{2J_1 + 1}}, \tag{5.45}$$

and the symmetries of the Clebsch-Gordan coefficients (cf. Exercise 5.5), we obtain the identity

$$\langle \alpha J_1 \| a_j \| \beta J_2 \rangle = (-1)^{J_1 - J_2 - j} \langle \beta J_2 \| a^\dagger_j \| \alpha J_1 \rangle^*. \tag{5.46}$$

Similar relationships to Equation (5.46) are obtained for other operators. Let $\hat{Z}^J$, with components $\{\hat{Z}^{JM} := (\hat{X}_{JM})^\dagger\}$, denote the Hermitian adjoint of a covariant spherical tensor $\hat{X}_J$. Then lower the indices of $\hat{Z}^{JM}$ to obtain the components,

$$\hat{X}^\ddagger_{JM} := \hat{Z}_{JM} = (-1)^{J-M} (\hat{X}_{J,-M})^\dagger, \tag{5.47}$$

of the so-called *covariant adjoint*, $\hat{X}^\ddagger_J$, of $\hat{X}_J$ (discussed further in Appendix A.7). As in the derivation of Equation (5.46), we then obtain

$$\langle \alpha J_1 \| \hat{X}^\ddagger_J \| \beta J_2 \rangle = (-1)^{J_1 - J_2 - J} \langle \beta J_2 \| \hat{X}_J \| \alpha J_1 \rangle^*. \tag{5.48}$$

Exercises

5.5 Use the symmetry relation

$$(j_1 m_1 j_2 m_2 | j m) = (-1)^{j - j_1 - m_2} \sqrt{\frac{2j + 1}{2j_1 + 1}} \, (j m \, j_2, -m_2 | j_1 m_1) \qquad (5.49)$$

for Clebsch-Gordan coefficients to derive Equation (5.46).

5.6 Show that, if the covariant fermion annihilation operators were defined by $a_{jm} := (-1)^{j+m} a^{j,-m}$ instead of by $a_{jm} := (-1)^{j-m} a^{j,-m}$, Equation (5.46) would become

$$\langle \alpha J_1 \| a_j \| \beta J_2 \rangle = (-1)^{J_1 - J_2 + j} \langle \beta J_2 \| a_j^\dagger \| \alpha J_1 \rangle^*.$$

5.4 One-body unitary transformations

Throughout this book, we encounter numerous groups and subgroups of one-body unitary transformations. In particular, as a result of the Schur-Weyl duality relationship between the symmetric group and the one-body unitary group, discussed in depth in Section 5.11, the shell-model Hilbert space of any nucleus carries a fully antisymmetric irrep of the group of all one-body unitary transformations.

If $\{|\varphi_\nu\rangle\}$ is an orthonormal basis of single-particle states for a one-nucleon Hilbert space, $\mathbb{H}^{(1)}$, then the states of any other orthonormal basis, $\{|\psi_\nu\rangle\}$, for $\mathbb{H}^{(1)}$ have expansions,

$$|\psi_\nu\rangle = \sum_\mu |\varphi_\mu\rangle U_{\mu\nu}. \qquad (5.50)$$

The orthogonality relationship $\langle \psi_\nu | \psi_{\nu'} \rangle = \delta_{\nu\nu'}$ then implies that the $U_{\mu\nu}$ coefficients satisfy the equation

$$\sum_\mu U_{\mu\nu}^* U_{\mu\nu'} = \delta_{\nu\nu'}, \qquad (5.51)$$

i.e., they are the elements of a unitary matrix. Such a transformation of an orthonormal basis of single-particle states to a new orthonormal basis, in which

$$|\varphi_\nu\rangle \to |\psi_\nu\rangle = \hat{U}|\varphi_\nu\rangle := \sum_\mu |\varphi_\mu\rangle U_{\mu\nu}, \qquad (5.52)$$

is said to be a *one-body unitary transformation*.

The group of all unitary transformations of $\mathbb{H}^{(1)}$ is infinite dimensional and its $\{U_{\mu\nu}\}$ matrices are likewise of infinite dimension. Thus, to avoid the subtleties and technicalities of infinite-dimensional groups, it will be convenient to suppose, whenever necessary, that $\mathbb{H}^{(1)}$ is finite-dimensional. However, we put no restriction on its size. Such an approximation is implicitly adopted whenever one resorts to a computer to diagonalise a matrix or numerically solve a differential equation.

5.4.1 *The Pauli and Schwinger representations of u(2)*

Consider first the simple unitary transformations of the two-dimensional Hilbert space spanned by two spin states $\{\varphi_m; m = \pm 1/2\}$ of a single nucleon. Because any unitary matrix g can be expressed as an exponential $g = e^{iX}$, with $X = X^\dagger$ a Hermitian matrix, it follows that the Lie algebra u(2) of the group U(2) is the set of all 2×2 Hermitian matrices. In particular, u(2) contains the three spin matrices

$$s_z = \tfrac{1}{2} \begin{pmatrix} 1 & 0 \\ 0 & -1 \end{pmatrix}, \quad s_x = \tfrac{1}{2} \begin{pmatrix} 0 & 1 \\ 1 & 0 \end{pmatrix}, \quad s_y = \tfrac{1}{2} \begin{pmatrix} 0 & -i \\ i & 0 \end{pmatrix}, \tag{5.53}$$

and the identity matrix

$$I_2 = \begin{pmatrix} 1 & 0 \\ 0 & 1 \end{pmatrix}. \tag{5.54}$$

An arbitrary u(2) matrix is a linear combination of the matrices $\{s_z, s_x, s_y, I_2\}$ with real coefficients. The matrices $\{s_z, s_x, s_y, I_2\}$, which to within factors of $1/2$ are the well-known Pauli matrices,[13] are a basis for the u(2) Lie algebra. The structure of this algebra is given by the standard commutation relationships

$$[s_x, s_y] = is_z, \quad [s_y, s_z] = is_x, \quad [s_z, s_x] = is_y,$$
$$[I_2, s_x] = [I_2, s_y] = [I_2, s_z] = 0. \tag{5.55}$$

The complex linear combinations of the above spin matrices (elements of the complex extension of su(2))

$$s_0 = s_z, \quad s_\pm = s_x \pm is_y, \tag{5.56}$$

satisfy the commutation relations

$$[s_0, s_\pm] = \pm s_\pm, \quad [s_+, s_-] = 2s_0, \tag{5.57}$$

and the corresponding operators act on the spin states in the standard way; i.e,

$$\hat{s}_0 \varphi_m = m\varphi_m, \tag{5.58}$$

$$\hat{s}_+ \varphi_{-1/2} = \varphi_{1/2}, \quad \hat{s}_+ \varphi_{1/2} = 0, \tag{5.59}$$

$$\hat{s}_- \varphi_{1/2} = \varphi_{-1/2}, \quad \hat{s}_- \varphi_{-1/2} = 0. \tag{5.60}$$

The spin operators are one-body operators and are realised in the language of second quantization by the expressions

$$\hat{s}_0 = \tfrac{1}{2}(a^\dagger_{1/2} a^{1/2} - a^\dagger_{-1/2} a^{-1/2}), \quad \hat{s}_+ = a^\dagger_{1/2} a^{-1/2}, \quad \hat{s}_- = a^\dagger_{-1/2} a^{1/2}. \tag{5.61}$$

---

[13]The standard Pauli matrices are given by $\sigma_x = 2s_x$, $\sigma_y = 2s_y$, $\sigma_z = 2s_z$; cf., for example, Messiah A. (1966), *Quantum Mechanics*, Vol. 2 (North Holland, Amsterdam), sixth printing, p. 544.

The identity matrix, I_2, of the u(2) algebra, has a realisation as the number operator (cf. Exercise (5.3))

$$\hat{n} = a^\dagger_{1/2} a^{1/2} + a^\dagger_{-1/2} a^{-1/2}. \tag{5.62}$$

It is easily ascertained that these expressions of the u(2) operators, known as the Schwinger realisation,[14] satisfy the u(2) commutation relations

$$[\hat{s}_0, \hat{s}_\pm] = \pm\hat{s}_\pm, \quad [\hat{s}_+, \hat{s}_-] = 2\hat{s}_0, \quad [\hat{n}, \hat{s}_0] = 0, \quad [\hat{n}, \hat{s}_\pm] = 0. \tag{5.63}$$

In general, single-particle states have spatial and isospin as well as spin degrees of freedom. Basis states can then be labelled by two sets of quantum numbers $\{\varphi_{\nu m}\}$, where ν is a combined set of spatial and isospin quantum numbers and $m = \pm 1/2$ is a spin quantum number. To accommodate this generalisation, the u(2) spin operators are expressed in the more general form

$$\hat{s}_0 = \tfrac{1}{2} \sum_\nu (a^\dagger_{\nu\,1/2} a^{\nu\,1/2} - a^\dagger_{\nu,-1/2} a^{\nu,-1/2}),$$
$$\hat{s}_+ = \sum_\nu a^\dagger_{\nu\,1/2} a^{\nu,-1/2}, \quad \hat{s}_- = \sum_\nu a^\dagger_{\nu,-1/2} a^{\nu\,1/2}, \tag{5.64}$$
$$\hat{n} = \sum_\nu (a^\dagger_{\nu,\,1/2} a^{\nu,\,1/2} + a^\dagger_{\nu,-1/2} a^{\nu,-1/2}).$$

The remarkable property of the Schwinger representation is that it extends the action of u(2), and the corresponding U(2) group, to the whole many-nucleon Fock space. A corresponding result holds for all one-body unitary transformations.

5.4.2 *Representation of the unitary transformations of* $\mathbb{H}^{(1)}$

If $\mathbb{H}^{(1)}$ is of dimension d, then the Lie algebra u(d) of the group of unitary transformations of $\mathbb{H}^{(1)}$ is the set of Hermitian $d \times d$ matrices. These matrices are naturally expanded in a basis of elementary matrices $\{E_{\mu\nu}\}$, where $E_{\mu\nu}$ is the matrix with elements

$$(E_{\mu\nu})_{\alpha\beta} = \delta_{\alpha\mu}\delta_{\beta\nu}. \tag{5.65}$$

These matrices have commutation relations

$$[E_{\mu\nu}, E_{\kappa\lambda}] = \delta_{\nu\kappa} E_{\mu\lambda} - \delta_{\mu\lambda} E_{\kappa\nu}. \tag{5.66}$$

It can be ascertained (cf. Exercise 5.9) that the one-body operators

$$\hat{E}_{\mu\nu} := a^\dagger_\mu a^\nu \tag{5.67}$$

satisfy the parallel commutation relations

$$[\hat{E}_{\mu\nu}, \hat{E}_{\kappa\lambda}] = \delta_{\nu\kappa} \hat{E}_{\mu\lambda} - \delta_{\mu\lambda} \hat{E}_{\kappa\nu}. \tag{5.68}$$

---

[14]Schwinger J. (1965), in *Quantum Theory of Angular Momentum*, edited by L.C. Biedenharn and H. Van Dam (Academic Press, New York), p. 229.

Thus, the map $E_{\mu\nu} \to \hat{E}_{\mu\nu}$ from matrices to operators is said to define a *realisation* of the u(d) Lie algebra. In this realisation, a matrix X of the u(d) Lie algebra is represented as a Hermitian one-body operator

$$\hat{X} := \sum_{\mu\nu} X_{\mu\nu} a_\mu^\dagger a^\nu, \quad \text{with } X_{\mu\nu}^* = X_{\nu\mu}. \tag{5.69}$$

A corresponding unitary matrix $U = e^{iX}$ of the group U(d) of one-body unitary transformations is similarly realised as the operator

$$\hat{U} := \exp\left(i \sum_{\mu\nu} X_{\mu\nu} a_\mu^\dagger a^\nu\right). \tag{5.70}$$

Thus, the elements of the Lie algebra u(d) are represented as Hermitian bilinear products of the fermion operators.

Exercises

5.7 Show that, if $[A, B] = AB - BA$ and $\{A, B\} = AB + BA$, any linear operators A, B, and C, satisfy both of the following identities

$$[AB, C] = A[B, C] + [A, C]B, \tag{5.71}$$
$$[AB, C] = A\{B, C\} - \{A, C\}B. \tag{5.72}$$

5.8 Show that the u(2) operators of the Schwinger representation (5.61) satisfy the commutation relations of Equation (5.63).

5.9 Show that

$$[a_\mu^\dagger a^\nu, a_\kappa^\dagger a^\lambda] = [a_\mu^\dagger a^\nu, a_\kappa^\dagger] a^\lambda + a_\kappa^\dagger [a_\mu^\dagger a^\nu, a^\lambda] \tag{5.73}$$

and that

$$[a_\mu^\dagger a^\nu, a_\kappa^\dagger] = \delta_\kappa^\nu\, a_\mu^\dagger, \quad [a_\mu^\dagger a^\nu, a^\lambda] = -\delta_\mu^\lambda\, a^\nu. \tag{5.74}$$

Hence, derive Equation (5.68).

5.10 Show that if $\{a_i^\dagger\}$ and $\{\alpha_i^\dagger\}$ are two sets of single-fermion creation operators for different (not necessarily orthonormal) bases of a single-particle Hilbert space and

$$|\varphi_1\rangle = \alpha_1^\dagger \alpha_2^\dagger \alpha_3^\dagger |-\rangle, \quad |\varphi_2\rangle = a_1^\dagger a_2^\dagger a_3^\dagger |-\rangle, \tag{5.75}$$

are two three-particle Slater determinants then, by using commutation relations such as

$$\langle -|\alpha^3 \alpha^2 \alpha^1 a_1^\dagger a_2^\dagger a_3^\dagger |-\rangle = \langle -|\alpha^3 \alpha^2 [\alpha^1, a_1^\dagger a_2^\dagger a_3^\dagger]|-\rangle \tag{5.76}$$

recursively, show that the overlap of the two determinants is given by

$$\langle \varphi_1 | \varphi_2 \rangle = \det(M), \tag{5.77}$$

where M is the matrix of single-particle overlaps with entries

$$M_{ij} = \{\alpha^i, a_j^\dagger\}. \tag{5.78}$$

249

5.5 Isospin in nuclear physics

Nuclei are said to be *isobaric* if they comprise different numbers of neutrons and protons but have the same total number of nucleons. Comparisons of the spectra of isobaric nuclei (e.g., Figure 1.28) suggest that the strong interactions between pairs of nucleons in states of the same permutation symmetry are essentially the same regardless of whether the two nucleons are neutrons, protons, or one of each. The body of nuclear data supports this ansatz. Such a symmetry of the two-nucleon interaction implies a corresponding symmetry of the nuclear Hamiltonian called *isospin symmetry*.

5.5.1 *Relationship with exchange symmetry*

The concept of isospin was introduced by Heisenberg[15] to explore the symmetry implications of the similar properties of neutron and protons. As we now show, the parallel between two kinds of nucleon (neutron and proton) and two orientations of a particle of spin $1/2$ (spin-up and spin-down) that underlies the isospin concept, proves to have powerful implications for the construction of shell-model coupling schemes.

In parallel with the Schwinger realisation of an su(2) algebra for describing the intrinsic spin states of nuclei, a parallel su(2)-isospin algebra is defined by the set of operators

$$\hat{T}_+ = \sum_\mu a^\dagger_{\mu n} a^{\mu p}, \quad \hat{T}_0 = \tfrac{1}{2} \sum_\mu (a^\dagger_{\mu n} a^{\mu n} - a^\dagger_{\mu p} a^{\mu p}), \quad \hat{T}_- = \sum_\mu a^\dagger_{\mu p} a^{\mu n}, \quad (5.79)$$

where μ labels all quantum numbers of a single-nucleon state other than its isospin. These operators satisfy the usual commutation relations of an su(2) algebra, namely

$$[\hat{T}_0, \hat{T}_\pm] = \pm \hat{T}_\pm, \quad [\hat{T}_+, \hat{T}_-] = 2\hat{T}_0. \quad (5.80)$$

When acting on any state $|\psi^{N,Z}\rangle$ of neutron number N and proton number Z, the $\hat{T}_0$ operator gives

$$\hat{T}_0 |\psi^{N,Z}\rangle = \tfrac{1}{2}(N - Z)|\psi^{N,Z}\rangle. \quad (5.81)$$

Thus, the state $|\psi^{N,Z}\rangle$ is an eigenstate of $\hat{T}_0$ with a well-defined value of $T_0 = \tfrac{1}{2}(N - Z)$. The $\hat{T}_+$ operator changes the state $|\psi^{N,Z}\rangle$ to a state $|\psi^{N+1,Z-1}\rangle$ by replacing a proton with a neutron and $\hat{T}_-$ does the opposite.

If the spinor wave functions $\varphi_{1/2, \pm 1/2}$ represent single-neutron and single-proton isospin states, respectively, then pairs of these wave functions combine to give two-nucleon wave functions of isospin $T = 1$ and 0. A $T = 1$ multiplet comprises the

---

[15]Heisenberg W. (1932), Z. *Phys.* **77**, 1.

$\{T_0 = 0, \pm 1\}$ triplet of isospin wave functions

$$\Phi_{T=1,T_0=1} = \varphi_{1/2,\,1/2}(1)\varphi_{1/2,\,1/2}(2),$$
$$\Phi_{T=1,T_0=0} = \tfrac{1}{\sqrt{2}}\left(\varphi_{1/2,\,-1/2}(1)\varphi_{1/2,\,1/2}(2) + \varphi_{1/2,\,1/2}(1)\varphi_{1/2,\,-1/2}(2)\right), \qquad (5.82)$$
$$\Phi_{T=1,T_0=-1} = \varphi_{1/2,\,-1/2}(1)\varphi_{1/2,\,-1/2}(2),$$

which represent a two-neutron pair, a symmetric neutron-proton pair, and a two-proton pair, respectively. The singlet isospin-zero wave function with $(T_0 = T = 0)$ is the antisymmetric neutron-proton function

$$\Phi_{T=0,T_0=0} = \tfrac{1}{\sqrt{2}}\left(\varphi_{1/2,\,-1/2}(1)\varphi_{1/2,\,1/2}(2) - \varphi_{1/2,\,1/2}(1)\varphi_{1/2,\,-1/2}(2)\right). \qquad (5.83)$$

These results are the first indications of a remarkable relationship between the su(2)-isospin and nucleon-exchange symmetry of a nuclear state, which proves to be vital in the construction of totally antisymmetric combinations of space, spin and isospin wave functions. For two-nucleon states, they show that it is necessary to combine $T = 0$ isospin wave function with symmetric two-nucleon space-spin wave functions, which are allowed for a neutron-proton pair but not for a pair of like nucleons (cf. Figure 1.28). Similarly, it is necessary to combine $T = 1$ isospin wave functions with antisymmetric two-nucleon space-spin wave functions, which are accessible to all three two-nucleon pairs. Thus, the physical significance of isospin is greatly enhanced by the discovery that the important space-spin permutation symmetry of a two-nucleon wave function is uniquely defined by the value of the isospin quantum number T. For a many-nucleon system there is a similar correspondence between isospin and space-spin exchange symmetry. This correspondence is an example of the Schur-Weyl duality theorem (described in Section 5.11.1).

The observation that the strong component of the nucleonic interaction is almost the same in two-nucleon states of the same space-spin symmetry implies that it is close to being invariant under isospin transformations. Thus, to a good approximation, an isospin transformation is a symmetry of the nuclear Hamiltonian. The Coulomb interaction between pairs of protons splits the degeneracy of isospin multiplets of states. However, because it is much weaker than the strong interaction and, perhaps more significantly, because antisymmetry requirements imply that states of different isospin must necessarily also have space-spin wave functions of different permutation symmetries, the Coulomb interaction is expected to cause relatively little mixing of states of different isospin symmetry. Thus, isospin is believed to be a rather good quantum number. Experimental data[16] attest to the goodness of isospin in nuclei.

It is worth noting that, even if isospin were not a good quantum number, it would still be useful for classifying and labelling basis states in shell model calculations. The fact that isospin is a rather good, even if imperfect, quantum number

---

[16]See, for example: Bertsch G.F. and Mekjian A. (1972), *Ann. Rev. Nucl. Sci* **22**, 25; Raman S., Walkiewicz T.A. and Behrens H. (1975), *At. Data Nucl. Data Tables* **16**, 451; Shlomo S. (1978), *Rep. Prog. Phys.* **41**, 957; and Auerbach N. (1983), *Phys. Repts.* **98**, 273.

makes it much more significant. For example, the inclusion of the small isospin symmetry-breaking terms in the nuclear Hamiltonian as first-order perturbations, would amount to using the subgroup chain $SU(2)_T \supset U(1)_T$ as a dynamical symmetry with good (T, T_0) quantum numbers. Moreover, the fact that many of the particles of nature belong to sets which can be regarded as multiplets of states of an approximate isospin dynamical symmetry suggests that isospin has a deep and fundamental significance.[17]

5.5.2 *Coulomb energies and the isobaric multiplet mass equation*

The Coulomb contribution to the energy of a nuclear state is easily estimated, e.g., using a uniformly-charged-sphere model for the electrostatic potential (cf. Exercises 5.14 and 5.15). By such methods, Wigner[18] derived the celebrated *Isobaric Multiplet Mass Equation* (IMME),

$$M(T_0) = a + bT_0 + cT_0^2, \tag{5.84}$$

which has been verified[19] to remarkable accuracy and with parameter values close[20,21] to those of a simple uniformly-charged-sphere model.

The IMME follows from an assumption that interactions in nuclei are predominantly two-body. Any two-body interaction which annihilates a pair of nucleons of maximum isospin $T = 1$ and recreates a pair also with maximum $T = 1$ isospin, can change the isospin of a many-nucleon state by at most $T = 2$; in technical terms, a general two-body nuclear interaction is an isospin tensor of maximum rank $T = 2$. It is then easy to show, using first-order perturbation theory (cf. Exercise 5.15), that the Coulomb correction to the energy of a nucleus can depend at most quadratically on T_0. Thus, if T_0^3 terms were observed in the masses of isobaric sequences, it would imply either the need for higher-order perturbation theory or the inclusion of 3-body forces. In fact, when the Coulomb component of the energy differences between isobaric nuclei is taken into account, it is observed that the energy levels of sequences of isobaric nuclei fall into *isospin multiplets*.

5.5.3 *Isospin multiplets*

An example of an isospin multiplet is given in Figure 1.28, which shows the low-lying energy levels of ^{42}Ca, ^{42}Sc and ^{42}Ti. The three nuclei are seen to have $T = 1$ states in common. The members of each $T = 1$ multiplet have the same excitation energies to within a few percent. The ground-state energy differences in an isobaric multiplet are primarily due to the Coulomb interaction energy differences between

[17]Miller G.A., Nefkens B.M.K. and Šlaus I. (1990), *Phys. Repts.* **194**, 1.

[18]Wigner E.P. (1957), in *Proc. of the Robert A. Welch Conferences on Chemical Research*, Vol. 1, edited by W.O. Milligan (Robert A. Welch Foundation, Houston, Texas), p. 67.

[19]Britz J., Pape A. and Anthony M.S. (1998), *At. Data Nucl. Data Tables* **69**, 125.

[20]Anthony M.S. and Pape A. (1984), *Phys. Rev.* **C30**, 1286.

[21]Pape A., Anthony M.S. and Georgiadis A. (1988), *Phys. Rev.* **C38**, 1952.

the isobars (cf. Exercise 5.14). However, the issue of Coulomb energies in isobaric multiplets still lacks a precise theoretical understanding.[16]

Examples of such multiplets are widespread in nuclei with $N \sim Z$. Particularly dramatic examples of $T = 1/2$ doublets are seen in the spectra of so-called *mirror nuclei* for which the neutron number of one nucleus is equal to the proton number of the other. Figure 5.1 shows the energy-level spectra of $^{45}_{23}\text{V}_{22}$ and $^{45}_{22}\text{Ti}_{23}$, with ground states adjusted for their Coulomb energy differences. It can be seen that

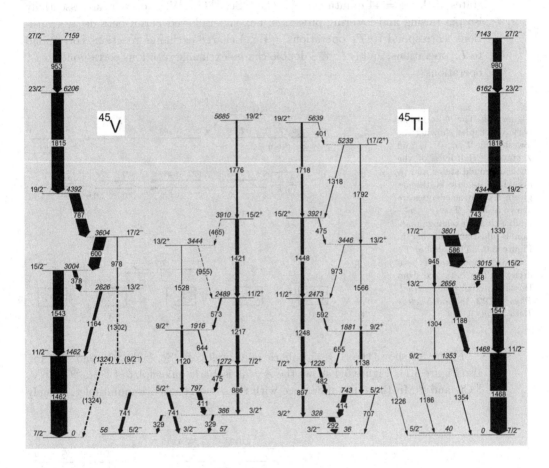

Figure 5.1: Energy levels of the nuclei $^{45}_{23}\text{V}_{22}$ and $^{45}_{22}\text{Ti}_{23}$. Electromagnetic transitions are shown with experimental intensities indicated by arrow widths. (The figure is from Bentley M.A. *et al.* (2006), *Phys. Rev.* **C73**, 024304.)

every energy level of one nucleus is paired with a nearly equal energy level of the other. The spectra are almost mirror images of one another. While the sequences of energy levels within each nucleus of the pair clearly exhibit interesting collective behaviour, the point we emphasise is the implication that the structures of the mirror pairs of states are essentially identical except for the substitution of a proton with a neutron, or vice versa; in other words, the results shown are consistent with the members of an isospin doublet being connected by isospin raising and lowering operators $\hat{T}_\pm$ that leave the space-spin degrees of freedom of the nucleons virtually unaltered.

Figure 5.2 shows features of the $A = 44$ isobars which are explained in terms of $T = 0$, 1, and 2 multiplets. It suggests that the beta-decay transitions between states of the $A = 44$ quintuplet of ^{44}Ca, ^{44}Sc, ^{44}Ti, ^{44}V, and ^{44}Cr, are essentially isospin raising and lowering processes; i.e., superallowed Fermi beta decay transitions, correspond to $\hat{T}_+$ operations, $(^3\text{He,t})$ charge exchange reactions correspond to $\hat{T}_-$ operations, and (π^+, π^-) double-charge-exchange reactions correspond to $\hat{T}_-^2$ operations.

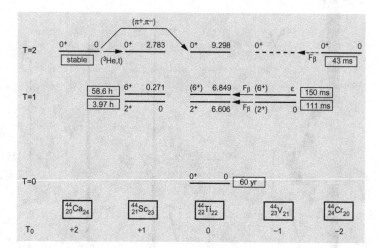

Figure 5.2: The $A = 44$ isobaric quintuplet showing the lowest-lying $T = 0$, 1, and 2 states. Half-lives of the nuclear ground states and β-decaying isomeric states are given. Excitation energies are shown in MeV. The (π^+, π^-) double-charge exchange reaction is illustrated further in Figure 5.6. The symbol F_β indicates superallowed beta decay. (The data are from Dossat C. *et al.* (2007), *Nucl. Phys.* **A792**, 18, and *Nuclear Data Sheets*.)

Figure 5.3 shows the $A = 45$ isobaric quadruplet of ^{45}Sc, ^{45}Ti, ^{45}V, and ^{45}Cr (cf. also Figure 5.1). Figure 5.4 shows the $A = 46$ isobaric quintuplet of ^{46}Sc, ^{46}Ti, ^{46}V, ^{46}Cr, and ^{46}Mn (note the difference with the $A = 44$ isobaric quintuplet, namely

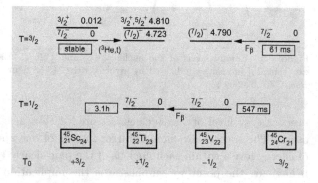

Figure 5.3: The $A = 45$ isobaric quadruplet showing the lowest-lying $T = 1/2$ (cf. Figure 5.1 for more detail) and $T = 3/2$ states. See the caption to Figure 5.2 for an explanation of details. (The data are from Dossat C. *et al.* (2007), *Nucl. Phys.* **A792**, 18, and *Nuclear Data Sheets*.)

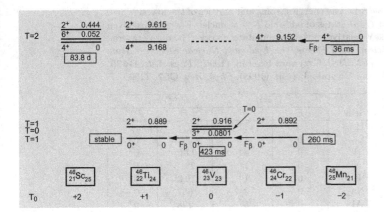

Figure 5.4: The $A = 46$ isobaric quintuplet showing the lowest-lying $T = 0$, 1, and 2 states (cf. also Figures 5.2 and 5.5). See the caption to Figure 5.2 for an explanation of details. (The data are from Dossat C. *et al.* (2007), *Nucl. Phys.* **A792**, 18, and *Nuclear Data Sheets*.)

that the $T = 1$ states are nearly degenerate with a $T = 0$ state in ^{46}V). The $A = 46$, $T = 1$ states form triplets of levels as shown in Figure 5.5.

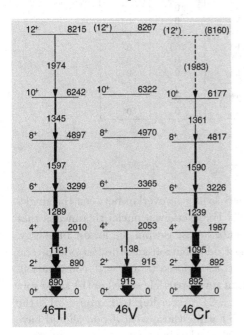

Figure 5.5: Isobaric triplets of energy levels in ^{46}Ti, ^{46}V, and ^{46}Cr. The E2 transitions observed between the levels, with their energies, are shown in keV. The widths of the arrows indicate observed experimental intensities. The numbers accompanying the arrows are the energies, in keV, of the corresponding gamma rays. (The figure is from Garrett P.E. *et al.* (2001), *Phys. Rev. Lett.* **87**, 132502.)

Isospin multiplets of states are observed in many isobaric ($N + Z = \text{constant}$) sequences in light nuclei. In even-even nuclei, the lowest energy states are invariably those of minimum isospin, i.e., $T = T_0 = \frac{1}{2}(N - Z)$. In odd-odd $N = Z$ nuclei, states of isospin $T = 0$ and $T = 1$ are generally commensurate in energy. The energies, spins and parities of the lowest $T = 0$ and $T = 1$ states in odd-odd $N = Z$ nuclei are shown in Table 5.1. In heavy nuclei, states of minimum possible isospin, i.e., $T = T_0 = \frac{1}{2}(N - Z)$, dominate the low-energy spectra. The expectation is that the

Table 5.1: The energies, spins and parities of the lowest $T = 0$ and $T = 1$ states of odd-odd $N = Z$ nuclei. Energies are given in keV relative to the ground-state energy. The data are from *Nuclear Data Sheets*, the *Evaluated Nuclear Structure Data File (ENSDF)*, Grzywacz R. *et al.* (1995), *Phys. Lett.* **B429**, 247, and Uusitalo J. *et al.* (1998), *Phys. Rev.* **C57**, 2259.

	$T = 0$		$T = 1$	
Nucleus	E	J^π	E	J^π
^{2}H	0	1^+		
^{6}Li	0	1^+	3563	0^+
^{10}B	0	3^+	1740	0^+
^{14}N	0	1^+	2313	0^+
^{18}F	0	1^+	1042	0^+
^{22}Na	0	3^+	657	0^+
^{26}Al	0	5^+	228	0^+
^{30}P	0	1^+	677	0^+
^{34}Cl	146	3^+	0	0^+
^{38}K	0	3^+	130	0^+
^{42}Sc	611	1^+	0	0^+
^{46}V	801	3^+	0	0^+
^{50}Mn	229	5^+	0	0^+
^{54}Co	197	7^+	0	0^+
^{58}Cu	0	1^+	203	0^+
^{62}Ga	571	1^+	0	0^+
^{66}As	837	1^+	0	0^+
^{70}Br	934	2^+	0	0^+
^{74}Rb	478	2^+	0	0^+
^{78}Y	< 500	5^+	0	0^+

lowest energy states are those for which there is largest overlap between the single-particle wave functions of the valence shell. For even-even nuclei this means that states of maximal space-spin symmetry, hence lowest isospin, lie lowest in energy. For odd-odd $N = Z$ nuclei, however, the last odd neutron and the last odd proton can often maximise their overlap in a $T = 1$, as opposed to a $T = 0$, combination. For example, a neutron and a proton, both in single-particle states of the same spin j, can have a combined spin of $J = 0$, 1, ..., $2j$. According to Exercise 5.12, the states with J even should have isospin $T = 1$ while those with J odd should have isospin $T = 0$. Now, a two-nucleon state in which the single-nucleon wave functions have maximum overlap is either the one in which the spins are parallel ($J = 2j$) or antiparallel ($J = 0$). Which isospin is energetically preferred is difficult to predict in general; it is even more difficult to predict when there are many valence nucleons. However, we can expect that the lowest energy $T = 1$ states of an odd-odd nucleus will have $J^\pi = 0^+$ as would be the case for an even-even $N = Z$ nucleus. We may also expect the lowest $T = 0$ states of such nuclei to have $J \neq 0$ values. The data shown in Table 5.1 are consistent with these observations.

5.5.4 *Isobaric analog states in the continuum*

Dramatic evidence of isospin symmetry is also given by the isobaric analog states observed in many medium and heavy nuclei.[22] An example of an isobaric analog state is shown for the ^{120}Sn(^{3}He,t)^{120}Sb reaction in Figure 1.84. The target nucleus, ^{120}Sn, contains more neutrons than the ^{120}Sb nucleus created from it by the replacement of one neutron by a proton in the (^{3}He,t) reaction. Isobaric analog states invariably show up as narrow (quasi-bound) resonance states in a continuous background of other states. This suggests that their structures are very different from those of neighbouring states. They have a natural interpretation if one supposes that the primary result of the (^{3}He,t) reaction is to simply replace a neutron in the target nucleus with a proton without otherwise altering the state of the target.

Recall that, if the nuclear Hamiltonian $\hat{H}$ were really invariant under isospin transformations, then the isospin operators would commute with $\hat{H}$,

$$[\hat{H}, \hat{T}_0] = [\hat{H}, \hat{T}_\pm] = 0, \tag{5.85}$$

and the states $\hat{T}_\pm |\psi^{N,Z}\rangle$ would either vanish or also be eigenstates of $\hat{H}$. The extremely narrow states populated in (p,n) or (^{3}He,t) charge-exchange reactions on the ground state of an $N > Z$ nucleus would then have a natural interpretation as the states generated by the operation of the $\hat{T}_-$ operator on the ground state, i.e.,

$$|\psi_{\text{IAS}}^{N-1,Z+1}\rangle \propto \hat{T}_- |\psi_{\text{g.s.}}^{N,Z}\rangle. \tag{5.86}$$

Similarly, it is understood that a (π^+, π^-) double charge-exchange reaction should populate a double isobaric analog state

$$|\psi_{\text{DIAS}}^{N-2,Z+2}\rangle \propto \hat{T}_-^2 |\psi_{\text{g.s.}}^{N,Z}\rangle. \tag{5.87}$$

The population, by the ^{44}Ca$(\pi^+, \pi^-)^{44}$Ti reaction, of the $T = 2$ double isobaric analog state (DIAS) of the ^{44}Ca ground state, which appears at 9.33 MeV in ^{44}Ti, is shown in Figure 5.6 (cf. also Figure 5.2).

The observation of isobaric analog states as narrow width resonance states in a continuous background of other states is evidence that states of different isospin mix relatively weakly.

5.5.5 *Isospin symmetry and gamma-decay selection rules*

Some of the best evidence for isospin symmetry in nuclei comes from gamma-ray selection rules.[23,24] Consider, for example, the $\mu = 0$ component of the electric

---

[22] Jänecke J. (1969), in *Isospin in Nuclear Physics*, edited by D.H. Wilkinson (North Holland, Amsterdam), chap. 8.

[23] Trainor L.E.H. (1952), *Phys. Rev.* **85**, 962.

[24] Wilkinson D.H., ed. (1969), *Isospin in Nuclear Physics* (North Holland, Amsterdam).

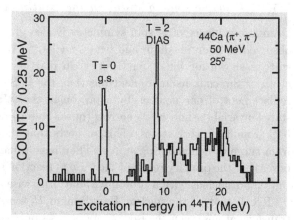

Figure 5.6: The population of the double isobaric analog state (DIAS) in ^{44}Ti (cf. Figure 5.2) via the double-charge-exchange reaction ^{44}Ca(π^+, π^-)^{44}Ti. (The figure is from Baer H.W. *et al.* (1991), *Phys. Rev.* **C43**, 1458.)

dipole, E1, operator

$$\hat{\mathfrak{M}}(E1,0) = e \sum_{i=1}^{Z} \hat{z}_i, \tag{5.88}$$

where e is the proton charge and z_i is the z-coordinate for the i'th proton.[25] This operator can be expressed as

$$\hat{\mathfrak{M}}(E1,0) = e \sum_{k=1}^{A} (\tfrac{1}{2} - \hat{T}_{0k}) \hat{z}_k, \tag{5.89}$$

where $\hat{T}_{0k}$ is the component of the isospin operator for nucleon k, with eigenvalue $\tau = +1/2$ when the k'th nucleon is a neutron and $\tau = -1/2$ when the k'th nucleon is a proton. It follows that the $\hat{\mathfrak{M}}(E1)$ operator has isoscalar and isovector components given, respectively, by

$$\hat{\mathfrak{M}}(E1,0; T=0) = \tfrac{1}{2} e \sum_{k=1}^{A} \hat{z}_k = \tfrac{1}{2} e A \hat{z}_{\text{cm}},$$

$$\hat{\mathfrak{M}}(E1,0; T=1) = -e \sum_{k=1}^{A} \hat{T}_{0k} \hat{z}_k, \tag{5.90}$$

where z_{cm} is a centre-of-mass coordinate. The isoscalar operator, $\hat{\mathfrak{M}}(E1,0; T=0)$, being proportional to a centre-of-mass coordinate, cannot excite the nucleus; it can only change its centre-of-mass state. Furthermore, from the basic rules of vector addition, the action of an isovector operator, such as $\hat{\mathfrak{M}}(E1,0; T=1)$, on a $T=0$

---

[25] For the reasons discussed in the section "Notations and conventions" following the Preface to this book, it is convenient to put a caret over a variable when it is to be interpreted as an operator. Thus, if a particle with (x, y, z) coordinates has dipole moment given by ez, the corresponding dipole operator, $e\hat{z}$, acts on a wave function, ψ, for the particle to map it to a new function, $\psi' = e\hat{z}\psi$, which has values $\psi'(x, y, z) = ez\psi(x, y, z)$.

state can only lead to $T = 1$ final states. It follows that the dipole gamma-ray decay of, for example, a $J^\pi = 1^-$, $T = 0$ state to a $J^\pi = 0^+$, $T = 0$ final state is forbidden if isospin is a good quantum number. Experimentally, E1, $\Delta T = 0$ transitions are observed[26] to be ∼10 times slower than E1, $\Delta T = 1$ transitions indicating that isospin mixing is weak for the states involved. An example is shown in Figure 5.7.

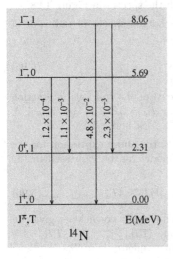

Figure 5.7: Example of allowed, $\Delta T = 1$, versus forbidden, $\Delta T = 0$, E1 transitions. Transition strengths are given as $B(\text{E1})$ values in W.u. (cf. Appendix B). (The data are from Ajzenberg-Selove F. (1991), *Nucl. Phys.* **A523**, 1.)

Exercises

5.11 Let

$$\Psi_{j_1 j_2 J M}(1, 2) = \sum_{m_1 m_2} (j_2 m_2\, j_1 m_1 | J M)\psi_{j_1 m_1}(1)\psi_{j_2 m_2}(2) \qquad (5.91)$$

denote a space-spin wave function for a neutron-proton pair of nucleons, of spins j_1 and j_2, coupled to total spin J. Show, for $j_1 \neq j_2$, that the wave function

$$\Psi^S_{j_1 j_2 J M} = \frac{1}{\sqrt{2}}\left[\Psi_{j_1 j_2 J M} + (-1)^{j_1 + j_2 - J}\Psi_{j_2 j_1 J M}\right] \qquad (5.92)$$

is symmetric under space-spin exchange and should therefore be assigned isospin $T = 0$. Conversely, show that the wave function

$$\Psi^A_{j_1 j_2 J M} = \frac{1}{\sqrt{2}}\left[\Psi_{j_1 j_2 J M} - (-1)^{j_1 + j_2 - J}\Psi_{j_2 j_1 J M}\right] \qquad (5.93)$$

is antisymmetric and should be assigned isospin $T = 1$. Hint: Use the symmetry property

$$(j_1 m_1\, j_2 m_2 | J M) = (-1)^{j_1 + j_2 - J}(j_2 m_2\, j_1 m_1 | J M) \qquad (5.94)$$

of the Clebsch-Gordan coefficients.

---

[26]See, for example, Endt P.M. (1979), *At. Data Nucl. Data Tables* **23**, 3.

5.12 Show that, for $j_1 = j_2$ and all other quantum numbers being equal, like nucleons have only even J values whereas a neutron-proton pair can have both even and odd J values. Hence, show that a neutron-proton pair with $j_1 = j_2$ has isospin $T = 1$ if J is even and $T = 0$ if J is odd.

5.13 Assuming that the lowest states for ^{42}Ca, ^{42}Sc and ^{42}Ti have two nucleons outside a closed-shell core in a $\left(1\mathrm{f}_{7/2}\right)^2$ configuration, determine the set of possible $J^\pi T$ states for these nuclei. Compare your results with the spectra in Figure 1.28.

5.14 Show that the Coulomb energy of a uniformly charged sphere of radius $R = R_0 A^{1/3}$ and electrostatic potential

$$V_c(r) = 2\pi\rho(R^2 - \tfrac{1}{3}r^2) \tag{5.95}$$

is given by $E_c = 3Z^2 e^2/5R_0 A^{1/3}$ and, with removal of the spurious interaction energy of the charge of each proton with itself, by $E_c = 3Z(Z-1)e^2/5R_0 A^{1/3}$. Estimate the Coulomb energies for the nuclei ^{42}Ca, ^{42}Sc and ^{42}Ti.

5.15 Show that, with the substitution $T_0 = (N - Z)/2$, the above expression for the Coulomb energy of a uniformly-charged sphere model can be expressed in the form

$$E_c = \frac{3e^2}{20R_0 A^{1/3}} \left[A(A-2) - 4(A-1)T_0 + 4T_0^2\right]. \tag{5.96}$$

5.16 Use the Wigner-Eckart theorem and the Clebsch-Gordan coefficients given in Appendix A.2, to show that the expectation value $\langle TT_0| \sum_{T'=1}^2 V^{T'} |TT_0\rangle$ of a tensor operator of maximum isospin $T = 2$ is at most quadratic in T_0.

5.6 Nucleon-nucleon interactions

The basic interactions between nucleons are inferred from the large body of two-nucleon data that have been observed, e.g., scattering phase shifts and the properties of the deuteron. One can also attempt to understand these interactions in terms of a fundamental theory of the strong interactions. Such an understanding is of paramount importance for determining if the largely separate developments of nuclear and particle physics are compatible with one another.

The generally accepted theory of the strong interactions, which bind quarks into baryons and mesons, is QCD (Quantum Chromodynamics). However, to derive nucleon-nucleon interactions directly from QCD is a challenge that, to date, has only been qualitatively successful.[27,28] On the other hand, quantitative success has been achieved within a field theory of nucleons interacting via the exchange of mesons[29] and within the framework of chiral effective field theory.[30,31,32] The development

---

[27] Bhaduri R.K. (1988), *Models of the Nucleon from Quarks to Soliton* (Addison-Wesley, Redwood City, California).

[28] Ishii N., Aoki S. and Hatsuda T. (2007), *Phys. Rev. Lett.* **99**, 022001.

[29] Machleidt R. (2001), *Phys. Rev.* **C63**, 024001.

[30] Entem D.R. and Machleidt R. (2002), *Phys. Lett.* **B524**, 93.

[31] Epelbaum E. *et al.* (2002), *Phys. Rev.* **C66**, 064001.

[32] Entem D.R. and Machleidt R. (2003), *Phys. Rev.* **C68**, 041001(R).

of nucleon-nucleon interaction theory has been reviewed by Machleidt and Šlaus[33] and, more recently, by Epelbaum and colleagues.[34]

5.6.1 *The bare nucleon-nucleon interaction*

The first step towards a fundamental theory of the nucleon-nucleon interaction was made by Yukawa[35] in 1935 and is a milestone of 20th century physics. In 1925, Born, Heisenberg, and Jordan[36] had applied their methods of quantization to the electromagnetic field. A precise description of electromagnetic interactions between charged particles mediated by the exchange of virtual photons followed and, by 1932, the field of quantum electrodynamics was firmly established.[37] What Yukawa did was to predict that the strong interaction between nucleons must similarly be mediated by the exchange of a particle that he called a *meson*. He showed by simple arguments, based on a solution of the Klein-Gordon equation, that the exchange of a meson of mass m would produce an interaction between a pair of nucleons equivalent to that of a potential of the form

$$V(r) = -g^2 \frac{e^{-\mu r}}{r}, \qquad (5.97)$$

where g is an adjustable coupling constant and $\mu = mc/\hbar$. Mesons had not been observed at that time. The first particle to be observed that showed promise as a candidate for Yukawa's meson,[38,39] called the mu meson at the time (now known to be a lepton and called the muon), proved to have too small a mass to be the meson of Yukawa's theory; moreover, it did not interact strongly with nucleons. However, in 1947, the pion (pi meson) was discovered[40] that did interact strongly and gave the observed range of the nuclear force.

With some adjustments, the basic Yukawa potential proved to be qualitatively successful in describing low-energy nucleon-nucleon scattering data. The primary adjustments were to include a spin-dependence and a tensor force, that come from a more complete field-theoretic treatment of the *one-pion exchange potential* (OPEP), and an isospin dependence, that comes from the different charge states of the pion. Another adjustment was to add the short-range repulsive interactions needed to explain the high-energy scattering data. Thus, many semi-phenomenological two-nucleon interactions were proposed and fitted the observed two-nucleon scattering phase-shift data and the properties of the deuteron with varying degrees of success.

[33]Machleidt R. and Šlaus I. (2001), *J. Phys. G: Nucl. Part. Phys.* **27**, R69.

[34]Epelbaum E., Hammer H.W. and Meissner U.G. (2008) ArXiv:0811.1338v1 [nucl-th].

[35]Yukawa H. (1935), *Proc. Phys. Math. Soc. (Japan)* **17**, 48, (reprinted in Beyer R.T., ed. (1949), *Foundations of Physics* (Dover Publications, New York)).

[36]Born *et al.*, *op. cit.* Footnote 7 on Page 238.

[37]Fermi E. (1932), *Rev. Mod. Phys.* **4**, 87.

[38]Neddermeyer S.H. and Anderson C.D. (1937), *Phys. Rev.* **51**, 884.

[39]Street J.C. and Stevenson E.C. (1937), *Phys. Rev.* **52**, 1003.

[40]Lattes C.M.G., Occhialini G.P.S. and Powell C.F. (1947), *Nature* **160**, 453.

The earliest two-nucleon potentials[41] were basically of the OPEP type, with adjustable parameters, plus a repulsive *hard core* interaction at ~ 0.45 fm. A potential that later became popular was the Reid soft-core potential.[42] This potential included short-range repulsive components of the Yukawa type that come from the exchange of heavy mesons. Modern *high precision* potentials[43] fit the two-nucleon data very accurately. Some, like the Argonne potential, are designed for easy use in nuclear many-body calculations. The (charge-dependent) CD-Bonn potential is special in as much as it is based on a full field-theoretic treatment of all single exchange processes of mesons with masses below the nucleon mass; it also includes two fictitious (so-called sigma) mesons with adjusted parameters to simulate two-meson exchange processes. A more fundamental, but still remarkably accurate, nucleon-nucleon potential, derived from chiral effective field theory, is the Idaho potential.[44,45] These several high-precision potentials are able to describe the observed two-nucleon data accurately and predict the *off-energy-shell* behaviour of nucleon-nucleon scattering.

Note that the scattering of two particles is said to take place *on the energy shell* when the relationships between the energies and momenta of the particles are those of free particles. However, in many-particle nuclei, pairs of nucleons interact in the fields of other nucleons; the relationships between their energies and momenta are then not those of free particles and the interactions are said to take place off the energy shell.

All of the above-mentioned potentials are between pairs of nucleons. However, it has been found[46,47,48] that these two-nucleon interactions alone underbind nuclei heavier than ^{2}H; they also overestimate the equilibrium density of nuclear matter. This suggests that three- and possibly four-body interactions are important in nuclei. Estimates have been made of the three-body interactions between nucleons,[49,50] based on multiple-pion exchange forces with parameters fitted in calculations of the binding energies and low-excited states of light nuclei with $A \leq 8$. It has been found that much-improved fits can be obtained over those obtainable

---

[41] Two-nucleon potentials in popular use in the early 1960's were those of Gammel J. and Thaler R. (1957), *Phys. Rev.* **107**, 1337, Hamada T. and Johnston I.D. (1962), *Nucl. Phys.* **34**, 382, and the Yale potential of Lassila K.E. *et al.* (1962), *Phys. Rev.* **126**, 881. The Hamada-Johnston potential is discussed in detail in the book of Heyde K.L.G. (1990), *The Nuclear Shell Model* (Springer-Verlag, Berlin), pp. 83-84.

[42] Reid R.V. (1968), *Ann. Phys. (NY)* **50**, 411.

[43] Modern potentials include the Paris potential (Lacombe M. *et al.* (1980), *Phys. Rev.* **C21**, 861), the Nijmegen potential (Stoks V.G.J. *et al.* (1994), *Phys. Rev.* **C49**, 2950), the Argonne potential (Wiringa R.B., Stoks V.G.J. and Schiavilla R. (1995), *Phys. Rev.* **C51**, 38), and the CD-Bonn potential (Machleidt, *op. cit.* Footnote 29 on Page 260).

[44] Entem and Machleidt, *op. cit.* Footnote 30 on Page 260.

[45] Entem and Machleidt, *op. cit.* Footnote 32 on Page 260.

[46] Pudliner B.S. *et al.* (1997), *Phys. Rev.* **C56**, 1720.

[47] Nogga A., Kamada H. and Glöckle W. (2000), *Phys. Rev. Lett.* **85**, 944.

[48] Wiringa R.B. *et al.* (2000), *Phys. Rev.* **C62**, 014001.

[49] Coon S.A. and Han H.K. (2001), *Few-body Systems* **30**, 131.

[50] Pieper S.C. *et al.* (2001), *Phys. Rev.* **C64**, 014001.

with just two-body interactions. On the other hand, two-nucleon potentials derived from inverse-scattering theory, which allow adjustment of the off-energy-shell components but have no connection to QCD or meson-exchange theory, have proved to be remarkably successful in application to the properties of $A \leq 6$ nuclei.[51]

5.6.2 Nucleon scattering phase shifts

The interaction between two nucleons depends on their relative orbital angular momentum L, their combined intrinsic spin S, their total angular momentum J, and, to a lesser extent, on whether the two nucleons are a pair of protons, a pair of neutrons, or one of each. In spectroscopic notation, the angular momentum state of a pair is described as a $^{2S+1}L_J$ state; it is also standard practice to denote the $L = 0, 1, 2, 3, 4, \ldots$ states by the letters $S, P, D, F, G, \ldots$, respectively.[52]

A spatial state of two nucleons of relative orbital angular momentum L has exchange symmetry $(-1)^L$ (see Exercise 5.17), i.e., it is symmetric if L is even and antisymmetric if L is odd, whereas a spin state is symmetric if $S = 1$ and antisymmetric if $S = 0$. Thus, in spectroscopic notation, a pair of like nucleons, whose combined space and spin wave functions must be antisymmetric, can be in a singlet S state (i.e., $^1S_0$), a triplet P state (i.e., $^3P_0$, $^3P_1$, $^3P_2$), a singlet D state (i.e., $^1D_2$), etc. As explained in the previous section, such two-nucleon states are said to have isospin $T = 1$. A neutron-proton pair is not subject to the space-spin antisymmetry constraint; in addition to $T = 1$ states, it also has $T = 0$ states that are symmetric under space-spin exchange. The $T = 0$ states include a triplet S state (i.e., $^3S_1$), a singlet P state (i.e., $^1P_1$), a triplet D state (i.e., $^3D_1$, $^3D_2$, $^3D_3$), etc. Possible values of L, S and T for a pair of nucleons are shown in Table 5.2.

Table 5.2: Exchange symmetries of a nucleon pair.

nn or pp	L	S	T
$^1S_0, {}^1D_2, \cdots$	even	0	1
$^3P_0, {}^3P_1, \cdots$	odd	1	1
np			
$^1S_0, {}^1D_2, \cdots$	even	0	1
$^3P_0, {}^3P_1, \cdots$	odd	1	1
$^3S_1, {}^3D_1, \cdots$	even	1	0
$^1P_1, {}^1F_3, \cdots$	odd	0	0

The scattering of nucleons due to their interaction is characterised by a scattering matrix, $S^{(\pi J)}$, which is diagonal relative to the parity $\pi = (-1)^L$ and total angular momentum J. It is also diagonal in the orbital angular momentum, L, when $\pi =$

---

[51]Shirokov A.M. et al. (2005), J. Phys. G: Nucl. Part. Phys. **31**, S1283.

[52]The letters S, P, D, and F stand for Sharp, Principal, Diffuse and Fundamental and refer to the appearance of spectral lines as seen in the early days of atomic spectroscopy. After F the letters follow an alphabetical sequence.

$(-1)^J$ and $L = J$. Its elements are then expressed simply in terms of energy-dependent phase shifts, δ_{LJ}, by

$$S^{(\pi=(-1)^J,J)} = e^{2i\delta_{JJ}}.$$ (5.98)

However, when $\pi = (-1)^{J+1}$ there is a coupling between the $L = J \pm 1$ channels (e.g., the $^3S_1$ and $^3D_1$ channels) due to the so-called *tensor force*. This component of the force is proportional to

$$S_{12} = \frac{3}{r^2}(\sigma_1 \cdot \mathbf{r})(\sigma_2 \cdot \mathbf{r}) - \sigma_1 \cdot \sigma_2,$$ (5.99)

where $\sigma_1 \cdot \mathbf{r}/r$ is the component of the Pauli spin (whose components are represented in quantum mechanics by the Pauli spin matrices $(\sigma_x, \sigma_y, \sigma_z)$) for nucleon 1 in the direction of nucleon 2. It is invariant under SU(2) rotations in the combined space of spin and spatial wave functions, but is not invariant under SO(3) rotations of just the spatial wave functions. Thus, it conserves the total angular momentum, J, and the parity, π, but not the orbital angular momentum, L. Consequently, for the $L \neq J$ channels, the scattering matrices are two-dimensional and expressible in the form[53]

$$S^{((-1)^{+1},J)} = \begin{pmatrix} e^{i\delta_{J-1,J}} & 0 \\ 0 & e^{i\delta_{J+1,J}} \end{pmatrix} \begin{pmatrix} \cos 2\varepsilon_J & i\sin 2\varepsilon_J \\ i\sin 2\varepsilon_J & \cos 2\varepsilon_J \end{pmatrix} \begin{pmatrix} e^{i\delta_{J-1,J}} & 0 \\ 0 & e^{i\delta_{J+1,J}} \end{pmatrix}.$$ (5.100)

Some pp scattering phase shifts and $T = 1$, np phase shifts, with fits obtained for the CD-Bonn potential, are shown in Figure 5.8. Some $T = 0$, np scattering phase shifts with fits obtained for the CD-Bonn potential are shown in Figure 5.9.

Because a positive phase shift signifies an attractive interaction whereas a negative phase shift signifies a repulsive interaction, it is apparent that the $^1S_0$ interaction is substantially more attractive at low energies than the ^3P interactions. This is understood by noting that the probability $|\psi_A(\mathbf{r} = 0)|^2$ for two protons to be at zero distance from one another is precisely zero in a spatially antisymmetric state. Consequently, for a zero-range interaction, there would be no interaction between particles in spatially antisymmetric states. In contrast, the probability amplitude $|\psi_S(\mathbf{r} = 0)|^2$ for a low-energy spatially-symmetric wave function is not generally zero. Thus, in general, for short-range interactions and low relative momenta (de Broglie wavelength large compared to the range of the two-proton interaction), the interaction is expected to be stronger in spatially-symmetric states than in antisymmetric states in accord with the observed phase shifts. Note, however, that when the relative orbital angular momentum is non-zero there is also a centrifugal repulsion that prevents nucleon pairs from approaching one another too closely. This further reduces the ^3P interactions relative to the $^1S_0$ interaction. It also explains why the $^1D_2$ interaction is only weakly attractive at low relative momenta.

---

[53]Stapp H.P., Ypsilantis T.J. and Metropolis N. (1957), *Phys. Rev.* **105**, 302.

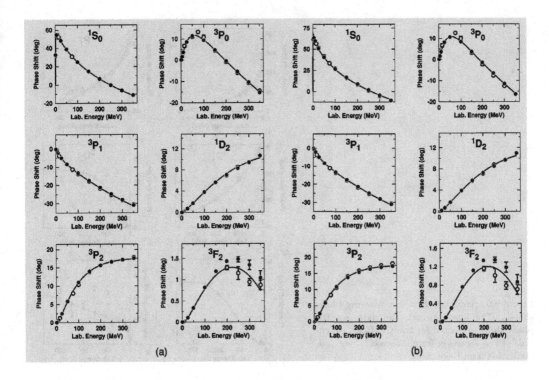

Figure 5.8: Some $T = 1$ phase shifts: (a) for pp scattering and (b) for np scattering. Most, but not quite all, of the differences between the two are due to the extra Coulomb interaction between a pair of protons. The lines through the data are the phase shifts predicted for the CD-Bonn potential. (The figure is a composite of figures from Machleidt, *op. cit.* Footnote 29 on Page 260.)

The above interpretation of low-energy phase shifts explains why, at higher energies (hence higher relative momenta and shorter de Broglie wavelengths), the $^1S_0$ interaction becomes less attractive and eventually repulsive. The implications are that the nucleon-nucleon interactions are strongly repulsive at short distances.

Phase shifts for nn scattering are difficult to measure. They are predicted by the CD-Bonn potential to be very similar to pp phase shifts when corrections are made for the Coulomb interaction between the charged particles. Additional, albeit small, differences are expected in meson exchange theory due to the different masses of the neutron and the proton.

Exercise

5.17 Given that the parity of a single-particle wave function of angular momentum l has parity $\pi = (-1)^l$ (see Section 5.8.2), show that a spatial state of two nucleons, with relative angular momentum L, has exchange symmetry given by $(-1)^L$. [Hint: show that the centre-of-mass component of the two-nucleon wave functions is invariant under exchange.]

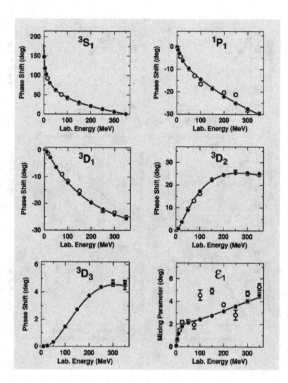

Figure 5.9: Some np, $T = 0$, scattering phase shifts. The lines through the data are the phase shifts predicted for the CD-Bonn potential. (The figure is from Machleidt, *op. cit.* Footnote 29 on Page 260.)

5.7 Effective interactions and operators for the shell model

The free-nucleon interactions as determined above are generally much too strong at short distances for direct use in the shell model. For example, if the actual eigenstates for a nucleus with realistic inter-nucleon interactions were to be expanded in terms of basis states convenient for shell-model calculations, they would inevitably be much too slowly convergent for practical purposes. Indeed, an interaction with an infinitely repulsive hard-core would have matrix elements with infinite values. Nevertheless, shell-model calculations with phenomenological interactions have proved to be remarkably successful in numerous applications in giving good results in spaces of modest dimensions. With the benefit of hindsight, it is easy to understand that this is because the strongly attractive components of the interactions more than compensate for the strongly repulsive short-range interactions. This is illustrated in a simple example, given below, for a zero-order effective interaction. However, to derive the actual effective interactions that should be used in shell-model calculations, from observed or predicted free-nucleon interactions, is a non-trivial task about which there is a huge literature. Moreover, it is a subject that is currently developing rapidly. Thus, it is impractical to present here more than a superficial outline of some of the promising methods. For example, we omit a discussion of the

Green's function Monte Carlo method.[54,55] Moreover, our treatment is essentially restricted to two-nucleon interactions between free nucleons and how they generate effective many-nucleon interactions. Complementary and different perspectives on the treatment of effective interactions for the shell model can be found in several recent review articles.[56,57,58,59,60]

5.7.1 A zero-order effective interaction

Consider two nucleons with $L = J = 0$ in the field of other nucleons (approximated by a harmonic oscillator potential) interacting via a potential of the Hamada-Johnston type with a hard repulsive core plus a one-pion exchange interaction given by

$$v(r) = \begin{cases} \infty & \text{for } r < 0.5 \text{ fm,} \\ -3 \frac{e^{-r/1.43}}{(r/1.43)} \text{MeV} & \text{for } r > 0.5 \text{ fm.} \end{cases} \tag{5.101}$$

This potential is shown in Figure 5.10. The figure also shows (dashed line) the

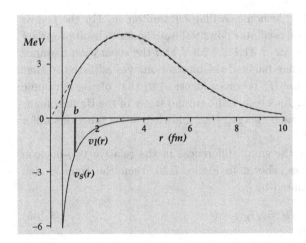

Figure 5.10: The relative radial wave function for the ground state of two nucleons in a harmonic oscillator shell-model potential with the addition of the short-range part, v_s, of the two-nucleon interaction. The separation distance, b, between the long- and short-range parts of the interaction is such that the wave function, shown as a full line, becomes identical to that (dashed line) of the unperturbed harmonic oscillator Hamiltonian for $r > b$.

radial part of the relative wave function for non-interacting nucleons in an $L = J = 0$ harmonic oscillator ground state.

We now enquire as to how the relative wave function shown will change as a result of the interaction. The technical problem is that, because harmonic oscillator wave functions are non-vanishing in the region $r < 0.5$ where the potential is singular, it

---

[54]Pudliner *et al.*, *op. cit.* Footnote 46 on Page 262.
[55]Pieper S.C., Varga K. and Wiringa R.B. (2002), *Phys. Rev.* **C66**, 044310.
[56]Brown B.A. (2001), *Prog. Part. Nucl. Phys.* **47**, 517.
[57]Otsuka T. *et al.* (2001), *Prog. Part. Nucl. Phys.* **47**, 319.
[58]Dean D. *et al.* (2004), *Prog. Part. Nucl. Phys.* **53**, 419.
[59]Caurier E. *et al.* (2005), *Rev. Mod. Phys.* **77**, 427.
[60]Coraggio L. *et al.* (2009), *Prog. Part. Nucl. Phys.* **62**, 135.

is impossible to diagonalise a Hamiltonian which includes the potential $\hat{v}$, such as

$$\hat{H} = \hat{H}_{\text{HO}} + \hat{v}, \tag{5.102}$$

in a harmonic oscillator basis because the matrix elements of $\hat{v}$ would be infinite. However, it is easy to solve the Schrödinger equation for such a Hamiltonian and compute the relative two-nucleon wave function, in an interval $0.5 \le r \le b$, with b as yet to be determined, in a basis of wave functions with boundary conditions adjusted so that they vanish at $r = 0.5$ and have the same logarithmic derivative (i.e, the same ratio $\psi^{-1}(r)d\psi(r)/dr$) as the harmonic oscillator ground-state wave function at $r = b$. Figure 5.10 shows the ground-state wave function for the Hamiltonian

$$\hat{H}_s = \hat{H}_{\text{HO}} + \hat{v}_s, \tag{5.103}$$

obtained by adding just the short-range component

$$v_s(r) = \begin{cases} v(r) & \text{for } r < b, \\ 0 & \text{for } r \ge b, \end{cases} \tag{5.104}$$

of the interaction potential to the harmonic oscillator Hamiltonian. For the results shown in Figure 5.10, the harmonic oscillator potential was chosen to be appropriate for a mass $A = 125$ nucleus with $\hbar\omega = 41A^{-1/3}$ MeV and the separation distance b, between the short- and long-range parts of the interaction, was adjusted so that the ground-state wave function for $\hat{H}_s$ becomes identical to that of the harmonic oscillator wave function for $r > b$. As a result, the ground states of the Hamiltonians $\hat{H}_{\text{HO}}$ and $\hat{H}_s$ have precisely the same energy, $\frac{3}{2}\hbar\omega$, and their wave functions differ only in a minor way for $r < b$.[61]

If one is prepared to neglect the small differences in the relative two-nucleon wave functions at short distances, shown in Figure 5.10, then the Hamiltonian $\hat{H}_v := \hat{H}_{\text{HO}} + \hat{v}$ can be approximated by

$$\hat{H}_l := \hat{H}_{\text{HO}} + \hat{v}_l, \tag{5.105}$$

where

$$v_l(r) = \begin{cases} 0 & \text{for } r < b, \\ v(r) & \text{for } r \ge b. \end{cases} \tag{5.106}$$

The potential v_l is then non-singular and could be used as a qualitative approximation to the residual effective interaction in a shell-model calculation. However, effective interactions and effective operators representing other physical observables can be defined rigorously without the need for such approximations.

---

[61] The idea of separating the two-nucleon interaction into long- and short-range parts in essentially this way, as a step towards deriving an effective interaction for use in the nuclear shell model, was proposed by Moszkowski S.A. and Scott B.L. (1960), *Ann. Phys.* (*NY*) **11**, 65; Moszkowski S.A. and Scott B.L. (1961), *Ann. Phys.* (*NY*) **14**, 107.

5.7.2 Principles of effective operator theory

Any theory of a complex physical system is inevitably an *effective theory* in the sense that it describes a model system in which much of the detailed structure of the real system is suppressed for the purposes of understanding the properties of the system of immediate concern. Indeed, even the best of theories, such as electromagnetic theory, general relativity, and quantum mechanics, are expected to break down in limiting situations. In spite of its inevitable limitations, a good theory can nevertheless be very precise within its domain of validity.

(i) *Definition of effective operators*

For present purposes, we start from a realistic model of a nucleus as a system of A nucleons with a Hilbert space $\mathbb{H}^{(A)}$, a Hamiltonian $\hat{H}$, and a set of Hermitian operators $\{\hat{X}_\alpha\}$ representing the observables of the nucleus. We assume that $\mathbb{H}^{(A)}$ is the infinite-dimensional shell-model space and that $\hat{H}$ incorporates the kinetic energies of the nucleons and potential energies corresponding to the bare nucleon interactions as described above.

As already noted, this model is intractable for A greater than two or three basically because, to handle its complexity, we need the symmetries and properties of a spherical harmonic-oscillator basis for $\mathbb{H}^{(A)}$, and face the problem that the expansion of eigenstates of $\hat{H}$ are much too slowly convergent (if at all) in such a basis. Thus, we need an effective isomorphic model defined by a similarity transformation, $\hat{S}^{-1}$, that maps a subset of eigenstates, $\{|\psi_\nu\rangle\}$, of $\mathbb{H}^{(A)}$ to a new set, $\{|\nu\rangle := \hat{S}^{-1}|\psi_\nu\rangle\}$ that have rapidly converging expansions on a harmonic-oscillator shell-model basis. By definition, a similarity transformation is invertible. As a result, if $\mathbb{H}^{(A)}$ and a set of operators, $\{\hat{X}^{(\alpha)}\}$, on $\mathbb{H}^{(A)}$ satisfy the equations

$$\hat{H}|\psi_\nu\rangle = E_\nu|\psi_\nu\rangle, \quad \hat{X}^{(\alpha)}|\psi_\nu\rangle = \sum_\mu |\psi_\mu\rangle X_{\mu\nu}^{(\alpha)}, \tag{5.107}$$

then the transformed operators

$$\hat{\mathcal{H}} = \hat{S}^{-1}\hat{H}\hat{S}, \quad \hat{\mathcal{X}}^{(\alpha)} = \hat{S}^{-1}\hat{X}^{(\alpha)}\hat{S}, \tag{5.108}$$

satisfy the similar equations

$$\hat{\mathcal{H}}|\nu\rangle = E_\nu|\nu\rangle, \quad \hat{\mathcal{X}}^{(\alpha)}|\nu\rangle = \sum_\mu |\mu\rangle X_{\mu\nu}^{(\alpha)}. \tag{5.109}$$

Thus, the transformed operators are effective operators which describe a more tractable effective shell model of the nucleus with properties that mimic those of the original intractable model.

Note that a similarity transformation does not preserve the Hermitian adjoint relationships of the operators it transforms unless it is also unitary. For example, if the original Hamiltonian is Hermitian, i.e., $\hat{H}^\dagger = \hat{H}$, then the Hermitian adjoint of the transformed Hamiltonian,

$$\hat{\mathcal{H}}^\dagger = \hat{S}^\dagger \hat{H} \left(\hat{S}^{-1} \right)^\dagger, \tag{5.110}$$

will be equal to $\hat{\mathcal{H}} = \hat{S}^{-1} \hat{H} \hat{S}$ if and only if $\hat{S}^\dagger = \hat{S}^{-1}$ as it is when $\hat{S}$ is unitary. Fortunately, the eigenstates, $\{|\nu\rangle\}$, of $\hat{\mathcal{H}}$ remain well-defined regardless of whether or not $\hat{\mathcal{H}}$ is Hermitian. The only concern is that, when $\hat{\mathcal{H}}$ is not Hermitian, they do not form an orthogonal set. This can be inconvenient because standard shell-model technology has been developed for use with orthonormal basis states. Fortunately, it is possible to transform an arbitrary basis to an orthonormal basis, e.g., by the standard Gram-Schmidt procedure.[62] The corresponding transformation of $\hat{\mathcal{H}}$ then gives a Hermitian effective Hamiltonian and effective operators with conserved Hermiticity relationships. The above correspondence can then be expressed in the useful form

$$\langle \mu | \nu \rangle = \langle \psi_\mu | \psi_\nu \rangle, \quad \langle \mu | \hat{\mathcal{H}} | \nu \rangle = \langle \psi_\mu | \hat{H} | \psi_\nu \rangle, \quad \langle \mu | \hat{\mathcal{X}}_\alpha | \nu \rangle = \langle \psi_\mu | \hat{X}_\alpha | \psi_\nu \rangle. \tag{5.111}$$

(ii) *A simple example*

A simple example of a unitary transformation that effectively suppresses the short-range correlations in the model space due to strongly repulsive short-range interactions is given by a unitary transformation that relates a basis of radial wave functions for the spherical harmonic oscillator shell-model basis to their counterparts for a modified harmonic oscillator, as described in Section 4.2.3. Recall that the radial wave function of a modified oscillator can be strongly reduced in magnitude at small values of the radius by the addition of a repulsive η/r^2 potential to the standard spherical harmonic oscillator potential. The radial wave functions for $1s$ states, related by such a unitary transformation. are shown in Figure 5.11.

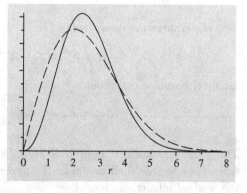

Figure 5.11: The dashed line shows the radial wave functions for a spherical harmonic oscillator $1s$ state. The full line shows a corresponding wave function for a modified oscillator wave function of slightly higher frequency.

An approach to the theory of effective interactions, using more general unitary

[62] Arfken, *op. cit.* Footnote 71 on Page 226.

transformations, has been developed by Feldmeier, Roth, and colleagues,[63,64] and is known as the *unitary correlation operator method* (UCOM). In particular, cluster models based on unitary transformations expressible in the form $e^{\hat{S}}$, where $\hat{S}$ is an antihermitian operator, were initially pursued by da Providencia and Shakin[65] and later developed into the so-called Unitary-Model-Operator Approach (UMOA) by Suzuki and Okamoto.[66,67,68] A related coupled-cluster method (CCM)[69] has also been developed in a modern context by Bishop and others.[70,71,72,73]

5.7.3 *Non-singular two-nucleon interactions*

Derivations of effective interactions for use in the shell model are conventionally tackled in two steps. The first step derives an effective two-nucleon interaction for the nucleus that would be appropriate for a shell model with a large (generally impractically large) but finite shell-model space. The rationale for starting in this way is based on an implicit assumption that, if there were only two-body interactions between free nucleons, then the induced many-body effective interactions would be small if the effective space were sufficiently large and sufficiently relevant to explicitly include the many-nucleon correlations in the model space. The second step derives corresponding effective interactions appropriate for the relatively small spaces in which practical many-nucleon shell-model calculations can be carried out.

Proceeding in two steps separates the major hurdles to be overcome in deriving many-nucleon effective interactions. The first step, which reduces bare two-nucleon interactions to well-behaved interactions, can be tackled in a space of just two nucleons. The complex second step of deriving many-nucleon effective interactions can then be tackled starting with well-behaved two nucleon interactions. The first step has traditionally been based on Brueckner-Bethe-Goldstone theory, but simpler and more promising methods have been pursued in recent years.

(i) *The Brueckner-Bethe-Goldstone approach*

The approach to a shell-model effective interaction initiated by Brueckner[74] is based on the observation that, even though the free-nucleon interaction may be highly singular and have a repulsive hard core, the scattering of pairs of nucleons from

---

[63]Feldmeier H. *et al.* (1998), *Nucl. Phys.* **A632**, 61.

[64]Roth R. *et al.* (2004), *Nucl. Phys.* **A745**, 3.

[65]da Providencia J. and Shakin C.M. (1964), *Ann. Phys. (NY)* **30**, 95.

[66]See Suzuki K. and Okamoto R. (1994), *Prog. Theor. Phys.* **92**, 1045, for a review.

[67]Fujii S., Okamoto R. and Suzuki K. (2004), *Phys. Rev.* **C69**, 034328.

[68]Fujii S., Okamoto R. and Suzuki K. (2009), *Phys. Rev. Lett.* **103**, 182501.

[69]Coester F. and Kümmel H. (1960), *Nucl. Phys.* **17**, 477.

[70]Bishop R.F. *et al.* (1993), *J. Phys. G: Nucl. Part. Phys.* **19**, 1163.

[71]Guardiola R. *et al.* (1996), *Nucl. Phys.* **A609**, 218.

[72]Heisenberg J.H. and Mihaila B. (1999), *Phys. Rev.* **C59**, 1440.

[73]Kowalski K. *et al.* (2004), *Phys. Rev. Lett.* **92**, 132501.

[74]Brueckner, *op. cit.* Footnote 3 on Page 235.

any initial state to any final state in a nucleus is well defined. Thus, Brueckner proposed that, to a first approximation, the effective interaction between nucleons in the shell model should be replaced by the reaction matrix for pairs of nucleons when they are surrounded by other nucleons which, because of the Pauli principle, prevent them from scattering into occupied single-particle states.

The scattering of free particles is described in terms of the so-called *t-matrix* of scattering theory defined by the equation

$$\hat{t}|\phi\rangle = \hat{v}|\psi\rangle, \tag{5.112}$$

where $|\phi\rangle$ denotes an initial non-interacting state of energy E of two non-interacting nucleons, $\hat{v}$ is the interaction, and $|\psi\rangle$ denotes the corresponding interacting state from which scattering cross sections can be derived. The operator, $\hat{t}$, which depends on the energy E, is then a solution of the *Lippman-Schwinger* equation[75]

$$\hat{t}(E) = \hat{v} - \hat{v}\frac{1}{\hat{T} - E}\hat{t}(E), \tag{5.113}$$

where $\hat{T}$ is the relative kinetic energy operator for the two particles.[76] The corresponding G matrix for the scattering of two nucleons, when surrounded by other nucleons that generate a mean-field potential $u(r)$, is given by the *Bethe-Goldstone* equation,[77]

$$\hat{G}(E) = \hat{v} - \hat{v}\frac{\hat{Q}}{\hat{H}_0 - E}\hat{G}(E), \tag{5.114}$$

where now $\hat{H}_0 = \hat{T} + \hat{u}$ is an independent-particle Hamiltonian for the two nucleons and $\hat{Q}$ denotes an operator that excludes states of the two nucleons in the summation over intermediate states that are occupied by the spectator nucleons.

The Brueckner G-matrix corresponds, in the diagrammatic language of many-body perturbation theory, to an infinite sum of a well-defined class of so-called *ladder diagrams*. Thus, it clearly defines what other terms in a perturbation expansion are needed to obtain complete many-body effective shell-model interactions.[78]

Unfortunately, the Bethe-Goldstone equation is an order of magnitude harder to solve than the Lippman-Schwinger equation. This is due to the exclusion of a subspace of intermediate scattering states that are occupied by other nucleons, and because the scattering takes place in the field of these nucleons which cannot be calculated until the effective interaction is known. Thus, the Bethe-Goldstone

[75]Lippman B.A. and Schwinger J. (1950), *Phys. Rev.* **79**, 469.

[76]The Lippman-Schwinger equation is a formal integro-differential equation in which a sum (or appropriate contour integral) over intermediate scattering states is implied; it is expressed in the form

$$\langle f|\hat{t}(E)|i\rangle = \langle f|\hat{v}|i\rangle - \sum_k \langle f|\hat{v}|k\rangle(T_k - E)^{-1}\langle k|\hat{t}(E)|i\rangle.$$

[77]Bethe and Goldstone, *op. cit.* Footnote 4 on Page 235.

[78]For more details, see Ellis P.J. and Osnes E. (1977), *Rev. Mod. Phys.* **49**, 777.

equation should be solved self-consistently. Other difficulties are that the G matrix is energy dependent and non-Hermitian. Nevertheless, in spite of the huge challenges and complexity of the Brueckner-based approach, G-matrix elements and effective interactions based on them have been computed, albeit with many approximations, for the $2s1d$ and $2p1f$ shells by Kuo and Brown.[79,80] More accurately computed G-matrix elements and higher-order corrections to the effective interactions were subsequently computed by others.[81] Moreover, these G matrix elements and the effective interactions for practical shell-model spaces, derived from them as discussed below, have been used in many shell-model calculations.[82]

(ii) $V_{\text{low}-k}$ and the renormalisation group

A useful perspective on replacing bare-nucleon interactions with effective interactions, is obtained in a momentum representation. From this perspective, the basic problem with the bare-nucleon interactions is that they are too strongly repulsive at high relative momenta. In light of the fact that the relative momenta between nucleons in low-energy nuclear states are typically much smaller than those for which the high-momentum repulsions are of concern, it is natural to seek effective nuclear interactions in which the high-momentum components are suppressed.

A transformation, based on renormalisation group methods, that effectively integrates out the high-momentum components of the two-nucleon interaction was developed by Bogner *et al.*[83,84,85] The resulting interaction, known as $V_{\text{low}-k}$, accurately describes low-energy nucleon phase shift data. Moreover, $V_{\text{low}-k}$ potentials have been used with remarkable success in many nuclear structure calculations (see Section 5.7.6). A notable observation,[86] as shown in Figure 5.12, is that $V_{\text{low}-k}$ interactions derived from different bare-nucleon interactions are remarkably similar; this is understood to be a consequence of the fact that the different interactions fit the same low-momentum data.

(iii) $V_{\text{low}-k}$ and the similarity renormalisation group

A concern with the $V_{\text{low}-k}$ potential, as originally derived, was that a sharp cut-off of high-momentum states, employed in its derivation, can lead to slowly convergent results in some situations. Thus, an improved $V_{\text{low}-k}$ interaction with a smooth high-momentum cut-off was derived by a unitary transformation[87] (see Section 5.7.2 for a definition of effective interactions by unitary transforma-

---

[79] Kuo T.T.S. and Brown G.E. (1966), *Nucl. Phys.* **85**, 40.

[80] Kuo T.T.S. and Brown G.E. (1968), *Nucl. Phys.* **A114**, 241.

[81] E.g., Barrett B.R., Hewitt R.G.L. and McCarthy R.J. (1971), *Phys. Rev.* **C3**, 1137.

[82] See, for example, the review article by Dean *et al.*, *op. cit.* Footnote 58 on Page 267.

[83] Bogner S.K., Kuo T.T.S. and Coraggio L. (2001), *Nucl. Phys.* **A684**, 432c.

[84] Bogner S.K. *et al.* (2002), *Phys. Rev.* **C65**, 051301(R).

[85] Bogner S.K., Kuo T.T.S. and Schwenk A. (2003), *Phys. Repts.* **386**, 1.

[86] Coraggio *et al.*, *op. cit.* Footnote 60 on Page 267.

[87] Bogner S.K., Furnstahl R.J. and Perry R.J. (2007), *Phys. Rev.* **C75**, 061001(R).

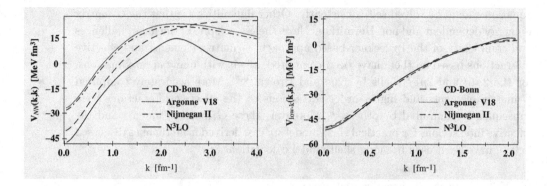

Figure 5.12: Figure (a) shows diagonal matrix elements of four bare-nucleon interactions for the $^1S_0$ channel in a momentum representation. Figure (b) shows the corresponding $V_{\text{low}-k}$ matrix elements with cut-off momentum of 2.1 fm^{-1}. (The figure is from Coraggio et al., op. cit. Footnote 60 on Page 267.)

tions), following so-called *similarity renormalisation group* (SRG) methods.[88,89,90]

The essence of the SRG method is to define a sequence of effective Hamiltonians, $\hat{H}_s$, indexed by a parameter s and defined by a corresponding sequence of unitary transformations of the original Hamiltonian,

$$\hat{H}_s := \hat{U}(s)\hat{H}\hat{U}^\dagger(s), \tag{5.115}$$

such that, as s increases, the coupling terms in $\hat{H}_s$ between high- and low-energy basis states are continuously decreased. The flow equation for the evolution of $\hat{H}_s$ with increasing s is given by

$$\begin{aligned}\frac{d\hat{H}_s}{ds} &= \frac{d\hat{U}(s)}{ds}\hat{H}\hat{U}^\dagger(s) + \hat{U}(s)\hat{H}\frac{d\hat{U}^\dagger(s)}{ds} \\ &= \frac{d\hat{U}(s)}{ds}\hat{U}^\dagger(s)\hat{H}_s + \hat{H}_s\hat{U}(s)\frac{d\hat{U}^\dagger(s)}{ds}.\end{aligned} \tag{5.116}$$

Thus, because

$$\frac{d\hat{U}(s)}{ds}\hat{U}^\dagger(s) + \hat{U}(s)\frac{d\hat{U}^\dagger(s)}{ds} = 0, \tag{5.117}$$

for a unitary transformation, the flow equation takes the form

$$\frac{d\hat{H}_s}{ds} = [\hat{\eta}_s, \hat{H}_s] \tag{5.118}$$

with an infinitesimal generator defined by the equation

$$\hat{\eta}_s := \frac{d\hat{U}(s)}{ds}\hat{U}^\dagger(s) = -\hat{U}(s)\frac{d\hat{U}^\dagger(s)}{ds}. \tag{5.119}$$

---

[88] Glazek S.D. and Wilson K.G. (1993), *Phys. Rev.* **D48**, 5863.
[89] Glazek S.D. and Wilson K.G. (1994), *Phys. Rev.* **D49**, 4214.
[90] Wegner F. (1994), *Ann. Phys. (Leipzig)* **3**, 77.

It follows that the evolution of $\hat{H}_s$, determined by $\hat{U}(s)$, is equivalently determined by $\hat{\eta}_s$. This expression of the flow is particularly useful because, as shown by Wegner,[91] if $\hat{\eta}_s$ is set equal to $[\hat{H}_d, \hat{H}_s]$, where $\hat{H}_d$ is the diagonal component of $\hat{H}$ (relative to a chosen basis), then the sum of the absolute squares of the non-diagonal elements of $\hat{H}_s$ monotonically decreases as s increases. Similarly, as shown by Glazek and Wilson,[92] the matrix elements coupling high- and low-momentum components of $\hat{H}$ are suppressed if $\hat{\eta}_s$ is set equal to $[\hat{T}_{\mathrm{rel}}, \hat{H}_s]$, where $\hat{T}_{\mathrm{rel}}$ is the relative kinetic energy of the two nucleons. Following this suggestion, Bogner et al.[93] used the flow equation

$$\frac{d\hat{H}_s}{ds} = [[\hat{T}_{\mathrm{rel}}, \hat{H}_s], \hat{H}_s], \qquad (5.120)$$

to derive an effective two-nucleon interaction, $V_{\mathrm{eff}}(s)$, with a degree of softening determined by the value of s.

Shell-model calculations with SRG effective two-nucleon interactions have been made by Bogner et al.[94] for light nuclei up to ^{7}Li for different values of s. For these calculations, truncation of the shell model space was inevitable and, accordingly, further renormalisation of the effective interaction was required for each nucleus. As shown in the following section, truncation of the many-nucleon shell-model space generates many-body effective interactions. Thus, the calculations of Bogner et al.[94] were preliminary in as much as they only included two-body effective interactions. Nevertheless, they succeeded in providing a rigorous first step towards deriving realistic effective interactions for shell-model calculations.

It is instructive to compare the SRG method with the Jacobi method for diagonalising a matrix. The Jacobi method for transforming a Hamiltonian to diagonal form is by a sequence of unitary transformations that progressively reduce off-diagonal matrix elements to zero. In contrast, the SRG sequence of transformations seeks only to find a unitary transformation that decouples a finite-dimensional subspace of low-energy states from the complementary space of high-energy states. With such an objective in mind, Anderson et al.[95] noted that, if $\hat{H}_s$ is expressed in block matrix form

$$\hat{H}_s = \begin{pmatrix} \hat{P}\hat{H}_s\hat{P} & \hat{P}\hat{H}_s\hat{Q} \\ \hat{Q}\hat{H}_s\hat{P} & \hat{Q}\hat{H}_s\hat{Q} \end{pmatrix}, \qquad (5.121)$$

where $\hat{P}$ denotes a projection operator to a model subspace of the Hilbert space and $\hat{Q}$ denotes the projection operator to the complementary orthogonal subspace,

[91] Wegner, op. cit. Footnote 90 on Page 274.
[92] Glazek and Wilson, op. cit. Footnote 89 on Page 274.
[93] Bogner et al., op. cit. Footnote 87 on Page 273.
[94] Bogner S.K. et al. (2008), Nucl. Phys. **A801**, 21.
[95] Anderson E. et al. (2008), Phys. Rev. **C77**, 037001.

the appropriate flow equation,

$$\frac{d\hat{H}_s}{ds} = [[\hat{G}_s, \hat{H}_s], \hat{H}_s], \tag{5.122}$$

with

$$\hat{G}_s = \begin{pmatrix} \hat{P}\hat{H}_s\hat{P} & 0 \\ 0 & \hat{Q}\hat{H}_s\hat{Q} \end{pmatrix}, \tag{5.123}$$

results in an evolution of the Hamiltonian, $\hat{H}_s$ with increasing s toward the diagonal form $\hat{G}_s$.

It is also interesting to note the suggestion by Sviratcheva et al.[96] that the SRG method can be used with $\hat{\eta}_s$ set equal to $[\hat{C}, \hat{H}_s]$, where $\hat{C}$ is the Casimir operator for a group corresponding to a given shell-model coupling scheme. Such a use of the flow equations would lead to a decomposition of the shell model into invariant subspaces of the group defining the coupling scheme. This would appear to be a potentially powerful result.

(iv) *Two-body potentials from inverse scattering fits*

The above approaches start with bare nucleon-nucleon interactions, developed from meson exchange theory and/or effective field theory, that achieve high precision in fitting two-nucleon data over a wide range of relative momenta. One can also define bare two-nucleon potentials, characterised by matrix elements in a truncated harmonic-oscillator basis, that directly fit the two-nucleon data. Such an approach was pursued by Elliott and colleagues[97] in deriving what are known as the *Sussex matrix elements* and, in terms of so-called *J-matrix inverse scattering* theory, by Shirokov and colleagues.[98,99]

Such potentials are ambiguous in as much as the two-nucleon data do not determine their off-energy shell properties. Thus, the off-energy shell characteristics are adjusted for ease of computation, or in shell-model fits to the spectra of very light nuclei. It is found that the results obtained are close to those with effective interactions derived from the more fundamental high-precision nucleon-nucleon potentials.

These results are significant because of their simplicity and, more importantly, because they show that a single effective two-nucleon interaction, derived directly from experimental data, can be used both for the two-nucleon data and for shell-model calculations in light nuclei.

---

[96]Sviratcheva K.D. *et al.* (2009) Talk available from the INT workshop website http://www.int.washington.edu/talks/WorkShops/int_09_1/.

[97]Elliott J.P. *et al.* (1968), *Nucl. Phys.* **A121**, 241.

[98]Shirokov A.M. *et al.* (2004), *Phys. Rev.* **C70**, 044005.

[99]Shirokov *et al.*, *op. cit.* Footnote 51 on Page 263.

(v) *Effective interactions from solution of the relative two-nucleon problem*

If it is possible to diagonalise a Hamiltonian in the full Hilbert space, it is also possible to determine a corresponding effective Hamiltonian on any convenient space and in any convenient basis that will accurately reproduce the properties of a chosen (e.g., lowest-energy) subset of states of the system. In particular, because it is possible to solve the Schrödinger equation numerically for two nucleons in a harmonic oscillator potential with any of the high-precision bare-nucleon interactions, it is also straightforward to derive matrix elements of a non-singular, and precisely-defined, effective two-nucleon interaction in a spherical harmonic-oscillator basis (with spin and isospin degrees of freedom). The effective operators and the space they act on are then an accurate representation of the chosen subset of states. To be useful, the wave functions for such a representation should resemble, as closely as possible, those of the accurately computed wave functions that they replace. This is naturally achieved by projection methods to a large-dimensional space of harmonic-oscillator wave functions.

We start by considering an A-nucleon Hamiltonian

$$\hat{H}^{(A)} := \sum_{i=1}^{A} \hat{T}_i + \sum_{i>j\geq 1} \hat{V}_{ij}, \qquad (5.124)$$

where $\hat{T}_i = -(\hbar^2/2M)\nabla_i^2$ is the kinetic energy of a nucleon and $\hat{V}_{ij}$ is the free two-nucleon interaction, and, for the moment, we suppose that free nucleons have only two-body interactions.

To take advantage of harmonic-oscillator symmetries, such as the separability of relative and centre-of-mass degrees of freedom, we want to derive an effective interaction appropriate for a shell-model subspace of spherical harmonic-oscillator states (combined with appropriate spin and isospin states). This is assisted by including a harmonic-oscillator potential in the centre-of-mass coordinates in the Hamiltonian of the nucleus. Thus, we consider the A-nucleon Hamiltonian

$$\hat{H}_{\omega}^{(A)} := \sum_{i=1}^{A} \hat{T}_i + \sum_{i>j\geq 1} \hat{V}_{ij} + \tfrac{1}{2}AM\omega^2 R_A^2, \qquad (5.125)$$

where $\mathbf{R}_A = \frac{1}{A}\sum_{i=1}^{A}\mathbf{r}_i$ is the centre-of-mass position vector for the A-nucleons. The addition of a centre-of-mass term greatly enhances the rate of convergence of the eigenfunctions of the Hamiltonian, in an oscillator basis, without affecting, in any way, the relative wave functions of the nucleus. From the identity

$$\sum_{i=1}^{A} r_i^2 = AR_A^2 + \frac{1}{A}\sum_{i>j\geq 1}|\mathbf{r}_i - \mathbf{r}_j|^2 \qquad (5.126)$$

(cf. Exercise 5.23), where $r_i^2 := \mathbf{r}_i \cdot \mathbf{r}_i$, it then follows that

$$\hat{H}_\omega^{(A)} = \sum_{i=1}^{A} \hat{h}_i + \sum_{i>\geq 1} \tilde{V}_{ij}, \tag{5.127}$$

where

$$\hat{h}_i := -\frac{\hbar^2}{2M}\nabla_i^2 + \frac{1}{2}M\omega^2 r_i^2, \quad \tilde{V}_{ij} := \hat{V}_{ij} - \frac{1}{2A}M\omega^2|\mathbf{r}_i - \mathbf{r}_j|^2. \tag{5.128}$$

Now observe that the Hamiltonian $\hat{H}_\omega^{(A)}$, being a one- plus two-body operator, is completely defined by its two-particle matrix elements, i.e., by the two-nucleon matrix elements of the Hamiltonian (for any two nucleons)[100]

$$\hat{H}_\omega^{(A,2)} := \hat{h}_1 + \hat{h}_2 + \tilde{V}_{12}. \tag{5.129}$$

Moreover, the two-nucleon eigenstates of $\hat{H}_\omega^{(A,2)}$ are easily determined by making a change of coordinates,

$$\mathbf{r} := \mathbf{r}_1 - \mathbf{r}_2, \quad \mathbf{R} := \tfrac{1}{2}(\mathbf{r}_1 + \mathbf{r}_2), \tag{5.130}$$

$$\hat{\mathbf{p}} := \tfrac{1}{2}(\hat{\mathbf{p}}_1 - \hat{\mathbf{p}}_2), \quad \hat{\mathbf{P}} := \hat{\mathbf{p}}_1 + \hat{\mathbf{p}}_2, \tag{5.131}$$

to obtain

$$\hat{H}_\omega^{(A,2)} = \hat{H}_\omega^{\text{rel}} + \hat{H}_\omega^{\text{cm}}, \tag{5.132}$$

with

$$\hat{H}_\omega^{\text{rel}} = \frac{1}{M}\hat{p}^2 + \frac{1}{4}M\omega^2 r^2 + \tilde{V}_{12} = \frac{1}{M}\hat{p}^2 + \frac{A-2}{4A}M\omega^2 r^2 + \hat{V}_{12}, \tag{5.133}$$

$$\hat{H}_\omega^{\text{cm}} = \frac{1}{4M}\hat{P}^2 + M\omega^2 R^2. \tag{5.134}$$

We can now project a finite number of lowest-energy states of $\hat{H}_\omega^{\text{rel}}$ onto an equal number of lowest-energy harmonic-oscillator states and so infer an effective interaction $\hat{\mathscr{V}}^{(2)}$ on these harmonic oscillator states such that the Hamiltonian

$$\hat{\mathcal{H}}_\omega^{\text{rel}} := \frac{1}{M}\hat{p}^2 + \frac{A-2}{4A}M\omega^2 r^2 + \hat{\mathscr{V}}^{(2)}, \tag{5.135}$$

has precisely the same energy spectrum as $\hat{H}_\omega^{\text{rel}}$. The desired effective interaction that reproduces the spectrum of the selected lowest-energy eigenstates of $\hat{\mathcal{H}}_\omega^{\text{rel}}$ can now be determined by the Lee-Suzuki method outlined in the following section.

Note that the projection of relative two-nucleon wave functions onto a harmonic oscillator basis naturally preserves all angular momentum, spin, parity, and isospin quantum numbers. Note also that, whereas the relative Hamiltonian,

---

[100]For convenience, we index the nucleons, $i = 1,\ldots,A$, even though the nucleons are indistinguishable, with the understanding that the Hamiltonian acts only on fully antisymmetric many-nucleon states. In the language of second quantisation, the Hamiltonian $\hat{H}_\omega^{(A,2)}$ is simply the restriction of $\hat{H}^{(A)}$ to the two-nucleon Hilbert space.

$\hat{H}_0^{\rm rel} = \frac{1}{M}\hat{p}^2 + \hat{V}_{12}$, without the added harmonic-oscillator potential, has unbound (scattering) states at low excitation energy and only bound states for a few low values of the relative angular momentum, the Hamiltonian $\hat{H}_\omega^{\rm rel}$ has no unbound states if $\omega > 0$. However, one can expect to find that, with increasing excitation energy, the energy eigenvalues of $\hat{H}_\omega^{\rm rel}$ rapidly approach the values $n(1-2/A)^{1/2}\hbar\omega$ that they would have if $\hat{V}_{12}$ were negligible. The extra binding energy for the two nucleons comes from the addition of a centre of mass field to the A-nucleon Hamiltonian.

5.7.4 *Effective operators for the many-nucleon shell-model*

Because many-nucleon shell-model calculations are impractical in the very large shell-model spaces for which the non-singular nucleon-nucleon interactions discussed in Section 5.7.3 are defined, it is necessary to determine corresponding effective interactions and operators for much smaller spaces. Many methods have been developed for this purpose. In earlier times, when computational resources were more limited than they are now, the methodology was based exclusively on many-body perturbation theory. Thus, sophisticated diagrammatic methods were introduced to keep track of the many terms in a perturbation expansion, and elegant linked cluster[101] and related *folded diagram* methods were developed[102,103] to streamline this process.[104]

Pioneering calculations, starting from G-matrix elements were carried out by Kuo and Brown.[105,106] Numerous contributions to the development of the subject followed as described and/or referenced in the review articles of Barrett and Kirson.[107] Kuo,[108] and the proceedings of the 1975 Tucson conference.[109] An introduction to the perturbation theory approach to effective operators in nuclei has been given by Ellis and Osnes.[110]

We present here an outline of some of the avenues being pursued within the formal structure laid out by Lee, Suzuki and Okamoto.[111,112,113] Because of its versatility, this formalism provides a unifying framework for the many developments.

---

[101]Goldstone J. (1957), *Proc. Roy. Soc. London* **A239**, 267.

[102]Morita T. (1963), *Prog. Theor. Phys.* **29**, 351.

[103]Brandow B.H. (1967), *Rev. Mod. Phys.* **39**, 771.

[104]See Kuo T.T.S. and Osnes E. (1990), *Folded-Diagram Theory of the Effective Interaction in Nuclei, Atoms and Molecules, Lecture Notes in Physics*, Vol. 364 (Springer-Verlag, Berlin), for a review of folded diagram theory.

[105]Kuo and Brown, *op. cit.* Footnote 79 on Page 273.

[106]Kuo and Brown, *op. cit.* Footnote 80 on Page 273.

[107]Barrett B.R. and Kirson M.W. (1973), *Adv. Nucl. Phys.* **6**, 219.

[108]Kuo T.T.S. (1974), *Ann. Rev. Nucl. Sci.* **24**, 101.

[109]Barrett B.R., ed. (1975), *Effective Interactions and Operators in Nuclei, Lecture Notes in Physics*, Vol. 40 (Springer-Verlag, Berlin).

[110]Ellis and Osnes, *op. cit.* Footnote 78 on Page 272.

[111]Lee S.Y. and Suzuki K. (1980), *Phys. Lett.* **91B**, 173.

[112]Suzuki K. and Lee S.Y. (1980), *Prog. Theor. Phys.* **64**, 2091.

[113]Suzuki and Okamoto, *op. cit.* Footnote 66 on Page 271.

(i) *The Lee-Suzuki formalism*

Let $\hat{H}$ denote a Hamiltonian on a Hilbert space $\mathbb{H}$ and let $\hat{P}$ and $\hat{Q}$, respectively, denote projections to a model subspace $\mathbb{H}_0 := \hat{P}\mathbb{H} \subset \mathbb{H}$ and to its orthogonal complement, $\hat{Q}\mathbb{H} = (\hat{I} - \hat{P})\mathbb{H}$. The Lee-Suzuki method[114,115,116] implicitly considers a subspace $\mathbb{W}_0 \subset \mathbb{H}$, of the same dimension as $\mathbb{H}_0$, spanned by a set, $\{|\psi_\nu\rangle\}$, of orthonormal eigenstates of $\hat{H}$ whose projections onto the model space $\mathbb{H}_0 = \hat{P}\mathbb{H}$ form a linearly-independent set. Each state, $|\psi_\nu\rangle$, in the set is then a sum of states,

$$|\psi_\nu\rangle = |\nu\rangle + \hat{Q}|\psi_\nu\rangle, \tag{5.136}$$

where $|\nu\rangle := \hat{P}|\psi_\nu\rangle$. Let $\hat{\eta}$ denote the operator on $\mathbb{H}$ that annihilates states in $\hat{Q}\mathbb{H}$ and maps states in $\hat{P}\mathbb{H}$ to states in $\hat{Q}\mathbb{H}$ such that

$$\hat{\eta}|\nu\rangle = \hat{Q}|\psi_\nu\rangle. \tag{5.137}$$

Then

$$|\psi_\nu\rangle = (\hat{I} + \hat{\eta})|\nu\rangle, \tag{5.138}$$

where $\hat{I}$ is the identity operator, $\hat{\eta}$ satisfies the equations

$$\hat{Q}\hat{\eta}\hat{P} = \hat{\eta}, \quad \hat{P}\hat{\eta}\hat{P} = \hat{Q}\hat{\eta}\hat{Q} = 0, \quad \hat{\eta}^2 = 0. \tag{5.139}$$

It follows that

$$e^{\pm\hat{\eta}} = \hat{I} \pm \hat{\eta}. \tag{5.140}$$

Thus, the operator $e^{\hat{\eta}}$ has an inverse, $e^{-\hat{\eta}}$, which means that the maps

$$|\psi_\nu\rangle = e^{\hat{\eta}}|\nu\rangle, \quad |\nu\rangle = e^{-\hat{\eta}}|\psi_\nu\rangle, \tag{5.141}$$

are similarity transformations of $\mathbb{H}$. Moreover, each of the states $|\nu\rangle$ in the model space, projected from an eigenstate $|\psi_\nu\rangle$ as described above, is an eigenstate of the Hamiltonian

$$\hat{\mathcal{H}} := e^{-\hat{\eta}}\hat{H}e^{\hat{\eta}} = (\hat{I} - \hat{\eta})\hat{H}(\hat{I} + \hat{\eta}) \tag{5.142}$$

These are remarkable results because they show that whereas the operator $e^{-\hat{\eta}}$ maps the orthonormal basis states, $\{|\psi_\nu\rangle\}$, for $\mathbb{W}_0$ to a basis, $\{|\nu\rangle\}$, for the model subspace $\mathbb{H}_0 \subset \mathbb{H}$, this same operator, being a similarity transformation, maps a complete basis for $\mathbb{H}$ to a new basis for $\mathbb{H}$. Moreover, it does so in such a way that the effective Hamiltonian relative to the transformed basis is block diagonal, i.e., it has no non-zero matrix elements between states of the model $\hat{P}\mathbb{H}$ space and those of

---

[114]Lee and Suzuki, *op. cit.* Footnote 111 on Page 279.
[115]Suzuki and Lee, *op. cit.* Footnote 112 on Page 279.
[116]Suzuki and Okamoto, *op. cit.* Footnote 66 on Page 271.

the $\hat{Q}\mathbb{H}$ space. Any operator $\hat{X}$ on $\mathbb{H}$ is mapped to a new operator, $\hat{\mathcal{X}} := e^{-\hat{\eta}}\hat{X}e^{\hat{\eta}}$, with the property that, if $\{|\varphi_\alpha\rangle\}$ is any basis for $\mathbb{H}$ for which

$$\hat{X}|\varphi_\alpha\rangle = \sum_\beta |\varphi_\beta\rangle X_{\beta\alpha}, \tag{5.143}$$

and $\{|\alpha\rangle := e^{-\hat{\eta}}|\varphi_\alpha\rangle\}$, then

$$\hat{\mathcal{X}}|\alpha\rangle = \sum_\beta |\beta\rangle X_{\beta\alpha}. \tag{5.144}$$

However, because the transformed basis is not orthonormal, care is needed in the calculation of matrix elements of such an effective operator.

For example, the model space $\hat{P}\mathbb{H}$ is spanned by the non-orthonormal basis $\{|\nu\rangle\}$. Matrix elements of a linear operator, $\hat{F}$, on $\hat{P}\mathbb{H}$ are then defined by the expansion

$$\hat{F}|\nu\rangle = \sum_\mu |\mu\rangle F_{\mu\nu}. \tag{5.145}$$

To evaluate such matrix elements when the basis is not orthonormal, it is useful to introduce a so-called *biorthogonal basis*. For a basis, $\{|\mu\rangle\}$, a biorthogonal basis, $\{|\tilde{\mu}\rangle\}$, is defined such that

$$\langle\tilde{\mu}|\nu\rangle = \delta_{\mu,\nu} \tag{5.146}$$

(see Exercise 5.21). We then have the identity

$$F_{\mu\nu} = \langle\tilde{\mu}|\hat{F}|\nu\rangle. \tag{5.147}$$

Thus, if $\hat{\mathcal{X}}$ is an effective operator on the space $\mathbb{H}_0$ corresponding to the operator $\hat{X}$ on $\mathbb{W}_0$ then

$$\langle\tilde{\mu}|\hat{\mathcal{X}}|\nu\rangle = \langle\psi_\mu|\hat{X}|\psi_\nu\rangle = X_{\mu\nu}. \tag{5.148}$$

Central to the Lee-Suzuki formalism is the identification of the operator $\hat{\eta}$. When the eigenstates, $\{|\psi_\nu\rangle\}$, are known and their overlaps with an orthonormal basis, $\{|i\rangle\}$, for $\hat{P}\mathbb{H}$ and an orthonormal basis, $\{|m\rangle\}$, for $\hat{Q}\mathbb{H}$ can be calculated, then $\hat{\eta}$ is determined as follows. With biorthogonal states for $\hat{P}\mathbb{H}$ defined such that $\langle\tilde{\mu}|\nu\rangle = \delta_{\mu,\nu}$, the operator $\hat{I}_P := \sum_\nu |\nu\rangle\langle\tilde{\nu}|$ is the identity operator on the model space $\hat{P}\mathbb{H}$. The matrix elements of $\hat{\eta}$ can then be expressed in the factored form

$$\langle m|\hat{\eta}|i\rangle = \sum_\nu \langle m|\hat{\eta}|\nu\rangle\langle\tilde{\nu}|i\rangle = \sum_\nu \langle m|\psi_\nu\rangle\langle\tilde{\nu}|i\rangle, \tag{5.149}$$

which identifies $\hat{\eta}$ as the operator

$$\hat{\eta} = \hat{Q}\sum_\nu |\psi_\nu\rangle\langle\tilde{\nu}|. \tag{5.150}$$

For any state $|j\rangle$ in the basis set $\{|i\rangle\}$, we also have the identity

$$\langle j|i\rangle = \sum_\nu \langle j|\nu\rangle\langle\tilde\nu|i\rangle = \delta_{j,i},$$ (5.151)

which implies that the matrix of overlaps, $\{\langle\tilde\nu|i\rangle\}$, is the inverse of the matrix of overlaps $\{\langle i|\nu\rangle\}$. Thus, from a knowledge of the overlaps $\langle m|\psi_\nu\rangle$ and $\langle j|\psi_\nu\rangle = \langle j|\nu\rangle$, we are able to evaluate the matrix elements $\hat\eta$, as given by Equation (5.149).

The value of the Lee-Suzuki method is that the operator $\hat\eta$ can also be derived by perturbative methods when the direct calculation, considered above, is not practicable, i.e., when the above expressions define the properties of the operator $\hat\eta$ but do not provide a way of deriving it.

Recall that the required operator, $\hat\eta$, is a linear operator from the model space, $\hat P\mathbb{H}$ to the excluded space. Hence, it satisfies the equations $\hat Q\hat\eta\hat P = \hat\eta$, $\hat Q\hat\eta\hat Q = \hat P\hat\eta\hat P = 0$, and $\hat\eta^2 = 0$. It also defines a similarity transformation, $e^{\hat\eta} = \hat I + \hat\eta$, and an effective Hamiltonian, $\hat{\mathcal{H}} = (\hat I - \hat\eta)\hat H(\hat I + \hat\eta)$. The all-important requirement that

$$\hat Q\hat{\mathcal{H}}\hat P = (\hat Q - \hat\eta)\hat H(\hat P + \hat\eta) = 0,$$ (5.152)

then ensures that $\hat{\mathcal{H}}$ is block diagonal, i.e., that the $\hat P\mathbb{H}$ and $\hat Q\mathbb{H}$ spaces are not coupled by the effective Hamiltonian.

For such an operator, $\hat\eta$, a (non-Hermitian) effective Hamiltonian for the model subspace is defined, in terms of the operator $\hat\eta$, by

$$\hat H_{\text{eff}}^{\text{NH}} := \hat P\hat{\mathcal{H}}\hat P = \hat P\hat H\hat P + \hat P\hat H\hat Q\hat\eta$$ (5.153)

and a complete set of effective operators for the model space are also defined. The challenge is then to find an operator, $\hat\eta$, for which Equation (5.152) is satisfied. However, before giving the standard solutions of this equation, we first give the subsequent transformation of the non-Hermitian $\hat H_{\text{eff}}^{\text{NH}}$ to an equivalent Hermitian effective Hamiltonian $\hat H_{\text{eff}}^{\text{H}}$.

(ii) *An equivalent Hermitian effective interaction*

As remarked above, a non-Hermitian effective interaction defined by a similarity transformation is readily transformed to a Hermitian effective interaction corresponding to a unitary transformation. This can be done explicitly for any solution of the Lee-Suzuki equations using a construction based on that given by Okubo.[117]

Consider first the similarity transformation

$$\hat S := \hat I + \hat\eta - \hat\eta^\dagger,$$ (5.154)

where $\hat\eta$ and its Hermitian adjoint, $\hat\eta^\dagger$, satisfy the equations,

$$\hat\eta\hat P = \hat Q\hat\eta = \hat\eta, \quad \hat P\hat\eta = \hat\eta\hat Q = 0, \quad \hat\eta^\dagger\hat Q = \hat P\hat\eta^\dagger = \hat\eta^\dagger, \quad \hat\eta^\dagger\hat P = \hat Q\hat\eta^\dagger = 0,$$ (5.155)

[117]Okubo S. (1954), *Prog. Theor. Phys.* **12**, 603.

as in Lee-Suzuki theory. It is seen that

$$\hat{S}^\dagger \hat{S} = \hat{I} + \hat{\eta}\hat{\eta}^\dagger + \hat{\eta}^\dagger\hat{\eta} \neq \hat{I}. \tag{5.156}$$

Thus, $\hat{S}$ is not unitary. However, $\hat{S}^\dagger \hat{S}$ is block diagonal in the sense that $\hat{P}\hat{S}^\dagger \hat{S}\hat{Q} = \hat{Q}\hat{S}^\dagger \hat{S}\hat{P} = 0$. We can therefore define a unitary transformation

$$\hat{U} := \hat{S}(\hat{S}^\dagger \hat{S})^{-\frac{1}{2}} = (\hat{I} + \hat{\eta} - \hat{\eta}^\dagger)(\hat{I} + \hat{\eta}\hat{\eta}^\dagger + \hat{\eta}^\dagger\hat{\eta})^{-\frac{1}{2}}, \tag{5.157}$$

$$\hat{U}^\dagger := (\hat{S}^\dagger \hat{S})^{-\frac{1}{2}}\hat{S}^\dagger = (\hat{I} + \hat{\eta}\hat{\eta}^\dagger + \hat{\eta}^\dagger\hat{\eta})^{-\frac{1}{2}}(\hat{I} - \hat{\eta} + \hat{\eta}^\dagger) = \hat{U}^{-1}, \tag{5.158}$$

and a Hermitian effective Hamiltonian, $\hat{\mathcal{H}}^H := \hat{U}^\dagger \hat{H}\hat{U}$. Moreover, because the operator

$$\hat{I} + \hat{\eta}\hat{\eta}^\dagger + \hat{\eta}^\dagger\hat{\eta} = (\hat{P} + \hat{\eta}^\dagger\hat{\eta}) + (\hat{Q} + \hat{\eta}\hat{\eta}^\dagger) \tag{5.159}$$

is block diagonal with sub-blocks given by the operators $(\hat{P} + \hat{\eta}^\dagger\hat{\eta})$ and $(\hat{Q} + \hat{\eta}\hat{\eta}^\dagger)$, on $\hat{P}\mathbb{H}$ and $\hat{Q}\mathbb{H}$, respectively, it follows that

$$(\hat{I} + \hat{\eta}\hat{\eta}^\dagger + \hat{\eta}^\dagger\hat{\eta})^{-\frac{1}{2}} = (\hat{P} + \hat{\eta}^\dagger\hat{\eta})^{-\frac{1}{2}} + (\hat{Q} + \hat{\eta}\hat{\eta}^\dagger)^{-\frac{1}{2}}, \tag{5.160}$$

and, hence, that

$$\hat{U} := (\hat{P} + \hat{\eta})(\hat{P} + \hat{\eta}^\dagger\hat{\eta})^{-\frac{1}{2}} + (\hat{Q} - \hat{\eta}^\dagger)(\hat{Q} + \hat{\eta}\hat{\eta}^\dagger)^{-\frac{1}{2}}, \tag{5.161}$$

$$\hat{U}^\dagger := (\hat{P} + \hat{\eta}^\dagger\hat{\eta})^{-\frac{1}{2}}(\hat{P} + \hat{\eta}^\dagger) + (\hat{Q} + \hat{\eta}\hat{\eta}^\dagger)^{-\frac{1}{2}}(\hat{Q} - \hat{\eta}). \tag{5.162}$$

The essential requirement for the determination of η is now that $\hat{Q}\hat{\mathcal{H}}^H\hat{P}$ and, hence $\hat{Q}\hat{U}^\dagger \hat{H}\hat{U}\hat{P}$, should be zero. Thus, it is required that

$$\hat{Q}\hat{\mathcal{H}}^H\hat{P} = (\hat{Q} + \hat{\eta}\hat{\eta}^\dagger)^{-\frac{1}{2}}(\hat{Q} - \hat{\eta})\hat{H}(\hat{P} + \hat{\eta})(\hat{P} + \hat{\eta}^\dagger\hat{\eta})^{-\frac{1}{2}} = 0, \tag{5.163}$$

which is satisfied if and only if $\hat{\eta}$ satisfies the equation

$$(\hat{Q} - \hat{\eta})\hat{H}(\hat{P} + \hat{\eta}) = 0. \tag{5.164}$$

This is again the fundamental equation of the Lee-Suzuki approach.

Now, if η satisfies this equation the effective Hamiltonian is Hermitian and of the block-diagonal form

$$\hat{\mathcal{H}}^H = \hat{P}\hat{\mathcal{H}}^H\hat{P} + \hat{Q}\hat{\mathcal{H}}^H\hat{Q} \tag{5.165}$$

with

$$\hat{P}\hat{\mathcal{H}}^H\hat{P} = (\hat{P} + \hat{\eta}^\dagger\hat{\eta})^{-\frac{1}{2}}(\hat{P} + \hat{\eta}^\dagger)\hat{H}(\hat{P} + \hat{\eta})(\hat{P} + \hat{\eta}^\dagger\hat{\eta})^{-\frac{1}{2}}, \tag{5.166}$$

$$\hat{Q}\hat{\mathcal{H}}^H\hat{Q} = (\hat{Q} + \hat{\eta}\hat{\eta}^\dagger)^{-\frac{1}{2}}(\hat{Q} - \hat{\eta})\hat{H}(\hat{Q} - \hat{\eta}^\dagger)(\hat{Q} + \hat{\eta}\hat{\eta}^\dagger)^{-\frac{1}{2}}. \tag{5.167}$$

These expressions can be simplified by use of the observation that, when Equation (5.164) is satisfied,

$$(\hat{P} + \hat{\eta})\hat{H}(\hat{P} + \hat{\eta}) = (\hat{I} - \hat{Q} + \hat{\eta})\hat{H}(\hat{P} + \hat{\eta}) = \hat{H}(\hat{P} + \hat{\eta}). \tag{5.168}$$

Thus, we can replace $(\hat{P} + \hat{\eta}^{\dagger})\hat{H}(\hat{P} + \hat{\eta})$ in Equation (5.166) with

$$(\hat{P} + \hat{\eta}^{\dagger})(\hat{P} + \hat{\eta})\hat{H}(\hat{P} + \hat{\eta}) = (\hat{P} + \hat{\eta}^{\dagger}\hat{\eta})\hat{H}(\hat{P} + \hat{\eta}) \tag{5.169}$$

to obtain

$$\hat{P}\hat{\mathcal{H}}^{\mathrm{H}}\hat{P} = (\hat{P} + \hat{\eta}^{\dagger}\hat{\eta})^{1/2}\hat{H}(\hat{P} + \hat{\eta})(\hat{P} + \hat{\eta}^{\dagger}\hat{\eta})^{-1/2}. \tag{5.170}$$

Similarly, it is shown that

$$\hat{Q}\hat{\mathcal{H}}^{\mathrm{H}}\hat{Q} = (\hat{Q} + \hat{\eta}\hat{\eta}^{\dagger})^{1/2}\hat{H}(\hat{Q} - \hat{\eta}^{\dagger})(\hat{Q} + \hat{\eta}\hat{\eta}^{\dagger})^{-1/2}. \tag{5.171}$$

Equation (5.170) shows that the Hermitian effective Hamiltonian for the model space is related to the non-Hermitian effective Hamiltonian of the Lee-Suzuki equations by the identity

$$\hat{H}_{\mathrm{eff}}^{\mathrm{H}} = \hat{P}\hat{\mathcal{H}}^{\mathrm{H}}\hat{P} = (\hat{P} + \hat{\eta}^{\dagger}\hat{\eta})^{1/2}\hat{H}_{\mathrm{eff}}^{\mathrm{NH}}(\hat{P} + \hat{\eta}^{\dagger}\hat{\eta})^{-1/2}. \tag{5.172}$$

Note, however, that the above simplification can only be used for operators, such as the Hamiltonian, for which the decoupling condition (5.164) applies. Thus, for arbitrary operators $\{\hat{X}^{(\alpha)}\}$ on $\mathbb{H}$, the effective model-space operators are given by[118,119]

$$\hat{P}\hat{\mathcal{X}}^{(\alpha)}\hat{P} = (\hat{P} + \hat{\eta}^{\dagger}\hat{\eta})^{-1/2}(\hat{P} + \hat{\eta}^{\dagger})\hat{X}^{(\alpha)}(\hat{P} + \hat{\eta})(\hat{P} + \hat{\eta}^{\dagger}\hat{\eta})^{-1/2}. \tag{5.173}$$

(iii) *The Bloch-Horowitz solution*

Equations (5.152) and (5.153) can be expressed in the form

$$\hat{Q}\hat{H}\hat{P} + \hat{Q}\hat{H}\hat{Q}\hat{\eta} = \hat{\eta}\hat{H}_{\mathrm{eff}}^{\mathrm{NH}}, \tag{5.174}$$

$$\hat{H}_{\mathrm{eff}}^{\mathrm{NH}} = \hat{P}\hat{H}\hat{P} + \hat{P}\hat{H}\hat{Q}\hat{\eta}. \tag{5.175}$$

equivalent to equations derived by Bloch, Horowitz, and Feshbach.[120,121,122] They have the potential to be solved self-consistently because, if a basis for the model space, $\{|\nu\rangle\}$, is defined by eigenstates of $\hat{H}_{\mathrm{eff}}^{\mathrm{NH}}$,

$$\hat{H}_{\mathrm{eff}}^{\mathrm{NH}}|\nu\rangle = E_{\nu}|\nu\rangle, \tag{5.176}$$

they imply that

$$(E_{\nu} - \hat{Q}\hat{H}\hat{Q})\hat{\eta}|\nu\rangle = \hat{Q}\hat{H}\hat{P}|\nu\rangle. \tag{5.177}$$

[118]Navratil P., Geyer H.B. and Kuo T. (1993), *Phys. Lett.* **B315**, 165.
[119]Suzuki K. and Okamoto R. (1995), *Prog. Theor. Phys.* **93**, 905.
[120]Bloch C. (1958), *Nucl. Phys.* **6**, 329.
[121]Bloch C. and Horowitz J. (1958), *Nucl. Phys.* **8**, 91.
[122]Feshbach H. (1962), *Ann. Phys. (NY)* **19**, 287.

Thus, they lead to the formal identities[123]

$$\hat{\eta}|\nu\rangle = \sum_\nu \frac{1}{E_\nu - \hat{Q}\hat{H}\hat{Q}}\hat{Q}\hat{H}\hat{P}|\nu\rangle, \tag{5.178}$$

$$\hat{H}_{\text{eff}}^{\text{NH}}|\nu\rangle = \left[\hat{P}\hat{H}\hat{P} + \sum_\nu \hat{P}\hat{H}\hat{Q}\frac{1}{E_\nu - \hat{Q}\hat{H}\hat{Q}}\hat{Q}\hat{H}\hat{P}\right]|\nu\rangle. \tag{5.179}$$

and, equivalently, the Bloch-Horowitz energy-dependent effective Hamiltonian,

$$\hat{H}_{\text{eff}}^{\text{NH}}(E) = \hat{P}\hat{H}\hat{P} + \sum_\nu \hat{P}\hat{H}\hat{Q}\frac{1}{E - \hat{Q}\hat{H}\hat{Q}}\hat{Q}\hat{H}\hat{P}. \tag{5.180}$$

This Hamiltonian has the property that

$$\hat{H}_{\text{eff}}^{\text{NH}}(E_\nu)|\nu\rangle = E_\nu|\nu\rangle. \tag{5.181}$$

It follows from these results that, if one could evaluate the energy-dependent effective Hamiltonian, $\hat{H}_{\text{eff}}^{\text{NH}}(E)$, and if one could find a set of d linearly-independent solutions to the eigenvalue equation (5.181), one would obtain an explicit, energy-independent, expression for the effective interaction in the form

$$\hat{H}_{\text{eff}}^{\text{NH}} = \hat{P}\hat{H}\hat{P} + \sum_\nu \hat{P}\hat{H}\hat{Q}\frac{1}{E_\nu - \hat{Q}\hat{H}\hat{Q}}\hat{Q}\hat{H}\hat{P}|\nu\rangle\langle\tilde{\nu}|, \tag{5.182}$$

where $\{|\tilde{\nu}\rangle\}$ is a basis for the model space, $\hat{P}\mathbb{H}$, that is biorthogonal to the $\{|\nu\rangle\}$ basis, as defined above.

To determine $\hat{H}_{\text{eff}}^{\text{NH}}(E)$ for a given value of E, when the matrix representation of the operator $(E - \hat{Q}\hat{H}\hat{Q})$ is too large to invert, necessitates replacing the right-hand side of Equation (5.180) by a perturbation expansion. Thus, like any perturbative method, it relies on a rapid convergence of the expansion. This is essential because, in practice, it may only be feasible to calculate a few terms. This limitation could be a problem. For if, as observed by Schucan and Weidenmüller,[124,125] the spectrum of the Hamiltonian $\hat{Q}\hat{H}\hat{Q}$ were to overlap with that of the effective Hamiltonian, $\hat{H}_{\text{eff}}^{\text{NH}}(E)$, then the energy denominator $E_\nu^n - \hat{Q}\hat{H}\hat{Q}$ in Equation (5.180) would take small, possibly vanishing, values and the iterative procedure would be unlikely to converge. This happens when so-called *intruder states*, i.e., states that are normally associated with higher-lying shells, show up in the low-energy domain. Such a lack of convergence was, in fact, observed by many researchers and alternative procedures were sought.[126,127,128,129,130]

---

[123]These identities are described as being *formal* because they ignore the possibility that the operator $(E_\nu - \hat{Q}\hat{H}\hat{Q})$ may have no inverse for some values of E_ν.

[124]Schucan T.H. and Weidenmüller H.A. (1972), *Ann. Phys. (NY)* **73**, 108.

[125]Schucan T.H. and Weidenmüller H.A. (1973), *Ann. Phys. (NY)* **76**, 483.

[126]Kirson M.W. (1971), *Ann. Phys. (NY)* **66**, 624.

[127]Kirson M.W. (1974), *Ann. Phys. (NY)* **82**, 345.

[128]Barrett and Kirson, *op. cit.* Footnote 107 on Page 279.

[129]Babu S. and Brown G.E. (1973), *Ann. Phys. (NY)* **78**, 1.

[130]Krenciglowa E.M. and Kuo T.T.S. (1974), *Nucl. Phys.* **A235**, 171.

(iv) *The Krenciglowa-Kuo solution*

The Krenciglowa-Kuo approach starts with the Hamiltonian $\hat{H}$ expressed as a sum of two terms, $\hat{H} = \hat{H}_0 + \hat{V}$, with $\hat{H}_0\hat{P} := E_0\hat{P}$ so that $\hat{H}_0$ acts on the model space as E_0 times the identity operator and $\hat{V}$ is the residual interaction. The result is conveniently expressed in terms of a so-called Q-box interaction,[131] defined by

$$\hat{Q}(E) := \hat{P}\hat{V}\hat{P} + \hat{P}\hat{V}\hat{Q}\frac{1}{E - \hat{Q}\hat{H}\hat{Q}}\hat{Q}\hat{V}\hat{P}. \tag{5.183}$$

The Q-box is a powerful concept because, to the extent that it is well-defined and can be determined accurately for a desired range of values of E, it enables many problems in effective interaction theory to be solved. (Note that it will inevitably have singularities at values of E equal to eigenvalues of $\hat{Q}\hat{H}\hat{Q}$.)

The Krenciglowa-Kuo solution is obtained by writing $\hat{H}_{\text{eff}}^{\text{NH}} := E_0\hat{P} + \hat{R}$, and re-expressing Equations (5.174) and (5.175) in terms of the *effective residual interaction*, $\hat{R}$, as

$$(E_0 - \hat{Q}\hat{V}\hat{Q})\hat{\eta} = \hat{Q}\hat{V}\hat{P} - \hat{\eta}\hat{R}, \tag{5.184}$$

$$\hat{R} = \hat{P}\hat{V}\hat{P} + \hat{P}\hat{V}\hat{Q}\hat{\eta}. \tag{5.185}$$

Equation (5.184) is then expressed in the form

$$\hat{\eta} = \frac{1}{E_0 - \hat{Q}\hat{H}\hat{Q}}[\hat{Q}\hat{V}\hat{P} - \hat{\eta}\hat{R}] \tag{5.186}$$

and iterated with respect to $\hat{\eta}$, keeping $\hat{R}$ fixed, to obtain the solution for $\hat{\eta}$ (in terms of $\hat{R}$)

$$\hat{\eta} = \frac{1}{E_0 - \hat{Q}\hat{H}\hat{Q}}\left[\hat{Q}\hat{V}\hat{P} - \frac{1}{E_0 - \hat{Q}\hat{H}\hat{Q}}\left(\hat{Q}\hat{V}\hat{P} - \frac{1}{E_0 - \hat{Q}\hat{H}\hat{Q}}\hat{\eta}\hat{R}\right)\hat{R}\right]$$

$$= -\sum_{m=0}^{\infty}\left(\frac{-1}{E_0 - \hat{Q}\hat{H}\hat{Q}}\right)^{m+1}\hat{Q}\hat{V}\hat{P}\hat{R}^m. \tag{5.187}$$

A corresponding equation for $\hat{R}$, obtained from Equation (5.184), is then given by[132]

$$\hat{R} = \hat{P}\hat{V}\hat{P} + \hat{P}\hat{V}\hat{Q}\frac{1}{E_0 - \hat{Q}\hat{H}\hat{Q}}\hat{Q}\hat{V}\hat{P} - \hat{P}\hat{V}\hat{Q}\sum_{m=1}^{\infty}\left(\frac{-1}{E_0 - \hat{Q}\hat{H}\hat{Q}}\right)^{m+1}\hat{Q}\hat{V}\hat{P}\hat{R}^m,$$

$$= \hat{Q}(E_0) + \sum_{m=1}^{\infty}\hat{Q}_m(E_0)\hat{R}^m, \tag{5.188}$$

---

[131]Kuo T.T.S., Lee S.Y. and Ratcliff K.F. (1971), *Nucl. Phys.* **A176**, 65.

[132]Related expressions were also derived by Brandow, *op. cit.* Footnote 103 on Page 279 and Des Cloizeaux J. (1960), *Nucl. Phys.* **20**, 321.

where

$$\hat{Q}_m(E_0) := -\hat{P}\hat{V}\hat{Q}\left(\frac{-1}{E_0 - \hat{Q}\hat{H}\hat{Q}}\right)^{m+1}\hat{Q}\hat{V}\hat{P}. \tag{5.189}$$

Solutions for $\hat{\eta}$, $\hat{Q}_m(E_0)$ and $\hat{R}$ can, in principle, be obtained iteratively. Krenciglowa and Kuo determined that, when such an iterative solution converges, it gives an effective interaction corresponding to eigenstates of the full Hamiltonian having maximum overlaps with the model states. Unfortunately, it appears that their perturbation expansion, while having improvements over the Bloch-Horowitz treatment, continues to have convergence problems when there are intruder states in the low-energy domain.

(v) *The Lee-Suzuki solution*

A perturbation expansion suggested by Lee and Suzuki[133] can be obtained by expressing Equations (5.184) and (5.184) in the recursive form

$$(E_0 - \hat{Q}\hat{V}\hat{Q})\hat{\eta}_n = \hat{Q}\hat{V}\hat{P} - \hat{\eta}_{n-1}\hat{R}_n, \tag{5.190}$$

$$\hat{R}_n = \hat{P}\hat{V}\hat{P} + \hat{P}\hat{V}\hat{Q}\hat{\eta}_n. \tag{5.191}$$

Defining

$$\hat{e} := E_0 - \hat{Q}\hat{V}\hat{Q}, \tag{5.192}$$

then leads to the perturbative expansions

$$\hat{\eta}_n = \frac{1}{\hat{e}}[\hat{Q}\hat{V}\hat{P} - \hat{\eta}_{n-1}\hat{R}_n]$$
$$= \frac{1}{\hat{e}}\hat{Q}\hat{V}\hat{P} - \frac{1}{\hat{e}^2}\hat{Q}\hat{V}\hat{P}\hat{R}_n + \frac{1}{\hat{e}^3}\hat{Q}\hat{V}\hat{P}\hat{R}_{n-1}\hat{R}_n - \ldots \tag{5.193}$$

and

$$\hat{R}_n = \hat{P}\hat{V}\hat{P} + \hat{P}\hat{V}\hat{Q}\frac{1}{\hat{e}}\hat{Q}\hat{V}\hat{P} - \hat{P}\hat{V}\hat{Q}\frac{1}{\hat{e}^2}\hat{Q}\hat{V}\hat{P}\hat{R}_n$$
$$+ \hat{P}\hat{V}\hat{Q}\frac{1}{\hat{e}^3}\hat{Q}\hat{V}\hat{P}\hat{R}_{n-1}\hat{R}_n - \ldots . \tag{5.194}$$

The latter expansion is expressed in terms of the Q-box interaction, of Equations (5.183) and (5.189), by

$$\hat{R}_n = \hat{Q} + \hat{Q}_1\hat{R}_n + \hat{Q}_2\hat{R}_{n-1}\hat{R}_n + \hat{Q}_3\hat{R}_{n-2}\hat{R}_{n-1}\hat{R}_n + \ldots . \tag{5.195}$$

Hence, we obtain

$$\left(1 - \hat{Q}_1 - \hat{Q}_2\hat{R}_{n-1} - \hat{Q}_3\hat{R}_{n-2}\hat{R}_{n-1} - \ldots\right)\hat{R}_n = \hat{Q} \tag{5.196}$$

---

[133]Lee and Suzuki, *op. cit.* Footnote 111 on Page 279.

and the Lee-Suzuki recursion relation

$$\hat{R}_n = \frac{1}{1 - \hat{Q}_1 - \sum_{m=2}^{n-1} \hat{Q}_m \prod_{k=n-m+1}^{n-1} \hat{R}_k} \hat{Q}. \tag{5.197}$$

This is a seemingly complicated expression for deriving the effective interaction $\hat{R}$. Moreover, like the Krenciglowa-Kuo approach, it depends on the accuracy to which one can calculate the Q-box interaction. However, given the Q-box interaction, it has been shown[134] to converge an order of magnitude faster than the Krenciglowa-Kuo solution. This is important because one can only hope to include the first few terms of either expansion. The reason the Lee-Suzuki method avoids the divergence problems associated with intruder states can be traced to the fact that it converges to the lowest-energy states of the original Hamiltonian, even when some of these lowest-energy states have largest overlaps with states that are not included in the model space.

(vi) *The Andreozzi solutions*

Andreozzi's solution[135] of the Lee-Suzuki equations starts with the decoupled P- and Q-space effective Hamiltonians given, respectively, by

$$\hat{p}(\hat{\eta}) := \hat{P}\mathcal{H}\hat{P} = \hat{P}\hat{H}\hat{P} + \hat{P}\hat{H}\hat{Q}\hat{\eta}, \tag{5.198}$$

$$\hat{q}(\hat{\eta}) := \hat{Q}\mathcal{H}\hat{Q} = \hat{Q}\hat{H}\hat{Q} - \hat{\eta}\hat{P}\hat{H}\hat{Q}. \tag{5.199}$$

The fundamental equation, (5.152), for $\hat{\eta}$ is then expressed in two alternative ways by

$$\hat{\eta}\hat{p}(\hat{\eta}) = \hat{Q}\hat{H}\hat{P} + \hat{Q}\hat{H}\hat{Q}\hat{\eta}, \tag{5.200}$$

$$\hat{q}(\hat{\eta})\hat{\eta} = \hat{\eta}\hat{P}\hat{H}\hat{P} - \hat{Q}\hat{H}\hat{P}. \tag{5.201}$$

Thus, Andreozzi devised iterative solutions of these equations equivalent to iterating one or other of the equations

$$\hat{\eta}_n = [\hat{Q}\hat{H}\hat{P} + \hat{Q}\hat{H}\hat{Q}\hat{\eta}_{n-1}]\frac{1}{p(\hat{\eta}_{n-1})}, \tag{5.202}$$

$$\hat{\eta}_n = \frac{1}{q(\hat{\eta}_{n-1})}[\hat{\eta}_{n-1}\hat{P}\hat{H}\hat{P} - \hat{Q}\hat{H}\hat{P}]. \tag{5.203}$$

The Andreozzi expansions are particularly useful when the $\hat{p}(\hat{\eta})$ and $\hat{q}(\hat{\eta})$ effective Hamiltonians can be inverted. It is then possible to avoid the use of perturbation theory. It will also be noted that the first of these equations requires only inversion of the P-space effective Hamiltonian, $\hat{p}(\hat{\eta})$, which by choice of a P space in which shell-model calculations are feasible, will naturally be invertible. However, in situations in which the Q-space effective Hamiltonian is also invertible, it appears that the second equation is much more rapidly convergent. Like the Lee-Suzuki algorithm,

---

[134]Lee and Suzuki, *op. cit.* Footnote 111 on Page 279.

[135]Andreozzi F. (1996), *Phys. Rev.* **C54**, 684.

on which the second of the above iterative solutions is based, Andreozzi claims that his equations also converge to the lowest-energy solutions and have no convergence problems.

5.7.5 *The no-core shell model (NCSM)*

The traditional approach to the calculation of effective shell-model operators is to define a model space with a closed-shell core with extra-core nucleons restricted to a valence space. However, for light nuclei, it is possible to consider an active shell-model space in which all the nucleons participate. The derivation of effective interactions for such a no-core shell model (NCSM) is greatly simplified primarily by avoiding the use of perturbation theory and the problems of non-convergence associated with core-excitations.

We first outline a simple idealised version of the *ab initio NCSM* of Navrátil, Vary, and Barrett,[136] for the purpose of identifying the principles and assumptions on which it is based. The model makes use of the Lee-Suzuki methods of Sections 5.7.4(i) and (ii) (but not the perturbative solution of 5.7.4(v)) and the principles underlying the cluster approximations of the *Unitary-Model-Operator Approach* (UMOA) of Suzuki and Okamoto.[137,138] In practice, additional approximations, as mentioned below, are needed to enable a practical version of the NCSM to be implemented.

(i) *An idealised NCSM*

Let

$$\mathbb{H}^{(A)} := \mathbb{H}^{(1)} \wedge \mathbb{H}^{(1)} \wedge \cdots \wedge \mathbb{H}^{(1)} \qquad (5.204)$$

denote the Hilbert space of antisymmetric tensor products of states in an infinite-dimensional Hilbert space, $\mathbb{H}^{(1)}$, of single-nucleon states.[139] The Hilbert space, $\mathbb{H}^{(1)}$, is spanned, for example, by single-particle harmonic-oscillator basis states combined with spin and isospin states. The objective is then to seek an effective Hamiltonian, $\hat{\mathcal{H}}^{(A)}_{n,\omega}$, appropriate for a finite-dimensional subspace,

$$\mathbb{H}^{(A)}_{n,\omega} := \mathbb{H}^{(1)}_{n,\omega} \wedge \mathbb{H}^{(1)}_{n,\omega} \wedge \cdots \wedge \mathbb{H}^{(1)}_{n,\omega}, \qquad (5.205)$$

of $\mathbb{H}^{(A)}$, where $\mathbb{H}^{(1)}_{n,\omega} \subset \mathbb{H}^{(1)}$ is a subspace of spherical harmonic-oscillator states (coupled to spin and isospin states) of frequency ω and no more than n harmonic-oscillator quanta.

---

[136] See Navrátil P., Vary J.P. and Barrett B.R. (2000), *Phys. Rev.* **C62**, 054311, and other references quoted therein.

[137] Suzuki K. and Okamoto R. (1986), *Prog. Theor. Phys.* **75**, 1388.

[138] Suzuki K. and Okamoto R. (1986), *Prog. Theor. Phys.* **76**, 127.

[139] The symbol $\wedge$, known as a *wedge product*, or *exterior product* as it is also called, is simply a generalisation to a higher-dimensional vector space of the familiar antisymmetric vector product. Such an antisymmetric Hilbert space is spanned by states with Slater-determinant wave functions.

As in Section 5.7.3(v), we start with the A-nucleon Hamiltonian

$$\hat{H}_\omega^{(A)} := \sum_{i=1}^{A} \hat{T}_i + \sum_{i>j=1}^{A} \hat{V}_{ij} + \frac{1}{2} A M \omega^2 R_A^2$$

$$= \sum_{i=1}^{A} \hat{h}_i + \sum_{i>j=1}^{A} \left[\hat{V}_{ij} - \frac{1}{2A} M \omega^2 |\mathbf{r}_i - \mathbf{r}_j|^2 \right], \qquad (5.206)$$

where $\hat{h}_i := -\frac{\hbar^2}{2M} \nabla_i^2 + \frac{1}{2} M \omega^2 r_i^2$. This Hamiltonian is obtained by adding a harmonic-oscillator potential in the centre-of-mass coordinates to localise the centre-of-mass of the nucleus in space, without affecting the internal (relative) wave functions of the nucleons and, thereby, making it possible to define translationally-invariant effective interactions.

An effective Hamiltonian, $\hat{\mathcal{H}}_{n,\omega}^{(A)}$, on the subspace $\mathbb{H}_{n,\omega}^{(A)} \subset \mathbb{H}^{(A)}$ which represents a subset of low-lying eigenstates of the Hamiltonian $\hat{H}_\omega^{(A)}$ on $\mathbb{H}^{(A)}$ is conveniently expressed as a sum of terms,

$$\hat{\mathcal{H}}_{n,\omega}^{(A)} = \hat{h} + \hat{\mathscr{V}}_n^{(2)} + \hat{\mathscr{V}}_n^{(3)} + \cdots + \hat{\mathscr{V}}_n^{(A)}, \qquad (5.207)$$

where $\hat{\mathscr{V}}_n^{(a)}$ is an a-body operator defined as follows.

Recall, as shown in the context of second quantization, that operators on an A-nucleon Hilbert space have natural extensions to other (totally antisymmetric) many-nucleon Hilbert spaces. For example, the single-particle Hamiltonian, $\hat{h}$, is defined for a nucleons by $\hat{h} = \sum_{i=1}^{a} \hat{h}_{i=1}$, for any finite value of $a > 0$. Similarly, the Hamiltonian $\hat{H}_\omega^{(A)}$ defines an a-nucleon Hamiltonian

$$\hat{H}_\omega^{(A,a)} := \sum_{i=1}^{a} \hat{h}_i + \sum_{i>j=1}^{a} \left[\hat{V}_{ij} - \frac{1}{2A} M \omega^2 |\mathbf{r}_i - \mathbf{r}_j|^2 \right], \qquad (5.208)$$

for any $a \geq 2$. By definition, an N-body operator is identically zero on an a-nucleon space if $a < N$. Thus, we define the a-nucleon Hamiltonian

$$\hat{\mathcal{H}}_{n,\omega}^{(A,a)} := \hat{h} + \hat{\mathscr{V}}_n^{(2)} + \hat{\mathscr{V}}_n^{(3)} + \cdots + \hat{\mathscr{V}}_n^{(a)}, \qquad (5.209)$$

as the effective Hamiltonian on the Hilbert space $\mathbb{H}_{n,\omega}^{(a)}$ corresponding to the Hamiltonian $\hat{H}_\omega^{(A,a)}$ on $\mathbb{H}^{(a)}$. This, definition for $a = 2, 3, 4, \ldots$ then defines the sequence of a-body interactions of the Hamiltonian $\hat{\mathcal{H}}_{n,\omega}^{(A)}$ of Equation (5.207) such that, by construction, the Hamiltonian, $\hat{\mathcal{H}}_{n,\omega}^{(A,A)} \equiv \hat{\mathcal{H}}_{n,\omega}^{(A)}$, with $a = A$, is the effective Hamiltonian for the A-nucleon Hilbert space $\mathbb{H}_{n,\omega}^{(A)}$. We now show that it is potentially possible (with adequate computer facilities) to derive the Hermitian effective Hamiltonian, $\hat{\mathcal{H}}_{n,\omega}^{(A)}$, without using perturbation theory by calculating the effective interactions, $\hat{\mathscr{V}}_n^{(a)}$, one at a time.

The following algorithm is based on the assumption that, if the free-nucleon interactions include only two-body interactions, the sequence of terms in Equation

(5.207) will likewise include only two-body interactions in the $n \to \infty$ limit. Conversely, we expect that, as the effective shell-model space is reduced in size, i.e., for more modest values of n, higher many-body effective interactions will be needed to compensate for the space truncation. The hope is that, for a sufficiently large value of n, the expansion

$$\hat{\mathcal{H}}_{n,\omega}^{(A)} \equiv \hat{\mathcal{H}}_{n,\omega}^{(A,A)} = \hat{h} + \mathscr{V}_n^{(2)} + \mathscr{V}_n^{(3)} + \cdots \tag{5.210}$$

can be terminated, after a relatively small number of terms to a satisfactory level of approximation. An important objective of no-core shell-model calculations is therefore to determine the number of terms needed in this sequence for given values of A and n. The following strategy is based on the presumption that, for free two-nucleon interactions, there is a sequence of integers $n_2 \geq n_3 \geq n_4 \geq \ldots$ such that only two-body effective interactions are needed to obtain results to an acceptable level of accuracy, if $n \geq n_2$, that at most three-body effective interactions are needed, if $n \geq n_3$, and so on. We discuss concerns with this presumption and its practical limitations in Subsection (iv).

The Schrödinger equation for the two-nucleon Hamiltonian, $\hat{H}_\omega^{(A,2)}$ is accurately solvable for any finite number of low-lying states as outlined in Section 5.7.3(v). Thus, its solutions can be projected onto a two-nucleon Hilbert space $\mathbb{H}_{n_2,\omega}^{(2)}$ and the Hermitian two-nucleon effective Hamiltonian, $\hat{\mathcal{H}}_{n_2,\omega}^{(A)} = \hat{h} + \mathcal{V}_{n_2}^{(2)}$, appropriate for this two-nucleon space can be determined for any finite value of n_2. The ideal strategy is to choose n_2 to be large enough so that, in accordance with the above conjecture, the three-body term, $\hat{\mathcal{V}}_{n_2}^{(3)}$, is negligible and, to the extent possible, n_2 is small enough that three-nucleon calculations can be carried out with the Hamiltonian $\hat{\mathcal{H}}_{n_2,\omega}^{(A,3)} :=$ $\hat{h} + \mathcal{V}_{n_2}^{(2)}$. A set of eigenstates of $\hat{\mathcal{H}}_{n_2,\omega}^{(A,3)}$ can then be projected down to a smaller space $\hat{\mathcal{H}}_{n_3,\omega}^{(A,3)}$, with $n_3 \leq n_2$, and a corresponding three-body effective interaction, $\mathcal{V}_{n_3}^{(3)}$, thereby derived. In principle, this process can be continued as far as computational resources permit or until convergence of the effective interaction sequence is obtained for a value of n that is sufficiently small for A-nucleon calculations to be possible. Truncating the series (5.207) to an a-body effective Hamiltonian is an a-body *cluster approximation*. The validity of a few-body cluster approximation in a given situation is something that one can aspire to determine in trial calculations.

If one wanted to start with bare three-nucleon, as well as bare two-nucleon interactions, they should be introduced in the solution of the three-nucleon problem for a suitably selected set of eigenstates in the full Hilbert space. In principle, this could be tackled by Fadeev methods.[140] Similarly, if one wanted to include bare four-nucleon interactions, they should be introduced in an accurate solution of the four-nucleon problem for a suitably selected set of eigenstates, e.g., by the methods of Yakubovsky[141,142]

---

[140]Fadeev L.D. (1960), *Zh. Eksperim. i Teor. Fiz.* **39**, 1459, (English translation: Fadeev L.D. (1961), *Soviet Phys. JETP* **12**, 1014).

[141]Yakubovsky O.A. (1967), *Sov. J. Nucl. Phys.* **5**, 937.

[142]Cf. calculations by Kamada H. and Glöckle W. (1992), *Nucl. Phys.* **A548**, 205.

Observe that it appropriate to choose the projection of a many-nucleon wave function onto a smaller space such that it preserves all quantum numbers arising from symmetries, such as angular momentum, spin, parity and isospin, that are compatible with a spherical harmonic oscillator basis. In particular, because the centre-of-mass component of a nuclear wave function in a spherical harmonic oscillator basis can be solved exactly, the projection is naturally required to conserve the translational invariance of the effective interaction.

The idealised NCSM, as presented here, is undoubtedly impractical for more than a few very light nuclei as it stands, primarily because the dimension of the model Hilbert space, $\mathbb{H}_{n,\omega}^{(A)}$, as defined by Equation (5.205), increases incrementally with $n \to n+1$ in steps that are much too large. Nevertheless, it might be considered that an accurate determinantion of $\hat{\mathcal{H}}_{n,\omega}^{(A,a)}$ for a as large as three or four, would already give considerable information about the rate of convergence of its expansion for various values of ω and n and whether or not reliable effective shell-model interactions could ever be derived by such a method. In fact, there are good reasons to suppose that such information might be misleading for the reasons considered in Subsection (iv).

(ii) *The ab initio NCSM of Navrátil, Vary, and Barrett*

In the practical NCSM of Navrátil, Vary, and Barrett,[143] the model Hilbert space, $\mathbb{H}_{N_{\max},\omega}^{(A)}$, is defined as the space spanned by A-nucleon harmonic-oscillator states of up to $N_{\max}$ harmonic-oscillator quanta above the valence space of fewest oscillator quanta. The dimension of the space $\mathbb{H}_{N_{\max},\omega}^{(A)}$ then increases much less rapidly with unit increases in $N_{\max}$. However, this modification requires adjustments and approximations to the idealised model presented above.

Several calculations have been carried out for nuclei of mass $A \leq 14$ in the two- and three-body cluster approximations to the Navrátil-Vary-Barrett ab initio NCSM.[144,143,145,146] The results are qualitatively successful in reproducing observed properties, such as energy levels, binding energies, and elastic electron scattering form factors. It was also found that significantly improved results in small shell-model spaces are obtained when three-body effective interactions are included.[146] The feasibility of calculations starting from three- as well as two-nucleon bare interactions has been explored by Marsden et al.,[147] and ab initio NCSM with such interactions have been reported by Návratil and colleagues.[148,149] However, it would appear that no attempts have been made to date to check the magnitude of a-body effective interactions for $a > 3$.

---

[143]Navrátil et al., op. cit. Footnote 136 on Page 289.

[144]Navrátil P., Vary J.P. and Barrett B.R. (2000), *Phys. Rev. Lett.* **84**, 5728.

[145]Navrátil P. et al. (2001), *Phys. Rev. Lett.* **87**, 172502.

[146]Navrátil P. and Ormand W.E. (2002), *Phys. Rev. Lett.* **88**, 152502.

[147]Marsden D.C.J. et al. (2002), *Phys. Rev.* **C66**, 044007.

[148]Navrátil P. and Ormand W.E. (2003), *Phys. Rev.* **C68**, 034305.

[149]Hayes A.C., Navrátil P. and Vary J.P. (2003), *Phys. Rev. Lett.* **91**, 012502.

(iii) *Effective transitions in the NCSM*

The calculation of effective transition operators, within the framework of the NCSM has been considered by Stetcu *et al.*[150] A concern that shows up in these calculations is a failure to get the large effective charges for E2 transitions that are needed for a shell-model description of experimental data. This result has been interpreted as a need for two-body effective E2 operators.[150] We discuss this interpretation in the following subsection.

(iv) *Insights gained from the NCSM and some concerns*

A major concern with the idealised NCSM algorithm, as expressed in Subsection (i), is that it makes use of a very specific expansion of the effective interaction as a sum of a-body operators with these operators defined in a very specific way. While the choice is convenient and legitimate, it is not unique and probably not the choice that would lead to the most rapidly convergent results. It is undoubtedly legitimate in the sense that, for free two-nucleon interactions, there must surely exist some large values of $n_2 \geq n_3 \geq n_4 \geq \ldots$ such that the presumption, on which the idealised NCSM is based, is valid. However, the real concerns are: (i) that the required values are too large for practical applications of the NCSM; and (ii) that an apparent convergence of the sequence (5.209) might be misleading. For example, a calculation of $\mathscr{V}_n^{(2)}$, $\mathscr{V}_n^{(3)}$, and $\mathscr{V}_n^{(4)}$, with $n_4 = 3$ might well appear to have converged to an acceptable degree of accuracy because of a negligibly small $\mathscr{V}_n^{(4)}$. However, such a result does not guarantee that subsequent terms in the sequence will remain small. In particular, one should be concerned that, in approaching a nucleus with highly-deformed states, many higher shell-model configurations are needed which, if not included in the active shell-model space, must be compensated for by effective interactions that were not needed in nuclei with at most four nucleons.

The most rapidly convergent expression of the effective Hamiltonian needed to describe a given set of energy levels of a mass-A nucleus on a specified shell-model space would no doubt be given by a sequence of least-squares fits to the data with Hamiltonians of the type expressed by Equation (5.209), for $a = 1, 2, 3, \ldots$, and corresponding fits to transition data. From phenomenological shell-model successes, one knows that good shell-model descriptions of the data can be obtained in this way with a restricted to small numbers. For example, Brown and Wildenthal[151,152] have determined that remarkably good fits to the spectra of sd-shell nuclei can be obtained with just two-body effective interactions. However, it is not obvious that the sequence of a-body effective Hamiltonians and other operators that would be obtained by idealised NCSM methods, assuming they could be done, would reproduce those of a most rapidly convergent fit.

[150]Stetcu I. *et al.* (2005), *Phys. Rev.* **C71**, 044325.
[151]Brown B.A. and Wildenthal B.H. (1988), *Ann. Rev. Nucl. Part. Sci.* **38**, 29.
[152]See also Brown B.A. and Richter W.A. (2006), *Phys. Rev.* **C74**, 0343150.

For example, the failure reported by Stetcu *et al.*,[150] to get the large effective charges for E2 transitions that are needed for a shell-model description of experimental data in NCSM calculations can be understood from this perspective. It was suggested in their paper that the NCSM effective E2 operator must have a significant two-body component whereas, in a fit to the data, a much smaller two-body effective E2 operator is needed if the one-body E2 operator is renormalised with an effective charge.

A notable fact, to be discussed more fully in Volume 2, is that the microscopic collective model,[153] with a shell model space spanned by subspaces of harmonic oscillator shells that are large enough to admit realisations of observed quadrupole deformations of nuclei, reproduces experimentally observed quadrupole moments and E2 transition rates with bare (unrenormalised one-body) E2 operators. An examination of the wide-scale prominence of quadrupole collective states in the low-energy spectra of nuclei also leads one to believe that their occurrence does not depend on the details of the nucleon-nucleon interactions; it is understood to be primarily a consequence of an adiabatic decoupling of the low-energy collective and higher-energy intrinsic dynamics. There are also indications that collective dynamics are *emergent phenomena*[154] in the sense that they are not readily predicted from first principles. Nevertheless, once they have been observed to occur experimentally, they have a simple explanation in microscopic terms (as will be described in Volume 2). Thus, one should attempt to design the effective shell-model theory to take the successes of the collective models into account. Such a perspective is consistent with the recent observations by Dytrych *et al.*[155,156] of the emergence of collective model dynamical symmetries in no-core shell model calculations.

5.7.6 *Realistic effective interactions in medium and heavy nuclei*

Computing effective interactions from bare nucleon-nucleon interactions for medium and heavy nuclei poses challenges that cannot reasonably be tackled by NCSM methods. The main problem is the explosion in size of the configuration spaces associated with the nuclear shells as one goes up in energy. This makes it necessary, in most cases, to restrict the active shell-model space to a relatively small number of shell-model configurations for the nucleons outside of a closed-shell core. Thus, large renormalisations of the effective interactions are needed.

The essential approach, as discussed in Section 5.7.3, is the two-step process in which the first step is a unitary transformation of the free two-nucleon interaction to a non-singular two-nucleon interaction. This step should provide a valid effective interaction for a shell-model on a very large, but finite, Hilbert space. In fact, the shell-model space appropriate for such an effective interaction should be large

---

[153]Rowe, *op. cit.* Footnote 66 on Page 222.
[154]Anderson, *op. cit.* Footnote 1 on Page vii.
[155]Dytrych T. *et al.* (2007), *Phys. Rev. Lett.* **98**, 162503.
[156]Dytrych T. *et al.* (2008), *J. Phys. G: Nucl. Part. Phys.* **35**, 123101.

enough to make the introduction of effective three-body interactions, at this point, unnecessary. The second step is then a perturbative derivation of the effective interaction appropriate for the necessary restriction to a much smaller space for which shell-model calculations become feasible. This two-step process can be carried out by combinations of the methods outlined above in Sections 5.7.3 and 5.7.4.

The first major steps in deriving effective shell-model interactions for the $2s1d$ and $2p1f$ shell nuclei calculations, in the pioneering calculations of Kuo and Brown,[157,158] followed this strategy. They started with the Brueckner G-matrix and added first-order core-polarization corrections following the methods of Bertsch.[159]

With the many subsequent developments, outlined in Sections 5.7.3 and 5.7.4, many of the problems encountered in the early calculations can now be avoided and, with modern computing power, the outlook is very promising. In recent years, it has proved to be more reliable to start with a V_{low-k} two-nucleon interaction, as described in Section 5.7.3, and add core-polarisation corrections using perturbation theory. This approach has been developed and reviewed by Coraggio, and colleagues.[160] One could also start with effective two-nucleon interactions obtained by numerical solution of the relative two-nucleon problem as described in Section 5.7.3(v) and possibly implement the Lee-Suzuki renormalisation procedure by the methods of Andreozzi (see Section 5.7.4(vi).

The description of a wide range of nuclear spectra with V_{low-k} interactions is impressive;[161,162,163] see, for example, Figure 6.1. However, to appreciate the significance of these results, it is important to learn how the results change when perturbative corrections to higher order and three-body interactions are included in the calculations.

5.7.7 *Importance of effective interaction theory*

Given that the shell model provides the basic formal framework for understanding nuclei in terms of interacting nucleons, one cannot stress too strongly the fundamental importance of effective interaction theory for establishing its foundations.

Recall that the nuclear shell model makes two essential assumptions: first, that the building blocks of nuclei are interacting nucleons, and second, that the effective interactions between nucleons are not so strong as to completely destroy the shell structure of the underlying independent-particle model. A justification of these assumptions in terms of elementary particle theory is therefore important for the unity of physics and, in particular, to determine if the hierarchical paradigm of physics applies in the nuclear domain. In proceeding from QCD to a derivation

---

[157]Kuo and Brown, *op. cit.* Footnote 79 on Page 273.
[158]Kuo and Brown, *op. cit.* Footnote 80 on Page 273.
[159]Bertsch G.F. (1965), *Nucl. Phys.* **74**, 234.
[160]Coraggio *et al.*, *op. cit.* Footnote 60 on Page 267.
[161]Coraggio L. *et al.* (2002), *Phys. Rev.* **C66**, 021303(R).
[162]Covello A. *et al.* (2007), *Prog. Part. Nucl. Phys.* **59**, 401.
[163]Coraggio L. *et al.* (2007), *Phys. Rev.* **C76**, 061303(R).

of the free two-nucleon interaction, it is presumed that nucleons and mesons are bound multi-quark systems. Nevertheless, considerable success has been achieved in calculating the interactions of nucleons at energies below their intrinsic excitation energies by regarding the nucleons and mesons as elementary particles. The shell model goes a step further; it dispenses with the mesons and describes nuclei as systems of nucleons with potential interactions. There are clearly times when it is necessary to probe more deeply, as for example, in describing nuclear beta decay. However, if it turned out that there were no underlying shell structure in nuclei and, even worse, if it were necessary to include the quarks and mesons as the elementary building blocks of nuclei, then nuclear structure theory would still be in a very primitive state.

Another objective of effective interaction theory is to understand, if only in qualitative terms, how the effective interactions needed for shell model applications come about and to gain a sense of how effective interactions and operators relate to the more fundamental operators that they represent. It is, after all, only too easy to interpret the results of a model calculation too literally.

Although effective interaction theory is already remarkably successful in addressing major concerns and in substantiating the shell model as the standard model of nuclear structure theory, a number of important questions remain to be answered. For example, why is it that two-body effective interactions work as well as they do in medium and heavy nuclei? It is reasonable to expect that the repulsive short-range interactions between nucleons should not induce many-body effective interactions because of the relative improbability that a significant number of nucleons will ever be simultaneously within the range of these interactions at nuclear densities. However, it is less clear why the restriction of a huge multi-shell model space to a single valence shell should not generate strong many-body effective interactions. Thus, one would particularly like to know better what effective theory actually does have to say about the rate of convergence of many-body effective interactions. Large computer calculations will, undoubtedly, be able to address this question in the near future. Calculations that have been done for p-shell nuclei[164,165] suggest that three-body effective forces are significant for the multi-shell spaces used for these nuclei. Thus, one might expect them to be even more significant in single-shell calculations for heavier nuclei. One can also address this question in model studies. For example, it is easy to imagine, as we shall discuss further in Volume 2, that the particles we describe in shell-model studies are not really nucleons but quasi-particles. Thus, an extra-closed-shell nucleon can polarize the core and, as it were, be followed by a tidal wave as it moves about the core and yet still behave as a particle, albeit one with modified (effective or "dressed") properties. Algebraic models can also be enlightening. For example, a realistic shell-model description of the properties of a rigidly-deformed nucleus requires an enormous shell-model

---

[164]Navrátil and Ormand, *op. cit.* Footnote 146 on Page 292.
[165]Hayes *et al.*, *op. cit.* Footnote 149 on Page 292.

space. Yet an SU(3) model, with an effective one-body quadrupole-moment, can rather accurately describe the projection of rigid-rotor model states onto a model consisting of a single harmonic-oscillator shell.

Finally, it should be emphasised that, however complete the theory of effective operators for the nuclear shell model eventually becomes, there will surely always be a place for phenomenological and symmetry-based interactions in the interpretation of nuclear data. In particular, it would not appear to be very enlightening to attempt a description of superdeformed rotational bands from first principles. As mentioned above (Section 5.7.5), it is probable that collective dynamics are emergent phenomena that would be realised, for more or less any reasonable attractive interactions, as a consequence of a general adiabaticity of collective motion in which all particles move slowly in unison in fluid-like flows that minimise their relative kinetic and interaction energies.

Exercises

5.18 Show that if a Hermitian Hamiltonian, represented by a matrix H, can be brought to block diagonal form

$$H \rightarrow U^\dagger H U = \begin{pmatrix} H_1 & 0 \\ 0 & H_2 \end{pmatrix}, \qquad (5.211)$$

then the submatrix H_1 has identical eigenvalues to a subset of eigenvalues of H.

5.19 Let $\{|\psi_\nu\rangle\}$ denote a low-energy set of eigenstates of the Hamiltonian

$$\hat{H} := \sum_{i=1}^{2} \left(\frac{\hat{p}_i^2}{2M} + \frac{1}{2} M\omega^2 r_i^2 \right) + \hat{V} \qquad (5.212)$$

and let $\hat{P}$ be an operator that projects each of these states onto a space of low-energy harmonic-oscillator states of the same angular momenta and other symmetries of the Hamiltonian. Show that, provided the projected states, $\{|\nu\rangle := \hat{P}|\psi_\nu\rangle\}$, form a linearly-independent set, the map

$$\hat{S} : |\nu\rangle \rightarrow |\psi_\nu\rangle \qquad (5.213)$$

is well-defined, has an inverse, and is a similarity transformation.

5.20 Given an arbitrary basis, $\{|\nu\rangle\}$, for a Hilbert space, $\mathbb{H}_0$, show how to derive a transformation that maps it to an orthonormal basis. [Hint: For an arbitrary basis, the overlap matrix with elements $\{\langle \mu|\nu\rangle\}$, is not a diagonal matrix. Define the unitary transformation that brings it to diagonal form and then the transformation that brings it to a unit matrix.]

5.21 Given an arbitrary basis, $\{|\nu\rangle\}$, for a Hilbert space, $\mathbb{H}_0$, show that the states of a biorthogonal basis, $\{|\tilde{\mu}\rangle\}$, i.e., states that satisfy the equation

$$\langle \tilde{\mu}|\nu\rangle = \delta_{\mu\nu}, \qquad (5.214)$$

are given by

$$|\bar{\mu}\rangle = \sum_{\nu} S^*_{\mu\nu}|\nu\rangle, \tag{5.215}$$

where S is the inverse of the overlap matrix N, i.e., $[S^{-1}]_{\mu\nu} = \langle\mu|\nu\rangle$.

5.22 Show that a zero-range interaction potential $V(\mathbf{r}_1 - \mathbf{r}_2) = V_0\delta(\mathbf{r}_1 - \mathbf{r}_2)$ has no effect on the scattering states of two nucleons unless they are in a relative S state.

5.23 Show that

$$\frac{1}{A}\sum_{i<j=1}^{A}|\mathbf{r}_i - \mathbf{r}_j|^2 = \sum_{i}^{A} r_i^2 - \frac{1}{A}\sum_{i,j=1}^{A}\mathbf{r}_i \cdot \mathbf{r}_j = \sum_{i}^{A} r_i^2 - AR^2. \tag{5.216}$$

5.8 Harmonic oscillator states and symmetries

It is instructive to reflect that, whereas a breakthrough in atomic physics came with the quantum-mechanical solution of the central force problem with a Coulomb potential, a parallel breakthrough in nuclear physics resulted from the quantum-mechanical solution for nucleons in a spherical harmonic oscillator potential.

In retrospect, these breakthroughs were highly fortuitous because, apart from their extensions to include centrifugal-like potentials, the Kepler and spherical harmonic oscillator are the only central-force problems that are analytically solvable.[166,167] They are analytically solvable because, in addition to being rotationally invariant, each of them has an su(1,1) spectrum generating algebra (SGA).

In fact, as we show in this section, the spherical harmonic oscillator has an exceedingly rich algebraic structure and a large variety of dynamical symmetries, many of which are exploited repeatedly throughout this volume. In addition to providing a simple basis for the nuclear shell model, the spherical harmonic oscillator has the invaluable property that its basis wave functions factor into products of relative and centre-of-mass wave functions. We have seen in Section 5.7 that this makes it possible, for example, to conserve translational invariance in deriving effective shell-model interactions. In subsequent sections of this chapter, it is shown that the many dynamical symmetries of the harmonic-oscillator shell model also provide an arsenal of coupling schemes that diagonalise important nuclear Hamiltonians and define solvable submodels of the shell model. In Volume 2, we shall be particularly concerned with the use of the special dynamical symmetries of the harmonic-oscillator shell model for the realisation of macroscopic collective models of the nucleus in microscopic terms.

---

[166]Čížek J. and Paldus J. (1977), *Int. J. Quantum Chem.* **12**, 875.
[167]Cooke T.H. and Wood J.L. (2002), *Amer. J. Phys.* **70**, 945.

5.8.1   *The simple harmonic oscillator (SHO)*

The SHO provides a prototypical example of the algebraic treatment of a quantum mechanical model. It is governed by the Hamiltonian

$$\hat{H} = \frac{1}{2M}\hat{p}^2 + \frac{1}{2}M\omega^2\hat{x}^2, \tag{5.217}$$

where $\hat{x}$ and $\hat{p}$ are mutually canonical position and momentum observables that, in quantum mechanics, satisfy the commutation relations

$$[\hat{x}, \hat{p}] = i\hbar\hat{I}, \tag{5.218}$$

and $\hat{I}$ is the identity operator. Thus, the observables, $\{\hat{x}, \hat{p}, \hat{I}\}$, span a Lie algebra; it is known as the *first Heisenberg-Weyl algebra*.

(i) *The Heisenberg-Weyl algebra, hw(1), as an SGA for the SHO*

The Heisenberg-Weyl algebra, hw(1), its representation, and its use as a spectrum generating algebra (SGA) for the SHO, are reviewed in Section 3.7.1. It is shown there that raising and lowering operators for this algebra, which satisfy the boson commutation relations

$$[c, c^\dagger] = \hat{I}, \quad [c, \hat{I}] = [c^\dagger, \hat{I}] = 0, \tag{5.219}$$

are given by

$$c^\dagger = \sqrt{\frac{M\omega}{2\hbar}}\left(\hat{x} - \frac{i}{M\omega}\hat{p}\right), \quad c = \sqrt{\frac{M\omega}{2\hbar}}\left(\hat{x} + \frac{i}{M\omega}\hat{p}\right). \tag{5.220}$$

When expressed in terms of these operators, the Hamiltonian becomes

$$\hat{H} = \tfrac{1}{2}\hbar\omega(c^\dagger c + cc^\dagger) = \hbar\omega(c^\dagger c + \tfrac{1}{2}\hat{I}). \tag{5.221}$$

The eigenstates of $\hat{H}$ then provide an orthonormal basis for a unitary irrep of the hw(1) algebra for which

$$\hat{H}|n\rangle = (n + \tfrac{1}{2})\hbar\omega|n\rangle, \quad n = 0, 1, 2, \ldots, \tag{5.222}$$

$$c^\dagger|n\rangle = \sqrt{n+1}\,|n+1\rangle, \quad c|n\rangle = \sqrt{n}\,|n-1\rangle, \quad \hat{I}|n\rangle = |n\rangle. \tag{5.223}$$

These basis states are defined by the equations

$$c|0\rangle = 0, \quad |n\rangle = \frac{1}{\sqrt{n!}}\,(c^\dagger)^n|0\rangle, \quad n = 0, 1, 2, \ldots. \tag{5.224}$$

(ii) *An su(1,1) algebra as an SGA for the SHO*

The SHO Hamiltonian has a second SGA, su(1,1), spanned by the bilinear combinations, $\{\hat{p}^2, \hat{x}^2, \hat{x}\hat{p} + \hat{p}\hat{x}\}$, of hw(1) elements. This algebra has raising and lowering operators, $\hat{S}_\pm$, and a Cartan weight operator, $\hat{S}_0$, given by the bilinear combinations of the hw(1) raising and lowering operators,

$$\hat{S}_+ = \tfrac{1}{2}c^\dagger c^\dagger, \quad \hat{S}_- = \tfrac{1}{2}cc, \quad \hat{S}_0 = \tfrac{1}{4}(c^\dagger c + cc^\dagger) = \tfrac{1}{2}(c^\dagger c + \tfrac{1}{2}\hat{I}), \qquad (5.225)$$

which satisfy the commutation relations

$$[\hat{S}_-, \hat{S}_+] = 2\hat{S}_0, \quad [\hat{S}_0, \hat{S}_+] = \hat{S}_+, \quad [\hat{S}_0, \hat{S}_-] = -\hat{S}_-. \qquad (5.226)$$

Comparison with Equation (5.221), reveals that the SHO Hamiltonian is an element of this su(1,1) algebra,

$$\hat{H} = 2\hbar\omega\hat{S}_0. \qquad (5.227)$$

Hence, it satisfies the equations of motion

$$[\hat{H}, \hat{S}_0] = 0, \quad [\hat{H}, \hat{S}_+] = 2\hbar\omega\,\hat{S}_+, \quad [\hat{H}, \hat{S}_-] = -2\hbar\omega\,\hat{S}_-. \qquad (5.228)$$

Thus, by determining the representations of the su(1,1) algebra on the Hilbert space of the SHO, we again determine the spectrum of $\hat{H}$.

The su(1,1) Lie algebra, as defined above, has two harmonic series irreps within the space of the SHO (cf. Section 3.7.3). Lowest-weight states for these irreps are given by the ground state, $|0\rangle$, and first-excited state, $|1\rangle$, respectively, of the Hamiltonian, (5.217). These states satisfy the equations

$$\hat{S}_-|0\rangle = \hat{S}_-|1\rangle = 0, \quad \hat{S}_0|0\rangle = \tfrac{1}{4}|0\rangle, \quad \hat{S}_0|1\rangle = \tfrac{3}{4}|1\rangle. \qquad (5.229)$$

However, both lowest-weight states can be raised indefinitely with the $\hat{S}_+$ raising operator. Thus, similar to the irrep of the Heisenberg-Weyl algebra, the su(1,1) irreps within the space of the SHO have no highest-weight states and are of infinite dimension. In accordance with the general expression of a unitary su(1,1) irrep with a lowest- but no highest-weight state given in Section 3.7.3, it is convenient to characterise such an irrep by its lowest weight which is defined to be an eigenvalue, λ, of $2\hat{S}_0$. For the irrep with lowest weight state $|0\rangle$, we then have $\lambda = \tfrac{1}{2}$ and, for the irrep with lowest weight state $|1\rangle$, $\lambda = \tfrac{3}{2}$.

Because the su(1,1) raising operator, $\hat{S}_+$, increases by two the number of quanta of any state it acts on, it follows that the subset, $\{|n\rangle; n = 0, 2, 4, \dots\}$, of SHO states with even values of n are a basis for the su(1,1) irrep $\lambda = \tfrac{1}{2}$ whereas the subset, $\{|n\rangle; n = 1, 3, 5, \dots\}$ with odd values of n are a basis for the irrep $\lambda = \tfrac{3}{2}$.

The representation theory of SU(1,1) and its Lie algebra, su(1,1), are discussed in more detail in Sections 3.7.3 and 4.2.2. An orthonormal basis for a unitary su(1,1) irrep with lowest weight λ is defined generally as a set of states $\{|\lambda\nu\rangle; \nu = 0, 1, 2, \dots\}$ such that $2\hat{S}_0|\lambda\nu\rangle = (\lambda + 2\nu)|\lambda\nu\rangle$. Thus, if the basis states, $\{|n\rangle\}$, of the SHO are relabelled by su(1,1) quantum numbers $\{|\lambda\nu\rangle; \nu = 0, 1, 2, \dots\}$ then, for n even, we have the correspondence $|n\rangle \equiv |\lambda = \tfrac{1}{2}, \nu = n/2\rangle$ and, for n odd, $|n\rangle \equiv |\lambda = \tfrac{3}{2}, \nu = (n-1)/2\rangle$. Equivalently, $|\lambda\nu\rangle \equiv |n\rangle$ with $n = \lambda + 2\nu - \tfrac{1}{2}$.

From the expressions given in terms of the $c^\dagger$ and c operators by Equations (5.223) and (5.224), it follows that, for $n = 0, 1, \ldots, \infty$,

$$
\begin{aligned}
\hat{S}_0|n\rangle &= \tfrac{1}{2}(n + \tfrac{1}{2})|n\rangle, \\
\hat{S}_+|n\rangle &= \tfrac{1}{2}\sqrt{(n+2)(n+1)}|n+2\rangle, \\
\hat{S}_-|n\rangle &= \tfrac{1}{2}\sqrt{(n-1)n}|n-2\rangle, \quad n \geq 1.
\end{aligned}
\tag{5.230}
$$

These are special $(\lambda = \tfrac{1}{2}, \tfrac{3}{2})$ cases of the general expressions

$$
\begin{aligned}
\hat{S}_0|\lambda\nu\rangle &= \tfrac{1}{2}(\lambda + 2\nu)|\lambda\nu\rangle, \\
\hat{S}_+|\lambda\nu\rangle &= \sqrt{(\lambda+\nu)(\nu+1)}\,|\lambda, \nu+1\rangle, \\
\hat{S}_-|\lambda\nu\rangle &= \sqrt{(\lambda+\nu-1)\nu}\,|\lambda, \nu-1\rangle,
\end{aligned}
\tag{5.231}
$$

derived for the harmonic series of unitary su(1,1) irreps in Section 3.7.3 (and more generally in Section 4.2.3). With the identification $\hat{H} = 2\hbar\omega\hat{S}_0$, we also obtain the energy spectrum for the SHO

$$
E_{\lambda\nu} = (\lambda + 2\nu)\hbar\omega,
\tag{5.232}
$$

which, with $n = \lambda + 2\nu - \tfrac{1}{2}$, gives the same result as Equation (5.222).

While su(1,1) is an SGA for the SHO Hamiltonian, it is not a full SGA for the SHO because the SHO Hilbert space is a sum of subspaces for two distinct su(1,1) irreps. Moreover, the su(1,1) Lie algebra does not contain any observables of odd degree in the c and $c^\dagger$ operators. However, it is useful to note that the latter operators satisfy the commutation relations

$$
\begin{aligned}
[\hat{S}_0, c^\dagger] &= \tfrac{1}{2}c^\dagger, & [\hat{S}_0, c] &= -\tfrac{1}{2}c, \\
[\hat{S}_-, c^\dagger] &= c, & [\hat{S}_-, c] &= 0, \\
[\hat{S}_+, c^\dagger] &= 0, & [\hat{S}_+, c] &= -c^\dagger,
\end{aligned}
\tag{5.233}
$$

which show that they transform under su(1,1) as a basis for a two-dimensional irrep with highest and lowest $2\hat{S}_0$ weights ± 1, respectively. Thus, they are components of an su(1,1) tensor (see Section 3.7.3) with $\lambda = -1$. These su(1,1) tensor operators are particularly useful because their matrix elements are already known from the hw(1) irreps.

(iii) Dynamical groups for the SHO

The observables of the hw(1) and su(1,1) algebras characterise different dynamical degrees of freedom of the SHO. This is seen explicitly by considering their expectation values in time-evolving states of the general form

$$
|\psi(t)\rangle := \exp\left(-\frac{i}{\hbar}\hat{H}t\right)|\psi(0)\rangle.
\tag{5.234}
$$

By expressing $\hat{x}$ and $\hat{p}$ in terms of creation and annihilation operators,

$$\hat{x} = \sqrt{\frac{\hbar}{2M\omega}}\,(c^\dagger + c), \quad \hat{p} = \mathrm{i}\sqrt{\frac{M\hbar\omega}{2}}\,(c^\dagger - c), \qquad (5.235)$$

and observing (see Exercise 5.26), for example, that

$$\begin{aligned}\langle\psi(0)|e^{\frac{\mathrm{i}}{\hbar}\hat{H}t}c^\dagger e^{-\frac{\mathrm{i}}{\hbar}\hat{H}t}|\psi(0)\rangle &= \langle\psi(0)|c^\dagger e^{\mathrm{i}\omega t}|\psi(0)\rangle,\\ \langle\psi(0)|e^{\frac{\mathrm{i}}{\hbar}\hat{H}t}c e^{-\frac{\mathrm{i}}{\hbar}\hat{H}t}|\psi(0)\rangle &= \langle\psi(0)|c e^{-\mathrm{i}\omega t}|\psi(0)\rangle,\end{aligned} \qquad (5.236)$$

it follows that the expectation values, $x(t) := \langle\psi(t)|\hat{x}|\psi(t)\rangle$ and $p(t) := \langle\psi(t)|\hat{p}|\psi(t)\rangle$, of the hw(1) observables have time evolutions given by

$$x(t) = x(0)\cos\omega t + \frac{1}{M\omega}p(0)\sin\omega t, \qquad (5.237)$$

$$p(t) = p(0)\cos\omega t - M\omega x(0)\sin\omega t. \qquad (5.238)$$

These are the standard harmonic vibrational modes of the SHO.

Making a similar expansion,

$$\hat{x}^2 = \frac{\hbar}{M\omega}\,(2\hat{S}_0 + \hat{S}_+ + \hat{S}_-), \qquad (5.239)$$

$$\hat{p}^2 = M\hbar\omega\,(2\hat{S}_0 - \hat{S}_+ - \hat{S}_-), \qquad (5.240)$$

it is determined that the expectation values, $\langle\hat{x}^2\rangle_t := \langle\psi(t)|\hat{x}^2|\psi(t)\rangle$ and $\langle\hat{p}^2\rangle_t := \langle\psi(t)|\hat{p}^2|\psi(t)\rangle$, of the su(1,1) observables have time evolutions given by

$$\frac{M\omega}{\hbar}\langle\hat{x}^2\rangle_t = \langle 2\hat{S}_0\rangle_0 + \langle\hat{S}_+ + \hat{S}_-\rangle_0\cos 2\omega t + \mathrm{i}\langle\hat{S}_+ - \hat{S}_-\rangle_0\sin 2\omega t, \qquad (5.241)$$

$$\frac{1}{M\hbar\omega}\langle\hat{p}^2\rangle_t = \langle 2\hat{S}_0\rangle_0 - \langle\hat{S}_+ + \hat{S}_-\rangle_0\cos 2\omega t - \mathrm{i}\langle\hat{S}_+ - \hat{S}_-\rangle_0\sin 2\omega t. \qquad (5.242)$$

These are so-called harmonic *breathing-mode* vibrations.

It is also of interest to consider the dynamical groups of transformations, HW(1) and SU(1,1), generated by their respective hw(1) and su(1,1) Lie algebras. For example, the group element $\exp\hat{X} \in$ HW(1), with $\hat{X} = \alpha c^\dagger - \alpha^* c + \mathrm{i}\gamma\hat{I}$, effects the transformation

$$c^\dagger \to e^{\hat{X}}c^\dagger e^{-\hat{X}} = c^\dagger - \alpha^*\hat{I}, \qquad (5.243)$$

$$c \to e^{\hat{X}}c e^{-\hat{X}} = c - \alpha\hat{I}. \qquad (5.244)$$

Such a transformation preserves the commutation relation $[c, c^\dagger] = \hat{I}$. The corresponding translations of the phase-space observables,

$$\hat{x} \to \hat{x} - \sqrt{\frac{\hbar}{2M\omega}}\,(\alpha^* + \alpha)\hat{I}, \qquad (5.245)$$

$$\hat{p} \to \hat{p} - \mathrm{i}\sqrt{\frac{M\hbar\omega}{2}}\,(\alpha^* - \alpha)\hat{I}, \qquad (5.246)$$

likewise preserve their commutation relations. They can be considered as translations in the $(\hat{x}, \hat{p})$ phase space.

Similar to the above transformations, as shown in Section 3.10, the elements of the group SU(1,1) effect the transformations

$$c^\dagger \to \alpha c^\dagger + \beta c, \tag{5.247}$$

$$c \to \beta^* c^\dagger + \alpha^* c, \tag{5.248}$$

with $|\alpha|^2 - |\beta|^2 = 1$. These transformations also preserve the $(\hat{x}, \hat{p})$ commutation relations and are known *general-linear canonical transformations* of the phase space.

5.8.2 *The spherical harmonic oscillator*

The spherical harmonic oscillator is the natural extension of the SHO to three dimensions. It is governed by a Hamiltonian

$$\hat{H} = \frac{1}{2M}\hat{\mathbf{p}}^2 + \frac{1}{2}M\omega^2\hat{\mathbf{r}}^2, \tag{5.249}$$

where $\hat{\mathbf{r}}^2 \equiv \hat{\mathbf{r}} \cdot \hat{\mathbf{r}} = \hat{x}^2 + \hat{y}^2 + \hat{z}^2$ and $\hat{\mathbf{p}}^2 = \hat{p}_x^2 + \hat{p}_y^2 + \hat{p}_z^2$.

(i) *An hw(3) algebra as an SGA for the spherical harmonic oscillator*

The operators $\{\hat{x}, \hat{y}, \hat{z}, \hat{p}_x, \hat{p}_y, \hat{p}_z, \hat{I}\}$ span a Heisenberg-Weyl Lie algebra, hw(3), which is an SGA for the spherical harmonic oscillator with non-vanishing commutation relations

$$[\hat{x}, \hat{p}_x] = i\hbar\hat{I}, \quad [\hat{y}, \hat{p}_y] = i\hbar\hat{I}, \quad [\hat{z}, \hat{p}_z] = i\hbar\hat{I}. \tag{5.250}$$

The spherical harmonic oscillator Hamiltonian, being a sum of three SHO Hamiltonians, can be expressed in the form

$$\hat{H} = \hbar\omega\big(c_x^\dagger c_x + c_y^\dagger c_y + c_z^\dagger c_z + \tfrac{3}{2}\hat{I}\big) \tag{5.251}$$

with

$$c_x^\dagger = \sqrt{\frac{M\omega}{2\hbar}}\left(\hat{x} - \frac{i}{M\omega}\hat{p}_x\right), \quad c_x = \sqrt{\frac{M\omega}{2\hbar}}\left(\hat{x} + \frac{i}{M\omega}\hat{p}_x\right), \quad \text{etc.} \tag{5.252}$$

Thus, $\hat{H}$ has a set of eigenstates labelled by a triple of integers (n_x, n_y, n_z) with eigenvalues given by

$$\hat{H}|n_x n_y n_z\rangle = \big(n_x + n_y + n_z + \tfrac{3}{2}\big)\hbar\omega|n_x n_y n_z\rangle. \tag{5.253}$$

However, in practice, it is desirable to take advantage of the spherical symmetry and to express the eigenstates of $\hat{H}$ in terms of quantum numbers associated with a spherical rather than a Euclidean geometry.

A set of eigenfunctions for the spherical harmonic oscillator can be constructed by expressing its Hamiltonian in terms of spherical polar coordinates and solving the coupled equations that emerge for the radial and angular wave functions. A simpler algebraic approach uses a combination of an su(1,1) SGA to generate radial states and an so(3) symmetry algebra to generate complementary angular states.

(ii) *An su(1,1) algebra as an SGA for the spherical harmonic oscillator*

An SGA for the spherical harmonic-oscillator Hamiltonian is spanned by operators

$$
\begin{aligned}
\hat{S}_+ &:= \tfrac{1}{2}(c_x^\dagger c_x^\dagger + c_y^\dagger c_y^\dagger + c_z^\dagger c_z^\dagger), \\
\hat{S}_- &:= \tfrac{1}{2}(c_x c_x + c_y c_y + c_z c_z), \\
\hat{S}_0 &:= \tfrac{1}{2}(c_x^\dagger c_x + c_y^\dagger c_y + c_z^\dagger c_z + \tfrac{3}{2}\hat{I}),
\end{aligned}
\tag{5.254}
$$

which also satisfy the commutation relations of Equation (5.226); $\hat{S}_\pm$ are likewise raising and lowering operators and $\hat{S}_0$ is a Cartan weight operator for an su(1,1) Lie algebra isomorphic to that of the SHO. Moreover, the spherical harmonic-oscillator Hamiltonian, $\hat{H} = 2\hbar\omega\hat{S}_0$, continues to be an element of the new su(1,1) algebra. The elements of this new realisation of su(1,1) are scalar (dot) products of vector operators. For example, we can define the vector operators $\mathbf{c}^\dagger := (c_x^\dagger, c_y^\dagger, c_z^\dagger)$ and $\mathbf{c} := (c_x, c_y, c_z)$ and write

$$
\hat{S}_+ = \tfrac{1}{2}\mathbf{c}^\dagger \cdot \mathbf{c}^\dagger, \quad \hat{S}_- = \tfrac{1}{2}\mathbf{c} \cdot \mathbf{c}, \quad \hat{S}_0 = \tfrac{1}{2}(\mathbf{c}^\dagger \cdot \mathbf{c} + \tfrac{3}{2}).
\tag{5.255}
$$

Consequently, *the elements of this su(1,1) algebra are O(3) invariant*, i.e., they are invariant under rotations and parity inversions, in which $\mathbf{c} \to -\mathbf{c}$. In particular, they commute with the angular-momentum operators.

Unitary irreps of the above realisation of the su(1,1) Lie algebra are given by the general expressions (5.231). As we show below, an infinite number of irreps, with lowest weights given by $\lambda = l + 3/2$, where $l = 0, 1, 2, \ldots$ is an angular momentum quantum number, occur in the space of the spherical harmonic oscillator.

(iii) *The angular momentum algebra as a symmetry algebra*

Any central-force Hamiltonian for a three-dimensonal space, such as that of the spherical harmonic oscillator, is rotationally invariant and has an SO(3) symmetry group. A Cartesian basis of infinitesimal generators of SO(3) is given by the angular momentum operators

$$
\begin{aligned}
\hbar\hat{L}_x &= (\hat{y}\hat{p}_z - \hat{z}\hat{p}_y) = -i\hbar(c_y^\dagger c_z - c_z^\dagger c_y), \\
\hbar\hat{L}_y &= (\hat{z}\hat{p}_x - \hat{x}\hat{p}_z) = -i\hbar(c_z^\dagger c_x - c_x^\dagger c_z), \\
\hbar\hat{L}_z &= (\hat{x}\hat{p}_y - \hat{y}\hat{p}_x) = -i\hbar(c_x^\dagger c_y - c_y^\dagger c_x).
\end{aligned}
\tag{5.256}
$$

These operators satisfy the commutation relations

$$[\hat{L}_x, \hat{L}_y] = \mathrm{i}\hat{L}_z, \quad \text{etc.,} \tag{5.257}$$

and span an $so(3) \sim su(2)$ Lie algebra. They also satisfy the equations

$$[\hat{H}, \hat{L}_x] = [\hat{H}, \hat{L}_y] = [\hat{H}, \hat{L}_z] = 0, \tag{5.258}$$

consistent with the rotational invariance of $\hat{H}$.

Irreps of the $so(3) \sim su(2)$ Lie algebra are derived in Section 3.7.2. An $so(3)$ irrep which generates a true irrep of the group SO(3), as opposed to a spinor irrep that is double-valued and thus a true SU(2) irrep, is characterised by a positive integer-valued (orbital) angular momentum quantum number, l, and a basis of states, $\{|lm; m = -l, \ldots, +l\}$. Thus, an $so(3)$ irrep (see Section 3.7.2) is defined by the equations

$$\hat{L}_0|lm\rangle = m|lm\rangle,$$
$$\hat{L}_\pm|lm\rangle = \sqrt{(l \mp m)(l \pm m + 1)}\,|lm \pm 1\rangle. \tag{5.259}$$

(iv) *A spherical-tensor basis for the hw(3) Lie algebra*

Raising operators for the hw(3) Lie algebra (interpreted as creation operators of harmonic-oscillator quanta) are given as components $\{c_m^\dagger; m = 0, \pm 1\}$ of an $l = 1$ spherical tensor (see Appendix A.3.2) by the combinations

$$c_0^\dagger = c_z^\dagger, \quad c_{\pm 1}^\dagger = \mp \frac{1}{\sqrt{2}}(c_x^\dagger \pm \mathrm{i}c_y^\dagger), \tag{5.260}$$

which satisfy the commutation relations

$$[\hat{L}_0, c_m^\dagger] = mc_m^\dagger, \quad [\hat{L}_\pm, c_m^\dagger] = \sqrt{(1 \mp m)(2 \pm m)}\,c_{m \pm 1}^\dagger, \tag{5.261}$$

with $m = 0, \pm 1$. Together with their Hermitian adjoints

$$c^0 = c_z, \quad c^{\pm 1} = \mp \frac{1}{\sqrt{2}}(c_x \mp \mathrm{i}c_y), \tag{5.262}$$

these operators satisfy the boson commutation relations

$$[c_m^\dagger, c_{m'}^\dagger] = [c^m, c^{m'}] = 0, \quad [c^m, c_{m'}^\dagger] = \delta_{m'}^m \hat{I}. \tag{5.263}$$

Lowering operators (annihilation operators of harmonic oscillator quanta) with lower indices are similarly defined as components of spherical tensors by

$$c_0 = c_z, \quad c_{\pm 1} = \mp \frac{1}{\sqrt{2}}(c_x \pm \mathrm{i}c_y). \tag{5.264}$$

This gives the relationship (see Appendix A.7.3)

$$c_m = (-1)^m c^{-m}, \quad m = 0, \pm 1. \tag{5.265}$$

305

The commutation relations of the su(1,1) operators with the hw(3) operators are now given by

$$[\hat{S}_0, c_m^\dagger] = \tfrac{1}{2} c_m^\dagger, \qquad\qquad [\hat{S}_0, c_m] = -\tfrac{1}{2} c_m,$$
$$[\hat{S}_-, c_m^\dagger] = c_m, \qquad\qquad [\hat{S}_-, c_m] = 0, \qquad\qquad (5.266)$$
$$[\hat{S}_+, c_m^\dagger] = 0, \qquad\qquad [\hat{S}_+, c_m] = -c_m^\dagger.$$

These equations show (see Section 5.8.1(ii)) that, for each $m = 0, \pm 1$, the operators $\{c_m, c_m^\dagger\}$ are components of an su(1,1) $\lambda = -1$ tensor operator. Thus, $c_1^\dagger \equiv c_{+1}^\dagger$ is simultaneously the highest-weight component of an $l = 1$ so(3) tensor operator and the highest-weight component of a $\lambda = -1$ su(1,1) tensor operator.

(v) *Basis states for the spherical harmonic oscillator*

Because the su(1,1) operators and the so(3) angular momentum operators commute with one another, they combine to form a direct sum Lie algebra su(1,1) + so(3) which is the Lie algebra of the direct product group SU(1,1) × SO(3). The next objective is then to combine radial states for su(1,1) irreps with angular-momentum states for so(3) irreps to form basis states for the spherical harmonic oscillator.

Let $\{|\lambda\nu lm\rangle\}$ denote states of the spherical harmonic oscillator for which ν labels an orthonormal basis for an su(1,1) irrep of lowest weight λ and m labels an orthonormal basis for an so(3) irrep of angular momentum l. The set $\{|\lambda\nu lm\rangle; \nu = 0, 1, 2, \ldots, m = -l, \ldots, +l\}$ is then an orthonormal basis for an su(1,1) + so(3) irrep, labelled by (λ, l). To determine which of these irreps is realised in the space of the spherical harmonic oscillator, we seek to identify the states $\{|\lambda 0 ll\rangle\}$ that are simultaneously of lowest su(1,1) weight and highest so(3) weight.

First observe that the spherical harmonic-oscillator ground state, $|0\rangle$, is the vacuum state for the boson operators, i.e., $c_x|0\rangle = c_y|0\rangle = c_z|0\rangle = 0$. Thus, it satisfies the equations

$$\hat{S}_-|0\rangle = 0, \quad \hat{S}_0|0\rangle = \tfrac{3}{4}|0\rangle, \qquad\qquad (5.267)$$
$$\hat{L}_x|0\rangle = \hat{L}_y|0\rangle = \hat{L}_z|0\rangle = 0, \qquad\qquad (5.268)$$

which imply that $|0\rangle$ is the state in the set $\{|\lambda\nu lm\rangle\}$ with $\lambda = 3/2$ and $\nu = l = m = 0$. Other states in the set include the subset

$$|l\rangle := |l + \tfrac{3}{2} 0 l l\rangle = \frac{1}{\sqrt{l!}} \, (c_1^\dagger)^l |0\rangle, \quad l = 0, 1, 2, \ldots \qquad (5.269)$$

which satisfy the equations (cf. Exercise 5.29)

$$\hat{L}_+|l\rangle = 0, \quad \hat{L}_0|l\rangle = l|l\rangle, \qquad\qquad (5.270)$$
$$\hat{S}_-|l\rangle = 0, \quad \hat{S}_0|l\rangle = \tfrac{1}{2}\big(l + \tfrac{3}{2}\big)|l\rangle. \qquad\qquad (5.271)$$

We now claim that the set of $\lambda = l + 3/2$ states

$$\{|\nu l m\rangle \equiv |l + \tfrac{3}{2}, \nu l m\rangle; \nu = 0, 1, 2, \ldots, l = 0, 1, 2, \ldots, m = -l, \ldots, +l\} \quad (5.272)$$

is a complete orthonormal basis for the spherical harmonic oscillator. This claim can be substantiated in many ways; e.g. from the observation that the su(1,1) states of a given angular momentum are in one-to-one correspondence with the corresponding solutions of the radial wave equation for the spherical harmonic oscillator.

Starting from the vacuum state, we now have a complete algebraic definition of the basis states, $\{|\nu l m\rangle\}$, for the spherical harmonic oscillator given by

$$|0ll\rangle = \frac{1}{\sqrt{l!}} (c_1^\dagger)^l |0\rangle, \quad l = 0, 1, 2, \ldots \quad (5.273)$$

$$\hat{L}_\pm |0lm\rangle = \sqrt{(l \mp m)(l \pm m + 1)} \, |0lm \pm 1\rangle, \quad (5.274)$$

$$\hat{L}_0 |0lm\rangle = m|0lm\rangle, \quad (5.275)$$

$$\hat{S}_+ |\nu l m\rangle = \sqrt{(l + \tfrac{3}{2} + \nu)(\nu + 1)} \, |\nu + 1, lm\rangle, \quad (5.276)$$

$$\hat{S}_- |\nu l m\rangle = \sqrt{(l + \tfrac{1}{2} + \nu)\nu} \, |\nu - 1, lm\rangle, \quad (5.277)$$

$$\hat{S}_0 |\nu l m\rangle = \tfrac{1}{2}(l + \tfrac{3}{2} + 2\nu)|\nu l m\rangle. \quad (5.278)$$

These states diagonalise the spherical harmonic-oscillator Hamiltonian and reproduce its complete angular-momentum spectrum, as illustrated in Figure 5.13.

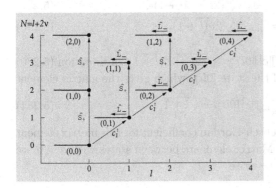

Figure 5.13: Energy levels of the spherical harmonic oscillator showing the values of the radial and angular momentum quantum numbers (ν, l) and the routes by which states are constructed from the ground state. The states $|l\rangle := |0ll\rangle$, which are simultaneously of highest so(3) weight and lowest su(1,1) weight, are generated by the repeated action of the $c_1^\dagger$ operators and correspond to the points on the lower diagonal (cf. Figure 1.15).

As well as being invariant under rotations, a central-force Hamiltonian is also invariant under inversion in which a position vectors, $\mathbf{r}$, is mapped to its negative. Thus, states of even or odd degree in the position coordinates are said to have positive or negative parity, respectively, and basis states, $\{|\nu l m\rangle\}$, have parity $\pi = (-1)^l$. The group of all orthogonal transformations of a three-dimensional Euclidean space which inversions and reflections, as well as rotations, is the group O(3).

Observe that the SU(1,1) and O(3) irreps, given by $\lambda = l + 3/2$ and $(l, \pi = (-1)^l)$, respectively, occur in uniquely-paired combinations. Thus, the radial component of a spherical harmonic-oscillator wave function, $R_{l\nu}$, is labelled by the same angular-momentum quantum number, l, as the spherical harmonic, Y_{lm}, with which it is combined. The l-dependence of the radial wave function is due to the presence of a centrifugal potential in the radial wave equation. Nevertheless, it is an example of a deep *duality relationship* between irreps of commuting groups that occurs frequently in nuclear physics. Such duality relationships are discussed further in Section 5.10, in subsequent sections of this chapter, and throughout the book.[168]

(vii) *Matrix elements*

Matrix elements of operators in the su(1,1) and so(3) Lie algebras are obtained immediately from Equations (5.274)–(5.278). Other basic matrix elements are expressed in terms of reduced matrix elements and SO(3) Clebsch-Gordan coefficients by using the Wigner-Eckart theorem (see Section A.3.3) which, for example, gives

$$\langle \nu_f l_f m_f | c_m^\dagger | \nu_i l_i m_i \rangle = (l_i m_i \, 1m | l_f m_f) \frac{\langle \nu_f l_f \| c^\dagger \| \nu_i l_i \rangle}{\sqrt{2l_f + 1}}. \tag{5.279}$$

Claim 5.1 *SO(3)-reduced matrix elements of the spherical harmonic oscillator boson operators are given by*

$$\langle \mu, l+1 \| c^\dagger \| \nu l \rangle = \delta_{\mu,\nu} \sqrt{(2l + 2\nu + 3)(l+1)}, \tag{5.280}$$

$$\langle \mu, l-1 \| c^\dagger \| \nu l \rangle = -\delta_{\mu,\nu+1} \sqrt{2(\nu+1)l}, \tag{5.281}$$

$$\langle \mu, l-1 \| c \| \nu l \rangle = -\delta_{\mu,\nu} \sqrt{(2l + 2\nu + 1)l}, \tag{5.282}$$

$$\langle \mu, l+1 \| c \| \nu l \rangle = \delta_{\mu,\nu-1} \sqrt{2\nu(l+1)}. \tag{5.283}$$

Proof: The special case, $\langle 0, l+1 \| c^\dagger \| 0l \rangle = \sqrt{(2l+3)(l+1)}$, of Equation (5.280) follows from the definition (5.273) of the state $|0ll\rangle$, which gives the matrix element

$$\langle 0, l+1, l+1 | c_1^\dagger | 0ll \rangle = \sqrt{l+1}, \tag{5.284}$$

and from the observation that the Clebsch-Gordan coefficient for this matrix element has the value $(ll\,11|l+1, l+1) = 1$. Matrix elements between states with $\nu > 0$ are obtained by using Equation (5.276) to write

$$|\nu lm\rangle = \frac{1}{\sqrt{(l + \nu + 1/2)\nu}} \hat{S}_+ |\nu - 1, lm\rangle. \tag{5.285}$$

Because $[c_m^\dagger, \hat{S}_+] = 0$, this gives the recursion relation

$$\langle \nu, l+1 \| c^\dagger \| \nu l \rangle = \frac{\langle \nu - 1, l+1 \| \hat{S}_- \hat{S}_+ c^\dagger \| \nu - 1, l \rangle}{\nu \sqrt{(l + \nu + 3/2)(l + \nu + 1/2)}}$$

$$= \sqrt{\frac{2l + 2\nu + 3}{2l + 2\nu + 1}} \langle \nu - 1, l+1 \| c^\dagger \| \nu - 1, l \rangle \tag{5.286}$$

---

[168] A review of duality relationships that occur in physics is given by Rowe D.J. *et al.* (preprint).

for which Equation (5.280) is the solution.

The special case of Equation (5.283),

$$\langle 0l\|c\|1, l-1\rangle = \sqrt{2l}\,, \tag{5.287}$$

is obtained from the identity

$$\langle 0ll|c_1|1, l-1, l-1\rangle = \frac{\langle 0ll|[c_1, \hat{S}_+]|0, l-1, l-1\rangle}{\sqrt{l + \frac{1}{2}}} = \frac{\langle 0ll|c_1^\dagger|0, l-1, l-1\rangle}{\sqrt{l + \frac{1}{2}}}. \tag{5.288}$$

The general case then follows by solution of the recursion relation

$$\langle \nu l\|c\|\nu + 1, l-1\rangle = \sqrt{\frac{\nu + 1}{\nu}}\langle \nu - 1, l\|c\|\nu, l-1\rangle, \tag{5.289}$$

derived from the observation that $[\hat{S}_-, c] = 0$ and the identity

$$\langle \nu l\|c\|\nu + 1, l-1\rangle = \frac{\langle \nu - 1, l\|\hat{S}_- c\,\hat{S}_+\|\nu, l-1\rangle}{(l + \nu + \frac{1}{2})\sqrt{\nu(\nu + 1)}}. \tag{5.290}$$

The remaining matrix elements of the boson operators are given by the symmetry relations for SO(3)-reduced matrix elements. Noting, from Equations (5.260) and (5.265), that $(c_{\pm 1}^\dagger)^\dagger = -c_{\mp 1}$, and $(c_0^\dagger)^\dagger = c_0$, it is determined that

$$(c_m^\dagger)^\dagger = (-1)^m c_{-m}. \tag{5.291}$$

Hence (cf. Appendix A.7.3), we obtain

$$\langle \nu l\|c\|\nu, l+1\rangle = -\langle \nu, l+1\|c^\dagger\|\nu l\rangle^*, \tag{5.292}$$

$$\langle \nu l\|c^\dagger\|\nu - 1, l+1\rangle = -\langle \nu - 1, l+1\|c\|\nu l\rangle^*. \tag{5.293}$$

This completes the proof of Claim 5.1. □

In using spherical harmonic-oscillator states as a basis for more general systems, it is useful to separate the radial and angular matrix elements. Because the angular wave functions are simply spherical harmonics, as they are for any three-dimensional central-force Hamiltonian, the reduced matrix elements of any spherical harmonic are given by the standard formula[169] (see Exercise A.6)

$$\langle l_3\|\hat{Y}_{l_2}\|l_1\rangle := \left[\frac{(2l_1 + 1)(2l_2 + 1)}{4\pi}\right]^{1/2} (l_1 0\, l_2 0 | l_3 0), \tag{5.294}$$

where $\langle l_3 m_3|\hat{Y}_{l_2 m_2}|l_1 m_1\rangle = \int_0^{2\pi} \int_0^\pi Y_{l_3 m_3}^*(\theta, \varphi) Y_{l_2 m_2}(\theta, \varphi) Y_{l_1 m_1}(\theta, \varphi)\, \sin\theta\, d\theta\, d\varphi.$

---

[169]This formula, or its equivalent, can be found in almost any book on angular momentum theory, e.g., Rose M.E. (1995), *Elementary Theory of Angular Momentum* (Dover Publications, New York; originally published by Wiley, 1957), Brink D.M. and Satchler G.R. (1994), *Angular Momentum* (Oxford University Press), third edn., or Edmonds A.R. (1996), *Angular Momentum in Quantum Mechanics* (Princeton University Press, Princeton, N.J.), paperback edition.

Matrix elements of even powers of r are determined from the identity

$$(a\hat{r})^2 = 2\hat{S}_0 + \hat{S}_+ + \hat{S}_-, \tag{5.295}$$

which is obtained from Equations (5.252)) and (5.254) with $a = \sqrt{M\omega/\hbar}$. By use of Equation (5.276)–(5.278) for the su(1,1) matrix elements, this expression gives the radial matrix elements

$$\langle \nu l | (a\hat{r})^2 | \nu l \rangle = l + \tfrac{3}{2} + 2\nu,$$
$$\langle \nu - 1, l | (a\hat{r})^2 | \nu l \rangle = \langle \nu l | (a\hat{r})^2 | \nu - 1, l \rangle = \sqrt{(l + \tfrac{1}{2} + \nu)\nu}. \tag{5.296}$$

Matrix elements of linear terms in r are obtained from the identity

$$ar Y_{10}(\theta, \varphi) = \sqrt{\frac{3}{4\pi}}\, az = \sqrt{\frac{3}{8\pi}}\, (c_0^\dagger + c_0), \tag{5.297}$$

which implies that $ar Y_{1m}(\theta, \varphi) = \sqrt{\frac{3}{8\pi}}\, (c_m^\dagger + c_m)$. By use of Claim 5.1 we have

$$\langle \nu, l+1 \| a\hat{r}\hat{Y}_1 \| \nu l \rangle = \sqrt{\frac{3}{8\pi}(2l + 2\nu + 3)(l+1)}, \tag{5.298}$$

$$\langle \nu + 1, l-1 \| a\hat{r}\hat{Y}_1 \| \nu l \rangle = -\sqrt{\frac{3}{4\pi}(\nu+1)l}, \tag{5.299}$$

$$\langle \nu, l-1 \| a\hat{r}\hat{Y}_1 \| \nu l \rangle = -\sqrt{\frac{3}{8\pi}(2l + 2\nu + 1)l}, \tag{5.300}$$

$$\langle \nu - 1, l+1 \| a\hat{r}\hat{Y}_1 \| \nu l \rangle = \sqrt{\frac{3}{4\pi}\nu(l+1)}. \tag{5.301}$$

Hence, by factoring out the matrix elements of the spherical harmonic Y_1, given by Equation (5.294), we obtain the radial matrix elements

$$\langle \nu - 1, l+1 | a\hat{r} | \nu l \rangle = \langle \nu l | a\hat{r} | \nu - 1, l+1 \rangle = \sqrt{\nu}, \tag{5.302}$$
$$\langle \nu, l-1 | ar | \nu l \rangle = \langle \nu l | ar | \nu, l-1 \rangle = \sqrt{l + \nu + \tfrac{1}{2}}. \tag{5.303}$$

The $a\hat{r}$ matrix elements can also be derived by the factorisation method described in Section 4.4.3.[170]

Other matrix elements, e.g., matrix elements of inverse powers of r, can be obtained by the general methods of Section 4.4. Radial matrix elements for a general potential energy function, for example, can also be calculated by direct integration over the radial wave functions, where it is noted that the general expression of a wave function $\psi_{\nu lm}$ for a spherical harmonic-oscillator state, $|\nu lm\rangle$, is given in spherical coordinates as a product,

$$\psi_{\nu lm}(r, \theta, \varphi) = \frac{a^{1/2}}{r} \mathcal{R}_\nu^{l+3/2}(ar) Y_{lm}(\theta, \varphi), \tag{5.304}$$

---

[170]Rowe, *op. cit.* Footnote 32 on Page 187.

of a radial wave function and a spherical harmonic. The radial wave function in this expression is given by

$$\mathcal{R}_\nu^{l+3/2}(r) = (-1)^\nu \sqrt{\frac{2\nu!}{\Gamma(\nu + l + 3/2)}} \, r^{l+1} L_\nu^{(l+1/2)}(r^2) \, e^{-r^2/2}, \quad \nu = 0, 1, 2, \ldots,$$

(5.305)

where $L_\nu^{(l+1/2)}$ is a generalised Laguerre polynomial.[171]

Note that the radial wave function, $\mathcal{R}_\nu^\lambda$, is here labelled by its SU(1,1) representation labels, (λ, ν). This is useful for the algebraic calculation of matrix elements which apply to harmonic oscillators of any dimension (cf. application to the collective model in Chapter 4). However, it is common practice, in nuclear physics, to label harmonic oscillator wave functions with a radial quantum number $n = \nu + 1$, taking values $n = 1, 2, 3, \ldots$, instead of $\nu = 0, 1, 2, \ldots$. Thus, the lowest-energy states of the spherical harmonic oscillator are described as $1s$ states rather than $0s$ states. Similarly, spherical harmonic-oscillator wave functions are commonly expressed in the form

$$\phi_{nlm}(r, \theta, \varphi) = \psi_{n-1,lm}(r, \theta, \varphi) = a^{3/2} R_{nl}(ar) Y_{lm}(\theta, \varphi), \quad n = 1, 2, 3, \ldots, \quad (5.306)$$

with $R_{nl}(r) = \frac{1}{r} \mathcal{R}_{n-1}^{l+3/2}(r)$.

5.8.3 The U(3) symmetry group

Any central-force Hamiltonian is invariant under the orthogonal group O(3), which includes SO(3) rotations and parity inversions. However, the spherical harmonic-oscillator Hamiltonian has a larger symmetry group, U(3), corresponding to the equal-energy degeneracies of states shown in Figure 5.13. The group U(3) is realised as the subset of one-body unitary transformations which leave the spherical-harmonic-oscillator Hamiltonian invariant. Its Lie algebra, u(3), is spanned by Hermitian linear combinations of the operators

$$\hat{C}_{\alpha\beta} = c_\alpha^\dagger c^\beta,$$

(5.307)

where $c_\alpha^\dagger$ and c^β are defined by Equations (5.260) and (5.262) with α and β running over the values $0, \pm 1$.

A state $|N\rangle = \frac{1}{\sqrt{N!}} (c_1^\dagger)^N |0\rangle$, defined by Equation (5.269) for any integer $N \geq 0$, satisfies the equations

$$\hat{C}_{\alpha\beta}|N\rangle = 0, \quad \text{for } \alpha > \beta, \qquad \hat{C}_{\alpha\alpha}|N\rangle = \delta_{\alpha 1} N |N\rangle.$$

(5.308)

Thus, it is a highest-weight state for a U(3) irrep $\{N, 0, 0\}$ spanned by all the equal-energy states $\{|\nu lm\rangle\}$ with $l + 2\nu = N$. (For $\nu = 0$, l is equal to N and $|0ll\rangle \equiv |N\rangle$.)

---

[171]Abramowitz and Stegun, *op. cit.* Footnote 14 on Page 176.

The group U(3) has many subgroups, including SU(3) whose Lie algebra, su(3), is the subset of operators $\hat{X} := \sum_{\alpha,\beta} X_{\alpha\beta} c^\dagger_\alpha c^\beta$, where X is a Hermitian matrix with vanishing trace, $\text{tr}(X) = 0$. In addition, U(3) contains O(1), SO(3), and SO(2) subgroups and the basis states, $\{|\nu l m\rangle\}$ reduce the subgroup chain

$$
\begin{array}{cccc}
\text{U}(3) & \supset & \text{O}(3) & \supset & \text{SO}(2) \\
\{N,0,0\} & & (l, \pi = (-1)^l) & & m
\end{array}
\tag{5.309}
$$

with the quantum numbers shown.

It is useful to note that the basis states, $\{|\nu l m\rangle\}$, defined by the quantum numbers of the subgroup chain (5.309), are identical to those defined in Section 5.8.2 by the quantum numbers of the groups in the chain

$$
\begin{array}{ccccc}
\text{SU}(1,1) & \times & \text{O}(3) & \supset & \text{U}(1) \times \text{SO}(2) \\
\lambda = l + {}^3\!/\!_2 & & (l, \pi = (-1)^l) & & \nu \qquad\quad m
\end{array},
\tag{5.310}
$$

with $N = l + 2\nu$. Thus, although the groups U(3) and SU(1,1) do not commute with one another, it is found that the basis states $\{|\nu l m\rangle\}$ are simultaneously classified by the two distinct subgroups chains, (5.309) and (5.310). This is an example of a so-called *dual pair of subgroup chains* discussed in Section 5.10 and in subsequent sections. As shown in Section 5.8.2, the SU(1,1) × O(3) chain is particularly useful for calculating single-particle matrix elements in the $\{|\nu l m\rangle\}$ basis. However, the U(3) chain proves to be more useful in the classification and interpretation of *many-nucleon* harmonic-oscillator shell-model states in collective-model terms.

5.8.4 *The Sp(3, $\mathbb{R}$) dynamical group*

It turns out that there is a dynamical group, Sp(3, $\mathbb{R}$), for the spherical-harmonic-oscillator Hamiltonian that includes all the subgroups of the chains (5.309) and (5.310). As will be shown in Volume 2, this group plays a central role in revealing how collective models are embedded in the shell model.

The group Sp(3,$\mathbb{R}$) is fundamentally *the set of linear canonical transformations of the phase-space coordinates for a particle in three-dimensional space*. Its Lie algebra, sp(3, $\mathbb{R}$), has a simple realisation as an extension of the combined su(1,1) and u(3) Lie algebras. It is spanned by Hermitian linear combinations of the operators

$$
\hat{A}_{\alpha\beta} = c^\dagger_\alpha c^\dagger_\beta, \quad \hat{B}_{\alpha\beta} = c^\alpha c^\beta, \quad \hat{C}_{\alpha\beta} = \tfrac{1}{2}(c^\dagger_\alpha c^\beta + c^\beta c^\dagger_\alpha) = \hat{C}_{\alpha\beta} + \tfrac{1}{2}\delta_{\alpha,\beta}\hat{I}, \tag{5.311}
$$

where $\hat{C}_{\alpha\beta}$ is given by Equation (5.307) with α and β taking the values $0, \pm 1$.

Observe that the states $|000\rangle$ and $|011\rangle$ of the set $\{|\nu l m\rangle\}$ satisfy the equations

$$
\hat{B}_{\alpha\beta}|000\rangle = \hat{B}_{\alpha\beta}|011\rangle = 0, \tag{5.312}
$$

$$
\hat{C}_{\alpha\beta}|000\rangle = \hat{C}_{\alpha\beta}|011\rangle = 0, \quad \alpha > \beta, \tag{5.313}
$$

$$
\hat{C}_{\alpha\alpha}|000\rangle = \tfrac{1}{2}|000\rangle, \quad \hat{C}_{\alpha\alpha}|011\rangle = \left(\tfrac{1}{2} + \delta_{\alpha,1}\right)|011\rangle. \tag{5.314}
$$

Thus, they are lowest-weight states for sp(3,$\mathbb{R}$) irreps, denoted by $\langle\frac{1}{2},\frac{1}{2},\frac{1}{2}\rangle$ and $\langle\frac{3}{2},\frac{1}{2},\frac{1}{2}\rangle$, respectively. It transpires that all single-nucleon states of a given angular momentum l span a single irrep of the SU(1,1) × O(3) ⊂ Sp(3,$\mathbb{R}$) subgroup. Moreover, because all $\{|0ll\rangle\}$ states are obtained by raising either from $|000\rangle$ or $|011\rangle$ with multiples of the $\hat{\mathcal{A}}_{11}$ raising operator, there are no other sp(3,$\mathbb{R}$) lowest-weight states in the space of the spherical harmonic oscillator. Thus, the two irreps $\langle\frac{1}{2},\frac{1}{2},\frac{1}{2}\rangle$ and $\langle\frac{3}{2},\frac{1}{2},\frac{1}{2}\rangle$ contain all states of even and odd angular momentum, respectively; the former contains the states of even parity and the latter those of odd parity.

The Sp(3,$\mathbb{R}$) group contains many subgroups. In particular, it contains the subgroups of the chain

$$\begin{array}{ccccccc}
\text{Sp}(3,\mathbb{R}) & \supset & \mathcal{U}(3) & \supset & \text{SO}(3) & \supset & \text{SO}(2) \\
\pi & & \{N+1/2,\, 1/2,\, 1/2\} & & l & & m
\end{array}, \qquad (5.315)$$

where $\mathcal{U}(3)$ is the subgroup of Sp(3,$\mathbb{R}$) with infinitesimal generators $\{\hat{\mathcal{C}}_{\alpha\beta}\}$, defined in Equation (5.311); it is isomorphic to the U(3) group with infinitesimal generators $\{\hat{C}_{\alpha\beta}\}$. The group Sp(3,$\mathbb{R}$) also contains the subgroups of the chain

$$\begin{array}{ccccc}
\text{Sp}(3,\mathbb{R}) & \supset & \text{SU}(1,1) \times \text{O}(3) & \supset & \text{U}(1) \times \text{SO}(2) \\
\pi & & l+3/2 \qquad (l,\pi) & & \nu \qquad\quad m
\end{array}, \qquad (5.316)$$

where SU(1,1) is the subgroup of Sp(3,$\mathbb{R}$) with infinitesimal generators $\{\hat{S}_{\pm}, \hat{S}_0\}$ given by Equation (5.254). It follows that basis states for the spherical harmonic oscillator can be classified by the quantum numbers of these two subgroup chains. However, with the parity related to the angular momentum by $\pi = (-1)^l$ and $N = l+2\nu$, it is found that the states classified by the two subgroups chains, (5.315) and (5.316), are identical to one another and to the states $\{|\nu lm\rangle\}$ classified by the chains (5.309) and (5.310). These apparent coincidences may also be understood in terms of the duality relationships discussed in Section 5.10.

Exercises

5.24 Show that, for the SHO, the group SU(1,1) is a group of linear canonical transformations of the $(\hat{x}, \hat{p})$ observables; i.e., linear transformations that leave the Heisenberg commutation relations $[\hat{x}, \hat{p}] = i\hbar\hat{I}$ invariant.

5.25 If $\hat{A}$ and $\hat{B}$ are linear operators on a common Hilbert space and the combination of these two operators,

$$\hat{A}(\lambda) := e^{\lambda\hat{B}}\hat{A}e^{-\lambda\hat{B}}, \qquad (5.317)$$

is an analytic function of a real parameter λ, show that $\hat{A}(\lambda)$ has a Taylor expansion

$$\hat{A}(\lambda) = \hat{A} + \lambda[\hat{B}, \hat{A}] + \frac{1}{2}\lambda^2[\hat{B}, [\hat{B}, \hat{A}]] + \frac{1}{3!}\lambda^3[\hat{B}, [\hat{B}, [\hat{B}, \hat{A}]]] + \dots. \qquad (5.318)$$

Hence, derive the Baker-Campbell-Hausdorff identity

$$e^B A e^{-B} = A + [B, A] + \frac{1}{2!}[B, [B, A]] + \frac{1}{3!}[B, [B, [B, A]]] + \cdots. \qquad (5.319)$$

5.26 Use the Baker-Campbell-Hausdorff identity to show that

$$e^{\frac{i}{\hbar}\hat{H}t}c^\dagger e^{-\frac{i}{\hbar}\hat{H}t} = c^\dagger e^{i\omega t}, \quad e^{\frac{i}{\hbar}\hat{H}t}c e^{-\frac{i}{\hbar}\hat{H}t} = ce^{-i\omega t}. \tag{5.320}$$

5.27 Define $i\hat{X} = \hat{S}_+ - \hat{S}_-$, where $\hat{S}_\pm$ are SHO realisations of the su(1,1) raising and lowering operators, and show that $e^{i\alpha\hat{X}}\hat{x}e^{-i\alpha\hat{X}} = e^{-\alpha}\hat{x}$ and that $e^{i\alpha\hat{X}}\hat{p}e^{-i\alpha\hat{X}} = e^{\alpha}\hat{p}$. Show also, for the spherical harmonic oscillator, that $\hat{X}$ is an infinitesimal generator of radial scale transformations, i.e., $e^{i\alpha\hat{X}}\hat{r}e^{-i\alpha\hat{X}} = e^{-\alpha}\hat{r}$.

5.28 If $|m\rangle$ is an eigenstate of an angular momentum operator $\hat{L}_0$ with eigenvalue m, show that $\exp(i2\pi\hat{L}_0)|m\rangle$ is equal to $|m\rangle$ when $2m$ is an even integer and to $-|m\rangle$ when $2m$ is an odd integer. Hence, show that $\hat{L}_0$ can only be an infinitesimal generator of an SO(3) rotation if m is an integer. [Hint: by definition, an SO(3) rotation through angle 2π about any axis is equivalent to no rotation at all.]

5.29 Show that

$$[\hat{S}_-, (c_1^\dagger)^l] = l\,(c_1^\dagger)^{l-1}c^{-1}, \quad [\hat{S}_0, (c_1^\dagger)^l] = \tfrac{1}{2}l\,(c_1^\dagger)^l, \tag{5.321}$$

and, hence, that

$$\hat{S}_-(c_1^\dagger)^l|0\rangle = 0, \quad \hat{S}_0(c_1^\dagger)^l|0\rangle = \tfrac{1}{2}(l+\tfrac{3}{2})(c_1^\dagger)^l|0\rangle. \tag{5.322}$$

5.30 Show that the commutation relations of Equation (5.233) define a finite-dimensional non-unitary matrix irrep of su(1,1) (i.e., for which $S_+^\dagger \neq S_-$) with

$$S_0 = \begin{pmatrix} 1/2 & 0 \\ 0 & -1/2 \end{pmatrix}, \quad S_+ = \begin{pmatrix} 0 & -1 \\ 0 & 0 \end{pmatrix}, \quad S_- = \begin{pmatrix} 0 & 0 \\ 1 & 0 \end{pmatrix}. \tag{5.323}$$

Show also that, unlike the harmonic series unitary irreps of su(1,1), this irrep has both highest and lowest weights corresponding to eigenvalues ± 1 of $2S_0$. [We refer to the operators $\{c, c^\dagger\}$, that transform as basis vectors for this finite-dimensional irrep, as components of an su(1,1) tensor.] Consider the differences between the su(2) and su(1,1) algebras to interpret the significance of this result.

5.31 Use Equation (5.278) to show that

$$|\nu l m\rangle = \sqrt{\frac{\Gamma(l+\tfrac{3}{2})}{\Gamma(\nu+l+\tfrac{3}{2})\nu!}}\,\hat{S}_+^\nu|0lm\rangle. \tag{5.324}$$

5.32 Show that, because of the Pauli exclusion principle, it is possible to put at most $2(n+1)(n+2)$ neutrons and $2(n+1)(n+2)$ protons into a spherical harmonic-oscillator shell of states of n harmonic-oscillator quanta. Use Equation (5.295) to show that the mean-square radius for a nucleon in such a shell is given by

$$\langle r^2\rangle_n = (n+3/2)\frac{\hbar}{M\omega}. \tag{5.325}$$

Hence, determine the mean-square radii, in harmonic-oscillator units, for a few of the nuclei whose root-mean-square charge radii are shown in Figure 1.5. Use the values $\hbar c = 197$ MeV fm and $Mc^2 = 939$ MeV, to show that the harmonic oscillator frequency $\hbar\omega \approx 41A^{-1/3}$ MeV reproduces the observed root-mean-square radius of a medium to heavy mass-A nucleus.

5.9 Transformation to relative and centre-of-mass coordinates

The previous section shows how to calculate matrix elements for the states of a single particle in a spherical harmonic-oscillator potential. The next challenge is to calculate the matrix elements of two-body operators. This is straightforward in a harmonic-oscillator basis for two-body operators that are translationally invariant. A subsequent concern of this chapter will be to use the methods of second quantization and the representation theory of Lie algebras to calculate the matrix elements of one- and two-body operators in suitably-defined bases of many-nucleon states.

Evaluation of the matrix elements of two-body interactions, that depend only on the relative coordinates of two nucleons, is conventionally handled in a spherical harmonic-oscillator basis by means of a Talmi transformation[172] from particle coordinates to relative and centre-of-mass coordinates. The basic idea is to express an angular-momentum-coupled wave function, $\left[\psi_{\nu_1 l_1}(\mathbf{r}_1) \otimes \psi_{\nu_2 l_2}(\mathbf{r}_2)\right]_{LM}$, as a sum of coupled products of relative and centre-of-mass wave functions,

$$\left[\psi_{\nu_1 l_1}(\mathbf{r}_1) \otimes \psi_{\nu_2 l_2}(\mathbf{r}_2)\right]_{LM} = \sum_{\nu l, \mu \Lambda} \langle \nu l, \mu \Lambda : L | \nu_1 l_1, \nu_2 l_2 : L \rangle \left[\psi_{\nu l}(\mathbf{r}) \otimes \psi_{\mu \Lambda}(\mathbf{R})\right]_{LM},$$
(5.326)

where $\mathbf{r}$ and $\mathbf{R}$ are relative position and centre-of-mass vectors, respectively, for the two nucleons.

An algorithm for computing the coefficients of the Talmi transformation (5.326) was devised by Moshinsky[173] and tables of coefficients were computed by Brody and Moshinsky.[174] These coefficients are referred to variously as Talmi coefficients, Moshinsky brackets, or Brody-Moshinsky coefficients. Subsequently, numerous alternative algorithms have been developed for computing these coefficients and their generalisations. Explicit expressions have been determined by Bakri[175] and Buck and Merchant.[176] The following derivation is based on that of Buck and Merchant.

If $\mathbf{r}_1$ and $\mathbf{r}_2$ denote position vectors for two nucleons, corresponding relative and centre-of-mass vectors are defined[177] by an orthogonal transformation as

$$\mathbf{r} = \frac{1}{\sqrt{2}}(\mathbf{r}_1 - \mathbf{r}_2), \quad \mathbf{R} = \frac{1}{\sqrt{2}}(\mathbf{r}_1 + \mathbf{r}_2).$$
(5.327)

In fact, it is just as easy to consider an arbitrary real orthogonal transformation

$$\mathbf{r}_3 = \alpha \mathbf{r}_1 + \beta \mathbf{r}_2, \quad \mathbf{r}_4 = \beta \mathbf{r}_1 - \alpha \mathbf{r}_2,$$
(5.328)

---

[172] Talmi I. (1952), *Helv. Phys. Acta* **25**, 185.

[173] Moshinsky M. (1959), *Nucl. Phys.* **13**, 104.

[174] Brody T.A. and Moshinsky M. (1960), *Tables of Transformation Brackets* (Monografias del Instituto de Fisica, Universidad Nacional Autonoma de Mexico, Mexico).

[175] Bakri M.M. (1967), *Nucl. Phys.* **A96**, 115.

[176] Buck B. and Merchant A.C. (1996), *Nucl. Phys.* **A600**, 387.

[177] The frequently used definitions $\mathbf{r} = (\mathbf{r}_1 - \mathbf{r}_2)$ and $\mathbf{R} = \frac{1}{2}(\mathbf{r}_1 + \mathbf{r}_2)$ are not given by an orthogonal transformation and tend to obscure the simplicity of the transformation.

with $\alpha^2+\beta^2 = 1$. The corresponding raising and lowering operators are then related by

$$\mathbf{c}_3^\dagger = \alpha\mathbf{c}_1^\dagger + \beta\mathbf{c}_2^\dagger, \quad \mathbf{c}_4^\dagger = \beta\mathbf{c}_1^\dagger - \alpha\mathbf{c}_2^\dagger, \tag{5.329}$$

where, for convenience, we regard the triples of creation operators $(c_{ix}^\dagger, c_{iy}^\dagger, c_{iz}^\dagger)$ as components of a vector operator $\mathbf{c}_i^\dagger$.

Now, if

$$|\nu l m\rangle = \Psi_{\nu l m}(\mathbf{c}^\dagger)|0\rangle \tag{5.330}$$

is a spherical harmonic-oscillator state and

$$|\nu_i l_i, \nu_j l_j : LM\rangle = \Phi_{\nu_i l_i, \nu_j l_j : LM}(\mathbf{c}_i^\dagger, \mathbf{c}_j^\dagger)|0\rangle, \tag{5.331}$$

where

$$\Phi_{\nu_i l_i, \nu_j l_j : LM}(\mathbf{c}_i^\dagger, \mathbf{c}_j^\dagger) := \left[\Psi_{\nu_i l_i}(\mathbf{c}_i^\dagger) \otimes \Psi_{\nu_j l_j}(\mathbf{c}_j^\dagger)\right]_{LM}, \tag{5.332}$$

is a coupled two-particle spherical harmonic-oscillator state of angular momentum L, then the Talmi coefficients for the above orthogonal transformation are the overlaps

$$\langle \nu_3 l_3, \nu_4 l_4 : L|\nu_1 l_1, \nu_2 l_2 : L\rangle$$
$$= \langle 0|\Phi_{\nu_3 l_3, \nu_4 l_4 : LM}^\dagger(\mathbf{c}_3^\dagger, \mathbf{c}_4^\dagger)\Phi_{\nu_1 l_1, \nu_2 l_2 : LM}(\mathbf{c}_1^\dagger, \mathbf{c}_2^\dagger)|0\rangle$$
$$= \frac{(-1)^L}{\sqrt{2L+1}}\langle 0|\left[\Phi_{\nu_4 l_4, \nu_3 l_3 : L}(\mathbf{c}_4, \mathbf{c}_3) \otimes \Phi_{\nu_1 l_1, \nu_2 l_2 : L}(\mathbf{c}_1^\dagger, \mathbf{c}_2^\dagger)\right]_0|0\rangle. \tag{5.333}$$

We show in the following that these coefficients are given by

$$\langle \nu_3 l_3, \nu_4 l_4 : L|\nu_1 l_1, \nu_2 l_2 : L\rangle = \frac{\pi}{4}\left(\prod_{i=1}^4 \sqrt{\nu_i!\Gamma(\nu_i+l_i+3/2)}\right)$$
$$\times \left(\prod_{p=a}^d \frac{2l_p+1}{k_p!\Gamma(k_p+l_p+3/2)}\right)\sum_{l_a l_b l_c l_d}(-1)^{l_d}\alpha^{2k_a+l_a+2k_d+l_d}\beta^{2k_b+l_b+2k_c+l_c} \tag{5.334}$$
$$\times (l_a 0\, l_b 0|l_1 0)(l_c 0\, l_d 0|l_2 0)(l_a 0\, l_c 0|l_3 0)(l_b 0\, l_d 0|l_4 0)\begin{Bmatrix} l_a & l_b & l_1 \\ l_c & l_c & l_2 \\ l_3 & l_4 & L \end{Bmatrix},$$

where the values of k_a, k_b, k_c, and k_d are related to those of l_a, l_b, l_c, and l_d by solution of the equations

$$2k_a + l_a + 2k_b + l_b = 2\nu_1 + l_1, \tag{5.335}$$
$$2k_c + l_c + 2k_d + l_d = 2\nu_2 + l_2, \tag{5.336}$$
$$2k_a + l_a + 2k_c + l_c = 2\nu_3 + l_3, \tag{5.337}$$
$$2k_b + l_b + 2k_d + l_d = 2\nu_4 + l_4. \tag{5.338}$$

This expression is equivalent to that obtained by Buck and Merchant.[178] It is derived by the following sequence of steps. However, the reader may choose to skip this rather technical derivation, which is included primarily for reference purposes.

The following steps are designed to guide the reader through the derivation of Equation (5.334).

Step 1:

An essential simplification of the transformation to relative and centre-of-mass coordinates is achieved by working in a Bargmann coherent-state representation. For present purposes, we need only the observation that, in the Bargmann representation, a SHO state $|n\rangle = \Psi_n(c_z^\dagger)|0\rangle$, for which $\Psi_n(c_z^\dagger) = \frac{1}{\sqrt{n!}}(c_z^\dagger)^n$, is represented by the wave function $\Psi_n(z) = \frac{1}{\sqrt{n!}}z^n$.

The first step is then to show that the single-particle spherical-harmonic-oscillator states of an orthonormal basis, $\{|\nu l m\rangle\}$, can be expressed in the form

$$|\nu l m\rangle = \Psi_{\nu l m}(\mathbf{c}^\dagger)|0\rangle, \tag{5.339}$$

with Bargmann wave functions given by

$$\Psi_{\nu l m}(\mathbf{r}) = A_{\nu l}r^{2\nu+l}Y_{lm}(\theta,\varphi), \tag{5.340}$$

where Y_{lm} is a spherical harmonic, $\mathbf{r}$ is a vector with spherical polar coordinates, (r,θ,φ), and

$$A_{\nu l} := \sqrt{\frac{2\pi\sqrt{\pi}}{2^{2\nu+l}\Gamma(\nu+l+\,{}^{3}\!/_{2})\nu!}}. \tag{5.341}$$

Equation (5.339) is derived for $\nu=0$ from Equation (5.273) which gives

$$\Psi_{0ll}(\mathbf{c}^\dagger) = \frac{1}{\sqrt{l!}}(c_1^\dagger)^l, \tag{5.342}$$

with $c_1^\dagger = -\frac{1}{\sqrt{2}}(c_x^\dagger + ic_y^\dagger)$. Comparison with the expression

$$-\frac{1}{\sqrt{2}}(x+iy) = \sqrt{\frac{4\pi}{3}}\,rY_{11}(\theta,\varphi), \tag{5.343}$$

then shows that

$$\Psi_{0ll}(\mathbf{r}) = \frac{1}{\sqrt{l!}}r^l\left[\sqrt{\frac{4\pi}{3}}\,Y_{11}(\theta,\varphi)\right]^l. \tag{5.344}$$

---

[178]The expression of Buck and Merchant, *op. cit.* Footnote 176 on Page 315, differs from ours by a phase factor $(-1)^{\nu_1+\nu_2+\nu_3+\nu_4}$. The Buck-Merchant phase factor is obtained in the above derivation if the factor $(-1)^\nu$ is removed from Equation (5.305) and, hence, added to Equation (5.341.) The phases of single-particle wave functions are, in principle, arbitrary. However, the choice used here is such that the matrix elements given by Equations (5.298)–(5.301) are consistent with the standard representation matrices of SU(1,1) used to derive them.

Thus, from the formula for combining spherical harmonics with a common argument,

$$Y_{11} \times Y_{l-1,l-1} = \sqrt{\frac{3}{4\pi} \frac{l}{2l+1}} \, Y_{ll}, \tag{5.345}$$

it is determined that

$$\left[\sqrt{\frac{4\pi}{3}} \, Y_{11}\right]^l = \sqrt{\frac{4\pi l!}{(2l+1)!!}} \, Y_{ll} \tag{5.346}$$

and, hence, that

$$\Psi_{0lm}(\mathbf{r}) = \sqrt{\frac{4\pi}{(2l+1)!!}} \, r^l Y_{lm}(\theta,\varphi). \tag{5.347}$$

The creation operator, $\Psi_{\nu lm}(\mathbf{c}^\dagger)$, for a general state $|\nu lm\rangle$ is now determined from Equation (5.324), with $\hat{S}_+ = \frac{1}{2}\mathbf{c}^\dagger \cdot \mathbf{c}^\dagger$, to be given by

$$\Psi_{\nu lm}(\mathbf{c}^\dagger) = \sqrt{\frac{\Gamma(l+\frac{3}{2})}{\Gamma(\nu+l+\frac{3}{2})\nu!}} \frac{1}{2^\nu} (\mathbf{c}^\dagger \cdot \mathbf{c}^\dagger)^\nu \Psi_{0lm}(\mathbf{c}^\dagger). \tag{5.348}$$

It follows that

$$\Psi_{\nu lm}(\mathbf{r}) = \sqrt{\frac{\Gamma(l+\frac{3}{2})}{\Gamma(\nu+l+\frac{3}{2})\nu!}} \frac{1}{2^\nu} r^{2\nu} \Psi_{0lm}(\mathbf{r})$$

$$= \sqrt{\frac{4\pi \, \Gamma(l+\frac{3}{2})}{2^{2\nu}\Gamma(\nu+l+\frac{3}{2})\nu!(2l+1)!!}} \, r^{2\nu+l} Y_{lm}(\theta,\varphi). \tag{5.349}$$

Finally, with the substitution

$$\Gamma(l+\tfrac{3}{2}) = \frac{\sqrt{\pi}\,(2l+1)!!}{2^{l+1}}, \tag{5.350}$$

the square-root factor on the right-hand side of Equation (5.349) becomes the $A_{\nu l}$ coefficient of Equation (5.341).

Step 2:

The next step is to show that $\Psi_{\nu lm}(\mathbf{c}^\dagger)$ can also be expressed in the form

$$\Psi_{\nu lm}(\mathbf{c}^\dagger) = \frac{1}{A_{\nu l}(2\nu+l)!} \left[\frac{d^{2\nu+l}}{dr^{2\nu+l}} \int Y_{lm}(\theta,\varphi) \exp(\mathbf{r}\cdot\mathbf{c}^\dagger) \sin\theta \, d\theta \, d\varphi\right]_{r=0}, \tag{5.351}$$

where $\mathbf{r}$ is a real vector and $A_{\nu l}$ is given by Equation (5.341).

Equation (5.351) follows from the expansion

$$\exp(\mathbf{r}\cdot\mathbf{c}^\dagger) = \sum_{\nu lm} \Psi^*_{\nu lm}(\mathbf{r}) \, \Psi_{\nu lm}(\mathbf{c}^\dagger), \tag{5.352}$$

which, with $\Psi_{\nu lm}$ given by Equation (5.340), yields

$$\exp(\mathbf{r} \cdot \mathbf{c}^\dagger) = \sum_{\nu lm} A_{\nu l} r^{2\nu+l} Y_{lm}^*(\theta, \varphi)\, \Psi_{\nu lm}(\mathbf{c}^\dagger). \qquad (5.353)$$

The expansion (5.352) is a three-dimensional generalisation of the equality

$$\exp(xc^\dagger) = \sum_n \frac{xc^\dagger}{n!} = \sum_n \Psi_n^*(x)\Psi_n(c^\dagger), \qquad (5.354)$$

which holds, for the simple (one-dimensional) harmonic oscillator, when

$$\Psi_n(x) = \Psi_n^*(x) = \frac{x^n}{\sqrt{n!}}, \quad \Psi_n(c^\dagger) = \frac{(c^\dagger)^n}{\sqrt{n!}}. \qquad (5.355)$$

In fact, this equality is valid whenever the wave functions, $\{\Psi_n\}$, are an orthonormal basis for the SHO, as they are for the wave functions of Equation (5.355) with respect to the inner product

$$\langle m|n\rangle = \langle 0|\frac{c^m}{\sqrt{m!}}\frac{(c^\dagger)^n}{\sqrt{n!}}|0\rangle = \delta_{m,n}. \qquad (5.356)$$

This follows because any two orthonormal basis sets are related by a unitary transformation. Thus, under a unitary transformation

$$\Psi_n \to \sum_m \Psi_m U_{mn}, \qquad (5.357)$$

the sum

$$\sum_n \Psi_n^*(x)\Psi_n(c^\dagger) \to \sum_{mm'n} \Psi_{m'}^*(x)\Psi_m(c^\dagger) U_{m'n}^* U_{mn} = \sum_n \Psi_n^*(x)\Psi_n(c^\dagger) \qquad (5.358)$$

remains invariant.

An extension of these results to three-dimensional space gives

$$\exp(\mathbf{r} \cdot \mathbf{c}^\dagger) = \sum_n \Psi_n^*(\mathbf{r})\Psi_n(\mathbf{c}^\dagger), \qquad (5.359)$$

where n is a triple of integers $n := (n_x, n_y, n_z)$ and

$$\Psi_n(\mathbf{c}^\dagger) := \frac{(c_x^\dagger)^{n_x}(c_y^\dagger)^{n_y}(c_z^\dagger)^{n_z}}{\sqrt{n_x! n_y! n_z!}}. \qquad (5.360)$$

Again this result is valid in any orthonormal basis and, in particular, it is valid in the spherical basis in which n is replaced by $\{\nu lm\}$. Thus, we obtain Equation (5.352).

Step 3:

In proceeding to the third step, it is useful rewrite Equation (5.351) in the form

$$\Psi_{\nu l m}(\mathbf{c}^\dagger) = \frac{1}{A_{\nu l}(2\nu + l)!} \left[\frac{d^{2\nu+l}}{dr^{2\nu+l}} \int Y_{lm}(\bar{r}) \exp(\mathbf{r} \cdot \mathbf{c}^\dagger) \, d\bar{r} \right]_{r=0}, \tag{5.361}$$

where $\bar{\mathbf{r}} := \mathbf{r}/r$ is a unit vector corresponding to a point with spherical coordinates (θ, φ) and $d\bar{\mathbf{r}} := \sin\theta \, d\theta \, d\varphi$. With creation operators for single-particle states given by this equation, the desired overlap integral (5.333) then assumes the form

$$\langle \nu_3 l_3, \nu_4 l_4 : L | \nu_1 l_1, \nu_2 l_2 : L \rangle$$

$$= \frac{(-1)^L}{\sqrt{2L+1}} \left(\prod_{i=1}^{4} \frac{1}{A_{\nu_i l_i}(2\nu_i + l_i)!} \right)$$

$$\times \left[\frac{\partial^{2\nu_1+l_1}}{\partial s^{2\nu_1+l_1}} \frac{\partial^{2\nu_2+l_2}}{\partial t^{2\nu_2+l_2}} \frac{\partial^{2\nu_3+l_3}}{\partial u^{2\nu_3+l_3}} \frac{\partial^{2\nu_4+l_4}}{\partial v^{2\nu_4+l_4}} X(s,t,u,v) \right]_{s,t,u,v=0}, \tag{5.362}$$

where

$$X(s,t,u,v) = \int \left[[Y_{l_4}(\bar{\mathbf{v}}) \otimes Y_{l_3}(\bar{\mathbf{u}})]_L \otimes [Y_{l_1}(\bar{\mathbf{s}}) \otimes Y_{l_2}(\bar{\mathbf{t}})]_L \right]_0$$

$$\times J(\mathbf{s},\mathbf{t},\mathbf{u},\mathbf{v}) \, d\bar{\mathbf{s}} \, d\bar{\mathbf{t}} \, d\bar{\mathbf{u}} \, d\bar{\mathbf{v}}, \tag{5.363}$$

with

$$J(\mathbf{s},\mathbf{t},\mathbf{u},\mathbf{v}) = \langle 0 | \exp(\mathbf{v} \cdot \mathbf{c}_4 + \mathbf{u} \cdot \mathbf{c}_3) \exp(\mathbf{s} \cdot \mathbf{c}_1^\dagger + \mathbf{t} \cdot \mathbf{c}_2^\dagger) | 0 \rangle. \tag{5.364}$$

Thus, the next steps are to determine $J(\mathbf{s},\mathbf{t},\mathbf{u},\mathbf{v})$ and then $X(s,t,u,v)$.

Step 4:

It is easy to show (cf. Exercise 5.33) that

$$\langle 0 | \exp(\mathbf{a} \cdot \mathbf{c}_i) \exp(\mathbf{b} \cdot \mathbf{c}_j^\dagger) | 0 \rangle = \delta_{i,j} \exp(\mathbf{a} \cdot \mathbf{b}) \tag{5.365}$$

and, hence, with $\mathbf{c}_3^\dagger$ and $\mathbf{c}_4^\dagger$ given by Equation (5.329), that

$$J(\mathbf{s},\mathbf{t},\mathbf{u},\mathbf{v}) = \exp(\alpha \mathbf{u} \cdot \mathbf{s} + \beta \mathbf{v} \cdot \mathbf{s} + \beta \mathbf{u} \cdot \mathbf{t} - \alpha \mathbf{v} \cdot \mathbf{t}). \tag{5.366}$$

Thus, the plane wave expansion,

$$\exp(\mathbf{k} \cdot \mathbf{r}) = \exp(-\mathrm{i}(\mathrm{i}\mathbf{k} \cdot \mathbf{r}) = 4\pi \sum_l \mathrm{i}^l \sqrt{2l+1} \, j_l(\mathrm{i}kr) [Y_l(\bar{\mathbf{k}}) \otimes Y_l(\bar{\mathbf{r}})]_0, \tag{5.367}$$

gives

$$J(\mathbf{s},\mathbf{t},\mathbf{u},\mathbf{v}) = (4\pi)^4 \sum_{l_a l_b l_c l_d} \mathrm{i}^{l_a+l_b+l_c+l_d} [l_a][l_b][l_c][l_d]$$

$$\times j_{l_a}(\mathrm{i}\alpha us) j_{l_b}(\mathrm{i}\beta vs) j_{l_c}(\mathrm{i}\beta ut) j_{l_d}(-\mathrm{i}\alpha vt) [Y_{l_a}(\bar{\mathbf{u}}) \otimes Y_{l_a}(\bar{\mathbf{s}})]_0$$

$$\times [Y_{l_b}(\bar{\mathbf{v}}) \otimes Y_{l_b}(\bar{\mathbf{s}})]_0 [Y_{l_c}(\bar{\mathbf{u}}) \otimes Y_{l_c}(\bar{\mathbf{t}})]_0 [Y_{l_d}(\bar{\mathbf{v}}) \otimes Y_{l_d}(\bar{\mathbf{t}})]_0, \tag{5.368}$$

where

$$[l] := \sqrt{2l+1}. \tag{5.369}$$

Now the product $[Y_{l_a}(\bar{\mathbf{u}}) \otimes Y_{l_a}(\bar{\mathbf{s}})]_0 [Y_{l_b}(\bar{\mathbf{v}}) \otimes Y_{l_b}(\bar{\mathbf{s}})]_0$ can be recoupled and expressed (cf. Exercises 5.34 and 5.35) as

$$\begin{aligned}
&\left[Y_{l_a}(\bar{\mathbf{u}}) \otimes Y_{l_a}(\bar{\mathbf{s}})\right]_0 \left[Y_{l_b}(\bar{\mathbf{v}}) \otimes Y_{l_b}(\bar{\mathbf{s}})\right]_0 \\
&= \frac{1}{\sqrt{4\pi}} \sum_{l_1} (l_a 0\, l_b 0 | l_1 0) \left[Y_{l_1}(\bar{\mathbf{s}}) \otimes \left[Y_{l_a}(\bar{\mathbf{u}}) \otimes Y_{l_b}(\bar{\mathbf{v}})\right]_{l_1}\right]_0.
\end{aligned} \tag{5.370}$$

Similarly

$$\begin{aligned}
&\left[Y_{l_c}(\bar{\mathbf{u}}) \otimes Y_{l_c}(\bar{\mathbf{t}})\right]_0 \left[Y_{l_d}(\bar{\mathbf{v}}) \otimes Y_{l_d}(\bar{\mathbf{t}})\right]_0 \\
&= \frac{1}{\sqrt{4\pi}} \sum_{l_2} (l_c 0\, l_d 0 | l_2 0) \left[Y_{l_2}(\bar{\mathbf{t}}) \otimes \left[Y_{l_c}(\bar{\mathbf{u}}) \otimes Y_{l_d}(\bar{\mathbf{v}})\right]_{l_2}\right]_0.
\end{aligned} \tag{5.371}$$

Then, with the recoupling (cf. Exercise 5.34)

$$\begin{aligned}
&\left[Y_{l_2}(\bar{\mathbf{t}}) \otimes \left[Y_{l_c}(\bar{\mathbf{u}}) \otimes Y_{l_d}(\bar{\mathbf{v}})\right]_{l_2}\right]_0 \left[Y_{l_1}(\bar{\mathbf{s}}) \otimes \left[Y_{l_a}(\bar{\mathbf{u}}) \otimes Y_{l_b}(\bar{\mathbf{v}})\right]_{l_1}\right]_0 \\
&= \sum_J (-1)^{l_1 + l_2 - J} \frac{[J]}{[l_1][l_2]} \\
&\quad \times \left[\left[Y_{l_2}(\bar{\mathbf{t}}) \otimes Y_{l_1}(\bar{\mathbf{s}})\right]_J \otimes \left[Y_{l_a}(\bar{\mathbf{u}}) \otimes Y_{l_b}(\bar{\mathbf{v}})\right]_{l_1} \otimes \left[Y_{l_c}(\bar{\mathbf{u}}) \otimes Y_{l_d}(\bar{\mathbf{v}})\right]_{l_2}\right]_0
\end{aligned} \tag{5.372}$$

and (cf. Exercise 4.35)

$$\begin{aligned}
&\left[\left[Y_{l_a}(\bar{\mathbf{u}}) \otimes Y_{l_b}(\bar{\mathbf{v}})\right]_{l_1} \otimes \left[Y_{l_c}(\bar{\mathbf{u}}) \otimes Y_{l_d}(\bar{\mathbf{v}})\right]_{l_2}\right]_J \\
&= \frac{1}{4\pi} \sum_{l_3 l_4} [l_a][l_b][l_c][l_d][l_1][l_2] (l_a 0\, l_c 0 | l_3 0)(l_b 0\, l_d 0 | l_4 0) \\
&\quad \times \begin{Bmatrix} l_a & l_b & l_1 \\ l_c & l_d & l_2 \\ l_3 & l_4 & J \end{Bmatrix} \left[Y_{l_3}(\bar{\mathbf{u}}) \otimes Y_{l_4}(\bar{\mathbf{v}})\right]_J,
\end{aligned} \tag{5.373}$$

we obtain

$$\begin{aligned}
&\left[Y_{l_a}(\bar{\mathbf{u}}) \otimes Y_{l_a}(\bar{\mathbf{s}})\right]_0 \left[Y_{l_b}(\bar{\mathbf{v}}) \otimes Y_{l_b}(\bar{\mathbf{s}})\right]_0 \left[Y_{l_c}(\bar{\mathbf{u}}) \otimes Y_{l_c}(\bar{\mathbf{t}})\right]_0 \left[Y_{l_d}(\bar{\mathbf{v}}) \otimes Y_{l_d}(\bar{\mathbf{t}})\right]_0 \\
&= \frac{1}{(4\pi)^2} \sum_{l_1 l_2 l_3 l_4 J} (-1)^{l_1 + l_2 - J} [l_a][l_b][l_c][l_d][J] \begin{Bmatrix} l_a & l_b & l_1 \\ l_c & l_d & l_2 \\ l_3 & l_4 & J \end{Bmatrix} \\
&\quad \times (l_a 0\, l_b 0 | l_1 0)(l_c 0\, l_d 0 | l_2 0)(l_a 0\, l_c 0 | l_3 0)(l_b 0\, l_d 0 | l_4 0) \\
&\quad \times \left[\left[Y_{l_2}(\bar{\mathbf{t}}) \otimes Y_{l_1}(\bar{\mathbf{s}})\right]_J \otimes \left[Y_{l_3}(\bar{\mathbf{u}}) \otimes Y_{l_4}(\bar{\mathbf{v}})\right]_J\right]_0
\end{aligned} \tag{5.374}$$

and

$$
\begin{aligned}
J(\mathbf{s},\mathbf{t},\mathbf{u},\mathbf{v}) = (4\pi)^2 \sum_{\substack{l_a l_b l_c l_d \\ l_1 l_2 l_3 l_4 J}} & (-1)^{l_1+l_2-J}[l_a]^2\,[l_b]^2\,[l_c]^2\,[l_d]^2\,[J] \begin{Bmatrix} l_a & l_b & l_1 \\ l_c & l_d & l_2 \\ l_3 & l_4 & J \end{Bmatrix} \\
& \times i^{l_a+l_b+l_c+l_d}(l_a\,0\,l_b0|l_10)(l_c0\,l_d0|l_20)(l_a0\,l_c0|l_30)(l_b0\,l_d0|l_40) \\
& \times j_{l_a}(i\alpha us)j_{l_b}(i\beta vs)j_{l_c}(i\beta ut)j_{l_d}(-i\alpha vt) \\
& \times \Big[\big[Y_{l_2}(\bar{\mathbf{t}}) \otimes Y_{l_1}(\bar{\mathbf{s}})\big]_J \otimes \big[Y_{l_3}(\bar{\mathbf{u}}) \otimes Y_{l_4}(\bar{\mathbf{v}})\big]_J \Big]_0.
\end{aligned}
\tag{5.375}
$$

Step 5:

The integral in the expression for $X(s,t,u,v)$ is determined to have the value (cf. Exercise 5.37)

$$
\int \Big[\big[Y_{l_4}(\bar{\mathbf{v}}) \otimes Y_{l_3}(\bar{\mathbf{u}})\big]_L \otimes \big[Y_{l_1}(\bar{\mathbf{s}}) \otimes Y_{l_2}(\bar{\mathbf{t}})\big]_L \Big]_0
\tag{5.376}
$$

$$
\times \Big[\big[Y_{l_2}(\bar{\mathbf{t}}) \otimes Y_{l_1}(\bar{\mathbf{s}})\big]_J \otimes \big[Y_{l_3}(\bar{\mathbf{u}}) \otimes Y_{l_4}(\bar{\mathbf{v}})\big]_J \Big]_0 \, d\bar{\mathbf{s}}\, d\bar{\mathbf{t}}\, d\bar{\mathbf{u}}\, d\bar{\mathbf{v}} = \delta_{L,J}.
$$

Thus, we obtain

$$
\begin{aligned}
X(s,t,u,v) = (4\pi)^2 \sum_{l_a l_b l_c l_d} & (-1)^{l_1+l_2-L}[l_a]^2\,[l_b]^2\,[l_c]^2\,[l_d]^2\,[L] \begin{Bmatrix} l_a & l_b & l_1 \\ l_c & l_d & l_2 \\ l_3 & l_4 & L \end{Bmatrix} \\
& \times i^{l_a+l_b+l_c+l_d}(l_a0\,l_b0|l_10)(l_c0\,l_d0\,|l_20)(l_a0\,l_c0|l_30)(l_b0\,l_d0|l_40) \\
& \times j_{l_a}(i\alpha us)j_{l_b}(i\beta vs)j_{l_c}(i\beta ut)j_{l_d}(-i\alpha vt).
\end{aligned}
\tag{5.377}
$$

Step 6:

This step is to evaluate the expression

$$
Z = \left[\frac{\partial^{2\nu_1+l_1}}{\partial s^{2\nu_1+l_1}} \frac{\partial^{2\nu_2+l_2}}{\partial t^{2\nu_2+l_2}} \frac{\partial^{2\nu_3+l_3}}{\partial u^{2\nu_3+l_3}} \frac{\partial^{2\nu_4+l_4}}{\partial v^{2\nu_4+l_4}} X(s,t,u,v) \right]_{s,t,u,v=0}.
\tag{5.378}
$$

From the standard expression for a spherical Bessel function,[179]

$$
j_l(z) = \sqrt{\pi} \sum_{k=0}^{\infty} \frac{(-1)^k z^{2k+l}}{2^{2k+l+1} k!\, \Gamma(k+l+3/2)},
\tag{5.379}
$$

which is valid for complex as well as real values of its argument, it is determined, for example, that

$$
j_{l_a}(i\alpha us) = \frac{\sqrt{\pi}}{2} i^{l_a} \sum_{k_a=0}^{\infty} \frac{(\alpha us)^{2k_a+l_a}}{2^{2k_a+l_a} k_a!\, \Gamma(k_a+l_a+3/2)}.
\tag{5.380}
$$

---

[179]Cf. Equations 9.1.10 and 10.1.1 of Abramowitz and Stegun, *op. cit.* Footnote 14 on Page 176.

It follows that

$$
\left[\frac{\partial^{2\nu_1+l_1}}{\partial s^{2\nu_1+l_1}} \frac{\partial^{2\nu_2+l_2}}{\partial t^{2\nu_2+l_2}} \frac{\partial^{2\nu_3+l_3}}{\partial u^{2\nu_3+l_3}} \frac{\partial^{2\nu_4+l_4}}{\partial v^{2\nu_4+l_4}} j_{l_a}(i\alpha us) j_{l_b}(i\beta vs) j_{l_c}(i\beta ut) j_{l_d}(-i\alpha vt) \right]_{s,t,u,v=0}
$$

$$
= \frac{\pi^2}{16} i^{l_a+l_b+l_c-l_d} \frac{(2\nu_1+l_1)!(2\nu_2+l_2)!(2\nu_3+l_3)!(2\nu_4+l_4)!}{2^{2\nu_1+l_1+2\nu_2+l_2}} \tag{5.381}
$$

$$
\times \frac{\alpha^{2k_a+l_a+2k_d+l_d} \beta^{2k_b+l_b+2k_c+l_c}}{k_a! \, k_b! \, k_c! \, k_d! \, \Gamma(k_a+l_a+3/2) \Gamma(k_b+l_b+3/2) \Gamma(k_c+l_c+3/2) \Gamma(k_d+l_d+3/2)}
$$

with the values of k_a, k_b, k_c, and k_d given by solution of Equations (5.335)–(5.338). Thus, we obtain

$$
Z = \pi^4 (-1)^{l_1+l_2-L} (2\nu_1+l_1)!(2\nu_2+l_2)!(2\nu_3+l_3)!(2\nu_4+l_4)! \, 2^{-(2\nu_1+l_1+2\nu_2+l_2)}
$$

$$
\times \sum_{l_a l_b l_c l_d} (-1)^{l_a+l_b+l_c} [l_a]^2 \, [l_b]^2 \, [l_c]^2 \, [l_d]^2 \, [L] \begin{Bmatrix} l_a & l_b & l_1 \\ l_c & l_d & l_2 \\ l_3 & l_4 & L \end{Bmatrix} \tag{5.382}
$$

$$
\times (l_a 0 \, l_b 0 \, | l_1 0)(l_c 0 \, l_d 0 | l_2 0)(l_a 0 \, l_c 0 | l_3 0)(l_b 0 \, l_d 0 | l_4 0)
$$

$$
\times \frac{\alpha^{2k_a+l_a+2k_d+l_d} \beta^{2k_b+l_b+2k_c+l_c}}{k_a! \, k_b! \, k_c! \, k_d! \, \Gamma(k_a+l_a+3/2) \Gamma(k_b+l_b+3/2) \Gamma(k_c+l_c+3/2) \Gamma(k_d+l_d+3/2)}.
$$

Step 7:

The final step is to put everything together to obtain Equation (5.334), noting that the phase factor $(-1)^{l_1+l_2+l_a+l_b+l_c}$ simplifies by use of the identity

$$
2\nu_1 + l_1 + 2\nu_2 + l_2 = \sum_{p=a}^{d} (2k_p + l_p) \tag{5.383}
$$

to give $(-1)^{l_1+l_2+l_a+l_b+l_c} = (-1)^{l_d}$.

Exercises

5.33 Show that, for a simple harmonic oscillator with raising and lowering operators $c^\dagger$ and c and with a and b real or complex numbers,

$$
\langle 0 | e^{a\,c} e^{b\,c^\dagger} | 0 \rangle = e^{ab}. \tag{5.384}
$$

5.34 Use the identity

$$
U(l_b l_b l_a l_a : 0 l_1) = (-1)^{l_a+l_b-l_1} \frac{[l_1]}{[l_a][l_b]} \tag{5.385}
$$

for the Racah U-coefficient, where $[l] := \sqrt{2l+1}$, to show that

$$
\left[Y_{l_a}(\bar{\mathbf{u}}) \otimes Y_{l_a}(\bar{\mathbf{s}}) \right]_0 \left[Y_{l_b}(\bar{\mathbf{v}}) \otimes Y_{l_b}(\bar{\mathbf{s}}) \right]_0
$$

$$
= \sum_{l_1} \frac{[l_1]}{[l_a][l_b]} \left[\left[Y_{l_a}(\bar{\mathbf{u}}) \otimes Y_{l_b}(\bar{\mathbf{v}}) \right]_{l_1} \otimes \left[Y_{l_a}(\bar{\mathbf{s}}) \otimes Y_{l_b}(\bar{\mathbf{s}}) \right]_{l_1} \right]_0. \tag{5.386}
$$

5.35 Use the identities $(l_b 0\, l_a 0 | l_1 0) = (l_a 0\, l_b 0 | l_1 0)$,

$$\left[Y_{l_a}(\bar{\mathbf{s}}) \otimes Y_{l_b}(\bar{\mathbf{s}}) \right]_{l_1} = \frac{[l_a]\,[l_b]}{\sqrt{4\pi}\,[l_1]} (l_b 0\, l_a 0 | l_1 0)\, Y_{l_1}(\bar{\mathbf{s}}), \qquad (5.387)$$

and the results of the previous exercise to show that

$$\left[Y_{l_a}(\bar{\mathbf{u}}) \otimes Y_{l_a}(\bar{\mathbf{s}}) \right]_0 \left[Y_{l_b}(\bar{\mathbf{v}}) \otimes Y_{l_b}(\bar{\mathbf{s}}) \right]_0$$
$$= \frac{1}{\sqrt{4\pi}} \sum_{l_1} (l_a 0\, l_b 0 | l_1 0) \left[Y_{l_1}(\bar{\mathbf{s}}) \otimes \left[Y_{l_a}(\bar{\mathbf{u}}) \otimes Y_{l_b}(\bar{\mathbf{v}}) \right]_{l_1} \right]_0. \qquad (5.388)$$

5.36 Use the recoupling relation

$$\left[\left[Y_{l_a}(\bar{\mathbf{u}}) \otimes Y_{l_b}(\bar{\mathbf{v}}) \right]_{l_1} \otimes \left[Y_{l_c}(\bar{\mathbf{u}}) \otimes Y_{l_d}(\bar{\mathbf{v}}) \right]_{l_2} \right]_J$$
$$= \sum_{l_3 l_4} [l_1]\,[l_2]\,[l_3]\,[l_4] \begin{Bmatrix} l_a & l_b & l_1 \\ l_c & l_d & l_2 \\ l_3 & l_4 & J \end{Bmatrix}$$
$$\times \left[\left[Y_{l_a}(\bar{\mathbf{u}}) \otimes Y_{l_c}(\bar{\mathbf{u}}) \right]_{l_3} \otimes \left[Y_{l_d}(\bar{\mathbf{v}}) \otimes Y_{l_b}(\bar{\mathbf{v}}) \right]_{l_4} \right]_J \qquad (5.389)$$

and the results of Exercise 5.35 to derive Equation (5.373).

5.37 Show that

$$\int \left[\left[Y_{l_4}(\bar{\mathbf{v}}) \otimes Y_{l_3}(\bar{\mathbf{u}}) \right]_L \otimes \left[Y_{l_1}(\bar{\mathbf{s}}) \otimes Y_{l_2}(\bar{\mathbf{t}}) \right]_L \right]_0$$
$$\times \left[\left[Y_{l_2}(\bar{\mathbf{t}}) \otimes Y_{l_1}(\bar{\mathbf{s}}) \right]_J \otimes \left[Y_{l_3}(\bar{\mathbf{u}}) \otimes Y_{l_4}(\bar{\mathbf{v}}) \right]_J \right]_0 d\bar{\mathbf{s}}\, d\bar{\mathbf{t}}\, d\bar{\mathbf{u}}\, d\bar{\mathbf{v}}$$
$$= \delta_{L,J} \frac{1}{[L]^2} \int \left[\left[Y_{l_1}(\bar{\mathbf{s}}) \otimes Y_{l_2}(\bar{\mathbf{t}}) \right]_L \otimes \left[Y_{l_2}(\bar{\mathbf{t}}) \otimes Y_{l_1}(\bar{\mathbf{s}}) \right]_L \right]_0 d\bar{\mathbf{s}}\, d\bar{\mathbf{t}}$$
$$\times \int \left[\left[Y_{l_4}(\bar{\mathbf{v}}) \otimes Y_{l_3}(\bar{\mathbf{u}}) \right]_L \otimes \left[Y_{l_3}(\bar{\mathbf{u}}) \otimes Y_{l_4}(\bar{\mathbf{v}}) \right]_L \right]_0 d\bar{\mathbf{u}}\, d\bar{\mathbf{v}} \qquad (5.390)$$

and that

$$\int \left[\left[Y_{l_1}(\bar{\mathbf{s}}) \otimes Y_{l_2}(\bar{\mathbf{t}}) \right]_L \otimes \left[Y_{l_2}(\bar{\mathbf{t}}) \otimes Y_{l_1}(\bar{\mathbf{s}}) \right]_L \right]_0 d\bar{\mathbf{s}}\, d\bar{\mathbf{t}} = (-1)^L [L]. \qquad (5.391)$$

Hence derive Equation (5.376).

5.10 Dual pairs of group representations

Several examples are encountered in the preceding sections of this book in which the irreps of two commuting groups on a given space occur in paired relationships. For example, in Chapter 4, it is shown that the harmonic series of SU(1,1) irreps on the Hilbert space of the 5-dimensional Bohr model always occur in unique paired combinations with those of a commuting O(5) group. A similar pairing, shown in Section 5.8, occurs between the harmonic series irreps of the SU(1,1) group of radial transformations and those of a commuting O(3) group of rotations and inversions on the Hilbert space of the three-dimensional spherical harmonic oscillator. It is shown that the Hilbert space, $\mathcal{L}^2(\mathbb{R}^3)$, of the three-dimensional spherical harmonic oscillator has an orthonormal basis of states $\{|\nu l m\rangle\}$ which, for each value of l, form a basis for an irrep of the direct product group $SU(1,1) \times O(3)$. Thus, it is found that every state in $\mathcal{L}^2(\mathbb{R}^3)$ that belongs to an $O(3)$ irrep of angular momentum l simultaneously belongs to an $SU(1,1)$ irrep labelled by $\lambda = l + \frac{3}{2}$. Similarly, it is observed in Chapter 4 that every state of an $O(5)$ irrep of seniority v in the Bohr model Hilbert space, $\mathcal{L}^2(\mathbb{R}^5)$, simultaneously belongs to an $SU(1,1)$ irrep labelled by $\lambda = v + \frac{5}{2}$. These $SU(1,1) \times O(5)$ and $SU(1,1) \times O(3)$ duality relationships are shown in Chapter 4 and in Section 5.8 to be powerful aids both for the classification of states and in the calculation of matrix elements.

Several variations on the definition of dual pairs of representations exist in the literature.[180] The following definition is useful for present purposes.

Definition 5.1 Two groups of transformations, G_1 and G_2, of a Hilbert space, $\mathbb{H}$, are said to have dual representations if they commute with one another, so that they can be combined to form a direct product group $G_1 \times G_2$, and if they satisfy the following conditions. Let $\{|\rho\lambda\mu\kappa\nu\rangle\}$ denote a basis of states for $\mathbb{H}$, labelled by quantum numbers such that: λ and κ label irreps of G_1 and G_2, respectively; μ and ν label bases for these irreps; and ρ distinguishes multiple occurrences of the $G_1 \times G_2$ irrep (λ, κ). The representations of G_1 and G_2 on $\mathbb{H}$ are then said to form a dual pair if: (i) there are no multiple occurrences of equivalent $G_1 \times G_2$ irreps, so that the multiplicity index ρ is not needed, and (ii) an irrep λ of G_1 only occurs in combination with a unique irrep $\kappa(\lambda)$ of G_2 and vice-versa. In such a situation, the label κ also becomes redundant and it is sufficient to label basis states, $\{|\lambda\mu\nu\rangle\}$, by the quantum numbers λ, μ, and ν.

A special kind of pairing of group representations is observed in Section 5.5.1, where it is shown that the concept of isospin, defined in terms of a $U(2)$ group, provides a way of keeping track of particle exchange symmetries, defined in terms of the symmetric (particle-permutation) group. This is a simple example of the extraordinarily powerful Schur-Weyl duality theorem,[181,182] given in Section 5.11, which shows that the symmetric group and the group of one-body unitary transformations have dual representations on many-body Hilbert spaces. As a result the representation theory of the two groups is intertwined and powerful techniques, based on the use of Young diagrams, have been introduced to facilitate its application.

It is hard to imagine how the shell model might have developed without the Schur-Weyl theorem. Essential to the shell model is the construction of physically

---

[180]An overview of the literature on dual pairs of group representations and a review of their applications in physics is given by Rowe *et al.*, *op. cit.* Footnote 168 on Page 308.

[181]Weyl H. (1953), *Classical Groups, their Invariants and Representations* (Princeton University Press, Princeton, N.J.), second edn.

[182]Weyl H. (1950), *The Theory of Groups and Quantum Mechanics* (Dover Publications, New York), translation of *Gruppentheorie und Quantenmechanik*, published in 1931.

relevant basis states and the calculation of matrix elements. To this end, coupling schemes are devised to classify many-nucleon basis states with quantum numbers defined by the irreps of chains of subgroups. Among the subgroups in these chains are some that transform the spatial wave functions and others that transform the spin and isospin wave functions. However, without the Schur-Weyl theorem, one would not know how to combine these various wave functions to form totally anti-symmetric functions that respect the Pauli principle.

Important examples of dual pairs of group representations arise in applications of the pair-coupling models.[183] As shown in later sections of this chapter and in Chapter 6, pair-coupling models are solved in coupling schemes which diagonalise chains of subgroups involving groups with complicated representations. However, because the irreps of these groups are partnered with the irreps of simpler groups, analytical results for the models are often obtained.[184] This illustrates a common and valuable property of the dual pairing of group representations.

A family of dual pairs of representations, widely studied in physics and mathematics,[185,186,187,188] are those of the $\text{Sp}(n, \mathbb{R}) \times \text{O}(A)$ groups, which will be shown in Volume 2 to play an important role in the microscopic realisation of the nuclear collective model. Examples in this volume include the $\text{SU}(1,1) \times \text{O}(5)$ and $\text{SU}(1,1) \times \text{O}(3)$ dualities mentioned above. Many other examples are given by Howe,[189] Gelbart,[190] and in a recent review article.[191]

5.11 The unitary and symmetric groups

Nuclear wave functions are many-particle combinations of spatial, spin, and isospin wave functions. However, because the only admissible combinations are those that are totally antisymmetric with respect to exchange of identical fermions, they are much easier to construct than might be expected. The construction makes use of the powerful symmetry techniques that result from the Schur-Weyl duality theorem.

5.11.1 *Schur-Weyl duality*

Schur-Weyl duality is concerned with an N-particle Hilbert space

$$\mathbb{H}_d^N := \mathbb{H}_d \otimes \cdots \otimes \mathbb{H}_d \quad (N \text{ copies}) \tag{5.392}$$

---

[183] The first of these was shown by Helmers K. (1961), *Nucl. Phys.* **23**, 594.

[184] Cf., for example, the quasi-spin model of Kerman A.K. (1961), *Ann. Phys. (NY)* **12**, 300.

[185] Moshinsky M. and Quesne C. (1971), *J. Math. Phys.* **12**, 1772.

[186] Kashiwara M. and Vergne M. (1978), *Invent. Math.* **44**, 1.

[187] Howe R. (1979), *Proc. Sympos. Pure Math.* **33**, 275.

[188] Howe R. (1989), *Trans. Amer. Math. Soc.* **313**, 539.

[189] Howe R. (1985), *Lect. Appl. Math.* **21**, 179.

[190] Gelbart S. (1979), *Proc. Sympos. Pure Math.* **33**, 287.

[191] Rowe *et al.*, *op. cit.* Footnote 168 on Page 308.

which is a tensor product (sometimes denoted $\mathbb{H}_d^{\oplus N}$) of N copies of a single-particle Hilbert space $\mathbb{H}_d$ of dimension d. Thus, if $\nu = 1, \ldots, d$ indexes an orthonormal basis, $\{\psi_\nu\}$, of single-particle wave functions for $\mathbb{H}_d$, a corresponding basis for $\mathbb{H}_d^N$ is given by the N-particle wave functions

$$\Psi_{\nu_1 \nu_2 \ldots \nu_N} = \psi_{\nu_1} \otimes \psi_{\nu_2} \otimes \ldots \otimes \psi_{\nu_N}, \tag{5.393}$$

with $1 \leq \nu_i \leq d$. The notation means that, if $i = 1, \ldots, N$ indexes the particles and $\psi_\nu(i)$ denotes a wave function for particle i, the N-particle basis wave functions are simple products of the form

$$\Psi_{\nu_1 \nu_2 \ldots \nu_N}(1, 2, \ldots, N) = \psi_{\nu_1}(1) \, \psi_{\nu_2}(2) \, \ldots \, \psi_{\nu_N}(N). \tag{5.394}$$

Note that $\mathbb{H}_d^N$ is not here restricted to antisymmetric products; it contains all N-particle wave functions.

Let $U(d)$ denote the group of unitary $d \times d$ matrices and let $\hat{T}^{(1)}(g)$ denote the representation of a matrix $g \in U(d)$ as a transformation of the single-particle wave functions of $\mathbb{H}_d$, i.e.,

$$\hat{T}^{(1)}(g)\psi_\nu = \sum_\mu \psi_\mu \, g_{\mu\nu}, \quad g \in U(d). \tag{5.395}$$

This d-dimensional irrep of $U(d)$ is called the *standard* or *defining* representation. The Hilbert space $\mathbb{H}_d^N$ then carries a *tensor product* representation of $U(d)$,

$$\hat{T}^{(N)}(g)\Psi_{\nu_1 \nu_2 \ldots \nu_N} := \sum_\mu \Psi_{\mu_1 \mu_2 \ldots \mu_N} \, g_{\mu_1 \nu_1} g_{\mu_2 \nu_2} \cdots g_{\mu_N \nu_N}, \tag{5.396}$$

that is reducible if $N > 1$. The representations that can be constructed in this way, as tensor products or as subrepresentations of tensor products of the defining representation of a Lie group or its Lie algebra, form a particularly important class of representations that we describe here as *tensor representations*.

If $N > 1$, the space $\mathbb{H}_d^N$ also carries a reducible representation of the symmetric group S_N of permutations of the particles. For example, the permutation $\hat{P}_{12}$, which exchanges the 1 and 2 particle coordinates

$$[\hat{P}_{12}\Psi_{\nu_1 \nu_2 \nu_3 \ldots \nu_N}](1, 2, 3, \ldots, N) := \Psi_{\nu_1 \nu_2 \nu_3 \ldots \nu_N}(2, 1, 3, \ldots, N), \tag{5.397}$$

is equivalent to the transformation

$$\hat{P}_{12}\Psi_{\nu_1 \nu_2 \nu_3 \ldots \nu_N} := \Psi_{\nu_2 \nu_1 \nu_3 \ldots \nu_N}. \tag{5.398}$$

The above actions of S_N and $U(d)$ are seen to commute with one another. Thus, the tensor product space $\mathbb{H}_d^N$ carries a representation $\hat{\Gamma}_d^{(N)}$ of the direct product group $S_N \times U(d)$, defined for $P \in S_N$ and $g \in U(d)$ by

$$\hat{\Gamma}_d^{(N)}(P, g)\Psi_{\nu_1 \nu_2 \ldots \nu_N} := \sum_\mu \hat{P} \, \Psi_{\mu_1 \mu_2 \ldots \mu_N} \, g_{\mu_1 \nu_1} g_{\mu_2 \nu_2} \cdots g_{\mu_N \nu_N}. \tag{5.399}$$

The representation $\hat{\Gamma}_d^{(N)}$ is again reducible for $N > 1$. According to the Schur-Weyl theorem, it is a direct sum of uniquely paired irreps of S_N and $U(d)$.

Theorem 5.2 (Schur-Weyl)[192] *The groups S_N and $U(d)$ have dual representations (as defined in Section 5.10) on the space $\mathbb{H}_d^N$.*

This theorem has proved to be enormously powerful. In Weyl's hands,[193] it laid the foundations for the branch of mathematics known as *invariant theory*, which has led to a huge arsenal of techniques for the study of many groups in addition to the symmetric and unitary groups.[194] These techniques make it possible to evaluate the tensor products of many group irreps and determine branching rules that give the irreps of one group contained in an irrep of another.[195] The coupling schemes, used in the nuclear shell model and in the expression of the microscopic collective model in algebraic terms, rely heavily on these developments.

It was shown in Section 5.5.1 that the permutation symmetry of a two-nucleon state is uniquely defined by its isospin. In the following section, we construct the representations of S_N and $U(d)$ on the Hilbert space $\mathbb{H}_d^N$ and show that they satisfy the more general duality relationship claimed by the theorem.

5.11.2 *Dual $S_N \times U(d)$ representations*

The group $U(d)$ of $d \times d$ unitary matrices has the property that a matrix $g \in U(d)$ can be expressed as an exponential, $g = e^{iX}$, where X is a $d \times d$ Hermitian matrix. Thus, the Lie algebra, $u(d)$, of the group $U(d)$ is the set of Hermitian $d \times d$ matrices (see Section 3.4).

Let $C_{\mu\nu}$ denote a matrix that has the entry 1 at the intersection of row μ with column ν and 0 everywhere else, i.e.,

$$\left(C_{\mu\nu}\right)_{ij} = \delta_{\mu,i}\delta_{j,\nu}. \tag{5.400}$$

These matrices have commutation relations

$$[C_{\mu\nu}, C_{\kappa\lambda}] = \delta_{\nu\kappa}C_{\mu\lambda} - \delta_{\mu\lambda}C_{\kappa\nu} \tag{5.401}$$

---

[192]A first version of the Schur-Weyl theorem, was given in Schur's doctoral thesis: Schur I. (1901), *Uber eine Klasse von Matrizen, die sich einer gegebenen Matrix zuordnen lassen*, Ph.D. thesis, Berlin. It was developed and related to the Young tableau methods by Weyl, cf. for example, Weyl, *op. cit.* Footnote 182 on Page 325. A proof of the Schur-Weyl theorem can be found in Chapter V of Weyl's book and also in the review article of Rowe *et al.*, *op. cit.* Footnote 168 on Page 308.

[193]Weyl, *op. cit.* Footnote 181 on Page 325.

[194]In addition to Schur and Weyl, major participants in these developments were Young, see Robinson G. de B., ed. (1977), *The Collected Papers of Alfred Young* (University of Toronto Press), and Littlewood D.E. (1950), *The Theory of Group Characters and Matrix Representations of Groups* (Oxford University Press), second edn.

[195]The book, Wybourne B.G. (1970), *Symmetry Principles and Atomic Spectroscopy* (Wiley, New York), and the paper, King R.C. (1975), *J. Phys. A: Math. Gen.* **8**, 429, provide an entry to the large literature on the subject of branching rules.

and span the *complex extension* of the u(d) Lie algebra. A basis for u(d) is given, in terms of them, by the Hermitian linear combinations

$$C_{\mu\nu} + C_{\nu\mu}, \quad \mathrm{i}(C_{\mu\nu} - C_{\nu\mu}), \quad 1 \le \mu,\nu \le d. \tag{5.402}$$

An arbitrary Hermitian matrix $X \in \mathrm{u}(d)$ has a standard d-dimensional irrep as a linear operator, $\hat{X}^{(1)}$, on a d-dimensional (one-particle) Hilbert space $\mathbb{H}_d$ with orthonormal basis $\{\psi_\nu; \nu = 1, \ldots, d\}$, defined by

$$\hat{X}^{(1)}\psi_\nu := \sum_{\mu=1}^{d} \psi_\mu X_{\mu\nu}. \tag{5.403}$$

Thus, $\hat{X}^{(1)}$ is realised as the differential operator

$$\hat{X}^{(1)} = \sum_{\mu,\nu=1}^{d} X_{\mu\nu}\, \psi_\mu \frac{\partial}{\partial\psi_\nu}, \quad X \in \mathrm{u}(d). \tag{5.404}$$

If the basis vectors $\{\psi_\nu\}$ of $\mathbb{H}_d$ are interpreted as single-particle wave functions, this standard irrep has a natural extension to a tensor representation of u(d) on the N-particle Hilbert space, $\mathbb{H}_d^N$, defined by Equation (5.392) and given by[196,197]

$$\hat{X}^{(N)} := \sum_{i=1}^{N} \sum_{\mu,\nu=1}^{d} X_{\mu\nu}\, \psi_\mu(i) \frac{\partial}{\partial\psi_\nu(i)}. \tag{5.405}$$

Such tensor representations are generally reducible and expressible as direct sums of tensor irreps.

Useful basis states for an irrep of U(d) and its Lie algebra u(d) are given by the simultaneous eigenstates of the subset of commuting operators which represent the diagonal matrices of a Cartan subalgebra of u(d). We call these operators *Cartan operators*. For the tensor representations, they are the operators

$$\hat{C}_{\nu\nu} = \sum_{i=1}^{N} \psi_\nu(i) \frac{\partial}{\partial\psi_\nu(i)}, \quad (\textit{Cartan operators}). \tag{5.406}$$

---

[196]It may be noted that the expression for $\hat{X}$ given by Equation (5.405) is very similar to the extension of X to a many-body operator given by second quantization, i.e., $\hat{\mathcal{X}} := \sum_{\mu\nu} X_{\mu\nu}\, a_\mu^\dagger a^\nu$. However, unlike $\hat{X}$, the action of the operator $\hat{\mathcal{X}}$ is only defined on the subspace of $\mathbb{H}_d^N$ of fully antisymmetric states. Thus, the operator $\hat{X}$ applies more generally.

[197]Exercises 5.39 and 5.40 confirm that the operators $\{\hat{X}^{(N)}\}$ define a u(d) representation by ascertaining that, if the u(d) matrices X, Y, and Z satisfy the commutation relation $[X, Y] = \mathrm{i}Z$, the corresponding operators $\hat{X}^{(N)}$, $\hat{Y}^{(N)}$, and $\hat{Z}^{(N)}$ satisfy the commutation relation $[\hat{X}^{(N)}, \hat{Y}^{(N)}] = \mathrm{i}\hat{Z}^{(N)}$.

Corresponding raising and lowering operators are realised as[198]

$$\hat{C}_{\mu\nu} = \sum_{i=1}^{N} \psi_\mu(i) \frac{\partial}{\partial \psi_\nu(i)}, \quad \mu < \nu \quad (raising\ operators), \tag{5.407}$$

$$\hat{C}_{\mu\nu} = \sum_{i=1}^{N} \psi_\mu(i) \frac{\partial}{\partial \psi_\nu(i)}, \quad \mu > \nu \quad (lowering\ operators). \tag{5.408}$$

Although a tensor representation, as defined above, is generally reducible, the basis wave functions of Equation (5.394) are already eigenfunctions of the Cartan operators. For example, the $N = 6$, $d = 7$ wave function $\Psi_{231532} = \psi_2 \otimes \psi_3 \otimes \psi_1 \otimes \psi_5 \otimes \psi_3 \otimes \psi_2$ satisfies the equation

$$\hat{C}_{\nu\nu} \Psi_{231532} = [\delta_{\nu,1} + 2\delta_{\nu,2} + 2\delta_{\nu,3} + \delta_{\nu,5}] \Psi_{231532}. \tag{5.409}$$

In general, a wave function, Ψ, that satisfies the equations

$$\hat{C}_{\nu\nu} \Psi = [\lambda_1 \delta_{\nu,1} + \lambda_2 \delta_{\nu,2} + \cdots + \lambda_d \delta_{\nu,d}] \Psi, \quad \nu = 1, \ldots, d, \tag{5.410}$$

is said to have *weight* $\lambda = (\lambda_1, \lambda_2, \ldots, \lambda_d)$. Thus, the function Ψ_{231532} has weight $(1, 2, 2, 0, 1, 0, 0)$. However, for compactness of notation, it is common practice to omit the commas and trailing zeroes unless it would be ambiguous to do so. The function Ψ_{231532} is then said to have weight (12201). For the above-defined tensor representations, each component of a weight has a physical interpretation as the number of particles that occupy the corresponding single-particle state. Thus, the state with wave function Ψ_{231532} and weight (12201), has one particle with wave function ψ_1, two particles with wave function ψ_2, two particles with wave function ψ_3, and one particle with wave function ψ_5.

The weights of the states of a U(d) representation can be ordered. For example, if a weight (2341) is identified with the number 2341, it is natural to say that it is higher than the weight (1234). Note, however, that it is important to take account of trailing zeroes in ordering weights. For example, the weight (341) of U(4) should be identified with the number 3410 and, consequently, it is a higher weight than (2341). When weights are ordered in this way, every finite-dimensional irrep of U(d) has unique highest and lowest weights.[199]

The action of a raising operator on a state of a U(d) representation either annihilates the state or produces a state of higher weight. For example, whereas Ψ_{123}

---

[198]The ordering of the state labels, which determines which of the $\{\hat{C}_{\mu\nu}\}$ are raising and which lowering, is arbitrary. Thus, one could equally decide that the operators with $\mu > \nu$ are raising and those with $\mu < \nu$ are lowering.

[199]An important result of Lie group theory is that all finite-dimensional irreps of a simple or semi-simple Lie group and its Lie algebra (see Footnote 226 on Page 365 for definition) have both highest and lowest weights. Another important result of Lie group theory is that all irreps of a compact Lie group and its Lie algebra are finite-dimensional. The group U(d) is compact but it is neither simple nor semi-simple, because it has an Abelian subgroup, U(1) $\subset$ U(d), whose elements commute will all elements of U(d). Nevertheless, it has a well-defined Cartan subgroup and all its irreps are finite-dimensional with both highest and lowest weights.

is a state of weight (111), the state $\hat{C}_{12}\Psi_{123} = \Psi_{113}$, if non-vanishing, has weight (201). Thus, in addition to being a simultaneous eigenstate of the Cartan operators, a highest-weight state is one that is annihilated by all the raising operators. It follows that the components of the highest weight λ, for any U(d) tensor irrep, are integers that satisfy the inequality[200]

$$\lambda_1 \geq \lambda_2 \geq \lambda_3 \geq \cdots \geq \lambda_d \geq 0. \tag{5.411}$$

An orthonormal weight basis for a U(d) irrep, i.e., a basis of eigenstates of the Cartan operators, is also called a *Gel'fand-Tsetlin* basis in Section 5.11.7. Such bases have many useful properties.

A highest-weight wave function for a U(d) tensor irrep of a given highest weight is easy to construct. For example, a wave function, $\Psi_{\text{h.wt.}}^{(421)}$, with highest weight (421) for an $N = 7$, $d = 3$ irrep is required to satisfy the equations

$$\hat{C}_{\mu\nu}\Psi_{\text{h.wt.}}^{(421)} = 0, \quad \text{for } \mu < \nu,$$
$$\hat{C}_{11}\Psi_{\text{h.wt.}}^{(421)} = 4\Psi_{\text{h.wt.}}^{(421)}, \quad \hat{C}_{22}\Psi_{\text{h.wt.}}^{(421)} = 2\Psi_{\text{h.wt.}}^{(421)}, \quad \hat{C}_{33}\Psi_{\text{h.wt.}}^{(421)} = \Psi_{\text{h.wt.}}^{(421)}. \tag{5.412}$$

These equations imply that the wave function $\Psi_{\text{h.wt.}}^{(421)}$ is a linear combination of the subset of basis wave functions $\{\Psi_{\nu_1\nu_2\nu_3\nu_4\nu_5\nu_6\nu_7}\}$ for which four of the ν_i indices have the value 1, two have value 2, and one has value 3. Each of the basis wave functions in this subset has weight (421). To be the highest-weight wave functions for an irrep, it must be the combination that satisfies the equations

$$\sum_{i=1}^{7} \psi_\mu(i)\frac{\partial}{\partial\psi_\nu(i)}\Psi_{\text{h.wt.}}^{(421)} = 0, \quad \forall\,\mu < \nu. \tag{5.413}$$

Consider, for example, the highest-weight state for the $N = 3$ irrep of U(3) with highest weight $(1^3) \equiv (111)$. It must be a linear combination of wave functions of the form $\{\psi_1(i)\psi_2(j)\psi_3(k)\}$, where i, j, and k index any three distinct particles. The particular linear combination that is annihilated by the raising operators is given (to within a normalisation factor) by the Slater determinant

$$\Psi_{\text{h.wt.}}^{(111)}(1,2,3) = \begin{vmatrix} \psi_1(1) & \psi_1(2) & \psi_1(3) \\ \psi_2(1) & \psi_2(2) & \psi_2(3) \\ \psi_3(1) & \psi_3(2) & \psi_3(3) \end{vmatrix}. \tag{5.414}$$

Similarly, a highest-weight state for the $N = 7$ irrep of U(3) with highest weight

---

[200]As shown above for the tensor irreps, an ordered basis of Cartan operators, $\{\hat{C}_{\nu\nu}\}$, can be put into one-to-one correspondence with an ordered orthonormal basis of single-particle states. Hence, if the components of a weight do not satisfy the inequality (5.411) then a re-arrangement of the Cartan basis, to give a higher weight that does, can always be achieved by a U(d) transformation of one orthonormal single-particle basis to another. A similar argument applies to any U(d) highest weight.

(421) is given by

$$\Psi_{\text{h.wt.}}^{(421)}(1,2,3,4,5,6,7) = \Psi_{\text{h.wt.}}^{(111)}(1,2,3)\Psi_{\text{h.wt.}}^{(11)}(4,5)\Psi_{\text{h.wt.}}^{(1)}(6)\Psi_{\text{h.wt.}}^{(1)}(7), \qquad (5.415)$$

i.e., by the product of Slater determinants

$$\Psi_{\text{h.wt.}}^{(421)}(1,2,3,4,5,6,7) = \begin{vmatrix} \psi_1(1) & \psi_1(2) & \psi_1(3) \\ \psi_2(1) & \psi_2(2) & \psi_2(3) \\ \psi_3(1) & \psi_3(2) & \psi_3(3) \end{vmatrix} \begin{vmatrix} \psi_1(4) & \psi_1(5) \\ \psi_2(4) & \psi_2(5) \end{vmatrix} \psi_1(6)\psi_1(7). \qquad (5.416)$$

It can be seen that this wave function has four particles with wave function ψ_1, two particles with wave function ψ_2, and one particle with wave function ψ_3. Thus, it is of weight (421). Moreover, from its structure as a product of highest-weight determinantal functions, it is of highest weight.

The above construction gives a highest-weight state for any U(d) tensor irrep. Note, however, that there are many such highest-weight states for a given highest weight, λ, corresponding to the many possible permutations of the particle indices. For example, the wave function

$$\Psi_{\text{h.wt.}}^{(421)}(1,2,7,6,5,4,3) = \begin{vmatrix} \psi_1(1) & \psi_1(2) & \psi_1(7) \\ \psi_2(1) & \psi_2(2) & \psi_2(7) \\ \psi_3(1) & \psi_3(2) & \psi_3(7) \end{vmatrix} \begin{vmatrix} \psi_1(6) & \psi_1(5) \\ \psi_2(6) & \psi_2(5) \end{vmatrix} \psi_1(4)\psi_1(3) \qquad (5.417)$$

is evidently of U(d) highest-weight (421) but is different from that given by Equation (5.416). In fact, the set of all such wave functions of U(d) highest-weight (421), obtained by permutations of the particle indices, spans a representation of the symmetric group S$_7$ (see Exercise 5.43). According to the Schur-Weyl theorem, this S$_7$ representation is irreducible and is said to be dual to the U(d) irrep with highest weight (421). More generally, the Schur-Weyl theorem implies that the highest weight, λ, for any tensor irrep of U(d) on $\mathbb{H}_d^N$ defines a dual irrep of S$_N$ with $N = \sum_{i=1}^{d} \lambda_i$.

5.11.3 Classification of S$_N$ and U(d) irreps by Young diagrams

The duality relationship between the S$_N$ and integer U(d) irreps is clarified considerably by Young diagram techniques.[201]

As the preceeding $N = 7$, $d = 3$ example illustrates, both a tensor irrep of U(d) and the corresponding S$_N$ irrep are characterised by a partition λ of N into p non-zero parts, with $p \leq d$, such that Equation (5.411) is satisfied, i.e., by $\lambda = (\lambda_1, \lambda_2, \ldots, \lambda_p)$ with $\sum_\nu \lambda_\nu = N$ and $\lambda_1 \geq \lambda_2 \geq \cdots \geq \lambda_p > 0$. A sequence of integers, λ, that satisfies this condition is said to be an *ordered partition* of N and expressed symbolically by $\lambda \vdash N$. An ordered partition, λ, with p parts is said to have *depth* (often called *length*) $l(\lambda) = p$.

---

[201] A review of Young diagram and related Schur function techniques has been given, for example, by Wybourne B.G. (1970), *Symmetry Principles and Atomic Spectroscopy* (Wiley, New York).

To avoid ambiguity, it is conventional to distinguish the S_N and $U(d)$ irreps, labelled by a partition λ, by putting the former in parenthesis, (λ), and the latter in brace brackets, $\{\lambda\}$. The Schur-Weyl theorem can then be interpreted as the statement that the representation $\hat{\Gamma}_d^{(N)}$ of the direct product group $S_N \times U(d)$ carried by the tensor product space $\mathbb{H}_d^N$ is a direct sum of irreps given by the combination of S_N and $U(d)$ irreps

$$\hat{\Gamma}_d^{(N)} = \bigoplus_{\lambda \vdash N}^{l(\lambda) \leq d} \left((\lambda), \{\lambda\} \right). \tag{5.418}$$

The irreps of S_N and $U(d)$ corresponding to an ordered partition, $\lambda \vdash N$, are equivalently characterised by a so-called Young diagram, Y^λ, which is an array of left adjusted boxes with λ_1 boxes in the first row, λ_2 in the second row, $\ldots$, λ_p in the p'th row. For example, the partition $\lambda = (421)$ is identified with the Young diagram

$$Y^{(421)} := \quad\quad . \tag{5.419}$$

The boxes of a Young diagram can be regarded as containers for indices that label particles for S_N and single-particle wave functions for $U(d)$. Thus, we need different Young diagrams for the S_N and $U(d)$ indices. For example, for the highest-weight state with wave function given by Equation (5.416), we can put the particle-number indices $1, \ldots, 7$ into an S_7 diagram and the single-particle state indices into a $U(3)$ diagram as follows:

$$
\Psi_{\text{h.wt.}}^{(421)}(1,2,3,4,5,6,7) \sim \left(
\begin{array}{|c|c|c|c|}
\hline
1 & 4 & 6 & 7 \\
\hline
\end{array}
\;,\;
\begin{array}{|c|c|c|c|}
\hline
1 & 1 & 1 & 1 \\
\hline
\end{array}
\right), \tag{5.420}
$$

where the S_7 tableau is shown with rows: $1\,4\,6\,7$ / $2\,5$ / 3, and the $U(3)$ tableau with rows: $1\,1\,1\,1$ / $2\,2$ / 3,

where the columns of the tableau for this state correspond to the determinantal factors in the wave function (5.416):

$$
\begin{vmatrix} \psi_1(1) & \psi_1(2) & \psi_1(3) \\ \psi_2(1) & \psi_2(2) & \psi_2(3) \\ \psi_3(1) & \psi_3(2) & \psi_3(3) \end{vmatrix} \sim \left(\begin{array}{|c|} \hline 1 \\ \hline 2 \\ \hline 3 \\ \hline \end{array} \;,\; \begin{array}{|c|} \hline 1 \\ \hline 2 \\ \hline 3 \\ \hline \end{array} \right), \tag{5.421}
$$

$$
\begin{vmatrix} \psi_1(4) & \psi_1(5) \\ \psi_2(4) & \psi_2(5) \end{vmatrix} \sim \left(\begin{array}{|c|} \hline 4 \\ \hline 5 \\ \hline \end{array} \;,\; \begin{array}{|c|} \hline 1 \\ \hline 2 \\ \hline \end{array} \right), \tag{5.422}
$$

$$
\psi_1(6) \sim \left(\boxed{6} \;,\; \boxed{1} \right), \qquad \psi_1(7) \sim \left(\boxed{7} \;,\; \boxed{1} \right). \tag{5.423}
$$

Numbered diagrams of this kind are called standard *Young tableaux*. The pair of tableaux in Equation (5.420) are special because the first represents what we call a *leading state* of the S_7 irrep (421) and the second represents a state of highest weight for the U(3) irrep {421}, as constructed in Section 5.11.2. The leading state is defined, somewhat arbitrarily, to be one for which the particle indices, 1,2,...,7, are entered sequentially down columns, starting from the first column. The Young tableau for a U(d) highest weight state is one for which the integer i fills all boxes of row i. Other basis states are obtained by putting the numbers into the boxes in different ways, subject to the condition that all the particle indices must be distinct. However, simple rules must be followed to avoid getting an overcomplete set. For example, inspection of the wave function $\Psi_{\text{h.wt.}}^{(421)}$ shows that the numbers in corresponding columns of the S_7 and U(3) tableaux, respectively, give the particle and state indices of the single-particle wave functions of a Slater determinant. Interchanging their order, in either tableau, can at most change the sign of the wave function. To obtain a linearly-independent set of states, it is therefore appropriate to impose the rule that the numbers in any column must always increase strongly from top to bottom (increasing strongly simply means that no number is repeated whereas increasing weakly means not decreasing). A second, less obvious rule, is that, to obtain linearly-independent states, the numbers in any row of boxes should also increase (weakly in the case of a U(d) tableau and strongly in the case of S_N) from left to right. Note also that, while state indices may be repeated, the particles indices must all be distinct.

It is easy to check that the above rules work out in given situations. For example, the 4-dimensional space, $\mathbb{H}_{d=2}^{N=2}$, is spanned by three states of the $\lambda = (2)$ irrep of $S_2 \times$ U(2) and one state of the $\lambda = (11)$ irrep:

$$
\begin{array}{cc}
S_2 \quad \times \quad \text{U}(2) & \\[6pt]
\left(\boxed{1\,2} \ , \ \boxed{1\,1} \right) \sim & \psi_1(1)\psi_1(2), \\[10pt]
\left(\boxed{1\,2} \ , \ \boxed{1\,2} \right) \sim & \psi_1(1)\psi_2(2), \\[10pt]
\left(\boxed{1\,2} \ , \ \boxed{2\,2} \right) \sim & \psi_2(1)\psi_2(2), \\[10pt]
\left(\begin{array}{c}\boxed{1}\\[-2pt]\boxed{2}\end{array} \ , \ \begin{array}{c}\boxed{1}\\[-2pt]\boxed{2}\end{array} \right) \sim & \begin{vmatrix} \psi_1(1) & \psi_1(2) \\ \psi_2(1) & \psi_2(2) \end{vmatrix},
\end{array}
\tag{5.424}
$$

Similarly, the 8-dimensional space, for $\mathbb{H}_{d=2}^{N=3}$, is spanned by four states of the $\lambda = (3)$

irrep of $S_3 \times U(2)$ and four states of the $\lambda = (21)$ irrep:

$$
\left(\begin{array}{|c|c|c|} \hline 1 & 2 & 3 \\ \hline \end{array} \,,\, \begin{array}{|c|c|c|} \hline 1 & 1 & 1 \\ \hline \end{array} \right), \qquad
\left(\begin{array}{|c|c|c|} \hline 1 & 2 & 3 \\ \hline \end{array} \,,\, \begin{array}{|c|c|c|} \hline 1 & 1 & 2 \\ \hline \end{array} \right),
$$

$$
\left(\begin{array}{|c|c|c|} \hline 1 & 2 & 3 \\ \hline \end{array} \,,\, \begin{array}{|c|c|c|} \hline 1 & 2 & 2 \\ \hline \end{array} \right), \qquad
\left(\begin{array}{|c|c|c|} \hline 1 & 2 & 3 \\ \hline \end{array} \,,\, \begin{array}{|c|c|c|} \hline 2 & 2 & 2 \\ \hline \end{array} \right),
$$

$$
\left(\begin{array}{|c|c|} \hline 1 & 2 \\ \hline 3 \\ \cline{1-1} \end{array} \,,\, \begin{array}{|c|c|} \hline 1 & 1 \\ \hline 2 \\ \cline{1-1} \end{array} \right), \qquad
\left(\begin{array}{|c|c|} \hline 1 & 3 \\ \hline 2 \\ \cline{1-1} \end{array} \,,\, \begin{array}{|c|c|} \hline 1 & 1 \\ \hline 2 \\ \cline{1-1} \end{array} \right),
$$

$$
\left(\begin{array}{|c|c|} \hline 1 & 2 \\ \hline 3 \\ \cline{1-1} \end{array} \,,\, \begin{array}{|c|c|} \hline 1 & 2 \\ \hline 2 \\ \cline{1-1} \end{array} \right), \qquad
\left(\begin{array}{|c|c|} \hline 1 & 3 \\ \hline 2 \\ \cline{1-1} \end{array} \,,\, \begin{array}{|c|c|} \hline 1 & 2 \\ \hline 2 \\ \cline{1-1} \end{array} \right). \tag{5.425}
$$

5.11.4 Dimensions of S_N and $U(d)$ irreps

The dimensions of the S_N and $U(d)$ irreps, corresponding to a particular ordered partition $\lambda = (\lambda_1, \lambda_2, \ldots, \lambda_p)$, are given by the possible numberings of the corresponding Young tableaux. Thus, the dimensions are readily obtained by counting, in simple cases, or by the following formulae expressed in terms of so-called *hook lengths*, which were obtained by combinatorial methods.

Let (i, j) index the box in row i and column j of the Young diagram, Y^λ, for the partition λ. The corresponding hook length, $h_{i,j}$ is defined as the number of boxes in the "hook" in Y^λ that extends from the box (i, j) to the right and to the bottom. For example, the hook length for each of the boxes of $Y^{(431)}$ is given by the number in the corresponding box in the non-standard tableau

$$
\begin{array}{|c|c|c|c|} \hline 6 & 4 & 3 & 1 \\ \hline 4 & 2 & 1 \\ \cline{1-3} 1 \\ \cline{1-1} \end{array} \,. \tag{5.426}
$$

The dimension of the S_N irrep (λ) is then given by the formula[202]

$$
\dim(\lambda) = \frac{N!}{H^\lambda}, \tag{5.427}
$$

where H^λ is the product of hook lengths,

$$
H^\lambda := \prod_{(i,j) \in Y^\lambda} h_{i,j}. \tag{5.428}
$$

For example, the S_8 irrep (431) has dimension

$$
\dim(431) = \frac{8!}{6 \cdot 4 \cdot 3 \cdot 4 \cdot 2} = 70. \tag{5.429}
$$

---

[202]Frame J.S., Robinson G. de B. and Thrall R.M. (1954), *Can. J. Math.* **6**, 316.

The dimension of the U(d) irrep $\{\lambda\}$ is given by the formula[203]

$$\dim\{\lambda\} = \frac{N_d^{\{\lambda\}}}{H^\lambda}, \qquad (5.430)$$

where

$$N_d^{\{\lambda\}} := \prod_{(i,j)\in Y^\lambda} (d - i + j). \qquad (5.431)$$

For example, the U(4) irrep $\{431\}$ has dimension

$$\dim\{431\} = \frac{4\cdot 5\cdot 6\cdot 7\cdot 3\cdot 4\cdot 5\cdot 2}{6\cdot 4\cdot 3\cdot 4\cdot 2} = 175. \qquad (5.432)$$

For low-dimensional irreps, these formulae can be checked against the number of tableau for U(d) and S_N irreps (cf. Exercise 5.44).

5.11.5 *Unitary-unitary duality*

A second duality relationship follows directly from the Schur-Weyl theorem.

Suppose that a single-particle Hilbert space $\mathbb{H}_{mn}$ is a tensor product

$$\mathbb{H} = \mathbb{H}_m \otimes \mathbb{H}_n, \qquad (5.433)$$

where $\mathbb{H}_m$ and $\mathbb{H}_n$ are of dimension m and n, respectively. For example, $\mathbb{H}_m$ might be a Hilbert space of spatial wave functions and $\mathbb{H}_n$ a combined space of spin-isospin wave functions. The Hilbert space for N-fermions is then the antisymmetric subspace

$$\mathbb{H}_{AS}^N := \left[\mathbb{H}_m^N \otimes \mathbb{H}_n^N\right]_{AS} \subset \mathbb{H}_m^N \otimes \mathbb{H}_n^N, \qquad (5.434)$$

where $\mathbb{H}_m^N$ and $\mathbb{H}_n^N$ are, respectively, the tensor products of N copies of $\mathbb{H}_m$ and $\mathbb{H}_n$. Thus, while the spaces $\mathbb{H}_m^N$ and $\mathbb{H}_n^N$ need not be separately antisymmetric, only the antisymmetric combinations of these spaces form acceptable many-fermion states.

The classification of basis states for the antisymmetric subspace of states in the tensor product space $\mathbb{H}^N = \mathbb{H}_m^N \otimes \mathbb{H}_n^N$ is much simplified by use of the following duality theorem.

Theorem 5.3 (Unitary-unitary duality for fermions) *The groups* U(m) *and* U(n) *have dual representations on the antisymmetric subspace* $\mathbb{H}_{AS}^N$ *of the tensor-product space* $\mathbb{H}^N$.

A basic property of the representation theory of the symmetric groups is that the tensor product of two S_N irreps contains a fully antisymmetric irrep, (1^N), if and only if the two S_N irreps are conjugate to one another, as defined below. Thus, the unitary-unitary duality theorem amounts to the statement that to form fully

---

[203]Robinson G. de B. (1958), *Can. Math. Bull.* **1**, 21.

antisymmetric states, it is necessary to combine states of permutation symmetry (λ) from $\mathbb{H}_m^N$ with states of conjugate symmetry $(\tilde{\lambda})$ from $\mathbb{H}_n^N$. We express this relationship for S_N by writing

$$\left[(\kappa) \otimes (\lambda)\right]_{\mathrm{AS}} = \delta_{\kappa, \tilde{\lambda}}(1^N), \qquad (5.435)$$

where $\left[(\kappa) \otimes (\lambda)\right]_{\mathrm{AS}}$ denotes the fully antisymmetric component of the S_N tensor product $(\kappa) \otimes (\lambda)$.

The Young diagrams for conjugate irreps are related to one another by the exchange of rows and columns. They can also be seen as images of one another under reflection in a diagonal line passing through the top-left corner of each diagram. Thus, for example, the conjugate of the partition $\{\lambda\} = \{32^2 1\}$ is $\{\tilde{\lambda}\} = \{431\}$, viz.

$$\{\lambda\} = \{32^2 1\} \qquad\qquad \{\tilde{\lambda}\} = \{431\}$$

$$(5.436)$$

With these results we can now prove the above unitary-unitary duality theorem. First recall that, according to the Schur-Weyl theorem expressed by Equation (5.418), the Hilbert space $\mathbb{H}_m^N$ carries a reducible representation $\hat{\Gamma}_m^{(N)}$ of $S_N \times U(m)$ and the Hilbert space $\mathbb{H}_n^N$ carries a reducible representation $\hat{\Gamma}_n^{(N)}$ of $S_N \times U(n)$, where

$$\hat{\Gamma}_m^{(N)} = \bigoplus_{\kappa \vdash N}^{l(\kappa) \leq m} \left((\kappa), \{\kappa\}\right), \quad \hat{\Gamma}_n^{(N)} = \bigoplus_{\lambda \vdash N}^{l(\lambda) \leq n} \left((\lambda), \{\lambda\}\right), \qquad (5.437)$$

and $l(\lambda) = \tilde{\lambda}_1$ is the depth of the partition λ. Thus, the tensor product space $\mathbb{H}_m^N \otimes \mathbb{H}_n^N$ carries the representation of the direct product group $S_N \times U(m) \times U(n)$ given by

$$\hat{T}^{(N)} = \bigoplus_{\kappa \vdash N}^{l(\kappa) \leq m} \bigoplus_{\lambda \vdash N}^{l(\lambda) \leq n} \left((\kappa) \otimes (\lambda), \{\kappa\}, \{\lambda\}\right). \qquad (5.438)$$

It immediately follows that the subspace, $\mathbb{H}_{\mathrm{AS}}^N$, of antisymmetric states in $\mathbb{H}^N = \mathbb{H}_m^N \otimes \mathbb{H}_n^N$ carries the $S_N \times U(m) \times U(n)$ representation

$$\hat{T}_{\mathrm{AS}}^{(N)} = \bigoplus_{\kappa \vdash N}^{l(\kappa) \leq m} \bigoplus_{\lambda \vdash N}^{l(\lambda) \leq n} \left(\left[(\kappa) \otimes (\lambda)\right]_{\mathrm{AS}}, \{\kappa\}, \{\lambda\}\right). \qquad (5.439)$$

Therefore, by use of Equation (5.435) and the fact that the S_N irrep (1^N) is one-dimensional, it follows that $\mathbb{H}_{\mathrm{AS}}^N$ carries a reducible representation

$$\hat{\Gamma}_{\mathrm{AS}}^{(N)} = \bigoplus_{\substack{\lambda \vdash N}}^{\lambda_1 \leq m, l(\lambda) \leq n} \left(\{\lambda\}, \{\tilde{\lambda}\} \right). \tag{5.440}$$

of $\mathrm{U}(m) \times \mathrm{U}(n)$. This completes the proof of the theorem. $\quad\square$

The above unitary-unitary duality theorem is almost obvious when expressed in the language of second quantization. Let

$$\{a_{s\nu}^\dagger \,;\, s = 1, \ldots, m, \; \nu = 1, \ldots, n\} \tag{5.441}$$

denote a set of single-fermion creation operators, where s and ν index orthonormal bases of single-particle states of $\mathbb{H}_m$ and $\mathbb{H}_n$, respectively. By the methods of Section 5.4.2, it is observed that the infinitesimal generators of a fully antisymmetric irrep of the group $\mathrm{U}(mn)$ are Hermitian linear combinations of the operators

$$\hat{C}_{s\mu,t\nu}^{(mn)} = a_{s\mu}^\dagger a^{t\nu}. \tag{5.442}$$

It is also observed that infinitesimal generators of the $\mathrm{U}(m)$ and $\mathrm{U}(n)$ subgroups of $\mathrm{U}(mn)$ are given in terms of the partial traces of these operators,

$$\hat{C}_{st}^{(m)} = \sum_{\nu=1}^n \hat{C}_{s\nu,t\nu}^{(mn)} = \sum_{\nu=1}^n a_{s\nu}^\dagger a^{t\nu}, \tag{5.443}$$

$$\hat{C}_{\mu\nu}^{(n)} = \sum_{s=1}^m \hat{C}_{s\mu,s\nu}^{(mn)} = \sum_{s=1}^m a_{s\mu}^\dagger a^{s\nu}. \tag{5.444}$$

With respect to these $\mathrm{U}(m)$ and $\mathrm{U}(n)$ operators, it is now easily ascertained, for example, that the fully antisymmetric 7-fermion state

$$(a_{11}^\dagger a_{12}^\dagger a_{13}^\dagger)\,(a_{21}^\dagger a_{22}^\dagger a_{23}^\dagger)\, a_{31}^\dagger \,|0\rangle, \tag{5.445}$$

is of $\mathrm{U}(m)$ highest weight $\{3^2 1\}$ and $\mathrm{U}(n)$ highest weight $\{3 2^2\}$. Thus, it is a highest-weight state for a $\mathrm{U}(m) \times \mathrm{U}(n)$ irrep $(\{3^2 1\}, \{3 2^2\})$ with mutually conjugate $\mathrm{U}(m) \times \mathrm{U}(n)$ Young tableaux

$$|(\{3^2 1\}, \{3 2^2\})\mathrm{h.wt.}\rangle \sim \left(\begin{array}{|c|c|c|} \hline 1 & 1 & 1 \\ \hline 2 & 2 & 2 \\ \hline 3 \\ \cline{1-1} \end{array} \;,\; \begin{array}{|c|c|c|} \hline 1 & 1 & 1 \\ \hline 2 & 2 \\ \cline{1-2} 3 & 3 \\ \cline{1-2} \end{array} \right). \tag{5.446}$$

The labelling of these Young tableaux can be understood by associating the first factor $(a_{11}^\dagger a_{12}^\dagger a_{13}^\dagger)$ with the tableaux representing the state

$$(a_{11}^\dagger a_{12}^\dagger a_{13}^\dagger)|0\rangle \sim \left(\begin{array}{|c|c|c|} \hline 1 & 1 & 1 \\ \hline \end{array} \;,\; \begin{array}{|c|} \hline 1 \\ \hline 2 \\ \hline 3 \\ \hline \end{array} \right), \tag{5.447}$$

which spans the $U(m) \times U(n)$ irrep $(\{3\}, \{1^3\})$, the second factor $(a_{21}^\dagger a_{22}^\dagger a_{23}^\dagger)$ with state

$$(a_{21}^\dagger a_{22}^\dagger a_{23}^\dagger)|0\rangle \sim \left(\boxed{\begin{array}{|c|c|c|} \hline 2 & 2 & 2 \\ \hline \end{array}} \, , \, \begin{array}{|c|} \hline 1 \\ \hline 2 \\ \hline 3 \\ \hline \end{array} \right), \qquad (5.448)$$

and the third factor $a_{31}^\dagger$ with

$$a_{31}^\dagger |0\rangle \sim \left(\boxed{3} \, , \, \boxed{1} \right). \qquad (5.449)$$

There is a parallel unitary-unitary theorem for bosonic systems.

Theorem 5.4 (Unitary-unitary duality for bosons) *The groups* $U(m)$ *and* $U(n)$ *have dual representations on the symmetric subspace* $\mathbb{H}_S^N$ *of the tensor-product space* $\mathbb{H}^N$.

This theorem is a direct analogue of its fermionic counterpart. It follows from the observation that the tensor product of two S_N irreps contains the fully symmetric (identity) irrep, (N), if and only if the two S_N irreps are equivalent to one another. This fact is expressed by the equation

$$[(\kappa) \otimes (\lambda)]_S = \delta_{\kappa,\lambda}(N), \qquad (5.450)$$

where $[(\kappa) \otimes (\lambda)]_S$ denotes the fully symmetric component of the S_N tensor product $(\kappa) \otimes (\lambda)$. The proof of Theorem 5.4 follows in close parallel to that of Theorem 5.3 and leads to the conclusion that $\mathbb{H}_S^N$ carries the representation

$$\hat{\Gamma}_A^{(N)} = \bigoplus_{\lambda \vdash N}^{l(\lambda) \leq m} \left(\{\lambda\}, \{\lambda\} \right) \qquad (5.451)$$

of $U(m) \times U(n)$.

5.11.6 Tensor products of unitary representations

The tensor product of the Hilbert spaces for two $U(d)$ irreps is a new Hilbert space which, by definition, carries a tensor product representation of $U(d)$. Such tensor products occur frequently in nuclear physics. Thus, a frequently occurring problem is to determine the irreps contained in the expansion of such a tensor product. For the tensor irreps of the unitary groups, the coefficients $\Gamma_{\kappa\lambda}^\sigma$ in the expansion

$$\{\kappa\} \otimes \{\lambda\} = \bigoplus_\sigma \Gamma_{\kappa\lambda}^\sigma \{\sigma\}, \qquad (5.452)$$

which give the multiplicity of the irrep $\{\sigma\}$ occurring in the tensor product $\{\kappa\} \otimes \{\lambda\}$, are known as *Littlewood-Richardson coefficients*. The following is a simple prescription for determining these coefficients.

The dimension of the tensor product space is equal to $\dim\{\kappa\} \times \dim\{\lambda\}$. If the irreps $\{\kappa\}$ and $\{\lambda\}$ have Young diagrams with N_1 and N_2 boxes, respectively, the corresponding Young tableaux can be interpreted as representing states of N_1 and N_2 particles, respectively. The tensor product states are then states of $N_1 + N_2$ particles; this implies that the Young diagrams for all irreps $\{\lambda\}$ appearing in the above sum must have $N_1 + N_2$ boxes. As a consequence of Schur-Weyl duality, the partition $\{\kappa\}$ defines an irrep of the direct product group $S_{N_1} \times U(d)$. Similarly, $\{\lambda\}$ defines an irrep of $S_{N_2} \times U(d)$. The problem of making the decomposition (5.452) is then equivalent to finding the irreps of $S_{N_1+N_2} \times U(d)$ within the tensor product space. This problem was solved by Littlewood and Richardson.[204] The result is summarised in terms of a set of diagrammatic rules, known as the Littlewood-Richardson rule, as follows. First represent the tensor product of interest in terms of Young diagrams. Then fill the boxes of one Young diagram (it is most efficient to choose the smaller but, in principle, it could be either) with the numbers 1, 2, 3, ... such that the first row contains all 1's, the second row all 2's, etc. Next add the numbered boxes to those of the other Young diagram in numerical order such that:

(i) No two identical numbers are in the same column.

(ii) Counting the occurrences of numbers from right to left and from top to bottom, the number of 1's, at any step, is never less than the number of 2's, the number of 2s is never less than the number of 3's, etc.

(iii) The diagram generated corresponds to that of a regular partition of N_1+N_2 of depth not greater than d.

The set of all diagrams that can be generated in this way gives the decomposition of the tensor product into irreps.

Consider, for example, the U(3) tensor product $\{321\} \otimes \{21\}$. The diagrammatic rules for this product give

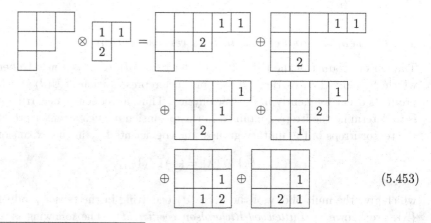

$$\text{(5.453)}$$

[204]Littlewood D.E. and Richardson A.R. (1934), *Philos. Trans. Roy. Soc. London* **A233**, 99.

or

$$\{321\} \otimes \{21\} = \{531\} \oplus \{52^2\} \oplus 2\{432\} \oplus 2\{3^3\}. \tag{5.454}$$

5.11.7 *Gel'fand-Tsetlin bases*

Young tableaux are defined in Section 5.11.3 as a way of labelling basis states for irreps of symmetric and unitary groups. It is easy to put the tableaux into one-to-one correspondence with those of a basis for either an S_N or $U(d)$ irrep. However, for practical purposes, one would like to have orthonormal basis states. This is achieved by identifying S_N tableaux with basis states that reduce the subgroup chain

$$S_N \supset S_{N-1} \supset S_{N-2} \supset \quad \cdots \quad \supset S_2, \tag{5.455}$$

and by identifying $U(d)$ tableaux with basis states that reduce the chain

$$U(d) \supset U(d-1) \supset U(d-2) \supset \quad \cdots \quad \supset U(1). \tag{5.456}$$

Such a basis is described as a *canonical* basis. We describe here the correspondence between Young tableaux for $U(d)$ and a canonical basis known as a *Gel'fand-Tsetlin* (GT) basis.[205] The algebraic properties of such a basis add many useful properties to the already powerful Young diagram methods.

Consider, for example, a state corresponding to the Young tableau

$$\begin{array}{|c|c|c|}
\hline
1 & 2 & 3 \\
\hline
3 & 4 \\
\cline{1-2}
4 \\
\cline{1-1}
\end{array} \tag{5.457}$$

of a $U(4)$ irrep $\{321\} \equiv \{3,2,1,0\}$. An element of the $U(3) \subset U(4)$ subgroup ignores any single-particle state, ψ_ν, with $\nu = 4$. Thus, if the state space for the $U(4)$ irrep $\{321\}$ is broken up into irreducible subspaces with respect to the subgroup $U(3) \subset U(4)$, then the state (5.457) belongs to the $U(3)$ subspace of $U(3)$ highest weight $\{31\}$ obtained by deleting all the 4's from the $U(4)$ tableau. Similarly, an element of the $U(2) \subset U(3) \subset U(4)$ subgroup ignores any single-particle state, ψ_ν, with $\nu = 3$ or 4. It follows that the state (5.457) likewise belongs to the irreducible $U(2)$ subspace of $U(2)$ highest weight $\{2\}$. Explicitly, the reduction is given directly from the diagram by deleting the numbered boxes sequentially from the highest down to show that the Young tableau of Equation (5.457) defines a state that reduces the subgroup chain $U(4) \supset U(3) \supset U(2) \supset U(1)$ with the sequence of

---

[205]Gel'fand I.M. and Tsetlin M.A. (1950), *Dokl. Akad. Nauk. USSR* **71**, 825.

highest weights shown by the chain of Young diagrams

$$\text{U}(4) \quad \supset \quad \text{U}(3) \quad \supset \quad \text{U}(2) \quad \supset \quad \text{U}(1)$$

$$\{3210\} \qquad \{310\} \qquad \{20\} \qquad \{1\} \tag{5.458}$$

A GT basis is conventionally labelled by a triangular array of regular partitions, called a *Gel'fand pattern*, denoting the irreps of the groups in the chain $\text{U}(d) \supset \text{U}(d-1) \supset \cdots$ to which a basis state belongs. Thus, for example, the basis state with Young tableau given by Equation (5.457) is identified with a GT state

$$\tag{5.459}$$

The equivalent Young tableaux and GT expressions of basis states both serve a useful purpose. The weight of a basis state is given by simply counting the number of ones, the number of twos, etc., in the tableau for the state. Thus, for example, the state (5.459) has weight $\nu = (1122)$. Observe also that the dimension of an irrep is readily determined, for simple irreps, by writing down all the tableaux for a given Young diagram or, equivalently, by writing down all the Gel'fand patterns for the corresponding highest weight. The latter is easiest to do once it is noted that, because of the rules for constructing Gel'fand patterns from tableaux, the numbers in a Gel'fand pattern must obey a so-called *betweenness* condition which is that every number in a given row must lie between or at the limits of the two closest numbers above it. So, if one row of an array is $\{\lambda_1 \lambda_2 \lambda_3 \cdots\}$, and the row immediately below it is $\{\kappa_1 \kappa_2 \kappa_3 \cdots\}$, then

$$\lambda_1 \geq \kappa_1 \geq \lambda_2 \geq \kappa_2 \geq \lambda_3 \geq \kappa_3 \geq \cdots . \tag{5.460}$$

5.11.8 U(d) ↓ U(n) × U($d - n$) and U(d) ↓ SU(d) branching rules

The above rules for constructing Gel'fand patterns immediately provide a simple algorithm for determining the U(d) ↓ U($d-1$) branching rules. If $\{\lambda\}$ denotes a U(d) irrep then the restriction of this irrep to U($d-1$) contains the U($d-1$) irrep $\{\kappa\}$ if and only the elements of $\{\kappa\}$ and $\{\lambda\}$ satisfy the betweenness condition (5.460). For example, the set of possible second rows of the Gel'fand patterns for the U(d) irrep $\{321\} \equiv \{3210\}$ (for $d > 3$) gives the branching rule

$$\text{U}(d) \downarrow \text{U}(d-1) \; : \; \{321\} \mapsto \{321\} \oplus \{32\} \oplus \{31^2\} \oplus \{2^21\}$$
$$\oplus \{31\} \oplus \{2^2\} \oplus \{21^2\} \oplus \{21\}. \tag{5.461}$$

The U(d) $\downarrow$ U($d-1$) branching rule obtained in this way can be expressed generally in the form

$$\text{U}(d) \downarrow \text{U}(d-1) \; : \; \{\lambda\} \downarrow \bigoplus_k \{\lambda/k\}, \tag{5.462}$$

where $\{k\}$, a one-rowed partition with k a positive integer, characterises an irrep of the group U(1) and $\{\lambda/k\}$ denotes an operation called *factorisation*. The precise meaning of 'factorisation of one U(d) irrep by another' is explained in the following section. However, in the present application, it is easy to see what it means in the context of the example given by Equation (5.461). Factoring a Young diagram for U(d) by $\{k\}$, simply means forming all the Young tableaux in which the integer d appears k times and then removing all the boxes which contain the integer d. For example, to determine $\{321\}/\{2\}$ for U(4), first observe that the number 4 can appear twice in the Young diagram for the irrep $\{321\}$ in the following boxes:

$$\tag{5.463}$$

Thus, removing the boxes containing the integer 4 leaves the tableaux corresponding to the irreps

$$\{31\}, \quad \{2^2\}, \quad \{21^2\}. \tag{5.464}$$

which are three of the terms in the branching rule (5.461).

The branching rule (5.462) extends naturally to give

$$\text{U}(d) \downarrow \text{U}(1) \times \text{U}(d-1) \; : \; \{\lambda\} \downarrow \bigoplus_k \left(\{k\}, \{\lambda/k\} \right), \tag{5.465}$$

where $k \geq 0$ is an integer. A further generalisation leads to the branching rule

$$\text{U}(d) \downarrow \text{U}(n) \times \text{U}(d-n) \; : \; \{\lambda\} \downarrow \bigoplus_\sigma \left(\{\sigma\}, \{\lambda/\sigma\} \right), \tag{5.466}$$

where $\{\sigma\}$ runs over all irreps of U(n) given by ordered partitions of positive integers.

Branching rules for restriction of a U(d) irrep to other subgroups of U(d) have been considered by many authors.[206,207] One important example is the branching

---

[206] A survey of branching rule methods and results has been given by King R.C. (1975), *J. Phys. A: Math. Gen.* **8**, 429.

[207] The above U(d) $\downarrow$ U(n) $\times$ U($d-n$) branching rules can be derived by use of the Schur-Weyl duality theorem; see, e.g., Rowe *et al.*, *op. cit.* Footnote 168 on Page 308.

rule for $U(d) \downarrow SU(d)$. On restriction to $SU(d)$, a $U(d)$ irrep remains irreducible. Thus, the $U(d) \downarrow SU(d)$ branching rule simply reflects the way an $SU(d)$ irrep is labelled. The standard labelling gives the branching rule

$$U(d) \downarrow SU(d); \quad \{\lambda_1 \lambda_2 \ldots \lambda_d\} \mapsto (\lambda_1 - \lambda_2, \lambda_2 - \lambda_3, \ldots, \lambda_{d-1} - \lambda_d). \qquad (5.467)$$

For example, the one-dimensional irreps, $\{0\}$ and $\{1^2\} \equiv \{11\}$, of $U(2)$ both restrict to the same $SU(2)$, spin zero, irrep (0). The two-dimensional $U(2)$ irreps $\{1\}$ and $\{21\}$ both restrict to the $SU(2)$, spin $j = 1/2$, irrep (1) (isomorphic to the spinor irrep $[1/2]$ of $SO(3)$). The three-dimensional $U(2)$ irreps $\{2\}$ and $\{31\}$ restrict to the $SU(2)$, spin $j = 1$, irrep (2). Thus, whereas the two-dimensional space of a single spin-$1/2$ particle carries the standard irrep $\{1\}$ of $U(2)$ and the standard, spin-$1/2$, irrep (1) of its subgroup $SU(2)$, the states of N such particles, which span a tensor irrep $\{\lambda_1, \lambda_2\}$ of $U(2)$ with $\lambda_1 + \lambda_2 = N$, are interpreted as having total spin $S = (\lambda_1 - \lambda_2)/2$ (see Exercise 5.49).

Branching rules for the restrictions $U(n) \downarrow O(n)$ and $U(2n) \downarrow USp(2n)$ are given in Sections 5.12.7 and 5.12.10.

5.11.9 *Factorisation as the inverse of multiplication*

The operation of factorisation, by manipulation of Young diagrams as explained above for the construction of Gel'fand patterns, is defined more generally as the inverse of the process of taking tensor products, described in Section 5.11.6. Thus, whereas the tensor product of two $U(d)$ irreps has the expansion

$$\{\kappa\} \otimes \{\sigma\} = \bigoplus_\lambda \Gamma^\lambda_{\kappa\sigma} \{\lambda\}, \qquad (5.468)$$

the quotient of two irreps, $\{\lambda\}$ and $\{\sigma\}$, is defined by

$$\{\lambda\}/\{\sigma\} \equiv \{\lambda/\sigma\} := \bigoplus_\kappa \Gamma^\lambda_{\kappa\sigma} \{\kappa\}, \qquad (5.469)$$

where the $\Gamma^\lambda_{\kappa\sigma}$ coefficients in the expansion of the tensor product, Equation (5.468), are the Littlewood-Richardson coefficients of Section 5.11.6. Thus, $\{\lambda\}/\{\sigma\}$ is the direct sum of all irreps whose tensor products with the irrep $\{\sigma\}$ contain $\{\lambda\}$.

Consider, for example, the quotient $\{321\}/\{21\}$ for $U(3)$ (whose irreps have at most three rows). There are four ways to obtain the irrep $\{321\}$ as a term in the tensor product of some other irrep with {21}; in graphical terms

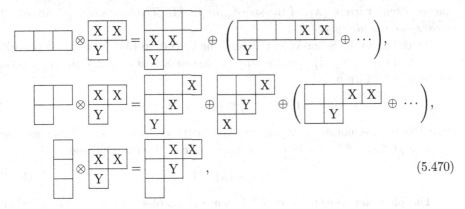

$$(5.470)$$

where the diagrams inside brackets on the right sides of these equations include only those for irreps other than {321}. These diagrammatic equations express the tensor product identities:

$$\{3\} \otimes \{21\} = \{321\} \oplus \big[\text{irreps other than}\{321\}\big],$$
$$\{21\} \otimes \{21\} = 2\{321\} \oplus \big[\text{irreps other than}\{321\}\big], \qquad (5.471)$$
$$\{1^3\} \otimes \{21\} = \{321\},$$

and imply that

$$\{321\}/\{21\} = \{3\} \oplus 2\{21\} \oplus \{1^3\}. \qquad (5.472)$$

5.11.10 *Space symmetry; plethysms*

A typical shell-model valence space is spanned by fully-antisymmetric states of nucleons with spin, isospin and a restricted set of spatial quantum numbers. Suppose, for example, that the restricted single-nucleon spatial states span a d-dimensional space, $\mathbb{H}_L$, that carries the standard irrep {1} of the so-called *space group*, U(d), of unitary transformations of these states. Similarly, the complementary four-dimensional space, $\mathbb{H}_{ST}$, spanned by the spin-isospin states of a nucleon carries the standard irrep {1} of the corresponding spin-isospin group U(4). The combined space-spin-isospin Hilbert space of a single nucleon is then the tensor product space $\mathbb{H} := \mathbb{H}_{ST} \otimes \mathbb{H}_L$ and, according to Theorem 5.3, the Hilbert space, $\mathbb{H}_{AS}^N$, of antisymmetric N-nucleon states in the valence shell is a sum of tensor product subspaces,

$$\mathbb{H}_{AS}^N = \bigoplus_{\lambda \vdash N} \mathbb{H}_{ST}^{\{\tilde{\lambda}\}} \otimes \mathbb{H}_L^{\{\lambda\}}, \qquad (5.473)$$

of which $\mathbb{H}_{ST}^{\{\tilde{\lambda}\}}$ carries a tensor irrep, $\{\tilde{\lambda}\}$, of the spin-isospin group $U(4)$, and $\mathbb{H}_L^{\{\lambda\}}$ carries a tensor irrep, $\{\lambda\}$, of the space group $U(d)$. The states of $\mathbb{H}_L^{\{\lambda\}}$ are said to have *space symmetry* $\{\lambda\}$.

According to the Schur-Weyl theorem, the $U(d)$ irrep $\{\lambda\}$, defined by a partition $\lambda \vdash N$ and carried by $\mathbb{H}_L^{\{\lambda\}}$, is the irreducible component of the N'th tensor power of the standard irrep, $\{1\}$,

$$\{1\}^N := \{1\} \otimes \{1\} \otimes \cdots \otimes \{1\}, \qquad (5.474)$$

that has S_N permutation symmetry (λ). Littlewood called such a symmetrised power a *plethysm*;[208] it is denoted here symbolically by the expression

$$\{1\} \, \textcircled{p} \, \{\lambda\} = \{\lambda\}. \qquad (5.475)$$

The plethysm concept is useful because it enables the principles underlying Schur-Weyl duality to be extended to more general situations. Suppose, for example, that, in addition to the irrep $\{1\}$ of $U(d)$, the d-dimensional Hilbert space $\mathbb{H}_L$ also carries an irrep $\{\kappa\}$ of a subgroup $U(n) \subset U(d)$. Then, the Hilbert space, $\mathbb{H}_L^{\{\lambda\}}$, for the tensor irrep $\{\lambda\}$ of $U(d)$ also carries a (generally reducible) representation of $U(n)$. This $U(n)$ representation is denoted by the plethysm $\{\kappa\} \, \textcircled{p} \, \{\lambda\}$. Thus, if the restriction to $U(n)$ of the irrep $\{1\}$ of $U(d)$ on $\mathbb{H}$ is expressed by the branching rule

$$U(d) \downarrow U(n) \; : \; \{1\} \downarrow \{\kappa\}, \qquad (5.476)$$

then, by definition of a plethysm, the restriction to $U(n)$ of the $U(d)$ irrep $\{\lambda\}$ is given by the branching rule

$$U(d) \downarrow U(n) \; : \; \{\lambda\} \downarrow \{\kappa\} \, \textcircled{p} \, \{\lambda\}. \qquad (5.477)$$

Branching rules for other subgroups of $U(d)$ can be derived by extensions of this method.[209]

To make use of a plethysm, one needs a procedure to calculate the coefficients in its expansion as a sum of $U(n)$ irreps, i.e.,

$$\{\kappa\} \, \textcircled{p} \, \{\lambda\} = \bigoplus_\nu P_{\kappa\lambda}^\nu \{\nu\}. \qquad (5.478)$$

Several computer codes exist for the determinations of these coefficients.[210,211,212]

Suppose, for example, that the single-nucleon spatial states are those of the $2s1d$ harmonic-oscillator shell. The space, $\mathbb{H}_L$, is the space spanned by the $l =$

[208]Littlewood D.E. (1950), *The Theory of Group Characters and Matrix Representations of Groups* (Oxford University Press), second edn.

[209]Wybourne B.G. (1970), *Symmetry Principles and Atomic Spectroscopy* (Wiley, New York).

[210]Wybourne B.G., SCHUR, An interaction program for calculating properties of Lie groups and symmetric functions, distributed by S. Christensen `http://smc.vnet.net/Christensen.html`.

[211]Draayer J.P. *et al.* (1989), *Comp. Phys. Comm.* **56**, 279.

[212]Carvalho M.J. and D'Agostino S. (2001), *Comp. Phys. Comm.* **141**, 282.

0 and 2 states of a single nucleon. It is 6-dimensional and carries the standard irrep $\{1\}$ of a U(6) space group. The states of this space have two spherical-harmonic-oscillator quanta and, as shown in Section 5.8.3, span the 6-dimensional irrep, $\{2\}$, of U(3), the symmetry group of the spherical-harmonic oscillator. It is then of importance for $2s1d$-shell-model calculations in a U(3)-coupled basis, to know what U(3) irreps there are in the $2s1d$ valence-shell space with space symmetry $\{\lambda\}$. In other words, one would like to know what irreps of U(3) are contained in the reducible representation of U(3) carried by the space $\mathbb{H}_L^{\{\lambda\}}$. The question is important because, by the unitary-unitary duality theorem, the SU(3)-coupled states belonging to irreps with space symmetry $\{\lambda\}$ must be coupled to spin-isospin states that belong to a U(4) irrep $\{\tilde{\lambda}\}$. The answer to the question is given by an expansion of the plethysm $\{2\} \textcircled{p} \{\lambda\}$. For example, using the plethysm code of Carvalho and D'Agostino,[213] we find that the U(3) irreps in the $2s1d$ shell, with space symmetries $\{4\}$, $\{31\}$, $\{22\}$, and $\{211\}$ are given by the plethysms

$$\{2\} \textcircled{p} \{4\} = \{8\} \oplus \{62\} \oplus \{422\} \oplus \{44\}, \tag{5.479}$$

$$\{2\} \textcircled{p} \{31\} = \{521\} \oplus \{71\} \oplus \{422\} \oplus \{62\} \oplus \{53\} \oplus \{431\}, \tag{5.480}$$

$$\{2\} \textcircled{p} \{22\} = \{422\} \oplus \{62\} \oplus \{521\} \oplus \{44\}, \tag{5.481}$$

$$\{2\} \textcircled{p} \{211\} = \{611\} \oplus \{521\} \oplus \{53\} \oplus \{332\} \oplus \{431\}. \tag{5.482}$$

These plethysms give corresponding U(6) $\downarrow$ U(3) branching rules in accord with Equation (5.477). Complete tables of the U(3) irreps in the harmonic oscillator sd shell for the range of possible space symmetries have been given by Elliott.[214].

As a second example, suppose that $\mathbb{H}_L \equiv \mathbb{H}_d$ is of dimension $d = 2l + 1$ and is spanned by single-nucleon spatial states of integer orbital angular momentum l. We then ask: "what values of the total angular momentum, L, can the N-nucleon states of space symmetry $\{\lambda\}$ have?" In fact, we can answer a more general question. "Suppose $\mathbb{H}_d$ is of dimension $d = 2j + 1$, where $2j$ is an even or odd integer, what values of the total angular momentum, J, can the N-nucleon states of symmetry $\{\lambda\}$ have?" In addressing this question, we note that $\mathbb{H}_d$ carries an SU(2) irrep $(2j)$ and, by simple extension, a U(2) irrep, $\{2j, 0\}$. Thus, the question can be rephrased: "What irreps of the group U(2) occur in the space of states, $\mathbb{H}_d^{\{\lambda\}}$?" The answer to this question is given by the expansion of the plethysm $\{2j\} \textcircled{p} \{\lambda\}$ as a direct sum of U(2) irreps followed by their restriction to SU(2). Thus, if $l = j = 2$ and $\{\lambda\} = \{32\}$, we seek to determine the SU(2) irreps that occur in the space $\mathbb{H}_d^{\{32\}}$. From the plethysm code of Carvalho and D'Agostino,[213] we obtain an expansion of the relevant $\{2j\}$ plethysm

$$\{4\} \textcircled{p} \{32\} = \{10, 10\} \oplus 2\{11, 9\} \oplus 4\{12, 8\} \oplus 3\{13, 7\} \oplus 4\{14, 6\}$$
$$\oplus 3\{15, 5\} \oplus 2\{16, 4\} \oplus \{17, 3\} \oplus \{18, 2\} \tag{5.483}$$

---

[213]Carvalho and D'Agostino, *op. cit.* Footnote 212 on Page 346.
[214]Elliott J.P. (1958), *Proc. Roy. Soc. London* **A245**, 128.

as a sum of U(2) irreps. Now, as discussed in Section 5.11.8, a U(2) irrep $\{\lambda_1, \lambda_2\}$ is spanned by the states of an SU(2) irrep $(\lambda_1 - \lambda_2)$. Thus, on restriction to SU(2), we obtain the irreps

$$\{4\} \, \circledP \, \{32\} \downarrow (0) \oplus 2(2) \oplus 4(4) \oplus 3(6) \oplus 4(8)$$
$$\oplus 3(10) \oplus 2(12) \oplus (14) \oplus (16). \tag{5.484}$$

which correspond to states of angular momentum

$$L = J = 0,\, 1^2,\, 2^4,\, 3^3,\, 4^4,\, 5^3,\, 6^2,\, 7,\, 8. \tag{5.485}$$

More details and a more precise definition of a plethysm can be found, for example, in the review article of Rowe *et al.*, *op. cit.* Footnote 168 on Page 308.

5.11.11 Casimir invariants of unitary Lie algebras

Casimir invariants are defined in Section 3.8 as combinations of elements of a Lie algebra (or its complex extension) that commute with all elements of the algebra.

The $u(d)$ Lie algebra is defined in Section 5.11.2 in terms of the matrices $\{C_{\mu\nu}; \mu, \nu = 1, \ldots, d\}$, defined by Equation (5.400). In a representation, in which these matrices are mapped to operators, $\{\hat{C}_{\mu\nu}\}$, the Casimir invariants are similarly represented by Casimir operators.

From the $u(d)$ commutation relations, one ascertains that the operators

$$\hat{\mathcal{K}}^{(1)}_{u(d)} := \sum_\nu \hat{C}_{\nu\nu}, \quad \hat{\mathcal{K}}^{(2)}_{u(d)} := \sum_{\mu,\nu} \hat{C}_{\mu\nu}\hat{C}_{\nu\mu}, \tag{5.486}$$

satisfy the identities

$$[\hat{\mathcal{K}}^{(1)}_{u(d)}, \hat{C}_{\mu\nu}] = 0, \quad [\hat{\mathcal{K}}^{(2)}_{u(d)}, \hat{C}_{\mu\nu}] = 0, \tag{5.487}$$

for all μ and ν. Thus, they represent first- and second-order Casimir invariants, respectively, for $u(d)$.

To determine the eigenvalue of a Casimir operator for any state of a $u(d)$ irrep $\{\lambda\}$, it is sufficient to consider the action of the operator on the highest-weight state $|\{\lambda\}\text{h.wt.}\rangle$ for the irrep. The components of a highest weight are defined by the equation

$$\hat{C}_{\nu\nu}|\{\lambda\}\text{h.wt.}\rangle = \lambda_\nu|\{\lambda\}\text{h.wt.}\rangle. \tag{5.488}$$

It follows that the eigenvalue of $\hat{\mathcal{K}}^{(1)}_{u(d)}$ is simply $\mathcal{K}^{(1)}_{u(d)}(\lambda) = \sum_\nu \lambda_\nu$. The action of

$\hat{\mathcal{K}}_{u(d)}^{(2)}$ on a highest-weight state is determined from the expansion

$$\hat{\mathcal{K}}_{u(d)}^{(2)} = \sum_{\mu < \nu} \left(\hat{C}_{\mu\nu} \hat{C}_{\nu\mu} + \hat{C}_{\nu\mu} \hat{C}_{\mu\nu} \right) + \sum_{\nu} \hat{C}_{\nu\nu} \hat{C}_{\nu\nu} \tag{5.489}$$

$$= \sum_{\mu < \nu} \left[2\hat{C}_{\nu\mu} \hat{C}_{\mu\nu} + \hat{C}_{\mu\mu} - \hat{C}_{\nu\nu} \right] + \sum_{\nu} \hat{C}_{\nu\nu} \hat{C}_{\nu\nu}, \tag{5.490}$$

where use has been made of the commutation relations of Equation (5.66). Then, because $\hat{C}_{\mu\nu} | \{\lambda\} \text{h.wt.} \rangle = 0$, for $\mu < \nu$, it is inferred (see Exercise 5.50) that the eigenvalue of $\hat{\mathcal{K}}_{u(d)}^{(2)}$ for the irrep $\{\lambda\}$ is

$$\mathcal{K}_{u(d)}^{(2)}(\lambda) = \sum_{\nu} \lambda_{\nu} (\lambda_{\nu} + d + 1 - 2\nu). \tag{5.491}$$

The second-order Casimir invariant for $su(d)$ can be derived from $\mathcal{K}_{u(d)}^{(2)}$. The important observation is that $su(d)$ is the subalgebra of trace-zero $u(d)$ matrices. Thus, a projection of $u(d)$ to its $su(d)$ subalgebra is achieved by mapping $C_{\mu\nu}$ to its traceless component

$$C'_{\mu\nu} := C_{\mu\nu} - \delta_{\mu,\nu} \frac{1}{d} \sum_{\nu} C_{\nu\nu}. \tag{5.492}$$

By this projection, we then obtain the $su(d)$ Casimir operator

$$\hat{\mathcal{K}}_{su(d)}^{(2)} := \sum_{\mu,\nu} \hat{C}'_{\mu\nu} \hat{C}'_{\nu\mu} = \hat{\mathcal{K}}_{u(d)}^{(2)} - \frac{1}{d} \left[\hat{\mathcal{K}}_{u(d)}^{(1)} \right]^2. \tag{5.493}$$

The eigenvalues of $\hat{\mathcal{K}}_{su(d)}^{(2)}$ can be derived immediately from those of $\hat{\mathcal{K}}_{u(d)}^{(2)}$. One simply observes that an $su(d)$ irrep can be labelled by a $u(d)$ highest weight, $\{\lambda'\}$, with the extra condition that $\sum_{\nu} \lambda'_{\nu} = 0$. Thus, we obtain

$$\mathcal{K}_{su(d)}^{(2)}(\lambda') = \sum_{\nu} \lambda'_{\nu} (\lambda'_{\nu} - 2\nu). \tag{5.494}$$

As an example, consider the $su(3)$ irrep whose highest-weight state is a highest-weight state for a $u(3)$ irrep $\{\lambda'_1, \lambda'_2, \lambda'_3\}$. This $su(3)$ irrep is customarily labelled by (λ, μ) with

$$\lambda := \lambda'_1 - \lambda'_2, \quad \mu := \lambda'_2 - \lambda'_3. \tag{5.495}$$

Thus, if we impose the condition that $\lambda'_1 + \lambda'_2 + \lambda'_3 = 0$, we obtain

$$\lambda'_1 = \frac{1}{3}(2\lambda + \mu), \quad \lambda'_2 = \frac{1}{3}(\mu - \lambda), \quad \lambda'_3 = -\frac{1}{3}(2\mu + \lambda), \tag{5.496}$$

and

$$\mathcal{K}_{su(3)}^{(2)}(\lambda, \mu) = \frac{2}{3} \left[\lambda^2 + \mu^2 + \lambda\mu + 3\lambda + 3\mu \right]. \tag{5.497}$$

Exercises

5.38 Determine the matrix given by the commutator $[X_{\mu\nu}, Y_{\mu'\nu'}]$, where $X_{\mu\nu}$ and $Y_{\mu\nu}$ are Hermitian matrices given by Equation (5.402), and show that it is equal to i times a Hermitian matrix.

5.39 Show that the operators

$$\hat{C}_{\mu\nu} = x_\mu \frac{\partial}{\partial x_\nu}, \quad \mu, \nu = 1, \ldots, d, \tag{5.498}$$

defined in terms of a set of variables $\{x_\nu; \nu = 1, \ldots, d\}$, satisfy the commutation relations

$$[\hat{C}_{\mu\nu}, \hat{C}_{\mu'\nu'}] = \delta_{\nu,\mu'} \hat{C}_{\mu\nu'} - \delta_{\mu,\nu'} \hat{C}_{\mu'\nu}. \tag{5.499}$$

5.40 Show that the operators

$$\hat{C}_{\mu\nu} = \sum_{i=1}^{N} x_{i\mu} \frac{\partial}{\partial x_{i\nu}}, \quad \mu, \nu = 1, \ldots, d, \tag{5.500}$$

defined in terms of a set of variables $\{x_{i\nu}; i = 1, \ldots, N, \nu = 1, \ldots, d\}$, also satisfy the commutation relations of Equation (5.499).

5.41 Use the Schur-Weyl theorem to deduce that the U(d) irreps $\{2\}$ and $\{1^2\}$ are of dimension $\frac{1}{2}d(d+1)$ and $\frac{1}{2}d(d-1)$, respectively.

5.42 Use the Schur-Weyl theorem and the symmetry property of SU(2) Clebsch-Gordan coefficients,

$$(j_1 m_1 j_2 m_2 | J, m_1 + m_2) = (-1)^{j_1+j_2-J}(j_2 m_2 j_1 m_1 | J, m_1 + m_2), \tag{5.501}$$

to show that the U($2j+1$) irreps $\{2\}$ and $\{1^2\}$, comprise states of angular momentum J even and odd, respectively.

5.43 Start from the observation that the actions of the symmetric group S_N and the unitary group U(d) on the Hilbert space, $\mathbb{H}_d^{(N)}$, commute with one another and show that the span of the states generated by the action of all S_N elements on a U(d) highest-weight state in this space carries an irrep of S_N.

5.44 Use Young tableaux to show that the irrep $(22) \times \{22\}$ of $S_4 \times$ U(3) is of dimension 12. Check your results by using the hook length formulae of Section 5.11.4.

5.45 Show that the Young tableau $\{4321\}$ is self-conjugate.

5.46 Determine the weights of the Gel'fand-Tsetlin states $\left| \begin{matrix} 2\,1\,0 \\ 2\,0 \\ 1 \end{matrix} \right\rangle$ and $\left| \begin{matrix} 2\,1\,0 \\ 1\,1 \\ 1 \end{matrix} \right\rangle$.

5.47 Derive the branching rule of Equation (5.461) from the general expression given by Equation (5.466).

5.48 Use the Littlewood-Richardson rule to compute the tensor products $\{3\} \otimes \{3\}$ and $\{21\} \otimes \{3\}$ of U(2) and U(3) irreps.

5.49 Let $\{a_{m\nu}^\dagger; m = -l, \ldots, +l, \nu = \pm 1/2\}$ denote a set of creation operators for neutrons of orbital angular momentum l and spin $1/2$. Show that the state

$$(a_{l,1/2}^\dagger a_{l-1,1/2}^\dagger \cdots a_{l-\lambda_1,1/2}^\dagger)(a_{l,-1/2}^\dagger a_{l-1,-1/2}^\dagger \cdots a_{l-\lambda_2,-1/2}^\dagger)|0\rangle, \tag{5.502}$$

where $|0\rangle$ is the vacuum state, has spin $S = (\lambda_1 - \lambda_2)/2$.

5.50 Starting from Equation (5.490), show that

$$\mathcal{K}^{(2)}_{\mathrm{u}(d)} = \sum_{\mu < \nu} (\lambda_\mu - \lambda_\nu) + \sum_\nu \lambda_\nu^2. \tag{5.503}$$

Hence, derive Equation (5.491).

5.12 Orthogonal and compact symplectic groups

Many groups feature in the development and application of nuclear models. After the symmetric and unitary groups, the next most important are the orthogonal and symplectic groups. Both the compact and non-compact symplectic groups have applications. However, in this chapter we are concerned only with the compact groups that are subgroups of a unitary group. Thus, when we say "symplectic", without qualification in this chapter, we are referring to a "compact symplectic", or "unitary-symplectic" group as it is also called. A treatment of the non-compact symplectic group is deferred to Volume 2.

Let $\mathrm{U}(d) = \mathrm{U}(d)_{\mathrm{space}}$ denote a group of unitary transformations of a d-dimensional Hilbert space, $\mathbb{H}_d$, of single-nucleon spatial wave functions, spanned by an orthonormal set $\{\psi_\nu\}$. Such spatial wave functions occur in combination with spin-isospin wave functions which, for a single nucleon, we index by $\tau = 1, \dots, 4$. The group $\mathrm{U}(d)$ is then a subgroup of a larger group, $\mathrm{U}(4d)$, of transformations of combined space, spin, and isospin wave functions which contains the subgroups

$$\mathrm{U}(4d) \supset \mathrm{U}(d) \times \mathrm{U}(4) \supset \mathrm{U}(d)_{\mathrm{space}} \times \mathrm{U}(2)_{\mathrm{spin}} \times \mathrm{U}(2)_{\mathrm{isospin}}. \tag{5.504}$$

Thus, infinitesimal generators for $\mathrm{U}(4d)$, $\mathrm{U}(d)$, and $\mathrm{U}(4)$, that span the complex extensions of Lie algebras $\mathrm{u}(4d)$, $\mathrm{u}(d)$, and $\mathrm{u}(4)$, are defined, respectively, (cf. Section 5.4) by Hermitian linear combinations of the operators

$$\hat{C}^{(4d)}_{\sigma\mu,\tau\nu} := a^\dagger_{\sigma\mu} a^{\tau\nu}, \quad \hat{C}^{(d)}_{\mu\nu} := \sum_\tau a^\dagger_{\tau\mu} a^{\tau\nu}, \quad \hat{C}^{(4)}_{\sigma\tau} := \sum_\nu a^\dagger_{\sigma\nu} a^{\tau\nu}, \tag{5.505}$$

where $a^\dagger_{\tau\nu}$ and $a^{\tau\nu}$ satisfy the anticommutation relations $\{a^{\sigma\mu}, a^\dagger_{\tau\nu}\} = \delta^\sigma_\tau \delta^\mu_\nu$ and are, respectively, creation and annihilation operators for nucleons with combined space, spin, and isospin wave functions.

The orthogonal subgroup, $\mathrm{O}(d)$, provides useful quantum numbers for the classification of spatial states by means of the subgroup chain

$$\mathrm{U}(d) \supset \mathrm{O}(d) \supset \mathrm{SO}(3), \tag{5.506}$$

where $\mathrm{SO}(3)$ is the orbital angular-momentum group. Similarly, the symplectic group $\mathrm{USp}(2d)$ provides useful quantum numbers for the classification of states of the spin-orbit subgroup $\mathrm{U}(2d) \subset \mathrm{U}(4d)$ by means of the chain

$$\mathrm{U}(2d) \supset \mathrm{USp}(2d) \supset \mathrm{SU}(2)_J, \tag{5.507}$$

where $SU(2)_J$ is the total angular-momentum group. For present purposes, these subgroups are defined in terms of the angular-momentum-zero two-nucleon-pair states that they leave invariant. We show that $O(d)$ leaves $L = 0$ two-nucleon pair states invariant and that $USp(2d)$ leaves $J = 0$ two-nucleon pair states invariant.

In working with these groups, there is a need for branching rules to decompose the representations obtained when the irreps of one group are restricted to a subgroup. For the most part, branching rules are described as the need arises. However, the $U(d) \downarrow O(d)$ branching rules are given special treatment in this section because they illustrate important general principles and the effective use of powerful Young-diagram techniques. The section concludes with a summary of corresponding results for the symplectic groups.

5.12.1 *The so(3) $\subset$ u(3) subalgebra*

The group $O(3)$ is initially defined as the subgroup of real linear transformations of the three-dimensional vector space, $\mathbb{R}^3$, that leaves invariant the norm,

$$r^2 = x^2 + y^2 + z^2, \tag{5.508}$$

of a vector $\mathbf{r} \simeq (x, y, z)$. A simple representation is obtained on the space spanned by the complex-valued functions on $\mathbb{R}^3$,

$$\psi_0(\mathbf{r}) = z, \quad \psi_{\pm 1}(\mathbf{r}) = \frac{1}{\sqrt{2}}(x \pm iy), \tag{5.509}$$

for which we also obtain the ($L = 0$) rotationally-invariant scalar product

$$\psi(\mathbf{r}) \cdot \psi(\mathbf{r}) := \sum_{m=-1}^{1} \psi_m(\mathbf{r})\, \psi_{-m}(\mathbf{r}) = r^2. \tag{5.510}$$

With $O(3)$ represented as the subgroup of $U(3)$ transformation of the above functions that leave their scalar product, $\psi \cdot \psi$, invariant, the $so(3) \subset u(3)$ subalgebra is the subset of $u(3)$ elements that have vanishing eigenvalues on these $L = 0$ pair-coupled functions. Thus, if a basis of $u(3)$ operators is chosen such that

$$\hat{C}_{\mu\nu}\psi_m = \delta_{\nu,m}\psi_\mu. \tag{5.511}$$

then

$$\hat{C}_{\mu\nu}\psi \cdot \psi := \psi_\mu\, \psi_{-\nu} + \psi_{-\nu}\, \psi_\mu, \tag{5.512}$$

it follows that the linear combinations, $\{\hat{X}_{\mu\nu} := \hat{C}_{\mu\nu} - \hat{C}_{-\nu,-\mu}\}$, of these operators satisfy the equations $\hat{X}_{\mu\nu}\psi \cdot \psi = 0$. Moreover, the linearly-independent set

$$\hat{L}_+ = \sqrt{2}(\hat{C}_{10} - \hat{C}_{0,-1}), \quad \hat{L}_0 = \hat{C}_{11} - \hat{C}_{-1,-1}, \quad \hat{L}_- = \sqrt{2}(\hat{C}_{01} - \hat{C}_{-1,0}), \tag{5.513}$$

is a basis for the complex extension of $so(3)$.

5.12.2 The so(d) ⊂ u(d) subalgebra

The above definitions generalise to any positive integer value of d. Let $O(d)$ denote the real linear transformations of the vector space, $\mathbb{R}^d$, for which a vector $\mathbf{r} \subset \mathbb{R}^d$ has Cartesian coordinates $\{x_1, x_2, \ldots, x_d\}$. When $d = 2n$ is even, we define a basis of complex-valued linear functions of $\mathbf{r} \in \mathbb{R}^d$ by

$$\psi_{\pm n}(\mathbf{r}) = \tfrac{1}{\sqrt{2}}(x_1 \pm \mathrm{i}x_2), \quad \psi_{\pm(n-1)}(\mathbf{r}) = \tfrac{1}{\sqrt{2}}(x_3 \pm \mathrm{i}x_4), \quad \cdots, \psi_{\pm 1}(\mathbf{r}) = \tfrac{1}{\sqrt{2}}(x_{2n-1} \pm \mathrm{i}x_{2n}), \tag{5.514}$$

and, when $d = 2n + 1$ is odd, we add the function

$$\psi_0(\mathbf{r}) = x_{2n+1}. \tag{5.515}$$

We then have the $SO(d)$-invariant scalar product

$$\psi(\mathbf{r}) \cdot \psi(\mathbf{r}) := \sum_{\nu=-n}^{n} \psi_\nu(\mathbf{r}) \psi_{\bar{\nu}}(\mathbf{r}) = r^2, \tag{5.516}$$

where $\psi_{\bar{\nu}} := \psi_{-\nu}$ and it is to be understood that ψ_0 is identically zero when $d = 2n$.

More generally, with the inclusion of spin and isospin degrees of freedom, we can define $SO(d)$-invariant pair-creation operators

$$a_\sigma^\dagger \cdot a_\tau^\dagger := \sum_{\nu=-d}^{d} a_{\sigma\nu}^\dagger a_{\tau\bar{\nu}}^\dagger. \tag{5.517}$$

It is then seen that, as for $SO(3)$, the subset of $U(d)$ operators which commute with these $O(d)$-invariant-pair operators are linear combinations of the operators, $\{\hat{X}_{\mu\nu} := \hat{C}_{\mu\nu}^{(d)} - \hat{C}_{\bar{\nu}\bar{\mu}}^{(d)}\}$. Thus, a basis for the complex extension of the so(d) Lie algebra is given (with arbitrary norms) by the operators

$$\hat{A}_{\mu\nu} = \hat{X}_{\mu\bar{\nu}}, \quad \mu > \nu \geq 0, \tag{5.518}$$

$$\hat{X}_{\mu\nu}. \quad \mu, \nu > 0, \tag{5.519}$$

$$\hat{B}_{\mu\nu} = \hat{X}_{\bar{\nu}\mu}, \quad \mu > \nu \geq 0, \tag{5.520}$$

with the understanding that $\nu = 0$ is excluded when d is even. The operators $\{\hat{A}_{\mu\nu}\}$ and $\{\hat{X}_{\mu\nu}; \mu > \nu\}$ are regarded as raising operators and the operators $\{\hat{B}_{\mu\nu}\}$ and $\{\hat{X}_{\mu\nu}; \mu < \nu\}$ are regarded as lowering operators. The remaining operators,

$$\hat{h}_\nu = \hat{C}_{\nu\nu}^{(d)} - \hat{C}_{\bar{\nu}\bar{\nu}}^{(d)}, \quad \nu > 0, \tag{5.521}$$

span a Cartan subalgebra for $so(n)$ which is a subalgebra of the Cartan subalgebra for $u(d)$ spanned by $\{\hat{C}_{\nu\nu}^{(d)}\}$. The corresponding root diagrams, and associated root vectors, are shown for so(4) and so(5) in Figure 5.14.

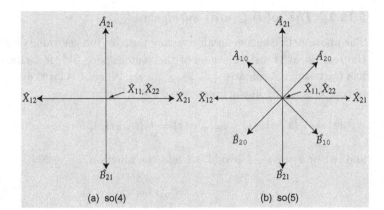

Figure 5.14: Root vectors for the so(4) and so(5) Lie algebras.

5.12.3 The labelling of SO(d) and O(d) irreps

As usual, we label an irrep of U(d) by its highest weight set in brace brackets, $\{\lambda\}$, and an SO(d) $\subset$ U(d) irrep by its highest weight set in square brackets, $[\kappa]$. The components of these weights are defined as follows. The natural ordering of the basis functions from highest to lowest is such that $\psi_n > \psi_{n-1} > \ldots \psi_{-n+1} > \psi_{-n}$, where $n := \lfloor d/2 \rfloor$.[215] Thus, to conform to the standard practice of labelling the components of a U(d) weight by a sequence $\{\lambda\} \equiv \{\lambda_1, \lambda_2, \ldots, \lambda_d\}$, we define λ_i to be an eigenvalue of the U(d) Cartan operator $\hat{C}^{(d)}_{\nu_i \nu_i}$, with

$$\left. \begin{array}{ll} \nu_i = n+1-i, & 1 \le i \le n, \\ \nu_i = n-i, & n < i \le 2n, \end{array} \right\} \quad \text{when} \quad d = 2n \text{ (even)}, \tag{5.522}$$

$$\nu_i = n+1-i, \quad 1 \le i \le 2n+1, \quad \text{when} \quad d = 2n+1 \text{ (odd)}. \tag{5.523}$$

Similarly, we define the component κ_i of an SO(d) weight κ to be an eigenvalue of the SO(d) Cartan operator $\hat{h}_{\nu_i}$, as defined by Equation (5.521), with $\nu_i = n+1-i$ for $i = 1, \ldots, n$, for $n = \lfloor d/2 \rfloor$. Thus, $\lambda_1, \lambda_2, \ldots$ are eigenvalues of $\hat{C}^{(d)}_{nn}, \hat{C}^{(d)}_{n-1,n-1}, \ldots$ and $\kappa_1, \kappa_2, \ldots$ are eigenvalues of $\hat{h}_n, \hat{h}_{n-1}, \ldots$.

For an O(d) irrep, an extra label is generally required. This is because O(d) contains elements, for which $\det(g) = -1$, in addition to the elements of its SO(d) subgroup for which $\det(g) = 1$. Thus, many irreps of O(d) occur in so-called *associated pairs*, $\hat{T}_1$ and $\hat{T}_2$, whose elements are related to one another by $\hat{T}_1(g) = \det(g)\hat{T}_2(g)$, for $g \in$ O(d).

Because the highest weight of an O(d) irrep is defined by the eigenvalues of a set of Cartan operators belonging to the Lie algebra of SO(d), associated pairs of O(d) irreps share common highest weights. It is then customary to distinguish distinct members of an associated pair of O(d) irreps, with common highest weight, by giving one an asterisk. Thus, the O(d) irreps $[\kappa]$ and $[\kappa]^*$ both restrict to a

---

[215]Note that $\lfloor x \rfloor$ denotes the largest integer less than or equal to x. Thus, $\lfloor d/2 \rfloor = d/2$ if d is even and $(d-1)/2$ if d is odd. Recall also that ψ_0 is identically zero when d is even.

common representation $[\kappa]$ of SO(d) but differ by a factor $\det(g) = \pm 1$ in their representation of an element $g \in$ O(d). For example, the U(d) irrep $\varepsilon := \{1^d\}$, is the fully antisymmetric one-dimensional irrep in which an element $g \in$ U(d) is represented simply by its determinant, i.e., $\varepsilon(g) = \det(g)$. Thus, on restriction to O(d), the U(d) irrep $\{1^d\}$ becomes the O(d) irrep $[0]^*$. For O(3), we would more commonly describe the irreps by angular momentum and parity, L^π. The O(3) irreps $[0]$ and $[0]^*$ would then correspond to $L^\pi = 0^+$ and 0^-, respectively.

It is important to be aware that, when d is even, some irreps of O(d) are reducible on restriction to SO(d). For example, the O(2) irrep $[2]$, which is two-dimensional and has basis wave functions $\{e^{\pm 2i\varphi}\}$, becomes a direct sum of two one-dimensional irreps, labelled by $m = \pm 2$, on restriction to SO(2). In this example, it is seen that the two wave functions, $e^{\pm 2i\varphi}$, are each invariant under SO(2) rotations (to within multiplication by a phase factor) but that the parity inversion operator maps $e^{2i\varphi} \to e^{-2i\varphi}$ and vice versa. It is also important to be aware that, when d is even, $[\kappa] \equiv [\kappa]^*$ for some irreps of O(d). Such irreps are said to be *self associated*. In fact, as we discuss further in Section 5.12.9, it is the self-associated irreps that are reducible on restriction to SO(d).

5.12.4 Highest-weight states for O(d) irreps

Because the chosen Cartan subalgebra for so(d), spanned by the operators $\{\hat{X}_{\mu\nu} := \hat{C}_{\mu\nu}^{(d)} - \hat{C}_{\bar\nu\bar\mu}^{(d)}\}$, is a subalgebra of the standard choice for u(d), it follows that a highest-weight state for every U(d) irrep is simultaneously a highest-weight state for an irrep of the O(d) $\subset$ U(d) subgroup. Thus, by the methods given in Section (5.11.2), we can immediately construct the highest-weight states for such O(d) irreps. We now show that a complete set of O(d) irreps can be defined in this way.

First observe that the highest-weight state, $|\phi\rangle$, for a U(d) irrep $\{\kappa\}$, with $\kappa_i = 0$ for all $i > \lfloor d/2 \rfloor$, is simultaneously a highest-weight state for an O(d) irrep with the same highest weight κ, i.e., the definition of the so(d) Cartan operators given by Equation (5.521) signifies that $\hat{C}_{\bar\nu\bar\nu}^{(d)}|\phi\rangle = 0$, for $\nu \geq 0$ and, hence, that

$$\hat{h}_\nu|\phi\rangle = \hat{C}_{\nu\nu}|\phi\rangle, \quad \nu > 0. \tag{5.524}$$

We label such an O(d) irrep by $[\kappa]$ without an asterisk. It remains then to define a highest-weight state for the associated O(d) irrep, $[\kappa]^*$.

As noted above, the U(d) irrep $\{1^d\}$ restricts to the O(d) irrep $[0]^*$. It follows that the highest-weight state for the U(d) irrep $\{1^d\} \otimes \{\kappa\}$ is simultaneously a highest-weight state for the O(d) irrep $[\kappa]^*$.

5.12.5 Particle-hole symmetry

A particle-hole symmetry, which relates irreps with distinct partition labels, arises from the fact that fermion holes have as much right to be regarded as fermions as do fermion particles. However, under unitary transformations, holes transform as the complex conjugates of particles, as can be seen by comparing the transformations

$$a^{\dagger}_{\tau\nu} \to \sum_{\mu} a^{\dagger}_{\tau\mu} U_{\mu\nu}, \tag{5.525}$$

$$a^{\tau\nu} \to \sum_{\mu} a^{\tau\mu} U^{*}_{\mu\nu}, \tag{5.526}$$

where $a^{\dagger}_{\tau\nu}$ and $a^{\tau\nu}$ are single-nucleon creation and annihilation operators. It follows that, whereas a product of fermion creation operators, $a^{\dagger}_{1\nu_p} a^{\dagger}_{1\nu_{p-1}} \ldots a^{\dagger}_{1\nu_1}$, transforms according to the irrep $\{1^p\}$, its Hermitian adjoint is a product of fermion annihilation operators $a^{1\nu_1} a^{1\nu_2} \ldots a^{1\nu_p}$ that transforms according to the contragredient (i.e., complex conjugate) of the irrep $\{1^p\}$, which we denote by $\{\overline{1^p}\}$.

The totally antisymmetric, d-fermion state

$$|1^d\rangle = a^{\dagger}_{1\nu_d} a^{\dagger}_{1\nu_{d-1}} \ldots a^{\dagger}_{1\nu_1} |0\rangle, \tag{5.527}$$

for which all $\tau = 1$ single-particle states are filled, spans the one-dimensional U(d) irrep $\{1^d\}$ for which

$$e^{\hat{X}} |1^d\rangle = \det(e^X) |1^d\rangle, \quad \text{for } X \in \mathrm{u}(d). \tag{5.528}$$

Therefore, because a partially-filled p-fermion-particle state can equally well be regarded as a $(d-p)$-fermion-hole state, i.e.,

$$a^{\dagger}_{1\nu_p} a^{\dagger}_{1\nu_{p-1}} \ldots a^{\dagger}_{1\nu_1} |0\rangle \equiv a^{1\nu_d} a^{1\nu_{d-1}} \ldots a^{1\nu_{p+1}} |1^d\rangle \tag{5.529}$$

the particle-hole relationship implies a symmetry between U(d) irreps in which $\{1^p\} \equiv \varepsilon \otimes \{\overline{1^{d-p}}\}$, where $\varepsilon(g) := \det(g)$ for any $g \in$ U(d).

More generally, a U(d) irrep $\{\lambda\}$ of is equivalent to an irrep $\varepsilon^{\lambda_1}\{\overline{\kappa}\}$ with $\{\kappa\}$ given by the Young diagram which, when rotated through 180°, complements the Young diagram for the partition $\{\lambda\}$ in the rectangle of width λ_1 and depth d. For example, as shown graphically by the equation

$$\{4^232\} \quad \equiv \quad \varepsilon^4 \otimes \{\overline{421}\}, \tag{5.530}$$

the U(5) irrep $\{4^232\}$ is identical to the irrep $\varepsilon^4\{\overline{421}\}$. This symmetry relationship is useful for determining U(d) $\downarrow$ O(d) branching rules, because O(d) can be viewed as a subgroup of real U(d) matrices and, for any $g \in$ O(d), $\varepsilon(g) = \pm 1$. Thus, for example, the U(5) irrep $\{4^232\}$ restricts to the same sum of O(5) irreps as the

U(5) irrep {421}. It is also useful for the classification of O(d) irreps, according to Littlewood's prescription.[216]

5.12.6 *Littlewood's classification of O(d) irreps*

It was determined in Section 5.12.3 that an O(d) irrep can be labelled by one or other of the symbols [κ] or [κ]*, where κ is an ordered partition of depth $p \leq d/2$. It was also determined that, if κ is an ordered partition of depth $p \leq d/2$, the highest-weight state of a U(d) irrep {κ} is simultaneously the highest-weight state for an O(d) irrep [κ]. Likewise, the highest-weight state of a U(d) irrep {1^d} $\otimes$ {κ} is simultaneously the highest-weight state for an O(d) irrep [κ]*. We now show, by use of the above particle-hole symmetry relationships, that highest-weight states for [κ]* irreps can be identified with those of U(d) highest-weight states in a more useful way.

Claim 5.5 *If κ denotes an ordered partition of depth $p = \tilde{\kappa}_1$ and κ' denotes an associated partition whose columns are identical to those of κ except for the first column which has depth $l\tilde{\kappa}'_1) = d - p$, then, if $p \leq d/2$, the highest-weight state for a U(d) irrep {κ} is simultaneously a highest-weight state for an O(d) irrep [κ] and {κ'} is simultaneously a highest-weight state for an O(d) irrep [κ]*.*

This claim is a generalisation of the above observation that the one-dimensional U(d) irrep

$$\{1^{d-p}\} \equiv \varepsilon \otimes \{\overline{1^p}\}, \tag{5.531}$$

restricts to the O(d) irrep [1^p]*, when $p \leq d/2$.

If $q = \kappa_1$, then the U(d) irrep {κ} can be regarded as the stretched component (i.e., the irreducible component of greatest highest weight) of the tensor product

$$\{\kappa\} \simeq \left[\{1^{\tilde{\kappa}_1}\} \otimes \{1^{\tilde{\kappa}_2}\} \otimes \cdots \{1^{\tilde{\kappa}_q}\}\right]_{\text{stretched}}. \tag{5.532}$$

Thus, for example, in terms of Young diagrams,

(5.533)

Similarly, {κ'}, as defined in the claim, can be expressed as the stretched tensor product

$$\{\kappa'\} = \left[\{1^{d-\tilde{\kappa}_1}\} \otimes \{1^{\tilde{\kappa}_2}\} \otimes \cdots \{1^{\tilde{\kappa}_q}\}\right]_{\text{stretched}}. \tag{5.534}$$

It follows, from Equation (5.531), that

$$\{\kappa'\} = \varepsilon \otimes \left[\{\overline{1^{\tilde{\kappa}_1}}\} \otimes \{1^{\tilde{\kappa}_2}\} \otimes \cdots \{1^{\tilde{\kappa}_q}\}\right]_{\text{stretched}} \tag{5.535}$$

[216]Littlewood, *op. cit.* Footnote 208 on Page 346.

and that its highest weight is also that of the O(d) irrep $[\kappa]^*$. □

For example, the Young diagram for the O(7) irrep $[\kappa] = [32]$ has a first column of depth $p = \tilde{\kappa}_1 = 2$. Extending it to depth $d - p = 5$ gives $[\kappa'] = [321^3] \equiv [32]^*$: viz.

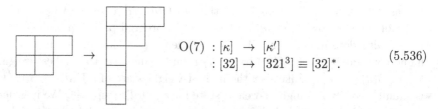

$$O(7) \; : [\kappa] \; \rightarrow \; [\kappa']$$
$$: [32] \; \rightarrow \; [321^3] \equiv [32]^*. \qquad (5.536)$$

The above results suggest that an alternative to labelling O(d) irreps by $[\kappa]$ and $[\kappa]^*$, is to label them, respectively, by $[\kappa]$ and $[\kappa']$, where $[\kappa'] \equiv [\kappa]^*$ is related to $[\kappa]$ as defined in the claim. This is the labelling scheme suggested by Littlewood.[217] Note that, to obtain Littlewood labels (without asterisks) for a complete set of O(d) irreps, only the depth of the first column need be allowed to exceed $d/2$. Moreover, it is always required that $\tilde{\kappa}_1 \geq \tilde{\kappa}_2$. The complete set of O($d$) irreps is then labelled by the regular partitions $[\kappa]$ subject to the restriction (see Exercise 5.51)

$$\tilde{\kappa}_1 + \tilde{\kappa}_2 \leq d. \qquad (5.537)$$

It is also seen that if d is even and $\tilde{\kappa}_1 = d/2$, then $[\kappa] \equiv [\kappa]^*$ and the irrep $[\kappa]$ is self-associated.[218]

5.12.7 U(d) ↓ O(d) *branching rules*

It has been determined, from the definition of O(d) weights given in Section 5.12.3, that the highest-weight state for a U(d) irrep $\{\lambda\}$ is also a highest-weight state for an irrep of the O(d) ⊂ U(d) subgroup. For the moment, we simply label the O(d) irrep with this highest weight by $[\lambda]$ with the recognition that, if $\tilde{\lambda}_1 + \tilde{\lambda}_2 > d$, this is not a standard labelling. Nevertheless, it identifies one O(d) irrep in the U(d) ↓ O(d) branching rule. The challenge is to determine the others.

We first consider the U(d) ↓ O(d) branching rule for the U(d) irrep $\{2\}$ with Hilbert space given by the subspace, $\mathbb{H}^{\{2\}} \subset \mathbb{H}^2$, of symmetric two-particle wave functions in the tensor product space $\mathbb{H}^2 := \mathbb{H} \otimes \mathbb{H}$, where $\mathbb{H}$ is a d-dimensional Hilbert space spanned by an orthonormal set of single-particle spatial wave functions, $\{\psi_\nu\}$. The highest-weight state for this U(d) irrep is also the highest-weight state for the O(d) irrep $[2]$ for which the carrier space, $\mathbb{H}_2^{[2]}$, is a subspace of $\mathbb{H}^{\{2\}}$. However, $\mathbb{H}^{\{2\}}$ also contains the O(d)-invariant wave function

---

[217]Littlewood, *op. cit.* Footnote 208 on Page 346.

[218]On restriction to SO($2n$), a self-associated O($2n$) irrep reduces to a direct sum of the two irreps of SO($2n$) having highest weights with components given, for common values of $\kappa_1 \geq \kappa_2 \geq \cdots \geq \kappa_n > 0$, by $[\kappa_1, \kappa_2, \ldots, \kappa_{n-1}, \kappa_n]$ and $[\kappa_1, \kappa_2, \ldots, \kappa_{n-1}, -\kappa_n]$.

$\Phi_0(1,2) := \frac{1}{\sqrt{d}} \sum_\nu \psi_\nu(1)\psi_{\bar{\nu}}(2)$. It follows that $\mathbb{H}^{\{2\}}$ is a direct sum of the two $O(d)$-invariant subspaces

$$\mathbb{H}^{\{2\}} = \mathbb{H}_2^{[2]} \oplus \mathbb{H}_2^{[0]}, \qquad (5.538)$$

where $\mathbb{H}_2^{[0]} \subset \mathbb{H}^{\{2\}}$ is the 1-dimensional subspace spanned by $\Phi_0(1,2)$ and $\mathbb{H}_2^{[2]}$ is its orthogonal complement. (The subscript N indicates that $\mathbb{H}_N^{[\kappa]}$ is an N-particle space.)

This example for the irrep $\{2\}$ serves as a prototype for a general identification of $O(d)$-invariant subspaces of the Hilbert space for a $U(d)$ irrep by contractions. The relevant contractions parallel those for constructing tensors of reduced rank by contraction with a metric tensor. The parallel follows from the observation that wave functions of the Hilbert space for an N-particle $U(d)$ irrep can be identified with symmetrised rank-N tensors. In accordance with the expansions given in Section 5.11.3, an arbitrary N-particle wave function, in the N-particle Hilbert space, $\mathbb{H}^{\{\lambda\}}$, for a $U(d)$ irrep $\{\lambda\}$ with $\lambda \vdash N$, is a rank-N tensor with components, $\{\Psi^{\nu_1 \nu_2 \cdots \nu_N}\}$, given by the coefficients in the expansion

$$\Psi(1,2,\cdots,N) := \sum_\nu \Psi^{\nu_1 \nu_2 \cdots \nu_N} \psi_{\nu_1}(1)\psi_{\nu_2}(2)\cdots\psi_{\nu_N}(N). \qquad (5.539)$$

For example, a general two-particle wave function, $\Psi \in \mathbb{H}^{\{2\}}$, is a symmetric rank-2 tensor with components, $\{\Psi^{\nu_1 \nu_2}\}$, given by the expansion

$$\Psi(1,2) = \sum_{\nu_1 \nu_2} \Psi^{\nu_1 \nu_2} \psi_{\nu_1}(1)\psi_{\nu_2}(2). \qquad (5.540)$$

Thus, the $O(d)$-scalar component of a general wave function, $\Psi \in \mathbb{H}^{\{2\}}$, is given by the projection

$$\mathbb{H}^{\{2\}} \to \mathbb{H}_2^{[0]} \;:\; \Psi \to \Psi^{[0]} = \sum_\nu \Psi^{\nu\bar{\nu}}\Phi_0 \qquad (5.541)$$

corresponding to the contraction,

$$\{\Psi^{\nu_1 \nu_2}\} \to \sum_{\nu_1 \nu_2} \Psi^{\nu_1 \nu_2} \delta_{\nu_1, \bar{\nu}_2}, \qquad (5.542)$$

of the tensor, $\{\Psi^{\nu_1 \nu_2}\}$, with two indices to a scalar with no uncontracted indices. The complementary orthogonal projection,

$$\mathbb{H}^{\{2\}} \to \mathbb{H}_2^{[2]} \;:\; \Psi \to \Psi^{[2]} = \Psi^{\{2\}} - \Psi^{[0]}, \qquad (5.543)$$

has the property that the contraction of $\Psi^{[2]}$ is zero, i.e., $\Psi^{[2]}$ has no $O(d)$-invariant component. The result is an expression of the Hilbert space $\mathbb{H}^{\{2\}}$ as a direct sum of two $O(d)$-invariant subspaces, $\mathbb{H}_2^{[2]} \oplus \mathbb{H}_2^{[0]}$, in accordance with Equation (5.538).

Because a two-particle wave function $\Psi \in \mathbb{H}^{\{1^2\}}$ is antisymmetric, the sum $\sum_\nu \Psi^{\nu\bar{\nu}} = \sum_\nu \Psi^{\bar{\nu}\nu}$ is zero which implies that $\mathbb{H}^{\{1^2\}}$ has no $O(d)$ scalar component. It follows that

$$\mathbb{H}^{\{1^2\}} = \mathbb{H}_2^{[1^2]}. \tag{5.544}$$

However, a wave function $\Psi \in \mathbb{H}_3^{\{3\}}$, being fully symmetric, has the contraction

$$\Psi(1,2,3) \to \frac{1}{\sqrt{3}} \left(\psi(1)\Phi^{[0]}(2,3) + \psi(2)\Phi^{[0]}(3,1) + \psi(3)\Phi^{[0]}(1,2) \right), \tag{5.545}$$

where $\psi(i) \propto \sum_\nu \Psi^{\mu\nu\bar{\nu}}\psi_\mu(i)$. These contracted wave functions transform in the same way as single-particle wave functions. It follows that

$$\mathbb{H}^{\{3\}} = \mathbb{H}_3^{[3]} \oplus \mathbb{H}_3^{[1]}. \tag{5.546}$$

The above results imply the branching rules

$$U(d) \downarrow O(d) : \{2\} \downarrow [2] \oplus [0], \tag{5.547}$$
$$: \{1^2\} \downarrow [1^2], \tag{5.548}$$
$$: \{3\} \downarrow [3] \oplus [1], \tag{5.549}$$

consistent with a general algorithm, devised by Littlewood,[219] which we now outline.

The principle underlying Littlewood's algorithm is the identification of all possible contractions of the basis states of a $U(d)$ irrep when viewed as components of an $O(d)$ tensor. The first step is the identification of all $U(d)$ irreps which, on restriction to $O(d)$, contain a copy of the identity irrep. The above considerations show that the irreps $\{0\}$ and $\{2\}$ are among this set. Littlewood showed that the complete set is given by the irreps contained in the plethysm (see Section 5.11.10)

$$D := \bigoplus_{k=0}^{\infty} \{2\} \circledP \{k\} = \{0\} \oplus \{2\} \oplus \{4\} \oplus \{2^2\} \oplus \{6\} \oplus \{42\} \oplus \{2^3\} \oplus \dots. \tag{5.550}$$

This set is known as the D series of irreps; it is the set of partitions of even numbers with only even parts. It follows that every $U(d)$ irrep in the D series can be fully contracted to an $O(d)$ identity irrep and, conversely, no irrep that is not in the D series can be fully contracted.

Observe that if, for any even-n integer in the range $0 \leq n \leq N$, the Hilbert space $\mathbb{H}^{\{\lambda\}}$, for a $U(d)$ irrep $\{\lambda\}$ with $\lambda \vdash N$, were to be expressed as a direct sum of coupled-tensor products of the form $\left[\mathbb{H}^{\{\kappa\}} \otimes \mathbb{H}^{\{\delta\}} \right]^{\{\lambda\}}$, with $\delta \vdash n$ labelling an irrep in the D series and $\kappa \vdash N - n$, then on restriction of the $U(d)$ irrep $\{\lambda\}$ to $O(d)$ there will be a term in the expansion in which the irrep $\{\delta\}$ is contracted to the $O(d)$ scalar irrep $[0]$ and an irrep $[\kappa]$ remains. It is then seen that these $[\kappa]$

---

[219]Littlewood, *op. cit.* Footnote 208 on Page 346.

irreps are those that appear in the expansion of the quotient $\{\lambda/\delta\} = \bigoplus_\kappa \Gamma_{\kappa\delta}^\lambda \{\kappa\}$. Thus, the $\mathrm{U}(d) \downarrow \mathrm{O}(d)$ branching rule is given by the expression

$$\mathrm{U}(d) \downarrow \mathrm{O}(d) \; : \; \{\lambda\} \downarrow \bigoplus_{\delta \in D} [\lambda/\delta] = \bigoplus_{\delta \in D} \bigoplus_\kappa \Gamma_{\kappa\delta}^\lambda [\kappa], \qquad (5.551)$$

where $\Gamma_{\kappa\delta}^\lambda$ is a Littlewood-Richardson coefficient. However, this equation needs some interpretation for several reasons.

One need for interpretation of Equation (5.551) arises because the labels for some $\mathrm{O}(d)$ irreps in this expression are non-standard. Another need arises because the contracted subspaces given by the above algorithm are not necessarily distinct when they share a common $\mathrm{O}(d)$ highest weight; thus, there may be overcounting. A remarkable addendum to the above algorithm is that the required adjustments to obtain meaningful branching rules in standard notation are achieved by so-called *modification rules*.

5.12.8 *Modification rules for non-standard O(d) labels*

Suitable modification rules have been given by Newell,[220] by Littlewood,[221] and by King.[222,223] According to King's rules, which are the simplest, each $\mathrm{O}(d)$ label $[\lambda]$ of depth greater than $d/2$ is to be replaced by an equivalent label

$$[\lambda] \to (-1)^{x-1} [\lambda - h]^*, \quad h = 2p - d, \qquad (5.552)$$

where $p = \tilde{\lambda}_1$ is the *depth* of the partition $\lambda = (\lambda_1, \ldots, \lambda_p)$ (i.e., the depth of the first column of the corresponding Young diagram). The Young diagram corresponding to the label $[\lambda - h]$ is constructed by removing a continuous boundary strip of length h from the Young diagram $[\lambda]$, starting from the foot of the first column and ending in column x. If the diagram that results is a not a regular Young diagram, then it does not correspond to an $\mathrm{O}(d)$ irrep and is discarded. For example, the non-standard partition $[43^22]$ for $\mathrm{O}(5)$ is of depth $p = 4$, so that $h = 3$. However, removal of a boundary strip of length 3 from the Young diagram for this partition results in the non-regular diagram

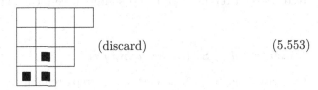 (discard) $\qquad (5.553)$

which does not correspond to any $\mathrm{O}(5)$ irrep. Thus, it is discarded.

[220]Newell M.J. (1951), *Proc. Roy. Irish Acad.* **A54**, 153.
[221]Littlewood D.E. (1958), *Can. J. Math.* **10**, 17.
[222]King R.C. (1971), *J. Math. Phys.* **12**, 1588.
[223]King, *op. cit.* Footnote 206 on Page 343.

If the diagram that emerges, following the above boundary strip removal, is a regular Young diagram, but does not correspond to a standard labelling for an O(d) irrep, the replacement must be repeated. For example, the non-standard partition [432^2] for O(5) is of depth $p = 4$ and, again, we have to remove a boundary strip of length 3 from the corresponding Young diagram. This boundary strip ends in column 2, and so we add a negative sign. The result is seen to be a regular Young diagram, but it still does not yield a standard label. However, by a repeated application of the modification rule we obtain

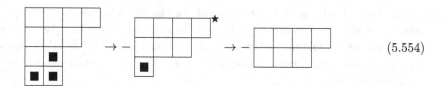

$$\tag{5.554}$$

or, in terms of partition labels,

$$[432^2] \rightarrow -[431]^* \rightarrow -[43]. \tag{5.555}$$

It is important to understand the meaning of the negative sign in front of the O(5) label [43] in Equation (5.555). The result of any branching rule must be a sum of irreps with only non-negative-integer multiplicities. Thus, if a term in the above algorithm should give a term with a negative coefficient, there must be other similar terms such that their sum has only non-negative coefficients. Thus, while the leading term [432^2] in the U(5)↓O(5) branching rule for the U(5) irrep {432^2} is equivalent to $-[43]$, there must also be a [43] term with a positive coefficient arising from the [{432^2}/{2^2}] contraction to cancel it.

Note that King's modification rules are consistent with Littlewood's alternative labelling of O(d) irreps. For example, application of King's modification rules restores the irrep label [431] to its alternative form, [43]*.

5.12.9 O(d) ↓ SO(d) *branching rules*

On restriction to SO(d), most O(d) irreps remain irreducible. In fact, for every partition λ of depth strictly less than $d/2$, both of the O(d) irreps [λ] and [λ]* restrict to a single SO(d) irrep [λ]. Exceptions are the self-associated O(d) irreps which occur when d is even. It is seen from the characterisation of associated O(d) irreps, given in Section 5.12.6, that an O($2m$) irrep [κ] = [$\kappa_1, \kappa_2, \ldots, \kappa_m$] is self-associated if and only if $\kappa_m \neq 0$. Such a self-associated O($2m$) irrep restricts to a direct sum of SO($2m$) irreps:[224,225]

$$O(2m) \downarrow SO(2m) : [\kappa_1, \ldots, \kappa_{m-1}, \kappa_m]$$
$$\downarrow [\kappa_1, \ldots, \kappa_{m-1}, \kappa_m] \oplus [\kappa_1, \ldots, \kappa_{m-1}, -\kappa_m]. \quad (5.556)$$

5.12.10 *The USp(2d) symplectic group*

The (compact) symplectic group USp($2d$), also called a unitary-symplectic group, is defined as the subgroup of the unitary group U($2d$) that leaves an antisymmetric scalar product invariant.

Suppose, for example, that a set of $2d = 2j + 1$ single-particle wave functions spans a space of combined spin and orbital angular momentum j, where $2j > 0$ is an odd integer. An antisymmetric scalar product of two sets of angular-momentum j wave functions, $\{\psi_{\sigma jm}\}$ and $\{\psi_{\tau jm}\}$, is then defined as the $J = 0$ combination

$$\psi_{\sigma j} \cdot \psi_{\tau j} := \sqrt{2j+1} \sum_{m=-j}^{j} (j, -m\, jm|00)\, \psi_{\sigma jm}\psi_{\tau j,-m}$$
$$= \sum_{m}(-1)^{j+m}\psi_{\sigma jm}\psi_{\tau j,-m}, \quad (5.557)$$

where σ and τ are isospin labels. If, to simplify the notation, we replace pair of indices jm by a single index $\mu \equiv (jm)$ and define

$$a^\dagger_{\sigma\mu} := a^\dagger_{\sigma jm}, \quad a^\dagger_{\sigma\bar{\mu}} := (-1)^{j+m}a^\dagger_{\sigma j,-m}, \quad (5.558)$$

then the creation operator for a $J = 0$ nucleon-pair is given by the dot product

$$a^\dagger_\sigma \cdot a^\dagger_\tau = \sum_\mu a^\dagger_{\sigma\mu}a^\dagger_{\tau\bar{\mu}}. \quad (5.559)$$

Corresponding annihilation operators, with lower indices, defined by

$$a_{\sigma\mu} := a_{\sigma jm}, \quad a_{\sigma\bar{\mu}} := (-1)^{j+m}a_{\sigma j,-m} = a^{\sigma jm}, \quad (5.560)$$

satisfy the anticommutation relations

$$\{a_{\sigma\bar{\mu}}, a^\dagger_{\tau\nu}\} = \delta_{\sigma,\tau}\delta_{\mu,\nu}, \quad \{a_{\sigma\mu}, a^\dagger_{\tau j\bar{\nu}}\} = -\delta_{\sigma,\tau}\delta_{\mu,\nu}. \quad (5.561)$$

Thus, in addition to the $J = 0$ nucleon-pair creation operator, we can also define the scalar products:

$$a_\sigma \cdot a_\tau := \sum_\nu a_{\sigma\nu}a_{\tau\bar{\nu}} = \sum_\nu a^{\tau\bar{\nu}}a^{\sigma\nu}, \quad a^\dagger_\sigma \cdot a_\tau := \sum_\nu a^\dagger_{\sigma\nu}a_{\tau\bar{\nu}}, \quad (5.562)$$

[224]Louck J.D. and Galbraith H.W. (1972), *Rev. Mod. Phys.* **44**, 540.

[225]Murnaghan F.D. (1938), *The Theory of Group Representations* (The Johns Hopkins Press, Baltimore).

where $a_\sigma \cdot a_\tau$ is a $J = 0$ nucleon-pair annihilation operator and $\hat{n}_\tau := a_\tau^\dagger \cdot a_\tau$, for example, is a nucleon number operator. A scalar (dot) product is similarly defined for any $2d = \sum_k (2j_k + 1)$.

In pairing-force theory, barred indices are used to denote time-reversed single-particle states (see Section 6.2.1). Thus, the maps

$$\hat{T} : a_{\tau\nu}^\dagger \to a_{\tau\bar\nu}^\dagger, \quad \hat{T} : a_{\tau\bar\nu}^\dagger \to -a_{\tau\nu}^\dagger. \tag{5.563}$$

are regarded as time-reversal operations. Note, that because $2j$ is odd, all single-particle states occur in time-reversed pairs with indices $(\nu, \bar\nu)$; there are no unpaired indices. It is then convenient to assign a positive value to one member of a pair and a negative value to its partner.

By definition, $\mathrm{USp}(2j + 1)$ is the subgroup of $\mathrm{U}(2j + 1)$ elements which leave the $J = 0$ scalar products of Equation (5.559) invariant. The Lie algebra $\mathrm{usp}(2d)$ of the group $\mathrm{USp}(2d)$ is then the subset of $\mathrm{u}(2d)$ operators which commute with the $a_\sigma^\dagger \cdot a_\tau^\dagger$ ($J{=}0$) scalars. It is spanned by Hermitian combinations of the operators

$$\hat{A}_{\mu\nu} = \hat{X}_{\mu\bar\nu}, \quad \mu \geq \nu > 0, \tag{5.564}$$

$$\hat{X}_{\mu\nu}. \qquad \mu, \nu > 0, \tag{5.565}$$

$$\hat{B}_{\mu\nu} = \hat{X}_{\bar\nu\mu}, \quad \mu \geq \nu > 0, \tag{5.566}$$

where

$$\hat{X}_{\mu\nu} := \hat{C}_{\mu\nu} - \hat{C}_{\bar\nu\bar\mu}, \quad \hat{C}_{\mu\nu} := \sum_\sigma a_{\sigma\mu}^\dagger a_{\sigma\bar\nu}. \tag{5.567}$$

The operators $\{\hat{A}_{\mu\nu}\}$ and $\{\hat{X}_{\mu\nu}; \mu > \nu\}$ are regarded as raising operators and the operators $\{\hat{B}_{\mu\nu}\}$ and $\{\hat{X}_{\mu\nu}; \mu < \nu\}$ are regarded as lowering operators. The remaining operators,

$$\hat{h}_\nu = \hat{C}_{\nu\nu} - \hat{C}_{\bar\nu\bar\nu}, \quad \nu > 0, \tag{5.568}$$

span a Cartan subalgebra for $\mathrm{usp}(2d)$ which is a subalgebra of the Cartan subalgebra for $\mathrm{u}(d)$ spanned by $\{\hat{C}_{\nu\nu}\}$.

This $\mathrm{usp}(2d)$ Lie algebra has much in common with $\mathrm{so}(d)$ but some essential differences. One difference results from the negative sign in the second term of Equation (5.563). Thus, for example, whereas raising operators for $\mathrm{so}(d)$ are given by $\hat{A}_{\mu\nu} = \hat{C}_{\mu\bar\nu} - \hat{C}_{\nu\bar\mu}$, the corresponding operators for $\mathrm{usp}(2d)$ are $\hat{A}_{\mu\nu} = \hat{C}_{\mu\bar\nu} + \hat{C}_{\nu\bar\mu}$ An important common feature is that, as for $\mathrm{so}(d) \subset \mathrm{u}(d)$, the Cartan subalgebra for $\mathrm{usp}(2d)$ is a subalgebra of that for $\mathrm{u}(2d)$. Thus, a highest-weight state for a $\mathrm{U}(2d)$ irrep is simultaneously a highest-weight state for a $\mathrm{USp}(2d)$ irrep.

A significant difference between the group $\mathrm{USp}(2d)$ and the group $\mathrm{O}(d)$ is that whereas $\mathrm{USp}(2d)$ is a *simple* Lie group,[226] $\mathrm{O}(d) \supset \mathrm{O}(1) \times \mathrm{SO}(d)$, is neither simple

---

[226]A Lie group, and its Lie algebra, are said to be *semi-simple* if the Lie algebra contains no first-order invariant, i.e., no element that commutes with all other elements. They are said to be *simple* if, in addition, the Lie algebra is not a direct sum of semi-simple Lie algebras. More precise

nor semi-simple. As a consequence, USp($2d$) irreps are completely identified by their highest weights; in contrast to those of O(d) which require a distinction to be made between two irreps $[\kappa]$ and $[\kappa]^*$ that have the same highest weight κ but transform differently under an O(1) inversion. An irrep of USp($2d$), labelled by its highest weight, is denoted by an ordered partition in angle brackets, $\langle \kappa \rangle = \langle \kappa_1, \kappa_2, \dots, \rangle$, where κ_ν is the eigenvalue of $\hat{X}_{\nu\nu}$ for $1 \leq \nu \leq d$.

5.12.11 U($2d$) $\downarrow$ USp($2d$) *branching rules*

The U($2d$) $\downarrow$ USp($2d$) branching rules are derived by use of contraction mappings in parallel with those used for U(d) $\downarrow$ O(d). Because the basic antisymmetric tensor, which is defined to be invariant under USp($2d$), transforms according to an irrep $\{1^2\}$ under U($2d$), the contractions now lead to the U($2d$) $\downarrow$ USp($2d$) branching rule for an irrep $\{\lambda\}$ as a sum of USp($2d$) representations obtained by factoring out from $\{\lambda\}$ all the U($2d$) irreps in the series

$$B = \bigoplus_{k=0}^{\infty} \{1^2\} \, \circledD \, \{k\} = \sum_{\delta \in D} \{\tilde{\delta}\} = \{0\} \oplus \{1^2\} \oplus \{1^4\} \oplus \{2^2 2^2\} \oplus \{1^6\} \oplus \{2^2 1^2\} \oplus \dots ,$$

$$(5.569)$$

where $\{k\}$ is again a completely symmetric (i.e., one-row) U(d) irrep. Each term in this series is the conjugate of an element δ in the D series of Equation (5.550). Thus, we now have the branching rule

$$\text{U}(2d) \downarrow \text{USp}(2d) \ : \ \{\lambda\} \downarrow \bigoplus_{\delta \in D} \langle \lambda/\tilde{\delta} \rangle = \bigoplus_{\delta \in D} \bigoplus_{\kappa} \Gamma^{\lambda}_{\kappa \tilde{\delta}} \langle \kappa \rangle . \qquad (5.570)$$

As for O(d), modification rules must be applied to interpret this branching rule. The modification rules given by King[227],[228] are again defined by a process of boundary strip removals to replace a non-standard USp($2d$) label, $\langle \lambda \rangle$, by a new label,

$$\langle \lambda \rangle \to (-1)^x \langle \lambda - h \rangle, \quad h = 2p - 2d - 2, \qquad (5.571)$$

where $p = \tilde{\lambda}_1$ is the *depth* of the partition $\lambda = (\lambda_1, \dots, \lambda_p)$, h is the length of the continuous boundary strip to be removed from the Young diagram, starting from the foot of the first column, and x is the column in which the boundary strip ends (see examples for O(d) in Section 5.12.8). This replacement is repeated until the result either corresponds to an irregular Young diagram, in which case it is discarded, or it is a standard label for a USp($2d$) irrep.

---

definitions are given, for example, by Wybourne, *op. cit.* Footnote 15 on Page 176.

[227]King, *op. cit.* Footnote 222 on Page 361.

[228]King, *op. cit.* Footnote 206 on Page 343.

5.12.12 Dimensions of O(d) and USp(d) irreps

The dimensions of O(d) and USp(d) irreps are given[229] by formulae similar to those for S$_N$ and U(d) irreps presented in Section 5.11.4.

The dimension of the O(d) irrep [λ] is given by

$$\dim_d[\lambda] = \frac{1}{H^\lambda} \prod_{i \geq j} (d + \lambda_i + \lambda_j - i - j) \prod_{i < j} (d - \tilde{\lambda}_i - \tilde{\lambda}_j + i + j - 2), \qquad (5.572)$$

where the products are taken over all pairs (i, j), with the restriction indicated, that label boxes in the Young diagram for the partition λ and H^λ is the product of hook lengths as defined by Equation (5.428). For example, the O(5) irrep [53] has dimension

$$\dim[53] = \frac{13 \cdot 10 \cdot 7 \cdot 2 \cdot 3 \cdot 5 \cdot 6 \cdot 4}{6 \cdot 5 \cdot 4 \cdot 2 \cdot 3 \cdot 2} = 455. \qquad (5.573)$$

The dimension of the USp(d) irrep $\langle \lambda \rangle$ is given by

$$\dim_d\langle \lambda \rangle = \frac{1}{H^\lambda} \prod_{i > j} (d + \lambda_i + \lambda_j - i - j + 2) \prod_{i \leq j} (d - \tilde{\lambda}_i - \tilde{\lambda}_j + i + j). \qquad (5.574)$$

For example, the USp(6) irrep $\langle 532 \rangle$ has dimension

$$\dim\langle 532 \rangle = \frac{13 \cdot 11 \cdot 8 \cdot 2 \cdot 3 \cdot 5 \cdot 7 \cdot 8 \cdot 4 \cdot 6}{7 \cdot 6 \cdot 4 \cdot 2 \cdot 4 \cdot 3 \cdot 2} = 5720. \qquad (5.575)$$

5.12.13 Groups dual to O(d) and USp(d)

Because the orthogonal and unitary symplectic groups are subgroups of one-body unitary transformations that, respectively, leave $L = 0$ and $J = 0$ two-particle states invariant, it follows that the creation and annihilation operators of such two-particle states are, respectively, infinitesimal generators of groups that commute with O(d) and USp($2d$). Thus, together with USp($2d$), the group generated by the $J = 0$ two-nucleon creation and annihilation operators forma pair of direct product groups, and together with O(d), the group generated by the $L = 0$ two-nucleon creation and annihilation operators form another pair of direct product groups. As we show explicitly in Chapter 6, these pairs of commuting groups have dual representations on many-nucleon Hilbert spaces in accord with Definition 5.1. Their duality relationships, developed in Chapter 6, play an essential role in revealing how the pair-coupling models relate to the shell-model coupling schemes of Section 5.13. In this section, we identify prototypes of such duality relationships which have extensions to larger multi-shell spaces.

---

[229]El Samra N. and King R.C. (1979), *J. Phys. A: Math. Gen.* **12**, 2317.

(i) *The $SU(2)$ quasi-spin and $USp(2j+1)$ duality relationship*

Within the space of nucleons of a single type (neutrons or protons but not both labelled by τ) in a single-j shell, there is a single $J = 0$ pair-creation operator and a single $J = 0$ pair-annihilation operator, with dot product defined by Equation (5.559), i.e.,

$$\hat{S}_+ := \tfrac{1}{2} a_\tau^\dagger \cdot a_\tau^\dagger, \quad \hat{S}_- := \tfrac{1}{2} a_\tau \cdot a_\tau, \tag{5.576}$$

The commutator of these operators generates a new $J = 0$ operator,

$$[\hat{S}_+, \hat{S}_-] = \tfrac{1}{2} \big(a_\tau^\dagger \cdot a_\tau + a_\tau \cdot a_\tau^\dagger \big) := 2\hat{S}_0. \tag{5.577}$$

The operators $\hat{S}_+, \hat{S}_-$ and $\hat{S}_0$ span an su(2) *quasi-spin algebra* which, as shown in Section 6.3, is the SGA (spectrum generating algebra) for the standard pairing model. The elements of this su(2) algebra are the infinitesimal generators of a corresponding $SU(2)$ quasi-spin group whose representations on the space of nucleons of a single type in a single-j shell are determined to be dual to those of $USp(2j+1)$, i.e., all the states with a common set of $USp(2j+1)$ quantum numbers span an irrep of $SU(2)$ and vice versa.

By considering the commutation relations of elements of the su(2) quasi-spin algebra with those of u(2j + 1), it is also shown, in Section 6.5, that whereas the complex extension of u(2j + 1) is spanned by a set of operators, $\{\hat{C}_{JM} := [a_{\tau j}^\dagger \otimes a_{\tau j}]_{JM}\}$ with angular momentum J taking all values from $0, \ldots, 2j$, the complex extension of the subalgebra, usp(2j + 1) $\subset$ u(2j + 1), is spanned by the subset with J taking only odd values.

(ii) *The $SO(5)$ and $USp(2j+1)$ duality relationship*

When both neutrons and protons are present, within the space of a single-j shell, the $(J = 0)$-coupled pair-creation and annihilation operators extend to the set

$$\hat{S}_+^{nn} := \tfrac{1}{2} a_n^\dagger \cdot a_n^\dagger, \quad \hat{S}_-^{nn} := -\tfrac{1}{2} a_n \cdot a_n, \tag{5.578}$$

$$\hat{S}_+^{pp} := \tfrac{1}{2} a_p^\dagger \cdot a_p^\dagger, \quad \hat{S}_-^{pp} := -\tfrac{1}{2} a_p \cdot a_p, \tag{5.579}$$

$$\hat{S}_+^{np} := \tfrac{1}{2} \big(a_n^\dagger \cdot a_p^\dagger + a_p^\dagger \cdot a_n^\dagger \big), \quad \hat{S}_-^{np} := -\tfrac{1}{2} \big(a_n \cdot a_p + a_p \cdot a_n \big), \tag{5.580}$$

whose commutation relations generate an so(5) Lie algebra. As shown in Section 6.6, this Lie algebra is an SGA for an isospin-conserving neutron-proton pairing model whose elements are the infinitesimal generators of an $SO(5)$ group[230] and whose representations are dual to those of $USp(2j+1)$ on the space of neutrons and protons in a single-j shell.

---

[230]We use the notation $SO(5)$ here, instead of SO(5), to avoid confusion with orthogonal subgroups of U(2j + 1).

(iii) *The $\mathcal{SO}(8)$ and $O(2l+1)$ duality relationship*

If $U(2l+1)$ is the group of unitary transformations of a space of single-particle spatial wave functions of orbital angular momentum l, then the standard representation of the subgroup $O(2l+1) \subset U(2l+1)$, as defined in Sections 5.12.2 and 5.12.3, is the set of $U(2l+1)$ transformations that leave the $L = 0$ pair-coupled products, $\psi_\sigma \cdot \psi_\tau$, invariant, where σ and τ run over the four spin-isospin states of a nucleon. Thus, the corresponding $L = 0$ pair-creation and annihilation operators,

$$\hat{\mathcal{A}}_{\sigma\tau} := a_\sigma^\dagger \cdot a_\tau^\dagger, \quad \hat{\mathcal{B}}_{\sigma\tau} := a_\sigma \cdot a_\tau, \tag{5.581}$$

commute with the infinitesimal generators of $O(2l + 1)$. As shown in Section 6.7, these pair operators generate an so(8) Lie algebra, which is the SGA for an $L = 0$ pairing model. Moreover, its elements are the infinitesimal generators of an $\mathcal{SO}(8)$ group, whose representations are dual to those of $O(2l+1)$ on the space of neutrons and protons in a single-l shell. This $\mathcal{SO}(8)$ group contains the $\mathcal{SU}(4)$ supermultiplet group (see Section 5.13.2) as a subgroup.

Exercises

5.51 Observe that if $\tilde{\kappa}_1 + \tilde{\kappa}_2 > d$ and $\tilde{\kappa}_1' = d - \tilde{\kappa}_1$, $\tilde{\kappa}_i' = \tilde{\kappa}_i$ for $i > 1$, then $\tilde{\kappa}_1' < \tilde{\kappa}_2'$ so that κ' is not a regular partition. Show that any $O(d)$ irrep labelled by an ordered partition $[\kappa]$ with $\tilde{\kappa}_1 + \tilde{\kappa}_2 \le d$ is either of the form $[\kappa]$ or $[\kappa']^*$, according as $\tilde{\kappa}_1 \le [d/2]$ or $> [d/2]$ and κ' is related to κ as defined in Claim 5.5.

5.52 Show that if l is an integer and

$$a_\sigma^\dagger \cdot a_\tau^\dagger = \sum_{m=-l}^{+l} (-1)^{l+m} a_{\sigma lm}^\dagger a_{\tau l,-m}^\dagger \tag{5.582}$$

then $a_\sigma^\dagger \cdot a_\tau^\dagger = -a_\tau^\dagger \cdot a_\sigma^\dagger$, whereas, if j is a half-odd integer and

$$a_\sigma^\dagger \cdot a_\tau^\dagger = \sum_{m=-j}^{+j} (-1)^{j+m} a_{\sigma jm}^\dagger a_{\tau j,-m}^\dagger \tag{5.583}$$

then $a_\sigma^\dagger \cdot a_\tau^\dagger = a_\tau^\dagger \cdot a_\sigma^\dagger$. Explain the difference between these results and the exchange symmetries of the corresponding scalar products, $\psi_\sigma \cdot \psi_\tau$, for $L = 0$ and $J = 0$ coupled wave functions.

5.53 Show that the $U(8)$ irrep $\{21^5\}$ restricts to the unmodified sum of $USp(8)$ irreps $\langle 21^5 \rangle \oplus \langle 1^5 \rangle \oplus \langle 21^3 \rangle \oplus \langle 21 \rangle \oplus \langle 1^3 \rangle \oplus \langle 1 \rangle$ and that, after modification to standard form, it gives the $U(8) \downarrow USp(8)$ branching rule

$$U(8) \downarrow USp(8) : \{21^5\} \downarrow \langle 21 \rangle \oplus \langle 1^3 \rangle \oplus \langle 1 \rangle. \tag{5.584}$$

5.13 The classification of shell-model basis states

Useful basis states and coupling schemes are labelled by the quantum numbers of physically meaningful dynamical groups and Lie algebras of observables. Such basis states are not eigenstates of a general Hamiltonian. However, they do diagonalise useful model Hamiltonians and give physical meaning to the content of eigenstates expressed in terms of corresponding coupling schemes. Moreover, they enable the powerful techniques of group theory to be utilised in the calculation of matrix elements of physical observables.

In this section, we consider basis states classified according to the so-called jj and LST coupling schemes. These coupling schemes take advantage of some important properties of group representations, e.g., the *duality* relationships between pairs of group actions, Young diagram techniques, and branching rules, as described in Sections 5.11 – 5.12.

5.13.1 *Nuclear jj coupling*

Single-nucleon states in the shell model have wave functions with space, spin, and isospin components. In jj coupling, the orbital angular momentum, l, and the spin, $s = 1/2$, of each nucleon are first coupled to form single-particle states of combined angular momentum $j = l \pm 1/2$ and parity $(-1)^l$. The single-nucleon wave functions are then combined to form many-nucleon wave functions of total angular momentum J, parity π, and isospin T that are antisymmetric under nucleon exchange. Thus, the first step of jj coupling is to construct fully antisymmetric basis states, with good J, π, and T within the shell-model subspaces of $(j)^n$ shell-model configurations. Basis states for general multi-j-shell configurations are then given by the angular-momentum- and isospin-coupled products of subshell configurations.

A natural basis for a $(j)^n$ configuration is given by states with quantum numbers that label irreps of the groups in the subgroup chain

$$
\begin{array}{ccc}
\mathrm{U}(2(2j+1)) \supset & \mathrm{U}(2j+1) \times & \mathrm{U}(2)_T \\
\{1^n\} & \{\lambda\} & (n,T)
\end{array}, \tag{5.585}
$$

where $\mathrm{U}(2(2j+1))$ is the group of unitary transformations of the $2(2j+1)$ single-particle states and λ is a partition of n. Because many-nucleon states are fully antisymmetric, the states of a $(j)^n$ configuration are required to carry the fully antisymmetric irrep $\{1^n\}$ of $\mathrm{U}(2(2j+1))$ with $n \leq 2(2j+1)$.

Let $\{a^\dagger_{\tau jm}, a^{\tau jm}\}$ denote a set of nucleon creation and annihilation operators, where $\tau = \pm 1/2$ is an isospin label and m runs over the range $m = -j, \ldots, +j$. When considering a single fixed value of j, it will be convenient to simplify the notation by suppressing the j index and abbreviating the indices of a single-nucleon by using nm or pm, respectively, to denote a neutron state or a proton state with angular momentum quantum numbers: $nm \equiv (\tau = 1/2, jm)$ and $pm \equiv (\tau = -1/2, jm)$.

Infinitesimal generators of $U(2(2j+1))$ are then given in the language of second quantization by Hermitian linear combinations of the operators

$$\hat{C}_{\tau m, \tau' m'} = a^\dagger_{\tau m} a^{\tau' m'};$$
(5.586)

i.e., by elements of the Lie algebra $u(2(2j+1))$.

Infinitesimal generators of the groups in the chain (5.585) are given as follows. The subalgebra $u(2j+1) \subset u(2(2j+1))$ is spanned by the Hermitian linear combinations of the operators

$$\hat{C}^{(2j+1)}_{mm'} = \sum_\tau a^\dagger_{\tau m} a^{\tau m'}.$$
(5.587)

A Cartan subalgebra for $u(2j+1)$ is spanned by the commuting operators

$$\{\hat{C}^{(2j+1)}_{mm}\,;\, m = -j, \ldots, +j\}.$$
(5.588)

The Lie algebra $u(2)_T$ of $U(2)_T$ is spanned by Hermitian linear combinations of the $su(2)_T$ isospin operators,

$$\hat{T}_+ = \sum_m a^\dagger_{nm} a^{pm}, \quad \hat{T}_0 = \tfrac{1}{2} \sum_m (a^\dagger_{nm} a^{nm} - a^\dagger_{pm} a^{pm}), \quad \hat{T}_- = \sum_m a^\dagger_{pm} a^{nm}, \quad (5.589)$$

and the $u(1)$ number operator,

$$\hat{n} = \sum_{\tau m} a^\dagger_{\tau m} a^{\tau m}.$$
(5.590)

A Cartan subalgebra for $u(2)_T$ is spanned by the commuting operators

$$\hat{n}_n = \tfrac{1}{2}\hat{n} + \hat{T}_0 = \sum_m a^\dagger_{nm} a^{nm}, \quad \hat{n}_p = \tfrac{1}{2}\hat{n} - \hat{T}_0 = \sum_m a^\dagger_{pm} a^{pm}, \quad (5.591)$$

the eigenvalues of which are, respectively, the neutron and proton numbers.

Bases for the j shell, which reduce the subgroup chain (5.585) are given by antisymmetric states, $\{|\lambda \kappa n T T_0\rangle\}$, for which κ indexes a basis for a $U(2j+1)$ irrep of highest weight $\{\lambda\}$, n is the nucleon number, and (T, T_0) are isospin quantum numbers. In fact, the label λ is redundant because it is uniquely defined by the $U(2)_T$ quantum numbers. This follows directly from the unitary-unitary duality theorem (cf. Section 5.11.5). Thus, a state of a j shell that belongs to a $U(2j+1)$ irrep of highest weight λ, simultaneously belongs to a $U(2)_T$ irrep of conjugate highest weight $\tilde{\lambda}$. For example, if $j = {}^5\!/_2$, then the state

$$a^\dagger_{n\,5/2} a^\dagger_{p\,5/2} a^\dagger_{n\,3/2} a^\dagger_{p\,3/2} a^\dagger_{n\,1/2} a^\dagger_{n,-1/2} |0\rangle$$
(5.592)

is a state of $n = 6$ and $T = 1$ or, equivalently, of $\tilde{\lambda}_1 = n_n = 4$ and $\tilde{\lambda}_2 = n_p = 2$. Hence, it is of $U(2)_T$ highest weight $\tilde{\lambda} = \{42\}$ and $U(2j+1)$ highest weight $\lambda = \{2^2 1^2\}$.

It remains to define a basis for the $U(2j+1)$ irrep $\{\lambda\}$ determined by n and T. There are two natural choices. One is the Gel'fand-Tsetlin basis, described in Section 5.11.7, which reduces the subgroup chain

$$U(d) \supset U(d-1) \supset U(d-2) \supset \cdots \supset U(1), \qquad (5.593)$$

with $d = 2j+1$. This basis is convenient for many purposes, largely because of its simplicity. It is particularly well-adapted to the binary bit structure of computers and the natural representation of shell model basis states in the occupation-number representation in which a single-particle state is assigned a value of 1 or 0 according as it is occupied or unoccupied. Such techniques have been developed with great success for huge shell-model calculations in the so-called *m scheme*.[231] However, if one is concerned with the interpretation of shell model results in terms of their dynamical structures and symmetry properties, it is appropriate to choose a basis of states with good angular-momentum quantum numbers that take advantage of the rotational invariance and other symmetries of a nuclear Hamiltonian.

A suitable basis, which reduces the subgroup chain

$$\begin{array}{ccccccc} U(2(2j+1)) \supset & U(2j+1) \times U(1)_T & \supset USp(2j+1) \supset & SU(2)_J \supset & U(1)_J \\ \{1^n\} & \{\lambda\} \equiv (nT) \quad T_0 & \rho \qquad (v,t) & \alpha \qquad J & M \end{array}, \qquad (5.594)$$

was introduced by Flowers[232], as an extension to include isospin of a subgroup chain used by Racah[233] in atomic physics. The groups $SU(2)_J \supset U(1)_J$ in this chain provide the total angular momentum quantum numbers, JM, for a nucleus. The many states of a given J and M in a $U(2j+1)$ irrep are then partially distinguished by the irrep labels, (v,t), of the unitary-symplectic group, $USp(2j+1)$. These quantum numbers must be supplemented, in general, by *multiplicity indices*, ρ and α, to provide a complete classification of shell-model basis states.

Because it commutes with the isospin group $U(2)_T$, the group $U(2j+1)$ leaves all isospin quantum numbers invariant. In addition, its subgroup, $USp(2j+1) \subset U(2j+1)$, leaves two-nucleon $J = 0$ states of $T = 1$ pairs invariant (cf. Section 5.12.11). Thus, $USp(2j+1)$ is the symmetry group of a useful pair-coupling model with states classified by *seniority*, v, and *reduced-isospin*, t, quantum numbers.[234] The physical meaning of these quantum numbers follows directly from a duality relationship between the $USp(2j+1)$ irreps and those of a complementary $\mathcal{SO}(5)$ dynamical group (see Section 5.12.13(ii)). The Lie algebra of the group $\mathcal{SO}(5)$ contains elements that create and annihilate nucleons in $J = 0$, $T = 1$ pairs. The $USp(2j+1)$-$\mathcal{SO}(5)$ duality relationship states that, because the $USp(2j+1)$ group leaves $J = 0$, $T = 1$ two-nucleon pairs invariant, the $USp(2j+1)$ quantum numbers

[231] Whitehead R.R. *et al.* (1977), *Adv. Nucl. Phys.* **9**, 123.

[232] Flowers B.H. (1952), *Proc. Roy. Soc. London* **A212**, 248.

[233] Racah G. (1943), *Phys. Rev.* **63**, 367.

[234] The seniority pair-coupling scheme was introduced in *LS* coupling, in atomic physics, by Racah, *op. cit.* Footnote 233; Flowers, *op. cit.* Footnote 232; and Racah G. and Talmi I. (1952), *Physica* **18**, 1097.

of a state are unchanged by adding or removing such two-nucleon pairs. Thus the
$USp(2j+1)$ quantum numbers of a state are identical to those of a $SO(5)$ lowest-
weight state that contains no such two-nucleon-pairs and whose nucleon number
and isospin are given, respectively, by v and t. This duality relationship, shown by
Helmers,[235] implies that shell-model states are equivalently classified by means of
the simpler subgroup chain

$$SO(5) \times SU(2)_J \supset U(2)_T \times U(1)_J \supset U(1)_T$$
$$(v,t) \quad J \quad (n,T) \quad M \quad T_0 \qquad (5.595)$$

The construction of basis states for a j shell by means of $USp(2j+1)$ and its dual
partner $SO(5)$ is described in Chapter 6 in the context of the pair-coupling model
that they diagonalise.

For j-shell configurations comprising only neutrons or only protons, the isospin
of a two-nucleon pair is automatically $T=1$ and for n such nucleons the isospin
takes its maximum value, $T=n/2$. Thus, the reduced isospin t is directly related
to the seniority by $t=v/2$ and becomes redundant. Basis states are then labelled
by the quantum numbers of the chain

$$U(2j+1) \supset USp(2j+1) \supset SU(2)_J \supset U(1)_J$$
$$\{1^n\} \quad v \quad J \quad M \qquad (5.596)$$

With the restriction to nucleons of a single type, the group $SO(5)$ also restricts to
a simpler group, $SU(2) \subset SO(5)$, known as the quasi-spin group.[236] The seniority
quantum number v is then simply related to the quasi-spin of an $SU(2)$ irrep. Thus,
the states of a single nucleon type in a single j shell are equivalently classified by
the dual subgroup chain

$$SU(2) \times SU(2)_J \supset U(1)_J \times U(1)$$
$$v \quad J \quad M \quad n \qquad (5.597)$$

The pair-coupling model, whose Hamiltonian is diagonalised by these subgroup
chains, is also described in Chapter 6.

In concluding this section, it is worth emphasising two important points. The
first is that isospin plays a valuable role in the classification of states regardless of
whether or not it is conserved. In fact, isospin is a relatively good symmetry in
nuclear physics (cf. Section 5.5). But, even if it were not, it would still be valuable
for the construction of fully antisymmetric basis states. The second is that, because
of the duality between the representations of the groups $U(2j+1)$ and $U(2)_T$, each
irrep of the direct product group $U(2j+1) \times U(2)_T$ occurs at most once; i.e., there
is no multiplicity of equivalent $U(2j+1) \times U(2)_T$ irreps. Similarly, because of the
duality between the representations of the groups $USp(2j+1)$ and $SO(5)$, each
irrep of the direct product group $USp(2j+1) \times SO(5)$ occurs at most once and

[235]Helmers, *op. cit.* Footnote 183 on Page 326.
[236]Kerman, *op. cit.* Footnote 184 on Page 326.

again no multiplicity index is needed. Thus, the concepts of isospin and the duality properties that result from the restriction to totally antisymmetric states are major assets in the formulation of theories that have their foundations in the nuclear shell model.

5.13.2 *Supermultiplet symmetry*

Although nuclear states are represented by antisymmetric combinations of space, spin, and isospin wave functions, the separate components of a wave function can have any exchange symmetry provided the other components supply the complementary symmetry to make the total combination antisymmetric. In supermultiplet theory, spin and isospin wave functions are combined first and a group U(4) of combined spin and isospin transformations is introduced to characterise the spin-isospin structure of states and keep track of their exchange symmetries.

As shown in a classic paper by Wigner,[237,238] the supermultiplet group provides extra labels, in addition to spin and isospin, to assist in the classification of many-nucleon states. This is particularly useful when the supermultiplet group is an approximate symmetry of the Hamiltonian and is invaluable in the microscopic shell-model expression of nuclear collective models.

A nucleon has four distinct spin-isospin states $|m_s m_t\rangle$:

$$|1\rangle := |{}^1\!/_2, {}^1\!/_2\rangle, \quad |2\rangle := |{}^1\!/_2, -{}^1\!/_2\rangle,$$
$$|3\rangle := |-{}^1\!/_2, {}^1\!/_2\rangle, \quad |4\rangle := |-{}^1\!/_2, -{}^1\!/_2\rangle. \tag{5.598}$$

Thus, if ν labels an orthonormal basis of spatial states for a nucleon, a complete orthonormal basis of single-nucleon states is labelled

$$|\sigma\nu\rangle = a^\dagger_{\sigma\nu}|0\rangle, \tag{5.599}$$

where the spin-isospin label σ runs from 1 to 4.

The supermultiplet group U(4) is the group of all unitary transformations of the spin-isospin component of a single-nucleon wave function. A set of infinitesimal generators of U(4) is given by the Hermitian linear combinations of the 16 operators

$$\hat{C}_{\sigma\tau} = \sum_\nu a^\dagger_{\sigma\nu} a^{\tau\nu}. \tag{5.600}$$

These operators span a u(4) Lie algebra which contains the su(2)$_S$ spin and su(2)$_T$ isospin algebras as subalgebras, i.e.,

$$\mathrm{u}(4) \supset \mathrm{su}(2)_\mathrm{S} + \mathrm{su}(2)_\mathrm{T}, \tag{5.601}$$

where su(2)$_S$ and su(2)$_T$ are spanned, respectively, by the spin and isospin operators

$$\hat{S}_+ = (\hat{C}_{13} + \hat{C}_{24}), \quad \hat{S}_0 = \tfrac{1}{2}(\hat{C}_{11} + \hat{C}_{22} - \hat{C}_{33} - \hat{C}_{44}), \quad \hat{S}_- = (\hat{C}_{31} + \hat{C}_{42}), \tag{5.602}$$

---

[237]Wigner E.P. (1937), *Phys. Rev.* **51**, 106.
[238]See also Hund F. (1937), *Z. Phys.* **105**, 202.

$$\hat{T}_+ = (\hat{C}_{12} + \hat{C}_{34}), \quad \hat{T}_0 = \tfrac{1}{2}(\hat{C}_{11} - \hat{C}_{22} + \hat{C}_{33} - \hat{C}_{44}), \quad \hat{T}_- = (\hat{C}_{21} + \hat{C}_{43}). \quad (5.603)$$

A Cartan subalgebra of u(4) is spanned by four commuting operators:

$$\hat{C}_{11}, \quad \hat{C}_{22}, \quad \hat{C}_{33}, \quad \hat{C}_{44}. \quad (5.604)$$

Thus, basis states of simultaneous eigenstates of these operators can be constructed for a U(4) irrep. The quadruplet, $w = (w_1, w_2, w_3, w_3)$, of eigenvalues of these operators is then the U(4) *weight* of the corresponding eigenstate. Note that, because

$$\hat{n} = \sum_\sigma \hat{C}_{\sigma\sigma} = \sum_{\sigma\nu} a^\dagger_{\sigma\nu} a^{\sigma\nu} \quad (5.605)$$

is the nucleon number operator, the U(4) weight $w = (w_1, w_2, w_3.w_4)$ of a state of n nucleons is an ordered partition of $n = \sum_\sigma w_\sigma$.

Useful basis states for an irrep of the supermultiplet group are ones which reduce the physically relevant subgroup chain

$$\begin{matrix} \text{U}(4) & \supset & \text{SU}(2)_S \times \text{SU}(2)_T & \supset & \text{U}(1)_S \times \text{U}(1)_T \\ \{\sigma\} \; \rho & & S \qquad\qquad T & & M_S \qquad\qquad T_0 \end{matrix}, \quad (5.606)$$

where the multiplicity label, ρ, distinguishes distinct states with common $\{\sigma\}$, S, T, M_S, and T_0.

The spin and isospin states contained within a U(4) irrep can be determined from the branching rules[239]

$$\text{U}(4) \downarrow \text{O}(4) \downarrow \text{SO}(4) \simeq \text{SU}(2) \times \text{SU}(2). \quad (5.607)$$

Consider, for example, the U(4) irrep $\{31^2\}$. The U(4) $\downarrow$ O(4) branching rule (see Sections 5.12.7 and 5.12.8) gives

$$\text{U}(4) \downarrow \text{O}(4) : \{31^2\} \downarrow [3]^* \oplus [21] \oplus [1]^*. \quad (5.608)$$

The O(4) irreps [3] and [1] remain irreducible on restriction to SO(4). But, according to Equation (5.556), the self-associated O(4) irrep [21] restricts to the sum of SO(4) irreps $[21] \oplus [2, -1]$. Thus, we obtain the branching rule

$$\text{U}(4) \downarrow \text{SO}(4) : \{31^2\} \downarrow [3] \oplus [21] \oplus [2, -1] \oplus [1]. \quad (5.609)$$

Finally, by relating the Cartan weight operators of SU(2) $\times$ SU(2) to those of SO(4), we obtain the corresponding U(4) $\downarrow$ SU(2) $\times$ SU(2) branching rule. The Cartan weight operators for SO(4), given by Equation (5.521), are

$$\hat{h}_1 = \hat{C}_{11} - \hat{C}_{44}, \quad \hat{h}_2 = \hat{C}_{22} - \hat{C}_{33}. \quad (5.610)$$

From Equations (5.602) and (5.603), we then have

$$\hat{S}_0 = \tfrac{1}{2}(\hat{h}_1 + \hat{h}_2), \quad \hat{T}_0 = \tfrac{1}{2}(\hat{h}_1 - \hat{h}_2), \quad (5.611)$$

---

[239]They can also be obtained from a generating function, given by Patera J. and Sharp R.T. (1980), *J. Phys. A: Math. Gen.* **13**, 397.

and the equivalent $O(4) \simeq SU(2) \times SU(2)$ labelling of an irrep

$$[\kappa_1, \kappa_2] \simeq \left(\tfrac{1}{2}(\kappa_1 + \kappa_2), \tfrac{1}{2}(\kappa_1 - \kappa_2)\right). \qquad (5.612)$$

Thus, we obtain the branching rule

$$U(4) \downarrow SU(2) \times SU(2) : \{31^2\} \downarrow \left(\tfrac{3}{2}, \tfrac{3}{2}\right) \oplus \left(\tfrac{3}{2}, \tfrac{1}{2}\right) \oplus \left(\tfrac{1}{2}, \tfrac{3}{2}\right) \oplus \left(\tfrac{1}{2}, \tfrac{1}{2}\right). \qquad (5.613)$$

Table 5.3 shows the (S, T) content of a few supermultiplet irreps. For the irreps shown, the multiplicity label ρ is not needed. A larger list, determined by a more lengthy procedure, has been given by Hamermesh.[240]

Table 5.3: The spin-isospin content of some U(4) supermultiplet irreps.

σ	$\dim(\sigma)$	(S, T)
$\{1\}$	4	$(^1/_2, ^1/_2)$
$\{2\}$	10	$(1, 1),\ (0, 0)$
$\{3\}$	20	$(^3/_2, ^3/_2),\ (^1/_2, ^1/_2)$
$\{n\}$	$\frac{1}{6}(n+1)(n+2)(n+3)$	$(\frac{n}{2}, \frac{n}{2}),\ (\frac{n}{2}-1, \frac{n}{2}-1),\ \ldots (0,0)$ or $(\frac{1}{2}, \frac{1}{2})$
$\{1^2\}$	6	$(1, 0),\ (0, 1)$
$\{1^3\}$	4	$(^1/_2, ^1/_2)$
$\{21\}$	20	$(^3/_2, ^1/_2),\ (^1/_2, ^3/_2),\ (^1/_2, ^1/_2)$
$\{31\}$	45	$(2, 1),\ (1, 2),\ (1, 1),\ (1, 0),\ (0, 1)$
$\{n1\}$	$\frac{1}{2}n(n+2)(n+3)$	$(\frac{n+1}{2}, \frac{n-1}{2}),\ (\frac{n-1}{2}, \frac{n-3}{2}),\ (\frac{n-3}{2}, \frac{n-5}{2}),\ \cdots,\ (1, 0)$ or $(\frac{3}{2}, \frac{1}{2})$ $(\frac{n-1}{2}, \frac{n+1}{2}),\ (\frac{n-3}{2}, \frac{n-1}{2}),\ (\frac{n-5}{2}, \frac{n-3}{2}),\ \cdots,\ (0, 1)$ or $(\frac{1}{2}, \frac{3}{2})$ $(\frac{n-1}{2}, \frac{n-1}{2}),\ (\frac{n-3}{2}, \frac{n-3}{2}),\ (\frac{n-5}{2}, \frac{n-5}{2}),\ \cdots,\ (1, 1)$ or $(\frac{1}{2}, \frac{1}{2})$
$\{2^2\}$	20	$(2, 0),\ (0, 2),\ (1, 1),\ (0, 0)$
$\{32\}$	60	$(^5/_2, ^1/_2),\ (^1/_2, ^5/_2),\ (^3/_2, ^3/_2),\ (^3/_2, ^1/_2),\ (^1/_2, ^3/_2),\ (^1/_2, ^1/_2)$
$\{31^2\}$	36	$(^3/_2, ^3/_2),\ (^3/_2, ^1/_2),\ (^1/_2, ^3/_2),\ (^1/_2, ^1/_2)$

A detailed review of supermultiplet theory has been given by Hecht and Pang.[241] A construction of the u(4) matrix elements in an $SU(2)_S \times SU(2)_T$ basis has been given by Hecht.[242]

---

[240]Hamermesh M. (1962), *Group theory and its applications to physical problems* (Addison-Wesley).

[241]Hecht K.T. and Pang S.C. (1969), *J. Math. Phys.* **10**, 1571.

[242]Hecht K.T. (1994), *J. Phys. A: Math. Gen.* **27**, 3445.

5.13.3 *LST and pair coupling*

In *LST* coupling, the many-particle spin-isospin wave functions are combined to form irreps of the Wigner supermultiplet U(4) group. They are then coupled to spatial single-particle wave functions to form totally antisymmetric combinations. Thus, if d denotes the number of states in an orthonormal basis of single-particle spatial wave functions, the set of totally antisymmetric n-nucleon wave functions spans an irrep $\{1^n\}$ of the group U(4d). A natural basis for this irrep is one that reduces the subgroup chain

$$
\begin{array}{ccc}
\mathrm{U}(4d) \supset \mathrm{U}(4) \times \mathrm{U}(d) \\
\{1^n\} \qquad \{\tilde{\lambda}\} \qquad \{\lambda\}
\end{array} \, ,
\tag{5.614}
$$

where it is noted that, because of unitary-unitary duality (cf. Section 5.11.5), the only U(4) irrep that can combine with a U(d) irrep $\{\lambda\}$ to form a fully antisymmetric U(4d) irrep is the irrep $\{\tilde{\lambda}\}$. For example, if $\{a^\dagger_{\sigma\nu}; \sigma = 1,\ldots,4, \ \nu = 1,\ldots,d\}$ denotes a set of nucleon creation operators, then the 6-particle state,

$$
|w\lambda\rangle = a^\dagger_{11} a^\dagger_{12} a^\dagger_{21} a^\dagger_{22} a^\dagger_{31} a^\dagger_{41} |0\rangle,
\tag{5.615}
$$

is simultaneously a highest-weight state for a U(4) irrep $\{w\} = \{\tilde{\lambda}\} = \{2^2 1^2\}$ and a highest-weight state for a Ud) irrep $\{\lambda\} = \{42\}$. Thus, it defines a U(4) × U(d) irrep

$$
\left(\{2^2 1^2\}, \{42\} \right) \quad \sim \quad \left(\begin{array}{c} \text{diagram} \end{array}, \begin{array}{c} \text{diagram} \end{array} \right).
\tag{5.616}
$$

The important observation is that the U(d) symmetry of a state (often called its *space symmetry*) is defined by its U(4) (supermultiplet) symmetry and vice versa.

Infinitesimal generators of U(d) are given by Hermitian linear combinations of the operators,

$$
\hat{C}^{(d)}_{\mu\nu} = \sum_{\sigma=1}^{4} a^\dagger_{\sigma\mu} a^{\sigma\nu}.
\tag{5.617}
$$

The operators $\{\hat{C}^{(d)}_{\mu\nu}\}$ span the complex extension of a u(d) Lie algebra. We can define the subset $\{\hat{C}^{(d)}_{\mu\nu}; \mu < \nu\}$ to be raising operators and the subset $\{\hat{C}^{(d)}_{\mu\nu}; \mu > \nu\}$ to be lowering operators. (Note, however, that the ordering of the indices which determines which is higher and which is lower is arbitrary and different choices may be appropriate in different situations.) The remaining operators $\{\hat{C}^{(d)}_{\nu\nu}\}$ then span a Cartan subalgebra.

A possible choice of basis states for a U(d) irrep is a Gel'fand-Tsetlin basis. However, it is more instructive to choose basis states with good orbital angular momentum quantum numbers. There are many possibilities. If l denotes the orbital angular momentum of a single particle, we consider an l^n configuration and a basis of n-nucleon states for a U(d) irrep, with $d = 2l + 1$, given by states which reduce the subgroup chain

$$\begin{matrix} \mathrm{U}(2l+1) \times \mathrm{SU}(2)_S \times \mathrm{SU}(2)_T \supset \mathrm{O}(2l+1) \times \mathrm{U}(1)_S \times \mathrm{U}(1)_T \supset \mathrm{SO}(3)_L \supset \mathrm{SO}(2)_L \\ \{\lambda\} \qquad S \qquad T \qquad \rho \quad [\kappa] \qquad M_S \quad T_0 \qquad \alpha \quad L \qquad M \end{matrix},$$
$$(5.618)$$

where the intermediate orthogonal group, O($2l + 1$), is defined as the subgroup of U($2l+1$) transformations that leave the states of two-nucleon $L = 0$ pairs invariant (cf. Section 5.12.2). It now turns out that the orthogonal group O($2l + 1$) is the symmetry group of a useful pair-coupling model.

An *LST* coupling scheme, based on the subgroup chain (5.618), was proposed by Bayman[243] as a natural parallel to the jj-coupling scheme of Flowers and Racah (cf. Section 5.13.1). Unaware of Bayman's scheme, Flowers and Spzikowski[244] proposed a pair-coupling scheme suitable for *LST* coupling, based on an extension of the supermultiplet U(4) group to an $\mathcal{SO}(8)$ group.[245] It turns out that these two coupling schemes are, in fact, equivalent as a result of the duality relationship between the two mutually commuting groups $\mathcal{SO}(8)$ and O($2l + 1$) discussed in Section 5.12.13(iii). Thus, an irrep $[\kappa]$ of O($2l + 1$) uniquely defines a dual irrep of $\mathcal{SO}(8)$, which we label by $[\bar{\kappa}]$, and a basis for the l^n configuration spaces is defined by the subgroup chain

$$\begin{matrix} \mathcal{SO}(8) \times \mathrm{SO}(3)_L \supset \mathrm{U}(4) \times \mathrm{SO}(2)_L \supset \mathrm{SU}(2)_T \times \mathrm{SU}(2)_S \supset \mathrm{U}(1)_T \times \mathrm{U}(1)_S \\ \alpha[\bar{\kappa}] \qquad L \qquad \rho \; \{\tilde{\lambda}\} \qquad M \qquad T \qquad S \qquad T_0 \qquad M_S \end{matrix}.$$
$$(5.619)$$

The construction of *LST*-coupled basis states which reduce this subgroup chain is described in the context of the pair-coupling model in Chapter 6.

It should be noted that there are variations on the $d = 2l + 1$ and a basis *LST* coupling scheme for multi-l shell spaces. For example, if $d = d_1 + d_2$ with $d_1 = 2l_1 + 1$ and $d_2 = 2l_2 + 1$, basis states can be classified by the subgroup chain

$$\mathrm{U}(d) \supset \mathrm{U}(d_1) \times \mathrm{U}(d_2) \supset \mathrm{O}(d_1) \times \mathrm{O}(d_2) \supset \mathrm{SO}(3)_{L_1} \times \mathrm{SO}(3)_{L_2} \supset \mathrm{SO}(3)_L \qquad (5.620)$$

or by

$$\mathrm{U}(d) \supset \mathrm{O}(d) \supset \mathrm{O}(d_1) \times \mathrm{O}(d_2) \supset \mathrm{SO}(3)_{L_1} \times \mathrm{SO}(3)_{L_2} \supset \mathrm{SO}(3)_L. \qquad (5.621)$$

The first of these chains is dual to the supermultiplet chain

$$\mathcal{SO}(8)_1 \times \mathcal{SO}(8)_2 \supset \mathrm{U}(4)_1 \times \mathrm{U}(4)_2 \supset \mathrm{SU}(2)_T \times \mathrm{SU}(2)_S \qquad (5.622)$$

[243]Bayman B.F. (1960), *Some Lectures on Groups and their Applications to Spectroscopy* (Nordita, Copenhagen).

[244]Flowers B.H. and Szpikowski S. (1964), *Proc. Phys. Soc.* **84**, 673.

[245]We use the notation $\mathcal{SO}(8)$ instead of SO(8) to distinguish it from the orthogonal subgroups of the U($2l + 1$) group in Equation (5.618).

and the second is dual to the chain

$$\mathcal{SO}(8)_1 \times \mathcal{SO}(8)_2 \supset \mathcal{SO}(8) \supset \mathrm{U}(4) \supset \mathrm{SU}(2)_T \times \mathrm{SU}(2)_S. \tag{5.623}$$

5.13.4 *LST coupling in SU(3) and Sp*$(3, \mathbb{R})$ *bases*

When the spatial components of the single-particle states span a major spherical harmonic-oscillator shell, i.e., the space of all states of given number of harmonic oscillator quanta, then a useful alternative to the bases defined, for example, by Equations (5.620) or (5.621), is given by states which reduce the subgroup chain

$$\mathrm{U}(d) \supset \mathrm{U}(3) \supset \mathrm{SU}(3) \supset \mathrm{SO}(3)_L \supset \mathrm{SO}(2)_L, \tag{5.624}$$

where $\mathrm{U}(3)$ is the symmetry group of the spherical harmonic oscillator (see Section 5.8.3). This coupling scheme will be deployed in Volume 2 in the development of Elliott's SU(3) model,[246],[247] a model that has been most influential in the construction of a shell-model theory of nuclear rotations.

It will also be shown in Volume 2 that a full microscopic multi-shell theory of nuclear quadrupole collective motion,[248],[249],[250],[251] including rotations with vorticity degrees of freedom, as well as monopole and quadrupole giant-resonance degrees of freedom, is given by an *LST* shell-model coupling scheme which involves the $\mathrm{Sp}(3, \mathbb{R})$ dynamical group of the spherical harmonic oscillator (see Section 5.8.4),

$$\mathrm{Sp}(3, \mathbb{R}) \supset \mathrm{U}(3) \supset \mathrm{SU}(3) \supset \mathrm{SO}(3)_L \supset \mathrm{SO}(2)_L. \tag{5.625}$$

5.13.5 *Competing coupling schemes*

Each of the above-defined coupling schemes diagonalises model Hamiltonians of interest. They all diagonalise a harmonic-oscillator shell-model Hamiltonian, $\hat{H}_{\mathrm{HO}}$. In addition, the jj-coupling scheme diagonalises single-particle spin-orbit, $\mathbf{l} \cdot \mathbf{s}$, and $J = 0$ pairing interactions. The technology associated with jj coupling is the simplest, most fully developed, and most widely used in shell-model calculations. The *LST* coupling schemes provide basis states with supermultiplet, space symmetry, and other symmetries. In addition to the harmonic-oscillator Hamiltonian, some versions of *LST* coupling diagonalise $L = 0$ pairing interactions, while those based on the $\mathrm{Sp}(3, \mathbb{R}) \supset \mathrm{U}(3)$ groups provide the appropriate bases for diagonalisation of microscopic collective-model Hamiltonians.

---

[246]Elliott, *op. cit.* Footnote 214 on Page 347.
[247]Elliott J.P. (1958), *Proc. Roy. Soc. London* **A245**, 562.
[248]Rosensteel and Rowe, *op. cit.* Footnote 64 on Page 222.
[249]Rosensteel and Rowe, *op. cit.* Footnote 65 on Page 222.
[250]Draayer J.P., Weeks K.J. and Rosensteel G. (1984), *Nucl. Phys.* **A413**, 215.
[251]Rowe, *op. cit.* Footnote 66 on Page 222.

In realistic situations, mixtures of interactions, which separately diagonalise different coupling schemes are needed and, as a result, the various coupling schemes are in competition with one another. Fortunately, it frequently transpires (or so it would appear) that the interactions favouring one or other coupling scheme tend to be dominant in any given situation. This suggests that different coupling schemes are appropriate in different situations.

Exercises

5.54 Show that a complete set of $U(2j+1) \times U(2)_T$ irreps for a nuclear j^6 configuration is given for any $j > \frac{5}{2}$ either by partition labels or by isospin, as follows:

$$\begin{aligned} \{1^6\} \times \{6\} &\sim (T=3), \quad \{21^4\} \times \{51\} \sim (T=2), \\ \{2^2 1^2\} \times \{42\} &\sim (T=1), \quad \{2^3\} \times \{33\} \sim (T=0). \end{aligned} \tag{5.626}$$

5.55 Give explicit expressions for the highest-weight states of each of the $U(2j+1) \times U(2)_T$ irreps with $n = 6$ listed in Equation (5.626).

5.56 Show that the infinitesimal generators of $U(4)$, defined by Equation (5.600), satisfy the commutation relations

$$[\hat{C}_{\sigma\tau}, \hat{C}_{\sigma'\tau'}] = \delta_{\sigma'\tau}\hat{C}_{\sigma\tau'} - \delta_{\sigma\tau'}\hat{X}_{\sigma'\tau}. \tag{5.627}$$

5.57 Show that the operators of Equations (5.602) and (5.603) obey SU(2) commutation relations.

5.58 Show that the $U(4)$ irrep of highest weight (2) is of dimension 10 and contains states of $(S, T) = (1, 1)$ and $(0, 0)$.

5.59 Show that the $u(d)$ operators of Equation (5.617) commute with the supermultiplet operators of Equation (5.600).

5.60 Show that the appropriate unitary group of one-body unitary transformations of the spatial wave functions for a system of nucleons in a $(2p, 1f)^n$ configuration is $U(10)$ and that for a $(3s, 2d, 1g)^n$ configuration it is $U(15)$.

5.14 Coefficients of fractional parentage (cfp)

Coefficients of fractional parentage (cfp) are often regarded as technical appendages to the shell model. There is also a tendency to separate their treatment from that of the role of symmetries in the solution of model problems. In fact, cfp are essential components of the shell model and are central to the application of algebraic (symmetry-based) methods to complex model problems.

Consider, for example, the solution of a model problem with a Hilbert space classified by basis states that reduce a chain of Lie groups

$$G_1 \supset G_2 \supset G_3. \tag{5.628}$$

If the Hamiltonian and other observables of the model are polynomials of elements of the Lie algebra $\mathbf{g}_1$ of G_1, then straightforward algebraic methods can be used to compute their matrix elements and solve the model problem. However, when the model Hamiltonian, for example, contains terms that mix the irreps of G_1, as is often the case, extra information is needed to compute its matrix elements.

A practical approach to the calculation of such matrix elements is to build up n-nucleon basis states by expressing them in terms of coupled tensor products of single-nucleon and $(n-1)$-nucleon states. The coefficients for these couplings are traditionally expressed in terms of cfp. This approach was developed in atomic spectroscopy[252,253,254,255] and first applied to the nuclear shell model by Edmonds and Flowers.[256] Reviews of traditional methods for computing cfp in the nuclear shell model have been given by de-Shalit and Talmi.[257,258]

The following treatment takes explicit advantage of the facts that many fermion states belong to antisymmetric irreps of a group of one-body unitary transformations and that the cfp for a many-fermion system are simply Clebsch-Gordan coefficients for this group.

5.14.1 *Cfp as SU(2)-reduced Clebsch-Gordan coefficients*

Consider sets of n-nucleon states comprising antisymmetric products of single-particle states from a single-particle space of dimension d. For example, in jj coupling, d would be equal to $2(2j + 1)$. Suppose that these n-nucleon states are classified by the quantum numbers of a subgroup chain

$$
\begin{array}{cccccc}
\mathrm{U}(d) & \supset & \mathrm{SU}(2) & \supset & \mathrm{U}(1) \\
\{1^n\} & \alpha & J & & M
\end{array},
\tag{5.629}
$$

where $\{1^n\}$, J, and M label irreps of subgroups in the chain; SU(2) is the angular-momentum group; and α denotes the quantum numbers supplied by the inclusion of extra groups in the chain and any other labels needed to define the orthonormal basis states of a particular coupling scheme. Given sets of (antisymmetric) n_1- and n_2-nucleon basis states, defined by such a subgroup chain, they can be coupled to form (antisymmetric) $(n = n_1+n_2)$-nucleon basis states,

$$
|n\alpha JM\rangle = \sum_{\substack{\alpha_1 J_1 M_1 \\ \alpha_2 J_2 M_2}} |n_1\alpha_1 J_1 M_1\rangle \otimes |n_2\alpha_2 J_2 M_2\rangle
\tag{5.630}
$$

$$
\times (n_2\alpha_2 J_2 M_2; n_1\alpha_1 J_1 M_1 | n\alpha JM),
$$

[252]Bacher R.F. and Goudsmit S. (1934), *Phys. Rev.* **46**, 948.

[253]Racah, *op. cit.* Footnote 233 on Page 371.

[254]Jahn H.A. (1951), *Proc. Roy. Soc. London* **A205**, 192.

[255]Jahn H.A. and van Wieringen H. (1951), *Proc. Roy. Soc. London* **A209**, 502.

[256]Edmonds A.R. and Flowers B.H. (1952), *Proc. Roy. Soc. London* **A214**, 515.

[257]de Shalit A. and Talmi I. (1963), *Nuclear Shell Theory* (Academic Press, New York).

[258]Talmi I. (1993), *Simple Models of Complex Nuclei* (Harwood Academic Publishers, Chur, Switzerland).

where

$$(n_2\alpha_2 J_2 M_2; n_1\alpha_1 J_1 M_1 | n\alpha J M)$$
$$:= \left[\langle n_2\alpha_2 J_2 M_2 | \otimes \langle n_1\alpha_1 J_1 M_1 |\right] |n\alpha J M\rangle \qquad (5.631)$$

is a U(d) Clebsch-Gordan coefficient. Embedded in Equation (5.630) is an expression of the fact that any fully antisymmetric ($n = n_1 + n_2$)-particle state has an expansion as a linear combination of tensor products of states that are separately antisymmetric with respect to subsets of particles, n_1 and n_2. Moreover, because of the Schur-Weyl theorem, the converse is also true. Taking care of the U(d) coupling properties automatically ensures that the combinations are fully antisymmetric.

Equation (5.630) is simplified, in parallel with the Wigner-Eckart theorem, by factoring out an SU(2) Clebsch-Gordan coefficient. This is done by making the expansion of $|n\alpha J M\rangle$ in terms of angular-momentum-coupled states

$$|n\alpha J M\rangle = \sum_{\alpha_1 J_1 \alpha_2 J_2} \left[|n_1\alpha_1 J_1\rangle \otimes |n_2\alpha_2 J_2\rangle\right]_{JM} [n_2\alpha_2 J_2; n_1\alpha_1 J_1|\}n\alpha J], \qquad (5.632)$$

where

$$[n_2\alpha_2 J_2; n_1\alpha_1 J_1|\}n\alpha J] := \sum_{M_1 M_2} (J_2 M_2\, J_1 M_1 | J M)$$
$$\times (n_2\alpha_2 J_2 M_2; n_1\alpha_1 J_1 M_1 | n\alpha J M). \qquad (5.633)$$

Thus, we obtain the more useful expression

$$[n_2\alpha_2 J_2; n_1\alpha_1 J_1|\}n\alpha J] := \langle n\alpha J M | \left[|n_1\alpha_1 J_1\rangle \otimes |n_2\alpha_2 J_2\rangle\right]_{JM}^*. \qquad (5.634)$$

The coefficient, $[n_2\alpha_2 J_2; n_1\alpha_1 J_1|\}n\alpha J]$, is known as a cfp. In other contexts, it would be called an SU(2)-reduced U(d) Clebsch-Gordan coefficient or an *isoscalar factor*. In practice, phase conventions are invariably used such that these cfp are real.

5.14.2 *Single-particle cfp as reduced matrix elements*

According to Equation (5.634), single-particle cfp are the overlaps

$$[(n-1)\alpha_1 J_1; \nu j|\}n\alpha J] = \langle \Psi_{n\alpha J M} | \left[\psi_{\nu j} \otimes \Psi_{(n-1)\alpha_1 J_1}\right]_{JM}\rangle^*. \qquad (5.635)$$

These cfp are identically zero for $n > d$, where d is the dimension of the space of single-nucleon states,

Claim 5.6 *For $n \le d$, the single-particle cfp defined by Equation (5.635) satisfy*

the identities:

$$[(n-1)\alpha_1 J_1; \nu j|\}n\alpha J] = \frac{1}{\sqrt{n(2J+1)}} \langle n\alpha J \| a_{\nu j}^\dagger \| (n-1)\alpha_1 J_1 \rangle^*, \quad (5.636)$$

$$= \frac{(-1)^{J+j-J_1}}{\sqrt{n(2J+1)}} \langle (n-1)\alpha_1 J_1 \| a_{\nu j} \| n\alpha J \rangle. \quad (5.637)$$

Proof: By construction, the wave function $\Psi_{(n-1)\alpha_1 J_1}$ is antisymmetric with respect to permutations of its $n-1$ nucleons. However, the n-nucleon tensor product, $[\psi_{\nu j} \otimes \Psi_{(n-1)\alpha_1 J_1}]_{JM}$, is not antisymmetric relative to permutations involving the n'th nucleon. In contrast, the state $[a_{\nu j}^\dagger \otimes |(n-1)\alpha_1 J_1\rangle]_{JM}$, in which the single-particle wave function $\psi_{\nu j}$ is replaced with the nucleon creation operator $a_{\nu j}^\dagger$, is fully antisymmetric. The latter has wave function

$$[a_{\nu j}^\dagger \otimes |(n-1)\alpha_1 J_1\rangle]_{JM} \equiv \frac{1}{\sqrt{n}}\left(1 - \sum_{k=1}^{n-1} \hat{P}_{nk}\right)[\psi_{\nu j} \otimes \Psi_{(n-1)\alpha_1 J_1}]_{JM}, \quad (5.638)$$

where $\hat{P}_{nk}$ is an operator that exchanges the n'th and k'th nucleons. Now, in the evaluation of the overlap in Equation (5.635), integration is carried out over all nucleon coordinates to obtain a single number (as opposed to a function of the coordinates). This overlap cannot depend on the labelling of the particles. Thus, it is invariant under the action of the exchange operator $\hat{P}_{nk}$ and

$$\left(1 + \sum_{k=1}^{n-1} \hat{P}_{nk}\right)\langle \Psi_{n\alpha JM}|[\psi_{\nu j} \otimes \Psi_{(n-1)\alpha_1 J_1}]_{JM}\rangle$$

$$= n\langle \Psi_{n\alpha JM}|[\psi_{\nu j} \otimes \Psi_{(n-1)\alpha_1 J_1}]_{JM}\rangle. \quad (5.639)$$

Moreover, because the wave function $\Psi_{n\alpha JM}$ is antisymmetric under the exchange of any two nucleons, it follows that

$$\left(1 + \sum_{k=1}^{n-1} \hat{P}_{nk}\right)\langle \Psi_{n\alpha JM}| = \langle \Psi_{n\alpha JM}|\left(1 - \sum_{k=1}^{n-1} \hat{P}_{nk}\right). \quad (5.640)$$

Therefore

$$\langle n\alpha JM|[a_{\nu j}^\dagger \otimes |(n-1)\alpha_1 J_1\rangle]_{JM} = \sqrt{n}\,\langle \Psi_{n\alpha JM}|[\psi_{\nu j} \otimes \Psi_{(n-1)\alpha_1 J_1}]_{JM}\rangle \quad (5.641)$$

and we obtain the identity

$$[(n-1)\alpha_1 J_1; \nu j|\}n\alpha J]^* = \frac{1}{\sqrt{n}}\langle n\alpha JM|[a_{\nu j}^\dagger \otimes |(n-1)\alpha_1 J_1\rangle]_{JM}. \quad (5.642)$$

From the general expression of the Wigner-Eckart theorem, cf. Equation (A.27),

$$\langle \beta J_3 M_3|[\hat{W}_{J_2} \otimes |\alpha J_1\rangle]_{J_3 M_3} = \frac{\langle \beta J_3 \| \hat{W}_{J_2} \| \alpha J_1\rangle}{\sqrt{2J_3+1}}, \quad (5.643)$$

it is now determined that

$$[(n-1)\alpha_1 J_1; \nu j|\}n\alpha J] = \frac{1}{\sqrt{n(2J+1)}}\langle n\alpha J\|a_{\nu j}^\dagger\|(n-1)\alpha_1 J_1\rangle^*, \qquad (5.644)$$

in accordance with the first equation of the Claim. From the identity

$$\langle \alpha JM|a_{\nu jm}^\dagger|\alpha_1 J_1 M_1\rangle = \langle \alpha_1 J_1 M_1|a^{\nu jm}|\alpha JM\rangle^*$$
$$= (-1)^{j+m}\langle \alpha_1 J_1 M_1|a_{\nu j,-m}|\alpha JM\rangle^* \qquad (5.645)$$

and the Clebsch-Gordan symmetry relation

$$\sqrt{2J+1}\,(JM\,j,-m|J_1 M_1) = (-1)^{J-J_1-m}\sqrt{2J_1+1}\,(J_1 M_1\,jm|JM), \qquad (5.646)$$

one also obtains the second equation of the Claim. □

In view of the close relationship between standard single-particle cfp and reduced matrix elements of nucleon creation and annihilation operators given by Claim 5.6, it is convenient to refer to both sets as single-particle cfp. However, when we wish to be specific, we refer to $[(n-1)\alpha_1 J_1; \nu j|\}n\alpha J]$ as a standard single-particle cfp.

5.14.3 *Some properties of the single-particle cfp*

(i) *Orthonormality relationships*

As Clebsch-Gordan coefficients, the standard cfp are the coefficients of a transformation from one orthonormal basis to another. Thus, they are elements of a unitary matrix and, for $n \leq d$, they satisfy the orthonormality relationship

$$\sum_{\nu j\beta J}\left[(n-1)\beta J; \nu j|\}n\alpha_1 J_1\right]^*\left[(n-1)\beta J; \nu j|\}n\alpha_2 J_2\right] = \delta_{\alpha_1,\alpha_2}\delta_{J_1,J_2}. \qquad (5.647)$$

The corresponding single-particle reduced matrices satisfy the relationship

$$\sum_{\nu j\beta J}\langle n\alpha_1 J_1\|a_{\nu j}^\dagger\|(n-1)\beta J\rangle\langle n\alpha_2 J_2\|a_{\nu j}^\dagger\|(n-1)\beta J\rangle^*$$
$$= n(2J+1)\delta_{\alpha_1,\alpha_2}\delta_{J_1,J_2}. \qquad (5.648)$$

(ii) *A recursion relation*

The following recursion relation is an adaptation of a relation derived by Zamick and Escuderos,[259] which corrected an inconvenient characteristic of an early method of Redmond[260] of generating non-orthogonal single-particle cfp.

---

[259]Zamick L. and Escuderos A. (2006), *Ann. Phys. (NY)* **321**, 987.

[260]Redmond P.J. (1954), *Proc. Roy. Soc. London* **A222**, 84. A review of the Redmond method has been given, for example, by Talmi, *op. cit.* Footnote 258 on Page 380.

Claim 5.7 *If d is the dimension of the space of single-nucleon states, then, for $n \leq d$, single-particle cfp satisfy a recursion relation given by*

$$\frac{1}{2J+1} \sum_{\alpha} \langle (n+1)\alpha J \| a_{\mu j_1}^\dagger \| n\beta_1 J_1 \rangle \langle (n+1)\alpha J \| a_{\nu j_2}^\dagger \| n\beta_2 J_2 \rangle^*$$

$$= \delta_{J_1,J_2} \delta_{\beta_1,\beta_2} \delta_{j_1,j_2} \delta_{\mu,\nu} - \sum_{\gamma J'} \frac{U(J_1 j_1 j_2 J_2 : JJ')}{\sqrt{(2J+1)(2J'+1)}} \tag{5.649}$$

$$\times \langle n\beta_2 J_2 \| a_{\mu j_1}^\dagger \| (n-1)\gamma J' \rangle^* \langle n\beta_1 J_1 \| a_{\nu j_2}^\dagger \| (n-1)\gamma J' \rangle.$$

Proof: Equation (5.649) is derived by equating the matrix elements of both sides of the identity,

$$\left[a_{\nu j_2} \otimes a_{\mu j_1}^\dagger \right]_{J'} = -\sqrt{2j_1+1}\, \delta_{\mu,\nu} \delta_{j_1,j_2} \delta_{J',0} \hat{I}$$

$$- (-1)^{j_1+j_2-J'} \left[a_{\mu j_1}^\dagger \otimes a_{\nu j_2} \right]_{J'}, \tag{5.650}$$

which is an expression of the anticommutation relation

$$\{ a_{\nu j_2 m_2}, a_{\mu j_1 m_1}^\dagger \} = (-1)^{j_2-m_2} \delta_{\mu,\nu} \delta_{j_1,j_2} \delta_{m_1,-m_2} \hat{I} \tag{5.651}$$

in angular-momentum-coupled form (cf. Equation (A.118)); $\hat{I}$ is the identity operator. From the general Equation (A.47), matrix elements for the terms in this identity are given by

$$\langle n\beta_2 J_2 \| \left[a_{\nu j_2} \otimes a_{\mu j_1}^\dagger \right]_{J'} \| n\beta_1 J_1 \rangle = \sum_{\alpha L} \frac{U(J_1 j_1 J_2 j_2 : LJ')}{\sqrt{2L+1}}$$

$$\times \langle n\beta_2 J_2 \| a_{\nu j_2} \| (n+1)\alpha L \rangle \langle (n+1)\alpha L \| a_{\mu j_1}^\dagger \| n\beta_1 J_1 \rangle, \tag{5.652}$$

$$\langle n\beta_2 J_2 \| \hat{I} \| n\beta_1 J_1 \rangle = \delta_{\beta_1,\beta_2} \delta_{J_1,J_2} \sqrt{2J_2+1}, \tag{5.653}$$

and

$$\langle n\beta_2 J_2 \| \left[a_{\mu j_1}^\dagger \otimes a_{\nu j_2} \right]_{J'} \| n\beta_1 J_1 \rangle = \sum_{\gamma L} \frac{U(J_1 j_2 J_2 j_1 : LJ')}{\sqrt{2L+1}}$$

$$\times \langle n\beta_2 J_2 \| a_{\mu j_1}^\dagger \| (n-1)\gamma L \rangle \langle (n-1)\gamma L \| a_{\nu j_2} \| n\beta_1 J_1 \rangle. \tag{5.654}$$

To manipulate these equations into the desired relationship, we need two identities satisfied by the Racah coefficients: the unitarity relationship

$$\sum_{J'} U(J_1 j_1 J_2 j_2 : JJ') U(J_1 j_1 J_2 j_2 : LJ') = \delta_{J,L} \tag{5.655}$$

and the sum rule[261]

$$\sum_{J'}(-1)^{J+L+J'}U(J_1j_1J_2j_2:JJ')\,U(J_1j_2J_2j_1:LJ')$$

$$=(-1)^{J_1+J_2+j_1+j_2}U(J_1j_1j_2J_2:JL). \qquad (5.656)$$

Applying a Racah transformation to Equations (5.652) - (5.654) leads to the equations

$$\sum_{J'}U(J_1j_1J_2j_2:JJ')\langle n\beta_2J_2\|[a_{\nu j_2}\otimes a^\dagger_{\mu j_1}]_{J'}\|n\beta_1J_1\rangle$$

$$=\sum_{\alpha}\frac{\langle n\beta_2J_2\|a_{\nu j_2}\|(n+1)\alpha J\rangle\langle(n+1)\alpha J\|a^\dagger_{\mu j_1}\|n\beta_1J_1\rangle}{\sqrt{2J+1}}, \qquad (5.657)$$

$$\sum_{J'}U(J_1j_1J_2j_2:JJ')\big[-\sqrt{2j_1+1}\,\delta_{\mu.\nu}\delta_{j_1,j_2}\delta_{J',0}\langle n\beta_2J_2\|\hat{I}\|n\beta_1J_1\rangle\big]$$

$$=(-1)^{J_1-j_1-J}\sqrt{2J+1}\,\delta_{\mu.\nu}\delta_{j_1,j_2}\delta_,\delta_{\beta_1,\beta_2}\delta_{J_1,J_2}, \qquad (5.658)$$

and

$$\sum_{J'}(-1)^{J'}U(J_1j_1J_2j_2:JJ')\langle n\beta_2J_2\|[a^\dagger_{\mu j_1}\otimes a_{\nu j_2}]_{J'}\|n\beta_1J_1\rangle$$

$$=\sum_{\gamma L}\frac{(-1)^{j_1+j_2+J_1+J_2-L-J}}{\sqrt{2L+1}}U(J_1j_1j_2J_2:JL) \qquad (5.659)$$

$$\times\langle n\beta_2J_2\|a^\dagger_{\mu j_1}\|(n-1)\gamma L\rangle\langle(n-1)\gamma L\|a_{\nu j_2}\|n\beta_1J_1\rangle.$$

Equating these matrix elements in accordance with Equation (5.650) leads to the result of the Claim. □

5.14.4 Applications of single-particle cfp

The matrix elements of any spherical-tensor operator can be expressed in terms of single-particle cfp. For example, by use of Equation (A.47), it follows that the

---

[261]This sum rule is more familiar when expressed in terms of $6j$-symbols as given, for example, by Brink D.M. and Satchler G.R. (1994), *Angular Momentum* (Oxford University Press), third edn, p. 143,

$$\sum_{J'}(-1)^{J+J'+L}(2J'+1)\begin{Bmatrix}J_1 & J_2 & J'\\ j_2 & j_1 & J\end{Bmatrix}\begin{Bmatrix}J_1 & J_2 & J'\\ j_1 & j_2 & L\end{Bmatrix}=\begin{Bmatrix}J_1 & j_1 & J\\ J_2 & j_2 & L\end{Bmatrix}.$$

one-body operator $\left[a^\dagger_{\mu j_1} \otimes a_{\nu j_2}\right]_{JM}$ has matrix elements

$$\langle n\alpha_f J_f \| \left[a^\dagger_{\mu j_1} \otimes a_{\nu j_2}\right]_J \| n\alpha_i J_i \rangle = \sum_{\beta J'} U(J_i j_2 J_f j_1 : J'J)$$

$$\times \frac{\langle n\alpha_f J_f \| a^\dagger_{\mu j_1} \| (n-1)\beta J'\rangle\langle (n-1)\beta J' \| a_{\nu j_2} \| n\alpha_i J_i \rangle}{\sqrt{2J'+1}}. \quad (5.660)$$

Two-particle cfp are similarly defined in terms of single-particle cfp by

$$\langle n\alpha J_f \| \left[a^\dagger_{\mu j_1} \otimes a^\dagger_{\nu j_2}\right]_J \| (n-2)\beta J_i \rangle = \sum_{\gamma J'} U(J_i j_2 J_f j_1 : J'J)$$

$$\times \frac{\langle n\alpha J_f \| a^\dagger_{\mu j_1} \| (n-1)\gamma J'\rangle\langle (n-1)\gamma J' \| a^\dagger_{\nu j_2} \| (n-2)\beta J_i \rangle}{\sqrt{2J'+1}}. \quad (5.661)$$

A knowledge of these coefficients enables the calculation of n-body matrix elements for any two-body interaction.

A two-body interaction, with two-nucleon matrix elements

$$V^J_{\nu_1 j_1 \nu_2 j_2, \nu_3 j_3 \nu_4 j_4} := \langle (\nu_1 j_2 \nu_2 j_2)JM|\hat{V}|(\nu_3 j_3 \nu_4 j_4)JM\rangle, \quad (5.662)$$

has the second-quantised expression

$$\hat{V} = \sum_{\substack{\nu_1\nu_2\nu_3\nu_4 \\ j_1 j_2 j_3 j_4 J}} \sqrt{2J+1}\, V^J_{\nu_1 j_1 \nu_2 j_2, \nu_3 j_3 \nu_4 j_4} \left[\hat{A}_{(\nu_1 j_1 \nu_2 j_2)J} \otimes \hat{B}_{(\nu_3 j_3 \nu_4 j_4)J}\right]_0, \quad (5.663)$$

where

$$\hat{A}_{(\nu_1 j_1 \nu_2 j_2)JM} := \frac{1}{\sqrt{1+(-1)^J \delta_{\nu_1 j_1, \nu_2 j_2}}} \left[a^\dagger_{\nu_1 j_1} \otimes a^\dagger_{\nu_2 j_2}\right]_{JM} \quad (5.664)$$

is the operator that creates the normalised two-nucleon state

$$|(\nu_1 j_2 \nu_2 j_2)JM\rangle := \hat{A}_{(\nu_1 j_1 \nu_2 j_2)JM}|0\rangle \quad (5.665)$$

and

$$\hat{B}_{(\nu_1 j_1 \nu_2 j_2)JM} := \frac{-1}{\sqrt{1+(-1)^J \delta_{\nu_1 j_1, \nu_2 j_2}}} \left[a_{\nu_1 j_1} \otimes a_{\nu_2 j_2}\right]_{JM}, \quad (5.666)$$

defined such that

$$\left(B_{(\nu_3 j_3 \nu_4 j_4)JM}\right)^\dagger = (-1)^{J-M} \hat{A}_{(\nu_3 j_3 \nu_4 j_4)J,-M}, \quad (5.667)$$

is its covariant adjoint.

Using Equation (A.64), we find that the matrix elements of $\hat{V}$ between n-nucleon states are given by

$$\langle n\alpha_f JM|\hat{V}|n\alpha_i JM\rangle = \sum_{\substack{\nu_1\nu_2\nu_3\nu_4 \\ j_1 j_2 j_3 j_4 J_1 \\ \beta J'}} V^{J_1}_{\nu_1 j_1 \nu_2 j_2, \nu_3 j_3 \nu_4 j_4} \frac{(-1)^{J+J_1-J'}}{2J+1}$$

$$\times \langle n\alpha_f J\|\hat{A}_{(\nu_1 j_1 \nu_2 j_2)J_1}\|(n-2)\beta J'\rangle \langle (n-2)\beta J'\|\hat{B}_{(\nu_3 j_3 \nu_4 j_4)J_1}\|n\alpha_i J\rangle. \quad (5.668)$$

From Equation (5.667), it is determined (cf. Equation (A.90)) that

$$\langle (n-2)\beta J'\|\hat{B}_{(\nu_3 j_3 \nu_4 j_4)J_1}\|n\alpha_i J\rangle$$

$$= (-1)^{J+J_1-J'} \langle n\alpha_i J\|\hat{A}_{(\nu_3 j_3 \nu_4 j_4)J_1}\|(n-2)\beta J'\rangle^*. \quad (5.669)$$

Thus, we obtain arbitrary matrix elements of a two-body interaction in terms of two-particle cfp given by

$$\langle n\alpha_f JM|\hat{V}|n\alpha_i JM\rangle = \sum_{\substack{\nu_1\nu_2\nu_3\nu_4 \\ j_1 j_2 j_3 j_4 J_1 \\ \beta J'}} V^{J_1}_{\nu_1 j_1 \nu_2 j_2, \nu_3 j_3 \nu_4 j_4} \frac{1}{2J+1}$$

$$\times \langle n\alpha_f J\|\hat{A}_{(\nu_1 j_1 \nu_2 j_2)J_1}\|(n-2)\beta J'\rangle \langle n\alpha_i J\|\hat{A}_{(\nu_3 j_3 \nu_4 j_4)J_1}\|(n-2)\beta J'\rangle^*. \quad (5.670)$$

5.14.5 *Calculation of single-particle cfp*

Tables of single-particle cfp for states of good seniority in a jj-coupling scheme have been computed by Bayman and Lande[262] by sequentially determining linear combinations of tensor product states $|[\psi_{\nu j} \otimes \Psi_{n-1,\alpha_1 J_1}]_{JM}\rangle$, for increasing values of n, that diagonalise the Casimir operators of the relevant symmetry groups.[263] Sophisticated methods which exploit the binary bit manipulations of computers have also been developed for use with the LST-SU(3) coupling scheme by Bahri and Draayer.[264] The following algorithm is an adaptation of the Bayman-Lande method for a general coupling scheme.

We first consider the construction of arbitrary single-particle cfp by use of the orthonormality equation, (5.647), and the recursion relation, (5.649). Let us suppose that single-particle cfp have been determined for states of nucleon number up to $n-1$ and that we now wish to determine the n-nucleon coefficients,

$$C_{\alpha p} := \left[(n-1)\beta J_1; \nu j|\}n\alpha J\right]^* = \frac{\langle n\alpha J\|a^\dagger_{\nu j}\|(n-1)\beta J_1\rangle}{\sqrt{n(2J+1)}}, \quad (5.671)$$

---

[262]Bayman B.F. and Lande A. (1966), *Nucl. Phys.* **77**, 1.

[263]Tables of cfp have also been computed by Towner I.S. and Hardy J.C. (1969), *At. Data Nucl. Data Tables* **6**, 153, Hubbard L.B. (1971), *At. Data Nucl. Data Tables* **9**, 85, and Shlomo S. (1972), *Nucl. Phys.* **A184**, 545.

[264]Bahri C. and Draayer J.P. (1994), *Comp. Phys. Comm.* **83**, 59.

where $\alpha = 1, \ldots, N$ indexes an orthonormal basis of antisymmetric n-nucleon states, $\{|n\alpha JM\rangle\}$, of a particular angular momentum JM, and $p \equiv \nu j\beta J_1$ indexes the coupled-tensor product states, $\{[a_{\nu j}^\dagger \otimes |(n-1)\beta J_1\rangle]_{JM}\}$.

Equations (5.647) and (5.649) can be expressed, respectively, as matrix equations; viz.

$$CC^\dagger = I_N, \quad \text{i.e.,} \quad \sum_p C_{\alpha p} C_{\alpha' p}^* = \delta_{\alpha,\alpha'} \tag{5.672}$$

and

$$C^\dagger C = X, \quad \text{i.e.,} \quad \sum_\alpha C_{\alpha p}^* C_{\alpha q} = X_{pq}, \tag{5.673}$$

where X_{pq} is defined by the right side of Equation (5.649) and, being expressed in terms of $(n-1)$-nucleon cfp, is presumed to be known. These equations have many solutions each corresponding to an arbitrary set of orthonormal basis states. One set of solutions is simply obtained by use of the following claim.

Claim 5.8   *The matrix X can be diagonalised by a unitary transformation and brought to the form*

$$UXU^\dagger = \begin{pmatrix} I_N & 0 \\ 0 & 0 \end{pmatrix}, \tag{5.674}$$

where I_N is the $N \times N$ identity matrix with $N = \mathrm{Tr}(X)$ and 0 signifies a submatrix with zero entries.

Proof: First observe, from the definition (5.635) of single-particle cfp and Claim 5.6, that the matrix elements of $X = C^\dagger C$ are given by

$$X_{p_1 p_2} = \sum_\alpha \langle \alpha|p_1\rangle^* \langle \alpha|p_2\rangle = \sum_\alpha \langle p_1|\alpha\rangle\langle \alpha|p_2\rangle, \tag{5.675}$$

where

$$|\alpha\rangle \equiv |\Psi_{n\alpha JM}\rangle, \quad |p\rangle \equiv |[\psi_{\nu j} \otimes \Psi_{(n-1)\alpha J'}]_{JM}\rangle. \tag{5.676}$$

Thus, $X_{p_1 p_2}$ can be viewed as a matrix element of a projection operator,

$$\hat{X} := \sum_\alpha |\alpha\rangle\langle \alpha|, \tag{5.677}$$

which projects states in the tensor product space $\mathbb{H}^1 \otimes \mathbb{H}_{AS}^{n-1}$ onto those in the fully antisymmetric n-nucleon subspace $\mathbb{H}_{AS}^n$. Moreover, because X is Hermitian, it can be diagonalised by a unitary transformation and, to satisfy the identity $XX = X$ required of a projection operator, it must have a diagonal form as given by the claim. $\qquad\square$

As a result of the claim, Equation (5.673) implies that

$$UC^\dagger CU^\dagger = \begin{pmatrix} I_N & 0 \\ 0 & 0 \end{pmatrix},$$

(5.678)

and we obtain the solution

$$C = (I_N\ 0)U.$$

(5.679)

For many purposes, shell-model calculations can be carried out in any jj-coupled basis and the above-defined cfp serve such purposes. However, to obtain a physical understanding of the shell-model states, it is desirable to have them expressed in terms of basis states corresponding to relevant coupling schemes.

The cfp corresponding to a preferred coupling scheme can be determined as follows. First, define an operator

$$\hat K := \mathcal{K}_1 + \mathcal{K}_2 + \cdots,$$

(5.680)

to be a sum of Casimir operators for the groups in the subgroup chain, (5.629), that define the coupling scheme of interest. Next, evaluate the matrix elements, $K_{\alpha\alpha'} = \langle n\alpha JM|\hat K|n\alpha' JM\rangle$, as bilinear products of the now-determined single-particle cfp, $\{C_{\alpha p}\}$, by use of Equations (5.660) and (5.668). It then only remains to make unitary transformations of these coefficients, such that the matrix K becomes diagonal, to obtain the cfp relative to the desired coupling scheme. This can be done sequentially for all values of n to give a complete set for any shell-model space of interest.

Exercises

5.61 Let $a_m^\dagger := a_{jm}^\dagger$ denote the creation operator for a nucleon of spin $j = 3/2$. Use a table of Clebsch-Gordan coefficients, as necessary, to show that $n = 2$ and $n = 3$ nucleon states $|nJM\rangle$ of angular momentum JM are given by

$$|200\rangle = \frac{1}{\sqrt 2}\left(a_{\frac12}^\dagger a_{-\frac12}^\dagger - a_{\frac32}^\dagger a_{-\frac32}^\dagger\right)|0\rangle,$$

(5.681)

$$|220\rangle = -\frac{1}{\sqrt 2}\left(a_{\frac12}^\dagger a_{-\frac12}^\dagger + a_{\frac32}^\dagger a_{-\frac32}^\dagger\right)|0\rangle,$$

(5.682)

$$|3\tfrac32\tfrac32\rangle = a_{\frac32}^\dagger a_{\frac12}^\dagger a_{-\frac12}^\dagger|0\rangle.$$

(5.683)

Hence derive the reduced matrix elements $\langle 3\,3/2\|a^\dagger\|20\rangle$ and $\langle 3\,3/2\|a^\dagger\|22\rangle$ and determine the cfp

$$[20; 3/2|\}3\,3/2] = \frac{1}{\sqrt 6}, \quad [22; 3/2|\}3\,3/2] = -\frac{5}{\sqrt 6}.$$

(5.684)

5.62 Show that if a two-body interaction is expressed in the form

$$\hat{V} = \sum_{\substack{\nu_1 \nu_2 \nu_3 \nu_4 \\ j_1 j_2 j_3 j_4 J}} \sqrt{2J+1}\, V^J_{\nu_1 j_1 \nu_2 j_2, \nu_3 j_3 \nu_4 j_4} \left[\hat{A}_{(\nu_1 j_1 \nu_2 j_2)J} \otimes \hat{B}_{(\nu_3 j_3 \nu_4 j_4)J} \right]_0, \quad (5.685)$$

then its two-nucleon matrix elements are given by

$$\langle (\nu_1 j_1 \nu_2 j_2)JM | \hat{V} | (\nu_3 j_3 \nu_4 j_4)JM \rangle = V^J_{\nu_1 j_1 \nu_2 j_2, \nu_3 j_3 \nu_4 j_4}. \quad (5.686)$$

5.15 Successes and limitations of the shell-model

The nuclear shell model is much more than a model of nuclear structure. It is the formal framework for a many-body theory of interacting nucleons in terms of which virtually all other microscopic models of nuclei have their ultimate expression. Nevertheless, it is based on assumptions and has its limitations.

A primary assumption is that the body of nuclear phenomena, at low energies, can be described in terms of interacting nucleons, without excitation of subnucleon degrees of freedom, within the framework of non-relativistic quantum mechanics.[265] Thus, in attempting to establish the microscopic shell-model foundations of nuclear physics, it is important to ascertain if the properties of nuclei are consistent with such assumptions and with effective interactions derived from experimental observations on two- and three-nucleon systems, as discussed in Section 5.7. Determining if this is the case is a formidable task.

Important contributions are made by the NCSM calculations (cf. References in Footnotes on page 292), which attempt to do the most complete shell-model calculations possible with the best effective interactions available. The calculations that have been done so far are remarkably successful in giving *ab initio* descriptions of some light nuclei. However, no-core shell-model calculations are limited by the rapid escalation, with increasing nucleon number, of the dimensions needed for reasonably reliable results. Moreover, any shell-model calculation is necessarily based on a selection of an effective model space which may be inappropriate for some low-energy states. With modern computers and the development of sophisticated techniques, it is nowadays possible to diagonalise matrices in determinantal bases of dimensionality up to 10^9 by means of the Lanczos algorithm.[266] Even so, it is doubtful that one could ever obtain a prediction of such emergent phenomena[267] as superdeformed bands, for example, from first principles.

---

[265]Some attempts have been made to include relativistic effects in the shell model; e.g., Celenza L.S., Pong W.S. and Shakin C.M. (1983), *Phys. Rev.* **C27**, 1799.

[266]For a survey of what can be done see, for example, the reviews of Brown B.A. (2002), *Nucl. Phys.* **A704**, 11c, and Caurier *et al.*, *op. cit.* Footnote 59 on Page 267.

[267]Emergent phenomena are complex behaviours that result from simple principles at a deeper level. A key paper on this topic is Anderson, *op. cit.* Footnote 1 on Page vii, in which the author makes the point that *"The ability to reduce everything to simple fundamental laws does not imply the ability to start from those laws and reconstruct the universe".*

Thus, the primary use of the shell model is not so much to predict the properties of nuclei from first principles as to provide the language and concepts for the interpretation of observed phenomena, to gain understanding of the circumstances in which particular phenomena arise, and to extrapolate into other mass regions to see if the interpretations are consistent.

A profitable use of the shell model is to start with a finite-dimensional effective shell-model space and derive a corresponding effective interaction by fitting the results of calculations to observed nuclear data. For example, one can adopt an active shell-model space for nuclei in the range $16 < A < 40$ consisting of configurations with a closed-shell ^{16}O core and valence-shell particles restricted to the $2s1d$ shell. For such a space, one can attempt to determine the one-, two-, and possibly three-body components of the effective Hamiltonian by fits to nuclei neighbouring ^{16}O with zero, one, two, and three extra-core nucleons. It is then possible to see how well the remaining $2s1d$-shell nuclei are described. Numerous calculations of this type, usually with only two-body interactions, have been completed for the $1p$ and $2s1d$ shells.[268],[269] In the latter calculation, Brown and Wildenthal adjusted the parameters of an effective Hamiltonian to fit the low-energy data of nuclei across the whole sd shell. They found that good fits could be obtained with a two-body effective interaction if the interaction strength was scaled according to an $e^{-0.3A}$ power law to take account of the changing volume of nuclei which was presumed to scale linearly with the nucleon number, A. They also found their fitted two-body effective interactions to be in general agreement with the two-body components of effective interactions derived from free-nucleon interactions. Similar successes have been achieved for the fp-shell nuclei by the Strasbourg group[270] and for intermediate and heavy mass nuclei, albeit with severely truncated shell-model spaces, by the Oak Ridge-Oslo and other groups.[271] The results lead one to believe that the shell model, with restriction to a single major shell, provides a reasonable effective-model approximation to a microscopic theory of a subset of states in light and singly-closed shell nuclei and can be used with a degree of confidence for the interpretation of these states. (Comments on the interpretation of observed states that are not described within the space of a single-shell are given below.)

Some of the best evidence in support of the shell model is that it gives simple and convincing physical interpretations in numerous situations.[272] Examples are the interpretation of the spectra of odd-mass nuclei in terms of an extra nucleon coupled to an even-mass core. It appears that the coupling is weak in some nuclei and strong in others. Thus, by examining the relative properties of adjacent even-

---

[268]For applications to $1p$-shell nuclei, see, for example, Cohen S. and Kurath D. (1965), *Nucl. Phys.* **73**, 1, and Cohen S. and Kurath D. (1967), *Nucl. Phys.* **A101**, 1.

[269]For applications to $2s1d$-shell nuclei, see, for example Brown and Wildenthal, *op. cit.* Footnote 151 on Page 293.

[270]For a review, see Caurier *et al.*, *op. cit.* Footnote 59 on Page 267.

[271]For a review, see Dean *et al.*, *op. cit.* Footnote 58 on Page 267.

[272]Many examples are given in the book by Talmi, *op. cit.* Footnote 258 on Page 380.

and odd-mass nuclei and interpreting the results in terms of a particle-plus-core shell model, the extra nucleon becomes an effective probe of what is going on inside the core. For example, the low-energy spectra of ^{209}Bi and ^{209}Pb, shown in Figure 1.13, are rather well described by the weak coupling of a nucleon to a closed-shell ^{208}Pb core. This suggests that a shell model with a closed-shell ^{208}Pb core might provide a meaningful description of the $N = 126$ isotones and $Z = 82$ isotopes; which indeed appears to be the case. There are many such examples.

Other sequences of isotopes or isotones have been identified which have simple shell-model interpretations in terms of more than one nucleon coupled to a core. For

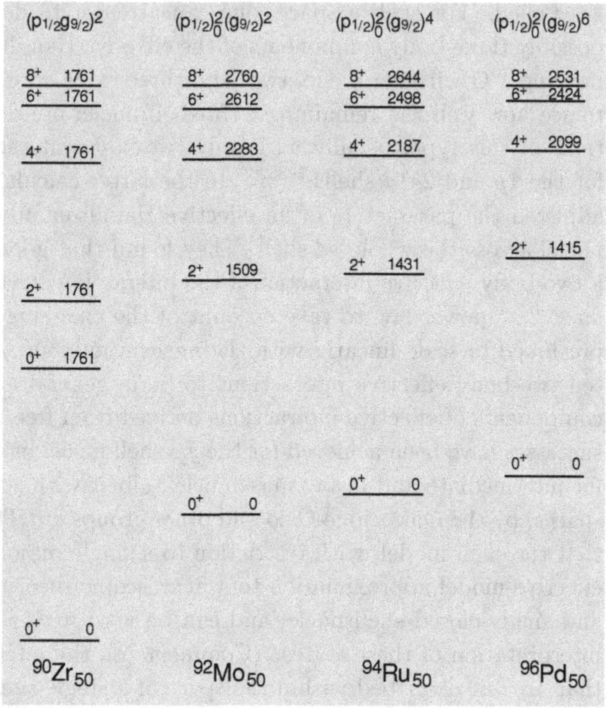

Figure 5.15: Energy levels of a sequence of $N = 50$ isotones. A simple shell model description of the states shown is given by considering a few active protons in just the $2p_{1/2}$ and $1g_{9/2}$ shells. (The data are from Nuclear Data Sheets.)

example, Figure 5.15 shows the energy-levels of a sequence of $N = 50$ isotones. In a simple shell model, all these isotones would have their 50 neutrons in closed shells including the $2p1f$ and $1g_{9/2}$ single-particle shells. Then, to a first approximation, one might suppose the protons in the ^{90}Zr$_{50}$ ground state to close the $2p1f$ shell and the extra protons in the succeeding $A > 40$ isotones to be occupying the $1g_{9/2}$ orbital. With this assumption, the spectra of the $Z > 40$ isotones have a simple interpretation in terms of a proton pair-coupling model (cf. Chapter 6) and the USp($2j+1$) coupling scheme of Section 5.13.1. From this explanation, the question arises as to why the closed-shell nucleus ^{90}Zr$_{50}$ exhibits similar sequences of states to the other isotones, albeit at higher excitation energies. It is then recalled that there is only a small gap between the $2p_{1/2}$ and $1g_{9/2}$ shells (cf. Figure 1.17). Thus, a natural interpretation is that the excited states of ^{90}Zr$_{50}$ correspond to the excitation of the two protons that occupy the $2p_{1/2}$ orbit in the closed-$2p1f$-shell ground state into a $(1g_{9/2})^2_J$ configuration, as in ^{92}Mo$_{50}$. One can then ask: why does the first excited 0^+ state not follow the same trend as its counterparts in the $Z > 40$ isotones? An immediate answer is that there must be a mixing of the $(2p_{1/2})^2_{J=0}$ and $(1g_{9/2})^2_{J=0}$ states which pushes them apart. This interpretation can be tested by further experimental observations and detailed shell-model calculations.[273,274,275,276] Pair-coupling submodels of the shell model, of particular relevance to singly-closed shell nuclei, are treated in depth in the following chapter.

The situation in the middle of doubly open-shell medium and heavy nuclei is the most challenging for the shell model. As discussed in Chapter 1, the data suggest that such nuclei have predominantly ellipsoidal shapes and exhibit rotational bands. They are the nuclei for which the collective rotational model is most successful. Thus, it is desirable to determine if the shell model can provide a microscopic description of these nuclei. Again considerable insight is obtained by considering the extra nucleon in an odd-mass nucleus as a probe of the even-even core. It turns out, as will be discussed in some depth in Volume 2, that the spectral properties of an odd-mass nucleus in the region of rotational nuclei is, generally, well described in terms of a strongly-coupled particle-core (Nilsson) model with a nucleon moving in the shell-model potential generated by a deformed rotational even-even core. The success of the model is again a success for the shell model. However, an examination of the wave function for the extra-core particle indicates that, while it is simply expressed in terms of a deformed harmonic oscillator (Nilsson model) basis, its expression in terms of standard spherical-shell-model single-particle basis wave functions requires a superposition of wave functions from many spherical harmonic oscillator shells. This is evidence that a full description of many-nucleon rotational states will likewise require shell-model states with components coming from many major oscillator shells. Similar evidence is provided by the examination of E2 transition rates in heavy nuclei and their interpretation in terms of the Nilsson model.[277] Fortunately, it turns out that the task of constructing such a shell-model theory of nuclear rotational states is not as formidable as might be supposed.

A successful many-nucleon description of rotational nuclei is provided by the *unified model* of Bohr and Mottelson (see Volume 2).[278,279] This model, which is a hybrid with both collective and many-nucleon degrees of freedom, has made a huge contribution to the understanding of nuclear collective phenomena. More-

[273] Talmi I. and Unna I. (1960), *Nucl. Phys.* **19**, 225.

[274] Ji X. and Wildenthal B.H. (1988), *Phys. Rev.* **C37**, 1256.

[275] Ji X. and Wildenthal B.H. (1988), *Phys. Rev.* **C38**, 2849.

[276] Sinatkas J. *et al.* (1992), *J. Phys. G: Nucl. Part. Phys.* **18**, 1377.

[277] Jarrio M., Wood J.L. and Rowe D.J. (1991), *Nucl. Phys.* **A528**, 409.

[278] Bohr and Mottelson, *op. cit.* Footnote 5 on Page 98.

[279] Bohr and Mottelson, *op. cit.* Footnote 6 on Page 98.

over, it gives clear indications of what shell-model configurations are needed for a meaningful description of nuclear rotations.

A major step towards a shell-model theory of nuclear rotations was achieved with Elliott's SU(3) coupling scheme.[280],[281] This coupling scheme is mentioned in Section 5.13.4 and will be considered in depth in Volume 2. Bands of states, with properties similar to those of the rigid-rotor model, emerge in the SU(3) coupling scheme when it is used with a schematic $Q \cdot Q$ interaction, where Q is the nuclear quadrupole-moment tensor restricted to a single major harmonic-oscillator shell. The SU(3) model has subsequently proved to be central to the development of a microscopic theory of quadrupole collective dynamics.

A microscopic model of quadrupole collective motion, which is a realisation of the macroscopic collective model as a submodel of the shell model, was sought for many years. A breakthrough came with the discovery that the simplest algebraic structure that contains both the many-nucleon quadrupole moments and the many-nucleon kinetic energy is the symplectic Lie algebra sp(3, $\mathbb{R}$) (cf. Sections 5.8.4 and 5.13.4).[282] Moreover, this Lie algebra contains infinitesimal generators of monopole and quadrupole vibrations, and of both irrotational and vortex-spin rotations. Thus, the symplectic shell model incorporates the dynamics of the Bohr collective model and, in addition, giant-resonance monopole and quadrupole vibrations, vortex-spin rotations, and the dynamics of Elliott's SU(3) model, which it contains as a submodel. A shell-model *LST* coupling scheme, which makes use of the subgroup chain

$$\text{Sp}(3, \mathbb{R}) \supset \text{SU}(3) \supset \text{SO}(3), \tag{5.687}$$

will be considered in Volume 2.

The primary importance of a microscopic collective model is that it enables the macroscopic collective-model perspectives of nuclear structure to be embraced within the formal framework of many-nucleon theory in shell-model terms. This identifies the essential approximations implicit in the model, the circumstances under which they are valid, and what needs to be done to remedy their deficiencies. Conversely, phenomenological models with microscopic realisations have the potential to provide physically meaningful interpretations of complicated shell-model states. However, it should be recognised that, while giving a model a microscopic expression opens up the possibility of asking deeper questions about the interpretation of some observed phenomena, the answers may not come easily. For example, finding efficient ways to handle the mixing of Sp(3, $\mathbb{R}$) irreps due to pairing and other residual interactions, proves to be a major challenge.

As often happens, valuable insights into the physics of a situation are obtained from the failures of a model. For example, for a closed-shell nucleus such as ^{16}O, the

---

[280]Elliott, *op. cit.* Footnote 214 on Page 347.

[281]Elliott, *op. cit.* Footnote 247 on Page 378.

[282]The historical developments that culminated in the symplectic model are reviewed in the article Rowe, *op. cit.* Footnote 66 on Page 222.

standard shell model predicts the lowest-energy excited states to be negative-parity states in which one nucleon is promoted from the occupied $1p$ single-particle shell to the $2s1d$ shell. In fact, the first excited state of ^{16}O, at 6.05 MeV, is of positive parity and is the lowest state of a band of states with rotational properties; cf. Figure 1.73. Examination of the properties of this band of states and shell-model calculations in large multishell spaces, indicate that the 6.05 MeV state is a so-called 4-particle–4-hole state and that it belongs predominantly to a $(1p)^{-4}(2s1d)^4$ configuration.[283,284,285,286] Subsequently, many states with unexpected deformations have been observed in the low-energy spectra of nuclei. Dramatic examples are given by the Sn isotopes (cf. Figure 1.76) which are traditionally viewed as singly-closed-shell nuclei with spherical ground states. However, their energy-level spectra exhibit rotational-like bands at quite low excitation energies. The phenomenon is known as *shape coexistence*.[287] A standard shell-model explanation of the appearance of such unexpected states in the low-energy domain is to characterise them as *intruder states*.[288] The reason they show up at low energies has a simple explanation[289] in shell-model terms as follows.

The need for one-particle–one-hole configurations in describing negative-parity excitations of even nuclei is expected. However, the widespread need for multi-particle-multi-hole configurations for the description of some low-energy positive parity states in closed- and singly-closed-shell nuclei might appear disturbing until it is recognised that the excitation of neutrons in an otherwise closed-neutron shell nucleus or the excitation of protons in an otherwise closed-proton shell nucleus creates a doubly open-shell situation and the potential for highly-correlated, low-energy, many-nucleon states. Moreover, the suggestion that multi-particle-multi-hole states generate low-lying rotational bands, is consistent with the predominance of rotational structure in doubly open-shell nuclei. Thus, the observation of shape-coexisting states and their interpretation in shell-model terms, is an indication that one should not expect a given valence shell-model space to be capable of describing all the low-energy states of a given nucleus.

In this book, we make the assumption that nuclear states can, in principle, be described in shell-model terms. However, we do not assume that all low-energy states can be described in terms of any given shell-model space. We recommend the following strategy. The first step towards identifying a suitable shell-model space for the description of some nuclear phenomenon is profitably tackled by systematically studying a wide range of experimental observations of the phenomenon and constructing phenomenological models to explain what is observed. The next step

---

[283]Morinaga, *op. cit.* Footnote 31 on Page 86.

[284]Talmi I. and Unna I. (1962), *Nucl. Phys.* **30**, 280.

[285]Brown G.E. and Green A.M. (1965), *Phys. Lett.* **15**, 168.

[286]Brown and Green, *op. cit.* Footnote 33 on Page 88.

[287]For a review of shape coexistence, see Wood *et al.*, *op. cit.* Footnote 60 on Page 221.

[288]Heyde K. *et al.* (1987), *Nucl. Phys.* **A466**, 189.

[289]Rowe *et al.*, *op. cit.* Footnote 34 on Page 92.

is to identify the models as submodels of the shell model. This identification is naturally achieved when a model can be expressed in terms of a dynamical group, or a spectrum generating algebra, which has representations on subspaces of the shell model. Of course, this may not be possible or it may be possible only with some adjustment of the model to make it compatible with the microscopic structure of the nucleus as a system of interacting nucleons. Assuming it can be done, the model can then be viewed as a submodel of the shell model with restricted dynamics which, in practice, means a shell model with a correspondingly truncated Hilbert space. The penultimate step is to determine if the phenomenon of interest can be described by a shell-model calculation in such a truncated space. The final step is to understand if the dynamical symmetries of the model are well-conserved by a realistic many-nucleon Hamiltonian; in other words, find out if the states inside the truncated Hilbert space are coupled relatively weakly to states outside of this space. If this is not the case, it means that the dynamical symmetries of the model are not well-conserved. One should then enquire if the success of the phenomenological model is fortuitous or not.[290] If one can follow through all of these steps, one will have learnt a lot about the physics of the nucleus.

The above strategy is based on the observation that it is generally much easier to substantiate a given interpretation of a phenomenon than to derive it. Similarly, while it is virtually impossible to predict the properties of complex nuclei from first principles, it may be possible to ascertain whether or not a model explanation of what is observed is consistent with a first-principles theory. Thus, the development of sophisticated shell-model codes that can work with many choices of basis states and coupling schemes is extremely important for the development of nuclear theory, even if their first-principles predictive powers are limited.

It is beyond the scope of this book to enter into the details of shell model technology and its numerous applications. The following are a few key resources that give a sense of the evolution of the subject:

- Elliott J.P. and Flowers B.H. (1955), *Proc. Roy. Soc. London* **A229**, 536;
- Elliott J.P. and Lane A.M. (1957), in *Handbuch der Physik*, Vol. 39, edited by S. Flügge (Springer-Verlag, Berlin);
- de Shalit and Talmi, *op. cit.* Footnote 257 on Page 380;
- Cohen S. and Kurath D. (1965), *Nucl. Phys.* **73**, 1; Cohen S. and Kurath D. (1967), *Nucl. Phys.* **A101**, 1;
- French J.B. *et al.* (1969), in *Advances in Nuclear Physics, Vol. 3*, edited by

---

[290]It can happen that states in a subspace of the shell model may, in fact, be quite strongly coupled to states outside of the subspace but, nevertheless, the subspace serves very well as an effective shell-model space (with suitably adjusted effective shell-model operators) for representing the essential physics of a given situation. In this situation, the dynamical symmetry of the model is an *effective dynamical symmetry* or *quasi-dynamical symmetry* for the model images of the states in question; cf. the review on quasi-dynamical symmetry by Rowe D.J. (2004), in *Int. Conf. on Computational and Group-Theoretical Methods in Nuclear Physics*, edited by J. Escher *et al.* (World Scientific, Singapore).

M. Baranger and E. Vogt (Plenum, New York), p. 193;

- Bertsch G.F. (1972), *The Practitioner's Shell Model* (North-Holland, Amsterdam);
- Whitehead *et al.*, *op. cit.* Footnote 231 on Page 371;
- Brussaard P.J. and Glaudemans P.W.M. (1977), *Shell-Model Applications in Nuclear Spectroscopy* (North-Holland, Amsterdam);
- Lawson R.D. (1980), *Theory of the Shell Model* (Oxford University Press);
- Brown and Wildenthal, *op. cit.* Footnote 151 on Page 293;
- Heyde K.L.G. (1990), *The Nuclear Shell Model* (Springer-Verlag, Berlin);
- Talmi, *op. cit.* Footnote 258 on Page 380;
- Martínez-Pinedo G. *et al.* (1997), *Phys. Rev.* **C55**, 187;
- Koonin S.E., Dean D.J. and Langanke K. (1997), *Phys. Repts.* **278**, 2;
- Honma M., Mizusaki T. and Otsuka T. (1995), *Phys. Rev. Lett.* **75**, 1284; Mizusaki T. *et al.* (1999), *Phys. Rev.* **C59**, R1846;
- Brown, *op. cit.* Footnote 56 on Page 267;
- Otsuka *et al.*, *op. cit.* Footnote 57 on Page 267
- Otsuka T., Tanihata I. and Sakurai H., eds. (2002), *Proceedings of The International Symposium Shell Model 2000, Nucl. Phys.*, Vol. A704;
- Dean *et al.*, *op. cit.* Footnote 58 on Page 267;
- Caurier *et al.*, *op. cit.* Footnote 59 on Page 267.

Chapter 6

Pair-coupling models

6.1 Introduction

The tendency of Fermi particles to pair has been recognised in several systems. It was understood in atomic physics many years ago and motivated Racah[1],[2] to introduce the seniority coupling scheme. Its importance in nuclear physics was also recognised and used by Mayer[3] in 1950 to explain why the ground states of even-even nuclei have spin zero while those of odd-mass nuclei have the spin of the last unpaired nucleon. Such considerations also led Flowers[4] and Racah and Talmi[5] in 1952 to introduce the seniority coupling scheme into nuclear physics. A major advance in the understanding of pairing phenomena came with the theory of superconductivity. For some time the close similarity between the properties of superfluid ^{4}He and superconducting metals was a mystery. It was understood that helium owed its superfluid properties to the bosonic nature of the ^{4}He atom. But the electrons in a metal are fermions! The connection was established by Cooper[6] who showed, in 1956, that fermions of opposite spin in a metallic lattice have an attractive effective interaction and form correlated boson-like pairs, subsequently referred to as *Cooper pairs*. Recent overviews of pairing in nuclei and its connections to non-nuclear systems has been given by Dean and Hjorth-Jensen[7] and Brink and Broglia.[8]

Evidence for $J = 0$ pair-coupling of nucleons in nuclei is given in Sections 1.4 and 1.5. Without exception, all doubly-even nuclei are observed to have $J = 0$ ground states. This suggests that identical nucleons have a predisposition to form $J = 0$ coupled pairs. Support for this suggestion is provided by the staggering of one-

---

[1]Racah G. (1942), *Phys. Rev.* **62**, 438.

[2]Racah, *op. cit.* Footnote 233 on Page 371.

[3]Mayer M.G. (1950), *Phys. Rev.* **78**, 16.

[4]Flowers, *op. cit.* Footnote 232 on Page 371.

[5]Racah G. and Talmi I. (1952), *Physica* **18**, 1097.

[6]Cooper, *op. cit.* Footnote 10 on Page 24.

[7]Dean D.J. and Hjorth-Jensen M. (2003), *Rev. Mod. Phys.* **75**, 607.

[8]Brink D.M. and Broglia R.A. (2005), *Nuclear superfluidity: pairing in finite systems* (Cambridge University Press, Cambridge).

nucleon separation energies. As illustrated in Figure 1.8 for the calcium isotopes, all isotopic sequences of nuclei exhibit an odd-even staggering in S_n (the one-neutron separation energy), and all isotonic sequences exhibit an odd-even staggering in S_p (the one-proton separation energy). Direct evidence for pairing is also provided by the favouring of $J = 0$ states of two nucleons in a j^2 configuration. This is illustrated in Figure 6.1 which shows that, for states of the ^{210}Po nucleus, it requires an energy of some 1.2 MeV to break the $J = 0$ coupling of the pair of protons interpreted as occupying $j = 9/2$ single-particle states outside of a closed-shell ^{208}Pb core. Such an energy gap, attributed to a $J = 0$ pairing interaction, is a dramatic feature of a wide range of doubly-even nuclei (cf. Figure 1.29).

Figure 6.1: Low-lying energy levels of ^{210}Po and ^{210}Bi (cf. Figures 1.26 and 1.27). The observed energy levels are compared to those calculated with a zero-range (delta), a $J = 0$ pairing interaction, and a realistic effective interaction. (The spectrum of ^{210}Bi for a zero-range force was computed by Talmi, *op. cit.* Footnote 258 on Page 380 and that for a realistic interaction by Coraggio *et al.*, *op. cit.* Footnote 163 on Page 295.)

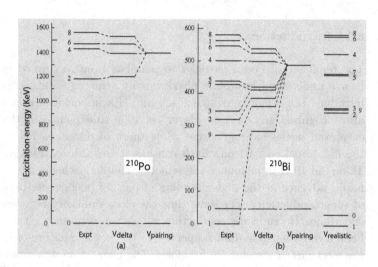

In general, pair-coupling of nucleons gives rise to a type of collectivity quite different from that of the shape-vibrations and rotations of the Bohr model. The perspective gained from studies of the two types of collective motion in nuclei is that shape vibrations and rotations are associated with long-range correlations between many nucleons and are most in evidence in doubly-open-shell nuclei. In contrast, pairing correlations are associated more with short-range nucleon interactions and tend to be dominant in singly-closed-shell nuclei. The kind of collectivity associated with pairing correlations is described as *superconductivity*, especially in macroscopic and mesoscopic systems.[9] The remarkable fact is that both kinds of collective motion in nuclei are characterised by dynamical symmetries and shell-model coupling schemes. In this chapter we explore the dynamical symmetries and coupling schemes associated with pairing interactions.

Insight into the interactions occurring within a two-proton pair and within a neutron-proton pair are obtained from a comparison of the low-lying energy levels of

---

[9]The theory of nuclear superconductivity was introduced into nuclear physics by Bohr *et al.*, *op. cit.* Footnote 15 on Page 33.

^{210}Po and ^{210}Bi, shown in Figure 6.1. Figure 6.1(a) also shows the spectrum of ^{210}Po modelled as a closed-shell ^{208}Pb core plus two protons in $1h_{9/2}$ single-particle states: first with a zero-range interaction, $V_{\text{delta}} = V_0\delta(|\mathbf{r}_1 - \mathbf{r}_2|)$, between them and, second, with a $J = 0$ pairing interaction. The delta interaction is seen to be quite successful in describing the two-proton spectrum. Even the simple pairing interaction, which lowers the energy of just the $J = 0$ two-proton pair state relative to the other $J \neq 0$ states, is qualitatively successful (note that the zero of the energy scale is set to be the energy of the ground state which, for ^{210}Po, is the $J = 0$ state). As the following shows, a Hamiltonian with just a pairing interaction has the big advantage of having a simple dynamical symmetry group which makes it easy to compute the pair-coupling spectra for sequences of singly-closed-shell isotones and isotopes. Moreover, it is shown that the $J = 0$ pair-coupling scheme admits more general *seniority conserving* interactions and is capable of giving results substantially better than those obtainable with a simple delta force. Thus, considerable insight into the characteristics of some major coupling schemes are gained from a consideration of the simple pair-coupling models.

Calculations for ^{210}Bi, modelled as a closed-shell ^{208}Pb core plus a proton in a $1h_{9/2}$ single-particle state and a neutron in a $2g_{9/2}$ single-particle state are shown in Figure 6.1(b). The figure shows that the delta and $J = 0$ pairing interactions are not as successful in approximating the neutron-proton interaction in ^{210}Bi as they are in approximating the two-proton interaction in ^{210}Po, particularly as regards the $J = 1$ state. This suggests that the $J = 0$ pairing interaction is a less dominant component of the two-nucleon interaction when neutrons and protons are both present. Two primary differences between neutron-proton and two-proton interactions in heavy nuclei are apparent: (i) the neutron is occupying a different single-particle shell relative to the proton, and (ii) the antisymmetry requirements for a two-proton pair and a neutron-proton pair are different. For example, whereas antisymmetry forbids two $j = 9/2$ protons from coupling their angular momenta to $J = 9$ and 1, such alignment (anti-alignment) with maximal overlaps of their wave functions, and hence a lowering of the two-nucleon $J = 1, 9$ energies, is possible for a neutron-proton pair.

Recall that, while two protons can occur only in isospin $T = 1$ pair combinations, a neutron-proton pair can have both $T = 1$ and $T = 0$. Moreover, in an *LST* coupling scheme, a spatially symmetric two-nucleon pair must be complemented by $(S = 0, T = 1)$ or $(S = 1, T = 0)$ spin-isospin wave functions. This suggests that $J = 1$ states with large $L = 0$, $S = 1$, $T = 0$ components should also be lowered by short-range interactions.

For comparison, Figure 6.1 shows the results of the first successful shell-model calculation of the spectrum of ^{210}B with realistic effective interactions determined according to the methods of Section 5.7.6.[10] These calculations showed the necessity of including core excitations to obtain the $J^\pi = 1^-$ state below the 0^+ state.

---

[10]Coraggio *et al.*, *op. cit.* Footnote 163 on Page 295.

For all the above reasons, the $J = 0$ pair-coupling models are considered to be most appropriate for singly-open-shell nuclei whereas $L = 0$ pair-coupling may be important in doubly-open-shell nuclei, especially in light nuclei for which the neutrons and protons are filling the same single-particle shells.

6.2   Time-reversal and the su(2) quasi-spin algebra

It is important to take time reversal into account in a treatment of nucleon pairing forces because if two nucleons approach one another in time-reversed states then, for a time-reversal invariant interaction, such as the strong interaction is understood to be, the nucleons can only scatter into states that continue to be time reverses of one another.

6.2.1   *The time-reversal operation*

The time-reversal operation is equivalent to complex conjugation combined with rotation of spin wave functions through an angle π about the y-axis;[11] the direction of rotation is a matter of convention. Thus, the transformation of a spin wave function under time reversal is given by

$$\chi_m \to e^{i\pi \hat{s}_y} \chi_m = (-1)^{\frac{1}{2}+m} \chi_{-m}, \tag{6.1}$$

whereas a spherical harmonic transforms as

$$Y_{lm} \to Y_{lm}^* = (-1)^m Y_{l,-m} = (-1)^l e^{i\pi \hat{L}_y} Y_{lm}. \tag{6.2}$$

Thus, by adding an i^l phase factor to the definition of a spin-orbit coupled single-particle wave function,

$$\psi_{nljm}(r,\theta,\varphi) = R_{nl}(r)\, i^l \, [Y_l \otimes \chi]_{jm}, \tag{6.3}$$

its transformation under time reversal is given by

$$\psi_{nljm} \to \psi_{nlj\bar{m}} = e^{i\pi \hat{J}_y} \psi_{nljm} = (-1)^{j+m} \psi_{nlj,-m}. \tag{6.4}$$

Also important is the fact that, with this phase convention, the sign of a matrix element $\langle (j)^2 J=0|V|(j')^2 J=0\rangle$ of a short-range interaction is invariably the same for all j and j' (cf. Exercise 5.1).[12]

The definition (6.4) is consistent with the conventions already introduced for covariant spherical tensors and their contravariants. Recall (cf. Sections 5.3 and

---

[11]Messiah A. (1966), *Quantum Mechanics*, Vol. 1 (North Holland, Amsterdam), fourth printing, Ch. XV.

[12]An analysis of alternative phase conventions can be found in Appendix A of Rowe, *op. cit.* Footnote 4 on Page x.

A.7.2) that upper and lower indices for fermion operators are defined by

$$a^{\alpha jm} := (-1)^{j+m} a_{\alpha j, -m} := \left(a^{\dagger}_{\alpha jm}\right)^{\dagger}. \tag{6.5}$$

Thus, it is appropriate to define creation and annihilation operators for time-reversed single-particle states by

$$a^{\dagger}_{\alpha j \bar{m}} := (-1)^{j+m} a^{\dagger}_{\alpha j, -m}, \quad a_{\alpha j \bar{m}} := (-1)^{j+m} a_{\alpha j, -m} = a^{\alpha jm}. \tag{6.6}$$

Then, because $\{a^{\alpha jm}, a^{\dagger}_{\alpha jn}\} = \delta_{m,n}$, we obtain the convenient expressions for the fermion anticommutation relations

$$\{a_{\alpha j \bar{m}}, a^{\dagger}_{\beta j' m'}\} = \delta_{\alpha, \beta} \delta_{j, j'} \delta_{m, m'}, \quad \{a_{\alpha jm}, a^{\dagger}_{\beta j' \bar{m}'}\} = -\delta_{\alpha, \beta} \delta_{j, j'} \delta_{m, m'}. \tag{6.7}$$

These definitions are consistent with those of Section 5.12.10. They imply that Hermitian adjoint relationships are given by

$$\left(a^{\dagger}_{\alpha jm}\right)^{\dagger} = a_{\alpha j \bar{m}}, \quad \left(a_{\alpha jm}\right)^{\dagger} = -a^{\dagger}_{\alpha j \bar{m}}, \tag{6.8}$$

and that applying the time-reversal operation twice reverses signs, i.e.,

$$a^{\dagger}_{\alpha j \bar{\bar{m}}} = -a^{\dagger}_{\alpha jm}, \quad a_{\alpha j \bar{\bar{m}}} = -a_{\alpha jm}. \tag{6.9}$$

6.2.2 The su(2) quasi-spin algebra

We now consider a pairing model for which the interactions between nucleons are non-zero only when the nucleons are in $J = 0$ states. Thus, if $i \equiv (\alpha_i j_i)$ and $k \equiv (\alpha_k j_k)$ index sets of single-nucleon shells in jj coupling, a $J = 0$ pair-coupling model Hamiltonian is of the form

$$\hat{H} := \sum_i \varepsilon_i \hat{n}_i - \sum_{ik} G_{ik} \hat{S}^i_+ \hat{S}^k_-, \tag{6.10}$$

where

$$\hat{n}_i = \sum_{m=-j_i}^{j_i} a^{\dagger}_{im} a^{im} = \sum_{m>0} (a^{\dagger}_{im} a_{i\bar{m}} - a^{\dagger}_{i\bar{m}} a_{im}) \tag{6.11}$$

is a number operator and

$$\hat{S}^i_+ := \sum_{m>0} a^{\dagger}_{im} a^{\dagger}_{i\bar{m}}, \quad \hat{S}^i_- := \left(\hat{S}^i_+\right)^{\dagger} = \sum_{m>0} a_{i\bar{m}} a_{im} \tag{6.12}$$

are, respectively, creation and annihilation operators for a $J = 0$ coupled and time-reversal-invariant two-nucleon pair (see Exercise 6.5).[13]

---

[13]The theory of superconductivity of Bardeen *et al.*, *op. cit.* Footnote 11 on Page 24, which is treated in Section 6.8, makes use of generalised fermion operators, $\{\alpha^{\dagger}_{jm}\}$, that are linear combinations of nucleon creation and annihilation operators. To conserve angular momentum, these operators are restricted to linear combinations of operators $\{a^{\dagger}_{\alpha jm}\}$ and $\{a_{\beta jm}\}$ having common values of j and m. Thus, in pairing theory, it is useful to work with creation and annihilation operators labelled by lower (covariant) indices.

The eigenvalue equations for Hamiltonians, such as that of Equation (6.10), are relatively easy to solve (analytically in simple cases) because the operators $\hat{S}^i_{\pm}$ are raising and lowering operators of su(2) Lie algebras,[14] i.e., they satisfy the commutation relations

$$[\hat{S}^i_+, \hat{S}^k_-] = \delta_{ik} 2\hat{S}_0, \quad [\hat{S}^i_0, \hat{S}^k_{\pm}] = \pm\delta_{ik}\hat{S}^k_{\pm}, \tag{6.13}$$

where

$$\hat{S}^i_0 := \tfrac{1}{2}\sum_m (a^\dagger_{im} a_{i\bar{m}} - 1) = \tfrac{1}{2}(\hat{n}_i - \Omega_i), \tag{6.14}$$

and

$$\Omega_i := \tfrac{1}{2}(2j_i + 1) \tag{6.15}$$

denotes the maximum number of $J = 0$ pairs that can be put into the j_i shell. Thus, the Hamiltonian for a $J = 0$ pairing model is expressible entirely in terms of the infinitesimal generators of a set of $\mathcal{SU}(2)$ groups known as *quasi-spin* groups.[15]

Recall that the standard, angular-momentum, su(2) algebra has irreps with basis states $\{|JM\rangle; M = -J, \ldots, +J\}$ labelled by JM quantum numbers and that the operators, $\{\hat{J}_0, \hat{J}_{\pm}\}$, of this algebra act according to the equations (see Section 3.7.2)

$$\hat{J}_0|JM\rangle = M|JM\rangle,$$
$$\hat{J}_{\pm}|JM\rangle = \sqrt{(J \mp M)(J \pm M + 1)}\,|J, M \pm 1\rangle. \tag{6.16}$$

Similarly, the quasi-spin su(2) algebra has irreps with basis states $\{|s\nu\rangle; \nu = -s, \ldots, +s\}$ labelled by quasi-spin quantum numbers, $s\nu$, on which $\hat{S}_0$ and $\hat{S}_{\pm}$ act according to the equations

$$\hat{S}_0|s\nu\rangle = \nu|s\nu\rangle,$$
$$\hat{S}_{\pm}|s\nu\rangle = \sqrt{(s \mp \nu)(s \pm \nu + 1)}\,|s, \nu \pm 1\rangle. \tag{6.17}$$

Exercises

6.1 Show that if a position vector $\mathbf{r}$ is assigned spherical polar coordinates (r, θ, φ) then the delta function $\delta(\mathbf{r}_1 - \mathbf{r}_2)$ is expressible in the form

$$\delta(\mathbf{r}_1 - \mathbf{r}_2) = \frac{\delta(r_1 - r_2)}{r_1^2}\sum_{lm} Y_{lm}(\theta_1, \varphi_1)\, Y^*_{lm}(\theta_2, \varphi_2). \tag{6.18}$$

---

[14]Recall that we use lower case type to distinguish a Lie algebra from its group; e.g., su(2) is the Lie algebra of the group SU(2).

[15]We use the notation $\mathcal{SU}(2)$ instead of SU(2) to avoid confusion with the number-conserving SU(2) groups. The concept of quasi-spin was introduced into nuclear physics by Kerman, *op. cit.* Footnote 184 on Page 326; cf. also Wada Y., Takano F. and Fukuda N. (1958), *Prog. Theor. Phys.* **19**, 597; Anderson P.W. (1958), *Phys. Rev.* **112**, 1900.

Then use the fact that the time reverse of a number is just complex conjugation to show that the two-body matrix elements of a delta-function potential $V(\mathbf{r}_1 - \mathbf{r}_2) = V_0 \delta(\mathbf{r}_1 - \mathbf{r}_2)$ satisfy

$$\langle \mu \bar{\mu} | V | \nu \bar{\nu} \rangle = V_0 F \sum_{lm} |\langle \mu | Y_{lm} | \nu \rangle|^2, \qquad (6.19)$$

where F is positive-definite and proportional to a radial integral.

6.2 Show that the su(2) commutation relations of Equation (6.13) follow from the definitions of the operators, $\hat{S}^i_{\pm}$, $\hat{S}^i_0$, and the anticommutation relationships (6.7).

6.3 Use Equation (6.17) to show that a state with quasi-spin quantum numbers $|s\nu\rangle$ can be expressed in terms of its lowest-weight state, $|s, -s\rangle$, by

$$|s\nu\rangle = \sqrt{\frac{(s - \nu)!}{(2s)!(s + \nu)!}} \, \hat{S}_+^{s+\nu} |s, -s\rangle. \qquad (6.20)$$

6.4 Consider the fermion-pair creation operators

$$c^\dagger = \frac{1}{\sqrt{2s}} \hat{S}_+, \quad c = \frac{1}{\sqrt{2s}} \hat{S}_-. \qquad (6.21)$$

Let $|p\rangle = |s\nu\rangle$ be a state of a quasi-spin irrep with $p = s + \nu$. Show that

$$[c, c^\dagger] |p\rangle = \left(1 - \frac{p}{2s}\right) |p\rangle. \qquad (6.22)$$

Hence show that, when s is large, c and $c^\dagger$ behave as boson operators, with commutation relations

$$[c, c^\dagger] \approx 1, \qquad (6.23)$$

when their actions are restricted to states of $p \ll s$. They are referred to as *quasi-boson* operators.

6.5 Show that, because the time-reversal operation on a single-fermion creation operator changes the sign of the operator when applied twice with the phase convention chosen in the text, the quasi-spin step-up operator, $\hat{S}^j_+$, is a creation operator for a time-reversal-invariant two-nucleon pair.

6.3 The single-shell $J = 0$ pairing model

Within the space of states for nucleons of a single type (i.e., only neutrons or only protons) in a single-j shell, the i and k indices take a single value and the pairing Hamiltonian (6.10) reduces to

$$\hat{H} = \varepsilon \hat{n} - G \hat{S}_+ \hat{S}_-, \qquad (6.24)$$

where $\hat{n} = 2\hat{S}_0 + \Omega$, with $\Omega = \frac{1}{2}(2j + 1)$, is the nucleon number operator for the j shell. Because of its rotational invariance, $\hat{H}$ has eigenstates $\{|JMs\nu\rangle\}$ labelled by angular-momentum quantum numbers JM, quasi-spin quantum numbers $s\nu$, and such other quantum numbers as are needed. From Equation (6.17), it follows directly that these states have energy eigenvalues given by

$$E(s,\nu) = n\varepsilon - G(s+\nu)(s-\nu+1), \tag{6.25}$$

where $n = 2\nu + \Omega$ is the number of nucleons in the shell.

To determine the model spectrum for each n it is necessary to consider the range of values that the quantum numbers J and s can take. First observe that a classification of states by means of quasi-spin quantum numbers is equivalent to specifying the number of nucleons that occur in a state in $J = 0$ coupled pairs and, conversely, the number of nucleons, v, called *seniority*,[16] that are coupled to $J \neq 0$. The lowest-weight state for a quasi-spin irrep is a state of $\nu = -s$ that is annihilated by the $\hat{S}_-$ operator. This state contains no $J = 0$ pairs and its seniority v is equal to its nucleon number. Other states of a quasi-spin irrep are constructed by adding $J = 0$ coupled pairs to a lowest-weight state; they have the same seniority as the lowest-weight state of the irrep. Hence, the quasi-spin quantum numbers $s\nu$ are related to the nucleon number and seniority by

$$2\nu = n - \Omega, \quad 2s = \Omega - v, \quad 2\Omega = 2j + 1. \tag{6.26}$$

These relationships are most easily understood by consideration of an example.

The j-shell vacuum state $|0\rangle$, being a state of seniority zero, is the lowest-weight state for a quasi-spin irrep with $s = \Omega/2 = (2j+1)/4$. The single-nucleon, $n = 1$, states $\{a_{jm}^\dagger|0\rangle; m = -j, \ldots, +j\}$ are likewise lowest-weight quasi-spin states; they have seniority $v = 1$ and quasi-spin $s = (\Omega - 1)/2$. Among the two-nucleon states $|(j)^2 JM\rangle$, the $J = 0$ state, proportional to $\hat{S}_+|0\rangle$, is a member of the quasi-spin irrep whose lowest-weight state is the zero-seniority vacuum state. It has seniority $v = 0$ and quasi-spin $s = \Omega/2$. The remaining two-nucleon states, $|(j)^2 JM\rangle$ with $J \neq 0$, all have seniority $v = 2$ and, therefore, quasi-spin $s = \Omega/2 - 1$. With the observation that all states of a given quasi-spin irrep have common angular momentum quantum numbers JM (because the quasi-spin operators are $J = 0$ tensors and commute with the angular momentum operators), the pattern of states becomes recognisable. It is shown in Figure 6.2 for the $j = 7/2$ shell.

The range of values that J can take for a given value of j and nucleon number can be determined by the 'peeling-off' procedure (cf. Section 2.3) or more routinely by use of a plethysm program, as described in Section 5.11.10 (see also Box 6.I).

The energy eigenvalues of Equation (6.25), when expressed in terms of nucleon number and seniority, are of the form

$$E(n,v) = n\varepsilon - \tfrac{1}{4}G(n-v)(2\Omega - n - v + 2). \tag{6.27}$$

The spectrum of $E(n,v)$ energies is shown for $j = 9/2$ ($\Omega = 5$) with $\varepsilon = 0$ in Figure 6.3. For even nuclei the lowest levels are seniority zero and two states,

---

[16]See remarks on the origins of seniority in Section 6.1. For other references and a review, see the book by Talmi, *op. cit.* Footnote 258 on Page 380.

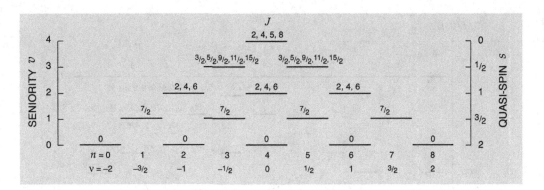

Figure 6.2: The states of nucleons of a single type (and hence maximum isospin $T = n/2$) for the $j = 7/2$ shell. Each horizontal line, with the given values of the nucleon number n, quasi-spin s, and seniority v, denotes a set of states with the angular momenta shown above it. As discussed in Section 6.5, the states associated with each such line span an irrep of a $USp(2j+1)$ group (equal to $USp(8)$ in the present case) and the set of all such $USp(2j+1)$ irreps of a given nucleon number make up a $U(2j+1)$ (equal to $U(8)$) irrep.

6.I Angular momentum states in a $(j)^n$ configuration

The states of a single particle of angular momentum j span an $SU(2)$ irrep. They also span a $U(2)$ irrep labelled by the one-rowed partition $\{2j\}$. (Note that the two-dimensional representation $\{1\}$ of $U(2)$ has angular momentum $j = 1/2$.) The totally antisymmetric states of an n-nucleon configuration $(j)^n$ are then given by the plethysm $\{2j\} \circledD \{1^n\}$. For example, using the plethysm code of Carvalho and D'Agostino[a] (cf. Section 5.11.10), we obtain for the $(9/2)^4$ configuration

$$\{9\} \circledD \{1^4\} = 2\{18, 18\} + 2\{20, 16\} + \{21, 15\} + 3\{22, 14\} + \{23, 13\} + 3\{24, 12\}$$
$$+ \{25, 11\} + 2\{26, 10\} + \{27, 9\} + \{28, 8\} + \{30, 6\}. \qquad (6.\text{I}.\text{i})$$

Now, recalling that a $U(2)$ irrep $\{\lambda_1, \lambda_2\}$ has $SU(2)$ angular momentum $J = (\lambda_1 - \lambda_2)/2$ (cf. Exercise 6.6), we determine that the fully antisymmetric states of the $(9/2)^4$ configuration have the angular momentum values

$$J = 0^2, 2^2, 3, 4^3, 5, 6^3, 7, 8^2, 9, 10, 12. \qquad (6.\text{I}.\text{ii})$$

The four-nucleon configuration can have seniority $v = 0$, 2, and 4. States of $v = 0$ have angular momentum $J = 0$ and, for a $j = 9/2$ shell, states of $v = 2$ can have $J = 2, 4, 6$, or 8. It follows that the remaining states of angular momenta

$$J = 0, 2, 3, 4^2, 5, 6^2, 7, 8, 9, 10, 12 \qquad (6.\text{I}.\text{iii})$$

have seniority $v = 4$. The angular momenta of the $j = 9/2$ shell states of other seniorities, derived in a similar way, are shown in Figure 6.3.

---

[a]Carvalho and D'Agostino, *op. cit.* Footnote 212 on Page 346.

while for odd nuclei they are seniority one and three states. The figure shows two important properties of the pair-coupling scheme: namely an odd-even staggering

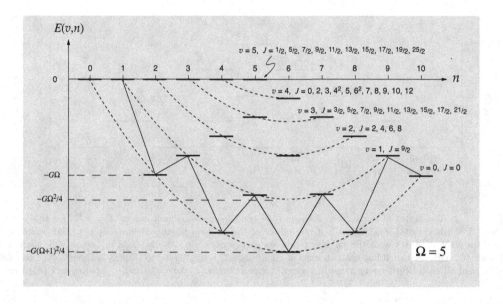

Figure 6.3: The energy levels for nucleons of a single type in a $j = 9/2$ shell with a $J = 0$ pairing interaction and $\varepsilon = 0$. For even nuclei, the $J = 0$ ground state has seniority $v = 0$ and the lowest excited $J \neq 0$ states have seniority $v = 2$. For odd nuclei, the lowest energy level is a seniority $v = 1$ multiplet of states and the next lowest level is a seniority $v = 3$ multiplet.

of the ground-state energies and an energy gap between the ground and lowest multiplet of excited states.

The odd-even staggering of the ground-state energies gives rise to a corresponding staggering of the separation energy of the last nucleon, as observed experimentally. This effect occurs in the model because the last nucleon in an odd nucleus is not part of a $J = 0$ pair.

Single-nucleon separation energies, defined in Equation (1.4), are given in the model for n even by

$$S_n = -E(n,0) + E(n-1,1) = \tfrac{1}{2}G(2\Omega + 1) - \tfrac{1}{2}(n-1)G - \varepsilon, \qquad (6.28)$$

and for n odd by

$$S_n = -E(n,1) + E(n-1,0) = -\tfrac{1}{2}(n-1)G - \varepsilon. \qquad (6.29)$$

These expressions show the odd-even staggering observed, for example, in Figure 1.8 for the Ca isotopes, as the $1f_{7/2}$ orbital is filled by neutrons. Neutron separation energies are shown for the multi-j-shell example of the tin isotopes in Figure 6.4.

Comparing the theoretical expression, (6.28) or (6.29), with the data, one finds that the slopes of the (parallel) S_n curves for n even and n odd provide a measure

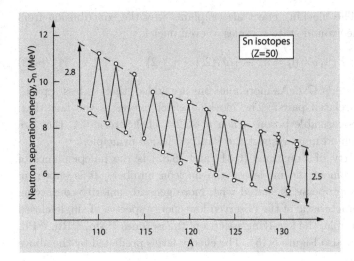

Figure 6.4: One-neutron separation energies, S_n, for the tin isotopes. The lines are drawn simply to emphasise the trends; they are not theoretical fits. Note the near constant odd-even staggering (2.8 MeV at $A \sim 110$ and 2.5 MeV at $A \sim 128$). The data shown do not follow the curved dependence on nucleon number seen in Figure 6.3 which shows results restricted to a single-j shell with the single-particle energy set equal to zero. (The data are from Audi G., Wapstra A.H. and Thibault C. (2003), *Nucl. Phys.* **A729**, 337.)

of the pair-coupling constant; i.e.,

$$G = -2\frac{dS_n}{dn}. \tag{6.30}$$

Likewise the magnitude of the staggering, when equated to $\frac{1}{2}G(2\Omega+1)$, is a measure of the effective number, $2\Omega_{\text{eff}}$, of degenerate single-particle states involved in the pair coupling.

The effects of gaps between single-particle levels show up in the experimental separation energy curves as discontinuities. Thus, for the Ca isotopes, one sees discontinuities at $A = 40$ and $A = 48$, the first corresponding to closing of the sd shell and the second corresponding to closing of the $f_{7/2}$ shell. In Figure 6.4 one sees a small discontinuity around $A = 115$ which can be attributed to the isolation of the $3s_{1/2}$ subshell (cf. Figure 1.30). Such discontinuities highlight the need to consider model spaces with (non-degenerate) multiple-j shells (cf. following section).

The energy level spectrum for an even nucleus is given in the model by

$$E(n,v) - E(n,0) = \tfrac{1}{4}Gv(2\Omega - v + 2). \tag{6.31}$$

The energy gap, $\Delta E = \Omega G$, between the ground ($v = 0, J = 0$) and first excited ($v = 2, J \neq 0$) states is the energy required to break a $J = 0$ pair. The energy level spectrum predicted by the model for an odd nucleus is

$$E(n,v) - E(n,1) = \tfrac{1}{4}G(v - 1)(2\Omega - v + 1). \tag{6.32}$$

The energy, $(\Omega - 1)G$, required to break a pair in an odd nucleus is seen to be slightly less than the energy gap ΩG in an even nucleus. This is a so-called *blocking effect* which results because the effective number of $(jm; j\bar{m})$ pairs available to form a correlated pair is reduced from Ω to $\Omega - 1$ when one state is occupied by an odd (unpaired) nucleon. The blocking effect also explains why the contribution from the pairing force to the ground-state energies of even nuclei

$$E(n, v{=}0) = n\varepsilon - \tfrac{1}{4}nG(2\Omega - n + 2) \qquad (6.33)$$

does not fall linearly as $-\tfrac{1}{2}nG\Omega$. As more nucleons are added, there is less "space" available for other correlated pairs. The blocking effect expresses the fact that, although nucleon pairs resemble bosons when $n \ll 2\Omega$, cf. Exercise 6.4, they are really pairs of fermions and must respect the Pauli exclusion principle.

A significant property of Equations (6.31) and (6.32) is the independence of the energy-level spectrum of the model on the nucleon number. It is shown in Section 6.5.5 that this property is shared with more general seniority-conserving interactions and is characteristic of the observed low-energy spectra of single-closed shell nuclei; cf., for example, the low-lying spectra of the isotones ^{92}Mo, ^{94}Ru, ^{96}Pd, shown in Figure 6.5 (cf. also Figure 5.15). The energy levels predicted by the above

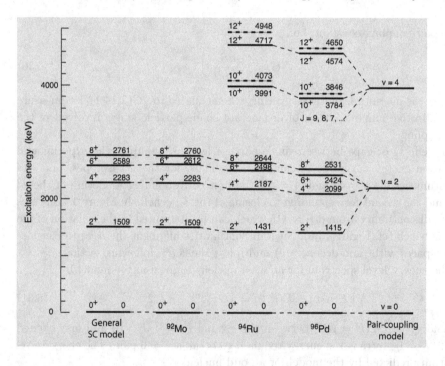

Figure 6.5: Energy levels of ^{92}Mo, ^{94}Ru, and ^{96}Pd taken from *Nuclear Data Sheets*. Only the highest spin states are known for the seniority $v = 4$ states (of which there are none in ^{92}Mo). The model energies are those of an $(N = 50, Z = 40)$ closed-shell plus valence-shell protons in the $1g_{9/2}$ subshell. The energy levels of the simple $J = 0$ pairing model are shown in the last column. The first column gives the results of the most general seniority-conserving (SC) model adjusted to fit the spectrum of ^{92}Mo (cf. Section 6.5.5). When adjusted to fit the seniority-two levels of ^{94}Ru and ^{96}Pd, the SC model also predicts the positions of the seniority-four $J^\pi = 10^+$ and 12^+ levels of these nuclei shown as dashed lines. (The SC model results are from Rosensteel G. and Rowe D.J. (2003), *Phys. Rev.* **C67**, 014303.)

$J = 0$ pair-coupling model are presented in the last column of this figure. They provide a qualitative explanation of the observed states and support the suggestion that the energy levels shown correspond to states of $(j = {}^9\!/_2)^n$ configurations. They also support the essential result of the pair-coupling model that, to a good approximation, the states are characterised by seniority quantum numbers. In fact, much stronger evidence for this conclusion comes from the remarkably good fit obtained with a general seniority-conserving model Hamiltonian given in Figure 6.5. General seniority-conserving models are discussed in Section 6.5.

Exercises

6.6 Let X_{ij} denote a 2×2 matrix with elements $\left(X_{ij}\right)_{kl} = \delta_{k,i}\delta_{l,j}$. Show that the matrices

$$S_+ := X_{12}, \quad S_- := X_{21}, \quad S_0 := \frac{1}{2}(X_{11} - X_{22}), \tag{6.34}$$

satisfy the standard su(2) commutation relations

$$[S_+, S_-] = 2S_0, \quad [S_0, S_\pm] = \pm S_\pm. \tag{6.35}$$

Let $\{\lambda_1, \lambda_2\}$ denote a U(2) irrep in which the matrices, $\{X_{ij}\}$, are represented by operators, $\{\hat{X}_{ij}\}$, and for which a state $|\varphi\rangle$ satisfies the equations

$$\hat{X}_{12}|\varphi\rangle = 0, \quad \hat{X}_{11}|\varphi\rangle = \lambda_1|\varphi\rangle, \quad \hat{X}_{22}|\varphi\rangle = \lambda_2|\varphi\rangle. \tag{6.36}$$

Show that $|\varphi\rangle$ is a highest-weight state for an SU(2) irrep of angular momentum $J = \frac{1}{2}(\lambda_1 - \lambda_2)$.

6.7 Estimate, from the results shown in Figure 6.4, the values of the pairing-force coupling constant G and the effective angular momentum, j_{eff}, needed to describe the neutron separation energies of the Sn isotopes in a single-j-shell pairing model.

6.8 Use Equation (6.20) to show that if $|n, 0\rangle$ is the ground state of an even-n nucleus in a j^n configuration and $|n + 1, 1j\rangle$ is a $J = j$, seniority $v = 1$, state of the neighbouring odd $n + 1$ nucleus, then in the single-shell pairing model,

$$|n, 0\rangle = \sqrt{\frac{(\Omega - n/2)!}{\Omega!(n/2)!}}\,(\hat{S}_+)^{n/2}|0\rangle, \tag{6.37}$$

$$|n + 1, 1jm\rangle = \sqrt{\frac{(\Omega - n/2 - 1)!}{(\Omega - 1)!(n/2)!}}\,(\hat{S}_+)^{n/2}a^\dagger_{jm}|0\rangle, \tag{6.38}$$

where $\Omega = (2j + 1)/2$.

6.9 Use the fact that $a^\dagger_{jm}(\hat{S}_+)^{n/2}|0\rangle := (\hat{S}_+)^{n/2}a^\dagger_{jm}|0\rangle$ to show that the spectroscopic factor, defined in terms of a reduced matrix (see Section A.3.3) by

$$S_j(n) := \frac{|\langle n + 1, 1j\|a^\dagger_j\|n0\rangle|^2}{2j + 1}, \tag{6.39}$$

has the value

$$S_j(n) = \frac{2j + 1 - n}{2j + 1}. \tag{6.40}$$

6.4 Multishell $J = 0$ pairing models

We now consider solutions of the pair-coupling Hamiltonian (6.10) in a multi-j-shell configuration. For nucleons of a single type (all neutrons or all protons) and a pairing interaction for which the coupling constants are all equal (cf. Section 6.2.1), i.e., $G_{ik} = G$, the Hamiltonian is of the form

$$\hat{H} = \sum_i \varepsilon_i \hat{n}_i - G\hat{S}_+ \hat{S}_-, \tag{6.41}$$

where $\hat{n}_i = \sum_m a_{j_i m}^\dagger a_{j_i \bar{m}}$ and

$$\hat{S}_\pm = \sum_i \hat{S}_\pm^i, \quad \hat{S}_0 = \sum_i \hat{S}_0^i. \tag{6.42}$$

Thus, the summed quasi-spin operators are also elements of a multishell quasi-spin su(2) algebra and satisfy the same standard commutation relations,

$$[\hat{S}_+, \hat{S}_-] = 2\hat{S}_0, \quad [\hat{S}_0, \hat{S}_\pm] = \pm\hat{S}_\pm, \tag{6.43}$$

as the corresponding single-shell operators as defined in Equation (6.12).

We first consider the simple two-nucleon case and then show, by Richardson-Gaudin methods, that $\hat{H}$ is, in general, the Hamiltonian of an integrable model.

6.4.1 *Two nucleons in many non-degenerate levels*

Any two-nucleon state, $|[j_1 j_2]JM\rangle$, satisfies the equation

$$\hat{S}_-|[j_1 j_2]JM\rangle = 0, \quad \text{when } J \neq 0, \tag{6.44}$$

and is an eigenstate of the pairing Hamiltonian (6.41) with energy given by the sum of single-particle energies according to the equation

$$\hat{H}|[j_1 j_2]JM\rangle = (\varepsilon_{j_1} + \varepsilon_{j_2})|[j_1 j_2]JM\rangle, \quad \text{when } J \neq 0. \tag{6.45}$$

Although it is not necessary, we suppose for simplicity that $j_i \neq j_k$ when $i \neq k$ and restrict consideration to solutions of the eigenvalue equation for $J = 0$ pair states.

Let

$$\hat{A}(\alpha) := \sum_i \alpha_i \hat{S}_+^i, \tag{6.46}$$

denote the $J = 0$ pair-creation operator for a two-nucleon $J = 0$, seniority zero, eigenstate of $\hat{H}$, expressed to within a convenient norm factor, N_α, by

$$|\alpha\rangle = N_\alpha \hat{A}(\alpha)|0\rangle. \tag{6.47}$$

The energy, E_α, of this state is then given, relative to the vacuum energy, $E_0 = 0$, by the equation of motion

$$[\hat{H}, \hat{A}(\alpha)]|0\rangle = (E_\alpha - E_0)\hat{A}(\alpha)|0\rangle = E_\alpha \hat{A}(\alpha)|0\rangle. \tag{6.48}$$

The left-hand side of this equation, evaluated from the identities

$$[\hat{n}_k, \hat{S}_+^i]|0\rangle = 2\delta_{k,i}\hat{S}_+^i|0\rangle, \tag{6.49}$$

$$[\hat{S}_+\hat{S}_-, \hat{S}_+^i]|0\rangle = -2\hat{S}_+\hat{S}_0^i|0\rangle = \Omega_i\hat{S}_+|0\rangle, \tag{6.50}$$

is given by

$$[\hat{H}, \hat{A}(\alpha)]|0\rangle = \sum_{ik} [2\varepsilon_k\delta_{i,k} - G\Omega_k]\,\alpha_k\hat{S}_+^i|0\rangle, \tag{6.51}$$

where $\Omega_k = \frac{1}{2}(2j_k + 1) = 2s_k$ (see Equation (6.26)). Thus, E_α is an eigenvalue of the equation

$$\sum_k (2\varepsilon_k\delta_{k,i} - G\Omega_k)\alpha_k = E_\alpha\alpha_i. \tag{6.52}$$

Equation (6.52) can be solved by matrix diagonalisation. However, a more insightful solution,[17] is obtained by rewriting it in the form

$$(2\varepsilon_i - E_\alpha)\alpha_i = G\sum_k \Omega_k\alpha_k. \tag{6.53}$$

Because of the norm factor N_α in Equation (6.47), the normalisation of the $\{\alpha_i\}$ coefficients is arbitrary. We can therefore choose it, for convenience, by setting

$$\sum_k \Omega_k\alpha_k = 1. \tag{6.54}$$

The eigenvectors of Equation (6.53) are then given by

$$\alpha_i = \frac{G}{2\varepsilon_i - E_\alpha}, \tag{6.55}$$

and the eigenvalues, E_α, are solutions of the equation

$$G\sum_i \frac{\Omega_i}{2\varepsilon_i - E_\alpha} = 1. \tag{6.56}$$

The squared norm factor, N_α^2, for the state $|\alpha\rangle$, is now given by

$$\langle\alpha|\alpha\rangle = N_\alpha^2 \sum_k \Omega_k|\alpha_k|^2 = 1. \tag{6.57}$$

Figure 6.6 shows a plot of the function

$$f(E) := \sum_i \frac{\Omega_i}{2\varepsilon_i - E} \tag{6.58}$$

---

[17]Högaasen-Feldman J. (1961), *Nucl. Phys.* **28**, 258.

for an arbitrary set of single-particle energies. Thus, the values of the pair energies, $\{E_\alpha\}$, are readily identified as the values of E for which $f(E) = 1/G$. More precise

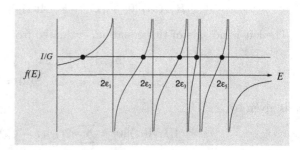

Figure 6.6: Graphical solution of Equation (6.56). The energies of two-particle $J = 0$ states are the values of E for which the function $f(E) = \sum_j \Omega_j/(2\varepsilon_j - E)$ takes the value $1/G$.

solutions to these equations are obtained numerically, cf. Exercise 6.10.

Figure 6.6 shows that the $J = 0$ two-particle states have energies that are equal to $\{2\varepsilon_i\}$ when G is zero and decrease as G increases; the lowest-energy solution decreases most. Excited $J = 0$ states have energies between the unperturbed energies and, when different single-particle orbitals are degenerate in energy, a corresponding number of $J = 0$ states become *trapped* at the unperturbed energies. Note that, when all the single-particle energies are degenerate, the energies of all but one pair state are trapped at the unperturbed energy; the lowest pair energy (the only one that is not trapped) is then that of the state $\hat{S}_+|0\rangle$.

Figure 6.7 shows the results of applying the above multishell-pairing model to the two-neutron-hole nucleus ^{206}Pb. The single-hole energies ε_j for the calculation

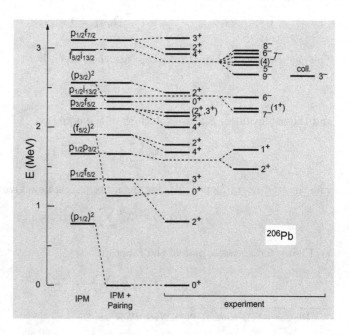

Figure 6.7: Energy levels of the two-neutron-hole nucleus ^{206}Pb described by the multishell-pairing model with single neutron-hole energies given in Exercise 6.12 and a coupling constant $G = 0.1472$ MeV. The first column shows independent-particle model energies obtained by simply adding the bare two neutron-hole energies. The second column shows the lowering of the $J = 0$ pairs by the pairing interaction. The energies are shown relative to the energy of the 0^+_1 ground state.

414

were taken from experimental data for ^{207}Pb (Figure 1.13) and the pair-coupling constant was adjusted to the value $G = 0.1472$ MeV. It is seen that this simple model accounts for all the observed states of ^{206}Pb below 3 MeV, except for a 3^- state at 2.65 MeV, which is presumably an analog of the collective (particle-hole) 3^- state of ^{208}Pb (cf. Figures 1.13 and 1.52). The model also gives a good description of some higher-lying states (not shown), such as $i_{13/2}^{-2}$ states with $J^\pi = 10^+$, 12^+ and $p_{1/2}^{-1}h_{9/2}^{-1}$ states with $J^\pi = 4^+$, 5^+ at 3.96, 4.03, 4.01, and 4.12 MeV, respectively.

6.4.2 *The Richardson equations*

Richardson[18,19,20,21,22] showed that the Högaasen-Feldman solution of the multishell-pairing model for a single pair can be generalised to accommodate any number of pairs.

Let N denote the number of shells and $\{|vJM\rangle\}$ a subset of states which satisfy the equations

$$\hat{S}_-^i|vJM\rangle = 0, \quad \hat{n}_i|vJM\rangle = v_i|vJM\rangle, \quad i = 1,\ldots,N; \qquad (6.59)$$

we shall refer to $v := (v_1, v_2, \ldots)$ as the *multishell seniority*. The states $\{|vJM\rangle\}$ are eigenstates of the Hamiltonian $\hat{H}$, Equation (6.41), and have energy eigenvalues given by

$$\hat{H}|vJM\rangle = \mathcal{E}(v)|vJM\rangle, \quad \mathcal{E}(v) = \sum_i \varepsilon_i v_i. \qquad (6.60)$$

These states contain no zero-coupled pairs and are the lowest-weight states for irreps of a multishell quasi-spin group $\mathcal{SU}(2)_1 \times \mathcal{SU}(2)_2 \times \cdots \times \mathcal{SU}(2)_N$ with *multishell quasi-spin* $s(v) := (s_1(v), s_2(v), \ldots, s_N(v))$, where

$$s_i(v) := \tfrac{1}{2}(\Omega_i - v_i) = s_i(0) - \tfrac{1}{2}v_i. \qquad (6.61)$$

Thus, they satisfy the equations

$$\hat{S}_0^i|vJM\rangle = -s_i(v)|vJM\rangle. \qquad (6.62)$$

The lowest-weight states $\{|vJM\rangle\}$ are easy to enumerate as combinations of single-shell lowest-weight states.

By use of the results of the previous section, the eigenstates $\{|vJM\rangle\}$ can be extended to include a set, $\{\hat{A}(\alpha)|vJM\rangle\}$, of one-pair states with energies $\mathcal{E}(v) + E_\alpha$,

---

[18]Richardson R.W. (1963), *Phys. Lett.* **3**, 277.

[19]Richardson R.W. (1963), *Phys. Lett.* **5**, 82.

[20]Richardson R.W. and Sherman N. (1964), *Nucl. Phys.* **52**, 221.

[21]Richardson R.W. and Sherman N. (1964), *Nucl. Phys.* **52**, 253.

[22]Many more references are cited in the review article of Dukelsky J., Pittel S. and Sierra G. (2004), *Rev. Mod. Phys.* **76**, 643.

where the pair-creation operator $\hat{A}(\alpha)$ is given, with a convenient normalisation, by

$$\hat{A}(\alpha) = \sum_i \frac{1}{2\varepsilon_i - E_\alpha} \hat{S}_+^i. \tag{6.63}$$

These states are eigenstates of the multi-j-shell pairing Hamiltonian, (6.41), for the $n = 2$ seniority-zero states. It is also seen that, in the limit in which the quasi-boson approximation (cf. Equation (6.23) is valid, a set of eigenstates can be constructed of the form

$$\left[\prod_\alpha^{\text{occ}} \hat{A}(\alpha) \right] |vJM\rangle, \tag{6.64}$$

where α labels a set of so-called *occupied* pair states. In an investigation of the adjustment required to take account of the fermion-pair nature of the $\{\hat{A}(\alpha)\}$ operators for these multi-pair states, Richardson discovered the remarkable result that the correction terms could be made to vanish by requiring the $\{E_\alpha\}$ energies to satisfy a modified equation. The possibility that a Hamiltonian such as $\hat{H}$ has eigenstates of the form (6.64) is known as a *Bethe ansatz*.[23]

With the benefit of hindsight, Richardson's equations are easily derived. It is first shown, by the methods of Section 6.4.1, that the identity

$$[\hat{H}, \hat{A}(\alpha)] - E_\alpha \hat{A}(\alpha) = \hat{S}_+ \left(1 + 2G \sum_i \frac{\hat{S}_0^i}{2\varepsilon_i - E_\alpha} \right) \tag{6.65}$$

is valid for any value of E_α. It then follows, from this equation and Equation (6.60), that

$$\hat{H}\hat{A}(\alpha)|vJM\rangle = \left[\mathcal{E}(v) + E_\alpha \right] \hat{A}(\alpha)|vJM\rangle$$
$$+ \left(1 - 2G \sum_i \frac{s_i(v)}{2\varepsilon_i - E_\alpha} \right) \hat{S}_+ |vJM\rangle \tag{6.66}$$

and that, when E_α satisfies the equation

$$1 - 2G \sum_i \frac{s_i(v)}{2\varepsilon_i - E_\alpha} = 0, \tag{6.67}$$

the state $\hat{A}(\alpha)|vJM\rangle$ becomes an eigenstate of $\hat{H}$, i.e.,

$$\hat{H}\hat{A}(\alpha)|vJM\rangle = \left[\mathcal{E}(v) + E_\alpha \right] \hat{A}(\alpha)|vJM\rangle, \tag{6.68}$$

consistent with the results of Section 6.4.1.

---

[23]The Bethe ansatz was introduced, Bethe H.A. (1931), *Z. Phys.* **71**, 205, in the study of critical points and phase transitions of magnetic systems in the Heisenberg model in terms of interacting spins on a lattice. The Heisenberg and other such models are reviewed, for example, in the book by Baxter R.J. (1982), *Exactly solved models in statistical mechanics* (Academic Press, London).

Now consider the identity

$$[\hat{H}, \hat{A}(\alpha)\hat{A}(\beta)] = \hat{A}(\alpha)[\hat{H}, \hat{A}(\beta)] + \hat{A}(\beta)[\hat{H}, \hat{A}(\alpha)] + [[\hat{H}, \hat{A}(\alpha)], \hat{A}(\beta)]. \quad (6.69)$$

From the above expression for $[\hat{H}, \hat{A}(\alpha)]$, it is determined that

$$[[\hat{H}, \hat{A}(\alpha)], \hat{A}(\beta)] = 2G\hat{S}_+ \sum_i \frac{\hat{S}_+^i}{(2\varepsilon_i - E_\alpha)(2\varepsilon_i - E_\beta)}. \quad (6.70)$$

Provided $E_\alpha \neq E_\beta$, the right-hand side of this equation can be reexpressed as

$$[[\hat{H}, \hat{A}(\alpha)], \hat{A}(\beta)] = \frac{2G\hat{S}_+}{E_\alpha - E_\beta} \sum_i \frac{(2\varepsilon_i - E_\beta) - (2\varepsilon_i - E_\alpha)}{(2\varepsilon_i - E_\alpha)(2\varepsilon_i - E_\beta)} \hat{S}_+^i$$

$$= \frac{2G}{E_\alpha - E_\beta} \hat{S}_+ [\hat{A}(\alpha) - \hat{A}(\beta)]. \quad (6.71)$$

It then follows that

$$\hat{H}\hat{A}(\alpha)\hat{A}(\beta)|vJM\rangle = [\mathcal{E}(v) + E_\alpha + E_\beta]\hat{A}(\alpha)\hat{A}(\beta)|vJM\rangle \quad (6.72)$$

$$+ \left(1 - \sum_i \frac{2Gs_i(v)}{2\varepsilon_i - E_\alpha} + \frac{2G}{E_\beta - E_\alpha}\right)\hat{S}_+\hat{A}(\beta)|vJM\rangle$$

$$+ \left(1 - \sum_i \frac{2Gs_i(v)}{2\varepsilon_i - E_\beta} + \frac{2G}{E_\alpha - E_\beta}\right)\hat{S}_+\hat{A}(\alpha)|vJM\rangle.$$

Thus, if the energies, E_α and E_β are chosen such that the second and third terms of this equation vanish, the state $\hat{A}(\alpha)\hat{A}(\beta)|vJM\rangle$ becomes an eigenstate of $\hat{H}$.

Generalisation of this result shows that a many-pair state $\prod_\alpha^{occ} \hat{A}(\alpha)|vJM\rangle$ is an eigenstate of $\hat{H}$, with energy given by

$$\hat{H}\prod_\alpha^{occ} \hat{A}(\alpha)|vJM\rangle = \left[\mathcal{E}(v) + \sum_\alpha^{occ} E_\alpha\right]\prod_\alpha^{occ} \hat{A}(\alpha)|vJM\rangle, \quad (6.73)$$

if all the occupied pair states have distinct E_α energies and these energies are solutions of the Richardson equations

$$1 - \sum_i \frac{2Gs_i(v)}{2\varepsilon_i - E_\alpha} + \sum_{\beta \neq \alpha}^{occ} \frac{2G}{E_\beta - E_\alpha} = 0. \quad (6.74)$$

It is interesting to note that the Richardson solution of the pair-coupling model was largely overlooked for many years,[24] until it was discovered[25,26] in an application in condensed matter physics to be relevant to the superconductivity properties

---

[24] A few applications were made, for example, by: Richardson, *op. cit.* Footnote 19 on Page 415; Bang J. and Krumlinde J. (1970), *Nucl. Phys.* **A141**, 18; Hasegawa M. and Tazaki S. (1987), *Phys. Rev.* **C35**, 1508; Hasegawa M. and Tazaki S. (1993), *Phys. Rev.* **C47**, 188.

[25] Dukelsky J. and Sierra G. (1999), *Phys. Rev. Lett.* **83**, 172.

[26] Sierra G. *et al.* (2000), *Phys. Rev.* **B61**, R11890.

of ultra-small metallic grains.[27] It is also interesting to note an intimate relationship between Richardson's solution of the pair-coupling model and a different family of solvable models known as the Gaudin magnet.[28] The connection[29,30,31] came through a proof of the integrability of the pairing model by Cambiaggio et al.[32] (see Box 6.II). A useful review of the Richardson-Gaudin models has been given by Dukelsky et al.[33]

6.4.3 Application of the Richardson equations

At first sight it would appear that, because of the $E_\beta \neq E_\alpha$ constraint, the Richardson equations only apply to a subset of energy eigenstates. In fact, as the following example illustrates, this is not a limitation. However, the Richardson equations do suffer from singularities which pose challenges for their solution.[34]

(i) *The single-j-shell pairing Hamiltonian*

Insights into the nature of the solutions of the Richardson equations are obtained by applying them to the, simply solvable, single-j-shell pairing Hamiltonian (6.24), $\hat{H} = -G\hat{S}_+\hat{S}_-$, with $\varepsilon = 0$. The seniority-zero eigenstates of this Hamiltonian (cf. Section 6.3) are of the form $(\hat{S}_+)^{n/2}|0\rangle$ and have energies (cf. Equation (6.27) with $\varepsilon = v = 0$ and $\Omega = 2s$)

$$E(s,n) = -\frac{1}{4}Gn(4s - n + 2). \tag{6.75}$$

For the one-pair $(n = 2)$ state, the Richardson equation

$$1 + \frac{2Gs}{E} = 0 \tag{6.76}$$

has solution $E \equiv E(s,2) = -2Gs$, in agreement with Equation (6.75). The two-pair equations

$$\begin{aligned}
1 + \frac{2Gs}{E_1} + \frac{2G}{E_2 - E_1} &= 0, \\
1 + \frac{2Gs}{E_2} + \frac{2G}{E_1 - E_2} &= 0,
\end{aligned} \tag{6.77}$$

---

[27]A review article on this topic has been given by von Delft J. and Ralph D.C. (2001), *Phys. Repts.* **345**, 61.

[28]Gaudin M. (1976), *J. Phys. (Paris)* **37**, 1087.

[29]Dukelsky J., Esebbag C. and Schuck P. (2001), *Phys. Rev. Lett.* **87**, 066403.

[30]von Delft J. and Poghossian R. (2002), *Phys. Rev.* **B66**, 134502.

[31]Links J. *et al.* (2002), *J. Phys. A: Math. Gen.* **35**, 6459.

[32]Cambiaggio M.C., Rivas A.M.F. and Saraceno M. (1997), *Nucl. Phys.* **A624**, 157.

[33]Dukelsky *et al.*, *op. cit.* Footnote 22 on Page 415.

[34]For discussions of the solution of the Richardson equations, cf., for example, the articles: Rombouts S., Van Neck D. and Dukelsky J. (2004), *Phys. Rev.* **C69**, 061303(R); Domínguez F., Esebbag C. and Dukelsky J. (2006), *J. Phys. A: Math. Gen.* **39**, 11349; and Dussel G.G. *et al.* (2007), *Phys. Rev.* **C76**, 011302(R).

6.II Richardson-Gaudin models

Gaudin[a] showed that, if $\{\hat{S}^i; i = 1, \ldots, N\}$ is a set of su(2) spin (or quasi-spin) vectors with the usual scalar product, $\hat{S}^i \cdot \hat{S}^j = \frac{1}{2}(\hat{S}^i_+ \hat{S}^j_- + \hat{S}^i_- \hat{S}^j_+) + \hat{S}^i_0 \hat{S}^j_0$, then the Hermitian operators

$$\hat{H}_i = \sum_{j \neq i} \frac{\hat{S}^i \cdot \hat{S}^j}{\varepsilon_i - \varepsilon_j}, \quad i = 1, \ldots, N, \tag{6.II.i}$$

form a complete mutually commuting set. More generally, Cambiaggio, Rivas, and Saraceno[b] discovered the commuting set

$$\hat{R}_i = \hat{S}^i_0 - G \sum_{j \neq i} \frac{\hat{S}^i \cdot \hat{S}^j}{\varepsilon_i - \varepsilon_j}, \quad i = 1, \ldots, N, \tag{6.II.ii}$$

and observed that these operators commute with the the the pairing Hamiltonian $\hat{H}$ of Equation (6.41). In fact, the pairing Hamiltonian $\hat{H}$ of Equation (6.41) is seen to be related to the sum

$$\sum_i 2\varepsilon_i \hat{R}_i = \sum_i 2\varepsilon_i \hat{S}^i_0 - G\hat{S}_+ \hat{S}_- - G\hat{S}_0(\hat{S}_0 - 1) + G \sum_i \hat{S}^i \cdot \hat{S}^i. \tag{6.II.iii}$$

Thus, the $\{\hat{R}_i\}$ operators have eigenstates, in common with $\hat{H}$, given by

$$|\Psi\rangle = \prod_{\alpha=1}^{\mathrm{occ}} \left(\sum_i \frac{\hat{S}^i_+}{2\varepsilon_i - E_\alpha} \right) |0\rangle, \tag{6.II.iv}$$

with $\{E_\alpha\}$ energies that solve the equations

$$1 - \sum_i \frac{2Gs_i}{2\varepsilon_i - E_\alpha} + \sum_{\beta \neq \alpha}^{\mathrm{occ}} \frac{2G}{E_\beta - E_\alpha} = 0. \tag{6.II.v}$$

Moreover, if r_i denotes an $\{E_\alpha\}$-dependent eigenvalue of $\hat{R}_i$, it follows from Equation (6.II.iii) that

$$\sum_i 2\varepsilon_i r_i = \sum_\alpha^{\mathrm{occ}} E_\alpha - \sum_i 2\varepsilon_i s_i - G\left(\sum_i s_i\right)^2 + G \sum_i s_i^2 + 2Gn_p \sum_i s_i - Gn_p(n_p - 1), \tag{6.II.vi}$$

where n_p is the number of occupied pair states. It was shown, by Sierra,[c] that

$$r_i = -s_i \left[1 + G \sum_{j \neq i} \frac{s_j}{\varepsilon_i - \epsilon_j} - 2G \sum_\alpha^{\mathrm{occ}} \frac{1}{2\varepsilon_i - E_\alpha} \right]. \tag{6.II.vii}$$

From these eigenvalues, one can also derive Equation (6.II.vi).

These results show the pair-coupling model to be an integrable system with good quantum numbers given by the eigenvalues of the $\{\hat{R}_i\}$ operators. Other integrable models that follow from the Richardson-Gaudin methods have been reviewed by Dukelsky et al.[d]

---

[a]Gaudin, *op. cit.* Footnote 28 on Page 418.
[b]Cambiaggio *et al.*, *op. cit.* Footnote 32 on Page 418.
[c]Sierra G. (2000), *Nucl. Phys.* **B572**, 517.
[d]Dukelsky *et al.*, *op. cit.* Footnote 22 on Page 415.

are also readily solved to give

$$E_1 = e_1 - e_2, \quad E_2 = e_1 + e_2, \tag{6.78}$$

with

$$e_1 = -G(2s - 1), \quad (e_2)^2 = -G^2(2s - 1). \tag{6.79}$$

Thus, for $s > \frac{1}{2}$, the pair energies are

$$E_1 = -G(2s - 1) - iG\sqrt{2s - 1}, \quad E_2 = -G(2s - 1) + iG\sqrt{2s - 1} \tag{6.80}$$

and $E_1 + E_2 \equiv E(s, 4) = -2G(2s - 1)$, again in accord with Equation (6.75). For more pairs, the equations must be solved numerically. For example, for $s = 2$, corresponding to $j = \frac{7}{2}$, the three-pair energies are given by

$$E_1 = -(1.69 + 2.51i)G, \quad E_2 = -(1.69 - 2.51i)G, \quad E_3 = -2.63G, \tag{6.81}$$

consistent with the exact result $E_1 + E_2 + E_3 = -6G$, and the four-pair energies are given by

$$\begin{aligned} E_1 &= -(1.73 + 0.89i)G, \ E_2 = -(1.73 - 0.89i)G, \\ E_3 &= -(0.27 + 2.50i)G, \ E_4 = -(0.27 - 2.50i)G, \end{aligned} \tag{6.82}$$

consistent with the exact results $E_1 + E_2 + E_3 + E_4 = -4G$

In general, all pair energies that are not real occur with complex conjugate partners so that the sum of the pair energies, which is what is observed, remains real.

(ii) *The energy-level spectrum of* ^{204}Pb

With the Richardson equations, the calculation of the $n = 2$ spectrum of the pairing Hamiltonian (6.41), appropriate for ^{206}Pb and shown in Figure 6.7, can be extended to other n values. We consider an application of the pair-coupling model to ^{204}Pb. However, whereas in calculating the spectrum of ^{206}Pb, we took the single-particle energies from the spectrum of ^{207}Pb and regarded the strength of the $J = 0$ pair-coupling interaction G as an adjustable parameter, we now retain the same value for G but take the energies of the $n = 2$ seniority-two parent states from the observed values for ^{206}Pb. The spectrum obtained in this way is shown in Figure 6.8.

For clarity, the predicted $J = 0$ seniority-zero levels below 3 MeV are shown in the first column. The predicted seniority-two levels associated with different configurations are shown in the succeeding six columns. The experimentally observed energy levels, most naturally associated with the predicted model levels, are shown as dashed lines in the corresponding columns. The remaining experimental energy levels are shown in the right-most column. Some of the latter energy levels can be associated with seniority-two levels predicted to lie at higher energies. The remainder are interpreted, in the model, as states of seniority four.

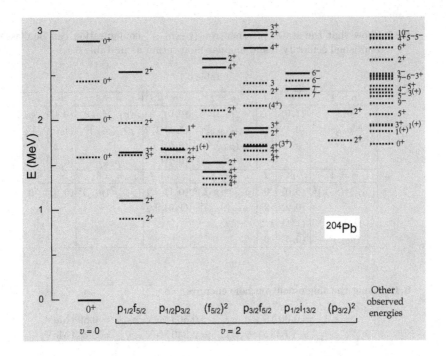

Figure 6.8: Energy levels of the four-neutron-hole nucleus ^{204}Pb described by the multishell-pairing model with the following parameters: single neutron-hole energies as given in Exercise 6.12; seniority-two parent-state energies taken from the observed ^{206}Pb spectrum; and a coupling constant $G = 0.1472$ MeV. The energies are shown relative to the energy of the 0_1^+ ground state. Experimental energy levels, from *Nuclear Data Sheets*, are shown as dashed lines. Calculated energy levels are shown as solid lines.

The detailed comparison between the model and the data is not impressive. Note, however, that none of the parameters was adjusted to fit the data; better agreement could have been obtained if the strength of the coupling constant G were weakened to reduce the energy gap between the ground and excited states by 100-200 keV. It is, nevertheless, notable that all the predicted energy levels below 3 MeV are observed. On the negative side, there is no clear separation in energy between the seniority two and four states which can therefore be expected to exhibit mixing.

Exercises

6.10 Show that an iterative procedure for finding a zero of the function $F(E) = f(E) - 1/G$ of the variable E is given by the sequence of approximations $\{E_1, E_2, E_3, \ldots\}$, where E_1 is a first guess for a solution, and

$$E_{k+1} = E_k - \frac{F(E_k)}{F'(E_k)}, \qquad (6.83)$$

where $F'(E) = \dfrac{d}{dE} F(E)$. (This is Newton's method of successive approximations.)

6.11 Show that the states of a neutron $(p_{1/2}p_{3/2})^n$ configuration can be classified by multishell seniority v and angular momentum as in Table 6.1.

Table 6.1.

$n = 1$	v	J	$n = 2$	v	J	$n = 3$	v	J
$p_{1/2}$	$(1,0)$	$1/2$	$(p_{1/2})^2$	$(0,0)$	0	$(p_{1/2})^2p_{3/2}$	$(0,1)$	$3/2$
$p_{3/2}$	$(0,1)$	$3/2$	$p_{1/2}p_{3/2}$	$(1,1)$	$1,2$	$p_{1/2}(p_{3/2})^2$	$(1,0)$	$1/2$
			$(p_{3/2})^2$	$(0,0)$	0		$(1,2)$	$3/2, 5/2$
				$(0,2)$	2	$(p_{3/2})^3$	$(0,1)$	$3/2$
$n = 4$	v	J	$n = 5$	v	J	$n = 6$	v	J
$(p_{1/2})^2(p_{3/2})^2$	$(0,0)$	0	$(p_{1/2})^2(p_{3/2})^3$	$(0,1)$	$3/2$	$(p_{1/2})^2(p_{3/2})^4$	$(0,0)$	0
	$(0,2)$	2	$p_{1/2}(p_{3/2})^4$	$(1,0)$	$1/2$			
$p_{1/2}(p_{3/2})^3$	$(1,1)$	$1,2$						
$(p_{3/2})^4$	$(0,0)$	0						

6.12 Using the single-neutron hole energies

$$\varepsilon_{p_{1/2}} = 0.00\,\text{MeV}, \quad \varepsilon_{f_{5/2}} = 0.569\,\text{MeV}, \quad \varepsilon_{p_{3/2}} = 0.898\,\text{MeV},$$
$$\varepsilon_{i_{13/2}} = 1.633\,\text{MeV}, \quad \varepsilon_{f_{7/2}} = 2.340\,\text{MeV}, \quad \varepsilon_{h_{9/2}} = 3.414\,\text{MeV}, \tag{6.84}$$

derive the $J = 0$ pairing-model energies given in Figure 6.7. Derive also the multishell-pairing prediction for the ^{206}Pb energies of the states with $i_{13/2}^{-2}$ and $p_{1/2}^{-1}h_{9/2}^{-1}$ configurations.

6.13 For the coupling constant $G = 0.1472$ and the single-particle energies of Exercise 6.12, as used in calculating the results shown in Figure 6.6, derive the corresponding pair amplitudes $\{\alpha_i\}$, defined by Equation (6.55).

6.5   The shell model in jj coupling; singly-closed shell nuclei

The pair-coupling models described in the previous sections provide valuable insights into the nuclear coupling schemes favoured by a major component of the nuclear interaction. In particular, they show that basis states, $\{|sv\alpha JM\rangle\}$, for the Fock space $\mathbb{F}^{(2j+1)}$ of single-j-shell configurations can be classified by means of the quasi-spin and angular-momentum quantum numbers of the chain of groups

$$\begin{array}{cccc} \mathcal{SU}(2) \times \text{SU}(2)_J & \supset & \mathcal{U}(1) \times \text{U}(1)_J, \\ s & J & \nu & M \end{array} \tag{6.85}$$

where an index, α, is needed to label the possible multiplicity of SU(2)$_J$ irreps of a given quasi-spin. We show in this section that, by establishing the correspondence between the states classified by this chain and those of a standard shell-coupling scheme, one gains a much more complete theory.

6.5.1 The group $U(2j+1)$ and its $USp(2j+1)$ subgroup

Our first objective is to show that the shell-model coupling scheme, which diagonalises a pairing-model Hamiltonian for nucleons of a single type in a single-j shell, is defined by the subgroup chain, $U(2j+1) \supset USp(2j+1)$.

A single-j-shell pairing-model Hamiltonian is a polynomial in the ($J=0$)-coupled quasi-spin operators

$$\hat{S}_+ = \sum_{m>0} a^\dagger_{jm} a^\dagger_{j\bar{m}}, \quad \hat{S}_- = \sum_{m>0} a_{j\bar{m}} a_{jm},$$
$$2\hat{S}_0 = \sum_{m>0} (a^\dagger_{jm} a_{j\bar{m}} + a_{jm} a^\dagger_{j\bar{m}}) = \hat{n} - \Omega, \tag{6.86}$$

where $\hat{n}$ is the nucleon number operator for the j shell and $\Omega = (2j+1)/2$.

We know (Section 5.11) that the set of all fully-antisymmetric states of a j^n configuration, comprising nucleons of a single type, is the Hilbert space for an irrep $\{1^n\}$ of the group, $U(2j+1)$, whose infinitesimal generators are Hermitian linear combinations of the nucleon-number-conserving operators

$$\hat{C}_{JM} = [a^\dagger_j \otimes a_j]_{JM}; \quad J = 0, \ldots, 2j. \tag{6.87}$$

Thus, to define basis states for a j^n configuration which diagonalise a pairing Hamiltonian, we seek the symmetry subgroup of all elements of $U(2j+1)$ that leave the above-defined su(2) quasi-spin operators invariant. By definition (Section 5.12.10), this is the subgroup, $USp(2j+1) \subset U(2j+1)$, for which the infinitesimal generators (elements of the usp($2j+1$) algebra) are the subset of u($2j+1$) elements that commute with the quasi-spin operators. They are identified as follows.

First observe that the single-nucleon operators satisfy the equations

$$[\hat{S}_+, a_{jm}] = -a^\dagger_{jm}, \quad [\hat{S}_-, a^\dagger_{jm}] = -a_{jm}, \quad [\hat{S}_+, a^\dagger_{jm}] = [\hat{S}_-, a_{jm}] = 0,$$
$$[\hat{S}_0, a^\dagger_{jm}] = \tfrac{1}{2} a^\dagger_{jm}, \quad [\hat{S}_0, a_{jm}] = -\tfrac{1}{2} a_{jm}. \tag{6.88}$$

Thus, the operators $(a^\dagger_{jm}, -a_{jm})$ are, respectively, the $\pm^1/_2$ components of a quasi-spin tensor.[35] It follows that the commutation relations of the quasi-spin operators with those of $U(2j+1)$ are given by

$$[\hat{S}_+, \hat{C}_{JM}] = -[a^\dagger_j \otimes a^\dagger_j]_{JM}, \quad [\hat{S}_-, \hat{C}_{JM}] = -[a_j \otimes a_j]_{JM},$$
$$[\hat{S}_0, \hat{C}_{JM}] = 0. \tag{6.89}$$

Because the operators $[a^\dagger_j \otimes a^\dagger_j]_{JM}$ and $[a_j \otimes a_j]_{JM}$ are non-zero for all values of J in the range $0 \leq J \leq 2j$ (cf. Exercise 6.15), it follows that the subset of u($2j+1$)

---

[35] Recall from Equation (6.17) that quasi-spin-$^1/_2$ states transform according the equations

$$\hat{S}_0|^1/_2, \nu\rangle = \nu|^1/_2, \nu\rangle, \quad \hat{S}_+|^1/_2, -^1/_2\rangle = |^1/_2, ^1/_2\rangle, \quad \hat{S}_-|^1/_2, ^1/_2\rangle = |^1/_2, -^1/_2\rangle.$$

Thus, with the standard phase conventions, $(a^\dagger_{jm}, -a_{jm})$ are the $\pm^1/_2$ components of a quasi-spin $^1/_2$ tensor.

operators, $\{\hat{C}_{JM}\}$, that commute with $\hat{S}_+$ (also $\hat{S}_-$ and $\hat{S}_0$) is the subset with J odd. Thus, the $\mathrm{usp}(2j+1)$ subalgebra of $\mathrm{u}(2j+1)$ is spanned by the Hermitian linear combinations of the operators $\{\hat{C}_{JM}; J \text{ odd}\}$.

From the above observations, it follows that the $J = 0$ pair-coupling model for either neutrons or protons, with $\mathrm{USp}(2+1)$ as its symmetry group, has an expression as a submodel of the jj-coupled shell model, as described in Section 5.13.1, in which states of each j^n configuration are classified by the quantum numbers n, v, J, and M of the subgroup chain

$$\begin{array}{ccccc}
\mathrm{U}(2j+1) & \supset & \mathrm{USp}(2j+1) & \supset & \mathrm{SU}(2)_J & \supset & \mathrm{U}(1)_J. \\
\{1^n\} & & \langle 1^v \rangle & \alpha & J & & M
\end{array} \tag{6.90}$$

The association of the pair-coupling model with a standard shell-model coupling scheme is significant for the development of nuclear structure theory because it means that the pair-coupling models, with interactions formed from nucleon number-conserving combinations of quasi-spin operators, have an immediate extension to more general quasi-spin- and, hence, seniority-conserving models in which the interactions include $(J{=}0)$-coupled combinations of $\mathrm{USp}(2j+1)$ operators. It also facilitates the identification of the components of a realistic interaction that are neglected in restricting a shell-model calculation to a pair-coupled basis; this makes it possible to consider ways to take them into account.

The quantum numbers of the subgroup chains (6.85) and (6.90) and the relationships between them are now obtained from the following duality properties and branching rules as mentioned in Section 5.12.13.

6.5.2 *Dual subgroup chains*

Because the groups $\mathrm{USp}(2j+1)$ and $\mathcal{SU}(2)$ commute with one another and the groups $\mathrm{U}(2j+1)$ and $\mathcal{U}(1)$ commute with one another, states of the pair-coupling model for nucleons of a single type in a single-j shell can be classified simultaneously by the quantum numbers of the two subgroup chains: (6.85) and (6.90). This does not provide extra quantum numbers because, consistent with the observations in Section 6.3, their values are related by the equations

$$n = \Omega + 2\nu, \quad v = \Omega - 2s, \quad \Omega = \tfrac{1}{2}(2j+1). \tag{6.91}$$

But, it does provide a richer algebraic structure that can be used to extend the pair-coupling model.

The identities of Equation (6.91) are direct results of powerful duality relationships (discussed in Sections 5.10 and 5.13.1);[36] e.g., the relationship, $2\hat{S}_0 \equiv \hat{n} - \Omega$, implies that the $\mathrm{U}(2j+1)$ irrep, $\{1^n\}$, is uniquely defined by the eigenvalue, 2ν, of $2\hat{S}_0$. Similarly, the irreps of $\mathrm{USp}(2j+1)$ are defined by those of the commuting quasi-spin group $\mathcal{SU}(2)$ according to the following theorem.

---

[36]Rowe *et al.*, *op. cit.* Footnote 168 on Page 308.

Theorem 6.1 *The groups $USp(2j+1)$ and $SU(2)$ have dual representations on $\mathbb{F}^{(2j+1)}$, the space of neutrons or protons (but not both) in a single-j shell.*[37]

This theorem is proved by the identification of all states of the single-j shell configuration that are simultaneously of highest $USp(2j+1)$ weight and lowest $SU(2)$ weight and observing that the irreps of these two groups all occur in multiplicity-free combinations and are such that an irrep of one group uniquely defines the irrep of the other group with which it is associated. First observe that a complete set of $U(2j+1)$ highest-weight states in the space $\mathbb{F}^{(2j+1)}$ is given by the states $\{|n\rangle := \hat{Z}_n|0\rangle; n = 0, \ldots, 2j+1\}$, where

$$\hat{Z}_0 := 1, \qquad \hat{Z}_n := a^\dagger_{jj}a^\dagger_{j,j-1}\ldots a^\dagger_{j,j-n+1}, \quad \text{for } n > 1. \tag{6.92}$$

This follows, if one chooses the operators $\{a^\dagger_{jm}a_{jm'}; m > m'\}$ to be raising operators for $U(2j+1)$, because in such a state the n nucleons occupy the single-particle states of highest-m values available to them. It is also seen that each $U(2j+1)$ highest-weight state, $|n\rangle$, is annihilated by the su(2) quasi-spin-lowering operator, $\hat{S}_-$, if and only if $n \leq (2j+1)/2$. Thus, the subset of these states that are simultaneously of $SU(2)$ lowest weight is the set

$$\{|v\rangle = \hat{Z}_v|0\rangle; 0 \leq v \leq \Omega\}. \tag{6.93}$$

Now, observe that, with a Cartan subalgebra for usp$(2j+1)$ that is a subalgebra of a u$(2j+1)$ Cartan subalgebra, a $U(2j+1)$ highest-weight state is also of $USp(2j+1)$ highest weight. Thus, the states $\{|v\rangle; v \leq \Omega\}$ are simultaneously of $USp(2j+1)$ highest weight and $SU(2)$ lowest weight. It is conceivable that there are other $USp(2j+1)$ highest-weight states within a $U(2j+1)$ irrep $\{1^n\}$ which are simultaneously of $SU(2)$ lowest weight. However, from the derivation of the $U(2j+1) \downarrow USp(2j+1)$ branching rules, given in Section 5.12.11, it is inferred that the n-nucleon $USp(2j+1)$ highest-weight states are of the form

$$|n, p\rangle \propto \left(\hat{S}_+\right)^p|n - 2p\rangle, \tag{6.94}$$

where $|n - 2p\rangle$ is the $(n - 2p)$-nucleon $U(2j+1)$ highest-weight state, for integer values of $p \leq n/2$. Because $\hat{S}_+$ is an $SU(2)$ raising operator, none of these states can be of $SU(2)$ lowest-weight unless $p = 0$. Thus, the set $\{|v\rangle; v \leq \Omega\}$ is a complete set of states that are simultaneously of highest $USp(2j+1)$ weight and of lowest $SU(2)$ weight. This result shows that each of the states $\{|v\rangle; v \leq \Omega\}$ defines a unique multiplicity-free irrep of the direct product group $USp(2j+1) \times SU(2)$ and that every irrep of $SU(2)$ in the shell-model space $\mathbb{F}^{(2j+1)}$ is paired with a unique irrep of $USp(2j+1)$. A single seniority label, v, thereby serves to index the irreps of both $USp(2j+1)$ and $SU(2)$ that occur in the space of a single-j shell. $\qquad\square$

---

[37]This duality theorem was proved, as part of a more general, symplectic-symplectic duality theorem, by Helmers, *op. cit.* Footnote 183 on Page 326. Helmers' theorem includes both Theorems 6.1 and 6.4 with the observation that $SU(2)$ is isomorphic to $USp(2)$ and $SO(5)$ is isomorphic to $USp(4)$ (see also Section 5.12.13).

The substance of this theorem is most readily understood in an example, Starting from a state $|v\rangle$ of $n = v$ neutrons or protons, regarded as the lowest-weight state for an $\mathcal{SU}(2)$ irrep of quasi-spin $s = \frac{1}{4}(2j + 1 - 2v)$, we can now identify all the $\mathrm{USp}(2j + 1)$ irreps that occur within the space of a single-j shell. For example, the set of $\mathrm{USp}(2j + 1)$ irreps for $j = {}^7/_2$, labelled by v, is given by the entries in Table 6.2 (cf. also Figure 6.2).

Table 6.2: The set of $\mathrm{USp}(2j + 1)$ irreps for nucleons of a single type in a $j = {}^7/_2$ shell. A $\mathrm{USp}(2j + 1)$ irrep is labelled by v which takes the values shown in the (n, v) entries for states of nucleon number n.The states of all equivalent $\mathrm{USp}(2j + 1)$ irreps in a horizontal row, i.e., those labelled by a common value of v, span an irrep of the direct product group $\mathrm{USp}(2j + 1) \times \mathcal{SU}(2)$. The states of all $\mathrm{USp}(2j + 1)$ irreps in a column, i.e., those labelled by a common value of n, span a $\mathrm{U}(2j + 1)$ irrep.

v	s :				(n, v)			
4	0 :				$(4, 4)$			
3	$\frac{1}{2}$:			$(3, 3)$		$(5, 3)$		
2	1 :		$(2, 2)$		$(4, 2)$		$(6, 2)$	
1	$\frac{3}{2}$:	$(1, 1)$		$(3, 1)$		$(5, 1)$		$(7, 1)$
0	2 : $(0, 0)$		$(2, 0)$		$(4, 0)$		$(6, 0)$		$(8, 0)$
	ν : -2	$-\frac{3}{2}$	-1	$-\frac{1}{2}$	0	$\frac{1}{2}$	1	$\frac{3}{2}$	2

6.5.3 *Duality and branching rules*

The $\mathrm{USp}(2j + 1) \leftrightarrow \mathcal{SU}(2)$ duality relationship has an intimate association with the $\mathrm{U}(2j + 1) \downarrow \mathrm{USp}(2j + 1)$ branching rules (cf. Section 5.12.10). This relationship was anticipated in establishing a correspondence between seniority and quasi-spin in Section 6.3. It is now exhibited explicitly, for the fully antisymmetric irreps of $\mathrm{U}(2j + 1)$, by arrays such as that of Table 6.2.

The set of all states of n nucleons of a single type in a single-j shell span a fully antisymmetric irrep, $\{1^n\}$, of $\mathrm{U}(2j + 1)$. Thus, the set of $\mathrm{USp}(2j + 1)$ irreps $\{\langle 1^v \rangle\}$ contained in the $\mathrm{U}(2j + 1)$ irrep $\{1^n\}$ corresponds to the set of v values in the n-nucleon column of the array, given for $j = {}^7/_2$ in Table 6.2. They correspond to the branching rules

$$
\begin{aligned}
\mathrm{U}(2j + 1) \downarrow \mathrm{USp}(2j + 1) &: \{0\} \downarrow \langle 0 \rangle & v &= 0, \\
&: \{1\} \downarrow \langle 1 \rangle & v &= 1, \\
&: \{1^2\} \downarrow \langle 1^2 \rangle \oplus \langle 0 \rangle & v &= 0, 2, \qquad (6.95) \\
&: \{1^3\} \downarrow \langle 1^3 \rangle \oplus \langle 1 \rangle & v &= 1, 3, \\
&: \{1^4\} \downarrow \langle 1^4 \rangle \oplus \langle 1^2 \rangle \oplus \langle 0 \rangle & v &= 0, 2, 4,
\end{aligned}
$$

from which the pattern is readily recognised and observed to be in accord with the general branching rule for $U(2j+1)$ irreps of type $\{1^n\}$

$$U(2j+1) \downarrow USp(2j+1) : \{1^n\} \downarrow \langle 1^n \rangle \oplus \langle 1^{n-2} \rangle \oplus \cdots \oplus \langle 1 \rangle \text{ or } \langle 0 \rangle, \quad (6.96)$$

for $n \leq (2j+1)/2$, and by the branching rule

$$U(2j+1) \downarrow USp(2j+1) : \{1^n\} \downarrow \langle 1^{2\Omega-n} \rangle \oplus \langle 1^{2\Omega-n-2} \rangle \oplus \cdots \oplus \langle 1 \rangle \text{ or } \langle 0 \rangle, \quad (6.97)$$

for $n > (2j+1)/2$, as obtained by the methods of Section 5.12.10.

6.5.4 Implications of duality for transition matrix elements

The above-defined duality relationships and the associated quasi-spin structure of operators of the $u(2j+1)$ Lie algebra imply useful relationships between the matrix elements of these operators. Recall that the single-particle creation and annihilation operators, $\{a_{jm}^\dagger\}$ and $\{-a_{jm}\}$, are components of quasi-spin-$1/2$ tensors. It follows that the elements of the $u(2j+1)$ Lie algebra, being bilinear combinations of these operators, are linear combinations of quasi-spin tensors of rank zero and one. Thus, because the elements $\{\hat{C}_{JM}; J \text{ odd}\}$ of the $usp(2j+1) \subset u(2j+1)$ subalgebra commute with the $su(2)$ quasi-spin operators, it is apparent that they are quasi-spin scalars. From the commutation relations of Equation (6.89), it also follows that the operators, $\{\hat{C}_{JM}; J \text{ even}\}$, are $s = 1$ quasi-spin tensors (see Appendix A.3 for a discussion of $SU(2)$ spherical tensors).

A consequence of the above tensorial properties is that many matrix elements of the $\{\hat{C}_{JM}\}$ operators are related to one another.[38] An excellent introduction to quasi-spin tensors and their matrix elements is given in the 1965 Summer School Lectures of Macfarlane.[39] Suppose that states $\{|s\nu JM\rangle\}$ of a single-j shell are labelled by quasi-spin, $s\nu$, and angular-momentum, JM, quantum numbers. Then, from the Wigner-Eckart theorem, it follows that

$$\langle s_f \nu_f J_f M_f | \hat{C}_{JM} | s_i \nu_i J_i M_i \rangle = \delta_{\nu_f, \nu_i} (s_i \nu_i \, 10 | s_f \nu_i)(J_i M_i \, JM | J_f M_f)$$

$$\times \frac{\langle s_f J_f ||| \hat{C}_J ||| s_i J_i \rangle}{\sqrt{(2s_f+1)(2J_f+1)}}, \quad \text{for } J \text{ even}, \quad (6.98)$$

where the reduced matrix elements are here reduced with respect to both angular momentum and quasi-spin. Note that, because of the relationship, $2\nu = n - \Omega$, given by Equation (6.26), and because $\hat{C}_{JM}$ is a nucleon number-conserving operator, these matrix elements vanish unless ν_f is equal to ν_i. Similarly, because $usp(2j+1)$

---

[38]These relationships were shown in papers by Arima A. and Kawarada H. (1964), *J. Phys. Soc. Japan* **19**, 1768; Watanable H. (1964), *Prog. Theor. Phys.* **32**, 106; and Lawson R.D. and Macfarlane M.H. (1965), *Nucl. Phys.* **66**, 80. Their implications for electromagnetic transition rates in the seniority coupling scheme were investigated notably in the paper of Lawson R.D. (1981), *Z. Phys.* **A303**, 51, and the review article of Blomqvist J. (1984), in *International Review of Nuclear Physics*, Vol. 2 (World Scientific, Singapore), pp. 1–32.

[39]Macfarlane M.H. (1966), *Lectures in Theoretical Physics*, Vol. **8c** (University of Colorado Press, Boulder).

operators have seniority $s = 0$, it follows that for J odd, their matrix elements simplify to

$$\langle s_f \nu_f J_f M_f | \hat{C}_{JM} | s_i \nu_i J_i M_i \rangle = \delta_{s_f, s_i} \delta_{\nu_f, \nu_i} (J_i M_i \, JM | J_f M_f)$$

$$\times \frac{\langle s_i J_f \| \hat{C}_J \| s_i J_i \rangle}{\sqrt{(2s_f + 1)(2J_f + 1)}}, \quad \text{for } J \text{ odd.} \quad (6.99)$$

These relationship are useful for reducing the number of matrix elements that need to be calculated in a variety of model situations. When used together with simple pair-coupling models, they also provide directly observable predictions. For example, when restricted to the states of nucleons of a single type occupying a single-j configuration, the effective electric quadrupole operators, $\hat{Q}_{2M}$, are proportional to the $s = 1$, $\{\hat{C}_{2M}\}$ operators. Thus, their matrix elements are given by Equation (6.98). Moreover, their reduced E2 transition rates are simply proportional to the squares of Clebsch-Gordan coefficients; e.g., for a transition between states of common quasi-spin quantum numbers, $s\nu$,

$$B(E2; s\nu J_i \to s\nu J_f) \propto (s\nu \, 10 | s\nu)^2 = \frac{\nu^2}{s(s+1)} = \frac{(n - \Omega)^2}{4s(s+1)}, \quad (6.100)$$

where $\Omega = (2j + 1)/2$. A notable characteristic of this expression is that it vanishes close to the mid-point of a shell. This effect is observed.

Figure 6.9 shows measured values of $\sqrt{B(E2)}$ for the E2 transitions between the first excited 10^+ and 8^+ states of the even $N = 82$ isotones and from the first

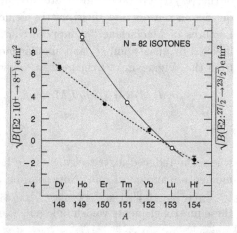

Figure 6.9: Square roots of the reduced E2 transition rates from the first-excited 10^+ state to the first-excited 8^+ state in the even $N = 82$ isotones and from the first-excited $27/2^-$ state to the first-excited $23/2^-$ state in the odd $N = 82$ isotones, shown as functions of the number of protons added to the closed-shell $^{146}_{64}$Gd nucleus. The lines through the data points are simply smooth curves. The choice of negative values for the square roots for $n = 7, 8$ is explained in the text. (The figure is from McNeill J.H. et al. (1989), Phys. Rev. Lett. **63**, 860.)

excited $27/2^-$ and $23/2^-$ states in the odd $N = 82$ isotones. The figure suggests a change in sign of the quadrupole transition matrix element in the mid-shell region in accord with Equation (6.100). Thus, the results shown are qualitatively consistent with a single-j-shell pairing model with $\Omega \sim 6$. This value corresponds to the value $j = {}^{11}/_2$ of the single proton $h_{11/2}$ orbital.

Similar results have been obtained for the Sn isotopes. Figure 6.10 shows such a plot for the Sn isotopes with the values for the odd isotopes scaled so that they lie on a common curve with those for the even isotopes. The turning point occurs

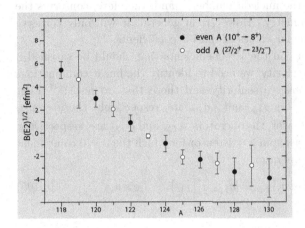

Figure 6.10: Square roots of the reduced E2 transition rates from the first-excited 10^+ state to the first-excited 8^+ state in the even Sn isotopes and from the first excited $27/2^-$ state to the first excited $23/2^-$ state in the odd Sn isotopes shown as functions of the nuclear mass number. The odd-A values are multiplied by 0.514. (The figure is from Lozeva R.L. *et al.* (2008), *Phys. Rev.* **C77**, 064313.)

now for A$\sim$ 123. Assuming the ^{114}Sn isotope to be a closed-shell nucleus, A$\sim$ 123 corresponds to an isotope with $n \sim 11$ neutrons in the valence shell. Such a turning point would correspond, in the single-j-shell pairing model to an unreasonably large value of j. However, Figure 1.31 shows that the $h_{11/2}$, $d_{3/2}$ and $s_{1/2}$ single-particle orbitals are all being filled simultaneously. Thus, the sign change in the transition matrix elements occurs at close to the effective nucleon number of a half-filled shell.

6.5.5 *General seniority-conserving single-j-shell models*

The quasi-spin algebra can be extended to a much larger fermion-pair algebra, as discussed in Section 7.7.5, which includes the creation and annihilation operators,

$$\hat{\mathcal{A}}_{JM} := \frac{1}{\sqrt{2}} \, [a_j^\dagger \otimes a_j^\dagger]_{JM}, \quad \hat{\mathcal{B}}_{JM} := \frac{1}{\sqrt{2}} \, [a_j \otimes a_j]_{JM}, \tag{6.101}$$

for all even angular-momentum-J nucleon-pair states; as already noted, these operators vanish identically if J is odd. This algebra is then an SGA (spectrum generating algebra) for a general two-body Hamiltonian for nucleons of a single type in a single-j shell of the form

$$\hat{H} := \epsilon \hat{n}_j - \sum_{J}^{\text{even}} \sqrt{2J+1} \, V^J \hat{Z}_J, \tag{6.102}$$

where V^J is the matrix element of a two-body interaction

$$V^J := \langle j^2; JM | \hat{V} | j^2; JM \rangle. \tag{6.103}$$

and

$$\hat{Z}_J := [\hat{A}_J \otimes \hat{B}_J]_0 \qquad (6.104)$$

Such a Hamiltonian conserves the nucleon number n and, therefore, conserves the dual $U(2j+1)$ dynamical symmetry. However, in general, it will only conserve $USp(2j+1)$ seniority for particular values of the V^J coefficients.

To determine the constraints on the V^J coefficients that should be satisfied if the interaction is to conserve seniority, we need to identify the linear combinations of the $\{\hat{Z}_J\}$ operators that conserve seniority and those that do not.[40,41] First observe that, because the operators $a^\dagger_{jm}$ and $-a_{jm}$ are, respectively, the $\nu = \pm 1/2$ components of a quasi-spin-$1/2$ tensor, the operators $\hat{A}_{JM}$ and $\hat{B}_{JM}$ are, respectively, the $\nu = \pm 1$ components of a quasi-spin $s = 1$ tensor for which the $\nu = 0$ component (to within a sign factor) is the operator

$$\hat{C}_{JM} := \tfrac{1}{2}\big([a^\dagger_j \otimes a_j]_{JM} + [a_j \otimes a^\dagger_j]_{JM}\big), \quad J \text{ even}, \qquad (6.105)$$

defined by the

$$[\hat{S}_-, \hat{A}_{JM}] = -\sqrt{2}\,\hat{C}_{JM}. \qquad (6.106)$$

The $\hat{Z}_J := [\hat{A}_J \otimes \hat{B}_J]_0$ operators are linear combinations of operators of quasi-spin 0, 1, or 2.

Claim 6.2 *Let $\mathcal{W}$ denote the linear space spanned by the $\hat{Z}_J$ operators, defined by Equations (6.101) and (6.104) for a given value of j. Let $\mathcal{W}_0$ and $\mathcal{W}_2$ denote subspaces of $\mathcal{W}$ spanned, respectively, by $(2\hat{I} - \hat{M})\hat{Z}_J$ and $(\hat{I} + \hat{M})\hat{Z}_J$, where $\hat{I}$ is the identity operator, $\hat{M}\hat{Z}_J := \sum_{J'} M_{JJ'}\hat{Z}_{J'}$ with*

$$M_{JJ'} = M_{J'J} = 2\sqrt{(2J+1)(2J'+1)}\,W(jjjj; JJ'), \qquad (6.107)$$

and $W(jjjj; JJ')$ is a Racah coefficient. Then $\mathcal{W}_0$ and $\mathcal{W}_2$ are subspaces of two-body operators which, to within multiples of the identity operator and terms linear in the number operator, are components of quasi-spin scalars and quasi-spin $s = 2$ tensors, respectively. Moreover, the operators

$$\hat{P}_0 := \frac{1}{3}(2\hat{I} - \hat{M}), \quad \hat{P}_2 := \frac{1}{3}(\hat{I} + \hat{M}), \qquad (6.108)$$

are, respectively, $\mathcal{W} \to \mathcal{W}_0$ and $\mathcal{W} \to \mathcal{W}_2$ projection operators.

---

[40] The conditions that an interaction must satisfy to conserve seniority were identified by French J.B. (1960), *Nucl. Phys.* **15**, 393, who first discovered most of the results that we review here. They were explored further by Arima A. and Kawarada H. (1964), *J. Phys. Soc. Japan* **19**, 1768; Lawson R.D. and Macfarlane M.H. (1965), *Nucl. Phys.* **66**, 80; Talmi I. (1971), *Nucl. Phys.* **A172**, 1; Shlomo S. and Talmi I. (1972), *Nucl. Phys.* **A198**, 81; Talmi, *op. cit.* Footnote 258 on Page 380; and others.

[41] We follow here the methods of Rosensteel G. and Rowe D.J. (2003), *Phys. Rev.* **C67**, 014303. Note that the latter authors included a factor of $1/4$ in their expression for $\hat{H}$. However, it makes no difference to their results, which scale with the interaction strength.

To prove the claim, we start from the observation that the $\nu = 0$ components of the $s = 0, 1, 2, J = 0$ tensors, formed from $(J=0)$-coupled bilinear products of the $s = 1$, J-even tensors, $\hat{A}_J$ and $\hat{B}_J$, are the operators given, respectively, by

$$
\begin{aligned}
\hat{X}_J^0 &= [\hat{A}_J \otimes \hat{B}_J]_0 - [\hat{C}_J \otimes \hat{C}_J]_0 + [\hat{B}_J \otimes \hat{A}_J]_0, \\
\hat{X}_J^1 &= [\hat{A}_J \otimes \hat{B}_J]_0 - [\hat{B}_J \otimes \hat{A}_J]_0, \\
\hat{X}_J^2 &= [\hat{A}_J \otimes \hat{B}_J]_0 + 2[\hat{C}_J \otimes \hat{C}_J]_0 + [\hat{B}_J \otimes \hat{A}_J]_0.
\end{aligned}
\tag{6.109}
$$

The right-hand sides of these equations can be normal ordered[42] and recoupled by means of the identities (cf. Exercises 6.17 - 6.21)

$$
[\hat{B}_J \otimes \hat{A}_J]_0 = \hat{Z}_J + \frac{\sqrt{2J+1}}{\Omega}\,(\hat{n} - \Omega)
\tag{6.110}
$$

$$
[\hat{C}_J \otimes \hat{C}_J]_0 = \sum_{J'}^{\text{even}} M_{JJ'}\,\hat{Z}_{J'} + \frac{\sqrt{2J+1}}{2\Omega}\,\hat{n} - \delta_{J,0}\big(\hat{n} - \tfrac{1}{2}\Omega\big),
\tag{6.111}
$$

where J, J' range over the even integers from 0 to $(2j - 1)$. This gives

$$
X_J^1 = \frac{\sqrt{2J+1}}{\Omega}\,(\Omega - \hat{n}),
\tag{6.112}
$$

and to within constants and terms linear in $\hat{n}$, which clearly conserve seniority,

$$
X_J^0 = (2 - \hat{M})\hat{Z}_J + \dots,
\tag{6.113}
$$

$$
X_J^2 = 2(1 + \hat{M})\hat{Z}_J + \dots.
\tag{6.114}
$$

Equations (6.113), (6.114), and he identity $\hat{P}_0 + \hat{P}_2 = \hat{I}$, imply that $\mathcal{W} = \mathcal{W}_0 \oplus \mathcal{W}_2$ is a direct sum of non-overlapping subspaces. Now, because the matrix M is real and symmetric, the operator $\hat{M}$ is fully diagonalisable. Therefore the eigenvectors of $\hat{M}$ must belong to either $\mathcal{W}_0$ or $\mathcal{W}_2$. It follows that any operator $\hat{Y} \in \mathcal{W}_0$ must satisfy the equation $(\hat{M} + 1)\hat{Y} = 0$ and any operator $\hat{Y} \in \mathcal{W}_2$ must satisfy the equation $(\hat{M} - 2)\hat{Y} = 0$. Thus, the eigenvalues of the operator $\hat{M}$ can only take the values -1 or $+2$ and, as claimed, the operators $\hat{P}_0 := \frac{1}{3}(2\hat{I} - \hat{M})$ and $\hat{P}_2 := \frac{1}{3}(\hat{I} + \hat{M})$ are, respectively, $\mathcal{W} \to \mathcal{W}_0$ and $\mathcal{W} \to \mathcal{W}_2$ projection operators. $\qquad\square$

Claim 6.3 *The number of linearly-independent combinations of the $\hat{Z}_J$ operators, for a given value of j, that do not conserve seniority is $\lfloor (2j - 3)/6 \rfloor$, where $\lfloor x \rfloor$ denotes the integer part of x.*

---

[42] A normal-ordered combination of operators belonging to an irrep of a Lie algebra is an expression in which no lowering operator appears to the left of a raising operator. Thus, for example, the operator $[\hat{A}_J \otimes \hat{B}_J]_0$ is normal ordered but $[\hat{B}_J \otimes \hat{A}_J]_0$ is not. However, the latter can be re-expressed as a normal-ordered expansion, as in Equation (6.110), by use of the commutation relations. Normal ordering is particularly useful in comparing operators which look to be different when expressed in arbitrary ways. It is also useful for evaluating the expectation value of an operator with respect to a lowest-weight state. For example, the lowest-weight expectation value $\langle 0|[\hat{B}_J \otimes \hat{A}_J]_0|0\rangle$ obtained by use of the normal-ordered expansion (6.110) is immediately seen to be $-\sqrt{2J+1}$. [For more details, see Section 7.5].

Because the eigenvalues of a projection operator can only be 1 or 0, it follows that the dimensions of the spaces $\mathcal{W}_0$ and $\mathcal{W}_2$ are given by

$$d_0 = \text{Tr}\hat{P}_0 = \frac{2}{3}\Big[\Omega - \sum_{J}^{\text{even}}(2J+1)\,W(jjjj;JJ)\Big], \tag{6.115}$$

$$d_2 = \text{Tr}\hat{P}_2 = \frac{1}{3}\Big[\Omega + 2\sum_{J}^{\text{even}}(2J+1)\,W(jjjj;JJ)\Big]. \tag{6.116}$$

Evaluation of these expressions gives[43]

$$d_2 = \lfloor (2j+3)/6 \rfloor, \quad d_0 + d_2 = \Omega = (2j+1)/2. \tag{6.118}$$

However, it is important to note that the operator $\hat{S}_+\hat{S}_-$, proportional to $\hat{Z}_0$, also conserves quasi-spin and seniority. To within constants and terms linear in the number operators, $\hat{Z}_0$ is a linear combination of the operators $\hat{X}_0^0$ and $\hat{X}_0^2$. Thus, there are only $d_2 - 1$ linearly-independent combinations of $\hat{Z}_J$ operators that do not conserve seniority. (This claim and a different proof of it was given by Talmi, *op. cit.* Footnote 258 on Page 380.) ☐

The claim shows that the number of seniority non-conserving linearly-independent combinations of the $\hat{Z}_J$ operators is zero for $j = 3/2, 5/2, 7/2$, one for $j = 9/2, 11/2, 13/2$, and two for $j = 15/2, 17/2, 19/2$. Thus, for $j \leq 7/2$, all interactions conserve seniority and for $9/2 \leq j \leq 13/2$, only one constraint on the values of the V^J matrix elements is needed to ensure that seniority is conserved. This constraint is obtained by requiring that, to within a term in $\hat{Z}_0$, a seniority-conserving two-body interaction $\hat{V}$ should be a quasi-scalar; i.e., an element of $\mathcal{W}_0$. Thus, it must satisfy the equation $\hat{P}_2(\hat{V} - \lambda\hat{Z}_0) = 0$ for some value of λ. The parameter, λ, can be fixed by requiring the $J = 0$ component of $\hat{P}_2(\hat{V} - \lambda\hat{Z}_0)$ to be zero. The seniority non-conserving component of a general interaction, $\hat{V}$, is then the projection

$$\hat{V}_{SNC} = \hat{P}_2(\hat{V} - \lambda\hat{Z}_0). \tag{6.119}$$

Using the above expression for $\hat{P}_2$, one determines that the condition for a given interaction to conserve seniority within a j shell, for $j = 9/2, 11/2,$ and $13/2$, is that the respective linear combinations of two-body matrix elements,

$$v^{(\frac{9}{2})} = 65V^2 - 315V^4 + 403V^6 - 153V^8, \tag{6.120}$$

$$v^{(\frac{11}{2})} = 1020V^2 - 3519V^4 + 637V^6 + 4403V^8 - 2541V^{10}, \tag{6.121}$$

$$v^{(\frac{13}{2})} = 1615V^2 - 4275V^4 - 1456V^6 + 3196V^8 + 5145V^{10} - 4225V^{12}, \tag{6.122}$$

---

[43]The identity

$$\frac{1}{3}\Big[\Omega + 2\sum_{J}^{\text{even}}(2J+1)\,W(jjjj;JJ)\Big] = \lfloor (2j+3)/6 \rfloor \tag{6.117}$$

was discovered empirically by Rosensteel and Rowe, *op. cit.* Footnote 41 on Page 430. This result was subsequently investigated by Zamick L. and Escuderos A. (2005), *Phys. Rev.* **C71**, 054308, and others quoted by them.

should vanish. More precisely, it is determined that the seniority non-conserving component, V_{SNC}, of an interaction V has two-body matrix elements

$$V^J_{SNC} = \langle j^2; JM | \hat{V}_{SNC} | j^2, JM \rangle = \frac{1}{k^{(j)}_J} v^{(j)}, \tag{6.123}$$

where $v^{(j)}$ is given by Equations (6.120) - (6.122) and the inverse scale factors, $\{k^{(j)}_J\}$, are listed in Table 6.3.

Table 6.3: Values of the inverse scale factors, $\{k^{(j)}_J\}$, appearing in Equation (6.123).

J	$j = \frac{9}{2}$	$j = \frac{11}{2}$	$j = \frac{13}{2}$
2	1980	15015	16380
4	$-5148/7$	$-180180/23$	$-55692/5$
6	$25740/31$	$437580/7$	$-188955/4$
8	-2860	$437580/37$	$1322685/47$
10	$-$	$-278460/11$	$151164/7$
12	$-$	$-$	$-406980/13$

For example, the two-body matrix elements defined by the energy levels of ^{148}Dy, and used by Lawson[44] to compute the energy levels of the nucleus ^{149}Ho, regarded as a system of 3 protons in an $h_{11/2}$ shell outside of a closed-shell core, has two-particle matrix elements and seniority-non-conserving matrix elements as shown in Table 6.4.

Table 6.4: Two-particle matrix elements, V^J (in units of MeV), inferred from the spectrum of ^{148}Dy, compared to those of its seniority-non-conserving component, V^J_{SNC}.

J	2	4	6	8	10
V^J	1.677	2.427	2.731	2.832	2.919
V^J_{SNC}	-0.003	0.005	-0.001	-0.003	0.002

Note that, to prove a given interaction is seniority-conserving in a single-j-shell configuration with $j \leq {}^{13}/_2$, it is sufficient to show that a single matrix element between states of different seniority is zero if it would be non-zero for a generic interaction. Similarly, for ${}^{15}/_2 \leq j \leq {}^{19}/_2$, it is sufficient to show that two matrix elements between states of different seniority are zero if they would both be zero only if seniority were conserved. This method was used in the derivations of Equations (6.120) - (6.122) referenced in Footnote 40 on Page 430.

---

[44]Lawson R.D. (1981), *Z. Phys.* **A303**, 51.

6.5.6 Applications of a seniority-conserving model in jj coupling

Candidates for a description in terms of protons occupying a single-j shell are given by the low-lying states of the $N = 50$ isotones, ^{92}Mo, ^{94}Ru and ^{96}Pd, the $N = 82$ isotones, ^{148}Dy, ^{150}Er, ^{152}Yb, ^{154}Hf, ..., and the $N = 126$ isotones ^{210}Po, ^{212}Rn, ^{214}Ra, ... The $N = 50$ isotones are modelled as ^{90}Zr closed shells plus 2, 4, and 6 protons, respectively, occupying the $1g_{9/2}$ single-particle shell. The $N = 82$ isotones are modelled as ^{146}Gd closed shells plus 2, 4, 6, 8, and 10 protons, respectively, occupying the $1h_{11/2}$ proton shell. The $N = 126$ isotones are modelled as ^{208}Pb closed shells plus 2, 4, and 6 protons, respectively, occupying the $1h_{9/2}$ proton shell. The observed low-energy states of the $N = 50$ isotones are shown in Figures 6.5 and 6.11. These figures also show the description of the states of these isotones with two-body seniority-conserving interactions, in which the strength parameters of the interaction are adjusted to fit the excitation energies of the seniority-two, $J^\pi = 2^+$, 4^+, and 8^+ states. Figure 6.11 shows that, with such an interaction, the energy levels of the remaining states observed, namely the seniority-two, $J^\pi = 6^+$ and the seniority-four, $J^\pi = 10^+$, and 12^+ states, are given remarkably accurately. In fact, a fit with unconstrained (seniority-mixing) two-body interactions would result in a negligibly-improved fit to the data. Similarly, results are shown for the ^{212}Rn and ^{214}Ra isotones in Figure 6.11.

Note the remarkable similarity between the low-energy spectra of two different sequences of isotones that is naturally explained in the jj-coupling scheme with seniority-conserving interactions in a single-j shell.

6.5.7 Seniority-conserving Hamiltonians in multishell spaces

The single-shell seniority coupling scheme is readily extended to a multishell space for nucleons of a single type in $j_1^{n_1} j_2^{n_2} j_3^{n_3} \cdots$ configurations. Basis states for such configurations are given by the coupled tensor products of the subshell bases

$$|nv\alpha(J_1 J_2 \cdots)JM\rangle := \Big[|n_1 v_1 \alpha_1 J_1\rangle \otimes |n_2 v_2 \alpha_2 J_2\rangle \otimes \cdots \Big]_{JM}, \qquad (6.124)$$

where $n := (n_1, n_2, \cdots)$ denotes a partition of the nucleons among the subshells, $v := (v_1, v_2, \cdots)$ is the multishell seniority, $\alpha_1, \alpha_2, \cdots$ are multiplicity indices, and JM are the total angular-momentum quantum numbers. Such basis states are commonly used in shell-model calculations for singly-closed-shell nuclei.

Whereas seniority is conserved in a single-j shell by a Hamiltonian of the type

$$\hat{H} = \varepsilon \hat{n}_j - G \hat{S}_+^j \hat{S}_-^j + \hat{V}_j, \qquad (6.125)$$

where $\hat{V}_j$ is a USp($2j + 1$) quasi-scalar two-body interaction (i.e., a scalar quasi-spin operator to within terms in the nucleon number operator), the most general Hamiltonian, with two-body interactions, that conserves multishell seniority is of

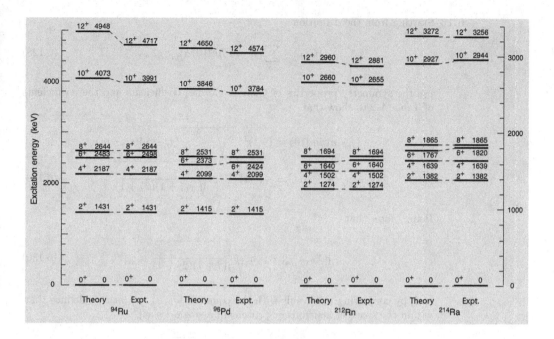

Figure 6.11: Energy levels of ^{94}Ru, ^{96}Pd, ^{212}Rn, and ^{214}Ra, taken from *Nuclear Data Sheets*. The observed levels were computed with seniority-conserving interactions for 4 and 6 protons, respectively, in the $1g_{9/2}$ shell for ^{94}Ru and ^{96}Pd and 4 and 6 protons, respectively, in the $1h_{9/2}$ shell for ^{212}Rn and ^{214}Ra . The strength parameters of the seniority-conserving interactions were adjusted to fit the $J^\pi = 0^+$, 2^+, 4^+ and 8^+ energy levels. (The model results are from Rosensteel and Rowe, *op. cit.* Footnote 41 on Page 430.)

the form

$$\hat{H} = \sum_j \varepsilon_j \hat{n}_j - \sum_{jj'} G_{jj'} \hat{S}_+^j \hat{S}_-^{j'} + \sum_j \hat{V}_j. \qquad (6.126)$$

Because of the multishell-pairing interaction, such a Hamiltonian does not conserve the partition of the nucleon numbers, $\{n_j\}$, occupying the different shells.

Exercises

6.14 Starting from the definitions $\hat{S}_+ := \sum_{m>0} a^\dagger_{jm} a^\dagger_{j\bar{m}}$ and $\hat{S}_- := (\hat{S}_+)^\dagger$, show that $\hat{S}_- = \sum_{m>0} a^{j\bar{m}} a^{jm}$. Then use the identities

$$a^\dagger_{j\bar{m}} = (-1)^{j+m} a^\dagger_{j,-m}, \quad a^{jm} = (-1)^{j+m} a_{j,-m}, \quad a^{j\bar{m}} = -a_{jm}, \qquad (6.127)$$

(obtained from Equations (5.40) and (5.558)) to show that $\hat{S}_- = \sum_{m>0} a_{j\bar{m}} a_{jm}$.

6.15 Use the symmetry $(j_1 m\, j_2 n | JM) = (-1)^{j_1+j_2-J} (j_1, -m\, j_2, -n | J, -M)$, for Clebsch-Gordan coefficients, to show that $\mathcal{A}_{JM}$ and $\mathcal{B}_{JM}$, defined by Equation (6.101), vanish if J is an odd integer.

6.16 Starting from the definition

$$\hat{C}_{10} := \sum_m (j, -m \, jm|10) \, a_{jm}^\dagger a_{j,-m}, \tag{6.128}$$

use the symmetry properties of Clebsch-Gordan coefficients and the coefficients of Table A.2 to show that

$$(j, -m \, jm|10) = (-1)^{j+m}\sqrt{\frac{3}{2j+1}} \, (j, -m \, 10|j, -m)$$

$$= -(-1)^{j+m}m\sqrt{\frac{3}{j(j+1)(2j+1)}}. \tag{6.129}$$

Hence, show that

$$[\hat{C}_{10}, a_{jm}^\dagger] = -m\sqrt{\frac{3}{j(j+1)(2j+1)}} \, a_{jm}^\dagger, \tag{6.130}$$

and, by comparing this result with the equation $[\hat{J}_0, a_{jm}^\dagger] = m \, a_{jm}^\dagger$, deduce that, within the space of neutrons or protons in a single-j shell,

$$\hat{J}_k = -\sqrt{\tfrac{1}{3}j(j+1)(2j+1)} \, \hat{C}_{1k}. \tag{6.131}$$

6.17 Show that the two-nucleon states of a single-nucleon type in a j^2 configuration separate into two subsets of states with quantum numbers

s	ν	v	J	
$\frac{1}{4}(2j+1)$	$-\frac{1}{4}(2j-3)$	0	0	(6.132)
$\frac{1}{4}(2j-3)$	$-\frac{1}{4}(2j-3)$	2	$2, 4, \ldots, 2j-1$	

Compare your results with those of Figure 6.2.

6.18 Use the standard expressions (cf. Section A.7.2),

$$a_{j\bar{m}} := (-1)^{j+m} a_{j,-m}, \quad \{a_{j\bar{m}}, a_{jn}^\dagger\} = \delta_{mn}, \tag{6.133}$$

to derive the identities:

$$[a_j^\dagger \otimes a_j]_0 = \frac{1}{\sqrt{2j+1}}\sum_m a_{jm}^\dagger a_{j\bar{m}} = \frac{1}{\sqrt{2j+1}}\hat{n}, \tag{6.134}$$

$$\{a_j^\dagger, a_j\}_{J0} := \sum_{mn}(jn \, jm|J0)\{a_{jm}^\dagger, a_{jn}\} = \sqrt{2j+1}\,\delta_{J,0}, \tag{6.135}$$

where $\hat{n} := \sum_m a_{jm}^\dagger a_{j\bar{m}}$ is the nucleon number operator. Show also that

$$\{a_j, a_j^\dagger\}_J = (-1)^{2j-J}\{a_j^\dagger, a_j\}_J = -\sqrt{2j+1}\,\delta_{J,0}. \tag{6.136}$$

6.19 Use the recoupling relations

$$\left[[a_j^\dagger \otimes a_j^\dagger]_J \otimes a_j\right]_j = \sum_{J'} U(jjjj:J'J)\left[a_j^\dagger \otimes [a_j^\dagger \otimes a_j]_{J'}\right]_j, \qquad (6.137)$$

$$\left[a_j \otimes [a_j^\dagger \otimes a_j^\dagger]_J\right]_j = \sum_{J'} U(jjjj:JJ')\left[[a_j \otimes a_j^\dagger]_{J'} \otimes a_j^\dagger\right]_j, \qquad (6.138)$$

to show that

$$\left[[a_j^\dagger \otimes a_j^\dagger]_J, a_j\right]_j = U(jjjj:0J)\left(\{a_j^\dagger, a_j\}_0 - (-1)^J \{a_j, a_j^\dagger\}_0\right)a_j^\dagger$$

$$= -\sqrt{\frac{2J+1}{2j+1}}\,\left(1 + (-1)^J\right)a_j^\dagger \qquad (6.139)$$

and, hence, that

$$[\mathcal{B}_J \otimes \mathcal{A}_J]_0 = [\mathcal{A}_J \otimes \mathcal{B}_J]_0 + \frac{\sqrt{2J+1}}{\Omega}\,(\hat{n} - \Omega), \qquad (6.140)$$

where $\Omega = (2j+1)/2$.

6.20 Show that $\hat{\mathcal{C}}_{JM}$, as defined by Equation (6.105), can also be expressed as

$$\hat{\mathcal{C}}_{JM} = [a_j^\dagger \otimes a_j]_{JM} - \frac{1}{2}\sqrt{2j+1}\,\delta_{J,0} \qquad (6.141)$$

and that

$$\left[\hat{\mathcal{C}}_J \otimes \hat{\mathcal{C}}_J\right]_0 = \left[[a_j^\dagger \otimes a_j]_J \otimes [a_j^\dagger \otimes a_j]_J\right]_0 - \left(\hat{n} - \tfrac{1}{4}(2j+1)\right)\delta_{J,0}. \qquad (6.142)$$

6.21 Show, for J even, that

$$\left[[a_j^\dagger \otimes a_j]_J \otimes [a_j^\dagger \otimes a_j]_J\right]_0 = \left[[a_j^\dagger \otimes a_j]_J \otimes a_j^\dagger \otimes a_j\right]_0$$

$$= \left[a_j^\dagger \otimes [a_j^\dagger \otimes a_j]_J \otimes a_j\right]_0 + \left[[a_j^\dagger \otimes a_j]_J, a_j^\dagger]_j \otimes a_j\right]_0, \qquad (6.143)$$

that

$$\left[a_j^\dagger \otimes [a_j^\dagger \otimes a_j]_J \otimes a_j\right]_0 = \sum_{J'} U(jjjj:J'J)\left[[a_j^\dagger \otimes a_j^\dagger]_J \otimes [a_j \otimes a_j]_J\right]_0, \qquad (6.144)$$

and that

$$\left[[a_j^\dagger \otimes a_j]_J, a_j^\dagger\right]_{jm} = \sqrt{\frac{2J+1}{2j+1}}\,a_{jm}^\dagger. \qquad (6.145)$$

Hence, derive Equation (6.111).

6.22 Show that the operators $\hat{P}_0$ and $\hat{P}_2$ of Equation (6.108) are $\mathcal{W} \to \mathcal{W}_0$ and $\mathcal{W} \to \mathcal{W}_2$ projection operators, respectively; i.e., show that they satisfy the identities

$$\hat{P}_0^2 = \hat{P}_0, \quad \hat{P}_2^2 = \hat{P}_2. \qquad (6.146)$$

6.6 $J = 0,\ T = 1$ pairing in jj coupling; open-shell nuclei

One can apply the coupling scheme of the previous section to the neutrons and protons separately. This may be appropriate when the neutrons and protons are occupying well-separated shells. However, when the neutrons and protons have valence shells in common, it is appropriate to consider the more general irreps of the $U(2j + 1) \supset USp(2j + 1) \supset SU(2)_J$ subgroup chain in the isospin formalism.

This section shows that, when both neutrons and protons are occupying a common single-j shell, the group $USp(2j + 1)$ has a dual $SO(5)$ group, which contains both the quasi-spin group, $SU(2)$, and the isospin group, $U_T(2)$, as subgroups. Thus, $SO(5)$ emerges as the dynamical group for an isospin-invariant pairing model for which $USp(2j + 1)$ remains a symmetry group. Moreover, as for the $SU(2)$ quasi-spin model, the $SO(5)$ model can be extended to a more general model with an $SO(5) \times USp(2j + 1)$ dynamical group which preserves the quantum numbers of the shell-model $U(2)_T \times U(2j + 1) \supset USp(2j + 1) \supset SU(2)_J$ coupling scheme.

It is interesting to note that the 1952 classification of shell-model states by the $USp(2j+1)$ coupling scheme in an isospin formalism,[45] already prepared the way for an isospin-invariant pair-coupling model. In fact, it led Helmers[46] in 1961 to identify a group, $USp(4) \simeq SO(5)$, that he described as the *commutator group* of Flower's symplectic group, which proved to be the dynamical group of the isospin-invariant pairing model, introduced shortly afterwards by Flowers and Szpikowski.[47,48] Thus, we now have an isospin-invariant pairing model with an $SO(5)$ dynamical group, and hence an so(5) spectrum generating algebra, and a $USp(2j+1)$ symmetry group, whose states provide a basis for the standard $U(2j + 1) \supset USp(2j + 1)$ coupling scheme of the shell model.

In reality, both $USp(2j + 1)$-conserving and non-conserving components of the nuclear interaction are important in doubly-open-shell nuclei. In fact, the observed properties of doubly open-shell nuclei (see Section 1.7) suggest that rotational correlations are generally expected to be much more important than pairing correlations in such nuclei. As a result, there is a competition between different coupling schemes associated with other subgroup chains. Brief comments on competing chains are given in Sections 6.6.7 and 6.7.9.

6.6.1 *The $U(2)_T$ isospin group*

Recall that, for nucleons of a single type, the Pauli principle permits at most one nucleon to occupy a given single-particle jm state. Thus, the states of a neutron or a proton j^n configuration span a single 'one-column' $U(2j+1)$ irrep $\{1^n\}$. However,

---

[45]Flowers, *op. cit.* Footnote 232 on Page 371.

[46]Helmers, *op. cit.* Footnote 183 on Page 326.

[47]Flowers B.H. and Szpikowski S. (1964), *Proc. Phys. Soc.* **84**, 193.

[48]Contributions to the development of charge-independent pairing models were also made by Ichimura M. (1964), *Prog. Theor. Phys.* **32**, 757.

a neutron and a proton can both occupy the same jm state. Thus, when neutrons and protons are both present in a j^n configuration, the U($2j+1$) irreps available are of the more general type corresponding to Young diagrams with one or two columns. As discussed in Section 5.13.1, the depths of the columns of the Young diagram for such an irrep $\{\lambda\}$ are the elements $\{\tilde{\lambda}_1, \tilde{\lambda}_2\}$ of a partition $\tilde{\lambda} \vdash n$ conjugate to $\{\lambda\}$ which labels a dual U(2)$_T$ irrep. Such a U(2)$_T$ irrep has nucleon number n and isospin T given by

$$n = \tilde{\lambda}_1 + \tilde{\lambda}_2, \quad T = (\tilde{\lambda}_1 - \tilde{\lambda}_2)/2. \tag{6.147}$$

Thus, the states of a nucleus of a j^n configuration with isospin T belong to a U($2j+1$) irrep $\{\lambda\}$ characterised by a Young diagram whose first and second columns consist of $\tilde{\lambda}_1 = n/2 + T$ and $\tilde{\lambda}_2 = n/2 - T$ boxes, respectively.

The one-to-one correspondence between the irreps of the U(2)$_T$ isospin group and those of U($2j + 1$) is an expression of the unitary-unitary duality relationship noted in Section 5.13.1. Thus, whereas for nucleons of a single type in a single-j shell the group dual to U($2j + 1$) is the group U(1), whose infinitesimal generator is the nucleon number operator, the group dual to U($2j + 1$) for a system of neutrons and protons is the larger group U(2)$_T$ whose infinitesimal generators comprise the elements of the isospin su(2)$_T$ algebra as well as the nucleon number operator.

6.6.2   *The group $\mathcal{SO}(5)$ dual to $USp(2j + 1)$*

We consider a system of neutrons and protons occupying a single-j shell. The Fock space, $\mathbb{F}^{(2(2j+2))}$, for this system is generated by a set of creation and annihilation operators, $\{a^\dagger_{\tau j m}, a_{\tau j m}\}$, where τ is equal to $\pm 1/2$ for a neutron and proton, respectively. However, to simplify the notation, it is convenient to suppress the index j, which is here restricted to a single value, and to denote the neutron and proton operators, respectively, by

$$a^\dagger_m \equiv a^\dagger_{1/2 j m}, \quad a_m \equiv a_{1/2 j m}, \quad b^\dagger_m \equiv a^\dagger_{-1/2 j m}, \quad b_m \equiv a_{-1/2 j m}. \tag{6.148}$$

The USp($2j + 1$) group is then defined (see Section 5.12.10) as the subgroup of all U($2j + 1$) transformations that commute with the fermion scalar products

$$a^\dagger \cdot a^\dagger, \quad b^\dagger \cdot b^\dagger, \quad a^\dagger \cdot b^\dagger + b^\dagger \cdot a^\dagger, \tag{6.149}$$

where, for example,

$$a^\dagger \cdot a^\dagger := \sum_m a^\dagger_m a^\dagger_{\bar{m}} = \sqrt{2j+1}\,[a^\dagger \otimes a^\dagger]_0. \tag{6.150}$$

Now, because USp($2j + 1$) is a subgroup of unitary transformations, its elements must also commute with the Hermitian adjoints of these $J = 0$ pair creation operators. Moreover, its elements must commute with the whole Lie algebra of $J = 0$

fermion-pair operators consisting of the three su(2) quasi-spin subalgebras, whose complex extensions are spanned by the operators

$$\hat{S}_+^n = \sum_{m>0} a_m^\dagger a_{\bar{m}}^\dagger, \quad \hat{S}_-^n = \sum_{m>0} a_{\bar{m}} a_m, \quad \hat{S}_0^n = \tfrac{1}{2} \sum_{m>0} \left(a_m^\dagger a_{\bar{m}} + a_m a_{\bar{m}}^\dagger\right), \quad (6.151)$$

$$\hat{S}_+^p = \sum_{m>0} b_m^\dagger b_{\bar{m}}^\dagger, \quad \hat{S}_-^p = \sum_{m>0} b_{\bar{m}} b_m, \quad \hat{S}_0^p = \tfrac{1}{2} \sum_{m>0} \left(b_m^\dagger b_{\bar{m}} + b_m b_{\bar{m}}^\dagger\right), \quad (6.152)$$

$$\hat{S}_+ = \sum_{m>0} (a_m^\dagger b_{\bar{m}}^\dagger + b_m^\dagger a_{\bar{m}}^\dagger), \quad \hat{S}_- = \sum_{m>0} \left(a_{\bar{m}} b_m + b_{\bar{m}} a_m\right),$$

$$\hat{S}_0 = \tfrac{1}{2} \sum_{m>0} \left(a_m^\dagger a_{\bar{m}} + a_m a_{\bar{m}}^\dagger + b_m^\dagger b_{\bar{m}} + b_m b_{\bar{m}}^\dagger\right) = \hat{S}_0^n + \hat{S}_0^p, \quad (6.153)$$

and the su(2)$_T$ isospin subalgebra

$$\hat{T}_+ = \sum_m a_m^\dagger b_{\bar{m}}, \quad \hat{T}_- = \sum_m b_m^\dagger a_{\bar{m}},$$

$$\hat{T}_0 = \tfrac{1}{2} \sum_m \left(a_m^\dagger a_{\bar{m}} - b_m^\dagger b_{\bar{m}}\right) = \hat{S}_0^n - \hat{S}_0^p. \quad (6.154)$$

Together, the Hermitian linear combinations of these operators span an so(5) Lie algebra[49] with the root diagram shown in Figure 6.12.

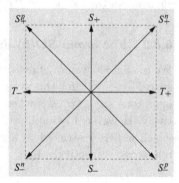

Figure 6.12: Root diagram for the generalised quasi-spin algebra, so(5), isomorphic to usp(4). The Cartan subalgebra is spanned by two elements, with zero roots: $\hat{S}_0$ and $\hat{T}_0$. A quasi-spin su(2) subalgebra for neutrons is spanned by the operators $\{\hat{S}_+^n, \hat{S}_0^n, \hat{S}_-^n\}$ and that for protons by $\{\hat{S}_+^p, \hat{S}_0^p, \hat{S}_-^p\}$. The isospin algebra u(2)$_T$ is spanned by the operators $\{\hat{T}_+, \hat{T}_0, \hat{T}_-, \hat{S}_0\}$.

Theorem 6.4 *The groups $USp(2j + 1)$ and $SO(5)$ have dual representations on the fermion Fock space, $\mathbb{F}^{(2(2j+1))}$, of neutron and proton states in single-j-shell configurations.*

This theorem is a special case of Helmer's theorem.[50] The proof[51] presented here is similar to that given for Theorem 6.1. It proceeds by the identification of all irreps of the direct product group USp2j + 1) × $SO(5)$ on the shell-model space of interest and the observation that their properties accord with the duality

[49] The irreps of this so(5) Lie algebra have been constructed by Hecht K.T. (1965), *Nucl. Phys.* **63**, 177; Hecht K.T. and Elliott J.P. (1985), *Nucl. Phys.* **A438**, 29; and Hecht K.T. (1987), *Nucl. Phys.* **A475**, 276.

[50] Helmers, *op. cit.* Footnote 183 on Page 326.

[51] Rowe *et al.*, *op. cit.* Footnote 168 on Page 308.

requirements. This is achieved by identifying every state that is simultaneously of
USp($2j + 1$) highest weight and $\mathcal{SO}(5)$ lowest weight in the j-shell model space.

Note that a lowest-weight state of $\mathcal{SO}(5)$ is also a lowest-weight state of its
U(2)$_T$ subgroup. Thus, as USp($2j + 1$) is a subgroup of U($2j + 1$), the proof starts
with the observation that the set of all U($2j + 1$) highest-weight states in the space
$\mathbb{F}^{(2(2j+1))}$ that are simultaneously of U(2)$_T$ lowest weight is the set

$$\{|\lambda\rangle := \hat{Z}_0^{\{\lambda\}}|0\rangle; \tilde{\lambda}_2 \leq \tilde{\lambda}_1 \leq 2j + 1\}, \tag{6.155}$$

where

$$\hat{Z}_0^{\{\lambda\}} := \left(b_j^\dagger b_{j-1}^\dagger \cdots b_{j+1-\tilde{\lambda}_1}^\dagger\right)\left(a_j^\dagger a_{j-1}^\dagger \cdots a_{j+1-\tilde{\lambda}_2}^\dagger\right). \tag{6.156}$$

These states are those corresponding to two-column Young tableaux of depth $2j+1$
or less. A complete set of states, that are both of U($2j + 1$) highest weight and of
$\mathcal{SO}(5)$ lowest weight in $\mathbb{F}^{(2(2j+1))}$, is then identified as the subset of states

$$\mathcal{W} := \{|\lambda\rangle := \hat{Z}_0^{\{\lambda\}}|0\rangle; \tilde{\lambda}_2 \leq \tilde{\lambda}_1 \leq \tfrac{1}{2}(2j + 1)\}, \tag{6.157}$$

which contain no nucleon in a single-particle state with $m < 0$ and are, therefore,
annihilated by the $\mathcal{SO}(5)$ operators, $\hat{S}_-^n$, $\hat{S}_-^p$, $\hat{S}_-$, and $\hat{T}_-$.

Now, because a highest-weight state for a u($2j + 1$) irrep is an eigenstate of all
elements of its Cartan subalgebra, and because usp($2j+1$) has a Cartan subalgebra
that is a subalgebra of that for u($2j + 1$) (see Section 5.12.10), it follows that a
highest-weight state of a u($2j + 1$) irrep is simultaneously a highest-weight state
for an irrep of the usp($2j + 1$) $\subset$ u($2j + 1$) subalgebra. Thus, the states of the
set $\mathcal{W}$ are simultaneously of highest weight relative to USp($2j + 1$) and of lowest
weight relative to $\mathcal{SO}(5)$. However, it is conceivable that there are other USp($2j+1$)
highest-weight states that are simultaneously of $\mathcal{SO}(5)$ lowest weight. The following
claim shows that there are no others.

Claim 6.5 *Every highest-weight state for a USp($2j + 1$) irrep in $\mathbb{F}^{(2(2j+1))}$, that
is simultaneously of $\mathcal{SO}(5)$ lowest weight, is a U($2j + 1$) highest-weight state.*

To prove this claim, we again start with the observation that a lowest-weight
state of $\mathcal{SO}(5)$ is also a lowest-weight state of its U(2)$_T$ subgroup. Then, if such
a U(2)$_T$ lowest-weight state belongs to a U(2)$_T$ irrep $\{\tilde{\lambda}\}$ it must, by the unitary-
unitary duality theorem (see Section 5.11.5), also belong to a dual U($2j + 1$) irrep
$\{\lambda\}$. Thus, we need only look for states of USp($2j + 1$) highest weight and $\mathcal{SO}(5)$
lowest-weight state among the states of the U($2j + 1$) irreps with highest-weight
states in the set $\mathcal{W}$. Suppose that $|\lambda\kappa^\mu\rangle$ is such a state, which belongs to a U($2j+1$)
irrep of highest weight λ and is of USp($2j + 1$) highest weight $\kappa^{(\mu)}$, where μ is its
U($2j + 1$) weight. The claim is then proved by showing that such a state can only
be of $\mathcal{SO}(5)$ lowest weight if $\mu = \lambda$ and, hence, if $|\lambda\kappa^{\{\lambda\}}\rangle$ is identically equal to the
state $|\lambda\rangle \in \mathcal{W}$.

Observe that the state $|\lambda\kappa^{\{\lambda\}}\rangle \equiv |\lambda\rangle = \hat{Z}_0^{\{\lambda\}}|0\rangle \in \mathcal{W}$ is created by an operator, $\hat{Z}_0^{\{\lambda\}}$, which is a component of a $\mathrm{USp}(2j+1) \times \mathrm{U}(2)_T$ tensor that is of highest weight relative to both $\mathrm{U}(2j+1)$ and $\mathrm{USp}(2j+1)$ and of lowest weight relative to $\mathrm{U}(2)_T$. Recall also that the $\mathrm{USp}(2j+1)$ irreps given by the $\mathrm{U}(2j+1) \downarrow \mathrm{USp}(2j+1)$ branching rules (see Section 5.12.7) are obtained by factoring out $\mathrm{USp}(2j+1)$ scalars from the creation operators of $\mathrm{U}(2j+1)$ states. As shown above, the $\mathrm{USp}(2j+1)$ scalar combinations of the nucleon creation operators are generated by the $\mathrm{so}(5)$ raising operators, $\{\hat{S}_+^n, \hat{S}_+, \hat{S}_+^p\}$, which can be regarded as components of a $T=1$, $\mathrm{U}(2)_T$ tensor, $\hat{\mathcal{A}}$. Thus, we can create a highest-weight state, $\langle\kappa^{\{\mu\}}\rangle$, for a $\mathrm{USp}(2j+1)$ irrep $\langle\kappa^{\{\mu\}}\rangle$ within the space of the $\mathrm{U}(2j+1)$ irrep $\{\lambda\}$ by adding a $J=0$ coupled nucleon pair to the highest-weight state $|\mu\rangle = Z_0^{\{\mu\}}|0\rangle$ of a $\mathrm{U}(2j+1)$ irrep with two fewer nucleons than the irrep $\{\lambda\}$; viz.

$$|\lambda\kappa^{\{\mu\}}\rangle = \left[\hat{\mathcal{A}} \otimes \hat{Z}^{\{\mu\}}\right]_0^{\{\lambda\}}|0\rangle, \tag{6.158}$$

where the bracket signifies a $\mathrm{U}(2)_T \times \mathrm{USp}(2j+1)$ coupled product of the tensors $\hat{\mathcal{A}}$ and $\hat{Z}^{\{\mu\}}$. Because $\hat{\mathcal{A}}$ is a $\mathrm{USp}(2j+1)$ scalar and an $\mathrm{U}(2)_T$, $T=1$ tensor, the coupled tensor product, $\left[\hat{\mathcal{A}} \otimes \hat{Z}^{\{\mu\}}\right]^{\{\lambda\}}$, is given simply by $\mathrm{SU}(2)_T$ coupling. Other $|\lambda\kappa^{\{\mu\}}\rangle$ states are of the form

$$|\lambda\kappa^{\{\mu\}}\rangle = \left[\hat{\mathcal{A}} \otimes \hat{\mathcal{A}} \otimes Z^{\{\mu\}}\right]_0^{\{\lambda\}}|0\rangle, \tag{6.159}$$

and so on. The $\mathrm{U}(2j+1) \downarrow \mathrm{USp}(2j+1)$ branching rules imply that every $\mathrm{USp}(2j+1)$ highest-weight state within a $\mathrm{U}(2j+1) \times \mathrm{U}(2)_T$ irrep can be obtained in this way. However, because the components of the tensor $\hat{\mathcal{A}}$ are raising operators of the $\mathrm{so}(5)$ Lie algebra, none of the states $\{|\lambda\kappa^{\{\mu\}}\rangle\}$ with $\mu \neq \lambda$ can be an $\mathcal{SO}(5)$ lowest-weight state. This completes the proof of the claim. $\square$

Having identified the states that are of $\mathrm{USp}(2j+1)$ highest weight and $\mathcal{SO}(5)$ lowest weight, it is now ascertained that each of them defines a unique multiplicity-free irrep of $\mathrm{USp}(2j+1) \times \mathcal{SO}(5)$ and that these irreps are such that each $\mathrm{USp}(2j+1)$ irrep occurs in combination with a uniquely defined irrep of $\mathcal{SO}(5)$ in accordance with the conditions for duality (Section 5.10). This, complete the proof of the theorem. $\square$

The one-to-one correspondence between paired irreps of $\mathrm{U\hat{S}p}(2j+1)$ and $\mathcal{SO}(5)$ is as follows. An $\mathcal{SO}(5)$ irrep, in the space of a single-j shell, is traditionally labelled by quantum numbers, (v, t), referred to as the *seniority* and *reduced isospin* of the irrep defined, respectively, as the nucleon number and isospin of the lowest-weight states for the irrep. The paired $\mathrm{USp}(2j+1)$ irreps can likewise be labelled by these same (v, t) quantum numbers. The κ labels of a $\mathrm{USp}(2j+1)$ irrep $\langle\kappa\rangle$ then have the same relationship to the seniority and reduced isospin as the labels of a $\mathrm{U}(2j+1)$ irrep $\{\lambda\}$ have to the particle number and isospin (cf. Equation (6.147)), i.e.,

$$v = \tilde{\kappa}_1 + \tilde{\kappa}_2, \quad t = (\tilde{\kappa}_1 - \tilde{\kappa}_2)/2. \tag{6.160}$$

6.6.3 Classification of states

The dualities between the irreps of $U(2j+1)$ and $U(2)_T$ and between the irreps of $USp(2j+1)$ and $\mathcal{SO}(5)$ simplify the definition of basis states for the jj-coupled shell model. Orthonormal basis states for a single-j shell are initially defined and labelled by the quantum numbers (irrep labels) of the subgroup chain

$$
\begin{array}{cccccc}
U(2j+1) \times U(2)_T & \supset & USp(2j+1) \times U(1)_T & \supset & SU(2)_J & \supset & U(1)_J \\
\{\lambda\} \equiv (n,T) & \alpha & \langle\kappa\rangle \equiv (v,t) & T_0 & \rho & J & M
\end{array}, \quad (6.161)
$$

where α and ρ are multiplicity indices. As a result of the above duality relationships, it is now seen that identical basis states are defined by the chain

$$
\begin{array}{cccccc}
\mathcal{SO}(5) \times SU(2)_J & \supset & U(2)_T \times U(1)_J & \supset & U(1)_T \\
(v,t) & \rho\,J & \alpha\;(n,T) & M & T_0
\end{array}. \quad (6.162)
$$

These equivalent (dual) subgroup chains exhibit the explicit way in which the $\mathcal{SO}(5)$ np-pairing model is embedded in the jj-coupled shell model. Moreover, they prepare the way for generalisations of the $\mathcal{SO}(5)$ model to include more general shell-model interactions, as is possible for the $\mathcal{SU}(2)$ pairing model.

6.6.4 The ranges of the $\mathcal{SO}(5) \times USp(2j+1)$ quantum numbers

The ranges of the above quantum numbers for a given j shell are readily determined as illustrated for $j = \frac{3}{2}$ in Table 6.5.

Table 6.5: States of the $j = \frac{3}{2}$ shell for nucleon number $n = 1, \ldots, 4$. For $n = 5, \ldots, 8$, the states are mirror images of the $n = 4, \ldots, 1$ states, as can be seen in Figure 6.13. Lowest-n $USp(2j+1)$ irreps of an $\mathcal{SO}(5) \times USp(4)$ irrep are denoted by an asterisk in the last column.

$\{\lambda\}$	n	T	$\langle\kappa\rangle$	v	t	J	l.wt.
$\{00\}$	0	0	$\langle 00\rangle$	0	0	0	*
$\{10\}$	1	$\frac{1}{2}$	$\langle 10\rangle$	1	$\frac{1}{2}$	$\frac{3}{2}$	*
$\{20\}$	2	0	$\langle 20\rangle$	2	0	1, 3	*
$\{1^2\}$	2	1	$\langle 00\rangle$	0	0	0	
			$\langle 11\rangle$	2	1	2	*
$\{21\}$	3	$\frac{1}{2}$	$\langle 10\rangle$	1	$\frac{1}{2}$	$\frac{3}{2}$	
			$\langle 21\rangle$	3	$\frac{1}{2}$	$\frac{1}{2}$, $\frac{5}{2}$, $\frac{7}{2}$	*
$\{1^3\}$	3	$\frac{3}{2}$	$\langle 10\rangle$	1	$\frac{1}{2}$	$\frac{3}{2}$	
$\{2^2\}$	4	0	$\langle 00\rangle$	0	0	0	
			$\langle 11\rangle$	2	1	2	
			$\langle 22\rangle$	4	0	2, 4	*
$\{21^2\}$	4	1	$\langle 20\rangle$	2	0	1, 3	
			$\langle 11\rangle$	2	1	2	
$\{1^4\}$	4	2	$\langle 00\rangle$	0	0	0	

The list of possible $U(2j+1)$ irreps, denoted by $\{\lambda\}$, is simply the list of ordered partitions of the nucleon number n corresponding to Young diagrams with at most two columns. The values of the isospin, T, for these irreps are then given by Equation (6.147). The angular momentum states in a given $U(2j+1)$ irrep $\{\lambda\}$ are given by the plethysm $\{2j\} \textcircled{p} \{\lambda\}$, as shown in Box 6.I on Page 407; e.g., for $j = {}^3\!/_2$ and $n \leq 4$ irreps, the needed plethysms are given by

$$
\begin{aligned}
\{3\} \textcircled{p} \{1\} &= \{3\} & \sim \quad &(J = {}^3\!/_2), \\
\{3\} \textcircled{p} \{2\} &= \{6\} \oplus \{42\} & \sim \quad &(J = 3, 1), \\
\{3\} \textcircled{p} \{1^2\} &= \{51\} \oplus \{33\} & \sim \quad &(J = 2, 0), \\
\{3\} \textcircled{p} \{21\} &= \{81\} \oplus \{72\} \oplus \{54\} \oplus \{63\} & \sim \quad &(J = {}^7\!/_2, {}^5\!/_2, {}^1\!/_2, {}^3\!/_2), \\
\{3\} \textcircled{p} \{1^3\} &= \{63\} & \sim \quad &(J = {}^3\!/_2), \\
\{3\} \textcircled{p} \{2^2\} &= \{10, 2\} \oplus \{66\} \oplus 2\{84\} & \sim \quad &(J = 4, 0, 2^2), \\
\{3\} \textcircled{p} \{21^2\} &= \{93\} \oplus \{75\} \oplus \{84\} & \sim \quad &(J = 3, 1, 2), \\
\{3\} \textcircled{p} \{1^4\} &= \{66\} & \sim \quad &(J = 0).
\end{aligned}
\tag{6.163}
$$

Note that the angular momentum content of $U(2j+1) = U(4)$ irreps for $4 < n \leq 8$ are simply determined by particle-hole symmetry. For example, the $n = 8$ filled-shell state has the same angular momentum ($J = 0$) as the zero-particle ($n = 0$) state, and the $8 - n$ hole states have the same angular momenta as corresponding n-particle states. Thus, from the list given in Table 6.5, we can display all the $USp(2j+1) = USp(4)$ irreps occurring in the $j = {}^3\!/_2$ shell labelled by their $SO(5)$ quantum numbers, as shown in Figure 6.13. It can be seen in the figure that the plots for each set of equivalent $USp(4)$ irreps are in one-to-one correspondence with $SO(5)$ weight diagrams.

6.6.5 The so(5) Casimir invariant

The so(5) Lie algebra has an so(4) subalgebra (isomorphic to su(2) $\oplus$ su(2)) whose root vectors $\hat{S}^n_\pm$ and $\hat{S}^p_\pm$ are those of the neutron and proton quasi-spin algebras. Thus, to within an arbitrary factor, the Casimir operator, $\hat{\mathcal{C}}_{so4}$, for so(4) is

$$
\hat{\mathcal{C}}_{so4} := 2\big(\hat{S}^n_0\big)^2 + \hat{S}^n_+ \hat{S}^n_- + \hat{S}^n_- \hat{S}^n_+ + 2\big(\hat{S}^p_0\big)^2 + \hat{S}^p_+ \hat{S}^p_- + \hat{S}^p_- \hat{S}^p_+.
\tag{6.164}
$$

From the symmetry of this expression and that of the so(5) root diagram, one may conjecture that the Casimir operator for so(5) is obtained by adding a multiple of the operator $\hat{S}_+ \hat{S}_- + \hat{S}_- \hat{S}_+ + \hat{T}_+ \hat{T}_- + \hat{T}_- \hat{T}_+$ to $\hat{\mathcal{C}}_{so4}$. This conjecture is confirmed by checking that, with an appropriate value of the multiple, which turns out to be $\frac{1}{2}$, the resulting expression for $\hat{\mathcal{C}}_{so5}$ commutes with all elements of the so(5) Lie algebra. Thus, it is determined that

$$
\hat{\mathcal{C}}_{so5} := \hat{T} \cdot \hat{T} + \big(\hat{S}_0\big)^2 - 3\hat{S}_0 + 2\hat{S}^n_+ \hat{S}^n_- + 2\hat{S}^p_+ \hat{S}^p_- + \hat{S}_+ \hat{S}_-.
\tag{6.165}
$$

The eigenvalues of $\hat{\mathcal{C}}_{so5}$ for an irrep (v, t) are most easily evaluated for a lowest-weight state, on which the operators $\hat{S}^n_-$, $\hat{S}^p_-$, and $\hat{S}_-$ vanish. From the expansion

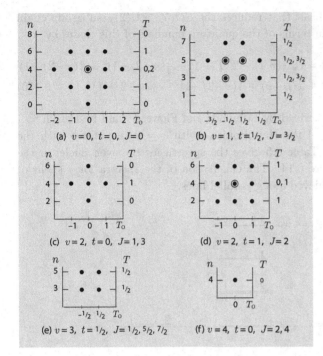

Figure 6.13: States of the $j = \frac{3}{2}$ shell classified by the quantum numbers $(nvtJTT_0)$ of the subgroup chain (6.162). Each diagram shows the states of a single $\mathcal{SO}(5) \times \mathrm{USp}(4)$ irrep. Subsets of states belonging to a single USp(4) irrep are shown as single dots. All the dots of a given diagram correspond to states of equivalent USp(4) irreps. Pairs of USp(4) irreps differing only by their isospin T quantum numbers are shown as dots with circles around them. For example, the two states at the centre of the weight diagram (a) have a complete set of quantum numbers in common except for their total isospin T. One state has isospin $T = 0$ and the other has isospin $T = 2$.

of $\hat{S}_0$, given by Equation (6.153), it is seen that a state of nucleon number n is an eigenstate of $\hat{S}_0$ with eigenvalue $\frac{1}{2}(n - 2j - 1)$. Thus, because $n = v$ and $T = t$ for a lowest-weight state, $\hat{C}_{\mathrm{so5}}$ has eigenvalues given for states of an $\mathcal{SO}(5)$ irrep (v,t) by

$$C_{\mathrm{so5}}(v,t) = t(t + 1) + \tfrac{1}{4}(v - 2j - 1)^2 - \tfrac{3}{2}(v - 2j - 1). \qquad (6.166)$$

6.6.6 The $\mathcal{SO}(5)$ pair-coupling model

Within the space of states of both neutrons and protons in a single-j shell, the $J = 0$ pairing Hamiltonian is of the form

$$\hat{H} = \varepsilon \, \hat{n} - G \, \hat{P}^\dagger \cdot \hat{P}, \qquad (6.167)$$

where $\hat{P}^\dagger \cdot \hat{P}$ is an isospin-invariant pairing interaction, that annihilates and recreates $J = 0, T = 1$ pair-coupled nucleons, and is defined in terms of the quasi-spin operators of Equations (6.151) - (6.154) by

$$\hat{P}^\dagger \cdot \hat{P} = 2\hat{S}_+^n \hat{S}_-^n + 2\hat{S}_+^p \hat{S}_-^p + \hat{S}_+ \hat{S}_-. \qquad (6.168)$$

It can be seen from the expression of the so(5) Casimir operator, given by Equation (6.165), that the pairing interaction, $\hat{P}^\dagger \cdot \hat{P}$, can be expressed in the form

$$\hat{P}^\dagger \cdot \hat{P} = \hat{C}_{\mathrm{so5}} - \hat{T} \cdot \hat{T} - \left(\hat{S}_0\right)^2 + 3\hat{S}_0. \qquad (6.169)$$

Thus, it is diagonal in the basis that reduces the $SO(5) \supset U(2)_T$ subgroup chain and has eigenvalues given, in terms of the quantum numbers of this chain, by

$$\langle vtanTT_0|\hat{P}^\dagger \cdot \hat{P}|vtanTT_0\rangle = \mathcal{C}_{so5}(v,t) - T(T+1) - \tfrac{1}{4}(n-2j-1)^2 + \tfrac{3}{4}(n-2j-1)$$
$$= t(t+1) - T(T+1) + \tfrac{3}{2}(n-v) + \tfrac{1}{4}(v-2j-1)^2 - \tfrac{1}{4}(n-2j-1)^2, \qquad (6.170)$$

consistent with expressions derived by Edmonds and Flowers[52] and by Parikh.[53]

Evaluating the energies of the Hamiltonian (with $\varepsilon = 0$) for states with the quantum numbers listed in Table 6.5 gives the spectra for the even nuclei in the $j = {}^3/_2$ shell shown in Figure 6.14. The calculation of the spectra for j shells of more physical interest is achieved along parallel lines.

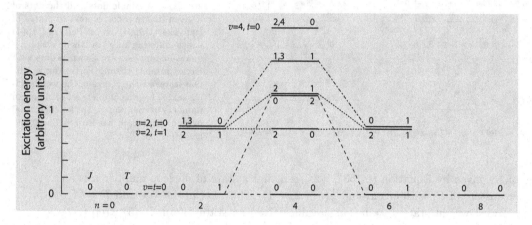

Figure 6.14: Energy levels of the $J = 0, T = 1$ isospin-invariant pairing model for even numbers of nucleons in a $j = {}^3/_2$ shell. Energy levels belonging to a common $SO(5)$ irrep and, thus, belonging to equivalent $USp(2j + 1)$ irreps are labelled by their seniority and reduced isospin quantum numbers, (v, t), and are connected by dashed or dotted lines.

6.6.7 General multishell seniority-conserving models

Because of the duality relationship between $SO(5)$ and $USp(2j + 1)$, it follows, as discussed in detail in Section 6.5.5 for nucleons of a single type, that seniority is conserved in a single-j shell by a Hamiltonian of the type

$$\hat{H} = \varepsilon\,\hat{n}_j - G\,\hat{P}_j^\dagger \cdot \hat{P}_j + \hat{V}_j, \qquad (6.171)$$

where $\hat{V}_j$ is a seniority-conserving interaction expressible as a bilinear combination of elements of the $usp(2j + 1)$ Lie algebra. Thus, a general Hamiltonian with two-

[52]Edmonds and Flowers, op. cit. Footnote 256 on Page 380.
[53]Parikh J.C. (1965), Nucl. Phys. **63**, 214.

body interactions that conserves multishell seniority is of the form

$$\hat{H} = \sum_j \varepsilon_j \hat{n}_j - \sum_{jj'} G_{jj'} \hat{P}_j^\dagger \cdot \hat{P}_{j'} + \sum_j \hat{V}_j. \tag{6.172}$$

Energy spectra for multishell-pairing Hamiltonians with unequal single-particle energies but a single coupling constant, i.e., $G_{jj'} = G$, and no $\hat{V}_j$ interactions, have been determined recently by Dukelsky et al.[54] using a generalisation, by Asorey et al.,[55] of the Richardson-Gaudin equations to Lie algebras of higher rank.

In concluding this section, we draw attention to the fact that, in a multishell situation, there are alternative and competing, coupling schemes which naturally diagonalise other components of the two-body interaction. For example, for two active j shells, the standard jj coupling scheme makes use of basis states that reduce the subgroup chain

$$U(2\Omega) \supset U(2\Omega_1) \times U(2\Omega_2) \supset USp(2\Omega_1) \times USp(2\Omega_2), \tag{6.173}$$

where $2\Omega_i = 2j_i + 1$ and $\Omega := \Omega_1 + \Omega_2$. This basis is favoured when the states of the two j shells are essentially uncoupled, e.g., because of large single-particle energy differences. The alternative possibility, which makes use of basis states that reduce the chain

$$U(2\Omega) \supset USp(2\Omega) \supset USp(2\Omega_1) \times USp(2\Omega_2), \tag{6.174}$$

is more appropriate when there are strong superconducting pairing interactions between the two shells.

Because of the above-described duality relations, reduction of the subgroups of Equations (6.173) and (6.174) is equivalent to reducing the, respective, subgroup chains

$$\mathcal{SO}(5)_1 \times \mathcal{SO}(5)_2 \supset U(2)_{T_1} \times U(2)_{T_2}, \tag{6.175}$$

$$\mathcal{SO}(5)_1 \times \mathcal{SO}(5)_2 \supset \mathcal{SO}(5) \supset U(2)_T. \tag{6.176}$$

The latter chains reveal that the transformation between the bases defined by Equations (6.173) and (6.174) is given by $\mathcal{SO}(5)$ Clebsch-Gordan coupling coefficients in a $U(2)_T \times U(2)_S$ basis.[56]

While jj-coupling schemes are ones for which shell-model calculations are most easily carried out, primarily because of the more ready availability of coefficients of fractional parentage in jj-coupling bases, it should also be recognised that, in doubly-open-shell nuclei, there will inevitably be strong mixing of states from different j-shell configurations due to deformation and rotational correlations.

[54]Dukelsky J. *et al.* (2006), *Phys. Rev. Lett.* **96**, 072503.
[55]Asorey M., Falceto F. and Sierra G. (2002), *Nucl. Phys.* **B622**, 593.
[56]Hecht K.T. (1993), *J. Phys. A: Math. Gen.* **26**, 329.

Exercises

6.23 Show that a state of nucleon number n is an eigenstate of the operator $\hat{S}_0$, given by Equation (6.153), with eigenvalue $\frac{1}{2}(n - 2j - 1)$.

6.24 Use the results of Table 6.5 to derive the U(4) ↓ USp(4) branching rules for the U(4) irreps shown in the table. Confirm that the results are consistent with those obtained, more generally, by the methods of Section 5.12.10.

6.7 $L = 0, T = 0, 1$ pairing in LST coupling; open-shell nuclei

Although the $J = 0$ pairing interaction of the $SO(5)$ model is isospin invariant, it applies only to $T = 1$ two-nucleon states. In fact, a short-range interaction also acts strongly on $L = 0$ states of both $T = 0$ and $T = 1$. To satisfy the fermion antisymmetry requirement, a state of two nucleons of common orbital angular momentum l is in an $S = T = 0$ or $S = T = 1$ spin-isospin state when its orbital angular momentum L is odd and in an $S = 1, T = 0$ or $S = 0, T = 1$ state, when L is even. Thus, an $L = 0$ pairing interaction lowers the energies of two-nucleon states of both $L = 0, S = 0, T = 1$ and $L = 0, S = 1, T = 0$. This is illustrated in Figure 6.15, for the states of two nucleons of angular momentum $l = 4$. Thus, for nuclei in

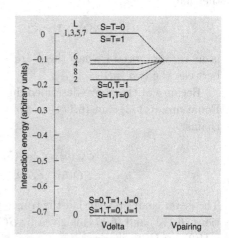

Figure 6.15: Two-nucleon interaction energy for a zero-range (delta-function) force and for an $L = 0$ pairing force when the two nucleons are in an $(l = 4)^2$ configuration coupled to orbital angular momentum L.

which both neutrons and protons are filling common valence shells, it makes sense to approximate the short-range component of the two-nucleon interaction by an $L = 0$ pairing interaction.

An $L = 0$ pairing model, based on an so(8) spectrum generating algebra, was proposed by Flowers and Szpikowski.[57] Somewhat earlier, although not widely publicised, it had been shown by Bayman[58] that an $L = 0$ pairing interaction is

---

[57]Flowers and Szpikowski, *op. cit.* Footnote 244 on Page 377.
[58]Bayman, *op. cit.* Footnote 243 on Page 377.

an invariant of an O(2l + 1) subgroup of U(2l + 1). However, although the two approaches were generally recognised to be complementary, the underlying duality relationships of their respective group structures was not pursued until later. It is shown in this section that, because of this duality, the $L = 0$ pairing Hamiltonian is diagonal in a standard LST coupling scheme based on the U(2l + 1) ⊃ O(2l + 1) ⊃ SO(3) subgroup chain. However, as is usually the case, other components of the two-nucleon interaction are also important, in general, and there ends up being a competition between alternative coupling schemes. For example, studies of ^{210}Bi, see Figure 6.1, indicate that $S = T = 0$ states of both maximum and minimum L values are favoured over those of intermediate L. Moreover, in doubly open-shell nuclei, the long-range components of the two-nucleon interaction, that give rise to quadrupole deformations and rotational structures, are understood to dominate over the short-range pairing interactions. As a result, the explicit effects of np-pairing correlations tend to be overshadowed and not easily seen.

6.7.1 *Single-nucleon operators and time reversal in LST coupling*

In parallel with the definitions for jj coupling, it is appropriate to label single-nucleon creation operators, $\{a^\dagger_{\sigma lm}\}$, in LST coupling with lm denoting the orbital angular momentum quantum numbers and σ indexing the spin and isospin and any other quantum numbers needed. The operator $a^\dagger_{\sigma lm}$ is then the m component of a covariant SO(3) spherical tensor of orbital angular momentum l. Together with corresponding annihilation operators, $\{a^{\sigma lm}\}$, defined by $a^{\sigma lm} := (a^\dagger_{\sigma lm})^\dagger$, these operators satisfy the anticommutation relations

$$\{a^{\sigma lm}, a^\dagger_{\tau l'm'}\} = \delta_{\sigma,\tau}\delta_{l,l'}\delta_{m,m'}. \tag{6.177}$$

The annihilation operators, $\{a^{\sigma lm}\}$, are components of contravariant tensors (see Section A.6). However, as in jj coupling, we can define covariant annihilation operators with lower indices by

$$a_{\sigma lm} := (-1)^{l-m}a^{\sigma l,-m}. \tag{6.178}$$

Such operators are needed in constructing coupled-tensor products of the form

$$(a^\dagger_{\sigma l} \otimes a_{\tau l'})_{LM} = \sum_{mm'}(l'm'\,lm|LM)a^\dagger_{\sigma lm}a_{\tau l'm'}. \tag{6.179}$$

Creation and annihilation operators for time-reversed single-nucleon states are defined, with a suitable choice of phase convention (see Section 6.2), in parallel with the definitions for jj coupling, by

$$a^\dagger_{\sigma l\bar{m}} := (-1)^{l+m}a^\dagger_{\sigma l,-m}, \quad a_{\sigma l\bar{m}} := (-1)^{l+m}a_{\sigma l,-m} = a^{\sigma lm}. \tag{6.180}$$

Then, because $\{a^{\sigma lm}, a^\dagger_{\tau l'n}\} = \delta_{\sigma,\tau}\delta_{l,l'}\delta_{m,n}$, we obtain the convenient expressions for the fermion anticommutation relations

$$\{a_{\sigma l \bar{m}}, a_{\tau l' m'}^{\dagger}\} = \{a_{\sigma l m}, a_{\tau' \bar{m}'}^{\dagger}\} = \delta_{\sigma, \tau} \delta_{l, l'} \delta_{m, m'}. \tag{6.181}$$

These definitions imply that Hermitian adjoint relationships are given by

$$\left(a_{\sigma l \bar{m}}\right)^{\dagger} = a_{\sigma l m}^{\dagger}, \quad \left(a_{\sigma l m}\right)^{\dagger} = a_{\sigma l m}^{\dagger}. \tag{6.182}$$

Note, however, that in contrast to the j-coupled single-particle operators, applying the time-reversal operation does not result in a sign change, i.e.,

$$a_{\sigma l \bar{m}}^{\dagger} = a_{\sigma l m}^{\dagger}, \quad a_{\sigma l \bar{m}} = a_{\sigma l m}. \tag{6.183}$$

6.7.2 *The $L = 0$ pairing interaction for states of a single l value*

Creation operators for two-nucleon $L = 0$ pair states are defined by

$$\hat{\mathcal{A}}_{\sigma \tau} := a_{\sigma}^{\dagger} \cdot a_{\tau}^{\dagger}, \tag{6.184}$$

where $a_{\sigma}^{\dagger} \cdot a_{\tau}^{\dagger}$ is the $L = 0$ scalar product

$$a_{\sigma}^{\dagger} \cdot a_{\tau}^{\dagger} := \sum_{m=-l}^{l} \sqrt{2l+1} \, (l, -m \, lm|00) \, a_{\sigma l m}^{\dagger} a_{\tau l, -m}^{\dagger} = \sum_{m} a_{\sigma m}^{\dagger} a_{\tau \bar{m}}^{\dagger}, \tag{6.185}$$

and we suppress the common l index so that

$$a_{\sigma m}^{\dagger} \equiv a_{\sigma l m}^{\dagger}, \quad a_{\tau \bar{m}}^{\dagger} \equiv (-1)^{l+m} a_{\sigma l, -m}^{\dagger}. \tag{6.186}$$

The expansion

$$\hat{\mathcal{A}}_{\sigma \tau} = \sum_{m} a_{\sigma m}^{\dagger} a_{\tau \bar{m}}^{\dagger} = \sum_{m} a_{\sigma \bar{m}}^{\dagger} a_{\tau m}^{\dagger} = -\hat{\mathcal{A}}_{\tau \sigma} \tag{6.187}$$

shows that, for neutrons and protons in a single l orbital, there are six linearly-independent two-nucleon $L = 0$ creation operators.

An $L = 0$ pairing interaction is now defined as

$$\hat{V} := -\tfrac{1}{4} G \sum_{\sigma \tau m n} a_{\sigma m}^{\dagger} a_{\tau \bar{m}}^{\dagger} a_{\tau n} a_{\sigma \bar{n}} = -\tfrac{1}{4} G \sum_{\sigma \tau} \hat{\mathcal{A}}_{\sigma \tau} \hat{\mathcal{B}}_{\sigma \tau}, \tag{6.188}$$

where

$$\hat{\mathcal{B}}_{\sigma \tau} := \left(\hat{\mathcal{A}}_{\sigma \tau}\right)^{\dagger} = \sum_{m} a_{\tau m} a_{\sigma \bar{m}} = -\hat{\mathcal{B}}_{\tau \sigma}. \tag{6.189}$$

An $L = 0$ pairing Hamiltonian is then given by

$$\hat{H} = \varepsilon \hat{n} - \tfrac{1}{4} G \sum_{\sigma \tau} \hat{\mathcal{A}}_{\sigma \tau} \hat{\mathcal{B}}_{\sigma \tau}, \tag{6.190}$$

where $\hat{n} = \sum_{\sigma m} a^\dagger_{\sigma m} a_{\sigma \bar{m}}$ is the nucleon number operator. The spectrum of such a Hamiltonian can be determined by the observation that, to within a constant, it is a quadratic in the elements of an so(8) spectrum generating algebra.

6.7.3 An so(8) spectrum generating algebra

The above-defined $\{\hat{A}_{\sigma\tau}\}$ and $\{\hat{B}_{\sigma\tau}\}$ operators satisfy the commutation relations

$$[\hat{A}_{\sigma\tau}, \hat{B}_{\sigma'\tau'}] = \sum_{mn}[a^\dagger_{\sigma m} a^\dagger_{\tau \bar{m}}, a_{\tau' n} a_{\sigma' \bar{n}}]$$

$$= \delta_{\sigma\sigma'}\hat{C}_{\tau\tau'} + \delta_{\tau\tau'}\hat{C}_{\sigma\sigma'} - \delta_{\sigma\tau'}\hat{C}_{\tau\sigma'} - \delta_{\tau\sigma'}\hat{C}_{\sigma\tau'}, \qquad (6.191)$$

where

$$\hat{C}_{\sigma\sigma'} := \sum_m a^\dagger_{\sigma m} a_{\sigma' \bar{m}} - \delta_{\sigma\sigma'}\tfrac{1}{2}(2l+1). \qquad (6.192)$$

The latter operators satisfy the commutation relations

$$[\hat{C}_{\sigma\tau}, \hat{C}_{\sigma'\tau'}] = \delta_{\tau\sigma'}\hat{C}_{\sigma\tau'} - \delta_{\tau'\sigma}\hat{C}_{\sigma'\tau} \qquad (6.193)$$

and their Hermitian linear combinations close on a u(4) Lie algebra.

The $\{\hat{C}_{\sigma\tau}\}$ operators are seen to be identical to the operators $\{\hat{C}_{\sigma\sigma'}\}$ of the u(4) supermultiplet algebra of Section 5.13.2 to within additive constants, i.e.,

$$\hat{C}_{\sigma\sigma'} = \hat{C}_{\sigma\sigma'} - \tfrac{1}{2}(2l+1)\delta_{\sigma\sigma'}. \qquad (6.194)$$

We therefore refer to the u(4) algebra spanned by these operators as a *shifted* supermultiplet algebra and, to avoid confusion, denote it by $\mathcal{U}(4)$ to distinguish it from the standard U(4) supermultiplet group. The important observation is that the irreps of the closely-related $\mathcal{U}(4)$ and U(4) groups are in one-to-one correspondence and are carried by identical Hilbert spaces; thus, a U(4) irrep with highest weight $\{\tilde{\lambda}\} \equiv \{\tilde{\lambda}_1, \tilde{\lambda}_2, \tilde{\lambda}_3, \tilde{\lambda}_4\}$ corresponds to a $\mathcal{U}(4)$ irrep with highest weight $\{\tilde{\lambda}_1 - d/2, \tilde{\lambda}_2 - d/2, \tilde{\lambda}_3 - d/2, \tilde{\lambda}_4 - d/2\}$, where $d = 2l + 1$. It is convenient to denote the latter, shifted highest weight, by $\tfrac{1}{2}d(\tilde{\lambda})$.

The commutation relations

$$[\hat{C}_{\sigma\tau}, \hat{A}_{\sigma'\tau'}] = \delta_{\tau\sigma'}\hat{A}_{\sigma\tau'} + \delta_{\tau\tau'}\hat{A}_{\sigma'\sigma}, \qquad (6.195)$$

$$[\hat{C}_{\sigma\tau}, \hat{B}_{\sigma'\tau'}] = -\delta_{\sigma\sigma'}\hat{B}_{\tau\tau'} - \delta_{\sigma\tau'}\hat{B}_{\sigma'\tau}, \qquad (6.196)$$

reveal that the combined set of operators $\{\hat{C}_{\sigma\tau}, \hat{A}_{\sigma\tau}, \hat{B}_{\sigma\tau}\}$ closes under commutation and spans an so(8) Lie algebra. This Lie algebra has a Cartan subalgebra spanned by the four commuting operators $\{\hat{C}_{\sigma\sigma}\}$. The remaining operators of the set are then identified as the (non-zero) root vectors of an so(8) Lie algebra. This so(8) Lie algebra, which is evidently a spectrum generating algebra for the $L = 0$ pairing Hamiltonian (6.190), has the useful property that it contains the (shifted) supermultiplet algebra, u(4), as a subalgebra.

It will be convenient to label an $\mathcal{SO}(8)$ irrep by $[\frac{1}{2}d(\tilde{\lambda})]$, if its $\mathcal{SO}(8)$ lowest-weight state, $|\phi\rangle$, defined by the equations

$$\hat{B}_{\sigma\tau}|\phi\rangle = 0, \quad \hat{C}_{\sigma\tau}|\phi\rangle = 0, \quad 1 \le \sigma < \tau \le 4, \tag{6.197}$$

has weight given by

$$\hat{C}_{\sigma\sigma}|\phi\rangle = (\lambda_\sigma - \tfrac{1}{2}d)|\phi\rangle, \quad \sigma = 1, \ldots, 4. \tag{6.198}$$

6.7.4 *The* $\mathrm{O}(2l+1) \subset \mathrm{U}(2l+1)$ *symmetry group dual to* $\mathcal{SO}(8)$

An important symmetry group of the $L = 0$ pairing interaction is the subgroup of one-body unitary transformations of the spatial wave functions that leave the $L = 0$ scalar products $\{a_\sigma^\dagger \cdot a_\tau^\dagger\}$ invariant. The identity of this subgroup becomes apparent if we regard the operators $\{a_{\sigma m}^\dagger; m = -l, \ldots, +l\}$, for each value of σ, as components of a $(2l + 1)$-component vector operator. The bilinear combinations, $\{a_\sigma^\dagger \cdot a_\tau^\dagger\}$, are then seen as a scalar product of such $(2l + 1)$-component vectors. Thus, as discussed in Section 5.12.2, the subgroup of $\mathrm{U}(2l + 1)$ that leaves such scalar products invariant is the orthogonal subgroup $\mathrm{O}(2l + 1)$.

In fact, all elements of the above so(8) spectrum generating algebra, defined in Section 6.7.3, are formed from scalar products of $(2l+1)$-component vectors and are similarly invariant under $\mathrm{O}(2l+1)$ transformations. It follows that the $L = 0$ pairing Hamiltonian (6.190) is $\mathrm{O}(2l + 1)$ invariant and diagonal in a basis that reduces the subgroup chain,

$$\begin{array}{cccc} \mathrm{U}(2l + 1) & \supset \mathrm{O}(2l + 1) & \supset \mathrm{SO}(3)_L & \supset \mathrm{SO}(2)_L \\ \{\lambda\} & [\kappa] & L & M_L \end{array}, \tag{6.199}$$

of the *LST* coupling scheme of Section 5.13.3.

Now recall, from Section 5.13.3, that the irreps of the $\mathrm{U}(4)$ supermultiplet group are dual to those of the $\mathrm{U}(2l+1)$ group in the space of states of an l^n configuration. This means that the fully antisymmetric many-nucleon states of an l^n configuration are obtained by combining the spatial states of a $\mathrm{U}(2l + 1)$ irrep $\{\lambda\}$ with spin-isospin states of a conjugate $\mathrm{U}(4)$ irrep $\{\tilde{\lambda}\}$.[59] A more precise expression of this duality is given by the branching rule

$$\mathrm{U}(4(2l + 1)) \downarrow \mathrm{U}(2l + 1) \times \mathrm{U}(4) : \{1^n\} \downarrow \bigoplus_{\lambda \vdash n} \{\lambda\} \times \{\tilde{\lambda}\}, \tag{6.200}$$

where $\mathrm{U}(4(2l+1))$ is the group of one-body unitary transformations of both the spatial, spin, and isospin wave functions of a nucleon with orbital angular momentum l, and $\{1^n\}$ is its fully antisymmetric n-nucleon irrep.

---

[59]Recall that an n-nucleon irrep $\{\lambda\}$ of $\mathrm{U}(N)$ is defined by an ordered partition, $\lambda \equiv \{\lambda_1, \lambda_2, \ldots, \lambda_N\}$, of n and characterised by a Young diagram having n boxes with λ_i boxes in the i'th row (cf. Section 5.11.3). A partition $\tilde{\lambda} \equiv \{\tilde{\lambda}_1, \tilde{\lambda}_2, \ldots\}$ of n, conjugate to λ, is defined such that $\tilde{\lambda}_i$ is the number of boxes in the i'th column of the Young diagram for λ, as illustrated in Equation (5.436).

We now show that the representations of the subgroup $O(2l + 1) \subset U(2l + 1)$, on the shell-model space under consideration, are dual to those of the group $\mathcal{SO}(8)$ whose infinitesimal generators are elements of the so(8) spectrum generating algebra identified above.[60]

Theorem 6.6   *The groups $O(2l + 1)$ and $\mathcal{SO}(8)$ have dual representations on the Fock space, $\mathbb{F}^{(4(2l+1))}$, of many-nucleon states in which each nucleon has a common orbital angular momentum l.*

The proof of this theorem is similar to that of Theorem 6.4. The strategy is to identify a complete set of states that are simultaneously of $O(2l + 1)$ highest weight and $\mathcal{SO}(8)$ lowest weight and observe that each of the states in this set defines a unique pair of $O(2l + 1)$ and $\mathcal{SO}(8)$ irreps in accord with the conditions for a duality relationship.

Observe that states of the form

$$|\lambda\rangle := \hat{Z}_0^{\{\lambda\}}|0\rangle, \tag{6.201}$$

where

$$\hat{Z}_0^{\{\lambda\}} = \left(a_{1l}^\dagger a_{1,l-1}^\dagger \cdots a_{1,l+1-\tilde{\lambda}_1}^\dagger\right) \cdots \left(a_{4l}^\dagger a_{4,l-1}^\dagger \cdots a_{4,l+1-\tilde{\lambda}_4}^\dagger\right), \tag{6.202}$$

for any partition, $\lambda \vdash n$, of length no greater than 4 and depth no greater than $2l + 1$, are of highest weight for a $U(2l + 1)$ irrep and of lowest weight for a dual irrep of the supermultiplet group, $U(4)$, with respect to the numerical ordering of the labels such that a single-particle state with label σm is said to be of lower $U(4)$ weight that one with label τn if $\sigma < \tau$ and higher $U(2l + 1)$ weight if $m > n$. Moreover, with $O(2l + 1)$ embedded in $U(2l + 1)$ such that $O(2l + 1)$ has a Cartan subalgebra that is a subalgebra of that for $U(2l + 1)$ (cf. Section 6.7.6), the states $\{|\lambda\rangle\}$ are also of $O(2l + 1)$ highest weight.

We can now identify the subset of states in the set $\{|\lambda\rangle\}$, defined by Equations (6.201) and (6.202), that are simultaneously of $\mathcal{SO}(8)$ lowest weight. It can be seen that $\mathcal{B}_{\sigma\tau}|\lambda\rangle$ will not vanish, if the product $\hat{Z}_0^{\{\lambda\}}$ of single-nucleon creation operators contains creation operators, $a_{\sigma m}^\dagger$ and $a_{\tau\bar{m}}^\dagger$, for which

$$m \geq l + 1 - \tilde{\lambda}_\sigma, \quad -m \geq l + 1 - \tilde{\lambda}_\tau, \tag{6.203}$$

for some m and some $\sigma \neq \tau$ (note that the operator $\mathcal{B}_{\sigma\tau}$ is zero if $\sigma = \tau$). The largest values of $\tilde{\lambda}_\sigma$ and $\tilde{\lambda}_\tau$ are $\tilde{\lambda}_1$ and $\tilde{\lambda}_2$. Thus, the condition for $\mathcal{B}_{\sigma\tau}|\lambda\rangle$ to vanish,

---

[60]Cf. Kota V.K.B. and Castilho Alcarás J.A. (2006), *Nucl. Phys.* **A764**, 181; Rowe D.J. and Carvalho M.J. (2007), *J. Phys. A: Math. Gen.* **40**, 471; and references therein. It is important to note, however, that in these references the correct distinction was not always made between the groups $\mathcal{O}(8)$ and $\mathcal{SO}(8)$. In reading these papers it should be understood that the group dual to $O(2l + 1)$ is $\mathcal{SO}(8)$ and not $\mathcal{O}(8)$, but that the groups $SO(2l + 1)$ and $\mathcal{SO}(8)$ are not dual to one another. The distinction is important because an orthogonal group $O(N)$, with N even, sometimes branches into two distinct irreps of its $SO(N)$ subgroup; cf. Section 5.12.3.

which is the condition for the state $|\lambda\rangle$ to be an $\mathcal{SO}(8)$ lowest-weight state, is that

$$\tilde{\lambda}_1 + \tilde{\lambda}_2 \leq 2l + 1. \tag{6.204}$$

There are conceivably other states that are simultaneously of $O(2l+1)$ highest weight and $\mathcal{SO}(8)$ lowest weight besides those in the set $\{|\lambda\rangle; \tilde{\lambda}_1 + \tilde{\lambda}_2 \leq 2l+1\}$. However, the following claim shows that there are no others.

Claim 6.7 *Every highest-weight state for an $O(2l+1)$ irrep in $\mathbb{F}^{(4(2l+1))}$ that is simultaneously of $\mathcal{SO}(8)$ lowest weight is a $U(2l+1)$ highest-weight state.*

To prove this claim, we start from the observation that a lowest-weight state of an $\mathcal{SO}(8)$ irrep is also, by the definition (6.197), a lowest-weight state of an irrep of the $\mathcal{U}(4) \subset \mathcal{SO}(8)$ subgroup and, hence, a U(4) lowest-weight state, where U(4) is the supermultiplet group and $\mathcal{U}(4)$ is the shifted supermultiplet group as indicated in Section 6.7.3.[61] Suppose the U(4) weight of this $\mathcal{SO}(8)$ lowest-weight state is $\tilde{\lambda}$. Then, according to the unitary-unitary duality theorem (see Section 5.11.5), this state must belong to a dual $U(2l+1)$ irrep $\{\lambda\}$. Thus, we first look for states, $|\lambda\kappa^{\{\mu\}}\rangle$, in a $U(2l+1) \times U(4)$ irrep, $(\{\lambda\}, \{\tilde{\lambda}\})$, that are of highest weight, $\kappa = \kappa^\mu$, for an $O(2l+1) \subset U(2l+1)$ irrep $[\kappa^{\{\mu\}}]$ and of highest weight for the U(4) irrep $\{\tilde{\lambda}\}$, where μ denotes the $U(2l+1)$ weight of the state $|\lambda\kappa^{\{\mu\}}\rangle$. The claim is then proved by showing that such a state is simultaneously of $\mathcal{SO}(8)$ lowest weight if and only if $\mu = \lambda$ and, hence, if and only if $|\lambda\kappa^{\{\lambda\}}\rangle$ is identically equal to $|\lambda\rangle$.

Observe that the operator, $\hat{Z}_0^{\{\lambda\}}$, which creates the state $|\lambda\kappa^{\{\lambda\}}\rangle \equiv |\lambda\rangle = \hat{Z}_0^{\{\lambda\}}|0\rangle$, is a component of an $O(2l+1) \times U(4)$ tensor that is of highest weight relative to $O(2l+1)$ and lowest weight relative to U(4). Also recall that the $O(2l+1)$ irreps given by the $U(2l+1) \downarrow O(2l+1)$ branching rules (see Section 5.12.7) are obtained by factoring out $O(2l+1)$ scalars from the creation operators of $U(2l+1)$ states. As shown above, the $O(2l+1)$ scalar combinations of the nucleon creation operators are generated by the so(8) raising operators, $\{\hat{A}_{\sigma\tau}\}$, which can be regarded as components of a U(4) tensor, $\hat{A}$. Thus, we can create a highest-weight state, $\langle\kappa^{\{\mu\}}\rangle$, for an $O(2l+1)$ irrep $\langle\kappa^{\{\mu\}}\rangle$ within the space of the $U(2j+1)$ irrep $\{\lambda\}$ by adding an $(L=0)$-coupled nucleon pair to the highest-weight state $|\mu\rangle = Z_0^{\{\mu\}}|0\rangle$ of a $U(2l+1)$ irrep with two fewer nucleons than the irrep $\{\lambda\}$; viz.

$$|\lambda\kappa^{\{\mu\}}\rangle = \left[\hat{A} \otimes \hat{Z}^{\{\mu\}}\right]_0^{\{\lambda\}}|0\rangle, \tag{6.205}$$

where the bracket signifies an $O(2l+1) \times U(4)$ coupled product of the tensors $\hat{A}$ and $\hat{Z}^{\{\mu\}}$. Because $\hat{A}$ is an $O(2l+1)$ scalar, the coupled tensor product, $\left[\hat{A} \otimes \hat{Z}^{\{\mu\}}\right]^{\{\lambda\}}$, is given simply by U(4) coupling. Other $|\lambda\kappa^{\{\mu\}}\rangle$ states are the form

$$|\lambda\kappa^{\{\mu\}}\rangle = \left[\hat{A} \otimes \hat{A} \otimes Z^{\{\mu\}}\right]_0^{\{\lambda\}}|0\rangle, \tag{6.206}$$

---

[61]Note that a given lowest-weight state of U(4) can also be considered a highest-weight with respect to a different ordering of its Cartan operators, i.e., a weight $\lambda = (\lambda_1\lambda_2\lambda_3\lambda_4)$ can be considered a highest weight with respect to the ordering $\lambda_1 \geq \lambda_2 \geq \lambda_3 \geq \lambda_4$ or a lowest weight weight with respect to the ordering $\lambda_4 \leq \lambda_3 \leq \lambda_2 \leq \lambda_1$.

and so on. From a knowledge of the $U(2l+1) \downarrow O(2l+1)$ branching rules, it is clear that every $O(2l+1)$ highest-weight state within a $U(2l+1) \times U(4)$ irrep can be obtained in this way. However, because the components of the tensor $\hat{\mathcal{A}}$ are raising operators of the so(8) Lie algebra, none of the states $\{|\lambda\kappa^{\{\mu\}}\rangle\}$ with $\mu \neq \lambda$ can be an $\mathcal{SO}(8)$ lowest-weight state. This completes the proof of the claim. $\square$

It is worth noting that the theorem has a natural generalisation to give a duality between the representations of $O(d)$, where $d = \sum_i(2l_i+1)$, and $\mathcal{SO}(8)$, which is useful when considering LST coupling over a shell-model space that includes many orbital angular momenta.

6.7.5 *Classification of states*

The dualities between the irreps of $U(2l+1)$ and the $U(4)$ supermultiplet group and between the irreps of $O(2l+1)$ and $\mathcal{SO}(8)$ indicate alternative ways of defining basis states for the LST-coupled shell model. A natural choice of basis is defined and labelled by the quantum numbers (irrep labels) of the subgroup chain

$$
\begin{array}{ccccccc}
U(2l+1) \times U(4) & \supset & O(2l+1) \times SU(2)_S \times SU(2)_T & \supset & SO(3)_L \\
\{\lambda\} \quad \{\tilde{\lambda}\} & \alpha & [\kappa] \qquad S \qquad\quad T & \rho & L
\end{array}, \quad (6.207)
$$

where α and ρ are multiplicity indices. However, the same basis states can be defined and labelled, equivalently, by the dual chain

$$
\begin{array}{ccccccc}
\mathcal{SO}(8) \times SO(3)_L & \subset & \mathcal{U}(4) & \supset & SU(2)_S \times SU(2)_T \\
[\frac{1}{2}d(\tilde{\kappa})] \quad \rho\, L & \alpha & \{\frac{1}{2}d(\tilde{\lambda})\} & & S \qquad\quad T
\end{array}, \quad (6.208)
$$

where $d = 2l+1$ and, as already noted, an $\mathcal{SO}(8)$ irrep is labelled by the $\mathcal{U}(4)$ irrep to which its lowest-weight state belongs.[62] These dual subgroup chains exhibit the explicit way in which the $L = 0$, $\mathcal{SO}(8)$ pairing model is embedded in the LST-coupled shell model. Moreover, they prepare the way for an extension of the $\mathcal{SO}(8)$ model to include more general shell-model interactions.

6.7.6 *Quantum number ranges from $U(d) \downarrow O(d)$ branching rules*

One can start by listing all the $U(d)$ irreps $\{\lambda\} \equiv \{\lambda_1, \lambda_2, \ldots, \lambda_r\}$, for $d = 2l+1$, corresponding to partitions of the number of nucleons in the shell. The admissible partitions are those with no more than d parts and for which no part λ_i exceeds four; otherwise there is no conjugate $U(4)$ irrep $\{\tilde{\lambda}\}$ with which it can be combined to form antisymmetric states.

---

[62]Labelling an irrep of a group by its lowest, instead of its highest, weight is unconventional but convenient in pairing-model contexts, because the lowest-weight state has fewest particles and is the simplest.

One can next use the algorithm given in Section 5.12.7 to determine the expansion coefficients in the branching rule

$$\mathrm{U}(d) \downarrow \mathrm{O}(d) \;:\; \{\lambda\} \downarrow \bigoplus_{\kappa} R_{\lambda\kappa}\,[\kappa]. \qquad (6.209)$$

These coefficients are shown for the U(3), $l = 1$, irreps of even nucleon number n in Table 6.6. Note that an $l = 1$ single-particle state has negative parity. Thus, the

Table 6.6: The non-zero $R_{\lambda\kappa}$ coefficients of the $\mathrm{U}(3) \downarrow \mathrm{O}(3) : \{\lambda\} \downarrow \bigoplus_{\kappa} R_{\lambda\kappa}\,[\kappa]$ branching rule for even-n irreps.

$[\kappa]$	$\{\lambda\}$										
	$\{0\}$ $\{4^3\}$	$\{2\}$ $\{4^22\}$	$\{1^2\}$ $\{43^2\}$	$\{4\}$ $\{4^2\}$	$\{31\}$ $\{431\}$	$\{2^2\}$ $\{42^2\}$	$\{21^2\}$ $\{3^22\}$	$\{42\}$	$\{41^2\}$ $\{3^2\}$	$\{321\}$	$\{2^3\}$
$[0]$	1	1		1		1		1			1
$[1^2] \equiv [1]^*$			1		1		1		1	1	
$[2]$		1		1	1	1		2		1	
$[31] \equiv [3]^*$					1			1	1		
$[4]$				1				1			

O(3) irreps for states of odd nucleon number have negative parity, whereas those for states of even nucleon number have positive parity. For example, the odd-nucleon number irreps, [1] and [3], have angular momentum and parity given, respectively, by $L^\pi = 1^-$ and 3^- whereas the even-number irreps, $[1^2] \equiv [1]^*$ and $[31] \equiv [3]^*$, have angular momentum and parity given, respectively, by $L^\pi = 1^+$ and 3^+.

Table 6.7 displays the results of Table 6.6 reorganised such that the columns show all the U(3) irreps that contain a given O(3) irrep $[\kappa]$.

Table 6.7: Columns of U(3) irreps, $\{\lambda\}$, given by partitions of even values of n, that contain a given O(3) irrep $[\kappa]$.

n	$[\kappa]$				
	$[0]$	$[1^2]$	$[2]$	$[31]$	$[4]$
0	$\{0\}$				
2	$\{2\}$	$\{1^2\}$	$\{2\}$		
4	$\{4\}, \{2^2\}$	$\{31\}, \{21^2\}$	$\{4\}, \{31\}, \{2^2\}$	$\{31\}$	$\{4\}$
6	$\{42\}, \{2^3\}$	$\{31^3\}, \{3^2\}, \{321\}$	$\{42\}^2, \{321\}$	$\{42\}, \{41^2\}, \{3^2\}$	$\{42\}$
8	$\{4^2\}, \{42^2\}$	$\{431\}, \{3^22\}$	$\{4^2\}, \{431\}, \{42^2\}$	$\{431\}$	$\{4^2\}$
10	$\{4^22\}$	$\{43^2\}$	$\{4^22\}$		
12	$\{4^3\}$				
L^π	0^+	1^+	2^+	3^+	4^+

The dual $\mathcal{SO}(8) \downarrow \mathcal{U}(4)$ branching rules

$$\mathcal{SO}(8) \downarrow \mathcal{U}(4) \ : \ [\tfrac{1}{2}d(\tilde{\kappa})] \downarrow \bigoplus_\lambda \mathcal{R}_{\lambda\kappa} \{\tfrac{1}{2}d(\tilde{\lambda})\} \qquad (6.210)$$

are now obtained immediately because

$$\mathcal{R}_{\lambda\kappa} = R_{\lambda\kappa}. \qquad (6.211)$$

This result follows from a comparison of the $\mathrm{U}(d) \times \mathrm{U}(4) \downarrow \mathrm{O}(d) \times \mathrm{U}(4)$ and $\mathrm{O}(d) \times \mathcal{SO}(8) \downarrow \mathrm{O}(d) \times \mathcal{U}(4)$ branching rules. The full l-shell Fock space is a direct sum of n-nucleon Hilbert spaces and carries a direct sum of $\mathrm{O}(N) \times \mathrm{U}(4)$ irreps given by

$$\mathrm{U}(4d) \ \downarrow \ \mathrm{U}(d) \times \mathrm{U}(4) \ \downarrow \ \ \mathrm{O}(d) \times \mathrm{U}(4)$$
$$\bigoplus_n \{1^n\} \downarrow \bigoplus_\lambda \{\lambda\} \times \{\tilde\lambda\} \downarrow \bigoplus_\lambda \bigoplus_\kappa R_{\lambda\kappa}[\kappa] \times \{\tilde\lambda\}, \qquad (6.212)$$

where λ is summed over all partitions of $n \le d$ of length no more than 4 and depth no more than d. Now, because of the $\mathrm{O}(d) \sim \mathcal{SO}(8)$ duality relationship, the l-shell Fock space also carries a direct sum $\bigoplus_\kappa [\kappa] \times [\tfrac{1}{2}d(\tilde\kappa)]$ of $\mathrm{O}(d) \times \mathcal{SO}(8)$ irreps. Thus, it carries a direct sum of $\mathrm{O}(d) \times \mathcal{U}(4)$ irreps given by

$$\mathrm{O}(d) \times \mathcal{SO}(8) \ \downarrow \ \ \mathrm{O}(d) \times \mathcal{U}(4)$$
$$\bigoplus_\kappa [\kappa] \times [\tfrac{1}{2}d(\tilde\kappa)] \downarrow \bigoplus_\lambda \bigoplus_\kappa \mathcal{R}_{\lambda\kappa}[\kappa] \times \{\tfrac{1}{2}d(\tilde\lambda)\}. \qquad (6.213)$$

Therefore, with the correspondence, $[\tilde\lambda] \sim [\tfrac{1}{2}d(\tilde\lambda)]$, between the $\mathrm{U}(4) \sim \mathcal{U}(4)$ irreps, it follows that Equations (6.212) and (6.213) can only be consistent if $\mathcal{R}_{\lambda\kappa} = R_{\lambda\kappa}$. Thus, tables listing the $\mathcal{U}(4)$ irreps contained in an $\mathcal{SO}(8)$ irrep are readily constructed; i.e., it is a simple matter of rearranging the table of $\mathrm{O}(d)$ irreps that are contained in a given $\mathrm{U}(d)$ irrep to obtain the columns of $\mathcal{U}(4)$ irreps in the corresponding $\mathcal{SO}(8)$ irreps. Thus, the results of Table 6.7 for $d = 3$, are rearranged to give the $\mathcal{SO}(8) \downarrow \mathcal{U}(4)$ branching rules shown in Table 6.8.

Table 6.8: The $\mathcal{U}(4)$ irreps contained in the even-n $\mathcal{SO}(8)$ irreps for $d = 3$. To simplify the table, a $\mathcal{U}(4)$ irrep $\{\tfrac{3}{2}(\tilde\lambda)\}$ is indicated simply by $\{\tilde\lambda\}$.

n	$[\frac{3}{2}(0)]$	$[\frac{3}{2}(2)]$	$[\frac{3}{2}(1^2)]$	$[\frac{3}{2}(21^2)]$	$[\frac{3}{2}(1^4)]$
0	$\{0\}$				
2	$\{1^2\}$	$\{2\}$	$\{1^2\}$		
4	$\{1^4\},\{2^2\}$	$\{21^2\},\{31\}$	$\{1^4\},\{21^2\},\{2^2\}$	$\{21^2\}$	$\{1^4\}$
6	$\{2^21^2\},\{3^2\}$	$\{31^3\},\{2^3\},\{321\}$	$\{2^21^2\}^2,\{321\}$	$\{2^21^2\},\{31^3\},\{2^3\}$	$\{2^21^2\}$
8	$\{2^4\},\{3^21^2\}$	$\{3^21\},\{3^22\}$	$\{2^4\},\{3^21\},\{3^21^2\}$	$\{3^21\}$	$\{2^4\}$
10	$\{3^22^2\}$	$\{3^31\}$	$\{3^22^2\}$		
12	$\{3^4\}$				

6.7.7 Spectra of the $L = 0$ pair-coupling model

We now show that the $L = 0$ pairing Hamiltonian is simply a linear combination of so(8) and u(4) Casimir operators and, as a consequence, its spectrum has a simple analytical expression.

A u(4) Casimir operator is defined in terms of the $\{\hat{\mathcal{C}}_{\sigma\tau}\}$ operators of Section 6.7.3 by

$$
\begin{aligned}
\hat{\mathcal{C}}_{\text{u4}} &:= \sum_{\sigma\tau} \hat{\mathcal{C}}_{\sigma\tau}\hat{\mathcal{C}}_{\tau\sigma} \\
&= \sum_{\sigma} \hat{\mathcal{C}}_{\sigma\sigma}\hat{\mathcal{C}}_{\sigma\sigma} + \sum_{\sigma>\tau} \left(2\hat{\mathcal{C}}_{\sigma\tau}\hat{\mathcal{C}}_{\tau\sigma} + \hat{\mathcal{C}}_{\tau\tau} - \hat{\mathcal{C}}_{\sigma\sigma}\right).
\end{aligned} \tag{6.214}
$$

The so(8) Casimir operator is similarly defined in terms of the operators of Section 6.7.3 by

$$
\begin{aligned}
\hat{\mathcal{C}}_{\text{so8}} &:= \sum_{\sigma\tau} \hat{\mathcal{C}}_{\sigma\tau}\hat{\mathcal{C}}_{\tau\sigma} + \sum_{\sigma>\tau} \left(\mathcal{A}_{\sigma\tau}\mathcal{B}_{\sigma\tau} + \mathcal{B}_{\sigma\tau}\mathcal{A}_{\sigma\tau}\right) \\
&= \hat{\mathcal{C}}_{\text{u4}} - 3\sum_{\sigma} \hat{\mathcal{C}}_{\sigma\sigma} + \sum_{\sigma\tau} \mathcal{A}_{\sigma\tau}\mathcal{B}_{\tau\sigma}.
\end{aligned} \tag{6.215}
$$

Thus, the Hamiltonian

$$
\hat{H} = -\tfrac{1}{4}G \sum_{\sigma\tau} \hat{\mathcal{A}}_{\sigma\tau}\hat{\mathcal{B}}_{\sigma\tau} \tag{6.216}
$$

can be expressed as

$$
\hat{H} = -\tfrac{1}{4}G \left[\hat{\mathcal{C}}_{\text{so8}} - \hat{\mathcal{C}}_{\text{u4}} + 3\sum_{\sigma} \hat{\mathcal{C}}_{\sigma\sigma}\right]. \tag{6.217}
$$

Now, when acting on a lowest-weight state $|f\rangle$ of a u(4) irrep for which

$$
\hat{\mathcal{C}}_{\sigma\sigma}|f\rangle = f_\sigma |f\rangle, \quad \hat{\mathcal{C}}_{\tau\sigma}|f\rangle = 0 \text{ for } \tau < \sigma, \tag{6.218}
$$

$\hat{\mathcal{C}}_{\text{u4}}$ has the eigenvalue

$$
\mathcal{C}_{\text{u4}}(f) = \sum_{\sigma} f_\sigma(f_\sigma + 5 - 2\sigma). \tag{6.219}
$$

Similarly, when acting on the lowest-weight state $|k\rangle$ of an so(8) irrep, for which

$$
\hat{\mathcal{C}}_{\sigma\sigma}|k\rangle = k_\sigma |k\rangle, \quad \mathcal{B}_{\tau\sigma}|k\rangle = 0, \tag{6.220}
$$

$\mathcal{C}_{\text{so8}}(k)$ has the eigenvalue

$$
\mathcal{C}_{\text{so8}}(k) = \sum_{\sigma} k_\sigma(k_\sigma + 2 - 2\sigma). \tag{6.221}
$$

Thus, when acting on a state of an O(d) irrep [κ] sitting inside a U(d) irrep $\{\lambda\}$ or, equivalently, when acting on a state of a $\mathcal{U}(4)$ irrep $\{\tfrac{1}{2}d(\tilde{\lambda})\}$ sitting inside an $\mathcal{SO}(8)$

irrep $[\frac{1}{2}d(\tilde{\kappa})]$, we have $f_\sigma \equiv \tilde{\kappa}_\sigma - \frac{1}{2}d$, $k_\sigma \equiv \tilde{\lambda}_\sigma - \frac{1}{2}d$, $\langle \hat{C}_{\sigma\sigma} \rangle \equiv \lambda_\sigma - \frac{1}{2}d$, and $\hat{H}$ has the eigenvalue

$$E_{\kappa\lambda} = -\frac{1}{4}G\sum_\sigma \left[(\tilde{\kappa}_\sigma - \frac{1}{2}d)(\tilde{\kappa}_\sigma - \frac{1}{2}d + 2 - 2\sigma) \right.$$
$$\left. - (\tilde{\lambda}_\sigma - \frac{1}{2}d)(\tilde{\lambda}_\sigma - \frac{1}{2}d + 2 - 2\sigma) \right]. \tag{6.222}$$

The excitation energies for $l = 1$ given by this equation are shown, for even-n nuclei, in Figure 6.16

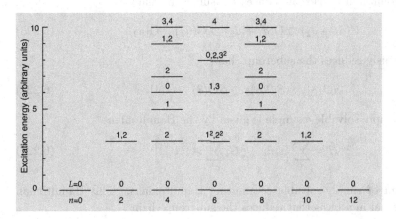

Figure 6.16: Excitation energies of the $L = 0$ pairing model for nucleons in an l^n configuration, with $l = 1$ and n even, given by Equation (6.222) with $G = 1$. The angular momentum L is shown for each energy level.

6.7.8 *General multishell $L = 0$ seniority-conserving models*

The $L = 0$ pairing Hamiltonian (6.190) for a single-l configuration has a natural extension to a Hamiltonian,

$$\hat{H} = \sum_l \varepsilon_l \hat{n}_l - \frac{1}{4}\sum_{ll'\sigma\tau} G_{ll'} \hat{\mathcal{A}}_{\sigma\tau}^{(l)} \hat{\mathcal{B}}_{\sigma\tau}^{(l')}, \tag{6.223}$$

that acts in a valence-shell space including multiple l orbitals; now $\hat{n}_l$, $\hat{\mathcal{A}}_{\sigma\tau}^{(l)}$, and $\hat{\mathcal{B}}_{\sigma\tau}^{(l')}$ are the number operator and $L = 0$ pair-creation and pair-annihilation operators, respectively, for nucleons in the l orbital. For arbitrary ε_l and $G_{ll'}$, the diagonalisation of this Hamiltonian involves coupling the states of several $\mathcal{SO}(8)$ irreps which must be done numerically. For a modest number of l orbitals, this is relatively straightforward.

In special cases, the Hamiltonian of Equation (6.223) is simply solvable. For example, if $\epsilon_l = \varepsilon$, and $G_{ll'} = G$ are l-independent, then

$$\hat{H} = \varepsilon \hat{n} - \frac{1}{4}G \sum_{\sigma\tau} \hat{\mathcal{A}}_{\sigma\tau} \hat{\mathcal{B}}_{\sigma\tau}, \tag{6.224}$$

where now

$$\hat{n} = \sum_l \hat{n}_l, \quad \hat{\mathcal{A}}_{\sigma\tau} = \sum_l \hat{\mathcal{A}}_{\sigma\tau}^{(l)}, \quad \hat{\mathcal{B}}_{\sigma\tau} = \sum_l \hat{\mathcal{B}}_{\sigma\tau}^{(l)}. \tag{6.225}$$

Therefore, $\hat{H}$ is again of the standard form given by Equation (6.190), except that now the dimension of the space of orbital wave functions is $d = \sum_l (2l + 1)$. Thus, as a result of the $\mathcal{SO}(8) \supset \mathcal{U}(4)$ duality relationships of Section 6.7.7, $\hat{H}$ remains diagonal in a U(d) $\supset$ O(d) basis. For example, if the orbital angular momentum l has two values, l_1 and l_2, and $d = d_1 + d_2$, with $d_i = 2l_i + 1$, then the Hamiltonian (6.224) is diagonal in a basis that reduces the subgroup chain

$$\mathrm{U}(d_1 + d_2) \supset \mathrm{O}(d_1 + d_2) \supset \mathrm{O}(d_1) \times \mathrm{O}(d_2) \tag{6.226}$$

and simultaneously reduces the subgroup chain

$$\mathcal{SO}(8)_1 \times \mathcal{SO}(8)_2 \supset \mathcal{SO}(8) \supset \mathcal{U}(4). \tag{6.227}$$

A second simply-solvable example is given by the Hamiltonian

$$\hat{H} = \sum_l \left[\varepsilon_l \hat{n}_l - \tfrac{1}{4} G_l \sum_{\sigma\tau} \hat{\mathcal{A}}_{\sigma\tau}^{(l)} \hat{\mathcal{B}}_{\sigma\tau}^{(l)} \right], \tag{6.228}$$

which is a sum of simply-solvable commuting Hamiltonians. This Hamiltonian (6.228) is diagonal in a basis that reduces the subgroup chain

$$\mathrm{U}(d_1 + d_2) \supset \mathrm{U}(d_1) \times \mathrm{U}(d_2) \supset \mathrm{O}(d_1) \times \mathrm{O}(d_2) \tag{6.229}$$

and simultaneously reduces the subgroup chain

$$\mathcal{SO}(8)_1 \times \mathcal{SO}(8)_2 \supset \mathcal{U}(4)_1 \times \mathcal{U}(4)_2 \supset \mathcal{U}(4), \tag{6.230}$$

where $\mathcal{SO}(8)_1$, $\mathcal{SO}(8)_2$ and $\mathcal{SO}(8)$ are the SO(8) groups dual to O(d_1), O(d_2) and O($d_1 + d_2$), respectively.

The simply-solvable pairing Hamiltonians are evidently those that diagonalise useful subgroup chains and, hence, are associated with shell-model coupling schemes. Thus, they are said to have good dynamical symmetries. However, other mixed-symmetry Hamiltonians may be more realistic. For example, a meaningful multishell-pairing Hamiltonian is one of the form

$$\hat{H} = \sum_l \varepsilon_l \hat{n}_l - \tfrac{1}{4} G \sum_{\sigma\tau} \hat{\mathcal{A}}_{\sigma\tau} \hat{\mathcal{B}}_{\sigma\tau}. \tag{6.231}$$

Such a Hamiltonian is solvable, in principle, by the generalised Richardson-Gaudin methods of Asorey et al.[63],[64]

---

[63] Asorey *et al.*, *op. cit.* Footnote 55 on Page 447.
[64] Lerma S. *et al.* (2007), *Phys. Rev. Lett.* **99**, 032501.

6.7.9   Concluding remark on $L = 0$ pairing

Although the $L = 0$ pairing interaction is a meaningful approximation to the short-range component of the two-nucleon interaction, as noted in the introduction to this chapter, experiment indicates that the short-range component is not the dominant component of the two-nucleon interaction in doubly open-shell nuclei. A longstanding objective is therefore to explore the competition between the long- and short-range components of the interactions in these nuclei.

An SU(3) model, with a schematic $Q \cdot Q$ approximation to the long-range component of the two-nucleon interaction, that is diagonal in a

$$U(N) \supset SU(3) \supset SO(3) \tag{6.232}$$

LST coupling scheme was introduced by Elliott[65] and, as will be shown in Volume 2, gives rise to shell-model states with the properties of rotational bands.

Thus, an instructive Hamiltonian with which to explore the superconducting pair correlations in rotational nuclei is one that includes both a pairing interaction and an Elliott $Q \cdot Q$ interaction. The diagonalisation of such a mixed-symmetry Hamiltonian is complicated if a $J = 0$ pairing interaction is used because this pairing interaction is diagonal in a jj coupling scheme whereas $Q \cdot Q$ is diagonal in *LST* coupling. In contrast, the $Q \cdot Q$ and $L = 0$ pairing interactions are both diagonalisable in *LST* coupling. Thus, they both preserve supermultiplet symmetry which results in considerable simplifications.[66]

It will also be shown in Volume 2 that a microscopic version of the collective model is numerically solvable in an

$$Sp(3, \mathbb{R}) \supset SU(3) \supset SO(3) \tag{6.233}$$

LST coupling scheme. Thus, the ability to handle pairing correlations in *LST* coupling is a step towards learning how to include such correlations in a more realistic treatment of nuclear collective structure.

Exercises

6.25 Show that, for an $l = 0$ shell, $U(2l+1)$ has irreps $\{0\}$, $\{1\}$, ..., $\{4\}$. Hence show that an s shell supports two $SO(8)$ irreps: $[\frac{1}{2}(0)]$ containing the even-nucleon states, and $[\frac{1}{2}(1)]$ containing the odd-nucleon states. Give the $\mathcal{U}(4)$ content of these irreps.

6.26 Show that the zero-nucleon, $n = 0$, state of an l shell, with $l > 0$, is the lowest-weight state for an $SO(8)$ irrep $[\frac{1}{2}d(0)]$, where $d = 2l + 1$, and that the $n = 2$ supermultiplet states have $SO(8) \supset \mathcal{U}(4)$ quantum numbers, $([\frac{1}{2}d(\tilde{\kappa})], \{\frac{1}{2}d(\tilde{\lambda})\})$, given by $([\frac{1}{2}d(0)], \{\frac{1}{2}d(1^2)\})$, $([\frac{1}{2}d(1^2)], \{\frac{1}{2}d(1^2)\})$, $([\frac{1}{2}d(2)], \{\frac{1}{2}d(2)\})$.

6.27 Use the energy formula of Equation (6.222) to deduce the energy spectrum of the $L = 0$ pairing Hamiltonian for two nucleons with $l = 2$.

---

[65]Elliott, *op. cit.* Footnote 214 on Page 347.
[66]Rosensteel G. and Rowe D.J. (2007), *Nucl. Phys.* **A797**, 94.

6.8   The BCS approximation

The BCS approximation has its origins in the theory of superconductivity.[67] The
form in which it is known in nuclear physics is due primarily to Bogolyubov[68] and
Valatin.[69] The theory was introduced into nuclear physics by Bohr, Mottelson, and
Pines[70] and developed by Belyaev,[71] Bayman,[72] and others.[73]

In this section, we apply the BCS approximation to a general pairing Hamiltonian, comprising an independent-particle energy and a pairing interaction, of the
form

$$\hat{H} = \sum_\nu \varepsilon_\nu a_\nu^\dagger a_{\bar\nu} - \sum_{\mu\nu>0} G_{\mu\nu} a_\mu^\dagger a_{\bar\mu}^\dagger a_{\bar\nu} a_\nu, \qquad (6.234)$$

where ν and $\bar\nu$, respectively, index a single-particle state and its time-reversed partner, as defined in Section 6.2.1. Thus, ν represents a set of indices, e.g., $\nu \equiv \alpha jm$.
A flexibility of interpretation of the single-particle basis facilitates the use of the
BCS approximation in a variety of situations. For example, the BCS approximation
is often used in conjunction with nuclear rotational models in which the angular
momentum is not a constant of the motion in the intrinsic frame of the rotor. It
will be presumed that phases are chosen such that the fermion operators satisfy the
anticommutation relations

$$\{a_{\bar\mu}, a_\nu^\dagger\} = \delta_{\mu\nu}, \quad \{a_\mu, a_{\bar\nu}^\dagger\} = -\delta_{\mu\nu}, \quad \{a_{\bar\mu}, a_\nu\} = \{a_{\bar\mu}^\dagger, a_\nu^\dagger\} = 0. \qquad (6.235)$$

6.8.1   *The BCS ground state as a quasi-particle vacuum*

The ground states of the pairing Hamiltonian for the seniority-zero states of a single-
j-shell configuration, cf. Section 6.3, have the simple form (to within a normalisation
factor)

$$|p0\rangle = (\hat{S}_+)^p|0\rangle = \left(\sum_{\nu>0} a_\nu^\dagger a_{\bar\nu}^\dagger\right)^{n/2}|0\rangle, \qquad (6.236)$$

for even $(n = 2p)$ nuclei, where $|0\rangle$ is a closed-shell core state, i.e., a vacuum state
relative to the j-shells under active consideration, and n is the number of valence-
shell nucleons. Solutions of the two-nucleon pairing Hamiltonian, in the form given
by Equation (6.46) in terms of a boson-like (see Exercise 6.4) Cooper pair creation

[67]Bardeen *et al.*, *op. cit.* Footnote 11 on Page 24.
[68]Bogolyubov, *op. cit.* Footnote 12 on Page 24.
[69]Valatin J.G. (1958), *Nuovo Cimento* **7**, 843.
[70]Bohr *et al.*, *op. cit.* Footnote 15 on Page 33.
[71]Belyaev S.T. (1959), *Mat. Fys. Medd. Dan. Vid. Selsk.* **31** (11).
[72]Bayman B.F. (1960), *Nucl. Phys.* **15**, 33.
[73]Early reviews of the subject were given in monographs by Lane A.M. (1964), *Nuclear Theory*
(Benjamin, New York and Amsterdam), and Rowe, *op. cit.* Footnote 4 on Page x.

operator, $\hat{A}(\alpha) = \sum_{j,m>0} z_j a_{jm}^\dagger a_{j\bar{m}}^\dagger$, suggest that the more general states,

$$|pz\rangle := \Big(\sum_{\nu>0} z_\nu a_\nu^\dagger a_{\bar{\nu}}^\dagger \Big)^p |0\rangle, \tag{6.237}$$

might provide reasonable approximations (to within a normalisation factors) for the ground states of a multishell-pairing Hamiltonian. Such a state can be optimised in a variational calculation in relatively simple situations; a method for doing this is described in Section 6.11. In the BCS approximation, a further approximation is made that results in more directly solvable equations, and is described as follows.

The BCS approximation, designed for application to systems of macroscopic particle number, makes use of the fact that when the number of particles is large the low-energy spectral properties of the pairing Hamiltonian are insensitive to the specific value of the particle number over a relatively narrow range of values. In fact, the results of the preceding sections show that, for pairing interactions, the number of particles need not be that large. Moreover, the spectral properties of several sequences of singly closed-shell isotopes and isotones of finite nuclei, for which the pairing models are applied, have remarkably similar properties; cf. Figures 6.5 and 6.11 and comment in Section 6.5.6. The BCS approximation capitalises on the fact that a superposition of different particle-number states, given by

$$|z\rangle := \sum_{p \geq 0} \frac{1}{p!} |pz\rangle = \exp \Big(\sum_{\nu>0} z_\nu a_\nu^\dagger a_{\bar{\nu}}^\dagger \Big) |0\rangle, \tag{6.238}$$

is the vacuum of a set of generalised fermion operators and, as a consequence, has many useful mathematical properties. Thus, it implicitly assumes the state $|z\rangle$, with suitably chosen (generally complex) values of $\{z_\nu\}$, to be a superposition of ground states of a relatively small number of neighbouring nuclei. BCS results should therefore be interpreted as averages over distributions of states of different particle number.

The distribution of states of different nucleon number contained in an expansion of the state $|z\rangle$ depends essentially on the overall magnitude of the $\{z_\nu\}$ coefficients. Consider, for example, the state $|x\rangle$ given by Equation (6.238) with $z = x$ real and ν restricted to a finite range $1 \leq \nu \leq N$. As determined in Exercise (6.28), this state has squared norm given by

$$\langle x|x \rangle = \sum_p \frac{N! x^{2p}}{p!(N-p)!}. \tag{6.239}$$

Thus, the probability that the state $|x\rangle$ is in a state of particle number $2p$ is

$$P_p(x) = \frac{N! x^{2p}}{p!(N-p)!\langle x|x \rangle}. \tag{6.240}$$

These probabilities are shown in Figure 6.17 for $N = 20$ and $x = 0.5$ and 1.0 (cf. also the comments on number fluctuation in Section 6.8.6). A partial suppression of the errors introduced by averaging over the particle number distribution are discussed in Section 6.8.7 and a more accurate number-projected BCS approximation is presented in Section 6.11.

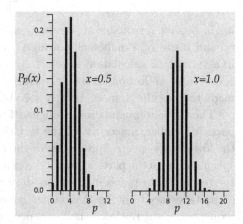

Figure 6.17: The probability $P_p(x)$ that the state $|x\rangle$ has particle number $2p$ when $N = 40$ for $x = 0.5$ and 1.0.

The fact that $|z\rangle$ is the vacuum of a set of generalised fermion operators follows from the observation (cf. Exercise (6.32)) that

$$\exp\left(\sum_{\nu>0} z_\nu a_\nu^\dagger a_{\bar\nu}^\dagger\right) a_\mu = (a_\mu - z_\mu a_{\bar\mu}^\dagger) \exp\left(\sum_{\nu>0} z_\nu a_\nu^\dagger a_{\bar\nu}^\dagger\right). \tag{6.241}$$

Thus, if the state $|0\rangle$ is annihilated by the fermion operators, $\{a_\mu\}$ then

$$(a_\mu - z_\mu a_{\bar\mu}^\dagger)|z\rangle = 0. \tag{6.242}$$

To take advantage of the vacuum property of BCS states, Bogolyubov and Valatin[74,75] introduced a transformation from the creation and annihilation operators of physical particles to *quasi-particle* operators, defined by

$$\alpha_\nu^\dagger := u_\nu a_\nu^\dagger + v_\nu a_{\bar\nu}, \quad \alpha_{\bar\nu}^\dagger := u_\nu a_{\bar\nu}^\dagger + v_\nu a_\nu, \tag{6.243}$$

with u_ν, v_ν coefficients which, in accord with usual practice, we assume to be real. From the Hermitian conjugation relationship, $(a_\nu^\dagger)^\dagger = a^\nu = a_{\bar\nu}$, which implies

$$(a_{\bar\nu})^\dagger = a_\nu^\dagger, \quad (a_\nu)^\dagger = -a_{\bar\nu}^\dagger, \tag{6.244}$$

we derive the corresponding expressions for the quasi-particle annihilation operators

$$\alpha_{\bar\nu} = u_\nu a_{\bar\nu} - v_\nu a_\nu^\dagger, \quad \alpha_\nu = u_\nu a_\nu - v_\nu a_{\bar\nu}^\dagger. \tag{6.245}$$

Thus, if the coefficients are normalised such that

$$u_\nu^2 + v_\nu^2 = 1, \tag{6.246}$$

[74]Bogolyubov, *op. cit.* Footnote 12 on Page 24.
[75]Valatin, *op. cit.* Footnote 69 on Page 462.

these quasi-particle operators obey standard fermion anticommutation relations, given for physical particles by Equation (6.235)[76] and for quasi-particles by the parallel expressions

$$\{\alpha_{\bar{\mu}}, \alpha_{\nu}^{\dagger}\} = \delta_{\mu,\nu}, \quad \{\alpha_{\mu}, \alpha_{\bar{\nu}}^{\dagger}\} = -\delta_{\mu,\nu}, \quad \{\alpha_{\bar{\mu}}, \alpha_{\nu}\} = \{\alpha_{\bar{\mu}}^{\dagger}, \alpha_{\nu}^{\dagger}\} = 0. \qquad (6.247)$$

It is now seen from Equation (6.242), that with $z_{\nu} := v_{\nu}/u_{\nu}$, the state $|z\rangle$ is the quasi-particle vacuum state, i.e.,

$$\alpha_{\nu}|z\rangle = 0, \quad \forall \nu. \qquad (6.248)$$

Moreover, from the identity

$$e^{z_{\nu} a_{\nu}^{\dagger} a_{\bar{\nu}}^{\dagger}} = 1 + z_{\nu} a_{\nu}^{\dagger} a_{\bar{\nu}}^{\dagger} = \frac{1}{u_{\nu}}(u_{\nu} + v_{\nu} a_{\nu}^{\dagger} a_{\bar{\nu}}^{\dagger}), \qquad (6.249)$$

it follows that the $|z\rangle$ is proportional to the normalised state

$$|\varphi(z)\rangle = \prod_{\nu>0}(u_{\nu} + v_{\nu} a_{\nu}^{\dagger} a_{\bar{\nu}}^{\dagger})|0\rangle, \qquad (6.250)$$

which is another standard expression of the quasi-particle vacuum state.

6.8.2 Bogolyubov-Valatin transformations as quasi-spin rotations

We now show that a Bololyubov-Valatin transformation, (6.243), can be expressed as a quasi-spin rotation, $a_{\nu}^{\dagger} \to e^{\hat{Z}(\theta)} a_{\nu}^{\dagger} e^{-\hat{Z}(\theta)}$, with

$$\hat{Z}(\theta) = \tfrac{1}{2} \sum_{\nu>0} \theta_{\nu}(\hat{S}_{+}^{\nu} - \hat{S}_{-}^{\nu}) \qquad (6.251)$$

where $\hat{S}_{+}^{\nu}\pm$ are the quasi-spin operators,

$$\hat{S}_{+}^{\nu} = a_{\nu}^{\dagger} a_{\bar{\nu}}^{\dagger}, \quad \hat{S}_{-}^{\nu} = a_{\bar{\nu}} a_{\nu}, \qquad (6.252)$$

defined in Section 6.2.2
From the commutation relations

$$[\hat{S}_{+}^{\nu}, a_{\mu}] = -\delta_{\mu\nu} a_{\nu}^{\dagger}, \quad [\hat{S}_{-}^{\nu}, a_{\mu}^{\dagger}] = -\delta_{\mu\nu} a_{\nu}, \qquad (6.253)$$

the quasi-particle operators are given by

$$\begin{aligned} \alpha_{\nu}^{\dagger} &:= e^{\hat{Z}(\theta)} a_{\nu}^{\dagger} e^{-\hat{Z}(\theta)} = \cos\left(\tfrac{1}{2}\theta_{\nu}\right) a_{\nu}^{\dagger} + \sin\left(\tfrac{1}{2}\theta_{\nu}\right) a_{\nu}, \\ \alpha_{\nu} &:= e^{\hat{Z}(\theta)} a_{\nu} e^{-\hat{Z}(\theta)} = \cos\left(\tfrac{1}{2}\theta_{\nu}\right) a_{\nu} - \sin\left(\tfrac{1}{2}\theta_{\nu}\right) a_{\nu}^{\dagger}. \end{aligned} \qquad (6.254)$$

Thus, a quasi-spin rotation reproduces the Bogolyubov-Valatin transformation of Equation (6.243) with

$$u_{\nu} = \cos\left(\tfrac{1}{2}\theta_{\nu}\right), \quad v_{\nu} = \sin\left(\tfrac{1}{2}\theta_{\nu}\right), \qquad (6.255)$$

---

[76]Note that $a_{\bar{\nu}}^{\dagger} = -a_{\nu}^{\dagger}$ and likewise $\alpha_{\bar{\nu}}^{\dagger} = -\alpha_{\nu}^{\dagger}$; cf. Equation (6.9).

from which we obtain the identities

$$u_\nu^2 = \tfrac{1}{2}(1 + \cos\theta_\nu), \quad v_\nu^2 = \tfrac{1}{2}(1 - \cos\theta_\nu). \tag{6.256}$$

Several insightful results follow from the interpretation of the Bogolyubov-Valatin transformation as a quasi-spin rotation. One important result is that the expression of the BCS quasi-particle vacuum state in the form

$$|\varphi(Z)\rangle = e^{\hat{Z}(\theta)}|0\rangle, \tag{6.257}$$

shows it to be a coherent-state of the direct product of multiple SU(2) quasi-spin groups.[77] Moreover, it is a minimum-uncertainly coherent state as discussed in Section 8.3. Consideration of the more general Hartree-Fock-Bogolyubov approximation in terms of coherent states is given in Chapter 7.

6.8.3 *The BCS equations*

The quasi-particle vacuum state, conveniently expressed in the form

$$|\varphi(Z)\rangle = \exp\sum_{\nu>0}[Z_\nu a_\nu^\dagger a_{\bar\nu}^\dagger - Z_\nu^* a_{\bar\nu} a_\nu]|0\rangle, \tag{6.258}$$

is optimised by minimising the expectation of the Hamiltonian, Equation (6.234), subject to the constraint that the mean value, $\langle\varphi(Z)|\hat{n}|\varphi(Z)\rangle$, of the number operator, $\hat{n} = \sum_\nu \hat{n}_\nu$, is equal to the desired nucleon number, n. This constraint is imposed by minimisation of the modified Hamiltonian, $\hat{H}' = \hat{H} - \lambda\hat{n}$, where λ is a Lagrange multiplier known as the *chemical potential*.

The expectation value $\langle\varphi(Z)|\hat{H}'|\varphi(Z)\rangle$ is a minimised if it satisfies the equations

$$\frac{\partial\langle\varphi(Z)|\hat{H}'|\varphi(Z)\rangle}{\partial Z_\nu} = \frac{\partial\langle\varphi(Z)|\hat{H}'|\varphi(Z)\rangle}{\partial Z_\nu^*} = 0. \tag{6.259}$$

Thus, the variational equations are expressed as

$$\langle\varphi|[\hat{H}', a_\nu^\dagger a_{\bar\nu}^\dagger]|\varphi\rangle = \langle\varphi|[\hat{H}', a_{\bar\nu} a_\nu]|\varphi\rangle = 0, \quad \forall\nu, \tag{6.260}$$

where $|\varphi\rangle$, without a Z argument, denotes the particular $|\varphi(Z)\rangle$ for which the variational equation is satisfied.

The commutator $[\hat{H}', a_\nu^\dagger a_{\bar\nu}^\dagger]$, for example, is given by

$$[\hat{H}', a_\nu^\dagger a_{\bar\nu}^\dagger] = 2(\varepsilon_\nu - \lambda)a_\nu^\dagger a_{\bar\nu}^\dagger - \sum_{\mu>0} G_{\mu\nu}a_\mu^\dagger a_{\bar\mu}^\dagger (a_{\bar\nu}a_\nu + a_{\bar\nu}a_\nu^\dagger). \tag{6.261}$$

---

[77]As defined in Chapter 8, a system of coherent states is a set of states $\{\hat{T}(g)|\phi\rangle, g \in G\}$, for any state $|\phi\rangle$ in the Hilbert space of a unitary irrep $\hat{T}$ of a group G. Exercises 8.1 and 8.10 show that, if $|\phi\rangle$ is a highest- or a lowest-weight state of a semi-simple Lie group, then the coherent states, $\{|\phi(g)\rangle\}$, are minimum-uncertainty states.

Expanding the fermion operators in terms of quasi-particle operators by the inverse Bogolyubov-Valatin transformation

$$a_\nu^\dagger = u_\nu \alpha_\nu^\dagger - v_\nu \alpha_{\bar\nu}, \quad a_\nu = u_\nu \alpha_\nu + v_\nu \alpha_{\bar\nu}^\dagger, \tag{6.262}$$

and using the fact that $\alpha_\nu|\phi\rangle = 0$, then gives

$$\langle\varphi|[\hat{H}', a_\nu^\dagger a_{\bar\nu}^\dagger]|\varphi\rangle = 2(\varepsilon_\nu - G_{\nu\nu}v_\nu^2 - \lambda)u_\nu v_\nu - \sum_{\mu>0} G_{\mu\nu}u_\mu v_\mu (u_\nu^2 - v_\nu^2). \tag{6.263}$$

Thus, the variational equation reduces to

$$2(\tilde{\varepsilon}_\nu - \lambda)u_\nu v_\nu = \Delta_\nu(u_\nu^2 - v_\nu^2), \tag{6.264}$$

where

$$\tilde{\varepsilon}_\nu := \varepsilon_\nu - G_{\nu\nu}v_\nu^2, \tag{6.265}$$

and

$$\Delta_\nu := \sum_{\mu>0} G_{\mu\nu}u_\mu v_\mu \tag{6.266}$$

is the so-called *gap parameter*. The equation $\langle\varphi|[\hat{H}', a_{\bar\nu}a_\nu]|\varphi\rangle = 0$ gives the same result.

Equation (6.264) is solved by squaring it and eliminating u_ν^4 and v_ν^4 with the square of the normalisation, $u_\nu^2 + v_\nu^2 = 1$, to obtain

$$u_\nu^2 v_\nu^2 = \frac{\Delta_\nu^2}{4[(\tilde{\varepsilon}_\nu - \lambda)^2 + \Delta_\nu^2]} = \frac{1}{4}\left[1 - \frac{(\tilde{\varepsilon}_\nu - \lambda)^2}{(\tilde{\varepsilon}_\nu - \lambda)^2 + \Delta_\nu^2}\right]. \tag{6.267}$$

It then follows that

$$u_\nu^2 = \frac{1}{2}\left[1 + \frac{\tilde{\varepsilon}_\nu - \lambda}{[(\tilde{\varepsilon}_\nu - \lambda)^2 + \Delta_\nu^2]^{1/2}}\right], \quad v_\nu^2 = \frac{1}{2}\left[1 - \frac{\tilde{\varepsilon}_\nu - \lambda}{[(\tilde{\varepsilon}_\nu - \lambda)^2 + \Delta_\nu^2]^{1/2}}\right]. \tag{6.268}$$

In terms of the quasi-spin rotation angles of Section 6.8.2, $u_\nu^2 = \frac{1}{2}(1 + \cos\theta_\nu)$ and $v_\nu^2 = \frac{1}{2}(1 - \cos\theta_\nu)$ with

$$\cos\theta_\nu = \frac{\tilde{\varepsilon}_\nu - \lambda}{[(\tilde{\varepsilon}_\nu - \lambda)^2 + \Delta_\nu^2]^{1/2}}. \tag{6.269}$$

Two further equations are needed to solve for the parameters Δ_ν and λ. One, obtained by substituting Equation (6.267) into (6.266), is the so-called gap equation

$$\Delta_\nu = \frac{1}{2}\sum_{\mu>0} G_{\mu\nu}\frac{\Delta_\mu}{[(\tilde{\varepsilon}_\mu - \lambda)^2 + \Delta_\mu^2]^{1/2}}. \tag{6.270}$$

The other is the number constraint

$$\langle\varphi|\hat{n}|\varphi\rangle = 2\sum_{\nu>0} v_\nu^2 = n. \tag{6.271}$$

Note that, if the pair coupling constants are all equal, i.e., $G_{\mu\nu} = G$, then the gap parameter $\Delta_\nu = \Delta$ takes the same value for all single-particle levels and the gap equation simplifies to

$$\tfrac{1}{2} G \sum_{\mu>0} \frac{1}{\left[(\tilde{\varepsilon}_\mu - \lambda)^2 + \Delta^2 \right]^{1/2}} = 1. \tag{6.272}$$

The above equations are readily solved with the help of a computer.[78]

6.8.4 *The independent-quasi-particle (IQP) Hamiltonian*

The BCS equations can be derived in other ways which give alternative perspectives and interpretations. The variational approach of the previous section focuses specifically on obtaining an approximate ground state for the pairing model Hamiltonian $\hat{H}'$. As we now show, the quasi-particle vacuum state is the ground state of an independent-quasi-particle (IQP) Hamiltonian,

$$\hat{H}_0 = W_0 + \sum_\nu \mathcal{E}_\nu \alpha_\nu^\dagger \alpha_{\bar{\nu}}. \tag{6.273}$$

Identifying this Hamiltonian enables it to be used to provide approximate (model) predictions for many spectroscopic properties of nuclei.

(i) *An equations-of-motion derivation of the IQP Hamiltonian*

The Hamiltonian $\hat{H}_0$ is easily derived within the framework of an equations-of-motion formalism.[79] In this derivation, the quasi-particle operators are, as usual, the generalised fermion operators, $\alpha_\nu^\dagger = u_\nu a_\nu^\dagger + v_\nu a_\nu$, $\alpha_\nu = u_\nu a_\nu - v_\nu a_\nu^\dagger$, of the Bogolyubov-Valatin transformation, Equations (6.243) and (6.245), and the equations-of-motion method is used to derive W_0, $\mathcal{E}_\nu$, and the parameters, u_ν, v_ν, in a mean-field approximation.

The derivation starts from the observation that $\mathcal{E}_\nu$ and $\hat{W}_0$ satisfy the equations

$$\{a_{\bar{\mu}}, [\hat{H}_0, \alpha_\nu^\dagger]\} = \delta_{\mu,\nu} u_\nu \mathcal{E}_\nu, \tag{6.274}$$

$$\{a_{\bar{\mu}}, [\hat{H}_0, \alpha_\nu]\} = \delta_{\mu,\nu} v_\nu \mathcal{E}_\nu, \tag{6.275}$$

$$\langle \varphi | \hat{H}_0 | \varphi \rangle = W_0, \tag{6.276}$$

where $|\varphi\rangle$ is a normalised quasi-particle vacuum state. The parameters of the Hamil-

---

[78]A simple computer program for the solution of the BCS equations for a constant coupling constant has been given, for example, in Heyde K.L.G. (1990), *The Nuclear Shell Model* (Springer-Verlag, Berlin), Section 9.8.

[79]The equations-of-motion formalism used here was introduced in Rowe D.J. (1968), *Rev. Mod. Phys.* **40**, 153, and developed in Rowe, *op. cit.* Footnote 4 on Page x. Its application to mean-field methods is reviewed in the following chapter; its application to vibrational excitations will be presented in Volume 2.

tonian, $\hat{H}_0$, are then defined by the equations

$$\langle\varphi|\{a_{\bar{\mu}}, [\hat{H}', \alpha_\nu^\dagger]\}|\varphi\rangle = \delta_{\mu,\nu} u_\nu \mathcal{E}_\nu, \qquad (6.277)$$

$$\langle\varphi|\{a_{\bar{\mu}}, [\hat{H}', \alpha_\nu]\}|\varphi\rangle = \delta_{\mu,\nu} v_\nu \mathcal{E}_\nu, \qquad (6.278)$$

$$\langle\varphi|\hat{H}'|\varphi\rangle = W_0, \qquad (6.279)$$

with $\hat{H}' = \hat{H} - \lambda\hat{n}$, where $\hat{H}$ is the original pairing-model Hamiltonian (6.234).

The left-hand sides of Equations (6.277) and (6.278) are linear combinations of the expectation values

$$\langle\varphi|\{a_{\bar{\mu}}, [\hat{H}', a_\nu^\dagger]\}|\varphi\rangle = \delta_{\mu,\nu}\left(\varepsilon_\nu - \lambda - G v_\nu^2\right) = \delta_{\mu,\nu}(\tilde{\varepsilon}_\nu - \lambda), \qquad (6.280)$$

$$\langle\varphi|\{a_{\bar{\mu}}, [\hat{H}', a_\nu]\}|\varphi\rangle = \delta_{\mu,\nu}\sum_{\sigma>0} G_{\nu\sigma} u_\sigma v_\sigma = \delta_{\mu,\nu}\Delta_\nu. \qquad (6.281)$$

Thus, Equations (6.277) and (6.278) combine to give the eigenvector equation, for the u_ν and v_ν coefficients of the Bogolyubov-Valatin transformation,

$$\begin{pmatrix} (\tilde{\varepsilon}_\nu - \lambda) & \Delta_\nu \\ \Delta_\nu & -(\tilde{\varepsilon}_\nu - \lambda) \end{pmatrix} \begin{pmatrix} u_\nu \\ v_\nu \end{pmatrix} = \mathcal{E}_\nu \begin{pmatrix} u_\nu \\ v_\nu \end{pmatrix} \qquad (6.282)$$

and their eigenvalues

$$\mathcal{E}_\nu = [(\tilde{\varepsilon}_\nu - \lambda)^2 + \Delta_\nu^2]^{1/2}. \qquad (6.283)$$

Once solutions have been found for the u_ν and v_ν coefficients, the quasi-particle vacuum energy is evaluated from the expression

$$W_0 = \langle\varphi|\hat{H}'|\varphi\rangle = \sum_{\nu>0}\left[2(\varepsilon_\nu - \lambda)v_\nu^2 - G_{\nu\nu}v_\nu^4 - \Delta_\nu u_\nu v_\nu\right]. \qquad (6.284)$$

Note that the vacuum state $|\varphi\rangle$ in the above equations is defined by quasi-particle operators which are known only after the equations have been solved. Thus, the BCS equations are non-linear. They are nevertheless solved rather easily by the iterative methods of self-consistent-field approximations.

(ii) *BCS theory as a generalised mean-field approximation*

The above equations-of-motion derivation of BCS theory can be seen as a self-consistent-field approximation, extended to include a pair field. This means that the BCS approximation is a special case of a more general Hartree-Fock-Bogolyubov approximation, as shown in Chapter 7.

In the extended mean-field approximation the Hamiltonian, $\hat{H}'$, is approximated by a Hamiltonian

$$\hat{H}_0 = \text{const.} + \hat{h} + \hat{\Delta}, \qquad (6.285)$$

where

$$\hat{h} := \sum_{\mu\nu} h_{\mu\nu} a_\mu^\dagger a_{\bar\nu} \qquad (6.286)$$

is a number-conserving Hamiltonian with matrix elements defined for an approximate ground state, $|\varphi\rangle$, by

$$h_{\mu\nu} := \langle \varphi | \{ a_{\bar\mu}, [\hat{H}', a_\nu^\dagger] \} | \varphi \rangle, \qquad (6.287)$$

and

$$\hat{\Delta} := - \sum_{\mu,\nu>0} \Delta_{\mu\nu} (a_\mu^\dagger a_{\bar\nu}^\dagger + a_{\bar\mu} a_\nu) \qquad (6.288)$$

is a so-called *pairing field* with matrix elements

$$\Delta_{\mu\nu} := \langle \varphi | \{ a_{\bar\mu}, [\hat{H}', a_\nu] \} | \varphi \rangle. \qquad (6.289)$$

A determination of $\tilde\varepsilon_\nu$, Δ, λ, and the state $|\varphi\rangle$ can be obtained by the following iterative self-consistent field procedure. The process can be initiated from a quasi-particle vacuum state, defined in terms of a Bogolyubov-Valatin transformation with randomly chosen u_ν and v_ν coefficients, subject to the condition that $2\sum_{\nu>0} v_\nu^2$ is equal to the desired nucleon number. One can then evaluate the matrix elements $h_{\mu\nu} = \delta_{\mu,\nu}(\tilde\varepsilon_\nu - \lambda)$ and $\Delta_{\mu\nu} := \delta_{\mu,\nu}\Delta_\nu$, as defined by Equations (6.280) and (6.281), and construct a first approximation for the Hamiltonian

$$\hat{H}_0 = \text{const.} + \sum_\nu (\tilde\varepsilon_\nu - \lambda) a_\nu^\dagger a_{\bar\nu} - \sum_{\mu,\nu>0} \Delta_\nu (a_\mu^\dagger a_{\bar\nu}^\dagger + a_{\bar\mu} a_\nu). \qquad (6.290)$$

An improved Bogolyubov-Valatin transformation is then obtained, by solution of Equation (6.282), which brings this approximation for $\hat{H}_0$ to the diagonal form, $\hat{H}_0 = W_0 + \sum_\nu \mathcal{E}_\nu \alpha_\nu^\dagger \alpha_{\bar\nu}$, in accordance with Equation (6.273). This improved Bogolyubov-Valatin transformation now defines an improved approximation for the quasi-particle state $|\varphi\rangle$. The process is then repeated until further iterations produce no change to whatever accuracy is required. Provided the iterative process converges, for some initial starting point, the system of equations will become *self-consistent*. Should the iterative solution not converge, it may be repeated from some other starting point. It should, in any case, be repeated from different initial starting points to find the lowest of perhaps several local minimum energy solutions.

When a solution exists with at least one non-zero value among a set of gap parameters, $\{\Delta_\nu\}$, the BCS equations are said to have a *superconducting solution*. However, it is possible that no superconducting solution to the BCS equations exists. The only solution will then be an independent-particle solution, for which $\hat{H}_0 = \text{const.} + \sum_\nu (\tilde\varepsilon_\nu - \lambda) a_\nu^\dagger a_{\bar\nu}$. Such situations are discussed in Section 6.8.8.

(iii) *The quasi-particle residual interaction*

The full pairing-model Hamiltonian $\hat{H}' = \hat{H} - \lambda\hat{n}$, with $\hat{H}$ defined by Equation (6.234), is expressed initially in terms of nucleon creation and annihilation operators. When re-expressed in terms of the above-defined quasi-particle operators, it takes the form

$$\hat{H}' = \hat{H}_0 + \hat{V}_{\text{res}} = \hat{W}_0 + \sum_\nu \mathcal{E}_\nu \alpha_\nu^\dagger \alpha_{\bar\nu} + \hat{V}_{\text{res}}, \tag{6.291}$$

where $\hat{V}_{\text{res}}$, the so-called *residual interaction*, is a Hermitian linear combination of the four quasi-particle operators in the set

$$\{\alpha^\dagger \alpha^\dagger \alpha^\dagger \alpha^\dagger, \quad \alpha^\dagger \alpha^\dagger \alpha^\dagger \alpha, \quad \alpha^\dagger \alpha^\dagger \alpha\alpha, \quad \alpha^\dagger \alpha\alpha\alpha, \quad \alpha\alpha\alpha\alpha\}. \tag{6.292}$$

This set contains only operators in which quasi-particle creation operators are to the left of quasi-particle annihilation operators. Such operators are said to be in *normal order* (see Footnote 42, on Page 431, and Section 7.5). Note that the residual interaction does not contain any bilinear terms of the type $\alpha^\dagger \alpha^\dagger$ and $\alpha\alpha$ because the Bogolyubov-Valatin transformation expresses the sum of all bilinear combinations of quasi-particle operators in the Hamiltonian, $\hat{H}'$, in the diagonal form, $\sum_\nu \mathcal{E}_\nu \alpha_\nu^\dagger \alpha_{\bar\nu}$ plus constant terms.

The normal-ordered expansion of operators is a valuable technique that is widely used in many-body theory, as discussed further in Chapter 7. For example, the normal-ordered expression of $\hat{V}_{\text{res}}$ ensures that it first contributes to the ground-state energy of $\hat{H}'$ as a second-order perturbation; i.e., the vacuum expectation of $\hat{V}_{\text{res}}$ is zero. It is clearly important to identify the residual interaction that is neglected in the basic BCS approximation if one wishes to go to higher order. In particular, it needs to be taken into account in the study of two-quasi-particle states. However, for a realistic calculation of two-quasi-particle states, one should also take account of other residual interactions in addition to those arising from the schematic pairing interaction. A further comment on this topic is given in the following section.

6.8.5 *Interpretation of the BCS approximation*

To make use of the BCS approximation in the description of nuclear structure, one must learn how to interpret its results. The variables that emerge from a BCS calculation, $\lambda, u_\nu, v_\nu, \mathcal{E}_\nu, \tilde{\varepsilon}_\nu$ and Δ_ν, have a natural interpretation. However, to avoid unrealistic conclusions, caution is needed because BCS states are not eigenstates of the nucleon number operator. For example, a gap parameter, Δ_ν, which plays a central role in BCS theory (as discussed explicitly in Section 6.8.8), is the quasi-particle vacuum expectation value of the operator

$$\hat{\Delta}_\nu = \sum_{\mu > 0} G_{\mu\nu} a_{\bar\mu} a_\mu. \tag{6.293}$$

Clearly, it would be inappropriate to suppose that the expectation value $\langle 0|\hat{\Delta}_\nu|0\rangle$ of $\hat{\Delta}_\nu$ in the exact ground state of the Hamiltonian $\hat{H}'$, which is an eigenstate of the nucleon number operator $\hat{n}$, could be approximated by the BCS value of Δ_ν or by any value other than zero. Nevertheless, the gap parameters are measures of the magnitudes of the pairing correlations and have physical interpretations. The following shows that a consistent interpretation of the BCS results emerges from the premise that its states are superpositions of eigenstates of $\hat{H}'$ of good nucleon number.

The BCS approximation is founded basically on three underlying assumptions. The first assumption, appropriate for the original formulation of BCS theory for a macroscopic system, is that the mean number of participating particles (regarded here as the number, n, of valence-shell nucleons) is relatively large. The second, is that the quasi-particle vacuum state is a superposition of the ground states of a range of nuclei of even particle number centred about a mean value $\bar{n} = \langle\varphi|\hat{n}|\varphi\rangle$. As shown in Section 6.8.6, the root-mean-square spread in the range of the nucleon number in the quasi-particle vacuum state is of the order $\sqrt{n}$. The third assumption, is that selected properties of nuclei in this range vary slowly and smoothly with n. This assumption is known to be valid for some sequences of isotopes and isotones (cf. Figures 1.33, 1.34 and for simple pairing model Hamiltonians, cf. Figures 6.5 and 6.11). The reliability of the model predictions depends on the extent to which these assumptions are valid.

The chemical potential is introduced in BCS theory in order that the ground-state expectation value of the Hamiltonian, $\hat{H}'$, for the nucleus of interest will be a local minimum relative to the ground states of neighbouring nuclei. If a smooth curve, drawn through a plot of the ground-state energies of even n-nucleon nuclei, were to have an energy dependence of the form

$$\langle n0|\hat{H}|n0\rangle \approx E_{\bar{n}} + (n - \bar{n})E'_{\bar{n}} + \tfrac{1}{2}(n - \bar{n})^2 E''_{\bar{n}} + \dots, \qquad (6.294)$$

for some n close to the value $\bar{n}$ of interest, then the appropriate value of λ would be given by $E'_{\bar{n}} = \partial\langle n0|\hat{H}|n0\rangle/\partial n$, so that $\partial\langle n0|\hat{H}'|n0\rangle/\partial n = 0$ when $n = \bar{n}$. However, for the energy expectation, $\langle n0|\hat{H}'|n0\rangle$, with such a value of λ, to be a local minimum, E''_{n_0} should be positive. This is usually the situation for open-shell nuclei for which the number of active particle-pairs, or hole-pairs in the latter half of the shell, is smallest at the closed-shell limits; cf. Figure 6.3. If it is not the situation, there is unlikely to be a superconducting solution towards which an iterative solution of the BCS equations will converge. Such situations are discussed in Section 6.8.8. We consider here situations in which a superconducting solution does exist. Thus, together with the BCS assumption that the excitation energies of the n-particle nuclei are (approximately) n-independent for n in the neighbourhood of the nucleus of interest and that the ground-state energy varies slowly with n, the energy levels of $\hat{H}'$ are, in effect, presumed to be symmetrical, as a function of n, as illustrated in Figure 6.18. We proceed to interpret the BCS results with

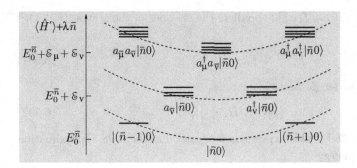

Figure 6.18: Schematic illustration of low-energy states of nuclei, interpreted in terms of the BCS model. The quasi-particle vacuum state is presumed to be a superposition of even-mass ground states. The one-quasi-particle states are interpreted as superpositions of low-energy states of odd-mass nuclei that are created by either adding or removing a single nucleon from the neighbouring even-mass nucleus.

the assumption that, to a first approximation, the quasi-particle vacuum state is a superposition of even-mass ground states and the one quasi-particle states are superpositions of odd-mass ground states.

With these assumptions, the expectation value

$$\langle \varphi | a_\nu^\dagger a_{\bar\nu} | \varphi \rangle = v_\nu^2 \tag{6.295}$$

is interpreted as the average occupancy of the single-particle state ν over the nucleon-number states $\{|n0\rangle\}$ that make up the vacuum state $|\varphi\rangle$. Assuming that the occupancy $\langle n0 | a_\nu^\dagger a_{\bar\nu} | n0 \rangle$ varies linearly with n for n close to $\bar n = \langle \varphi | \hat n | \varphi \rangle$, the occupancies of single-particle states in the n-nucleon ground state, $|\bar n 0\rangle$, are given in the BCS approximation by

$$\langle \bar n 0 | a_\nu^\dagger a_{\bar\nu} | \bar n 0 \rangle \approx \langle \varphi | a_\nu^\dagger a_{\bar\nu} | \varphi \rangle = v_\nu^2. \tag{6.296}$$

Similarly, the model predicts that

$$\langle \bar n 0 | a_{\bar\nu} a_\nu^\dagger | \bar n 0 \rangle \approx \langle \varphi | a_{\bar\nu} a_\nu^\dagger | \varphi \rangle = u_\nu^2. \tag{6.297}$$

A one-quasi-particle BCS state, $\alpha_\nu^\dagger | \varphi \rangle$, is interpreted as a superposition of number-eigenstates of $\hat H'$ of energy $E_0^{\bar n} + E_\nu$, in accord with the expansion

$$\alpha_\nu^\dagger | n0 \rangle = u_\nu a_\nu^\dagger | n0 \rangle + v_\nu a_{\bar\nu} | n0 \rangle. \tag{6.298}$$

In view of Equations (6.296) and (6.297), normalised states of an odd $\bar n \pm 1$ nucleus are approximated by

$$|(\bar n + 1)\nu\rangle \approx \frac{1}{u_\nu} a_\nu^\dagger | \bar n 0 \rangle, \quad |(\bar n - 1)\nu\rangle \approx \frac{1}{v_\nu} a_{\bar\nu} | \bar n 0 \rangle. \tag{6.299}$$

Thus, in the BCS approximation, single-nucleon transfer matrix elements are given by

$$\langle (\bar n + 1)\nu | a_\nu^\dagger | \bar n 0 \rangle \approx u_\nu, \quad \langle (\bar n - 1)\nu | a_{\bar\nu} | \bar n 0 \rangle \approx v_\nu. \tag{6.300}$$

These transition matrix elements are important in applications to nucleon transfer reactions (cf. Section 6.9.5). If the ground state $|\bar n 0\rangle$ has energy $E_0^{\bar n}$, then the BCS approximation $[\hat{H}', \alpha_\nu^\dagger]|n0\rangle \approx \mathcal{E}_\nu \alpha_\nu^\dagger |0\rangle$ predicts the energies of the states $|(\bar{n} \pm 1)\nu\rangle$ to be given by

$$E_\nu^{\bar{n}\pm1} \approx E_0^{\bar{n}} + \mathcal{E}_\nu \pm \lambda. \tag{6.301}$$

We now consider the single-particle energies,

$$\tilde{\varepsilon}_\nu = \langle \varphi | \{a_{\bar{\nu}}, [\hat{H}, a_\nu^\dagger]\} | \varphi \rangle, \tag{6.302}$$

defined by Equation (6.280). This expression is regarded as the BCS approximation for the single-particle energy

$$\tilde{\varepsilon}_\nu \approx \langle \bar{n}0 | \{a_{\bar{\nu}}, [\hat{H}, a_\nu^\dagger]\} | \bar{n}0 \rangle. \tag{6.303}$$

An expansion of this expression, by insertion of intermediate states, gives the general expression

$$\tilde{\varepsilon}_\nu \approx \sum_\kappa (E_\kappa^{\bar{n}+1} - E_0^{\bar{n}}) \, |\langle (\bar{n}+1)\kappa | a_\nu^\dagger | \bar{n}0 \rangle|^2$$
$$+ \sum_\kappa (E_0^{\bar{n}} - E_\kappa^{\bar{n}-1}) \, |\langle (\bar{n}-1)\kappa | a_{\bar{\nu}} | \bar{n}0 \rangle|^2, \tag{6.304}$$

where $\{|(\bar{n}\pm1)\kappa\rangle\}$ denote states of the $\bar{n}\pm1$ nuclei with energies $E_\kappa^{\bar{n}\pm1}$, respectively. Thus, $\tilde{\varepsilon}_\nu$ is a centroid of combined pick-up and stripping spectroscopic factors (cf. Section 6.9.5). In the BCS approximation, pick-up and stripping strengths for the operators $a_\nu^\dagger$ and $a_{\bar{\nu}}$ are concentrated in single states (cf. Equation (6.300)); i.e.,

$$\langle (\bar{n}+1)\kappa | a_\nu^\dagger | \bar{n}0 \rangle \approx \delta_{\kappa\nu} u_\nu, \quad \langle (\bar{n}-1)\kappa | a_{\bar{\nu}} | n0 \rangle \approx \delta_{\kappa\nu} v_\nu. \tag{6.305}$$

Insertion of these matrix elements and the energies of Equation (6.301) into Equation (6.304) then leads to the identity

$$\tilde{\varepsilon}_\nu - \lambda = (u_\nu^2 - v_\nu^2) \mathcal{E}_\nu, \tag{6.306}$$

consistent with an identical expression derived from Equations (6.268) and (6.283).

Similarly, when interpreted in terms of average matrix elements given by an expansion of $\langle (\bar{n}-2)0 | \{a_{\bar{\nu}}, [\hat{H}, a_\nu]\} | \bar{n}0 \rangle$, the gap parameter, $\Delta_\nu = \langle \varphi | \{a_{\bar{\nu}}, [\hat{H}, a_\nu]\} | \varphi \rangle$, is expected to satisfy the identity

$$\Delta_\nu = 2\mathcal{E}_\nu u_\nu v_\nu, \tag{6.307}$$

consistent with an identical expression derived directly from Equation (6.267).

As noted in the previous section, the single-particle energy $\tilde{\varepsilon}_\nu$ has an interpretation as an eigenvalue of a single-particle mean-field Hamiltonian, whereas the single-quasi-particle energy,

$$\mathcal{E}_\nu = \sqrt{(\tilde{\varepsilon}_\nu - \lambda)^2 + \Delta_\nu^2}, \tag{6.308}$$

has an interpretation as an eigenvalue of a generalised single-quasi-particle mean-field Hamiltonian which includes a number-non-conserving pair field.

We conclude this section by pointing out an insightful geometrical relation-ship between the single-particle and single-quasi-particle energies given by Equation (6.308). This identity shows that $\mathcal{E}_\nu$ is the length of the hypotenuse of a right-angled triangle for which the other two sides are of length $(\tilde{\varepsilon}_\nu - \lambda)$ and Δ_ν, as illustrated on the right-hand side of Figure 6.19. The angle θ_ν of this triangle

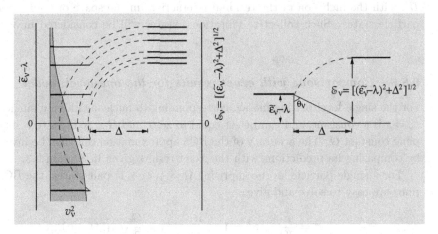

Figure 6.19: Possible single-particle energies $\{\tilde{\varepsilon}_\nu\}$ and the corresponding quasi-particle energies given by $\{\mathcal{E}_\nu = [(\tilde{\varepsilon}_\nu - \lambda)^2 + \Delta_\nu^2]^{1/2}\}$. In an independent-particle model, a single-particle state ν is either occupied or empty. In the BCS approximation, a single-particle state has an occupancy factor v_ν^2 ranging from zero to one. The occupancy factors are given in BCS theory by $v_\nu^2 = \frac{1}{2}(1 - \cos\theta_\nu)$ where $\cos\theta_\nu = (\tilde{\varepsilon}_\nu - \lambda)/\mathcal{E}_\nu$. For illustrative purposes, the gap parameter Δ_ν is assumed to take a single constant value Δ in the figure.

is the quasi-spin rotation angle, defined by Equation (6.269). Thus, if the single-particle energy levels $\{\tilde{\varepsilon}_\nu - \lambda\}$ of the mean-field approximation are plotted on a vertical axis, as on the left-hand side of the figure, there is a simple geometrical map from these single-particle energies to the corresponding quasi-particle energies $\{\mathcal{E}_\nu\}$. Moreover, as indicated by Equation (6.256), the occupancy factors v_ν^2, shown by the shaded area in the figure, are given in terms of the quasi-spin rotation angles by $v_\nu^2 = \frac{1}{2}(1 - \cos\theta_\nu)$.

An important property of the BCS approximation is that it reproduces the familiar odd-even staggering observed in the ground-state energies (cf. Figure 6.21). According to Equation (6.301), the magnitude of this staggering is given in the BCS approximation by

$$\frac{1}{2}\left(E_\nu^{\bar{n}+1} + E_\nu^{\bar{n}-1}\right) - E_0^{\bar{n}} = \mathcal{E}_\nu. \tag{6.309}$$

Thus, as Figure 6.19 illustrates, the odd-even staggering energy is predicted to be large when the gap parameters $\{\Delta_\nu\}$ are large.

The BCS approximation also predicts an energy gap of approximately twice the odd-even energy difference, between the ground and two-quasi-particle states of even nuclei. It should be noted, however, that the interpretation of excited states of even-mass nuclei as two-quasi-particle states, as in the basic BCS model, cannot realistically be considered a good approximation. In particular, it is expected that nuclei, for which the pair-coupling model is appropriate, exhibit low-lying collective states, notably quadrupole vibrational states, which can more realistically be described as coherent mixtures of states resulting from diagonalising the Hamiltonian $\hat{H}'$, with the inclusion of the residual interaction, in the space of many two-quasi-particle states. Such collective vibrational states will be considered in Volume 2.

6.8.6 Comparisons with exact results for the one-level model

For the single-level pairing model, corresponding to nucleons filling a single-j shell, $\varepsilon_\nu = \varepsilon$ is a constant, and can be set equal to zero, and $G_{\mu\nu}$ reduces to a single coupling constant G. The accuracy of the BCS approximation can then be investigated by comparing its predictions with the exact results given in Section 6.3.

For a single-particle level comprising $\Omega = \frac{1}{2}(2j + 1)$ pair states, the BCS equations are easy to solve and give

$$v^2 = \frac{n}{2\Omega}, \quad u^2 = 1 - \frac{n}{2\Omega}, \quad \Delta = \frac{G}{2}\sqrt{n(2\Omega - n)}. \tag{6.310}$$

(i) *Ground-state energy*

The BCS ground-state energy $\langle\varphi|H|\varphi\rangle = W_0 + \lambda n$, given by Equation (6.284), takes the value

$$\langle\varphi|H|\varphi\rangle = -\frac{1}{4}Gn\left(2\Omega - n + \frac{n}{\Omega}\right), \tag{6.311}$$

which may be compared with the exact result of Equation (6.27), for $v = 0$,

$$E(n, v{=}0) = -\tfrac{1}{4}Gn(2\Omega - n + 2). \tag{6.312}$$

The fractional error in the ground-state energy is seen to be $\sim n/\Omega^2$.

(ii) *Quasi-particle energies*

The quasi-particle energy $\mathcal{E} = [(\tilde{\varepsilon} - \lambda)^2 + \Delta^2]^{1/2}$, evaluated with

$$\lambda = \frac{\partial\langle\varphi|H|\varphi\rangle}{\partial n} = -\frac{1}{2}G\left(\Omega - n + \frac{n}{\Omega}\right),$$
$$\tilde{\varepsilon} = -Gv^2 = -G\frac{n}{2\Omega}, \tag{6.313}$$

is

$$\mathcal{E} = \tfrac{1}{2}G\Omega; \tag{6.314}$$

it may be compared with the exact expression,

$$\tfrac{1}{2}[E(n+1,1) + E(n-1,1)] - E(n,0) = \tfrac{1}{4}G(2\Omega+1), \tag{6.315}$$

for the energies of seniority-one states obtained from Equation (6.27). The fractional error is now $\sim 1/2\Omega$. For states of more than one quasi-particle, the residual interaction $\hat{V}_{\text{res}}$ has non-vanishing expectation values and should not be ignored.

(iii) Particle-number fluctuation

In the single-level pairing model, the BCS ground state $|\varphi\rangle$ is precisely a superposition of seniority-zero ground states of even nuclei and the one-quasi-particle states are precisely superpositions of seniority-one states of odd nuclei. Thus, the errors made by applying the BCS approximation to this model are entirely due to number fluctuations. The spread in particle number is defined by the variance

$$(\Delta n)^2 = \langle\varphi|(\hat{n} - \bar{n})^2|\varphi\rangle = \langle\varphi|(\hat{n}^2 - \bar{n}^2)|\varphi\rangle, \tag{6.316}$$

which is evaluated to be

$$(\Delta n)^2 = 4\Omega u_j^2 v_j^2 = 2\bar{n}\left(1 - \frac{\bar{n}}{2\Omega}\right). \tag{6.317}$$

Thus, the fractional uncertainty in particle number,

$$\frac{\Delta n}{\bar{n}} \le \left(\frac{2}{\bar{n}}\right)^{1/2}, \tag{6.318}$$

becomes negligible for large systems. However, it is not negligible in applications to finite nuclei and considerable advantage can be gained, as will be discussed in Section 6.11, by working with states of good nucleon number projected from BCS-like states.

6.8.7 Improved BCS approximations

(i) Lipkin-Nogami corrections for particle-number fluctuations

Some errors due to number fluctuation can be suppressed by a relatively simple adjustment of the BCS model as proposed by Lipkin and Nogami.[80,81] The idea is to replace the Hamiltonian $\hat{H}' = \hat{H} - \lambda\hat{n}$ of the standard BCS approximation, by a more general Hamiltonian

$$\hat{H}' = \hat{H} - f(\hat{n}), \tag{6.319}$$

where the function $f(\hat{n})$ is ideally chosen so that the BCS vacuum state has the desired mean nucleon number $\bar{n} = \langle\varphi|\hat{n}|\varphi\rangle$ and, in addition, the ground-state energies, relative to $\hat{H}'$, of nuclei with n in the neighbourhood of $\bar{n}$ are degenerate. In the standard BCS model, the Hamiltonian $\hat{H}'$ is adjusted so that its ground-state

[80]Lipkin H.J. (1960), Ann. Phys. (NY) 9, 272.
[81]Nogami Y. (1964), Phys. Rev. 134, B313.

energies, for neighbouring even-n nuclei, are degenerate to first order in $(n-\bar{n})$. By choosing

$$f(\hat{n}) = \lambda_1\hat{n} + \lambda_2\hat{n}^2, \qquad (6.320)$$

they can be made degenerate to order $(n-\bar{n})^2$. In the single-level pairing model, for which the exact ground-state energies are quadratic in n, one is able to get precise results in this way.

(ii) *Correction for blocking effects*

If the BCS quasi-particle vacuum state is expressed in the unnormalised form

$$|\varphi\rangle \propto \exp\Big(\sum_{\nu>0} \frac{v_\nu}{u_\nu} a_\nu^\dagger a_{\bar{\nu}}^\dagger\Big)|0\rangle, \qquad (6.321)$$

as in Equation (6.238), then a one quasi-particle state has the corresponding expression

$$\alpha_\mu^\dagger|\varphi\rangle \propto \exp\Big(\sum_{0<\nu\neq\mu} \frac{v_\nu}{u_\nu} a_\nu^\dagger a_{\bar{\nu}}^\dagger\Big) a_\mu^\dagger|0\rangle. \qquad (6.322)$$

In the standard BCS approximation, the coefficients u_j and v_j are derived so as to optimise the even n-particle ground state. However, a better description of odd nuclear states can be obtained by optimising the coefficients directly for the one quasi-particle states of interest. It is easy to see what the result would be. If a particular single-particle state μ is occupied by a physical particle, that state is no longer available to the other pair-coupled particles. The single-particle state μ is then said to be *blocked*. As a consequence, the paired particles are restricted to $(\Omega - 1)$ pair states and the (u_ν, v_ν) coefficients and energies should be adjusted accordingly.

While corrections for blocking effects[82] are relatively straightforward, there is a disadvantage to applying them in general because they introduce a different quasi-particle basis for each energy level, which destroys the simplicity of the BCS approximation. If blocking effects are a concern, one should probably go to the next step of working with number projected states (cf. Section 6.11). An in-depth study of blocking effects has been made by Wahlborn.[83]

(iii) *Corrections for spurious states*

If $\nu \equiv (jm)$, the BCS treatment gives $\Omega = \frac{1}{2}(2j+1)$ distinct two-quasi-particle $M = 0$ states, $\alpha_{jm}^\dagger \alpha_{j\bar{m}}^\dagger|\varphi\rangle$, whereas the exact solution allows only $\Omega - 1$ seniority-two states. The problem arises because, while $(\hat{n}-\bar{n})|\bar{n}0\rangle$ is identically zero, $(\hat{n}-\bar{n})|\varphi\rangle$ is not. Thus, the combination of two-quasi-particle states $\sum_{m>0} \alpha_{jm}^\dagger \alpha_{j\bar{m}}^\dagger|\varphi\rangle = \hat{S}_+|\varphi\rangle$

[82]Cf., for example, Soloviev V.G. (1961), *Mat. Fys. Skr. Dan. Vid. Selsk.* **1** (11), and Gallagher Jr. C.J. and Soloviev V.G. (1962), *Mat. Fys. Skr. Dan. Vid. Selsk.* **2** (2).
[83]Wahlborn S. (1962), *Nucl. Phys.* **37**, 554.

is spurious. In a number-projected BCS theory, this problem is avoided because the $n = \bar{n}$ component of the so-called spurious state does indeed vanish.

Spurious states are a characteristic of theories which break a fundamental symmetry of the Hamiltonian. They occur, for example, in the shell model, when translational invariance is violated, and in the Hartree-Fock theory of deformed nuclei, when rotational invariance is violated. As will be shown in Volume 2, the so-called *Quasi-Particle Random Phase Approximation* has the merit of identifying the spurious excitations of broken-symmetry states by giving them vanishing excitation energy.

6.8.8 *Normal and superconducting BCS solutions*

Consider a singly-closed shell nucleus for which the nucleon number is such that, without a pairing interaction, the active single-particle levels are either fully occupied or empty, as in a closed-subshell situation. There is then always a trivial solution to the BCS equations in which the quasi-particle vacuum is precisely the closed-subshell state. This is the solution for which

$$v_\nu = 1 \text{ or } 0 \quad \text{and} \quad u_\nu = 0 \text{ or } 1. \tag{6.323}$$

Such a solution is characterised by a zero value of the gap parameters,

$$\Delta_\nu = \sum_{\mu > 0} G_{\mu\nu} u_\mu v_\mu = 0, \tag{6.324}$$

and is described as a *non-superconducting solution*. In contrast, a non-trivial solution in which some of the gap parameters are non-zero is described as a *superconducting solution*. The single-particle occupancy factors, v_ν^2, for the two solutions are illustrated in Figure 6.20.

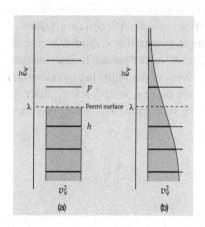

Figure 6.20: Single-particle energy levels, $\tilde{\varepsilon}_\nu$, and their occupancy factors, v_ν^2: (a) for a non-superconducting state, and (b) for a superconducting state with non-vanishing gap parameters. The energy ε_F for which all single-particle states with energies less than ε_F are fully occupied and all those above are empty is described as the *Fermi energy* or *Fermi surface*. Thus, figure (a) shows a state with a sharp Fermi surface and (b) one with a diffuse Fermi surface.

An interesting observation is that, for small values of the $G_{\mu\nu}$ coupling constants, a superconducting solution of the BCS equations may not exist. Suppose that, in the non-superconducting solution, h labels the highest-energy occupied single-particle state and p labels the lowest-energy unoccupied state. To satisfy the number equation for a superconducting solution, λ must have a value intermediate between ε_h and ε_p. Let us suppose, for simplicity, that $G_{\mu\nu} = G$ takes constant values. Then, if G is so small that, for such a value of λ,

$$\tfrac{1}{2} G \sum_{\nu>0} \frac{1}{|\tilde{\varepsilon}_\nu - \lambda|} < 1, \tag{6.325}$$

there can be no solution to the gap equation (6.272). This is presumed to be the situation for closed-shell nuclei for which the separation in energy between single-particle states above and below the Fermi surface is sufficiently large that the single-particle states below the Fermi surface are expected to be predominantly occupied and those above predominantly empty.

For a nucleus with a partly-filled subshell, a superconducting solution always exists, for any strength of the pairing interaction. However, such a situation only occurs when the other residual interactions between the particles are ignored. In particular, for a doubly-open shell nucleus, the experimental evidence suggests that deformation correlations, associated with rotational degrees of freedom, develop and compete with the pairing correlations. For example, the singly-closed shell Ni and Sn isotopes are interpreted as superconducting and spherical whereas many doubly-open-shell rare-earth nuclei appear to be both deformed and superconducting. In the latter situation, it is apparent that, for a realistic treatment, the deformation and pairing correlations need to be tackled simultaneously. For this problem, a combination of Hartree-Fock and BCS theory, known as Hartree-Bogolyubov theory, has been developed and is outlined in the following chapter.

In spite of the insights gained from the colourful picture painted by the BCS model, it is important to recognise that the sharp distinction between states in finite nuclei that are superconducting and others that are not is strictly an artefact of the model. Even within the framework of a simple pairing-force model, a precise treatment of the transition from one phase to the other, as the pair coupling constant is increased, is both continuous and smooth. However, as the following comparisons with precisely computed solutions to the two-level pairing model show, sharply-defined phase transitions, as predicted by the BCS approximation, are approached in the large particle-number limit.

Exercises

6.28 Show that, if

$$|x\rangle := \exp \left(x \sum_{\nu=1}^{N} a_\nu^\dagger a_{\bar{\nu}}^\dagger \right) |0\rangle, \tag{6.326}$$

then, for $N = 1$, the overlap $\langle x|x\rangle$ is given by $1 + x^2$ and, when N is any other positive integer,

$$\langle x|x \rangle = (1 + x^2)^N = \sum_p \binom{N}{p} x^{2p},$$ (6.327)

where $\binom{N}{p} = \dfrac{N!}{p!(N-p)!}$ is a binomial coefficient.

6.29 Use the definitions

$$a^{jm} := \left(a^\dagger_{jm} \right)^\dagger, \quad a_{jm} := (-1)^{j-m} a^{j,-m}, \quad a^\dagger_{j\bar{m}} = (-1)^{j+m} a^\dagger_{j,-m},$$ (6.328)

to show that

$$a^\dagger_{j\bar{m}} = -a^\dagger_{jm}, \quad a_{j\bar{m}} = -a_{jm}, \quad a_{j\bar{m}} = a^{jm}, \quad a^{j\bar{m}} = -a_{jm},$$ (6.329)

and that, if $a^\dagger_\nu = \sum_{jm} C^{jm}_\nu a^\dagger_{jm}$ with real coefficients, then

$$a^\dagger_{\bar{\nu}} = -a^\dagger_\nu, \quad (a_{\bar{\nu}})^\dagger = a^\dagger_\nu.$$ (6.330)

6.30 Use the results of the previous exercise and the anticommutation relations,

$$\{a^\mu, a^\dagger_\nu\} = \delta_{\mu\nu},$$ (6.331)

to show that

$$\{a_{\bar{\mu}}, a^\dagger_\nu\} = \delta_{\mu\nu}, \quad \{a_\mu, a^\dagger_{\bar{\nu}}\} = -\delta_{\mu\nu}.$$ (6.332)

6.31 Use the anticommutation relations of the previous exercise to derive the inverse of the Bogolyubov-Valatin transformation, Equation (6.243), as given by Equation (6.262).

6.32 Use the Baker-Campbell-Hausdorff identity

$$e^{\hat{A}} \hat{B} e^{-\hat{A}} = \hat{B} + [\hat{A}, \hat{B}] + \frac{1}{2!}[\hat{A}, [\hat{A}, \hat{B}]] + \dots$$ (6.333)

to show that, for any operator $\hat{A}$,

$$e^{\hat{A}} a_\mu = \left(a_\mu + [\hat{A}, a_\mu] + \frac{1}{2!}[\hat{A}, [\hat{A}, a_\mu]] + \dots \right) e^{\hat{A}}.$$ (6.334)

Hence, derive Equation (6.241).

6.33 Show that, if $|\varphi\rangle$ is the quasi-particle vacuum state then $\langle \varphi| a^\dagger_\mu a^\dagger_{\bar{\mu}} |\varphi \rangle = u_\mu v_\mu$ and, with $\hat{\Delta}_\nu$ defined by Equation (6.293), that

$$\langle \varphi| \hat{\Delta}_\nu |\varphi \rangle = \sum_{\mu>0} G_{\mu\nu} u_\mu v_\mu = \Delta_\nu.$$ (6.335)

6.34 Show that

$$\exp\left[\sum_{\nu>0} z_\nu a^\dagger_\nu a^\dagger_{\bar{\nu}} \right] = \prod_{\nu>0} (1 + z_\nu a^\dagger_\nu a^\dagger_{\bar{\nu}}).$$ (6.336)

6.35 Use the identity of Exercise 6.32 to derive the particle to quasi-particle transformation, $a^\dagger_\nu \to \alpha^\dagger_\nu = e^{\hat{Z}} a^\dagger_\nu e^{-\hat{Z}}$, given by Equation (6.254).

6.9 Implications and applications of BCS theory

BCS theory is successful in explaining many characteristics of nuclei in which pair correlations play an essential role. More importantly, it provides a formal framework within which more general theories, that take other correlations into account, can be developed.

6.9.1 Ground states of even nuclei

In the first instance, BCS theory provides a model description of the ground states of even singly-closed-shell nuclei. In particular, it gives criteria for when closed-shell configurations will be stable against break-down of their shell structures by the short-range forces between nucleons.

When the BCS predictions for ground-state energies are compared with those of exact calculations for the same pairing Hamiltonian, as done above for a single-shell model, the BCS approximation does rather well. However, it should be clearly understood that the BCS Hamiltonian was not designed to give realistic values for ground-state binding energies. The derivation of nuclear binding energies for systems of nucleons is of very considerable interest, but it is not usefully tackled with schematic interactions. BCS theory is much more relevant for understanding the energy differences between related nuclear states.

6.9.2 Odd-even mass differences

Because of the pairing interaction, the last unpaired nucleon in an odd-mass nucleus is much less strongly bound than in its even neighbours. One prediction of BCS theory is that if the valence-shell nucleon number n is even then the ground states of the odd $n \pm 1$ particle nuclei will have a common spin, j. Let E_0^n and $E_j^{n\pm1}$ denote the ground-state energies of these nuclei. The average odd-even mass difference, defined by

$$\mathcal{E}_j = \tfrac{1}{2}\left[E_j^{n+1} + E_j^{n-1}\right] - E_0^n, \tag{6.337}$$

is then predicted by BCS theory to be given by

$$\mathcal{E}_j = [(\bar{\varepsilon}_j - \lambda)^2 + \Delta_j^2]^{\frac{1}{2}}. \tag{6.338}$$

Thus, the average odd-even mass difference is predicted to have a value that always exceeds that of the BCS gap parameter; i.e., $\mathcal{E}_j > \Delta_j$. The large values of the odd-even mass differences observed (cf. Figure 6.4) and estimates of $(\bar{\varepsilon}_j - \lambda)$ given by Equation (6.306) (cf. Figure 6.19), indicate that, in singly-closed shell nuclei, the gap parameter is large compared to the values of $(\bar{\varepsilon}_j - \lambda)$ for the lowest-energy single-particle states; i.e., the measured odd-even energy differences are only marginally larger than Δ_j.

One-proton separation energies are shown for the $N = 82$ isotones in Figure 6.21 and one-neutron separation energies are shown for the $Z = 82$ (Pb) isotopes in Figure 6.22. Observe the very different slopes of of the dashed lines in the two

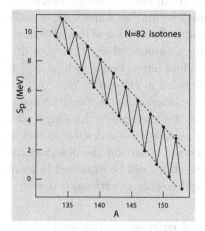

Figure 6.21: One-proton separation energies, S_p, for the $N = 82$ isotones. All lines are drawn to guide the eye. All uncertainties but one are smaller than the circles around the data points. Note that the extreme right-hand data point (for [153]Lu) corresponds to this nucleus being unbound with respect to one-proton emission. (Data taken from Audi G., Wapstra A.H. and Thibault C. (2003), *Nucl. Phys.* **A729**, 337.)

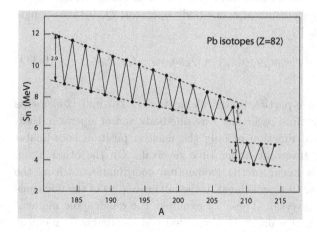

Figure 6.22: One-neutron separation energies, S_n, for the $Z = 82$ (Pb) isotopes. All lines are drawn to guide the eye. All uncertainties are smaller than the circles around the data points. (Data taken from Audi G., Wapstra A.H. and Thibault C. (2003), *Nucl. Phys.* **A729**, 337.)

figures which indicate the different strengths of the effective pairing interaction at $N = 82$ and $Z = 82$ (cf. Equation (6.30)). Note, also the discontinuity that occurs between [207]Pb and [209]Pb which is attributed to the "rapid" filling of the $3p_{1/2}$ subshell (cf. Figure 6.4 for the Sn isotopes).

6.9.3 *The energy gap and collective two quasi-particle states*

An important result of BCS theory is the prediction of an energy gap between the ground state of an even nucleus and its first excited state. In the BCS approximation, this energy gap is the energy required to create two quasi-particles, $2\mathcal{E}_j$, and is equal to twice the odd-even energy difference. To a first approximation, this prediction accords with the observed energy-level spectra of even singly-closed-shell nuclei; cf., for example, the spectra of the Sn isotopes shown in Figure 1.29 which, except for a 2^+ state, exhibit a distinct lack of states in the low-energy region.

To obtain the energy of the low-lying 2^+ state within the framework of BCS theory, it is necessary to take into account the residual interaction of the full expansion $\hat{H}' = W_0 + \sum_\nu E_\nu \alpha_\nu^\dagger \alpha_\nu + \hat{V}_{\rm res}$. Moreover, for a meaningful description of this state it is necessary to include more realistic residual interactions, such as $J \neq 0$ pairing interactions and other interactions not included in $\hat{H}' = \hat{H} - \lambda \hat{n}$.

For example, when a suitable Hamiltonian is diagonalised within the space of $J = 2$ two-quasi-particle states then, as shown for example by Kisslinger and Sorensen[84],[85] in the framework of the so-called *pairing-plus-quadrupole* model, a coherent mixture of two-quasi-particle states emerges at around the observed energy. Such coherent mixtures of two quasi-particle states will be discussed in more detail in Volume 2 and interpreted as collective quadrupole vibrational states.

6.9.4   *Electromagnetic matrix elements and pairing correlations*

Electromagnetic multipole moments and transitions are sometimes strongly modified by pairing. The matrix elements of a one-body operator, $\hat{Q}$, between single quasi-particle states are given by

$$\langle \varphi | \alpha_{\bar{\mu}} \hat{Q} \alpha_\nu^\dagger | \varphi \rangle = \sum_{\alpha\beta} \hat{Q}_{\alpha\beta} \langle \varphi | \alpha_{\bar{\mu}} a_\alpha^\dagger a_{\bar{\beta}} \alpha_\nu^\dagger | \varphi \rangle = Q_{\mu\nu} u_\mu u_\nu - Q_{\bar{\nu}\bar{\mu}} v_\mu v_\nu, \qquad (6.339)$$

where $Q_{\mu\nu} := \langle \mu | \hat{Q} | \nu \rangle$ is a single-particle matrix element of $\hat{Q}$. The matrix elements $Q_{\mu\nu}$ and $Q_{\bar{\nu}\bar{\mu}}$ are related by time reversal. Any one-body tensor operator that, in a Cartesian basis, is a real function of only the nucleon position coordinates will be both Hermitian and invariant under time reversal. On the other hand, a Hermitian operator that is linear in the momentum coordinates, such as the angular-momentum operators, and/or is linear in the intrinsic spins of the nucleons, will change sign under time reversal. Thus, for example, the quadrupole moment operators of a nucleus are positive under time reversal and the magnetic moment operators are negative. Note, however, that the components of a Hermitian tensor operator in a (complex) spherical tensor basis are not separately Hermitian. Instead, they satisfy the Hermiticity relationship

$$Q_{LM}^\dagger = (-1)^M Q_{L,-M}. \qquad (6.340)$$

Recall, for example, the familiar relationship, $Y_{lm}^* = (-1)^m Y_{l,-m}$, for spherical harmonics. Therefore, with the inclusion (by convention) of complex conjugation in the definition of the time-reversal operation (see Section 6.2), a Hermitian spherical

---

[84]Kisslinger L.S. and Sorensen R.A. (1963), *Rev. Mod. Phys.* **35**, 853.
[85]See also Bès D.R. and Sorensen R.A. (1969), *Adv. Nucl. Phys.* **2**, 129.

tensor operator is, respectively, positive or negative under time reversal, T, if

$$T(\hat{Q}_{LM}) = \tau \hat{Q}_{LM}^\dagger, \quad \tau = \pm 1. \tag{6.341}$$

Because the time reverse of a matrix element (being a complex number) is given simply by complex conjugation, it follows that

$$Q_{\bar{\nu}\bar{\mu}} = \langle \bar{\nu}|\hat{Q}|\bar{\mu}\rangle = T\langle \bar{\nu}|\hat{Q}|\bar{\mu}\rangle^* = \langle \mu|T(\hat{Q}^\dagger)|\nu\rangle = \tau Q_{\mu\nu}. \tag{6.342}$$

Thus, if $\hat{Q}$ is a component of a Hermitian spherical tensor, we obtain

$$\langle \varphi|\alpha_{\bar{\mu}}\hat{Q}\alpha_\nu^\dagger|\varphi\rangle = Q_{\mu\nu}(u_\mu u_\nu - \tau v_\mu v_\nu). \tag{6.343}$$

It is seen that the diagonal ($\mu = \nu$) one-quasi-particle matrix elements of a time-odd, e.g. magnetic-moment, operator are unchanged relative to single-particle matrix elements by the pairing correlations whereas off-diagonal matrix elements are of the same order of magnitude. In contrast, the matrix elements of a time-even, e.g. electric-quadrupole, operator may be substantially reduced.

For even nuclei, the opposite occurs. For example, the matrix element of $\hat{Q}$ between the ground state and a two-quasi-particle state is given in the BCS approximation by

$$\langle \varphi|\hat{Q}\alpha_\nu^\dagger\alpha_{\bar{\mu}}^\dagger|\varphi\rangle = \sum_{\alpha\beta} Q_{\alpha\beta}\langle \varphi|a_\alpha^\dagger a_{\bar{\beta}}\alpha_\nu^\dagger\alpha_{\bar{\mu}}^\dagger|\varphi\rangle = Q_{\mu\nu}(v_\mu u_\nu + \tau u_\mu v_\nu). \tag{6.344}$$

It is emphasised, however, that the strong mixing of two-quasi-particle configurations in low-lying collective states can very much enhance electromagnetic transition strengths in even nuclei over those predicted by a pure pairing model (this is discussed further in Volume 2).

6.9.5 *Transfer reactions and pairing correlations*

Experiments that are particularly sensitive to pairing correlations in nuclei are the stripping and pick-up reactions, cf. Section 1.3.2.[86,87]

---

[86]The theory of transfer reactions was initiated by Butler S.T. (1951), *Proc. Roy. Soc. London* **A208**, 559 and reviewed in his book Butler S.T. (1957), *Nuclear Stripping Reactions* (J. Wiley, New York). A classic review of the subject was presented by Macfarlane M.H. and French J.B. (1960), *Rev. Mod. Phys.* **32**, 567, who give a full list of references to the theory of direct nuclear reactions on which their treatment is based. A useful recent and comprehensive evaluation of spectroscopic factors for $A < 56$ is given by Lee J., Teang M.B. and Lynch W.G. (2007), *Phys. Rev.* **C75**, 064320

[87]The application of pairing theory to single-nucleon transfer reactions was initiated by Yoshida S. (1961), *Phys. Rev.* **123**, 2122, and later extended to two-nucleon stripping reactions by the same author, Yoshida S. (1962), *Nucl. Phys.* **33**, 685.

(i) *Single-nucleon transfer reactions*

The theory of single-nucleon-transfer reactions expresses pick-up and stripping cross sections in terms of *spectroscopic factors*

$$S(J_f; j, J_i) := \left| \langle \Psi_{J_f M} | \Phi_{J_f M}(j, J_i) \rangle \right|^2 = \frac{\left| \langle \Psi_{J_f} \| a_j^\dagger \| \Psi_{J_i} \rangle \right|^2}{2J_f + 1}, \tag{6.345}$$

where $\Psi_{J_f M}$ and $\Psi_{J_i M}$ are states of so-called *parent* and *daughter* nuclei, respectively; the state

$$\Phi_{J_f M}(j, J_i) = \sum_{M_i m} a_{jm}^\dagger | \Psi_{J_i M_i} \rangle \, (J_i M_i \, jm | JM) \tag{6.346}$$

is then the state of a nucleon of spin j coupled to the daughter nucleus. Thus, the spectroscopic factor $S(J_f; j, J_i)$ measures the extent to which a nucleon can be added to the state Ψ_{J_i} of the daughter nucleus to create the parent state Ψ_{J_f} or, conversely, the extent to which a nucleon can be removed from the parent to leave the daughter.

Suppose, for example, that the daughter state is that of an even nucleus in its ground state, represented in the BCS approximation by the quasi-particle vacuum, and that the parent is represented as a one quasi-particle state:

$$|J_i M_i\rangle = |0\rangle \approx |\varphi\rangle, \quad |\Phi_{jm}\rangle \approx \alpha_{jm}^\dagger |\varphi\rangle. \tag{6.347}$$

Then

$$S(j; j, 0) = |\langle \varphi | \alpha_{j\bar{m}} a_{jm}^\dagger | \varphi \rangle|^2 = u_j^2. \tag{6.348}$$

Thus, spectroscopic factors, when interpreted in terms of the BCS approximation, provide measurements of single-particle occupancies.

Figure 1.31 shows a plot of the occupation probabilities $v_j^2 = 1 - u_j^2$ of single-particle levels as measured[88] in (d,p) and (d,t) reactions on the Sn isotopes. They exhibit the filling of these levels with increasing mass number that one would expect. Moreover, they show that several levels do not fill sequentially, as in an independent-particle model, but in parallel as in a pair-coupling model.

(ii) *Two-nucleon transfer reactions*

For two-nucleon transfer reactions, it is more convenient to express the cross sections in terms of *spectroscopic amplitudes* rather than their squares. This is because more than one amplitude usually appears in the expression for a cross section and there are interference effects. Thus, following Yoshida,[89] we consider the spectroscopic

---

[88]Such spectroscopic factors, first measured by Cohen B.L. and Price R.E. (1961), *Phys. Rev.* **121**, 1441, were successfully described in terms of BCS theory by Kisslinger and Sorensen, *op. cit.* Footnote 84 on Page 484.

[89]Yoshida, *op. cit.* Footnote 87 on Page 485.

amplitudes

$$B(J_f; [j_1 j_2]J, J_i) := \frac{1}{\sqrt{(1 + \delta_{j_1, j_2})}} \langle \Psi_{J_f M_f} | \left[[a_{j_1}^\dagger \otimes a_{j_2}^\dagger]_J \otimes \Psi_{J_i} \right]_{J_f M_f}. \qquad (6.349)$$

The following results are then obtained in the BCS approximation:

(i) When the initial and final states are both quasi-particle vacuum states:

$$B(0; [jj]0, 0) = \sqrt{j + \tfrac{1}{2}} \, u_j v_j, \qquad (6.350)$$

where u_j and v_j are coefficients in the Bogolyubov-Valatin transformation. In defining these coefficients, Yoshida considered it appropriate to adopt the values of u_j and v_j for the initial and final nucleus, respectively. It should be understood, however, that the ambiguity about which coefficients to use leads to uncertainties that are within the inherent errors of the BCS approximation and, therefore, they are ignored in the results presented here.

(ii) When the initial and final states are vacuum and two-quasi-particle states, respectively:

$$B(J; [j_1 j_2]J, 0,) = u_{j_1} u_{j_2}. \qquad (6.351)$$

(iii) When the initial and final states are both one-quasi-particle states:

$$B(J_f; [j_1 j_2]J, J_i) = \delta_{J,0} \delta_{J_i, J_f} \delta_{j_1, j_2} \sqrt{j_1 + \tfrac{1}{2}} \, u_{j_1} v_{j_1} \qquad (6.352)$$

$$+ \sqrt{\frac{2J + 1}{(1 + \delta_{j_1, j_2})(2J_f + 1)}} \left[(-1)^{J_i + J_f - J} \delta_{j_1, J_f} \delta_{j_2, J_i} - \delta_{j_1, J_i} \delta_{j_2, J_f} \right] u_{J_f} v_{J_i}.$$

Examples of two-nucleon transfer are given in Figure 1.25 for the ^{210}Pb(p,t)^{208}Pb reaction and in Figure 1.32 for (p,t) and (t,p) reactions between Sn isotopes. These examples illustrate two different limits of the pair-coupling model when considered from the BCS perspective.

Schematic illustrations of the structures of the ground and first excited (4863 keV) 0^+ states of ^{208}Pb and the 0^+ ground state of ^{210}Pb are shown in Figure 6.23. With the presumption that ^{208}Pb is a rather good closed-shell nucleus, it is expected that the single-particle levels below the Fermi surface are predominantly occupied in the 0_1^+ ground state of ^{208}Pb, while those above are predominantly vacant, as illustrated in Figure 6.23(a). To a first approximation, the excited 0_2^+ state is presumed to be generated by promoting two neutrons across the Fermi level to create a correlated combination of two-particle-two-hole configurations relative to that of the ground state, as illustrated in Figure 6.23(b). Similarly, the dominant component of the ^{210}Pb ground state, is expected to be a state with two neutrons added to predominantly empty single-particle levels above the Fermi level.

Thus, if j_1 denotes the angular momentum of an *occupied* single-particle state and j_2 denotes that of an *unoccupied* single-particle state in the ostensibly closed-

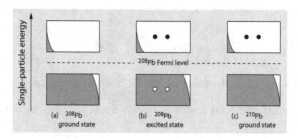

Figure 6.23: Schematic illustration of the occupancies of shell-model states in ^{208}Pb and ^{210}Pb shown as a function of single-particle energy by the widths of the grey areas. The ^{208}Pb ground state is considered to be in a closed-shell state with small pairing correlations. The ground state of ^{210}Pb is considered to have an additional correlated neutron pair in the lowest shell above the ^{208}Pb Fermi surface. The first excited 0^+ state of ^{208}Pb is considered to have both a correlated neutron pair in the shell above the Fermi surface and a corresponding correlated neutron-hole pair in the shell below.

shell ^{208}Pb ground state, we would expect to obtain the following information from the two-nucleon spectroscopic amplitudes between the nuclei:

(a) For the ^{210}Pb ground-state to ^{208}Pb ground-state transition

$$B(0_1; (j_1)^{-2}0, 0_1) \approx 0, \quad B(0_1; (j_2)^{-2}0, 0_1) \approx [u_{j_2}]^2 \mathcal{C}_{j_2}, \qquad (6.353)$$

where $|u_{j_2}|^2$ is a measure of the extent to which the j_2 level really is an unoccupied (particle) states in the ^{208}Pb ground state and $\mathcal{C}_{j_2}$ is the spectroscopic amplitude that would be expected if u_{j_2} were equal to 1.

(b) For the ^{210}Pb ground state to ^{208}Pb excited, 0_2^+ state, transition

$$B(0_2; (j_1)^{-2}0, 0_1) \approx [v_{j_1}]^2 \mathcal{D}_{j_1}, \quad B(0_2; (j_2)^{-2}0, 0_1) \approx 0, \qquad (6.354)$$

where $|v_{j_1}|^2$ is a measure of the extent to which the j_1 level really is an occupied (hole) state in the ^{208}Pb ground state and $\mathcal{D}_{j_1}$ is the spectroscopic amplitude that would be expected if v_{j_1} were were equal to 1.

Thus, if the (p,t) reaction picked up pairs equally from the upper and lower levels, the above BCS approximation would indicate that the cross sections to the ground and excited states should be of similar magnitudes (to within reaction theory differences). However, it is reasonable to suppose that pick up of j_1 pairs would be favoured over j_2 pairs; this would imply an enhancement of the cross section to the excited 0_2^+ state as is seen in Figure 1.25.

Schematic illustrations of the structures of the 0^+ states in the Sn isotopes are shown in Figure 6.24. A standard shell-model description of the low-energy states of these nuclei considers a distribution of neutrons over a suitably-chosen valence space outside of a doubly-closed-shell core. A more realistic description would have some neutrons in higher shells and some neutron holes in the otherwise occupied shells, as indicated in Figure 6.24. In the BCS approximation, the ground states of all the even Sn isotopes are quasi-particle vacuum states, as indicated in Figure 6.24(b). Excited 0^+ states are then interpreted as non-spurious $J = 0$ linear

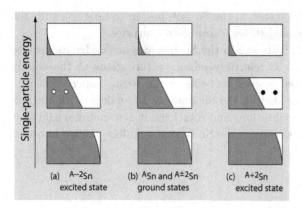

Figure 6.24: Schematic illustration of the neutron occupancies of shell-model states, e.g., in the ground and excited two-quasi-particle states of Sn isotopes, shown as a function of single-particle energy by the widths of the grey areas. Relative to the mass-A Sn isotopes, the two-quasi-particle states of the mass $A \pm 2$ Sn isotopes are, respectively, correlated two-particle and correlated two-hole states.

combinations of two-neutron quasi-particle states (cf. Section 6.8.7). Thus, in the BCS approximation, we consider an excited 0^+ two-quasi-particle state

$$|0_2^+\rangle = \sum_j C_j (\alpha_j^\dagger \otimes \alpha_j^\dagger)_0 |\varphi\rangle. \qquad (6.355)$$

We then obtain the spectroscopic amplitudes:
(a) For the transition from the $|0_1^+\rangle$ ground state of an Sn isotope of mass A to the $|0_1^+\rangle$ ground state of its mass $A \pm 2$ neighbour

$$B(0_1; (j)^2 0, 0_1) = \sqrt{j + \tfrac{1}{2}}\, u_j v_j. \qquad (6.356)$$

(b) For the transition from the $|0_1^+\rangle$ ground state of an Sn isotope of mass A to the $|0_2^+\rangle$ excited state of its mass $A + 2$ neighbour

$$B(0_2; (j)^2 0, 0_1) = C_j [u_j]^2. \qquad (6.357)$$

(c) For the transition from the $|0_1^+\rangle$ ground state of an Sn isotope of mass A to the $|0_2^+\rangle$ excited state of its mass $A - 2$ neighbour

$$B(0_2; (j)^2 0, 0_1) = C_j [v_j]^2. \qquad (6.358)$$

Although a calculation must be done to determine the C_j coefficients, a condition they must satisfy for the excited states to be orthogonal to the ground state can be inferred from the expression for the quasi-particle vacuum state, $|\varphi\rangle = \exp\left[\sum_\nu z_\nu a_\nu^\dagger a_{\bar\nu}^\dagger\right]|0\rangle$. This expression implies that, within the framework of the BCS approximation, the two-particle transfer operator $\sum_\nu z_\nu a_\nu^\dagger a_{\bar\nu}^\dagger$ should map the A-particle ground state, for which the BCS vacuum state is optimised, to just the $(A + 2)$–particle ground state. The implication is that, if the transfer operator for the (t,p) stripping reaction were, in fact, this operator, then the cross section for population of an excited 0^+ states in a (t,p) reaction or, conversely, in a (p,t) pick-up reaction would be zero. Thus, the observation of small two-nucleon transfer cross sections to excited states, in the Sn isotopes, in comparison with the cross sections for populating ground states, comes as no surprise.

The pairing correlations in ^{208}Pb and in the Sn isotopes are understood to be very different. They are viewed as relatively minor perturbations to the doubly closed-shell structure of ^{208}Pb as opposed to being of the strong superconducting type in the singly-closed valence shell of the Sn isotopes In analogy with the collective model concepts of shape vibrations and rotations, it is sometimes said that ^{208}Pb exhibits *pairing vibrations* whereas the Sn isotopes exhibit *pairing rotations* (see comments in Section 6.10.5).[90]

Exercises

6.36 Show that the electric multipole operators $\mathfrak{M}(E\lambda\mu) = e\sum_{i=1}^{Z} r_i^\lambda Y_{\lambda\mu}(\theta_i, \varphi_i)$, where $(r_i, \theta_i, \varphi_i)$ are spherical polar coordinates for a proton and $Y_{\lambda\mu}$ is a spherical harmonic, are even under time reversal.

6.37 Relate the spectroscopic factors, defined by Equation (6.345), to the single-particle coefficients of fractional parentage, defined in Section 5.14.

6.38 Show that

$$\frac{\langle \Psi_{J_f} \| [a_{j_1}^\dagger \otimes a_{j_2}^\dagger]_J \| \Psi_{J_i} \rangle}{\sqrt{2J_f + 1}} = \langle J_f M_f | \left[[a_{j_1}^\dagger \otimes a_{j_1}^\dagger]_J \otimes |J_i\rangle \right]_{J_f M_f}. \tag{6.359}$$

6.39 Use the recoupling methods of Section A.4 to show that if $|\Psi_{J_i M}\rangle$ and $|\Psi_{J_f M}\rangle$ are one-quasi-particle states then

$$\langle J_f M_f | \left[[a_{j_1}^\dagger \otimes a_{j_2}^\dagger]_J \otimes |J_i\rangle \right]_{J_f M_f} = I + II + III, \tag{6.360}$$

with

$$I = \delta_{J,0}\, \delta_{J_i,J_f}\, \delta_{j_1,j_2} \langle \varphi | [a_{j_1}^\dagger \otimes a_{j_1}^\dagger]_0 | \varphi \rangle,$$

$$II = \delta_{j_1,J_f}\, \delta_{j_2,J_i} U(J_i J_i J_f J_f : 0J) \langle J_f M_f | a_{J_f M_f}^\dagger | \varphi \rangle \langle \varphi | [a_{J_i}^\dagger \otimes |J_i\rangle]_0,$$

$$III = -\delta_{j_2,J_f}\, \delta_{j_1,J_i} (-1)^{J_i + J_f - J} U(J_i J_i J_f J_f : 0J)$$
$$\times \langle J_f M_f | a_{J_f M_f}^\dagger | \varphi \rangle \langle \varphi | [a_{J_i}^\dagger \otimes |J_i\rangle]_0. \tag{6.361}$$

Then use the identity

$$U(j_1 j_1 j_2 j_2 : 0J) = U(j_1 j_2 j_1 j_2 : 0J) = (-1)^{j_1 + j_2 - J} \sqrt{\frac{2J + 1}{(2j_1 + 1)(2j_2 + 1)}} \tag{6.362}$$

to derive Equation (6.352).

6.40 By linearly extrapolating the one-neutron separation energies for ^{209,211,213}Pb in Figure 6.22, estimate the mass number for which the odd-mass Pb isotopes become neutron unbound.

---

[90]A review of two-nucleon transfer reactions has been given by Broglia R.A., Hansen O. and Riedel C. (1973), in *Adv. Nucl. Phys.*, Vol. 6, edited by M. Baranger and E. Vogt (Plenum, New York), p. 287.

6.10 BCS and exact results for a two-level model

The pairing model of nucleons in a single level is simply solvable (Section 6.3) because of its su(2) algebraic structure (cf. Section 6.3). The SGA for a two-level pairing model, being a product of two SU(2) quasi-spin groups, is less simple but is readily handled with the help of a computer. Moreover, it is instructive because it is the simplest pairing model to exhibit a phase transition from a so-called *normal phase*, in which the nucleons predominantly occupy the lower of the two levels, and a *superconducting phase* in which they are distributed over both levels.

6.10.1   *The two-level model*

We consider a model[91] of a system of either neutrons or protons with $J = 0$ pairing interactions in a shell-model space comprising two single-particle levels of angular momentum j_1 and j_2 which, for simplicity, we set equal to a common value, $j_1 = j_2 = j$. The Hamiltonian chosen for this model is

$$\hat{H} = \varepsilon(\hat{S}_0^2 - \hat{S}_0^1) - G\hat{S}_+\hat{S}_-, \tag{6.363}$$

where $\varepsilon := \varepsilon_2 - \varepsilon_1$ is the energy difference of the two levels and $\hat{S}_\pm = \hat{S}_\pm^1 + \hat{S}_\pm^2$. The shell-model space is then the tensor product of two SU(2) quasi-spin spaces with $s_1 = s_2 = s = (2j+1)/4$ and the Hamiltonian is readily diagonalised. We also assume a nucleon number such that the lowest level is filled and the upper level is empty when $G = 0$. Thus, when $G = 0$, the ground state of the model is the state which reduces the subgroup chain

$$\begin{array}{cccc} \mathrm{SU(2)}_1 \times & \mathrm{SU(2)}_2 \supset & \mathrm{U(1)}_1 \times & \mathrm{U(1)}_2 \\ s_1 = s & s_2 = s & \nu_1 = s & \nu_2 = -s \end{array}. \tag{6.364}$$

However, in the limit when $G/\varepsilon \to \infty$, the Hamiltonian simplifies to $-G\hat{S}_+\hat{S}_-$ and is diagonal in a coupled quasi-spin basis

$$|S0\rangle = \sum_\nu |s\nu, s, -\nu\rangle \, (s, -\nu \, s\nu|S0), \quad S = 0, \ldots, 2s, \tag{6.365}$$

i.e., a basis labelled by the quantum numbers of the subgroup chain

$$\begin{array}{cccc} \mathrm{SU(2)}_1 \times & \mathrm{SU(2)}_2 \supset & \mathrm{SU(2)} \supset & \mathrm{U(1)} \\ s_1 = s & s_2 = s & S & M = 0 \end{array}. \tag{6.366}$$

A phase transition occurs from one coupling scheme to the other as G increases. For a small value of j the transition is found to be continuous and smooth. However, as the following results indicate, the transition takes place more suddenly as j is increased and becomes singular in the $j \to \infty$ limit.

---

[91]More details of this model can be found in the paper of Chen H., Brownstein J.R. and Rowe D.J. (1990), *Phys. Rev.* **C42**, 1422.

6.10.2 BCS solutions for the two-level model

The BCS equations are analytically solvable for the two-level pairing model (cf. Exercises 6.41 - 6.44). From the solutions, the critical value of the coupling constant is determined to be

$$G_{\mathrm{crit}} = \frac{\varepsilon}{2j},\qquad(6.367)$$

the gap parameter is

$$\Delta = \varepsilon\frac{2j+1}{4j}\left[\left(\frac{G}{G_{\mathrm{crit}}}\right)^2 - 1\right]^{\frac{1}{2}},\qquad(6.368)$$

the quasi-particle energy is

$$\mathcal{E} = \tfrac{1}{2}(2j+1)G,\qquad(6.369)$$

and the occupancy factors are given by

$$n_1 = v_1^2 = u_2^2 = \tfrac{1}{2}\left[1 + \frac{G_{\mathrm{crit}}}{G}\right],\quad n_2 = v_2^2 = u_1^2 = \tfrac{1}{2}\left[1 - \frac{G_{\mathrm{crit}}}{G}\right].\qquad(6.370)$$

The value of the gap parameter and the one- and two-quasi-particle energies are shown as functions of G/G_{crit} in Figure 6.25. The ratios of the occupancy factors given by the BCS approximation are shown in Figure 6.26. The BCS results in these figures are shown as dashed lines.

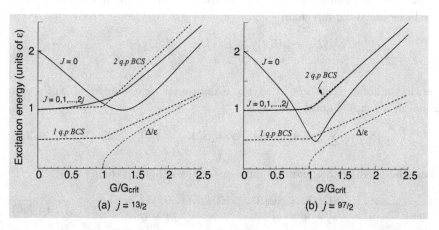

Figure 6.25: Accurately computed excitation energies of the first excited multiplet of states and the second $J = 0$ excited state of the two-level pairing model. Energies are shown in units of ε as functions of G/G_{crit}, where $G_{\mathrm{crit}} = \varepsilon/2j$: (a) for $j = {}^{13}/_2$ and (b) for $j = {}^{97}/_2$. When $G = 0$, the first excited multiplet of states, at $E = \varepsilon$, corresponds to the excitation of one particle to the higher level and the second multiplet, at $E = 2\varepsilon$, corresponds to the excitation of two particles. Also shown are the one-quasi-particle BCS energy, which represent states in the neighbouring odd-particle system, the two-quasi-particle energy, which is an approximation to the energiy of the first excited multiplet, and the value of the BCS gap parameter.

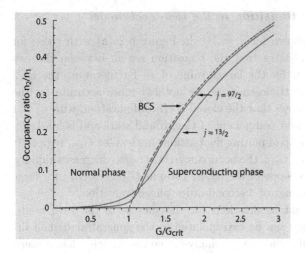

Figure 6.26: Occupancy ratios for the single-particle levels of the two-level pair-coupling model with equal quasi-spins as a function of the coupling constant G in units of $G_{\text{crit}} = \varepsilon/2j$. Results are shown for $j = {}^{13}\!/_{2}$ and $j = {}^{97}\!/_{2}$ and for a nucleon number equal to $2j+1$. Exact and BCS ($j \to \infty$) results are shown as continuous and dashed lines, respectively.

6.10.3 *Exact solutions of the two-level model*

Exact results (to within computer precision) of the the quasi-spin model, obtained by diagonalisation, are shown (as solid lines) in Figures 6.25 and 6.26. In the $G = 0$ limit, the first-excited energy level, shown in Figure 6.25, comprises a set of one-particle-one-hole states of angular momenta $J = 0, 1, 2, \ldots, 2j$, obtained by promoting a single particle from the lower to the upper level. In the language of seniority, these states have multishell seniority $(v_1, v_2) = (1, 1)$ and, in BCS language, they are two-quasi-particle states. Also, at $G = 0$, there is a multi-degenerate excited energy level comprising a set of two-particle-two-hole states with many values of J. These states have multishell seniority $(v_1, v_2) = (2, 2)$ and, in the BCS approximation, are represented as four-quasi-particle states. However, among these two-particle-two-hole states there is one special, multishell seniority $(0,0)$, $J = 0$ state formed by promoting a zero-coupled pair from the lower shell to the upper shell. This $J = 0$ state separates from the other two-particle-two-hole states and falls rapidly in energy as G increases. In fact, the excitation energy of this state is given accurately in first-order perturbation theory for relatively small values of G. This is the domain in which the coupling scheme corresponding to the subgroup chain (6.364) applies. However, as the energy of this excited $J = 0$ state approaches that of the closed-shell $J = 0$ state, i.e., as G approaches the critical value, $G_{\text{crit}} = \varepsilon/(2j)$, the pairing interaction causes these two states (and other multi-pair $J = 0$ excited states that are also collapsing) to mix strongly. Thus, the ground state rapidly changes its character, and level repulsion causes the orthogonal combinations of $J = 0$ states to rise again in energy. In the language of the BCS approximation, a phase transition to a superconducting phase is taking place. More precisely a transition to the alternative coupling scheme, corresponding to the subgroup chain (6.366), is occurring.

6.10.4 *The pairing phase transition in the two-level model*

A comparison of the results shown for $j = {}^{13}/_2$ in Figure 6.25(a) with those for $j = {}^{97}/_2$ in Figure 6.25(b) indicates that the transition region between the two phases is considerably narrower for the larger value of j. Furthermore, the BCS predictions for the two-quasi-particle energies are notably more accurate for the larger value of j. Figure 6.26 shows that the closed-shell configuration, with $n_1 = 1$ and $n_2 = 0$, remains the dominant component of the ground state and is markedly successful in resisting the build-up of pairing fluctuations in the $G < G_{\text{crit}}$ region. It also shows that, once G exceeds G_{crit} the occupancy ratio n_2/n_1 increases rapidly. In fact, the rapid change in the slope of the n_2/n_1 ratio as G passes G_{crit} for large values of j indicates the beginnings of a second-order phase transition.[92]

These results have important implications for the success of the shell model because, to the extent that they can be extrapolated to the general multishell situation, they suggest that shell structure, in singly-closed-shell nuclei, has a much greater degree of rigidity than one might have expected. They suggest that only shell-model configurations involving single-particle energy levels that are relatively close to one another are strongly mixed by the pairing interactions. Thus, when pairing interactions are dominant, the valence space of the shell model is expected to be relatively well-defined. However, in doubly-open-shell nuclei, it is not expected that the pairing interactions will be dominant but that deformation and associated deformation correlations between neutrons and protons will change the situation and cause significant mixing of major spherical-shell-model configurations.

The above results also illustrate the remarkable influence of symmetry in the structure of complex systems and, especially, their phase transitions. It appears that while neither of the two symmetry limits of the model has reason to be good for arbitrary values of ε and G, the system is clearly favouring one or the other. Thus, as we discover repeatedly, dynamical symmetries play a much more dominant role than could have reasonably been expected. Several examples of this dominance will be given in Volume 2.

Two characteristic features of the BCS approximation are evident in the comparisons: the first is that the BCS approximation gives accurate results for large values of j and large values of the particle number; the second is that the BCS approximation always predicts a sharp phase transition between the normal and superconducting phases, in contrast to the smooth phase transitions of the exact solutions for a finite system. Thus, the simple and easily solvable BCS approximation can be expected to give useful but only qualitatively reliable results for finite nuclear situations.

---

[92] A phase transition is said to be of second order if the essential parameters characterising the system change continuously across the critical point but some first derivatives change discontinuously. In Landau's theory, a second-order phase transition is associated with a continuous breaking of a symmetry; cf. Landau L.D. and Lifshitz E.M. (1980), *Statistical Physics* (Butterworth-Heinemann, Oxford), third edn.

6.10.5   *Analogy with shape vibrations and rotations*

The transition from a normal to a superconducting phase with increase of the pair coupling constant, seen in the above example, is analogous to the shape transition from a spherical vibrational phase to a deformed rotational phase in the collective model, as seen in Section 4.7.2. The condition for a phase transition to occur in a multi-level pairing model is that the model has (at least) two highly-symmetrical states: one state in which the single-particle levels are either fully occupied or empty and another state in which the particles are distributed over both levels and has the underlying symmetry of the BCS approximation. Thus, for low values of the coupling constant, the pairing model predicts a ground state with small pairing correlations and pair-excited states which, for large shells, have excitation energies approximately proportional to the number of pairs excited. The excited pairs and the pair-holes they leave behind are expected to be highly correlated and accordingly have similarities with the shape vibrational states of spherical nuclei. Conversely, for large values of the pair coupling constant, it is more appropriate to consider the participating levels as belonging to a common valence space. The BCS approximation, in which states of different nucleon number are mixed, then has analogies with a mean-field description of the rotational states of deformed nuclei in which states of different angular momenta are mixed (cf. comparisons of the Hartree-Fock-Bogolyubov and deformed Hartree-Fock approximations in the following chapter).

For the above reasons, the pair excitations of nuclei at and near doubly-closed shells, e.g., ^{208}Pb, are often described as *pairing vibrations*,[93] whereas the sequence of ground states that are strongly populated in two-nucleon transfer reactions in singly-closed shell nuclei are descried as *pairing rotations*.

Exercises

6.41 Show that, with the choice of $j_1 = j_2 = j$ in the above two-level pairing model, $v_1^2 = u_2^2$ and $v_2^2 = u_1^2$ in order for the mean nucleon number to be a constant, $\langle \hat{n} \rangle = 2j + 1$. Hence show that $\tilde{\varepsilon}_1 - \lambda = -(\tilde{\varepsilon}_2 - \lambda)$. Show also that, if the zero of the single-particle energy scale is defined by setting $\lambda = 0$, then, with $\tilde{\varepsilon} := \tilde{\varepsilon}_2 - \tilde{\varepsilon}_1$, we have $\tilde{\varepsilon}_2 = \tilde{\varepsilon}/2$ and $\tilde{\varepsilon}_1 = -\tilde{\varepsilon}/$.

6.42 Show that the gap equation (6.272), together with the expression (6.283) for a quasi-particle energy $\mathcal{E}_\nu$ in the above two-level model, gives

$$\mathcal{E}_1 = \mathcal{E}_2 = \mathcal{E} = \tfrac{1}{2}(2j + 1)G, \qquad (6.371)$$

and the critical value of the coupling constant

$$G_{\text{crit}} = \frac{\tilde{\varepsilon}}{2j + 1}. \qquad (6.372)$$

---

[93]Bès D.R. and Broglia R.A. (1966), *Nucl. Phys.* **80**, 289.

6.43 Use the identities $\tilde{\varepsilon}_\nu = \varepsilon_\nu - G v_\nu^2$ and

$$v_1^2 = \frac{1}{2}\left[1 + \frac{\tilde{\varepsilon}}{2\mathcal{E}}\right], \quad u_1^2 = \frac{1}{2}\left[1 - \frac{\tilde{\varepsilon}}{2\mathcal{E}}\right], \tag{6.373}$$

to show that

$$\tilde{\varepsilon} = \varepsilon + G(v_1^2 - u_1^2) \tag{6.374}$$

and that

$$\tilde{\varepsilon} = \frac{2j+1}{2j}\varepsilon \quad \text{for } G \geq G_{\text{crit}}. \tag{6.375}$$

6.44 From the above results for $\mathcal{E}$ and ε, show that

$$\frac{\Delta}{\varepsilon} = \frac{2j+1}{4j}\left[\left(\frac{G}{G_{\text{crit}}}\right)^2 - 1\right]^{\frac{1}{2}}. \tag{6.376}$$

6.11 The number-projected coherent-state (NPCS) model

The loss of nucleon number as a conserved quantity of the Hamiltonian in the BCS and HFB approximations is a serious deficiency in their applications to finite nuclei. However, it is restored with relative ease by number projection. Number projection was proposed as an improvement of the BCS approximation by Mottelson[94] and pursued by many.[95] Pairing models for states of definite particle number include the *generalised seniority* model,[96,97] the *broken-pair model*,[98] and the *coherent correlated-pair method*.[99] A closely-related approach in the theory of electronic structure is known as a theory of *antisymmetrised geminal power states*.[100,101]

The physical principles underlying the several techniques for restoring nucleon-number conservation have much in common. The goal is to conserve seniority as well as particle number as good quantum numbers and thereby restore two major symmetries of the pairing model.

Number projection from BCS states is, in principle, simple. For example, with the BCS quasi-particle vacuum state expressed, to within a normalisation factor, in the form $|z\rangle = \exp\left(\sum_{\nu>0} z_\nu a_\nu^\dagger a_{\bar\nu}^\dagger\right)|0\rangle$, as in Equation (6.238), its p-pair component

[94]Mottelson B.R. (1958), in *The Many-Body Problem* (Dunod, Paris), p. 283.

[95] For example, Dietrich K., Mang H.J. and Pradal J.H. (1964), *Phys. Rev.* **135**, B22; Lande A. (1965), *Ann. Phys.* (*NY*) **31**, 525; Ottaviani P.L. and Savoia M. (1969), *Phys. Rev.* **187**, 1306; Ottaviani P.L. and Savoia M. (1970), *Nuovo Cimento* **A67**, 630.

[96]Talmi I. (1971), *Nucl. Phys.* **A172**, 1.

[97]Talmi I. (1975), *Phys. Lett.* **B55**, 255.

[98]Cf. Lorazo B. (1969), *Phys. Lett.* **B29**, 150; Lorazo B. (1970), *Nucl. Phys.* **A153**, 255; Gambhir Y.K., Rimini A. and Weber T. (1969), *Phys. Rev.* **188**, 1573; and Gambhir Y.K., Rimini A. and Weber T. (1973), *Phys. Rev.* **C7**, 1454. The broken-pair model has been reviewed by Allaart K. *et al.* (1988), *Phys. Repts.* **169**, 209.

[99]Vary J.P. and Plastino A. (1983), *Phys. Rev.* **C28**, 2494.

[100]Coleman A.J. (1965), *J. Math. Phys.* **6**, 1425.

[101]Ortiz J.V., Weiner B. and Öhrn Y. (1981), *Int. J. Quantum Chem.* **S15**, 113.

is simply the state

$$|p(z)\rangle := \frac{1}{p!}\Big(\sum_{\nu>0} z_\nu a_\nu^\dagger a_{\bar\nu}^\dagger\Big)^p |0\rangle. \qquad (6.377)$$

Thus, it is straightforward to project all the states arising in BCS theory to their n-particle components. The more challenging problem is to reconstruct the BCS theory using states of well-defined particle number at each stage of its development. The approach of Dietrich, Mang, and Pradal,[95] for example, makes use of the contour integration formula,

$$f_p(z) = \frac{1}{2\pi i} \oint \frac{f(z)}{z^{p+1}} dz, \qquad (6.378)$$

to project out the component of degree p in z from any analytical function of z. They then replaced the various integrals, for which Bayman[102] introduced a saddle-point approximation to derive the BCS equations, by accurate expressions involving number-projected functions.

We present here an alternative number-projection technique that takes explicit advantage of the underlying coherent-state nature of the BCS approximation; we refer to it as the *number-projected coherent-state* (NPCS) model. In this model,[103] number projection is carried out simply and analytically in a coherent-state representation. A particularly desirable characteristic of this approach is that the number-projected wave functions are simple and well-known functions. Thus, it retains the essential simplicity of the BCS model.

6.11.1 *Coherent-state representation of pair-coupled states*

Collective model dynamics usually have a natural expression in terms of coherent state representations of associated dynamical groups of transformations that generate the collective motions of a model. Such representations associate an intrinsic wave function with a particular state, such as a lowest- (or highest-) weight state, which uniquely identifies a sequence (band) of collective states that span an irrep of the dynamical group. Thus, the different states of such a sequence have a common intrinsic structure and are distinguished by their collective wave functions.

Models which, to a first approximation, separate the dynamics of a system into decoupled collective and intrinsic parts are of considerable interest. The remarkable property of the pairing models is that they admit a simple and precise separation of the collective (pairing) and intrinsic degrees of freedom. This is because of the existence of dual pairs of commuting dynamical groups, one member of which is a dynamical group for the intrinsic structure and the other is a dynamical group for the "collective" pairing dynamics, as described in Sections 6.5, 6.6, and 6.7.

---

[102]Bayman, *op. cit.* Footnote 72 on Page 462.

[103]Rowe D.J., Song T. and Chen H. (1991), *Phys. Rev.* **C44**, R598; Chen H., Song T. and Rowe D.J. (1995), *Nucl. Phys.* **A582**, 181; Rowe D.J. (2001), *Nucl. Phys.* **A691**, 691.

For example, in the single-shell $J = 0$ pairing model, the dynamical group for the pairing dynamics is the $\mathcal{SU}(2)$ quasi-spin group and the complementary dynamical group for the intrinsic dynamics is the symplectic group, $USp(2j + 1)$.

The coherent-state approach applies to any pairing Hamiltonian. For illustrative purposes, we consider here the simple pair-coupling model Hamiltonian

$$\hat{H} := \sum_{i=1}^{N} \varepsilon_i \hat{n}_i - G \hat{S}_+ \hat{S}_- \qquad (6.379)$$

with a level-independent coupling constant, $G_{ik} = G$, and with $\hat{S}_\pm$ defined by

$$\hat{S}_\pm := \sum_i \hat{S}_\pm^i, \quad \hat{S}_+^i := \sum_{m>0} a_{j_i m}^\dagger a_{j_i \bar{m}}^\dagger, \quad \hat{S}_-^i := \sum_{m>0} a_{j_i \bar{m}} a_{j_i m}. \qquad (6.380)$$

(i) *Coherent-state wave functions for a single-j shell*

The zero-particle state $|0\rangle$, being a lowest-weight state for an irrep of a single-j-shell quasi-spin algebra, satisfies the equations

$$\hat{S}_-|0\rangle = 0, \quad \hat{S}_0|0\rangle = -s|0\rangle, \quad s = (2j + 1)/4. \qquad (6.381)$$

In a coherent-state representation, we regard the lowest-weight state, $|0\rangle$, as the intrinsic state for the irrep of quasi-spin s and assign to it an *intrinsic* wave function ξ_0. Any state $|\psi\rangle$ of the irrep with lowest-weight state $|0\rangle$ can be created by acting on $|0\rangle$ with some combination of the $\{\hat{S}_+\}$ operators. It is then observed that any such state is uniquely defined by its overlaps with the continuous set of states $\{|z\rangle := e^{z^* \hat{S}_+}|0\rangle\}$, where z is a real or complex variable; these states are known as $\mathcal{SU}(2)$ *coherent states*. Thus, any state $|\psi\rangle$ of zero seniority, $v = 0$, i.e., a state in the irrep with lowest-weight state $|0\rangle$, has a so-called *coherent-state wave function* defined in terms of these overlaps by

$$\Psi(z) := \xi_0 \langle z|\psi\rangle = \xi_0 \langle 0|e^{z\hat{S}_-}|\psi\rangle. \qquad (6.382)$$

For example, the p-pair ($2p$-particle) state of seniority $v = 0$, defined to within a known normalisation factor (cf. Exercise 6.45) by $|p, 0\rangle \propto [\hat{S}_+]^p|0\rangle$, has coherent-state wave function

$$\Psi_{p0}(z) \propto \xi_0 z^p. \qquad (6.383)$$

Coherent-state wave functions for a complete basis of single-j-shell states are similarly constructed. For example, if the lowest-weight state, $a_{jm}^\dagger|0\rangle$, for an $\mathcal{SU}(2)$ irrep is assigned the intrinsic wave function ξ_{jm}, then the $(2p+1)$-particle states of seniority $v = 1$ are given, again to within normalisation factors, by

$$\Psi_{p1jm} \propto \xi_{jm} z^p. \qquad (6.384)$$

In general, if a complete set of lowest-weight states, $\{|v\rho JM\rangle\}$, labelled by seniority v, angular momentum JM, and a multiplicity label ρ to distinguish distinct states with the same vJM, is represented by intrinsic wave functions $\{\xi_{v\rho JM}\}$, then an arbitrary state $|\psi\rangle$ in the single-j shell is represented by the wave function

$$\Psi(z) = \sum_{v\rho JM} \xi_{v\rho JM} \langle v\rho JM | e^{z\hat{S}_-} | \psi\rangle. \tag{6.385}$$

The coherent-state wave function for a particular p-pair, $(n = 2p + v)$-particle, state $|pv\rho JM\rangle \propto [\hat{S}_+]^p |v\rho JM\rangle$ is given by

$$\Psi pv\rho JM(z) = \mathcal{N}_{pv}\, \xi_{v\rho JM}\, z^p, \tag{6.386}$$

where $\mathcal{N}_{pv}$ is a normalisation constant that remains to be determined.

(ii) *Coherent-state wave functions for a multi-j shell*

Coherent-state representations of a multi-j shell state are defined in a natural extension of those for a single shell. The zero-particle state $|0\rangle$ is now a lowest-weight state for the N-fold direct product, $\mathcal{SU}(2) \times \mathcal{SU}(2) \times \cdots \times \mathcal{SU}(2)$ of $\mathcal{SU}(2)$ quasi-spin groups, where N is the number of shells. Thus, the state $|0\rangle$ satisfies the equations

$$\hat{S}_-^i |0\rangle = 0, \quad \hat{S}_0^i |0\rangle = -s_i |0\rangle, \quad s_i = (2j_i + 1)/4, \quad i = 1, \ldots, N, \tag{6.387}$$

and a state $|\psi\rangle$ in the irrep with lowest-weight state $|0\rangle$ has coherent-state wave function

$$\Psi(z) := \xi_0 \langle 0 | e^{\hat{S}_-(z)} | \psi\rangle, \tag{6.388}$$

where z now denotes a set of N variables, $z := (z_1, z_2, \ldots, z_N)$, and

$$\hat{S}_-(z) := \sum_i z_i \hat{S}_-^i. \tag{6.389}$$

Thus, a multishell seniority-zero state $\prod_i \left(\hat{S}_+^i \right)^{p_i} |0\rangle$ has a coherent-state wave function of the form

$$\Psi_{p0}(z) = \mathcal{N}_p\, \xi_0\, z_1^{p_1} z_2^{p_2} \ldots z_N^{p_N}. \tag{6.390}$$

Similarly, the coherent-state wave functions for basis states of multishell seniority $v := (v_1, v_2, \ldots, v_N)$ are of the form

$$\Psi_{pv\rho JM}(z) = \mathcal{N}_{pv}\, \xi_{v\rho JM}\, z_1^{p_1} z_2^{p_2} \ldots z_N^{p_N}. \tag{6.391}$$

6.11.2 *Coherent-state representation of quasi-spin operators*

In making use of the coherent-state representation, one needs to know how the various operators that appear in the pairing problem act on coherent-state wave functions. For example, if we are given the coherent-state wave function $\Psi(z) = \xi_0 \langle z | \psi\rangle$, for the state $|\psi\rangle$, we want to be able to calculate the corresponding wave functions for $\hat{S}_\pm |\psi\rangle$. This is achieved by the following standard procedure.

In a coherent-state representation in which a state $|\psi\rangle$ has wave function Ψ with values $\Psi(z) := \xi_0 \langle 0| e^{\hat{S}_-(z)} |\psi\rangle$, the state $\hat{S}_-^i |\psi\rangle$, for example, has coherent-state wave defined by the overlaps

$$\xi_0 \langle 0| e^{\hat{S}_-(z)} \hat{S}_-^i |\psi\rangle = \frac{\partial}{\partial z_i} \xi_0 \langle 0| e^{\sum_i z_i \hat{S}_-^i} |\psi\rangle = \frac{\partial}{\partial z_i} \Psi^s(z). \tag{6.392}$$

Thus, the state $\hat{S}_-^i |\psi\rangle$ has wave function given by the map

$$\hat{S}_-^i |\psi\rangle \rightarrow \frac{\partial}{\partial z_i} \Psi(z). \tag{6.393}$$

For the number operator $\hat{n}_i$, the general identity

$$e^A B = (B + [A, B] + \tfrac{1}{2}[A, [A, B]] + \cdots) e^A, \tag{6.394}$$

gives

$$\langle 0| e^{\hat{S}_-(z)} \hat{n}_i |\psi\rangle = \langle 0| [\hat{n}_i + 2z_i \hat{S}_-^i] e^{\hat{S}_-(z)} |\psi\rangle = 2z_i \frac{\partial}{\partial z_i} \langle 0| e^{\hat{S}_-(z)} |\psi\rangle. \tag{6.395}$$

Hence, we obtain

$$\hat{n}_i |\psi\rangle \rightarrow 2z_i \frac{\partial}{\partial z_i} \Psi(z). \tag{6.396}$$

For the operator $\hat{S}_+^i$, we similarly obtain

$$\langle 0| e^{\hat{S}_-(z)} \hat{S}_+^i |\psi\rangle = \langle 0| [\hat{S}_+^i - 2z_i \hat{S}_0^i - z_i^2 \hat{S}_-^i] e^{\hat{S}_-(z)} |\psi\rangle$$
$$= z_i \left(2s_i - z_i \frac{\partial}{\partial z_i} \right) \Psi(z). \tag{6.397}$$

Thus, the quasi-spin operators have a coherent-state representation, with quasi-spin s_i, given by

$$\hat{n}_i \rightarrow 2z_i \frac{\partial}{\partial z_i}, \quad \hat{S}_-^i \rightarrow \frac{\partial}{\partial z_i}, \quad \hat{S}_+^i \rightarrow z_i \left(2s_i - z_i \frac{\partial}{\partial z_i} \right), \quad \hat{S}_0^i \rightarrow z_i \frac{\partial}{\partial z_i} - s_i. \tag{6.398}$$

This representation is commonly known as a *Dyson representation*.[104]

6.11.3 *Wave functions for quasi-particle vacuum states*

Underlying the simplicity of the BCS approximation is the fact that a quasi-particle vacuum state, $|x\rangle$, as defined in unnormalised form by Equation (6.238), is itself a seniority-zero coherent state. This is evident from its expression in the form

$$|x\rangle = e^{\hat{S}_+(x)} |0\rangle, \quad \hat{S}_+(x) = \sum_i x_i \hat{S}_+^i. \tag{6.399}$$

---

[104]Dyson F.J. (1956), *Phys. Rev.* **102**, 1217.

Thus, the coherent-state wave functions for quasi-particle vacuum states, given by

$$\Psi(z,x) = \xi_0 \langle 0|e^{\hat{S}_-(z)}e^{\hat{S}_+(x)}|0\rangle = \xi_0\langle z|x\rangle, \tag{6.400}$$

are the overlaps of coherent states. As shown by Exercise (6.28), the overlap of the two coherent states in this wave function is given by

$$\langle z|x\rangle = \Phi^s(zx) := \prod_k (1 + z_k x_k)^{2s_k} \tag{6.401}$$

and we obtain

$$\Psi(z,z) = \xi_0 \Phi^s(zx). \tag{6.402}$$

6.11.4 Wave functions for number-projected seniority states

We now find that, because quasi-particle vacuum states are coherent states, the states of definite particle number projected from them have simple coherent-state wave functions. They are, in fact, standard orthogonal polynomials, known as *Schur functions*[105] (also known simply as *S*-functions). Such functions are widely used in group theory. The subset of *S*-functions needed for present purposes are, in fact, characters of fully antisymmetric representations of unitary groups. Thus, number projection is achieved simply by replacing one known function by another.

(i) *Number-projected seniority-zero coherent states*

The $2p$-particle state projected from a seniority-zero state $|x\rangle$, defined as a quasi-particle vacuum by Equation (6.238), is the p-pair state

$$|p(x)\rangle := \frac{1}{p!}[\hat{S}_+(x)]^p|0\rangle, \quad p \geq 0. \tag{6.403}$$

This state has coherent-state wave function,

$$\Psi_{p0}(z,x) = \xi_0 \Phi_p^s(zx) = \xi_0 \hat{\Pi}_p \Phi^s(zx), \tag{6.404}$$

where $\hat{\Pi}_p$ is a projection operator which, when operating on a polynomial function $F(z,x)$, picks out the component of the function of degree p in its z variables. Explicit expressions for Φ_p^s are derived as follows.

The definition (6.388) implies that

$$\Phi_p^s(z) = \frac{1}{p!}\langle 0|e^{\hat{S}_-(z)}[\hat{S}_+]^p|0\rangle. \tag{6.405}$$

Thus, $\Phi_0^s(z) = 1$ and, from Equation (6.398), we obtain the recursion relation

$$\Phi_{p+1}^s(z) = \frac{1}{p+1}\sum_i z_i \left(2s_i - z_i \frac{\partial}{\partial z_i}\right)\Phi_p^s(z). \tag{6.406}$$

---

[105]Macdonald I.G. (1979), *Symmetric Functions and Hall Polynomials* (Oxford University Press).

With a change of variables to the so-called *symmetric power sums*[106]

$$\phi_m^s = \sum_i 2s_i z_i^m, \quad m \geq 0, \tag{6.407}$$

for which

$$z_i \frac{\partial}{\partial z_i} = \sum_m z_i \frac{\partial \phi_m^s}{\partial z_i} \nabla_m^s = \sum_m m \phi_m^s \nabla_m^s, \tag{6.408}$$

where

$$\nabla_m^s = \frac{\partial}{\partial \phi_m^s}, \tag{6.409}$$

the recursion relation becomes

$$\Phi_{p+1}^s = \frac{1}{p+1}\Big[\phi_1^s - \sum_{m=1}^p m\phi_{m+1}^s \nabla_m^s\Big]\Phi_p^s. \tag{6.410}$$

Theorem 6.8 *The 2p-particle components of quasi-particle vacuum states have coherent-state wave functions that satisfy the recursion relation*

$$\Phi_p^s = \frac{1}{p}\sum_{m=1}^p (-1)^{m+1}\phi_m^s \Phi_{p-m}^s \tag{6.411}$$

whose solutions are the Schur functions

$$\Phi_p^s(z) = \frac{1}{p!}\det\begin{vmatrix} \phi_1^s & 1 & 0 & 0 & \cdots & 0 \\ \phi_2^s & \phi_1^s & 2 & 0 & \cdots & 0 \\ \phi_3^s & \phi_2^s & \phi_1^s & 3 & \cdots & 0 \\ & & \cdots & & & \\ \phi_p^s & \phi_{p-1}^s & \phi_{p-2}^s & \phi_{p-3}^s & \cdots & \phi_1^s \end{vmatrix}. \tag{6.412}$$

To prove this theorem, we first show that the function Φ_p^s satisfies the equation

$$m\nabla_m^s \Phi_p^s = (-1)^{m+1}\Phi_{p-m}^s, \quad \forall m > 0. \tag{6.413}$$

When $p = 1$, this equation reduces to $\nabla_m^s \Phi_1^s = \delta_{m,1}\Phi_0^s$, which is manifestly true because $\Phi_0^s = 1$ and $\Phi_1^s = \phi_1^s$. From Equation (6.410), we have the identity

$$\nabla_1^s \Phi_{p+1}^s = \frac{1}{p+1}\nabla_1^s\Big[\phi_1^s - \sum_{m=1}^p m\phi_{m+1}^s \nabla_m^s\Big]\Phi_p^s, \tag{6.414}$$

and, because $[\nabla_1^s, (\phi_1^s - \sum_{m=1}^p m\phi_{m+1}^s \nabla_m^s)] = 1$,

$$\nabla_1^s \Phi_{p+1}^s = \frac{1}{p+1}\Big[\Phi_p^s + \big(\phi_1^s - \sum_m m\phi_{m+1}^s \nabla_m^s\big)\nabla_1^s \Phi_p^s\Big]. \tag{6.415}$$

[106]Macdonald, *op. cit.* Footnote 105 on Page 501.

Thus, if Equation (6.413) is true, when $m = 1$, for some value of p then, for that value of p, Equation (6.415) gives

$$\nabla_1^s \Phi_{p+1}^s = \frac{1}{p+1} \Big[\Phi_p^s + \big(\phi_1^s - \sum_{m=1}^{p} m \phi_{m+1}^s \nabla_m^s \big) \Phi_{p-1}^s \Big], \qquad (6.416)$$

which, with Equation (6.410), implies that $\nabla_1^s \Phi_{p+1}^s = \Phi_p^s$ and that Equation (6.413) is also valid for all values of p when $m = 1$.

Now observe that, for $m > 1$, Equation (6.410) implies that

$$\begin{aligned}
\nabla_m^s \Phi_{p+1}^s &= \frac{1}{p+1} \nabla_m^s \big[\phi_1^s - \sum_n n \phi_{n+1}^s \nabla_n^s \big] \Phi_p^s \\
&= \frac{1}{p+1} \Big[\big(\phi_1^s - \sum_n n \phi_{n+1}^s \nabla_n^s \big) \nabla_m^s - \sum_n n (\nabla_m^s \phi_{n+1}^s) \nabla_n^s \Big] \Phi_p^s. \quad (6.417)
\end{aligned}$$

Thus, if Equation (6.413) is true for some value of p and any m, as it is for $p = 1$, then for such a value of p,

$$m \nabla_m^s \Phi_{p+1}^s = (-1)^{m+1} \Phi_{p+1-m}. \qquad (6.418)$$

Hence, Equation (6.413) is valid for all m and all p.

Equations (6.410) and (6.413) together now give the recursion relation (6.411) for which Equation (6.412) is the solution. $\qquad \square$

(ii) *Wave functions for number-projected coherent states of other seniority*

Coherent-state wave functions for number-projected coherent states of any seniority are obtained immediately from the above results.

Let $|v\rho JM\rangle$, with intrinsic wave function $\xi_{v\rho JM}$, be a normalised lowest-weight eigenstate of the multishell-pairing Hamiltonian (6.379) with multishell seniority $v := (v_1, v_2, \ldots, v_N)$. Such a state satisfies the equations

$$\begin{aligned}
\hat{S}_-^i |v\rho JM\rangle &= 0, \quad \hat{n}_i |v\rho JM\rangle = v_i |v\rho JM\rangle, \\
\hat{S}_0^i |v\rho JM\rangle &= -s_i(v)|v\rho JM\rangle, \quad s_i(v) = s_i - \tfrac{1}{2} v_i.
\end{aligned} \qquad (6.419)$$

Coherent states built on such a lowest-weight state, given by $e^{\hat{S}_+(x)}|v\rho JM\rangle$, then have properties similar to those built on the zero-particle state, except that the multishell seniority s is replaced by $s(v)$. In particular, the state $e^{\hat{S}_+(x)}|v\rho JM\rangle$ has the coherent-state wave function

$$\Psi_{v\rho JM}(x,z) = \xi_{v\rho JM} \langle v\rho JM | e^{\hat{S}_-(z)} e^{\hat{S}_+(x)} | v\rho JM \rangle = \xi_{v\rho JM} \Phi^{s(v)}(zx), \qquad (6.420)$$

where

$$\Phi^{s(v)}(zx) = \prod_i (1 + z_i x_i)^{2s_i(v)}. \qquad (6.421)$$

Moreover, the wave function for the number-projected p-pair component, $\frac{1}{p!}[\hat{S}_+(x)]^p|v\rho JM\rangle$, of the state $e^{\hat{S}_+(x)}|v\rho JM\rangle$, is given by

$$\Psi_{pv\rho JM}(z,x) = \xi_{v\rho JM}\Phi_p^{s(v)}(zx). \qquad (6.422)$$

6.11.5 *The BCS and number-projected BCS approximations*

The relationship between the standard and number-projected BCS approximations becomes transparent when viewed from the perspective of coherent-state theory. The important observation is that both the quasi-particle vacuum states and the corresponding number-projected vacuum states have simple and known coherent-state wave functions. Thus, in a coherent-state representation, the BCS approximation and the more accurate number-projected BCS approximation are of the same order of complexity. As is usual (but not necessary) in BCS theory, we restrict the coordinates $x = (x_1, x_2, \dots)$ of the coherent (quasi-particle vacuum) state in the following to real values. The generalisation to complex values is straightforward but, with a real value of the pair-coupling constant, G, it proves to be unnecessary.

(i) *The BCS approximation*

In the BCS approximation, the ground state of a $2p$-particle nucleus, in a shell-model space of multishell quasi-spin $s = (s_1, s_2, \dots, s_N)$, is approximated by the seniority-zero coherent state that minimises the energy expectation

$$E_0^s(x) := \frac{\langle x|\hat{H}|x\rangle}{\langle x|x\rangle}, \qquad (6.423)$$

subject to the constraint

$$\frac{\langle x|\hat{n}|x\rangle}{\langle x|x\rangle} = 2p. \qquad (6.424)$$

From the expressions of Section 6.11.2 for the Hamiltonian (6.379), it follows that

$$\langle z|\hat{H}|x\rangle = \langle z|\left[\sum_{i=1}^{N}\varepsilon_i\hat{n}_i - G\hat{S}_+\hat{S}_-\right]|x\rangle$$

$$= \sum_i\left[2\varepsilon_i z_i\frac{\partial}{\partial z_i} - G\sum_j z_i\left(2s_i - z_i\frac{\partial}{\partial z_i}\right)\frac{\partial}{\partial z_j}\right]\Phi^s(zx). \qquad (6.425)$$

From the expression $\Phi^s(zx) = \prod_k(1 + z_k x_k)^{2s_k}$, it is determined that

$$\frac{\partial}{\partial z_j}\Phi^s(zx) = \frac{2s_j x_j}{1 + z_j x_j}\Phi^s(zx), \qquad (6.426)$$

and that

$$z_i\left(2s_i - z_i\frac{\partial}{\partial z_i}\right)\frac{\partial}{\partial z_j}\Phi^s(zx) = \delta_{i,j}\frac{2s_iz_i^2x_i^2}{(1+z_ix_i)^2}\Phi^s(zx)$$
$$+\frac{4s_is_jz_ix_j}{(1+z_ix_i)(1+z_jx_j)}\Phi^s(zx). \qquad (6.427)$$

Now observe that, when Φ^s is divided by a factor $(1 + z_ix_i)$, its functional form is unchanged; what is changed is the multishell seniority, i.e., $s = (s_1, s_2, \ldots) \to s(i)$, where the components of $s(i)$ are $s_k(i) = s_k - \frac{1}{2}\delta_{i,k}$. Thus,

$$\frac{1}{1+z_jx_j}\Phi^s(zx) = \Phi^{s(j)} \qquad (6.428)$$

and, similarly,

$$\frac{1}{(1+z_ix_i)(1+z_jx_j)}\Phi^s(zx) = \Phi^{s(ij)}, \qquad (6.429)$$

where $s_k(ij) = s_k - \frac{1}{2}\delta_{i,k} - \frac{1}{2}\delta_{j,k}$. Thus, we obtain the simple expression

$$\langle z|\hat{H}|x\rangle = \sum_i 2s_iz_ix_i\left[2\varepsilon_i\Phi^{s(i)}(zx) - Gz_ix_i\Phi^{s(ii)}(zx)\right]$$
$$-G\sum_{ij}4s_is_jz_ix_j\Phi^{s(ij)}(zx) \qquad (6.430)$$

and the energy expectation

$$E_0^s(x) = \frac{\langle x|\hat{H}|x\rangle}{\langle x|x\rangle} = \frac{1}{\Phi^s(x^2)}\sum_i\left[4s_i\varepsilon_ix_i^2\Phi^{s(i)}(x^2) - 2Gs_ix_i^4\Phi^{s(ii)}(x^2)\right]$$
$$-G\sum_{ij}4s_is_jx_ix_j\Phi^{s(ij)}(x^2). \qquad (6.431)$$

This energy, $E_0^s(x)$, can be minimised as a function of the x parameters, with the constraint (6.424) to give yet another derivation of the BCS equations.

The u_i, v_i coefficients of BCS theory are obtained by considering the occupation probability of nucleons in the level i which is given in BCS theory by

$$\frac{\langle x|\hat{n}_i|x\rangle}{\langle x|x\rangle} = (2j_i + 1)v_i^2 = 4s_iv_j^2. \qquad (6.432)$$

From Equation (6.396), we have

$$\langle z|\hat{n}_i|x\rangle = 2z_i\frac{\partial}{\partial z_i}\Phi^s(zx) = \frac{4s_iz_ix_i}{1+z_ix_i}\Phi^s(zx). \qquad (6.433)$$

Thus, we obtain v_i^2 and, from the identity $u_i^2 + v_i^2 = 1$,

$$u_i^2 = \frac{1}{1+x_i^2}, \quad v_i^2 = \frac{x_i^2}{1+x_i^2}. \qquad (6.434)$$

(ii) *Number-projected states*

In the number-projected BCS approximation, the ground state of the 2p-particle nucleus, $|p(x)\rangle$, is the seniority-zero state that minimises the energy expectation

$$E^s_{p0}(x) = \frac{\langle p(x)|\hat{H}|p(x)\rangle}{\langle p(x)|p(x)\rangle} = \frac{\langle x|\hat{H}|p(x)\rangle}{\langle x|p(x)\rangle}. \tag{6.435}$$

From the definition of $|p(x)\rangle$ as the 2p-particle component of $|x\rangle$, we have

$$\langle z|\hat{H}|p(x)\rangle = \hat{\Pi}_p\langle z|\hat{H}|x\rangle, \tag{6.436}$$

where $\hat{\Pi}_p$ projects out the component of $\langle z|\hat{H}|x\rangle$ of degree p in the z variables. From the parallel definition, $\Phi^s_p = \hat{\Pi}_p\Phi^s$, and the observation, for example, that

$$\hat{\Pi}_p[z_iz_j\Phi^{s(ij)}(zx)] = z_iz_j\hat{\Pi}_{p-1}\Phi^{s(ij)}(zx) = z_iz_j\Phi^{s(ij)}_{p-1}(zx), \tag{6.437}$$

it also follows that $E^s_{p0}(x)$ has the explicit analytic expression

$$E^s_{p0}(x) = \frac{1}{\Phi^s_p(x^2)}\sum_i\left[4s_j\varepsilon_ix_i^2\Phi^{s(i)}_{p-1}(x^2) - 2Gs_ix_i^4\Phi^{s(ii)}_{p-2}(x^2)\right.$$
$$\left. -G\sum_{i,j}4s_is_jx_ix_j\Phi^{s(ij)}_{p-1}(x^2)\right]. \tag{6.438}$$

Thus, the value of x for which $E^s_{p0}(x)$ is minimised is easily determined with the help of a computer.

(iii) *One quasi-particle states*

In the BCS approximation the quasi-particle vacuum state is the normalised state

$$|\varphi(x)\rangle := \frac{|x\rangle}{\sqrt{\langle x|x\rangle}} = \frac{e^{\hat{S}_+(x)}|0\rangle}{\sqrt{\Phi^s(x^2)}} \tag{6.439}$$

and the one quasi-particle state $\alpha^\dagger_{jm}|\varphi(x)\rangle$ is the seniority-one state

$$|jm,\varphi(x)\rangle := \alpha^\dagger_{jm}|\varphi(x)\rangle = \frac{e^{\hat{S}_+(x)}|jm\rangle}{\sqrt{\Phi^{s(j)}(x^2)}}, \tag{6.440}$$

where $|jm\rangle := a^\dagger_{jm}|0\rangle$, and $x = (x_1, x_2, \dots)$ is chosen, as discussed above, to minimise the energy of the state $|\varphi(x)\rangle$. Thus, for example, one can use these expressions to derive the spectroscopic amplitude of the BCS approximation

$$\langle jm,\varphi(x)|a^\dagger_{jm}|\varphi(x)\rangle = \frac{\langle jm|e^{\hat{S}_-(x)}e^{\hat{S}_+(x)}|jm\rangle}{\sqrt{\Phi^{s(j)}(x^2)\Phi^s(x^2)}}$$
$$= \sqrt{\frac{\Phi^{s(j)}(x^2)}{\Phi^s(x^2)}} = \sqrt{\frac{1}{1+x_j^2}} = u_j. \tag{6.441}$$

In the corresponding NPCS approximation, the above quasi-particle states are replaced by the number-projected states

$$|p(x)\rangle := \frac{\Pi_p e^{\hat{S}_+(x)}|0\rangle}{\sqrt{\Phi_p^s(x^2)}}, \tag{6.442}$$

$$|jm, p(y)\rangle := \frac{\Pi_p e^{\hat{S}_+(y)}|jm\rangle}{\sqrt{\Phi_p^{s(j)}(y^2)}}, \tag{6.443}$$

where x and y are chosen independently to minimise the energies of the respective seniority zero and one states. The spectroscopic amplitudes are then given by

$$\langle jm, p(y)|a_{jm}^\dagger|p(x)\rangle = \frac{\langle jm|e^{\hat{S}_-(y)}\Pi_p e^{\hat{S}_+(x)}|jm\rangle}{\sqrt{\Phi_p^s(x^2)\Phi_p^{s(j)}(y^2)}}$$

$$= \frac{\Phi_p^{s(j)}(yx)}{\sqrt{\Phi_p^s(x^2)\Phi_p^{s(j)}(y^2)}}. \tag{6.444}$$

The latter results differ from their BCS counterparts both because the states have well-defined particle number and also because the number-conserving formalism makes it natural to adjust the values of the x and y variables separately to allow for the different seniorities of the states. Thus, the NPCS approximation enables full account to be taken of the blocking effects (see Section 6.8.7(ii) without creating non-orthogonality problems.

6.11.6 *Number-projected states of arbitrary seniority*

As observed in the previous section, the BCS vacuum and one quasi-particle states have well-defined multishell seniority. However, multiple quasi-particle BCS states do not, in general, have this property. For example, a particular linear combination of two-quasi-particle states is the so-called spurious, seniority-zero, state discussed in Section 6.8.7. In contrast, the NPCS model, along with other number-conserving models, does conserve the multishell seniority of a pairing Hamiltonian. It does so by taking advantage of the precise separation of the degrees of freedom in the pairing models, discussed in Section 6.11.1, which makes it possible to treat the determination of the intrinsic wave functions and the complementary coherent-state wave functions for the paired particles independently.[107] As we now show,

---

[107]One can think of the intrinsic states in the pair-coupling models as *parent* states which give rise to sequences of states by the addition of $J = 0$ coupled nucleon pairs. The challenge of separating the problems of determining the intrinsic and pair wave functions has motivated a large body of research which includes the so-called *quantized Bogolyubov-Valatin* approach, initiated by Suzuki T. and Matsuyanagi K. (1976), *Prog. Theor. Phys.* **56**, 1156, and methods which attempt to describe the intrinsic states and pair-coupled states by means of quasi-fermion and quasi-boson algebras, respectively; cf., for example, Geyer H.B. and Hahne F.J.W. (1980), *Phys. Lett.* **B90**, 6; and the review article of Klein A. and Marshalek E.R. (1991), *Rev. Mod. Phys.* **63**, 375.

the addition of more general multishell seniority-conserving interactions adds no complications to the NPCS model

The pairing Hamiltonian (6.379) can be expressed as a sum of two components

$$\hat{H} = \hat{H}_{\mathrm{int}} + \hat{H}_{\mathrm{P}} :$$ (6.445)

an intrinsic component,

$$\hat{H}_{\mathrm{int}} = \sum_i \varepsilon_j \hat{n}_j,$$ (6.446)

that acts only on parent (i.e., intrinsic) states, and a Hamiltonian for the pairing degrees of freedom,

$$\hat{H}_{\mathrm{P}} = \sum_i 2\varepsilon_j \hat{p}_j - G\hat{S}_+\hat{S}_-.$$ (6.447)

More realistic Hamiltonians, that also conserve multishell seniority, are given by an intrinsic Hamiltonian of the form

$$\hat{H}_{\mathrm{int}} = \sum_j \varepsilon_j \hat{v}_j + \sum_j \hat{V}_j,$$ (6.448)

where $\hat{V}_j$ is a $\mathrm{USp}(2j+1)$-scalar two-body interaction as discussed in Section 6.5.7.

(i) *Lowest-energy states of arbitrary seniority in the NPCS approximation*

Let $\{|v\rho JM\rangle\}$, with energies $\{E^s_{v\rho J}\}$, denote a set of eigenstates of a particular intrinsic Hamiltonian, $\hat{H}_{\mathrm{int}}$. As shown in Section 6.11.4, seniority v coherent states, analogous to those for the BCS vacuum state, are defined by $|xv\rho JM\rangle := e^{\hat{S}_+(x)}|v\rho JM\rangle$ and have coherent-state wave functions given by

$$\Psi_{v\rho JM}(x, z) = \xi_{v\rho JM}\Phi^{s(v)}(zx),$$ (6.449)

where $s_j(v) = s_j - \frac{1}{2}v_j$. In the NPCS approximation the lowest-energy state of any given nucleus, with $|v\rho JM\rangle$ as its intrinsic state, is approximated by a number-projected coherent state of the (unnormalised) form $|p(x)v\rho JM\rangle := \frac{1}{p!}[\hat{S}_+(x)]^p|v\rho JM\rangle$. Such states have coherent-state wave functions given by

$$\Psi_{pv\rho JM}(x, z) = \xi_{v\rho JM}\Phi^{s(v)}_p(zx).$$ (6.450)

The state $|p(x)v\rho JM\rangle$, defined above, has energy expectation value given by

$$E_{pv\rho J}(x) = E_{v\rho J} + E^{s(v)}_{p0}(x),$$ (6.451)

where, for the simple Hamiltonian $\hat{H}_{\mathrm{int}}$ of Equation (6.446),

$$E_{v\rho J} = \sum_i \varepsilon_i v_i.$$ (6.452)

Thus, the energies, $\{E_{pv\rho J}(x)\}$, can be minimised independently for each v as functions of the x variables by use of the same routine as used to determine the approximate seniority-zero ground state. They are all orthogonal to one another by virtue of the fact that their intrinsic states are orthogonal.

Equation (6.451) exhibits the fact that the energies of non-zero seniority states in the NPCS model have essentially the same dependence on the number of pairs, p, as the seniority-zero ground states. Thus, for example, the energy gap between the ground state and first-excited seniority-two state is expected to be similar to that for the $n = 2$ nucleus. For the Hamiltonian (6.379), a seniority-two state $[a_j^\dagger \otimes a_j^\dagger]_{JM}$, with $J \neq 0$ has energy $2\varepsilon_j$. Nevertheless, for non-zero G, an energy gap is predicted between the energy of this state and that of the one-pair, seniority-zero, ground state, whose energy is lowered by the pairing interaction. The emergence of an energy gap in this way is clearly visible in the spectra for the exactly solvable single-j-shell pairing Hamiltonian shown in Figure 6.3.

(ii) *Broken-pair excited states*

Excited states of a given multishell seniority can also be calculated within the NPCS model. For example, within the space spanned by seniority-zero states of the $(2p + 2)$-particle nucleus, one may seek states of the form $\sum_j c_j \hat{S}_+^j |p(x)\rangle$ by diagonalisation of the matrix

$$\langle x|\hat{S}_-^i \,\hat{H}\hat{S}_+^j |p(x)\rangle = \left[\hat{\Pi}_p \langle z|\hat{S}_-^i \,\hat{H}\hat{S}_+^j |x\rangle\right]_{z=x}. \tag{6.453}$$

Because

$$\langle x|\hat{S}_-^i \,\hat{H}\hat{S}_+^j |x\rangle = \left[\frac{\partial}{\partial z_i}\frac{\partial}{\partial x_j}\langle 0|e^{\hat{S}_-(z)}\hat{H}e^{\hat{S}_+(x)}|0\rangle\right]_{z=x}. \tag{6.454}$$

the required matrix elements are given simply by

$$\langle x|\hat{S}_-^i \,\hat{H}\hat{S}_+^j |p(x)\rangle = \hat{\Pi}_p\left[\frac{\partial}{\partial z_i}\frac{\partial}{\partial x_j}\langle z|\hat{H}|x\rangle\right]_{z=x}, \tag{6.455}$$

$$\langle x|\hat{S}_-^i \,\hat{S}_+^j |p(x)\rangle = \hat{\Pi}_p\left[\frac{\partial}{\partial z_i}\frac{\partial}{\partial x_j}\Phi^s(zx)\right]_{z=x}. \tag{6.456}$$

Note that, if the values of the $\{x_i\}$ variables in the above equations are chosen to be those for which the energy $\langle x|\hat{H}|p+1(x)\rangle$ is minimised, then the state $|p+1(x)\rangle$ will be expressible as

$$|p + 1(x)\rangle = \frac{1}{p+1}\hat{S}_+(x)|p(x)\rangle. \tag{6.457}$$

From the energy-minimisation variational principle, it then follows that the lowest-energy eigenstate obtained by diagonalisation of the matrix $\{\langle x|\hat{S}_-^i \,\hat{H}\hat{S}_+^j |p(x)\rangle\}$ will be the NPCS ground state $|p + 1(x)\rangle$. Thus, by the above diagonalisation, one obtains a set of orthogonal states. In the language of the broken-pair model,[108] the state $|p(x)\rangle$ is described as a *zero broken-pair state*. Conversely, the orthogonal linear combinations of the states $\{\hat{S}^i_+|p(x)\rangle; i = 1, \ldots, N\}$ are described as one broken-pair states.

States of two or more broken pairs can be derived in a similar way.

6.11.7 *Application to a two-level model*

For a modest number of j shells, the results of the BCS and NPCS pairing models can be compared with precisely computed results obtained by numerical diagonalisation. We consider a two-level model with $j_1 = j_2$ and a particle number such that the lower level is filled and the upper level is empty when $G = 0$. Figure 6.27 shows the occupancy ratios of the upper to lower level, for $j = 7/2$ and $j = 13/2$, as a function of G/G_{crit}, where G_{crit} is the critical value of G at which a phase transition occurs in BCS theory for $j \to \infty$. The figure shows that the BCS approximation

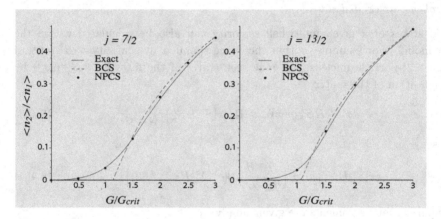

Figure 6.27: The ratio of the mean number of particles in the upper level to the mean number in the lower level for a two-level pairing Hamiltonian obtained accurately by numerical diagonalisation, and in the BCS and NPCS approximations. (Based on figures from Rowe D.J. (2001), *Nucl. Phys.* **A691**, 691.)

would indicate a sharp phase transition from one in which all the particles occupy the lower level to a *superconducting* phase in which the particles occupy both levels. Note, however, that the onset of the superconducting phase takes place continuously, although not smoothly, in the BCS approximation. Thus, the BCS model predicts a second-order phase transition. In contrast, the accurate results obtained by diagonalisation show a smooth transition from one phase to the other when the particle number is finite. Moreover, the figure shows the results of the NPCS model to be accurate to the level of precision shown in the figure.

The level of accuracy of the NPCS model can be seen more clearly in Figure 6.28,

---

[108] Allaart K. *et al.* (1988), *Phys. Repts.* **169**, 209.

which shows amplitudes for the ground and first-excited $J = 0$ states for $j = {}^{13}\!/_2$ and for two values of G. The amplitudes shown are the coefficients of the states in

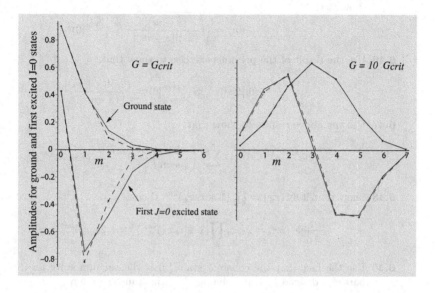

Figure 6.28: The C_m amplitudes in the expansion, $\sum_m C_m|m\rangle$, of the ground and first-excited $J = 0$ states of a two-level model with $j = {}^{13}\!/_2$; m denotes the number of $J = 0$ nucleon pairs in the upper level. The numerically precise results are shown as continuous lines and the NPCS results as dashed lines. (Based on a figure from Rowe D.J. (2001), *Nucl. Phys.* **A691**, 691.)

an expansion, $\sum_m C_m|m\rangle$, where m denotes the number of $J = 0$ nucleon pairs in the upper level. It is now seen that, even for $G = G_{\rm crit}$ (the value of G for which the BCS approximation is at its worst), the NPCS amplitudes are remarkably accurate. For larger and smaller values of G, the results are even more accurate.

An interesting observation is that, for large j and $G > G_{\rm crit}$, the C_n coefficients vary smoothly with n and acquire the shapes of harmonic-oscillator wave functions.[109]

---

[109]This was shown for the ground-state coefficients by Chen *et al.*, *op. cit.* Footnote 91 on Page 491.

Exercises

6.45 By iteration of the identity

$$\hat{S}_+|sm\rangle = \sqrt{(s+m)(s-m+1)}\,|s,m+1\rangle, \qquad (6.458)$$

show that

$$|sm\rangle = \sqrt{\frac{(s-m)!}{(2s)!(s+m)!}}\,[\hat{S}_+]^{s+m}|0\rangle. \qquad (6.459)$$

6.46 Use the result of the previous exercise to show that

$$\langle 0|[\hat{S}_-]^{s+m}[\hat{S}_+]^{s+m}|0\rangle = \frac{(2s)!(s+m)!}{(s-m)!}. \qquad (6.460)$$

6.47 Use the above results to show that

$$\langle 0|e^{z\hat{S}_-}e^{x\hat{S}_+}|0\rangle = \sum_{m=-s}^{s}\binom{2s}{s+m}(xz)^{s+m} = (1+zx)^{2s}. \qquad (6.461)$$

6.48 Show that if $\Phi^s(zx) = \prod_k(1+z_kx_k)^{2s_k}$ then

$$\frac{\partial}{\partial z_i}\Phi^s(zx) = 2s_ix_i\prod_k(1+z_kx_k)^{2s_i-\delta_{ki}} = 2s_ix_i\Phi^{s(i)}(zx). \qquad (6.462)$$

6.49 Use the fact that the coherent state $|x\rangle$ is the vacuum state for a set of quasi-particles defined by a Bogolyubov-Valatin transformation

$$\alpha_{jm}^\dagger = u_ja_{jm}^\dagger + v_ja_{jm}, \quad \alpha_{jm} = u_ja_{jm} - v_ja_{jm}^\dagger, \qquad (6.463)$$

with $x_j = v_j/u_j$ and $u_j^2 + v_j^2$, to show that

$$u_j^2 = \frac{1}{1+x_j^2}, \quad v_j^2 = \frac{x_j^2}{1+x_j^2}, \quad u_jv_j = \frac{x_j}{1+x_j^2}. \qquad (6.464)$$

6.50 Use the results of the previous exercise to show that

$$a_{jm}^\dagger|x\rangle = u_j\alpha_{jm}^\dagger|x\rangle. \qquad (6.465)$$

6.51 Use the above results to show that

$$\langle x|a_{j_im}a_{j_im}^\dagger|x\rangle = \frac{1}{1+x_i^2}\Phi^s(x^2) = \Phi^{s(i)}(x^2). \qquad (6.466)$$

6.52 By making the change of variables to power sums, show that the recursion relation (6.406) can be expressed in the more convenient form (6.410).

6.53 Confirm that Φ_p^s, given by Equation (6.412), satisfies the recursion relation (6.411).

Chapter 7

Mean-field approximations

7.1 Introduction

The mean-field method is basically an independent-particle approximation in which the interactions between nucleons are replaced by the interactions of each nucleon with a mean field. Prototypes of the mean-field method are given by the Hartree and Hartree-Fock approximations. In this chapter we also consider the Hartree-Fock-Bogolyubov approximation which is important for including pairing correlations and more general mean-field methods.

Recall that the essential strategy of the shell model (see Section 5.1) is to separate the many-nucleon Hamiltonian (with non-singular effective interactions) into two parts

$$\hat{H} := \hat{H}_0 + \hat{V}_{\text{res}}, \tag{7.1}$$

where $\hat{H}_0$ is an independent-particle Hamiltonian and $\hat{V}_{\text{res}} := \hat{H} - \hat{H}_0$ is the residual effective interaction. Hartree-Fock theory provides a way to define the Hamiltonian $\hat{H}_0$ as a first step in a many-body theory of nuclear ground-state binding energies in which the correlation energy due to the residual interaction is calculated in perturbation theory.[1] Another use (considered in the following chapter) is to obtain a semi-classical perspective of collective dynamics in nuclei.

In this chapter, we show that when Hartree-Fock theory is applied to open-shell nuclei, for example, it frequently generates nuclear states that break rotational invariance and appear in energy-degenerate sets with different rotational orientations. Thus, Hartree-Fock theory gives insightful indications of how deformed rotational states emerge in nuclei and the magnitudes of the deformations to be expected. When extended to include pairing fields in Hartree-Fock-Bogolyubov theory, there are other broken-symmetry solutions: some deformed and rotational, some spherical and superconducting, and others that are both deformed and superconducting.

---

[1] The many-body perturbation methods used for such purposes have been reviewed, for example, in the books of Brown G.E. (1967), *Unified Theory of Nuclear Models and Forces* (North-Holland, Amsterdam); Irvine J.M. (1972), *Nuclear Structure Theory* (Pergamon Press, Oxford); and Ring P. and Schuck P. (1980), *The Nuclear Many-Body Problem* (Springer-Verlag, New York).

These properties are exploited in the so-called *unified model*, to be discussed in Volume 2, to construct collective models with microscopic wave functions.

In contrast to most models of nuclear structure, whose states span Hilbert spaces, the states of mean-field models lie on geometrical surfaces. Both kinds of spaces have group theoretical properties. A typical shell-model Hilbert space carries a unitary irrep of a group G of one-body unitary transformations, whereas the geometrical space of the corresponding Hartree-Fock approximation is an *orbit* of this group. For example, the Hilbert space for an SO(3) irrep is a complex vector space whereas an SO(3) orbit within this Hilbert space might be the real two-dimensional surface of a sphere generated by rotating a normalised state-vector in the Hilbert space through all possible angles. Thus, whereas most of the models we consider have algebraic structures, the mean-field models have geometrical structures.

The states of a group orbit in a Hilbert space are known as *coherent states* of the corresponding group and the orbit itself is a *manifold of coherent states*. Such coherent-state manifolds have many remarkable properties and have been well studied. In particular, the coherent-state orbits of interest are known to have the properties of classical phase spaces and support classical (Poisson bracket) Hamiltonian dynamics. This is important for the theory of quantization. As a result, significant classical insights are gained from mean-field approximations, which are often described as *semi-classical* approximations.

This chapter describes the Hartree-Fock and Hartree-Fock-Bogolyubov self-consistent mean-field approximations as constrained quantum mechanics. i.e., with the time-evolution of states, according to the time-dependent Schödinger equation, constrained to remain coherent states at all times. Chapter 8 shows that constrained quantum mechanics defines maps from a quantum mechanical to a classical description of a many-particle system; i.e., a unitary irrep of a Lie algebra of one-body operators is mapped by Hartree-Fock theory to a classical representations with Poisson brackets replacing the commutator brackets of quantum mechanics. This provides a rationale for a classical interpretation of the results of a self-consistent mean-field calculation and the need for a so-called requantization of the subdynamics of interest. Quantization of a system is achieved by construction of a unitary representation of a spectrum-generating algebra or dynamical group for the system. A particularly powerful technique for the construction of such irreps is given by vector-coherent-state theory.[2] Thus, coherent-state theory provides a unifying framework both for quantizing a classical system and for dequantizing a quantal system.

---

[2] The vector-coherent-state method for constructing unitary irreps was introduced in the papers: Rowe D.J. (1984), *J. Math. Phys.* **25**, 2662; Rowe D.J., Rosensteel G. and Carr R. (1984), *J. Phys. A: Math. Gen.* **17**, L399; Rowe D.J., Rosensteel G. and Gilmore R. (1985), *J. Math. Phys.* **26**, 2787. A simple review of its early applications was given by Hecht K.T. (1987), *The Vector Coherent State Method and Its Application to Problems of Higher Symmetries*, Lecture Notes in Physics, Vol. 290 (Springer-Verlag, Berlin). A more mathematical review was given by Rowe D.J. and Repka J. (1991), *J. Math. Phys.* **32**, 2614.

7.2 The Hartree mean-field approximation

A physically intuitive introduction to mean-field theory, is given by the Hartree approximation for an independent-particle model (IPM) Hamiltonian, $\hat{H}_0$, for a system of atomic electrons.[3]

Consider the Hamiltonian

$$\hat{H} := \sum_i^Z \{\hat{T}_i + u_0(r_i)\} + \frac{1}{2} \sum_{ij}^Z v(\mathbf{r}_i, \mathbf{r}_j), \tag{7.2}$$

for the electrons of an atom, where $\hat{T}_i$ is the kinetic energy of the i'th electron, u_0 is the central Coulomb potential of the nucleus, and $v(\mathbf{r}_i, \mathbf{r}_j)$ is the interaction potential between two electrons at $\mathbf{r}_i$ and $\mathbf{r}_j$, respectively. With neglect of the interactions, the electrons are independent and have wave functions that satisfy the wave equation

$$\left\{-\frac{\hbar^2}{2M} \nabla^2 + u_0(r)\right\} \psi_i(\mathbf{r}) = \varepsilon_i \psi_i(\mathbf{r}). \tag{7.3}$$

With the inclusion of interactions, it may be supposed that each electron moves in an additional Coulomb potential due to all the other electrons. For electron i, this additional potential is

$$u_1(\mathbf{r}_i) := \sum_{j \neq i}^Z \int v(\mathbf{r}_i, \mathbf{r}_j) |\psi_j(\mathbf{r}_j)|^2 \, d^3\mathbf{r}_j. \tag{7.4}$$

The single-electron wave equation then becomes

$$\left\{-\frac{\hbar^2}{2M} \nabla_i^2 + u_0(r_i) + u_1(\mathbf{r}_i)\right\} \psi_i(\mathbf{r}_i) = \varepsilon_i \psi_i(\mathbf{r}_i). \tag{7.5}$$

The product of the lowest-energy solutions of this equation gives the Hartree many-electron wave function,

$$\Phi(\mathbf{r}_1, \mathbf{r}_2, \ldots, \mathbf{r}_Z) = \psi_1(\mathbf{r}_1)\psi_2(\mathbf{r}_2) \ldots \psi_Z(\mathbf{r}_Z), \tag{7.6}$$

as an eigenfunction of the independent-particle Hamiltonian

$$\hat{H}_0 = \sum_i \left\{-\frac{\hbar^2}{2M} \nabla_i^2 + u_0(r_i) + u_1(\mathbf{r}_i)\right\}, \tag{7.7}$$

with energy eigenvalue $E = \sum_i \varepsilon_i$. Note, however, that electrons are spin-$1/2$ fermions and obey the Pauli principle. This is taken into account by assigning at most one spin-up and one spin-down electron to each single-particle spatial state. Note also that, because the potential u_1 depends on the wave functions and the wave functions depend on u_1, the Hartree equation must be solved iteratively.

---

[3]Hartree D.R. (1928), *Proc. Camb. Phil. Soc.* **24**, 89.

In contrast to the situations for atomic electrons, the only central field for a nucleon in the nucleus is the field generated by all the other nucleons. Thus, the Hartree Hamiltonian for a nucleus is of the form

$$\hat{H}_0 = \sum_i^A \left\{ -\frac{\hbar^2}{2M} \nabla_i^2 + u(\mathbf{r}_i) \right\} , \tag{7.8}$$

with

$$u(\mathbf{r}_i) = \sum_{j \neq i}^A \int v(\mathbf{r}_i, \mathbf{r}_j) \, |\psi_j(\mathbf{r}_j)|^2 \, \mathrm{d}^3 \mathbf{r}_j . \tag{7.9}$$

Again, this is a self-consistent field equation because the potential u depends on the nucleon wave functions and the wave functions depend on u. However, it is known that, provided one starts with a reasonable first guess for u and provided the interactions are predominantly attractive and do not have singular short-range behaviour, the iterative solution of the Hartree equation converges.[4] Problems would arise if an attempt were made to perform a Hartree (or any other mean-field) calculation with interactions determined from nucleon-nucleon scattering experiments, which are strongly repulsive at short range. For such interactions, the mean-field equations are unlikely to converge. Thus, for application of mean-field theory, it is necessary to replace the so-called *bare* nucleon-nucleon interactions with non-singular effective interactions derived by the methods of Section 5.7.

In addition to the fact that Hartree theory neglects the correlations between particles, other than those generated by the mean field, its major deficiency is that it does not take account of the indistinguishability of like particles. Thus, it neglects the so-called *exchange forces*. A definition of single-particle states that takes exchange forces into account is provided by the following generalised mean-field equations.

Exercise

7.1 Show that although the Schrödinger equation for the Hartree Hamiltonian of Equation (7.7) can be solved by repeated solutions of linearly independent-particle equations until self-consistency is reached, it is fundamentally a non-linear many-particle equation.

7.3 Mean-field equations for indistinguishable fermions

To take the indistinguishability of nucleons into account it is convenient to use the language of second quantization. Recall that the fermion annihilation operators, $\{a^{jm}\}$, are the Hermitian adjoints of the fermion creation operators, $\{a_{jm}^\dagger\}$, and are labelled by upper indices to indicate that they transform contragrediently to

---

[4]Rowe D.J. and Rosensteel G. (1976), *Int. J. Theor. Phys.* **15**, 501.

the creation operators with lower indices. As discussed in Section 5.3, they are related by the definition

$$a^{jm} := (-1)^{j+m} a_{j,-m} \tag{7.10}$$

to annihilation operators, $\{a_{jm}\}$, which transform covariantly as the standard components of spherical tensors. With extension to the Hartree-Fock-Bogolyubov approximation in mind, it is also convenient to define operators that create and annihilate fermions in time-reversed single-particle states, as in the BCS approximation, by

$$a^\dagger_{j\bar{m}} := (-1)^{j+m} a^\dagger_{j,-m}, \quad a_{j\bar{m}} := (-1)^{j+m} a_{j,-m} = a^{jm}. \tag{7.11}$$

The combinations

$$\hat{n}_j = \sum_m a^\dagger_{jm} a^{jm} = \sum_m a^\dagger_{jm} a_{j\bar{m}} \tag{7.12}$$

are then (rotationally invariant) scalars. With these conventions, the standard fermion anticommutation relations are given by

$$\{a_{\bar{\mu}}, a^\dagger_\nu\} = \delta_{\mu,\nu}, \quad \{a_\mu, a^\dagger_{\bar{\nu}}\} = -\delta_{\mu,\nu}. \tag{7.13}$$

An IPM Hamiltonian $\hat{H}_0$ is a one-body operator and is expressible, in the language of second quantization (Section 5.2), in the form

$$\hat{H}_0 = \sum_{\mu\nu} \varepsilon_{\mu\nu} a^\dagger_\mu a_{\bar{\nu}}. \tag{7.14}$$

In contrast, a general nuclear Hamiltonian with non-singular two-body interactions is a sum of one- and two-body terms of the form

$$\hat{H} = \sum_\nu T_{\mu\nu} a^\dagger_\mu a_{\bar{\nu}} + \tfrac{1}{4} \sum_{\mu\mu'\nu\nu'} V_{\mu\mu'\nu\nu'} a^\dagger_\mu a^\dagger_{\mu'} a_{\bar{\nu}'} a_{\bar{\nu}}. \tag{7.15}$$

A mean-field approximation can be viewed as a map from a two-body Hamiltonian to a one-body Hamiltonian in which $\hat{H} \to \hat{H}_0$.

Observe that the single-particle matrix elements $\varepsilon_{\mu\nu}$ of the IPM Hamiltonian $\hat{H}_0$ are equal to the numerical values of the combined commutator-anti-commutator

$$\varepsilon_{\mu\nu} = \{a_{\bar{\mu}}, [\hat{H}_0, a^\dagger_\nu]\}. \tag{7.16}$$

Parallel commutator-anti-commutators for the Hamiltonian $\hat{H}$ give

$$\{a_{\bar{\mu}}, [\hat{H}, a^\dagger_\nu]\} = T_{\mu\nu} + \sum_{\mu'\nu'} V_{\mu\mu'\nu\nu'} a^\dagger_{\mu'} a_{\bar{\nu}'}. \tag{7.17}$$

The second term on the right-hand side of this equation is not a number, as required for an IPM Hamiltonian. However, its expectation value with respect to some

517

nuclear state $|\Psi\rangle$ is a number. Thus, by specification of a state $|\Psi\rangle$, the Hamiltonian H is mapped to an IPM Hamiltonian $\hat{H}_0 = \sum_{\mu\nu} \varepsilon_{\mu\nu} a_\mu^\dagger a_{\bar{\nu}}$ with

$$\varepsilon_{\mu\nu} = \langle\Psi|\{a_{\bar{\mu}}, [\hat{H}, a_\nu^\dagger]\}|\Psi\rangle = T_{\mu\nu} + \sum_{\mu'\nu'} V_{\mu\mu'\nu\nu'}\langle\Psi|a_{\mu'}^\dagger a_{\bar{\nu}'}|\Psi\rangle. \qquad (7.18)$$

Such a use of commutators and anti-commutators to pick out a component of a many-body operator is an application of a general equations-of-motion method.[5,6]

The physical content of the $\hat{H} \to \hat{H}_0$ mapping can be understood by writing $\hat{H}_0 = \hat{T} + \hat{v}$ with $\hat{v} = \sum_{\mu\nu} v_{\mu\nu} a_\mu^\dagger a_{\bar{\nu}}$ and

$$v_{\mu\nu} = \sum_{\mu'\nu'} V_{\mu\mu'\nu\nu'}\langle\Psi|a_{\mu'}^\dagger a_{\bar{\nu}'}|\Psi\rangle. \qquad (7.19)$$

This shows $\hat{v}$ to be the one-body potential generated by the interaction of a nucleon with all the other nucleons in the state $|\Psi\rangle$. Thus, $\hat{H}_0$ is a *mean-field* Hamiltonian.

The matrix ε can be brought to diagonal form,

$$\varepsilon_{\mu\nu} = \delta_{\mu\nu}\varepsilon_\nu, \qquad (7.20)$$

by a one-body unitary transformation to give the energies, $\{\varepsilon_\nu\}$, of a set of single-particle wave functions with a direct physical interpretation. Suppose, for example, that $|\Psi_0\rangle$ were the $J = 0$ ground state for an (even) A-particle nucleus of energy E_0. Then expanding the commutator-anti-commutator and inserting complete sets of states in appropriate places for the $A \pm 1$ particle nuclei gives

$$\begin{aligned}\varepsilon_\nu &= \langle\Psi_0|\{a_{\bar{\nu}}, [\hat{H}, a_\nu^\dagger]\}|\Psi_0\rangle \\ &= \sum_\alpha (E_\alpha^+ - E_0)|\langle\Psi_\alpha|a_\nu^\dagger|\Psi_0\rangle|^2 - \sum_\beta (E_\beta^- - E_0)|\langle\Psi_\beta|a_{\bar{\nu}}|\Psi_0\rangle|^2, \qquad (7.21)\end{aligned}$$

where α indexes states of the $A+1$ particle nucleus of energy E_α^+, β indexes states of the $A-1$ particle nucleus of energy E_β^-, and $E_0 = \langle\Psi_0|\hat{H}|\Psi_0\rangle$. The squared overlap $|\langle\Psi_\alpha|a_\nu^\dagger|\Psi_0\rangle|^2$ expresses the extent to which the state $|\Psi_\alpha\rangle$ is formed by adding a nucleon in the single-particle state ν to the A-particle ground state. Likewise $|\langle\Psi_\beta|a_{\bar{\nu}}|\Psi_0\rangle|^2$ expresses the extent to which the state $|\Psi_\beta\rangle$ is formed by removing a nucleon in the single-particle state ν from the A-particle ground state.[7] These are the quantities (spectroscopic factors) measured in stripping and pick-up reactions (cf. Sections 1.3.2 and 6.9.5). Thus, Equation (7.21) shows that the mean-field single-particle energies are energy-weighted centroids of combined pick-up and stripping spectroscopic factors.

It will be noted that, while the mean-field equations define single-particle states and their energies in a physically meaningful way in terms of the single-particle

---

[5]Rowe D.J. (1968), *Rev. Mod. Phys.* **40**, 153.

[6]Rowe, *op. cit.* Footnote 4 on Page x.

[7]A simpler relationship between independent-particle energies and single-particle removal energies is known as Koopman's theorem; Koopman T.A. (1933), *Physica* **1**, 104.

occupancies of the ground state of the nucleus, they do not automatically define many-particle ground states. However, this can be done in the Hartree-Fock *self-consistent field* approximation, in which the nuclear ground state $|\Psi_0\rangle$ is approximated by a Slater determinant.

Exercise

7.2 Give a definition of mean-field single-particle states for a nucleus according to the above prescription when there are three-body interactions and show that it involves both the one-body and two-body densities of the nuclear ground state.

7.4 Manifolds of Slater determinants

The many-nucleon wave functions of Hartree-Fock theory are antisymmetrised products of single-particle wave functions, i.e., Slater determinants, as introduced in Section 5.2.

Suppose the Hilbert space $\mathbb{H}^{(1)}$ of active single-particle wave functions is of dimension d and has an orthonormal basis $\mathcal{S} = \{\psi_\nu\}$.[8] As discussed in Section 5.2, a selected set of A basis wave functions in $\mathcal{S}$ defines a normalised Slater determinant and the set of all such determinants defines an orthonormal basis for the A-nucleon Hilbert space $\mathbb{H}^{(A)}$. However, there is a continuous infinity of orthonormal basis sets for $\mathbb{H}^{(1)}$ and a corresponding infinite number for $\mathbb{H}^{(A)}$. In Hartree-Fock theory, we consider the continuous set of all possible (normalised) A-nucleon Slater determinants although, in practice, one may choose to restrict to subsets (e.g., subsets having the desired number of neutrons and protons and restricted bases of single-particle wave functions).

It is shown in this section that the set of Slater determinants of A mutually orthonormal single-particle states can be viewed as a hypersurface in the A-nucleon Hilbert space. It is, in fact, a submanifold of the Hilbert space, $\mathcal{M}^{(A)} \subset \mathbb{H}^{(A)}$ and, as such, has many useful symmetries and geometrical properties. A closely-related manifold, designated $\overline{\mathcal{M}}^{(A)}$, which we now define, is of even more importance.

Recall that two states in a Hilbert space define the same physical state if they differ only by a phase factor. In particular, any linear operator $\hat{X}$ on a Hilbert space has the same expectation value $\langle\Psi|\hat{X}|\Psi\rangle = \langle\Psi'|\hat{X}|\Psi'\rangle$ with respect to any two state vectors $|\Psi\rangle$ and $|\Psi'\rangle$ that differ only by a phase factor. Thus, for many purposes it is useful to disregard the phase of a state. This is conveniently done when a physical state, $|\Psi\rangle$, (here assumed to be normalised) is characterised by a

---

[8]The Hilbert space of all single-nucleon wave functions is, a priori, of infinite dimension. However, in the shell model, it is projected to a possibly very large but nevertheless finite-dimensional *effective* subspace. This makes it possible to avoid the many subtleties associated with infinite-dimensional spaces. It is also common practice in applications of the shell-model and mean-field approximations, to restrict the calculations to relatively small spaces of so-called active single-particle states.

so-called *density*, ρ_Ψ, defined such that the expectation value of a linear operator $\hat{X}$ in the state with density ρ_Ψ is given by

$$\rho_\Psi(\hat{X}) := \langle \Psi | \hat{X} | \Psi \rangle. \qquad (7.22)$$

In density matrix theory as used, for example, in statistical mechanics,[9] such densities are often expressed as operators in the form

$$\rho_\Psi := | \Psi \rangle \langle \Psi | \qquad (7.23)$$

and the expectation value, $\rho_\Psi(\hat{X})$, is defined as the trace, $\mathrm{tr}(\rho_\Psi \hat{X})$. Thus, if $\{|\Psi_i\rangle\}$ is an orthonormal basis for the many-particle Hilbert space, we obtain

$$\rho_\Psi(\hat{X}) = \sum_i \langle \Psi_i | \Psi \rangle \langle \Psi | \hat{X} | \Psi_i \rangle = \sum_i \langle \Psi | \hat{X} | \Psi_i \rangle \langle \Psi_i | \Psi \rangle, \qquad (7.24)$$

consistent with Equation (7.22).

The densities for (normalised) state vectors in a Hilbert space, $\mathbb{H}$, are in one-to-one correspondence with the states of a *projective Hilbert space*,[10]

$$P\mathbb{H} \simeq \big\{ \rho_\Psi \,\big|\, |\Psi\rangle \in \mathbb{H}, \langle \Psi | \Psi \rangle = 1 \big\}. \qquad (7.25)$$

For the above reasons, it will useful, in discussing the properties of a manifold of Slater determinants, $\mathcal{M}^{(A)}$, to discuss also the properties of the associated manifold of densities

$$\overline{\mathcal{M}}^{(A)} := \big\{ \rho_\phi \,\big|\, |\phi\rangle \in \mathcal{M}^{(A)} \big\}. \qquad (7.26)$$

7.4.1   Slater determinants as particle-hole vacuum states

A normalised A-nucleon Slater determinant can expressed in the language of second quantization as

$$|\phi\rangle = \prod_{i=1}^{A} a_i^\dagger |0\rangle, \qquad (7.27)$$

where $\{a_i^\dagger; i = 1, \dots, A\}$ is a selected orthonormal set of A single-nucleon creation operators and $|0\rangle$ is the zero-particle vacuum state. The label i then indexes a set of single-particle states that are occupied in the many-nucleon state $|\phi\rangle$. Similarly, $\{a_m^\dagger, m = A+1, \dots\}$ denotes a complementary set of creation operators for unoccupied single-particle states which, together with the $\{a_i^\dagger\}$ operators, are the creation operators for a complete orthonormal basis of single-particle states.

---

[9]Feynman R.P. (1972), *Statistical Mechanics* (Benjamin, Reading, Mass.).

[10]A projective Hilbert space $P\mathbb{H}$ is the set of all normalised states of a Hilbert space $\mathbb{H}$ together with an equivalence relationship that identifies all states that differ only by a phase factor, i.e., $|\Psi\rangle \equiv e^{i\theta}|\Psi\rangle$. The defining equation, (7.25), is to be read: $P\mathbb{H}$ is isomorphic to the set of densities $\{\rho_\Psi\}$ subject to the constraint that $|\Psi\rangle$ is a normalised state in the Hilbert space $\mathbb{H}$.

A so-called *particle-hole* state is now created by promoting a nucleon from an occupied i state to an unoccupied m state, i.e.,

$$|mi^{-1}\rangle = a_m^\dagger a_{\bar{\imath}}|\phi\rangle. \tag{7.28}$$

Similarly, a two-particle-two-hole state is of the form

$$|mni^{-1}j^{-1}\rangle = a_m^\dagger a_n^\dagger a_{\bar{\imath}} a_{\bar{\jmath}}|\phi\rangle. \tag{7.29}$$

Thus, the state $|\phi\rangle$ is naturally regarded as a zero-particle-zero-hole state, i.e., a *particle-hole vacuum* state. Equivalently, one can define quasi-particle operators,

$$\alpha_m^\dagger := a_m^\dagger, \quad \alpha_{\bar{\imath}}^\dagger := a_{\bar{\imath}}, \tag{7.30}$$

as creation operators for *particles* and *holes*, respectively. The determinantal state $|\phi\rangle$ is then the vacuum state relative to these quasi-particle operators, i.e., it satisfies the equations

$$\alpha_{\bar{\imath}}|\phi\rangle = 0, \qquad \alpha_{\bar{m}}|\phi\rangle = 0. \tag{7.31}$$

Because of these relationships of states to the particle-hole vacuum, it is often convenient to describe occupied and unoccupied single-particle states as *hole states* and *particle states*, respectively.

7.4.2 *Slater determinants as coherent states*

Any orthonormal basis of single-nucleon wave functions for $\mathbb{H}^{(1)}$ can be transformed into any other by some element of the group $U(d)$ of one-body unitary transformations of $\mathbb{H}^{(1)}$, where d is the dimension of $\mathbb{H}^{(1)}$.[11] Thus, any normalised A-nucleon Slater determinant can be transformed into any other by an element of $U(d)$. This means that the group $U(d)$ is able to generate the whole manifold of normalised Slater determinants, $\mathcal{M}^{(A)}$, from any initial determinant. The group $U(d)$ is then said to act *transitively* on $\mathcal{M}^{(A)}$. In geometrical terms, $\mathcal{M}^{(A)}$ is said to be a *homogeneous space*; in fact, it is a special homogeneous space with many valuable properties.[12] One of its important geometrical property is that the associated manifold of densities, $\overline{\mathcal{M}}^{(A)}$, is a *symplectic* manifold and, consequently, it has the properties of a classical *phase space*. This property is vital for the development of time-dependent Hartree-Fock theory, as shown in the following chapter. In group theoretical terms, both $\mathcal{M}^{(A)}$ and $\overline{\mathcal{M}}^{(A)}$ are *group orbits*. The Slater determinants of $\mathcal{M}^{(A)}$ are also *coherent states* and $\overline{\mathcal{M}}^{(A)}$ is a set of *phase-equivalent coherent states*.

A group orbit is a manifold $\mathcal{M}$ that is generated by a group G of transformations such that an element $g \in G$ transforms any point $m \in \mathcal{M}$ into some other point denoted by $g \cdot m \in \mathcal{M}$. To qualify as being an orbit of the group G, there must also

---

[11]See Footnote 8 on Page 519.

[12]It is also a *symmetric space*, a *Grassmann manifold*, and a *Kähler manifold*.

exist some element $g \in G$ that will transform any point m of the manifold into any other. Thus, a group orbit is a set of points $\{g \cdot m \, ; \, g \in G\}$ that can all be reached by group transformations from any initial point m on the orbit. Recognising that a manifold is an orbit of a group, G, is valuable because it means that its properties are determined by the properties of G. A first step in the characterisation of a group orbit is to associate a point m on the manifold with an *isotropy subgroup* $G_0 \subset G$ (also called a *stability subgroup*), defined as the subset of elements of G that leave the point m invariant, i.e.,

$$G_0 = \{h \in G_0 \, | \, h \cdot m = m\}. \tag{7.32}$$

It follows that any two group elements g and g' transform m into the same point, i.e., $g \cdot m = g' \cdot m$, if and only if $g' = gh$ for some element $h \in G_0$. Thus, the group orbit is identified with the factor space G/G_0.

The concept of dividing a group G by a subgroup $G_0 \subset G$ to form a factor space G/G_0 amounts to putting group elements into packages of elements and regarding the packages as the elements of G/G_0. An element of a factor space G/G_0, called a *coset* gG_0, is then a subset of elements $\{gh \, ; \, h \in G_0\}$ with $g \in G$. Factoring a group by a subgroup can also be viewed as defining an equivalence relation that identifies all group elements as equivalent if they differ only by an element of G_0; they then belong to the same coset. This is appropriate for the definition of a homogenous space $\mathcal{M}$ because, if G_0 is the isotropy subgroup of a point $m \in \mathcal{M}$, then $gh \cdot m = g \cdot m$, for all $h \in G_0$. The point $g \cdot m$ is then in one-to-one correspondence with the coset gG_0.

The isotropy (stability) subgroup of elements of $U(d)$ that leave a Slater determinant invariant, to within a phase factor, is the direct product group $U(A) \times U(d-A)$, where $U(A)$ is the group of unitary transformations of the occupied single-particle states and $U(d - A)$ is the group of unitary transformations of the complementary subset of unoccupied single-particle states. (Note that a transformation of unoccupied states has no effect on a Slater determinant of occupied single-particle wave functions while a unitary transformation of occupied states can change a Slater determinant by at most a phase factor.) The corresponding manifold of densities (phase equivalent coherent states) is then isomorphic to the factor space

$$\overline{\mathcal{M}}^{(A)} \simeq U(d)/(U(A) \times U(d - A)). \tag{7.33}$$

In the context of Hartree-Fock theory, the invariance of a Slater determinant, to within a phase factor, under $U(A) \times U(d-A)$ will be used primarily at the algebraic level. If $\{\psi_i\}$ denotes the wave functions for the occupied single-particle states and $\{\psi_m\}$ denotes the remaining wave functions for the unoccupied single-particle states in an orthonormal basis for $\mathbb{H}^{(1)}$, then the infinitesimal generators of the subgroup $U(A)$ are Hermitian combinations of the operators $\{a_i^\dagger a_{\bar{i}'}\}$ whereas the infinitesimal generators of $U(d - A)$ are Hermitian combinations of the operators $\{a_m^\dagger a_{\bar{m}'}\}$. The infinitesimal generators of U(d) include, in addition, the Hermitian linear combinations of the particle-hole creation and annihilation operators $\{a_m^\dagger a_{\bar i}, a_i^\dagger a_{\bar m}\}$.

The particle-particle and hole-hole operators either annihilate the Slater determinant of occupied single-particle wave functions or leave it invariant; i.e.,

$$a_m^\dagger a_{\bar m'}|\phi\rangle = 0\,, \quad a_i^\dagger a_{\bar i'}|\phi\rangle = \delta_{i,i'}|\phi\rangle\,. \tag{7.34}$$

Thus, to within a phase factor $e^{i\theta}$, any Slater determinant can be transformed into any other in a common neighbourhood by a unitary transformation of the form

$$|\phi\rangle \to |\phi'\rangle = e^{i\theta}\exp\Big[\sum_{mi}(X_{mi}a_m^\dagger a_{\bar i} - X_{mi}^* a_i^\dagger a_{\bar m})\Big]|\phi\rangle\,; \tag{7.35}$$

i.e., *there is no need to involve the particle-particle and hole-hole operators.* This means that the real and imaginary parts of the $\{X_{mi}\}$ coefficients serve as local coordinates for a neighbourhood of the state $\rho_\phi \in \overline{\mathcal{M}}^{(A)}$, with ρ_ϕ sitting at the coordinate origin.

Exercises

7.3 Starting with the observation that a sphere is generated by rotation of some fixed vector in a 3-dimensional Euclidean space through all possible angles, show that a sphere is an SO(3) group orbit and that it can be identified with the homogeneous space SO(3)/SO(2).

7.4 Show that if $\mathcal{M}$ is a G-orbit and if $G_0 \subset G$ is the isotropy subgroup of a point $m \in \mathcal{M}$, then two elements, g_1 and g_2 of G are in the same coset of G/G_0 if and only if $g_1 \cdot m = g_2 \cdot m$.

7.5 The dimension of a factor space G/G_0, as a manifold, is defined to be the dimension of the space of tangent vectors at a point on the manifold and is given by

$$\dim(G/G_0) = \dim G - \dim G_0\,, \tag{7.36}$$

where the dimension of a Lie group is defined as the number of linearly independent infinitesimal generators of that group. Show that a unitary group U(n) has dimension n^2, and that $\mathcal{M}^{(A)} \sim \mathrm{U}(d)/(\mathrm{U}(A)\times \mathrm{U}(d-A))$ has dimension equal to twice the number of linearly-independent particle-hole creation operators.

7.5 Normal-ordered expansions

For many purposes, e.g., the formulation of Hartree-Fock theory, it is useful to define a so-called *normal ordering* of a sequence of fermion creation and annihilation operators. This is an ordering in which all annihilation operators are put to the right and all creation operators to the left of one another. We denote a normal-ordered product of an arbitrary sequence of fermion creation and annihilation operators $ABC\ldots$ by putting the sequence inside colons, i.e., $:ABC\cdots:$. This means that, if the sequence is not already in normal order it must be rearranged so that it is in normal order and multiplied by a factor ± 1 according as the rearrangement requires an even or an odd permutation of the operators; e.g.,

$$: a^\dagger_\mu a_{\bar\nu} := a^\dagger_\mu a_{\bar\nu}, \quad : a_{\bar\nu} a^\dagger_\mu := -a^\dagger_\mu a_{\bar\nu},$$
$$: a^\dagger_\mu a^\dagger_\nu := a^\dagger_\mu a^\dagger_\nu = -a^\dagger_\nu a^\dagger_\mu, \tag{7.37}$$
$$: a^\dagger_\mu a_{\bar\nu} a^\dagger_\lambda := -a^\dagger_\mu a^\dagger_\lambda a_{\bar\nu}. \tag{7.38}$$

Because of the property that the vacuum expectation of a normal-ordered operator (other than a constant) is always zero, it is useful to expand an arbitrary operator as a sum of normal-ordered operators; the vacuum expectation of the operator is then the constant term in the expansion. For example, the following are normal-ordered expansions:

$$a_{\bar\nu} a^\dagger_\mu = \delta_{\mu\nu} + : a_{\bar\nu} a^\dagger_\mu := \delta_{\mu\nu} - a^\dagger_\mu a_{\bar\nu},$$
$$a_{\bar\mu} a^\dagger_\nu a^\dagger_\lambda = \delta_{\mu\nu} a^\dagger_\lambda - \delta_{\mu\lambda} a^\dagger_\nu + a^\dagger_\nu a^\dagger_\lambda a_{\bar\mu}. \tag{7.39}$$

Normal-ordered products can also be defined relative to vacuum states other than the bare-particle state. As shown above, any Slater determinant $|\phi\rangle$, in which a subset of single-particle states are occupied, can be regarded as a *particle-hole* vacuum or, equivalently, as the vacuum of a set of quasi-particle operators, as defined by Equation (7.31). A normal-ordered expansion is then defined in terms of these quasi-particle operators.

Relative to any fermion-vacuum state $|\phi\rangle$, a pair of fermion operators AB has normal-ordered expansion

$$AB = \langle\phi|AB|\phi\rangle + : AB : . \tag{7.40}$$

A triple of fermion operators has expansion

$$ABC = \langle\phi|AB|\phi\rangle C + \langle\phi|BC|\phi\rangle A - \langle\phi|AC|\phi\rangle B + : ABC :, \tag{7.41}$$

and so on. Note that to put a sequence of operators into normal-order relative to a quasi-particle vacuum, the sequence must first be expanded in terms of the quasi-particle operators. For example, if

$$A = \sum_\mu (u_\mu \alpha^\dagger_\mu + v_\mu \alpha_\mu), \quad B = \sum_\nu (u_\nu \alpha_\nu - v_\nu \alpha^\dagger_\nu), \tag{7.42}$$

then

$$: AB := \sum_{\mu\nu} \left[u_\nu (u_\mu \alpha^\dagger_\mu + v_\mu \alpha_\mu) \alpha_\nu + v_\nu \alpha^\dagger_\nu (u_\mu \alpha^\dagger_\mu + v_\mu \alpha_\mu) \right]. \tag{7.43}$$

Replacing a pair of fermion operators in a product by its vacuum expectation value is referred to as a *contraction*. The rule in making a normal-ordered expansion is that a product of a sequence of fermion operators is equal to the sum of the normal-ordered products of the operators with all possible distinct contractions

524

and with each term in the sum multiplied by ± 1 according as it involves an even or an odd permutation of the operators. For example, the normal-ordered expansion of a two-body operator, relative to a Slater determinant $|\phi\rangle$, is:[13]

$$
\begin{aligned}
a_\mu^\dagger a_\nu^\dagger a_{\bar\nu'} a_{\bar\mu'} = {} & \langle\phi|a_\mu^\dagger a_{\bar\mu'}|\phi\rangle\langle\phi|a_\nu^\dagger a_{\bar\nu'}|\phi\rangle - \langle\phi|a_\mu^\dagger a_{\bar\nu'}|\phi\rangle\langle\phi|a_\nu^\dagger a_{\bar\mu'}|\phi\rangle \\
& + \langle\phi|a_\mu^\dagger a_{\bar\mu'}|\phi\rangle : a_\nu^\dagger a_{\bar\nu'} : + \langle\phi|a_\nu^\dagger a_{\bar\nu'}|\phi\rangle : a_\mu^\dagger a_{\bar\mu'} : \\
& - \langle\phi|a_\mu^\dagger a_{\bar\nu'}|\phi\rangle : a_\nu^\dagger a_{\bar\mu'} : - \langle\phi|a_\nu^\dagger a_{\bar\mu'}|\phi\rangle : a_\mu^\dagger a_{\bar\nu'} : \\
& + : a_\mu^\dagger a_\nu^\dagger a_{\bar\nu'} a_{\bar\mu'} : .
\end{aligned}
\tag{7.44}
$$

Note that, in this expansion, we have not included contractions of the type $\langle\phi|a_\mu^\dagger a_\nu^\dagger|\phi\rangle$ and $\langle\phi|a_{\bar\nu'} a_{\bar\mu'}|\phi\rangle$ because they are identically zero when the vacuum state $|\phi\rangle$ has a well-defined particle number. However, such terms must be included for a generalised quasi-particle vacuum state of the BCS or Hartree-Bogolyubov type (see Sections 6.8.4 and 7.7).

Exercises

7.6 Give the normal-ordered expansion of the operator

$$
a_{\bar\beta} a_{\bar\alpha} a_\mu^\dagger a_\nu^\dagger
\tag{7.45}
$$

relative to the bare particle vacuum.

7.7 Show that, if m and n index particle (unoccupied) states and i and j index hole (occupied) states, then, relative to the particle-hole vacuum,

$$
: a_m^\dagger a_i^\dagger a_{\bar\jmath} a_{\bar n} : = -a_m^\dagger a_{\bar\jmath} a_i^\dagger a_{\bar n} .
\tag{7.46}
$$

7.6 The Hartree-Fock approximation

Hartree-Fock theory can be viewed from several perspectives. In the first instance, it can be seen as a self-consistent mean-field theory. It can also be phrased in terms of a variational principle as an optimal independent-particle model (IPM) approximation. As such, its essential approximation is to neglect the effects of two-body correlations. However, Hartree-Fock theory need not be regarded as an end in itself. It can be viewed as providing the desired separation of a Hamiltonian into an independent-particle part and a residual interaction as required for a shell model calculation in an IPM basis. Even more generally, Hartree-Fock theory can be regarded as a map from the quantum mechanical to a classical description of a many-body system. This latter perspective is developed in Chapter 8.

---

[13]This is a special, simple case of Wick's theorem; Wick G.C. (1950), *Phys. Rev.* **80**, 268.

7.6.1 *Self-consistent field equations*

In the Hartree-Fock approximation[14,15] for a mass-A nucleus, the ground state, $|\Psi_0\rangle$, in the mean-field equations is replaced by a Slater determinant, $|\phi\rangle :=$ $\prod_{i=1}^{A} a_i^\dagger|0\rangle$, of A lowest-energy single-particle states. The corresponding Hartree-Fock IPM Hamiltonian then has matrix elements, defined by the mean-field expression, (7.18),

$$\varepsilon_{\mu\nu} = T_{\mu\nu} + \sum_{\mu'\nu'} V_{\mu\mu'\nu\nu'}\langle\phi|a_{\mu'}^\dagger a_{\bar\nu'}|\phi\rangle = T_{\mu\nu} + \sum_{i=1}^{A} V_{\mu i\nu i}. \tag{7.47}$$

A Hartree-Fock solution is obtained when the single-particle states, that make up the Slater determinant $|\phi\rangle$ and define the mean-field Hamiltonian $\hat{H}_0 = \sum_{\mu\nu}\varepsilon_{\mu\nu}a_\mu^\dagger a_{\bar\nu}$ generated by $|\phi\rangle$, diagonalise $\hat{H}_0$, i.e., are such that $\varepsilon_{\mu\nu} = \delta_{\mu,\nu}\varepsilon_\nu$.

Note that, because the single-particle states (corresponding to the creation operators $\{a_\nu^\dagger\}$) depend on the Hamiltonian which, in turn, depends on its lowest-energy occupied single-particle states, the Hartree-Fock equations have to be solved iteratively to ensure self-consistency. They are solved in the first round by diagonalisation of $\hat{H}_0$ with some first guess for its mean field. The A lowest-energy single-particle eigenfunctions of this guess for $\hat{H}_0$ are then used to construct an A-particle Slater determinant and a corresponding improved approximation to $\hat{H}_0$. Iterating this process, until there are no longer changes in the solutions, finally gives the required Hartree-Fock solution provided that the process converges.[16]

For an occupied single-particle state, in the Hartree-Fock approximation, the general mean-field expression (7.21) gives

$$\varepsilon_i = \langle\phi|\{a_{\bar i}, [\hat{H}, a_i^\dagger]\}|\phi\rangle = \langle\phi|[\hat{H}, a_i^\dagger]a_{\bar i}|\phi\rangle = E_0 - E_i, \tag{7.48}$$

and, for an unoccupied state, it gives

$$\varepsilon_m = \langle\phi|\{a_{\bar m}, [\hat{H}, a_m^\dagger]\}|\phi\rangle = \langle\phi|a_{\bar m}[\hat{H}, a_m^\dagger]|\phi\rangle = E_m - E_0. \tag{7.49}$$

In these equations, $E_0 := \langle\phi\hat{H}|\phi\rangle$ is the approximate ground-state energy of the even nucleus, to which the Hartree-Fock approximation is being applied, and E_i and E_m are the energies of states described, respectively, as one-hole and one-particle states in the neighbouring odd-mass nuclei. Equation (7.48), which relates the energy of an occupied single-particle state in the Hartree-Fock approximation to a separation energy, is known as Koopman's theorem.[17]

---

[14]Hartree, *op. cit.* Footnote 3 on Page 515.
[15]Fock V.A. (1930), *Z. Phys.* **61**, 126.
[16]Cf. Rowe and Rosensteel, *op. cit.* Footnote 4 on Page 516.
[17]Koopman, *op. cit.* Footnote 7 on Page 518.

7.6.2 The Hartree-Fock variational equations

Because no state of a system can have lower energy than its ground state, the lowest expectation value of the Hamiltonian $\langle\phi|\hat{H}|\phi\rangle$ that can be obtained within any space of normalised states is an upper bound on the ground-state energy for that Hamiltonian. Thus, the best approximation for the ground-state energy that can be obtained with states in a trial set is given by the state $|\phi\rangle$ with lowest-energy expectation value; this state satisfies the variational equation

$$\delta\langle\phi|\hat{H}|\phi\rangle = 0. \tag{7.50}$$

From the simple observation that a Slater determinant, $|\phi\rangle$, is defined by specification of an orthonormal set of occupied single-particle states, it immediately follows that any other determinantal state, $|\phi'\rangle$, is obtained by the one-body unitary transformation of $|\phi\rangle$, which transforms the single-particle states that are occupied in $|\phi\rangle$ into those that are occupied in $|\phi'\rangle$. Thus, we can write

$$|\phi'\rangle = e^{\hat{X}}|\phi\rangle\,, \tag{7.51}$$

where $\hat{X}$ is skew-Hermitian and $e^{\hat{X}}$ is an element of the group $U(d)$ of one-body unitary transformations. Then, because

$$\langle\phi|e^{-\epsilon\hat{X}}\hat{H}e^{\epsilon\hat{X}}|\phi\rangle = \langle\phi|\hat{H}|\phi\rangle + \epsilon\langle\phi|[\hat{H},\hat{X}]|\phi\rangle + \dots\,, \tag{7.52}$$

it follows that $|\phi\rangle$ has local minimum energy when

$$\langle\phi|[\hat{H},\hat{X}]|\phi\rangle = 0, \quad \forall\,\hat{X}\in u(d), \tag{7.53}$$

where $u(d)$ is the Lie algebra of $U(d)$. Moreover, the requirement that $\hat{X}$ should be skew-Hermitian can be relaxed because if $\hat{Y}$ is Hermitian, then $i\hat{Y}$ is skew-Hermitian. Thus, Equation (7.53) applies for $\hat{X}$ any one-body operator and the variational equation is expressed as

$$\langle\phi|[\hat{H},a_\alpha^\dagger a_{\bar\beta}]|\phi\rangle = 0, \quad \forall\,\alpha,\beta. \tag{7.54}$$

Evaluation of the expectation value of $[\hat{H},a_\alpha^\dagger a_{\bar\beta}]$, for the Hamiltonian of Equation (7.15) is straightforward. For the kinetic energy term,

$$\sum_{\mu\nu}T_{\mu\nu}\langle\phi|[a_\mu^\dagger a_{\bar\nu},a_\alpha^\dagger a_{\bar\beta}]|\phi\rangle = \sum_\mu T_{\mu\alpha}\langle\phi|a_\mu^\dagger a_{\bar\beta}|\phi\rangle - \sum_\nu T_{\beta\nu}\langle\phi|a_\alpha^\dagger a_{\bar\nu}|\phi\rangle\,. \tag{7.55}$$

Thus, if $|\phi\rangle$ is a Slater determinant of A single-particle states indexed by the label $\{i = 1,\dots,A\}$, then

$$\sum_{\mu\nu}T_{\mu\nu}\langle\phi|[a_\mu^\dagger a_{\bar\nu},a_\alpha^\dagger a_{\bar\beta}]|\phi\rangle = T_{\beta\alpha}\sum_i(\delta_{\beta i} - \delta_{\alpha i})\,. \tag{7.56}$$

527

With a similar treatment of the potential energy terms, we then obtain

$$\langle \phi | [\hat{H}, a_\alpha^\dagger a_{\bar{\beta}}] | \phi \rangle = \varepsilon_{\beta\alpha} \sum_i (\delta_{\beta i} - \delta_{\alpha i}), \qquad (7.57)$$

with

$$\varepsilon_{\beta\alpha} = T_{\beta\alpha} + \sum_i^{occ} V_{\beta i \alpha i}. \qquad (7.58)$$

It follows from Equation (7.57) that $\langle \phi | [\hat{H}, a_\alpha^\dagger a_{\bar{\beta}}] | \phi \rangle$ vanishes and the variational equation is satisfied when the matrix ε is diagonal.

It will be observed immediately that the matrix ε is identical to the matrix of the self-consistent field Hamiltonian of Equation (7.47). This is a remarkable result. It means that the Slater determinant of lowest-energy single-particle wave functions of the self-consistent field equations also satisfies the variational equation. This gives a double meaning to Hartree-Fock theory: the solutions of the self-consistent field equations give a best many-body ground state from the criterion of a variational principle, and they give single-particle states with a meaningful and natural self-consistent field interpretation.

It should be recognised that a given solution to the Hartree-Fock equations is not necessarily unique and there may be a better, i.e., lower energy, solution. It can be shown[18] that each iteration of a self-consistent field cycle gives a lower energy many-particle state. Thus, any Hartree-Fock solution, obtained by the self-consistent method, is at least a local minimum with respect to all considered variations. But, from the perspective of hills and valleys on the manifold of Slater determinants (cf. Section 8.4), there are generally many local-energy minima. This suggests that a Hartree-Fock calculation should be repeated with different initial guesses if one wishes to ensure that the solution obtained has the lowest energy of possibly many local minima. It also frequently happens that there are flat-bottomed valleys of equal-energy minima. Such minima, which require special treatment, are of considerable physical interest because they occur whenever the Hartree-Fock approximation breaks a symmetry of the Hamiltonian, as is often the case. They are discussed in Section 7.6.4.

7.6.3 *The residual interaction*

The explicit separation of a given Hamiltonian into a Hartree-Fock one-body Hamiltonian plus a residual interaction is achieved by a normal ordering of the Hamiltonian with respect to the Hartree-Fock Slater determinant.

If $|\phi\rangle := \prod_{i=1}^A a_i^\dagger |0\rangle$ is a Slater determinant of A single-particle states, then the

---

[18]Cf. Rowe and Rosensteel, *op. cit.* Footnote 4 on Page 516.

normal-ordered expansion of the kinetic energy,

$$\hat{T} = \sum_{\mu\nu} T_{\mu\nu} a_\mu^\dagger a_{\bar{\nu}} = \sum_{\mu\nu} T_{\mu\nu} \langle \phi | a_\mu^\dagger a_{\bar{\nu}} | \phi \rangle + \sum_{\mu\nu} T_{\mu\nu} : a_\mu^\dagger a_{\bar{\nu}} : , \qquad (7.59)$$

relative to $|\phi\rangle$ as particle-hole vacuum, gives

$$\sum_{\mu\nu} T_{\mu\nu} a_\mu^\dagger a_{\bar{\nu}} = \sum_i T_{ii} + \sum_{\mu\nu} T_{\mu\nu} : a_\mu^\dagger a_{\bar{\nu}} : . \qquad (7.60)$$

A similar expansion of the potential energy, making use of Equation (7.44), gives

$$\hat{V} = \tfrac{1}{4} \sum_{\mu\nu\mu'\nu'} V_{\mu\nu\mu'\nu'} a_\mu^\dagger a_\nu^\dagger a_{\bar{\nu}'} a_{\bar{\mu}'}$$

$$= \tfrac{1}{2} \sum_{ij} V_{ijij} + \sum_{\mu\nu i} V_{\mu i\nu i} : a_\mu^\dagger a_{\bar{\nu}} : + \tfrac{1}{4} \sum_{\mu\nu\mu'\nu'} V_{\mu\nu\mu'\nu'} : a_\mu^\dagger a_\nu^\dagger a_{\bar{\nu}'} a_{\bar{\mu}'} : . \quad (7.61)$$

It follows that $\hat{H}$ has the normal-ordered expansion

$$\hat{H} = E_0 + : \hat{H}_0 : + \hat{V}_{\text{res}} = \hat{W}_0 + \hat{H}_0 + \hat{V}_{\text{res}} , \qquad (7.62)$$

where

$$E_0 = \langle \phi | \hat{H} | \phi \rangle = \sum_i \left[T_{ii} + \tfrac{1}{2} \sum_j V_{ijij} \right], \qquad (7.63)$$

$$W_0 = E_0 - \langle \phi | \hat{H}_0 | \phi \rangle = -\tfrac{1}{2} \sum_{ij} V_{ijij}, \qquad (7.64)$$

$\hat{H}_0$ is the Hartree-Fock IPM Hamiltonian, and the residual interaction is

$$\hat{V}_{\text{res}} = \tfrac{1}{4} \sum_{\mu\nu\mu'\nu'} V_{\mu\nu\mu'\nu'} : a_\mu^\dagger a_\nu^\dagger a_{\bar{\nu}'} a_{\bar{\mu}'} : . \qquad (7.65)$$

7.6.4 Symmetry breaking in the Hartree-Fock approximation

A characteristic of Hartree-Fock theory is that the separation $\hat{H} = W_0 + \hat{H}_0 + \hat{V}_{\text{res}}$ does not ensure that $\hat{H}_0$ and $\hat{V}_{\text{res}}$ separately retain the symmetries of $\hat{H}$; in general, they do not. For example, the rotational invariance of $\hat{H}$ does not mean that $\hat{H}_0$ is rotationally invariant. This is because the manifold of Slater determinants is a space of trial wave functions of which only a few have angular momentum zero. Indeed, only in the special cases of closed-shell and closed-subshell configurations will there exist a suitable candidate for the nuclear ground state that is both a $J = 0$ state and a Slater determinant. Observe also that, if a determinantal state $|\phi\rangle$ does not have angular momentum zero, it is not rotationally invariant and neither, in general, is the corresponding Hartree-Fock single-particle Hamiltonian $\hat{H}_0$. In general, a symmetry of $\hat{H}$ is only a symmetry of $\hat{H}_0$ if it leaves $|\phi\rangle$ invariant (to within a phase factor).[19]

It is interesting to note that all unitary irreps of the group U(1) are one-dimensional. Thus, a Hartree-Fock state $|\phi\rangle$ that has good z-component of angular momentum, M, can only change by a phase factor $e^{iM\theta}$ under a U(1) rotation θ about the z-axis. Hence, its density and the corresponding Hartree-Fock Hamiltonian $\hat{H}_0$, have axes of symmetry and, as a consequence, axial symmetry will be conserved at each Hartree-Fock iteration. Thus, it is always possible to conserve axial symmetry in a Hartree-Fock calculation. In contrast, it is not generally possible to conserve spherical symmetry in a Hartree-Fock calculation because no single state is invariant under SU(2), even to within a phase factor, unless it has angular momentum $J = 0$.[20] It should also be understood that although it is possible to conserve axial symmetry, there may be lower-energy Hartree-Fock solutions that break this symmetry. Nevertheless, it is not surprising that axial symmetry is much more prevalent than full rotational invariance in the Hartree-Fock approximation.

Even more prevalent in Hartree-Fock approximations is time-reversal invariance. This is because the group generated by time-reversal has only two inequivalent irreps, both of which are one-dimensional. Thus, the Hartree-Fock Hamiltonian is time-reversal invariant for any Slater determinant that is either even or odd under time reversal. A remarkable fact is that, for a time-reversal invariant Hartree-Fock solution, the single-particle states of the approximation always occur in equal-energy time-reversed pairs in accordance with Kramer's *degeneracy theorem*.[21] For example, for an axially symmetric Hartree-Fock solution, the single-particle states can be labelled by the z-component of angular momentum, m. Under time-reversal m is mapped to $-m$ which cannot be equal to m because m is half an odd integer. Thus, for a time-reversal invariant Hartree-Fock solution, the time-reversed pairs of single-particle states have common single-particle energies. According to Kramer's theorem, this two-fold degeneracy persists when the axial symmetry of the Hartree-Fock solution is broken.

The origin of symmetry breaking in the Hartree-Fock approximation is understood from the observation that, while the set of Slater determinants for a many-particle system spans the Hilbert space for that system, the subset of Slater determinants having a given symmetry doesn't come close to spanning the subspace of the Hilbert space of that symmetry. For example, the number of linearly-independent $J = 0$ Slater determinants is a tiny fraction of the total number of

---

[19]Bar-Touv J. and Kelson I. (1965), *Phys. Rev.* **138**, B1035.

[20]One could, for example, conserve spherical symmetry in a highly restricted variational calculation in which one started with a closed-subshell, $J = 0$, Slater determinant and then restricted the class of variations to just the single-particle radial wave functions.

[21]Kramers degeneracy theorem states that the Hartree-Fock energy levels of atomic and other systems of electrons occur in doubly-degenerate pairs in the presence of purely electric fields (i.e., no magnetic fields). The theorem, enunciated by Kramers H.A. (1930), *Proc. Koninkl. Ned. Akad. Wetenschop.* **33**, 959, is discussed in many text books.

linearly-independent $J = 0$ states for any given nucleus. This is because most $J = 0$ states are linear combinations of Slater determinants that do not separately have well-defined values of J. Thus, it is not surprising that breaking a symmetry in the Hartree-Fock approximation generally results in a lower-energy solution. The breaking of point-group symmetries in the Hartree-Fock approximation has been explored by Dobaczewski et al.[22].

A breaking of fundamental symmetries can be interpreted as a failure of Hartree-Fock theory. However, on reflection, it appears that symmetry breaking is a strength rather than a weakness of the approximation. The breaking of a symmetry in the Hartree-Fock approximation is certainly an indication that the constraint to determinantal states is overly restrictive. But, there are strong indications that it is physically significant. In particular, it is frequently observed that the nuclear-matter-density distribution of a Hartree-Fock solution is most highly non-spherical when the corresponding nucleus exhibits a rotational spectrum. This suggests that the Hartree-Fock state might be interpreted as an approximation for the intrinsic state of a ground-state rotational band rather than for the ground state itself. Such an interpretation, developed in Section 8.2.5, arises naturally from a consideration of the semi-classical dynamics on the manifold of Slater determinants obtained by restriction of the quantum mechanics to this manifold.

7.6.5 *The restoration of broken symmetries*

A symmetry that is broken as a result of the Hartree-Fock approximation can be restored by taking linear combinations of equal-energy Hartree-Fock solutions.[23] For example, if a Hartree-Fock minimal-energy state, $|\phi\rangle$, for a rotationally-invariant Hamiltonian, $\hat{H}$, is not rotationally-invariant then, because

$$\langle\phi|\hat{R}(\Omega^{-1}\hat{H}\hat{R}(\Omega)|\phi\rangle = \langle\phi|\hat{H}\hat{R}(\Omega)|\phi\rangle, \qquad (7.66)$$

the state $\hat{R}(\Omega)|\phi\rangle$, obtained by rotating $|\phi\rangle$ through an angle Ω, has precisely the same energy expectation value as $|\phi\rangle$. A better approximation to an eigenfunction of the original Hamiltonian is then given by a linear combination of rotated states,

$$|\Psi\rangle = \int f(\Omega)\,\hat{R}(\Omega)|\phi\rangle\,d\Omega. \qquad (7.67)$$

In particular, if f is the constant function, $f(\Omega) = 1$, and $d\Omega$ is the usual volume element (invariant measure) for a rotation, then $|\Psi\rangle$ becomes rotationally invariant. Thus, it has angular momentum $J = 0$ and, for a doubly-even nucleus, gives a better approximation for the ground state.

Basis states of good angular momentum are constructed in this way by angular-momentum projection, which corresponds to particular choices for the function

---

[22]Dobaczewski J. *et al.* (2000), *Phys. Rev.* **C62**, 014311.
[23]Peierls R.E. and Yoccoz J. (1957), *Proc. Phys. Soc.* **70**, 381.

f in Equation (7.67). For example, if $|\phi\rangle$ is expanded in terms of good angular momentum states, $|\phi\rangle = \sum_{JK} C_{JK}|\phi_{\alpha_K JK}\rangle$, then

$$\hat{R}(\Omega)|\phi\rangle = \sum_{JKM} C_{JK}|\phi_{\alpha_K JM}\rangle \mathcal{D}^J_{MK}(\Omega), \qquad (7.68)$$

where $\mathcal{D}^J_{MK}(\Omega)$ is a Wigner rotation matrix. Thus, by using the identity

$$\int \mathcal{D}^{J*}_{MK}(\Omega)\,\mathcal{D}^{J'}_{M'K'}(\Omega)\,d\Omega = \frac{8\pi^2}{2J+1}\,\delta_{JJ'}\delta_{KK'}\delta_{MM'} \qquad (7.69)$$

and setting $f = \mathcal{D}^{J*}_{MK}$, we obtain the state

$$|\Psi_{KJM}\rangle := \int \mathcal{D}^{J*}_{MK}(\Omega)\,R(\Omega)|\phi\rangle\,d\Omega = \frac{8\pi^2}{2J+1}\,C_{JK}|\phi_{\alpha_K JM}\rangle. \qquad (7.70)$$

(Note that the label α_K, for the state $|\phi_{\alpha_K JK}\rangle$, is needed to exhibit the fact that states of different K may occur with a common value of J in different, although equivalent, irreps of SU(2).) Restoration of other lost symmetries can be achieved by projecting out states with good quantum numbers in a similar way.

An examination of what happens when a symmetry is broken reveals some general and useful principles. Suppose S is a symmetry group of the original Hamiltonian, $\hat{H}$, and that it has a representation, $\hat{U}$, as a subgroup of one-body unitary transformations of $\hat{H}$ (as do most symmetry groups of $\hat{H}$ of interest). If $|\phi\rangle$ is a Slater determinant, it follows that all states on the S orbit $\{|\phi(g)\rangle = \hat{U}(g)|\phi\rangle, g \in S\}$ are also Slater determinants. Moreover, these states have a common energy

$$\langle\phi|\hat{U}(g^{-1})\hat{H}\hat{U}(g)|\phi\rangle = \langle\phi|\hat{H}|\phi\rangle, \quad \forall g \in S. \qquad (7.71)$$

The complex linear span of the states of such an S orbit is an S-invariant subspace of the original many-particle Hilbert space to which the states $\{|\phi(g)\rangle, g \in S\}$ belong. Thus, if a Hartree-Fock lowest-energy state, $|\phi\rangle$, for the Hamiltonian $\hat{H}$ breaks the S symmetry, then better approximations to eigenstates of $\hat{H}$, which belong to irreps of the symmetry group S, are obtained and the symmetry is restored by diagonalising the Hamiltonian $\hat{H}$ in the S-invariant subspace spanned by the states $\{|\phi(g)\rangle, g \in S\}$. For simplicity, we refer to states obtained in this way as *projected Hartree-Fock* states, although, in fact, they are more general. Such a procedure is an effective way of requantizing the subdynamics associated with the broken symmetry group to regain a unitary representation of this group.

As illustrated in Section 6.11, a substantial gain in accuracy is achieved by an extension of a variational solution of pairing-force problems to include number projection. A similar gain in accuracy is achieved by restoring rotational and other symmetries in Hartree-Fock-based approximations. It may also be noted that lowest-energy states projected from a Slater determinant can, in general, be obtained with even lower energies if projected from a Slater determinant that differs somewhat from the Hartree-Fock minimum-energy determinant. Thus, for example, if a state $|J(\phi)\rangle$ of angular momentum J is projected from a determinant $|\phi\rangle$, then the particular state $|\phi\rangle$ that generates the lowest-energy state $|J(\phi)\rangle$ for a given value of J will generally differ somewhat from that of the Hartree-Fock minimum-energy solution and the extent of the difference will depend on the value of J. Indeed, one may expect that the deformation of the determinant appropriate for a given value of J should increase with increasing J due to centrifugal effects.

The determination of the state $|\phi\rangle$, that minimises the energy of a state $|J(\phi)\rangle$ of angular momentum J projected from the state $|\phi\rangle$, independently for each J is customarily described as *variation after projection* in contrast to the simpler procedure, described as *variation before projection*, in which all states are projected from the single minimum-energy Hartree-Fock determinant. Variation-after-projection calculations generally give more accurate results, but they are also more computationally intensive.

7.6.6   *The Skyrme-Levinson method*

An alternative way to restore a broken symmetry is to exploit it by creating a situation in which the symmetry is expected to be broken so that, when it is broken, it has an anticipated physical interpretation. Such an approach, outlined as follows, was proposed by Skyrme[24] and developed by Levinson.[25]

Consider a nucleus that has a ground-state band of rotational energies given by

$$E_J = E_0 + AJ(J+1), \qquad (7.72)$$

where A is a parameter. Then, if $\hat{H}$ is the rotationally invariant Hamiltonian for this nucleus, the related Hamiltonian

$$\hat{H}(A) = \hat{H} - A\mathbf{J}^2 \qquad (7.73)$$

has the same eigenstates as $\hat{H}$. However, the energies of all states of the ground-state band, relative to the Hamiltonian $\hat{H}(A)$, collapse to a common energy E_0 and form a multiply-degenerate ground state.

Application of the Hartree-Fock equations to the Hamiltonian $\hat{H}(A)$ is now expected to return a Slater determinant, $|\phi\rangle$, that is an approximation to some linear combination of states in the multiply-degenerate ground state. Moreover, all the equal-energy Hartree-Fock solutions in the set $\{\hat{R}(\Omega)|\phi\rangle; \Omega \in SO(3)\}$ are expected to approximate linear combinations of the multiply-degenerate ground state. Thus, to the extent that all the above assumptions are valid, the Hartree-Fock energy is an approximation for the energy E_0 and the angular-momentum states projected from $|\phi\rangle$ are approximations to states of the ground-state rotational band.

The parameter A might be taken from experiment. It can also be predicted

[24]Skyrme T.H.R. (1957), *Proc. Phys. Soc.* **A70**, 433.
[25]Levinson C.A. (1963), *Phys. Rev.* **132**, 2184.

self-consistently from the theory by adjusting its value to minimise the variance,

$$\sigma^2(A) := \langle\phi|\big[\hat{H}(A) - \langle\phi|\hat{H}(A)|\phi\rangle\big]^2|\phi\rangle, \qquad (7.74)$$

of the energy spread of the collapsed energy levels. This minimisation requires an extra level of self-consistency because the Hartree-Fock state depends on the value of A, whose value in turn depends on the Hartree-Fock state $|\phi\rangle$.

Using Equation (7.74) to determine $\sigma(A)$ would appear to present an extra challenge because of the presence of a quadratic term in the Hamiltonian. However, because the required expectation values in this equation are with respect to a Slater determinant, they are much more easily evaluated than would otherwise be the case. An application of the Skyrme-Levinson method has been made to the spectrum of ^{20}Ne by Kelson and Levinson.[26]

Exercises

7.8 Show that to satisfy the Hartree-Fock variational equation, it is not necessary to fully diagonalise the matrix ε; it is only necessary to block diagonalise it to ensure that the particle-hole matrix elements ε_{ph} and ε_{hp} are zero. (This can be done by showing that the right-hand side of Equation (7.57) vanishes when $\varepsilon_{ph} = \varepsilon_{hp} = 0$.)

7.9 Show that if $|\phi\rangle$ is a Hartree-Fock Slater determinant then

$$\langle\phi|V_{\text{res}}|\phi\rangle = \langle\phi|[V_{\text{res}}, a_p^\dagger a_{\bar{h}}]|\phi\rangle = 0, \quad \forall\, p, h, \qquad (7.75)$$

where p and h label unoccupied (particle) and occupied (hole) states, respectively. Because of this result, the Hartree-Fock solution is said to be *particle-hole stable*.

7.10 Show that the above derivations of Hartree-Fock theory naturally extend to Hamiltonians with three-body, or more, interactions.

7.7 The Hartree-Fock-Bogolyubov approximation

The Hartree-Fock-Bogolyubov (HFB) approximation is a generalisation of the independent-particle approximation of Hartree-Fock theory to incorporate the independent-quasi-particle ideas of the BCS model.[27]

7.7.1 *Generalised quasi-particles and the fermion-pair algebra*

In Hartree-Fock theory it is useful to describe unoccupied single-particle states as *particle* states and occupied single-particle states as *hole* states and to regard a minimum-energy Slater determinant as a *particle-hole vacuum* state. However, as seen in Chapter 6, the distinction between particle states and hole states becomes

---

[26]Kelson I. and Levinson C.A. (1964), *Phys. Rev.* **134**, B269.

[27]The generalisation was made by Baranger M. (1961), *Phys. Rev.* **122**, 992, and by Valatin, *op. cit.* Footnote 69 on Page 462.

blurred when the single-particle states in the neighbourhood of the Fermi surface are partially occupied as happens when pairing and other residual interactions are taken into account. One can then define quasi-particles as linear combinations of particle and hole operators in a generalisation of the Bogolyubov-Valatin transformation given in Section 6.8,

$$a_\nu^\dagger \to \alpha_\nu^\dagger = \sum_\mu (a_\mu^\dagger u_{\mu\nu} + a_\mu v_{\mu\nu}), \quad a_{\bar\nu} \to \alpha_{\bar\nu} = \sum_\mu (a_{\bar\mu} u_{\mu\nu}^* - a_{\bar\mu}^\dagger v_{\mu\nu}^*). \qquad (7.76)$$

With the definition of time-reversal given in Section 6.2.1, the corresponding transformations for quasi-particle operators for time-reversed states are given by

$$a_{\bar\nu}^\dagger \to \alpha_{\bar\nu}^\dagger = \sum_\mu (a_{\bar\mu}^\dagger u_{\mu\nu}^* + a_{\bar\mu} v_{\mu\nu}^*), \quad a_\nu \to \alpha_\nu = \sum_\mu (a_\mu u_{\mu\nu} - a_\mu^\dagger v_{\mu\nu}). \qquad (7.77)$$

The essential requirement of such quasi-particle operators is that they continue to satisfy the fermion anticommutation relationships

$$\{\alpha_{\bar\mu}, \alpha_\nu^\dagger\} = \delta_{\mu,\nu}, \quad \{\alpha_\mu, \alpha_\nu\} = \{\alpha_\mu^\dagger, \alpha_\nu^\dagger\} = 0, \qquad (7.78)$$

and preserve the Hermiticity relationship

$$(\alpha_{\bar\nu})^\dagger = \alpha_\nu^\dagger. \qquad (7.79)$$

Thus, the transformation matrices are required to satisfy the identity

$$\begin{pmatrix} u^\dagger & v^\dagger \\ -v^\dagger & u^\dagger \end{pmatrix} \begin{pmatrix} u & -v \\ v & u \end{pmatrix} = \begin{pmatrix} I & 0 \\ 0 & I \end{pmatrix}, \qquad (7.80)$$

where $I_{\mu\nu} = \delta_{\mu,\nu}$. From this identity, we obtain the inverses of the Bogolyubov-Valatin transformation:

$$a_\nu^\dagger = \sum_\mu (u_{\nu\mu}^* \alpha_\mu^\dagger - v_{\nu\mu}^* \alpha_\mu), \quad a_{\bar\nu} = \sum_\mu (u_{\nu\mu} \alpha_{\bar\mu} + v_{\nu\mu} \alpha_{\bar\mu}^\dagger), \qquad (7.81)$$

$$a_\nu = \sum_\mu (u_{\nu\mu}^* \alpha_\mu + v_{\nu\mu}^* \alpha_\mu^\dagger), \quad a_{\bar\nu}^\dagger = \sum_\mu (u_{\nu\mu} \alpha_{\bar\mu}^\dagger - v_{\nu\mu} \alpha_{\bar\mu}). \qquad (7.82)$$

7.7.2 Self-consistent field equations

From a mean-field perspective, HFB theory is defined by a map from the nuclear Hamiltonian,

$$\hat H' = \sum_{\mu\nu} (T_{\mu\nu} - \lambda\delta_{\mu,\nu}) a_\mu^\dagger a_{\bar\nu} + \tfrac{1}{4} \sum_{\mu\mu'\nu\nu'} V_{\mu\mu'\nu\nu'} a_\mu^\dagger a_{\mu'}^\dagger a_{\bar\nu'} a_{\bar\nu}, \qquad (7.83)$$

to an independent-quasi-particle Hamiltonian

$$\hat H_0' = W_0 + \sum_\nu \mathcal{E}_\nu \alpha_\nu^\dagger \alpha_{\bar\nu}. \qquad (7.84)$$

The Hamiltonian, $\hat{H}'$, used here differs from the Hamiltonian of Equation (7.15), used in Hartree-Fock theory, by the addition of a constraining term, i.e.,

$$\hat{H}' = \hat{H} - \lambda\hat{n}, \qquad (7.85)$$

where $\hat{n} = \sum_\nu a_\nu^\dagger a_{\bar{\nu}}$ is the nucleon number operator and λ is a Lagrange multiplier, often referred to as the *chemical potential*. (Note, the analogy between this equation and Equation (7.73).) The purpose of the constraining term is to ensure that the expectation value of the nucleon number relative to the HFB quasi-particle-vacuum state $|\phi\rangle$ is the nucleon number,

$$n = \langle\phi|\hat{n}|\phi\rangle, \qquad (7.86)$$

of the nucleus under investigation. Such a constraint is needed in HFB and BCS approximations in which states of different nucleon number are mixed.[28]

It can be seen that the independent-quasi-particle Hamiltonian, $\hat{H}_0'$, satisfies the equations of motion

$$\{a_{\bar{\mu}}, [\hat{H}_0', \alpha_\nu^\dagger]\} = \mathcal{E}_\nu\{a_{\bar{\mu}}, \alpha_\nu^\dagger\} = \mathcal{E}_\nu u_{\mu\nu}, \qquad (7.87)$$

$$\{a_{\bar{\mu}}, [\hat{H}_0', \alpha_\nu]\} = -\mathcal{E}_\nu\{a_{\bar{\mu}}, \alpha_\nu\} = \mathcal{E}_\nu v_{\mu\nu}. \qquad (7.88)$$

Thus, in parallel with the Hartree-Fock self-consistent-field methods of Section 7.6.1 and the BCS methods of Section 6.8, the desired quasi-particle operators and their energies are defined by requiring them to satisfy the generalised mean-field equations

$$\langle\varphi|\{a_{\bar{\mu}}, [\hat{H}', \alpha_\nu^\dagger]\}|\varphi\rangle = \mathcal{E}_\nu u_{\mu\nu}, \qquad (7.89)$$

$$\langle\varphi|\{a_{\bar{\mu}}, [\hat{H}', \alpha_\nu]\}|\varphi\rangle = \mathcal{E}_\nu v_{\mu\nu}, \qquad (7.90)$$

where $|\varphi\rangle$ is the quasi-particle vacuum state. Thus, one obtains the eigenvalue equations

$$\sum_{\nu'}\left[(\varepsilon_{\mu\nu'} - \lambda\delta_{\mu\nu'})u_{\nu'\nu} + \Delta_{\mu\nu'}v_{\nu'\nu}\right] = \mathcal{E}_\nu u_{\mu\nu}, \qquad (7.91)$$

$$\sum_{\nu'}\left[\Delta_{\mu\nu'}u_{\nu'\nu} - (\varepsilon_{\mu\nu'} - \lambda\delta_{\mu\nu'})v_{\nu'\nu}\right] = \mathcal{E}_\nu v_{\mu\nu}, \qquad (7.92)$$

where

$$\varepsilon_{\mu\nu} = \langle\varphi|\{a_{\bar{\mu}}, [\hat{H}, a_\nu^\dagger]\}|\varphi\rangle = T_{\mu\nu} + \sum_{\mu'\nu'\sigma} V_{\mu\mu'\nu\nu'}v_{\mu'\sigma}^* v_{\nu'\sigma}, \qquad (7.93)$$

$$\Delta_{\mu\nu} = \langle\varphi|\{a_{\bar{\mu}}, [\hat{H}, a_\nu]\}|\varphi\rangle = -\tfrac{1}{2}\sum_{\mu'\nu'\sigma} V_{\mu\bar{\nu}\mu'\bar{\nu}'}u_{\nu'\sigma}^* v_{\mu'\sigma}. \qquad (7.94)$$

---

[28]Such mixing of different nucleon number states is said to break the symmetry of the U(1) gauge group whose infinitesimal generator is the nucleon number operator.

Equations (7.91) and (7.92) can be combined and written in matrix form

$$\begin{pmatrix} (\varepsilon - \lambda I) & \Delta \\ \Delta & -(\varepsilon - \lambda I) \end{pmatrix} \begin{pmatrix} u_\nu \\ v_\nu \end{pmatrix} = \mathcal{E}_\nu \begin{pmatrix} u_\nu \\ v_\nu \end{pmatrix}, \tag{7.95}$$

where u_ν and v_ν are, respectively, the vectors with components $\{u_{\mu\nu}\}$ and $\{v_{\mu\nu}\}$.

Note that ε is Hermitian, i.e., $\varepsilon^*_{\nu\mu} = \varepsilon_{\mu\nu}$, and, provided $\hat{H}$ and the HFB approximation to $\hat{H}$ are time-reversal invariant, $\hat{\Delta}$ is also Hermitian (see Exercise 7.12). The HFB Hamiltonian matrix to be diagonalised in Equation (7.95) is then a Hermitian matrix and diagonalisable by a unitary transformation.

The above equations are solved iteratively. Starting with initial guesses for the single-particle energies, $\varepsilon_{\mu\nu} = \delta_{\mu,\nu}\varepsilon_\nu$, the chemical potential, λ, and the gap parameters, $\Delta_{\mu\nu} = \delta_{\mu,\nu}\Delta_\nu$, one derives the quasi-particle energies $\mathcal{E}_\nu$ and coefficients $(u_{\mu\nu}, v_{\mu\nu})$ by solving the matrix Equation (7.95). The value of λ is then adjusted so that the expectation of the number operator,

$$\langle \varphi | \hat{n} | \varphi \rangle = \sum_{\mu\nu} |v_{\mu\nu}|^2, \tag{7.96}$$

has its desired value. New values for the ε and Δ matrices are also determined by Equations (7.93) and (7.94). Thus, the process is repeated until subsequent iterations produce no change in the results. Note that, with real two-body matrix elements the parameters can all be assumed real.

There is no guarantee that a unique solution will be obtained or that the iterative method of solving the equations from any starting point will converge. Thus, it is important to repeat the calculations from a number of starting points. Fortunately, the experimental data make it possible to adopt sensible choices for the starting values of the parameters.

Because of the form of the above matrix equation, it can also be shown (Exercise 7.13) that, for every eigenvector (u_ν, v_ν) with $u_\nu = \{u_{\mu\nu}\}$, $v_\nu = \{v_{\mu\nu}\}$, and energy $\mathcal{E}_\nu$, there is another eigenvector $(-v_\nu, u_\nu)$ with energy $-\mathcal{E}_\nu$. Thus, if $\mathcal{E}_\nu > 0$, the first solution defines a quasi-particle creation operator $\alpha^\dagger_\nu = \sum_\mu (a^\dagger_\mu u_{\mu\nu} + a_\mu v_{\mu\nu})$ and the second defines the annihilation operator $\alpha_\nu = \sum_\mu (-a^\dagger_\mu v_{\mu\nu} + a_\mu u_{\mu\nu})$.

7.7.3 The full Hamiltonian in terms of quasi-particle operators

When expressed in normal order (cf. Section 7.5) relative to the quasi-particle vacuum state, the full Hamiltonian, (7.85), is of the form

$$\hat{H}' = \hat{H}'_0 + \hat{V}_{\text{res}}, \tag{7.97}$$

with

$$\hat{H}_0' := \langle\phi|\hat{H}'|\phi\rangle + \sum_{\mu\nu}(T_{\mu\nu} - \lambda\delta_{\mu,\nu}) : a_\mu^\dagger a_{\bar\nu} : + \sum_{\mu\nu\mu'\nu'} V_{\mu\nu\mu'\nu'}\Big[\langle\phi|a_\mu^\dagger a_{\bar\mu'}|\phi\rangle : a_\nu^\dagger a_{\bar\nu'} :$$

$$+\tfrac{1}{2}\big(\langle\phi|a_\mu^\dagger a_\nu^\dagger|\phi\rangle : a_{\bar\nu'}a_{\bar\mu'} : +\langle\phi|a_{\bar\nu'}a_{\bar\mu'}|\phi\rangle : a_\mu^\dagger a_\nu^\dagger : \big)\Big], \qquad (7.98)$$

and

$$\hat{V}_{\text{res}} := \sum_{\mu\nu\mu'\nu'} V_{\mu\nu\mu'\nu'} : a_\mu^\dagger a_\nu^\dagger a_{\bar\nu'}a_{\bar\mu'} : . \qquad (7.99)$$

When re-expressed in terms of quasi-particle operators, this expansion reduces to the form (as for the BCS expansion, Section 6.8.4(iii))

$$\hat{H}' = \langle\phi|\hat{H}'|\phi\rangle + \sum_\nu \mathcal{E}_\nu \alpha_\nu^\dagger a_{\bar\nu} + \hat{V}_{\text{res}}. \qquad (7.100)$$

7.7.4 *The group of Bogolyubov-Valatin transformations*

As defined in Section 7.7.1, Bogolyubov-Valatin transformations are linear transformations of a set of fermion creation and annihilation operators, $a_\nu^\dagger \to \alpha_\nu^\dagger$, $a_\nu \to \alpha_\nu^\dagger$ which preserve both their anticommutation relations and their Hermiticity properties. To identify the group of such transformations, consider its application to the Hermitian operators

$$A_\nu = \frac{1}{\sqrt{2}}(a_\nu^\dagger + a_{\bar\nu}), \quad B_\nu = \frac{i}{\sqrt{2}}(a_\nu^\dagger - a_{\bar\nu}), \quad \nu = 1,\dots,d. \qquad (7.101)$$

These operators span a real $2d$-dimensional vector space for which a symmetric scalar product of any pair of vectors in this space is defined by their anticommutator:

$$\{A_\mu, A_\nu\} = \{B_\mu, B_\nu\} = \delta_{\mu\nu}, \quad \{A_\mu, B_\nu\} = 0. \qquad (7.102)$$

It is known that the linear transformations of a set of real vectors that preserve their scalar products are those of an orthogonal group. It follows that the set of Bogolyubov-Valatin transformations is the orthogonal group O($2d$).

The infinitesimal generators of O($2d$) span a Lie algebra, so($2d$), known as the *fermion-pair algebra*. This Lie algebra comprises the Hermitian linear combinations of the bilinear operators $\{a_\mu^\dagger a_\nu^\dagger, (a_\mu^\dagger a_{\bar\nu} - a_{\bar\nu}a_\mu^\dagger), a_{\bar\mu}a_{\bar\nu}\}$. In addition to SO($2d$), the group O($2d$) also contains discrete subgroups of inversions and reflections.

It is clear that the above-defined fermion-pair operators cannot change a state with an even number of fermions into an odd-fermion state. Thus, the Fock space, $\mathcal{F}$, of all many-nucleon states with nucleons occupying the single-nucleon states of a selected d-dimensional space, separates into two irreducible subspaces, with respect to the dynamical group O($2d$). These irreducible subspaces contain the even- and odd-nucleon states, respectively. The lowest-weight state for the even-nucleon irrep is the zero-particle vacuum state, $|0\rangle$, and that for the odd-nucleon irrep is a one-quasi-particle state.

To obtain a complete spectrum-generating algebra and a full dynamical group for the many-nucleon Fock space, the fermion-pair algebra, so$(2d)$, can be augmented to a Lie algebra which contains the single-fermion operators $\{a_\nu^\dagger\}$ and $\{a_\nu\}$ in addition to the so$(2d)$ operators.[29] This Lie algebra is identified, from its root structure, as so$(2d+1)$.[30,31,32] Thus, the whole many-nucleon Fock space, $\mathcal{F}$, generated from a finite set of d single-particle states, carries an irrep of the dynamical group O$(2d+1)$. This group features in numerous articles relating to HFB theory.[33]

7.7.5 *The manifold of quasi-particle vacuum states*

It was shown in Section 7.4.2, that the Hartree-Fock manifold of Slater determinants is an orbit of the dynamical group, U(d), of one-body unitary transformations of the active many-nucleon configuration space (see Footnote 8 on Page 519). Thus, the underlying geometrical space of Hartree-Fock theory is a manifold of U(d) coherent states (see Section 7.4.2) which, to within phase factors, are elements of the factor space $\overline{\mathcal{M}}^{(A)} \simeq $ U$(d)/($U$(A) \times$ U$(d-A))$.

Having identified the group of Bogolyubov-Valatin transformation, we can now identify the corresponding manifold of quasi-particle vacuum states. Following the methods of Section 7.4.2, the above observations imply that this manifold is the O$(2d)$ group orbit containing the zero-particle vacuum state; i.e., it is a manifold of O$(2d)$ coherent states. Its geometrical properties are identified from the observation that the isotropy subgroup of O$(2d)$ transformations which leave the zero-particle vacuum state invariant is the number-conserving subgroup, U$(d) \subset$ O$(2d)$. Thus, to within phase factors, the geometric space of HFB theory is isomorphic to the factor space

$$\overline{\mathcal{M}}^{\mathcal{F}} \simeq \text{O}(2d)/\text{U}(d). \tag{7.103}$$

Because the Hartree-Fock transformations, which are elements of U(d), are a subset of the HFB transformations, it follows that $\overline{\mathcal{M}}^{(A)}$ is a submanifold of $\overline{\mathcal{M}}^{\mathcal{F}}$.

---

[29]As elements of the so$(2d+1)$ Lie algebra, the single-fermion operators satisfy the commutation relations $[a_\mu^\dagger, a_\nu] = a_\mu^\dagger a_\nu - a_\nu a_\mu^\dagger$, etc.

[30]Judd B.R. (1968), in *Group Theory and its Applications*, edited by M. Loebl (Academic Press, New York), p. 183.

[31]Cf., also discussion by Moshinsky M. and Quesne C. (1970), *J. Math. Phys.* **11**, 1631.

[32]The addition of the single-fermion creation and annihilation operators to the operators of so$(2d)$ also generates a *graded Lie algebra* or *superalgebra* as it is also called. In this superalgebra, commutation relations are used for the elements of so$(2d)$ and for mixed Lie brackets, such as $[a_\mu^\dagger a_\nu^\dagger, a_\sigma]$, while anti-commutation relations are used for the odd fermion operators. It is also notable that the matrices of the fundamental irrep of so$(2d+1)$ satisfy both the commutation relations of so$(2d+1)$ and the anticommutation relations of the corresponding superalgebra.

[33]See, for example, Nishiyama S. and Fukutome H. (1992), *J. Phys. G: Nucl. Part. Phys.* **18**, 317, and references therein.

7.7.6 The HFB variational equations

In complete parallel with the variational expression of the Hartree-Fock approximation, given in Section 7.6.2, the HFB approximation can be expressed in terms of a variational principle, in which the selected quasi-particle vacuum state, and the associated Bogolyubov-Valatin transformation, for a given Hamiltonian, $\hat{H}'$, are given by the quasi-particle vacuum state that minimises the expectation value of $\hat{H}'$. Because any quasi-particle vacuum state, $|\phi'\rangle$, can be expressed in terms of any other quasi-particle vacuum state, $|\phi\rangle$, by a transformation

$$|\phi'\rangle = e^{\hat{X}}|\phi\rangle, \quad X \in \text{so}(2d), \tag{7.104}$$

it follows that, for $|\phi\rangle$ to be of minimum-energy, it must satisfy the equation

$$\frac{d}{d\epsilon}\langle\phi|e^{-\epsilon\hat{X}}\hat{H}'e^{\epsilon\hat{X}}|\phi\rangle\big|_{\epsilon=0} = \langle\phi|[\hat{H}',\hat{X}]|\phi\rangle = 0, \quad \forall X \in \text{so}(2d). \tag{7.105}$$

From the expansion of $\hat{H}'$ given by Equation (7.100), it is seen that, in fact, $\langle\phi|[\hat{H}',\hat{X}]|\phi\rangle$ vanishes if $\hat{X}$ is any of the operators $\{a_\mu^\dagger a_\nu^\dagger, a_\mu a_\nu, a_\mu^\dagger a_\nu, a_\mu a_\nu^\dagger\}$. This shows that any quasi-particle vacuum state resulting from the self-consistent field equations satisfies the variational equation (7.105).

Exercises

7.11 Show that if

$$\alpha_\nu^\dagger = \sum_\mu (a_\mu^\dagger u_{\mu\nu} + a_\mu v_{\mu\nu}), \quad \alpha_{\bar\nu} = \sum_\mu (a_{\bar\mu} u_{\mu\nu}^* - a_{\bar\mu}^\dagger v_{\mu\nu}^*), \tag{7.106}$$

$$\alpha_{\bar\nu}^\dagger = \sum_\mu (a_{\bar\mu}^\dagger u_{\mu\nu}^* + a_{\bar\mu} v_{\mu\nu}^*), \quad \alpha_\nu = \sum_\mu (a_\mu u_{\mu\nu} - a_\mu^\dagger v_{\mu\nu}), \tag{7.107}$$

then the identities

$$a_\nu^\dagger = \sum_\mu \left[\{\alpha_{\bar\mu}, a_\nu^\dagger\}\alpha_\mu^\dagger - \{\alpha_{\bar\mu}^\dagger, a_\nu^\dagger\}\alpha_\mu\right], \tag{7.108}$$

$$a_{\bar\nu}^\dagger = \sum_\mu \left[-\{\alpha_\mu, a_{\bar\nu}^\dagger\}\alpha_{\bar\mu}^\dagger + \{\alpha_\mu^\dagger, a_{\bar\nu}^\dagger\}\alpha_{\bar\mu}\right], \tag{7.109}$$

imply that the inverse equations are given by

$$a_\nu^\dagger = \sum_\mu (u_{\nu\mu}^* \alpha_\mu^\dagger - v_{\nu\mu}^* \alpha_\mu), \quad a_{\bar\nu} = \sum_\mu (u_{\nu\mu}\alpha_{\bar\mu} + v_{\nu\mu}\alpha_\mu^\dagger), \tag{7.110}$$

$$a_{\bar\nu}^\dagger = \sum_\mu (u_{\nu\mu}\alpha_{\bar\mu}^\dagger - v_{\nu\mu}\alpha_\mu), \quad a_\nu = \sum_\mu (u_{\nu\mu}^*\alpha_\mu + v_{\nu\mu}^*\alpha_{\bar\mu}^\dagger). \tag{7.111}$$

7.12 Show that if $\hat{H}$ and $|\varphi\rangle$ are time-reversal invariant, then

$$\Delta_{\nu\mu}^* := \langle\varphi|\{a_{\bar\nu}, [\hat{H}, a_\mu]\}|\phi\rangle^* = \langle\varphi|\{a_{\bar\mu}, [\hat{H}, a_\nu]\}|\phi\rangle = \Delta_{\mu\nu}. \tag{7.112}$$

7.13 Use the symmetries of Equation (7.95) to show that for every eigenvector (u_ν, v_ν) with energy $\mathcal{E}_\nu$ there is another eigenvector $(-v_\nu, u_\nu)$ with energy $-\mathcal{E}_\nu$.

7.14 Show that, to satisfy the fermion anticommutation relations $\{\alpha_{\bar\mu}, \alpha_\nu^\dagger\} = \delta_{\mu,\nu}$ and $\{\alpha_{\bar\mu}^\dagger, \alpha_\nu^\dagger\} = 0$, the Bogolyubov-Valatin transformation matrices should satisfy the identities:

$$u^\dagger u + v^\dagger v = I, \quad u^\dagger v = v^\dagger u, \qquad (7.113)$$

where I is the identity matrix with elements $I_{\mu\nu} = \delta_{\mu,\nu}$.

7.15 Show that the commutator of $a_\mu^\dagger$ and $a_{\bar\nu}$ (but not the anticommutator) is an element of the so$(2d+1)$ Lie algebra.

7.16 Show that, if $\{a_{j,m}^\dagger\}$ are creation operators for a fermion of angular momentum j, there is no two-nucleon state of $J = 2j$ and and a two-nucleon state of $J = M = 2j - 1$ is given by $|J, M = J\rangle = a_{j,j}^\dagger a_{j,j-1}^\dagger |0\rangle$. Show also that

$$|J, J - 1\rangle = a_{j,j}^\dagger a_{j,j-2}^\dagger |0\rangle, \qquad (7.114)$$

$$|J, J - 2\rangle \propto \left[\sqrt{2j}\, a_{j,j-1}^\dagger a_{j,j-2}^\dagger + \sqrt{3(2j-2)}\, a_{j,j}^\dagger a_{j,j-3}^\dagger \right] |0\rangle, \qquad (7.115)$$

$$|J - 2, J - 2\rangle \propto \left[\sqrt{3(2j-2)},a_{j,j-1}^\dagger a_{j,j-2}^\dagger - \sqrt{2j}\, a_{j,j}^\dagger a_{j,j-3}^\dagger \right] |0\rangle. \qquad (7.116)$$

7.8 Constrained Hartree-Fock methods

The addition of a Lagrangian constraint term to the Hamiltonian of HFB theory is required to ensure that the states described by this theory have the desired mean value of the nucleon number equal to that of the nucleus under investigation. A similar technique can also be used to extract extra, more detailed, information from Hartree-Fock calculations. For example, by adding a Lagrangian constraint term, $-\alpha \hat{J}_x$, to the Hamiltonian of Hartree-Fock theory, so that $\hat{H} \to \hat{H} - \alpha \hat{J}_x$, where $\hat{J}_x$ is a component of the nuclear angular-momentum operator, it is possible to seek solutions of the Hartree-Fock equations with different expectation values of $\hat{J}_x$. The energy dependence of the solution on this expectation value then provides information about nuclear moments of inertia and the way they change with increasing angular momentum. One can similarly seek to map out the dependence of the nuclear energy on various deformation parameters. However, the fundamental principles underlying these methods are most usefully discussed in terms of an interpretation of the Hartree-Fock approximation as an example of a more general map from quantum to classical mechanics by use of coherent state methods. Thus, we defer further discussion of this interesting topic to the following chapter which considers such general coherent-state methods.

7.9 Applications of mean-field models

Mean-field models have a remarkable ability to provide a basic level of understanding of many nuclear structure phenomena, as will be shown in an exploration of the unified models in Volume 2. At first sight, it is surprising to find that phenomena associated more naturally with correlated-nucleon, rather than independent-particle, dynamics have first-order explanations in mean-field theory. In fact, the mean-field is generated by all the nucleons in the nucleus and is very much a collective construct. This observation underlies the the wide-ranging success of the Bohr-Mottelson unified model,[34] much of which will be covered in Volume 2. Particularly remarkable is the fact that, by the inclusion of a pairing field in the BCS and HFB models, mean-field theory has even been able to provide a first-order explanation of pair-coupling phenomena.[35,36] In doing so, they succeed in a way that is hard to improve upon, in handling the competition between seemingly incompatible degrees of freedom, e.g., in the description of deformed nuclei with superconducting rotational flows, and in describing the transitions between phases in which different competing dynamics dominate. Thus, although more detailed and more sophisticated methods are needed in particular situations, the overall perspective and unification of many nuclear models brought about by the mean-field methods is unique.

The following are some valuable attributes of the mean-field methods:

(i) They can be carried out in huge spaces of single-particle wave functions.

(ii) They predict the ground-state properties of A-particle nuclei, for a given Hamiltonian, with increasing accuracy as $1/A \to 0$. Thus, they are particularly useful for interpreting and predicting ground-state masses (binding energies), sizes (radii), shapes, and stabilities of a wide range of (relatively heavy) nuclei.

(iii) The time-dependent mean-field approximations (described in the following chapter), provide elegant and powerful techniques for a quantal description of normal-mode vibrational excitations. The small-amplitude time-dependent HF and HFB description of vibrational excitations turns out to be equivalent to the so-called *random phase approximation* (to be discussed in more depth in Volume 2).

(iv) The Hartree-Fock and HFB theories are prototypical examples of a more general classical approximation in which the quantal dynamics is restricted to an orbit of coherent states. Thus, they establish physical connections to the orbit methods used in the theory of group and Lie algebra representations[37] and in the theory of quantization.[38,39]

(v) Mean-field methods have the potential to be generalised in creative ways to yield important information about the most appropriate coupling schemes

---

[34]The unified model, initiated by A. Bohr, and developed by A. Bohr and B.R. Mottelson and many colleagues and visitors to the Niels Bohr Institute in Copenhagen, is the main theme of their book, Bohr and Mottelson, *op. cit.* Footnote 6 on Page 98.

[35]Bohr *et al.*, *op. cit.* Footnote 15 on Page 33.

[36]Belyaev, *op. cit.* Footnote 71 on Page 462.

[37]Kirillov A.A. (1976), *Elements of the Theory of Representations* (Springer-Verlag, Berlin).

[38]Souriau J.-M. (1966), *Comm. Math. Phys.* **1**, 374.

[39]Kostant B. (1970), in *Group Representations in Mathematics and Physics*, *Lecture Notes in Physics*, Vol. 6, edited by V. Bargmann (Springer-Verlag, Berlin).

and truncated spaces for shell-model calculations. One possibility is discussed in the concluding paragraph of this section.

7.9.1 *Effective interactions for mean-field calculations*

Mean-field approximations can be implemented in very large spaces. The minimal-energy state is nevertheless a single Slater determinant (or quasi-particle vacuum state). Thus, the effective interactions derived for many-particle shell model calculations restricted to relatively small valence-shell spaces are generally inappropriate for mean-field calculations. Effective interactions should be designed specifically for their intended use. The effective interaction derived for an A-nucleon Hilbert space, $\mathbb{H}_1^{(A)}$, spanned by a set of Slater determinants of single-nucleon wave functions from a large space, $\mathbb{H}_1$, as defined in Section 5.7.3, should be a good starting point.

Recall, from Section 5.7.5, that if two-body interactions were sufficient to describe the interaction between free nucleons, one would expect that by taking the space of single-particle states to be sufficiently large, the effective interaction for the corresponding A-nucleon space should likewise only require two-body interactions. However, it appears that mean-field calculations fail to reproduce the observed densities of nuclear-matter without the inclusion of three-body interactions. Moreover, it is not clear to what extent the required three-body interactions originate at a fundamental level as opposed to being generated by the mean-field restriction to a single-determinantal ground state. To obtain the ground-state properties of nuclei in a mean-field approximation, certainly major contributions to the effective interactions are to be expected from the suppressed coupling of the particle-hole vacuum to its many excited states. These contributions to the effective interactions will lower the ground-state energy of a nucleus by the so-called *correlation energy*.[40]

An important perspective on an approach to deriving an effective interaction for use in mean-field approximations is the observation that the basic interactions between nucleons in such a model are mediated by the mean fields that they generate. These fields are functionals of the nuclear density. Thus, to include three- and higher many-body contributions, it is appropriate to seek effective interactions between pairs of nucleons which depend on the local density of nuclear matter at the point of interaction. The idea of local-density-dependent effective interactions was explored by Negele[41] who proposed deriving them from nuclear matter calculations. Vautherin and Brink[42] showed that, when used in a mean-field calculation, an interaction due to Skyrme,[43,44] which included three-body terms, becomes equivalent to a two-body density-dependent interaction. A more general density-dependent

---

[40]Inclusion of the correlation energy is, in fact, the most challenging problem in the Brueckner, Bethe, Goldstone program (Brueckner, *op. cit.* Footnote 3 on Page 235; Bethe and Goldstone, *op. cit.* Footnote 4 on Page 235) for calculating the binding energy of nuclear matter.

[41]Negele J.W. (1970), *Phys. Rev.* **C1**, 1260.

[42]Vautherin D. and Brink D.M. (1972), *Phys. Rev.* **C5**, 626.

[43]Skyrme T.H.R. (1958/9), *Nucl. Phys.* **9**, 615.

[44]Skyrme T.H.R. (1958/9), *Nucl. Phys.* **9**, 635.

interaction was also proposed by Gogny.[45] The Skyrme and Gogny interactions, with adjustable parameters, subsequently became the most commonly used interactions in non-relativistic mean-field calculations. To quote from the review article of Bender *et al.*[46]

> The breakthrough came when the connection to the bare nucleon-nucleon force was abandoned and effective interactions tailored for use in mean-field calculations were directly adjusted to observables of finite nuclei. The rediscovery of Skyrme's interaction by Vautherin and Vénéroni (1969) and Vautherin and Brink (1972), the introduction of the Gogny force (Gogny, 1973), and finally the formulation of the relativistic mean-field model (Walecka, 1974; Boguta and Bodmer, 1977)[47,48] led to three "standard models" for the nuclear mean field which are widely used today and able to compete with the mic-mac[49] method on a quantitative level.

Determination of the relevant effective interaction for use in a mean-field calculation is related to deriving the expression for the ground-state energy in density functional theory.[50,51] A theorem of Hohenberg and Kohn[50] shows the ground-state energy of a system of atomic electrons in the Coulomb field of a nucleus to be a universal functional of the electron density. The theorem does not give an algorithm for deriving the functional. However, knowledge of its existence has prompted numerous studies of Coulomb many-body problems in terms of density functional theory that have met with remarkable success. It has also prompted a search for similar applications in nuclear physics,[52] based on an understanding that the Hohenberg-Kohn theorem also applies to nuclei.[53] It would then appear that a good approximation to such a density functional treatment of nuclear ground states would be given by a Hartree-Fock approximation for the density distribution and ground-state energy of a nucleus (presumably subject to corrections for symmetry breaking effects) with suitable density-dependent effective interactions.

---

[45]Gogny's original proposal, Gogny D. (1973), in *Proc. Int. Conf. on Nucl. Phys. (Munich)*, edited by J. de Boer and H.J. Mang (North-Holland, Amsterdam), was developed by Dechargé J. and Gogny D. (1980), *Phys. Rev.* **C21**, 1568.

[46]Bender M., Heenen P.-H. and Reinhard P.-G. (2003), *Rev. Mod. Phys.* **75**, 121.

[47]Walecka J.D. (1974), *Ann. Phys. (NY)* **83**, 491.

[48]Boguta J. and Bodmer A.R. (1977), *Nucl. Phys.* **A292**, 413.

[49]The so-called *mic-mac* method refers to a microscopic-macroscopic method for computing nuclear sizes, shapes, and binding energies by a combination of the macroscopic liquid-drop model with microscopic shell-model corrections. A technique for adding shell-model corrections to the liquid-drop model was originally proposed by Strutinski V.M. (1967), *Nucl. Phys.* **A95**, 420. The mic-mac model has been reviewed by Möller P. *et al.* (1995), *At. Data Nucl. Data Tables* **59**, 185.

[50]Hohenberg P. and Kohn W. (1964), *Phys. Rev.* **136**, B864.

[51]Kohn W. and Sham L. (1965), *Phys. Rev.* **140**, A1133.

[52]Cf. for example, Drut J., Furnstahl R. and Platter L. (2009), arXiv:0906.1463v2 [nucl-th].

[53]Bertsch G.F. (2007), *J. of Phys: Conf. Series* **78**, 012005.

7.9.2   Calculations of nuclear properties

The following results have been compiled by Bender et al.[54] from a number of mean-field calculations by different authors using different interactions and/or different parameters. They include: the Skyrme interactions SLy6 of Chabanat *et al.*,[55] SkI3 of Reinhard and Flocard,[56] and BSk1 of Samyn *et al.*;[57] the Gogny interaction D1S of Berger *et al.*;[58] and the relativistic mean-field parameterisations, NL-3 of Lalazissis *et al.*[59,60] and NL-Z2 of Bender *et al.*.[61]

(i) Nuclear masses

The most fundamental property of a nucleus is its mass or, equivalently, its binding energy. In particular, the binding energy of a nucleus determines the available energy that the nucleus can contribute to any reaction or decay in which it is involved. Thus, a first test of a global theory of nuclei is the ability of the theory to predict nuclear masses. Establishing reliable methods for predicting nuclear masses is not only important for understanding why known nuclei have the masses that they have, it is also important for predicting the values of the neutron and proton numbers for which new bound nuclei might be found.

Figure 7.1 shows the differences between experimental and calculated binding energies for nuclei with a singly-closed proton shell. Typical experimental binding energies are, e.g., ^{56}Ni (484 MeV), ^{208}Pb (1636 MeV). The results of the calculations shown achieve better than 1% agreement with measured values.

Systematic deviations are observed between the calculated and measured binding energies shown in Figure 7.1. The largest deviations are for nuclei farthest from the closed neutron shells, for which the calculated binding energies tend to be less than those observed. This suggests that there are additional correlations in the midshell regions that are not accounted for in the theory.

(ii) Nucleon separation energies

Quantities of major importance that are derived from nuclear masses are nucleon separation energies. When these go to zero, the drip line (proton or neutron) has been reached. Thus, separation energies lead to the delineation of the limits of nuclear stability and the boundaries to what is experimentally accessible. Figure 6.21 shows such a limit for the the proton-drip line at N = 82. Figure 7.2 shows two-neutron separation energies for the tin isotopes. The calculations predict that

---

[54]Bender *et al.*, *op. cit.* Footnote 46 on Page 544.
[55]Chabanat E. *et al.* (1998), *Nucl. Phys.* **A635**, 231, (erratum, *ibid.* **A643**, 441).
[56]Reinhard P.-G. and Flocard H. (1995), *Nucl. Phys.* **A584**, 467.
[57]Samyn M. *et al.* (2002), *Nucl. Phys.* **A700**, 142.
[58]Berger J.-F., Girod M. and Gogny D. (1984), *Nucl. Phys.* **A428**, 23c.
[59]Lalazissis G.A., König J. and Ring P. (1997), *Phys. Rev.* **C55**, 540.
[60]Lalazissis G.A., Raman S. and Ring P. (1999), *At. Data Nucl. Data Tables* **71**, 1.
[61]Bender M. *et al.* (1999), *Phys. Rev.* **C60**, 034304.

7 Mean-field approximations

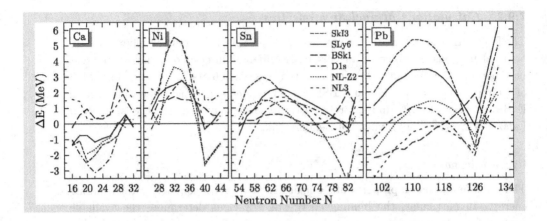

Figure 7.1: Deviations from experimental values of binding energies for singly-closed shell nuclei calculated using spherical mean-field approximations with various effective nucleon-nucleon interactions. Positive values correspond to under-binding with respect to experiment. (The figure is from Bender *et al.*, *op. cit.* Footnote 46 on Page 544.)

the neutron drip line for $Z = 50$ (which is far beyond experimentally known masses) is at $N \sim 126$, which is a closed neutron shell.

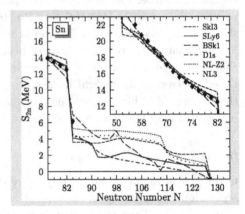

Figure 7.2: Two-neutron separation energies, S_{2n}, for the tin isotopes; a comparison between spherical mean-field calculations and experiment. The inset gives the comparison for $N \leq 82$. The main part of the figure shows the results of the various mean-field calculations out to the neutron drip line; all predict the $N = 126$ nucleus, i.e., ^{176}Sn, to be the most neutron-rich bound isotope of tin. (The figure is from Bender *et al.*, *op. cit.* Footnote 46 on Page 544.)

(iii) *Nuclear radii*

A second fundamental property of a nucleus is its size as determined by the root-mean-square of its radius. This is a manifestation of the density of the nuclear matter within the nucleus. As discussed in Section 1.2, observed rms radii of nuclei are fitted rather accurately by the expression $\left[\frac{5}{3} \langle r^2 \rangle_{\text{expt}} \right]^{\frac{1}{2}} \approx 1.1\, A^{1/3} + 0.65$ fm which implies that the volume of a nucleus is proportional to its nucleon number. This, in turn, implies that the density of nuclear matter in the interior of a nucleus saturates at a nucleon-number-independent value (see Figure 1.7). The resistance of nuclear matter to collapse under the basically attractive nucleon-nucleon interaction is due to the strong repulsive nature of this interaction at short distances and the Pauli exclusion principle. However, it turns out that the observed saturation properties of nuclei are only reproduced accurately in mean-field calculations if three-body, or equivalently, suitable density-dependent interactions are used. Thus, the success of a mean-field calculation in fitting observed nuclear radii is a sensitive test of the interaction used.

Figure 7.3 shows root-mean-squared radii in femtometres for singly-closed proton shell nuclei. The deviations are generally at the $\sim 2\%$ level; the calcium isotopes exhibit a systematic deviation indicating that they are slightly 'larger' than predicted.

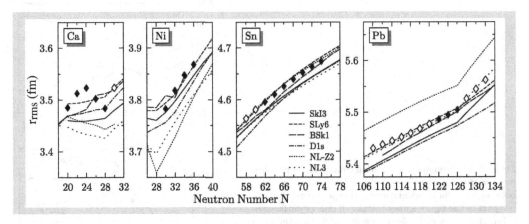

Figure 7.3: Root-mean-square radii, R_{rms}, in fm, for singly-closed proton shells; a comparison between spherical mean-field calculations and experiment. (The figure is from Bender *et al.*, *op. cit.* Footnote 46 on Page 544.)

(iv) *Nuclear deformations*

The lowest-energy self-consistent mean-field solutions for closed-shell nuclei are normally expected to be spherically symmetric. On the other hand, those for doubly-open-shell nuclei are generally deformed. Thus, deformation is associated with shell filling in accord with the nuclear Jahn-Teller effect as discussed, for example, by Nazarewicz.[62] When pairing fields are included, in HF-BCS or HFB calculations, the mean fields for singly-open-shell nuclei are likely to be spherical and superconducting.

As discussed in Section 7.6.4, the breaking of rotational symmetry in a mean-field calculation is an indication that the corresponding nucleus has a ground-state rotational band. Thus, self-consistent mean-field models provide valuable indica-

---

[62]Nazarewicz W. (1993), *Int. J. Mod. Phys.* **E2**, 51.

tions of where deformation might occur and what its magnitude might be.

It is also observed that doubly-closed and singly-closed-shell nuclei frequently exhibit excited rotational bands at low excitation energies. This phenomenon is known as *shape coexistence*.[63,64] Thus, it is of interest to study the energies of lowest-energy mean-field states under the constraint that each such state have a given value for a range of possible values of, say, a quadrupole moment. The presence of an excited deformed local-minimum energy would then suggest the existence of a corresponding low-energy excited rotational band.

Figure 7.4 shows HF-BCS energy surfaces as a function of a quadrupole moment,

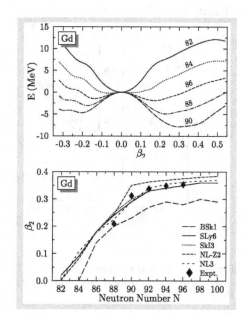

Figure 7.4: The shape deformation change in the gadolinium isotopes between $N = 82$ and $N = 100$; a comparison between mean-field calculations and experiment. The upper part of the figure shows calculated energy surfaces as a function of quadrupole deformation, β_2, for the even $N = 82$ to $N = 90$ isotopes. The lower part of the figure shows a comparison between experimental and predicted values of this deformation parameter. (The figure is from Bender *et al.*, *op. cit.* Footnote 46 on Page 544.)

β_2, for the gadolinium isotopes in the region where these isotopes undergo a rapid change from spherical shapes (at $N = 82$) to deformed shapes (at $N \sim 90$). The figure shows that the $N = 82$ gadolinium isotope, which is a singly-closed shell nucleus, has a minimum mean-field energy when $\beta_2 = 0$. However, this spherical minimum energy solution is replaced by a lower energy deformed ($\beta_2 > 0$) solution for the doubly-open shell isotopes with $N > 84$. The lower part of the figure compares the quadrupole deformations predicted by the mean-field calculations with experimentally deduced values.

---

[63]Heyde K. *et al.* (1983), *Phys. Repts.* **102**, 291.
[64]Wood *et al.*, *op. cit.* Footnote 60 on Page 221.

7.9.3 *A potential application of generalised mean-field methods*

Mean-field methods can, in principle, assist in optimising some of the choices to be made in designing a shell-model calculation within a truncated Hilbert space. For example, given a Hamiltonian, one needs to select a coupling scheme that will give the most accurate results for a truncated space of minimal dimensions. In addition, one needs to select basis states classified by the chosen coupling scheme to define the truncated Hilbert space. As we have seen in Chapter 5, a coupling scheme is defined by a subgroup chain. Thus, choosing subsets of basis states naturally corresponds to selecting irreps of subgroups in the chosen chain. Moreover, to each subgroup there corresponds a system of coherent states. Thus, mean-field methods can be used to assign a minimum-energy state to each subgroup in the chain. If this were done for each subgroup of the chains corresponding to potential coupling schemes it would clearly be of considerable assistance in making an intelligent choice of coupling scheme and a truncated basis for a shell-model calculation with a given Hamiltonian.

Chapter 8

Time-dependent mean-field theory; a classical approximation

The Hartree-Fock and HFB mean-field theories are special cases of more general mean-field theories in which quantum mechanics is constrained to suitable manifolds of coherent states. We show in this chapter that, when quantum mechanics is constrained to such submanifolds of a Hilbert space, the time-evolution of states is determined by classical Hamilton equations of motion.[1] Thus, mean-field theories and their generalisations are interpreted as classical approximations. This suggests ways in which they can be used and, conversely, how classical systems may be quantised. In particular, because of the close relationship between the classical and quantum mechanics of harmonic oscillators, the small-amplitude normal-mode vibrational states of time-dependent Hartree-Fock (TDHF) theory are shown to have an immediate interpretation in terms of low-energy vibrational states of nuclei.

8.1 The equations of motion of constrained quantum mechanics

The quantum dynamics of a system is determined by each of two fundamentally equivalent equations: the time-dependent Schrödinger equation,

$$\hat{H}|\psi(t)\rangle = i\hbar \frac{\partial}{\partial t}|\psi(t)\rangle, \qquad (8.1)$$

and the Heisenberg equations of motion for the expectation values of quantum-mechanical observables,

$$i\hbar \frac{d}{dt}\langle\psi(t)|\hat{X}|\psi(t)\rangle = \langle\psi(t)|[\hat{X}, \hat{H}]|\psi(t)\rangle. \qquad (8.2)$$

---

[1] The observation that quantum mechanics, when constrained to a space of Slater determinants, obey classical Hamilton equations was made, e.g., by Kerman A.K. and Koonin S.E. (1976), *Ann. Phys. (NY)* **100**, 332; Rowe D.J. and Basserman R. (1976), *Can. J. Phys.* **54**, 1941; Rowe D.J., Ryman A. and Rosensteel G. (1980), *Phys. Rev.* **A22**, 2362; and Kramers P. and Saraceno M. (1981), *Geometry of the Time-Dependent Variational Principle in Quantum Mechanics*, *Lecture Notes in Physics*, Vol. 140 (Springer-Verlag, Berlin). A more general procedure for deriving classical Hamilton dynamics from their quantal counterparts was reviewed by Bartlett S.D. and Rowe D.J. (2003), *J. Phys. A: Math. Gen.* **36**, 1683.

A small but important difference between these two equations arises because the expectation value of a linear operator, $\langle\psi|\hat{X}|\psi\rangle$, is independent of the phase of the state vector, $|\psi\rangle$, i.e., $\langle\psi|e^{-i\theta}\hat{X}e^{i\theta}|\psi\rangle = \langle\psi|\hat{X}|\psi\rangle$. Thus, Equation (8.2) is really an equation of motion for the density operator, ρ_ϕ, defined for the state vector, $|\phi\rangle$, such that

$$\rho_\phi(\hat{X}) := \langle\phi|\hat{X}|\phi\rangle \qquad (8.3)$$

for a linear operator $\hat{X}$ on the Hilbert space.

We use the latter equations of motion to derive the time evolution of states when constrained to a submanifold of the Hilbert space.[2] When the Hilbert space carries a unitary irrep of a dynamical group, G, a suitable submanifold consists of coherent states generated by this group with the proviso that the states of the submanifold are uniquely defined, to within phase factors, by the expectation values of a set of operators, $\{\hat{X}_\nu\}$, on $\mathbb{H}$, that span the representation of the Lie algebra, $\mathbf{g}$, of the group G. It will be shown that the time evolution of states constrained to such a submanifold is then well defined by the equations of motion

$$i\hbar\frac{d}{dt}\langle\psi(t)|\hat{X}_\nu|\psi(t)\rangle = \langle\psi(t)|[\hat{X}_\nu,\hat{H}]|\psi(t)\rangle, \quad \hat{X}_\nu \in \mathbf{g}, \qquad (8.4)$$

even when the operator $[\hat{X}_\nu,\hat{H}]$ is not an element of $\mathbf{g}$.

In Hartree-Fock theory, for which G is the group of one-body unitary transformations of the many-particle Hilbert space, the constraint submanifold of coherent states is the set of Slater determinants. These coherent states are particularly appropriate because a Slater determinant is uniquely defined by the expectation value of the one-body operators that span the Lie algebra of G and, as illustrated by Exercise 8.3, a Slater-determinant is a minimal-uncertainty state. Thus, because Hartree-Fock theory is a prototype of constrained quantum mechanics, we first consider the time evolution of states on the manifold of Slater determinants.

Let $|\phi(t)\rangle$ denote a determinantal state at time t and let i and m index single-particle states that are, respectively, occupied and unoccupied in this many-particle state. Then, from Equation (7.35), it follows that the state $|\phi(t+\delta t)\rangle$ to which $|\phi(t)\rangle$ evolves at time $t + \delta t$ can be expressed, to within a phase factor, by

$$|\phi(t + \delta t)\rangle = \exp\left[\delta t\sum_{mi}\left(C_{mi}a_m^\dagger a_{\bar{i}} - C_{mi}^*a_i^\dagger a_{\bar{m}}\right)\right]|\phi(t)\rangle + 0(\delta t^2). \qquad (8.5)$$

---

[2]Equations of motion for constrained quantum mechanics were first considered by Dirac P.A.M. (1930), *Cambridge Philos. Soc* **26**, 376 who derived them by considering extremal paths for the action integral $\int_{t_1}^{t_2}\langle\Phi(t)|\hat{H} - i\hbar\partial/\partial t|\phi(t)\rangle dt$ on the constraint hypersurface. As shown, for example, by Rowe D.J., Ryman A. and Rosensteel G. (1980), *Phys. Rev.* **A22**, 2362, this variational principle leads to a well-defined dynamics, if and only if the constraint hypersurface is a symplectic manifold, i.e., if and only if it has the properties of a classical phase space. The resulting dynamics is then expressed by classical Hamilton equations of motion. The algebraic approach adopted in this chapter, and developed in generality in Section 8.3, gives results identical to those of Dirac's variational principle, but is simpler and relates more directly to the underlying geometry of the submanifold.

It is seen that

$$\langle \phi(t + \delta t) | a_i^\dagger a_{\bar{m}} | \phi(t + \delta t) \rangle = \delta t \, C_{mi} + 0(\delta t^2) \tag{8.6}$$

and hence, by taking a time derivative, that

$$C_{mi} = \frac{d}{dt} \langle \phi(t) | a_i^\dagger a_{\bar{m}} | \phi(t) \rangle. \tag{8.7}$$

Thus, from Equation (8.4), the time-evolution of a TDHF state is given by

$$C_{mi} = -\frac{i}{\hbar} \langle \phi(t) | [a_i^\dagger a_{\bar{m}}, \hat{H}] | \phi(t) \rangle. \tag{8.8}$$

There are infinitely-many time-evolving paths on the manifold of Slater determinants and a challenge is to determine which paths are of interest. An application that has been explored is the description of heavy-ion scattering reactions for which it may be expected that the trajectories of the ions follow essentially classical paths but interact with one another in a quantum mechanical manner.[3] Another potentially useful application is to the identification of the relevant degrees of freedom involved in the structure of low-energy collective states of nuclei. We consider first the small-amplitude (normal-mode) solutions that have an immediate quantum-mechanical interpretation in terms of harmonic vibrations. Then, in Section 8.4, we consider the more general adiabatic dynamics of relevance when larger amplitudes and/or anharmonic dynamics are important.

Exercises

8.1 Show that if $\hat{A}$ and $\hat{B}$ are self-adjoint linear operators on a Hilbert space, $\mathbb{H}$, and if $|\phi\rangle$ is a member of a suitably-defined basis for $\mathbb{H}$ (e.g., a basis of states for which the expectation values $\langle \hat{A}^2 \rangle$, $\langle \hat{A}\hat{B} \rangle$, and $\langle \hat{B}^2 \rangle$ are finite),[4] then the identity $\langle \phi | (\hat{A} \pm i\lambda\hat{B})(\hat{A} \mp i\lambda\hat{B}) | \phi \rangle \geq 0$, for λ a real number, implies that

$$\langle \phi | \hat{A}^2 | \phi \rangle + \lambda^2 \langle \phi | \hat{B}^2 | \phi \rangle \pm i\lambda \langle \phi | [\hat{A}, \hat{B}] | \phi \rangle \geq 0. \tag{8.9}$$

Hence, by replacing $\hat{A}$ and $\hat{B}$ by $\hat{A} - \langle \phi | \hat{A} | \phi \rangle$ and $\hat{B} - \langle \phi | \hat{B} | \phi \rangle$, respectively, in this derivation, show that

$$\sigma_\phi^2(A) + \lambda^2 \sigma_\phi^2(B) \pm i\lambda \langle \phi | [\hat{A}, \hat{B}] | \phi \rangle \geq 0, \tag{8.10}$$

---

[3] A perspective of early activity in the field can be gained from the collection of contributions in Goeke K. and Reinhard P.-G., eds. (1982), *Proc. Int. Symposium, held at Bad Honnef, Germany, June 7–11, 1982, Lecture Notes in Physics*, Vol. 171 (Springer-Verlag, Berlin), and from the review article of Goeke K. *et al.* (1983), *Prog. Theor. Phys. Suppl.* **74**, 33. Later developments can be traced from articles in Bromley D.A., ed. (1985), *Treatise on Heavy-Ion Science*, Vol. 3 (Plenum Press, New York and London), and the review article of Feldmeier H. and Schnack J. (1997), *Prog. Part. Nucl. Phys.* **39**, 393.

[4] Even for the most familiar example, namely the $\hat{x} := x$ and $\hat{p} := -i\hbar d/dx$ operators of the Heisenberg-Weyl algebra, it is necessary to restrict their actions to smooth differentiable functions and to subsets of states in the Hilbert space for which $\langle \hat{x}^2 \rangle$ is well defined. A discussion of these technicalities is given, for example, in the book of Emch G.G. (1984), *Mathematical and Conceptual Foundations of 20th Century Physics* (North-Holland, Amsterdam).

where

$$\sigma_\phi^2(A) := \langle\phi|\hat{A}^2|\phi\rangle - \langle\phi|\hat{A}|\phi\rangle^2. \tag{8.11}$$

Now use the identity

$$[\sigma_\phi(A) - \lambda\sigma_\phi(B)]^2 - [\sigma_\phi^2(A) + \lambda^2\sigma_\phi^2(B)] + 2\lambda\sigma_\phi(A)\sigma_\phi(B) = 0 \tag{8.12}$$

to show that

$$[\sigma_\phi(A) - \lambda\sigma_\phi(B)]^2 + 2\lambda\left[\sigma_\phi(A)\sigma_\phi(B) \mp \tfrac{1}{2}\mathrm{i}\langle\phi|[\hat{A},\hat{B}]|\phi\rangle\right] \geq 0 \tag{8.13}$$

and, by setting $\lambda := \sigma_\phi(A)/\sigma_\phi(B)$, derive the uncertainty relation

$$\sigma_\phi(A)\sigma_\phi(B) \geq \tfrac{1}{2}\left|\langle\phi|[\hat{A},\hat{B}]|\phi\rangle\right|. \tag{8.14}$$

8.2 Derive the standard Heisenberg uncertainty relationship, $\Delta x \Delta p \geq \tfrac{1}{2}\hbar$, from the general expression given by Equation (8.14).

8.3 Show that if

$$\hat{Q}_{mi} = \frac{1}{\sqrt{2}}(\hat{a}_m^\dagger a_{\bar{\imath}} + \hat{a}_i^\dagger a_{\bar{m}}), \quad \hat{P}_{mi} = \frac{\mathrm{i}}{\sqrt{2}}(\hat{a}_m^\dagger a_{\bar{\imath}} - \hat{a}_i^\dagger a_{\bar{m}}), \tag{8.15}$$

where i and m index single-particle states that are, respectively, occupied and unoccupied in the Slater-determinant state $|\phi\rangle$, then

$$\sigma_\phi(Q_{mi})\sigma_\phi(P_{mi}) = \tfrac{1}{2}\left|\langle\phi|[\hat{Q}_{mi},\hat{P}_{mi}]|\phi\rangle\right|. \tag{8.16}$$

8.2 Small-amplitude normal-mode vibrations

Stationary-state mean-field theory gives an approximation for the ground state of a nucleus, or in the case of a broken symmetry, an approximation for an intrinsic state of a collective band of states. We now show that the solutions of the TDHF equations for small-amplitude *normal-mode* vibrations give predictions for low-energy vibrational excitations.

8.2.1 *Small-amplitude TDHF solutions*

We consider TDHF wave functions of the form

$$|\phi(t)\rangle = e^{-\frac{\mathrm{i}}{\hbar}E_0 t}e^{\epsilon\hat{X}(t)}|\phi\rangle, \tag{8.17}$$

with

$$\hat{X}(t) := \sum_{mi}\left[X_{mi}(t)a_m^\dagger a_{\bar{\imath}} - X_{mi}^*(t)a_i^\dagger a_{\bar{m}}\right], \tag{8.18}$$

for small-amplitude vibrations of a nucleus about a Hartree-Fock minimum-energy state $|\phi\rangle$. From the identity

$$\langle\phi(t)|a_i^\dagger a_{\bar{m}}|\phi(t)\rangle = \epsilon X_{mi}(t) + 0(\epsilon^2), \qquad (8.19)$$

and Equation (8.2),

$$i\hbar\frac{d}{dt}\langle\phi(t)|a_i^\dagger a_{\bar{m}}|\phi(t)\rangle = \langle\phi(t)|[a_i^\dagger a_{\bar{m}}, \hat{H}]|\psi(t)\rangle, \qquad (8.20)$$

we obtain, by equating the leading order terms in the arbitrarily small amplitude ϵ,

$$i\hbar\frac{d}{dt}X_{mi}(t) = \langle\phi|[[a_i^\dagger a_{\bar{m}}, \hat{H}], \hat{X}(t)]|\phi\rangle. \qquad (8.21)$$

Thus, with the expressed according to Equation (7.62) by

$$\hat{H} = E_0 + \sum_\nu \varepsilon_\nu : a_\nu^\dagger a_{\bar{\nu}} : + \tfrac{1}{4}\sum_{\mu\nu\mu'\nu'} V_{\mu\nu\mu'\nu'} : a_\mu^\dagger a_\nu^\dagger a_{\bar{\nu}'} a_{\bar{\mu}'} :, \qquad (8.22)$$

we obtain the equation for the $X_{mi}(t)$ coefficients

$$i\hbar\frac{d}{dt}X_{mi}(t) = (\varepsilon_m - \varepsilon_i)X_{mi}(t) + \sum_{nj} V_{mjin}X_{nj}(t) + V_{mnij}X_{nj}^*(t). \qquad (8.23)$$

Among the small-amplitude TDHF solutions we expect to find harmonic vibrations about the Hartree-Fock minimum-energy state. Thus, we seek solutions of the form

$$X_{mi}(t) = Y_{mi}e^{-i\omega t} + Z_{mi}^* e^{i\omega t}. \qquad (8.24)$$

Equation (8.23) then separates into a set of coupled eigenvalue equations:

$$\sum_{nj}\left[A_{minj}Y_{nj} + B_{minj}Z_{nj}\right] = \hbar\omega Y_{mi}, $$
$$\sum_{nj}\left[A_{minj}Z_{nj}^* + B_{minj}Y_{nj}^*\right] = -\hbar\omega Z_{mi}^*, \qquad (8.25)$$

where A is a Hermitian matrix and B is a symmetric matrix with components

$$A_{minj} := \langle\phi|[a_i^\dagger a_{\bar{m}}, [\hat{H}, a_n^\dagger a_{\bar{j}}]]|\phi\rangle = (\varepsilon_m - \varepsilon_i)\delta_{mi,nj} + V_{mjin},$$
$$B_{minj} := -\langle\phi|[a_i^\dagger a_{\bar{m}}, [\hat{H}, a_j^\dagger a_{\bar{n}}]]|\phi\rangle = V_{mnij}. \qquad (8.26)$$

After taking the complex conjugate of the second equation, these equations can be expressed in matrix form

$$\begin{pmatrix} A & B \\ B^* & A^* \end{pmatrix}\begin{pmatrix} Y \\ Z \end{pmatrix} = \hbar\omega\begin{pmatrix} Y \\ -Z \end{pmatrix}. \qquad (8.27)$$

8.2.2 *Small-amplitude vibrations of anharmonic oscillators*

The small-amplitude TDHF solutions clearly behave as a classical system of coupled harmonic oscillators. Thus, to interpret the results in quantum mechanical terms, it is appropriate to consider the corresponding results for a quantum system in terms of many harmonic-oscillator degrees of freedom.

Let $\hat{\mathcal{H}}$ denote a Hermitian Hamiltonian for a multi-dimensional harmonic oscillator and let $|\phi\rangle$ denote the vacuum state for a set of boson creation and annihilation operators, $\{c_\nu^\dagger, c_{\bar{\nu}}\}$, for this space that obey the standard commutation relations, e.g., $[c_{\bar{\mu}}, c_\nu^\dagger] = \delta_{\mu,\nu}$ and satisfy the variational equation

$$\langle\phi|[\hat{\mathcal{H}}, c_\nu^\dagger]|\phi\rangle = \langle\phi|[\hat{\mathcal{H}}, c_{\bar{\nu}}]|\phi\rangle = 0. \tag{8.28}$$

It then follows that the Hamiltonian, $\hat{\mathcal{H}}$, has an expansion given, up to quadratic terms in these boson operators, by

$$\hat{\mathcal{H}} := W_0 + \sum_{\mu\nu}\left(A_{\mu\nu}c_\mu^\dagger c_{\bar{\nu}} + \tfrac{1}{2}B_{\mu\nu}c_\mu^\dagger c_\nu^\dagger + \tfrac{1}{2}B_{\mu\nu}^*c_{\bar{\nu}}c_{\bar{\mu}}\right) + \cdots. \tag{8.29}$$

Small-amplitude vibrational solutions for $\hat{\mathcal{H}}$ are of the form

$$|\phi(t)\rangle = e^{-\frac{i}{\hbar}E_0 t}e^{\epsilon\hat{X}(t)}|\phi\rangle, \tag{8.30}$$

where, because $e^{\hat{X}(t)}$ is unitary, $\hat{X}(t)$ can be expressed as

$$\hat{X}(t) := \epsilon\left(\mathcal{O}_\lambda^\dagger e^{-i\omega_\lambda t} - \mathcal{O}_{\bar{\lambda}}e^{i\omega_\lambda t}\right). \tag{8.31}$$

If we seek approximate solutions with

$$\mathcal{O}_\lambda^\dagger := \sum_\nu\left(Y_\nu(\lambda)c_\nu^\dagger - Z_\nu(\lambda)c_{\bar{\nu}}\right), \quad \mathcal{O}_{\bar{\lambda}} := \sum_\nu\left(Y_\nu^*(\lambda)c_{\bar{\nu}} - Z_\nu^*(\lambda)c_\nu^\dagger\right), \tag{8.32}$$

then the equations of motion,

$$\langle\phi(t)|[c_\nu, \hat{\mathcal{H}}]|\psi(t)\rangle = i\hbar\frac{d}{dt}\langle\phi(t)|c_\nu|\phi(t)\rangle, \tag{8.33}$$

determine that the operators $\mathcal{O}_\lambda^\dagger$ and $\mathcal{O}_{\bar{\lambda}}$ should satisfy the equations

$$\langle\phi|[[c_{\bar{\nu}}, \hat{\mathcal{H}}], \mathcal{O}_\lambda^\dagger]|\phi\rangle = \langle\phi|[c_{\bar{\nu}}, [\hat{\mathcal{H}}, \mathcal{O}_\lambda^\dagger]]|\phi\rangle = \hbar\omega_\lambda\langle\phi|[c_{\bar{\nu}}, \mathcal{O}_\lambda^\dagger]|\phi\rangle, \tag{8.34}$$

$$\langle\phi|[[c_{\bar{\nu}}, \hat{\mathcal{H}}], \mathcal{O}_{\bar{\lambda}}]|\phi\rangle = \langle\phi|[c_{\bar{\nu}}, [\hat{\mathcal{H}}, \mathcal{O}_{\bar{\lambda}}]]|\phi\rangle = -\hbar\omega_\lambda\langle\phi|[c_{\bar{\nu}}, \mathcal{O}_{\bar{\lambda}}]|\phi\rangle, \tag{8.35}$$

where use has been made of the Jacobi identity (Exercise 8.4) in deriving these equations. Thus, the Y_ν and Z_ν coefficients are required to satisfy the equations

$$\sum_\nu\left(A_{\mu\nu}Y_\nu(\lambda) + B_{\mu\nu}Z_\nu(\lambda)\right) = \hbar\omega_\lambda Y_\mu(\lambda),$$
$$\sum_\nu\left(A_{\mu\nu}Z_\nu^*(\lambda) + B_{\mu\nu}Y_\nu^*(\lambda)\right) = -\hbar\omega_\lambda Z_\mu^*(\lambda), \tag{8.36}$$

which are of the same form as those given by Equation (8.25).

Manipulation of these equations gives the expressions

$$\langle\phi|[O_{\bar{\kappa}},[\hat{\mathcal{H}},O_\lambda^\dagger]]|\phi\rangle = \langle\phi|[[O_{\bar{\kappa}},\hat{\mathcal{H}}],O_\lambda^\dagger]|\phi\rangle = \hbar\omega_\lambda\langle\phi|[O_{\bar{\kappa}},O_\lambda^\dagger]|\phi\rangle, \qquad (8.37)$$

$$\langle\phi|[O_{\bar{\kappa}},[\hat{\mathcal{H}},O_{\bar{\lambda}}]]|\phi\rangle = \langle\phi|[[O_{\bar{\kappa}},\hat{\mathcal{H}}],O_{\bar{\lambda}}]|\phi\rangle = -\hbar\omega_\lambda\langle\phi|[O_{\bar{\kappa}},O_{\bar{\lambda}}]|\phi\rangle, \qquad (8.38)$$

from which it follows, as for the eigenstates of Hermitian matrices, that the solutions with positive $\hbar\omega_\lambda$ can be normalised to satisfy the identities

$$\langle\phi|[O_{\bar{\kappa}},O_\lambda^\dagger]|\phi\rangle = \delta_{\kappa,\lambda}, \quad \langle\phi|[O_{\bar{\kappa}},O_{\bar{\lambda}}]|\phi\rangle = \langle\phi|[O_\kappa^\dagger,O_\lambda^\dagger]|\phi\rangle = 0. \qquad (8.39)$$

In fact, in this example, they satisfy the boson commutation relations

$$[O_{\bar{\kappa}},O_\lambda^\dagger] = \delta_{\kappa,\lambda}, \quad [O_{\bar{\kappa}},O_{\bar{\lambda}}] = [O_\kappa^\dagger,O_\lambda^\dagger] = 0. \qquad (8.40)$$

When the Hamiltonian, $\hat{\mathcal{H}}$, is expressed in terms of these operators, assuming they form a complete set,[5] it takes the form

$$\hat{\mathcal{H}} := E_0 + \sum_\lambda \hbar\omega_\lambda O_\lambda^\dagger O_{\bar{\lambda}} + \cdots. \qquad (8.41)$$

Thus, with neglect of the higher-order terms, the ground state and one-phonon excited states of $\hat{\mathcal{H}}$ are defined by the equations

$$O_\lambda|0\rangle = 0, \quad |\lambda\rangle := O_\lambda^\dagger|0\rangle, \qquad (8.42)$$

for the solutions corresponding to positive values of $\hbar\omega_\lambda$. As is generally the case for harmonic-oscillator states, one can determine most of their properties directly from the boson operators without actually determining the states. Thus, for example, for any Hermitian operator,

$$\hat{Q} = \sum_\nu (Q_\nu c_\nu^\dagger + Q_\nu^* c_{\bar{\nu}}) + \ldots, \qquad (8.43)$$

that is at most quadratic in the $\{c_\nu^\dagger, c_{\bar{\nu}}\}$ operators, one can directly evaluate the matrix elements

$$\langle 0|\hat{Q}|\lambda\rangle = \langle 0|[\hat{Q},O_\lambda^\dagger]|0\rangle = [\hat{Q},O_\lambda^\dagger] = \sum_\nu (Q_\nu^* Y_\nu(\lambda) + Q_\nu Z_\nu(\lambda)). \qquad (8.44)$$

8.2.3 *Interpretation of the small-amplitude TDHF vibrations*

The above results reflect the observation that classical systems, in general, have harmonic small-amplitude normal modes of vibration about a stable equilibrium state (see the following section for a discussion of the stability conditions). Thus,

---

[5] As shown in the following, the operators $\{O_\lambda^\dagger\}$ and $\{O_{\bar{\lambda}}\}$ form an incomplete set of harmonic-oscillator raising and lowering operators when one or more of the energies, $\hbar\omega_\lambda$, vanishes or takes a complex value.

the above coupled-harmonic-oscillator relationships immediately provide an interpretation of the solutions of Equation (8.25)[6,7] as defining parallel excitation and de-excitation operators,

$$O_\lambda^\dagger := \sum_{mi} \left(Y_{mi}(\lambda) a_m^\dagger a_{\bar{i}} - Z_{mi}(\lambda) a_i^\dagger a_{\bar{m}} \right),$$
$$O_{\bar{\lambda}} := \sum_{mi} \left(Y_{mi}^*(\lambda) a_i^\dagger a_{\bar{m}} - Z_{mi}^*(\lambda) a_m^\dagger a_{\bar{i}} \right),$$
(8.45)

for the one-phonon quantal counterparts of classical normal-mode vibrations. These operators satisfy the equations

$$\langle \phi | [O_{\bar{\kappa}}, [\hat{H}, O_\lambda^\dagger]] | \phi \rangle = \hbar \omega_\lambda \langle \phi | [O_{\bar{\kappa}}, O_\lambda^\dagger] | \phi \rangle,$$
$$\langle \phi | [O_{\bar{\kappa}}, [\hat{H}, O_{\bar{\lambda}}]] | \phi \rangle = -\hbar \omega_\lambda \langle \phi | O_{\bar{\kappa}}, O_{\bar{\lambda}}] | \phi \rangle,$$
(8.46)

and the solutions with $\hbar \omega$ real and positive can be normalised (see Exercise 8.5) to satisfy the orthogonality relationships

$$\langle \phi | [O_{\bar{\kappa}}, O_\lambda^\dagger] | \phi \rangle = \sum_{mi} \left(Y_{mi}^*(\kappa) Y_{mi}(\lambda) - Z_{mi}^*(\kappa) Z_{mi}(\lambda) \right) = \delta_{\kappa, \lambda},$$
(8.47)

$$\langle \phi | [O_{\bar{\kappa}}, O_{\bar{\lambda}}] | \phi \rangle = \langle \phi | [O_\kappa^\dagger, O_\lambda^\dagger] | \phi \rangle = 0.$$
(8.48)

When the TDHF solutions are expressed in this way the operators $\{O_\lambda^\dagger\}$ and $\{O_\lambda\}$ have a natural interpretation as excitation and de-excitation operators of excited states of the nucleus from a so-called *correlated vacuum state*, $|0\rangle$, that is implicitly defined as the solution to the equations

$$O_\lambda |0\rangle = 0, \quad \forall O_\lambda \text{ for which } \omega > 0.$$
(8.49)

Thus, if the Z coefficients (often referred to as *backward-going amplitudes*) were all zero, the *uncorrelated* Hartree-Fock ground state, $|\phi\rangle$, would already be a solution of these equations. Conversely, non-zero Z coefficients imply the presence of ground-state correlations not present in the simple Hartree-Fock state $|\phi\rangle$. Note, that we describe the state $|0\rangle$ as an *implied* ground state because the harmonic-oscillator interpretation of the TDHF solutions predicts most properties of these states, that are of interest, by algebraic methods without explicitly deriving them. In particular, matrix elements of a one-body transition operator,

$$\hat{Q} = \sum_{\mu\nu} Q_{\mu\nu} a_\mu^\dagger a_{\bar{\nu}},$$
(8.50)

between the ground and excited one-phonon states, defined by

$$|\lambda\rangle := O_\lambda^\dagger |0\rangle, \quad \forall O_\lambda^\dagger \text{ for which } \omega > 0,$$
(8.51)

---

[6]Rowe D.J. (1966), *Nucl. Phys.* **80**, 209.
[7]Rowe, *op. cit.* Footnote 4 on Page x.

are evaluated from the expression

$$\langle 0|\hat{Q}|\lambda\rangle = \langle 0|[\hat{Q}, O_\lambda^\dagger]|0\rangle, \tag{8.52}$$

and are given, within the harmonic-oscillator approximation, by

$$\langle 0|\hat{Q}|\lambda\rangle = \langle \phi|[\hat{Q}, O_\lambda^\dagger]|\phi\rangle = \sum_{mi} \left(Y_{mi}(\lambda)Q_{im} + Z_{mi}(\lambda)Q_{mi}\right). \tag{8.53}$$

The above equations are identical to those of a theory of the elementary excitations of a many-body system known as the *random phase approximation* (RPA), when the latter is expressed in the *double-commutator equations-of-motion* formalism.[8] The properties of the RPA will be described in some detail in Volume 2, where it will be derived with a clear interpretation in mind. The above derivation by TDHF methods gives a different perspective of its significance.

8.2.4 *The Hartree-Fock stability condition*

The above small-amplitude solutions of the TDHF equations make the implicit assumption that the Hartree-Fock state, $|\phi\rangle$, is at least a local minimum-energy state and not merely a stationary state with respect to some displacements. Otherwise, $|\phi\rangle$ would not be a state of stable equilibrium and one would hardly expect to have solutions of the TDHF equations corresponding to small-amplitude vibrations about it. The implications of what would happen to the RPA, and hence the TDHF, solutions if the Hartree-Fock state were not one of stable equilibrium, were determined by Thouless.[9,10]

Consider the state

$$|\phi(\epsilon)\rangle := e^{\epsilon\hat{S}}|\phi\rangle, \tag{8.54}$$

where $|\phi\rangle$ is a Hartree-Fock particle-hole vacuum state and $\hat{S}$ is a skew-Hermitian one-body operator,

$$\hat{S} := \sum_{mi} \left(C_{mi}a_m^\dagger a_{\bar{i}} - C_{mi}^* a_i^\dagger a_{\bar{m}}\right). \tag{8.55}$$

The energy expectation for this state is given by

$$\langle\phi(\epsilon)|\hat{H}|\phi(\epsilon)\rangle = \langle\phi|\hat{H}|\phi\rangle + \epsilon\langle\phi|[\hat{H}, \hat{S}]|\phi\rangle + \tfrac{1}{2}\epsilon^2\langle\phi|[[\hat{H}, \hat{S}], \hat{S}]|\phi\rangle + \dots. \tag{8.56}$$

The Hartree-Fock equation,

$$\langle\phi|[\hat{H}, \hat{S}]|\phi\rangle = 0, \tag{8.57}$$

---

[8]Rowe, *op. cit.* Footnote 5 on Page 518.

[9]Thouless D.J. (1960), *Nucl. Phys.* **21**, 225.

[10]Thouless D.J. (1961), *Nucl. Phys.* **22**, 78.

ensures that the energy is stationary at $|\phi\rangle$. However, this equation alone does not ensure that the energy is a minimum. For its energy to be a minimum, the state $|\phi\rangle$ should satisfy the stability condition

$$\langle\phi|[[\hat{H},\hat{S}],\hat{S}]|\phi\rangle = \langle\phi|[\hat{S}^\dagger,[\hat{H},\hat{S}]]|\phi\rangle > 0 \tag{8.58}$$

for all $\hat{S}$ given by Equation (8.54). This condition is expressed in matrix form by

$$(C^\dagger \ \tilde{C})\begin{pmatrix} A & B \\ B^* & A^* \end{pmatrix}\begin{pmatrix} C \\ C^* \end{pmatrix} > 0, \quad \forall\,\{C_{mi}\}, \tag{8.59}$$

and is satisfied if and only if the matrix

$$M := \begin{pmatrix} A & B \\ B^* & A^* \end{pmatrix} \tag{8.60}$$

is positive definite.

As we discuss further in the following section, situations of particular interest are ones in which the matrix M has a zero eigenvalue. For example, if the Hamiltonian is invariant under translations, it commutes with the components of the centre-of-mass momentum operator, $\hat{P}_k$. Thus, if we set $\hat{S} = i\hat{P}_k$, we obtain

$$\langle\phi|[[\hat{H},\hat{S}],\hat{S}]|\phi\rangle = \langle\phi|[[\hat{H},\hat{P}_k],\hat{P}_k]|\phi\rangle = 0. \tag{8.61}$$

Thouless showed that, if the matrix M is positive definite, the TDHF equations can only have solutions with real values of ω. This inference follows immediately from the observation that any solution of the TDHF equations leads to the identity

$$(Y^\dagger \ Z^\dagger)\begin{pmatrix} A & B \\ B^* & A^* \end{pmatrix}\begin{pmatrix} Y \\ Z \end{pmatrix} = \hbar\omega \, (Y^\dagger \ Z^\dagger)\begin{pmatrix} Y \\ -Z \end{pmatrix}. \tag{8.62}$$

The left-hand side of this equation, being the expectation value of a Hermitian positive-definite matrix, has a positive real value. Moreover, the coefficient of $\hbar\omega$ on the right is real and so ω cannot be complex. Conversely, the existence of a TDHF solution with a complex value of ω implies that the matrix M has a negative eigenvalue and that the Hartree-Fock state is unstable.

8.2.5 Broken symmetries, rotations, and moments of inertia

A deeper understanding of Hartree-Fock theory is obtained by interpreting the manifold of Slater determinants as a classical phase space and the TDHF equations of motion as an expression of a classical Hamilton dynamics. Such an interpretation, developed in Section 8.3, extends the above interpretation of small-amplitude TDHF solutions to arbitrary solutions. Moreover, when TDHF theory is regarded as a classical approximation, the philosophical concern with broken symmetries ceases to be an issue. After all, one does not expect to see classical objects in zero-angular momentum eigenstates in the quantum mechanical sense. On the contrary, one would expect a classical object with zero translational momentum and zero angular momentum to be at rest and have a well-defined location and orientation in space. One would then say that the dynamics of the object was rotationally invariant if the energy of the object were the same for all its orientations. However, while the classical energy function, $\mathcal{H}$, is expected to be independent of the orientation of an object, it is also expected to have a non-vanishing rotational kinetic-energy term. Thus, we now consider what TDHF theory has to say about rotational moments of inertia.

A first objective is to determine the infinitesimal generators of angular momentum boosts. Observe that, with the substitution

$$\hat{Q}_\lambda := \left(\frac{\hbar}{2\omega_\lambda \mathcal{B}_\lambda}\right)^{\frac{1}{2}} (O_\lambda^\dagger + O_{\bar{\lambda}}), \quad \hat{P}_\lambda := i\left(\frac{\hbar\omega_\lambda \mathcal{B}_\lambda}{2}\right)^{\frac{1}{2}} (O_\lambda^\dagger - O_{\bar{\lambda}}), \qquad (8.63)$$

Equations (8.46)–(8.48) are equivalently expressed in terms of position- and momentum-like operators in the form

$$\langle\phi|[\hat{Q}_\kappa, \hat{P}_\lambda]|\phi\rangle = i\hbar\,\delta_{\kappa,\lambda}, \quad \langle\phi|[\hat{Q}_\kappa, \hat{Q}_\lambda]|\phi\rangle = \langle\phi|[\hat{P}_\kappa, \hat{P}_\lambda]|\phi\rangle = 0, \qquad (8.64)$$

$$\langle\phi|[\hat{P}_\kappa, [\hat{H}, \hat{P}_\lambda]]|\phi\rangle = i\hbar\omega_\lambda^2 \mathcal{B}_\lambda \langle\phi|[\hat{P}_\kappa, \hat{Q}_\lambda]|\phi\rangle = \hbar^2\omega_\lambda^2 \mathcal{B}_\lambda \delta_{\kappa,\lambda},$$
$$\langle\phi|[\hat{Q}_\kappa, [\hat{H}, Q_\lambda]]|\phi\rangle = -i\frac{\hbar}{\mathcal{B}_\lambda}\langle\phi|[\hat{Q}_\kappa, \hat{P}_\lambda]|\phi\rangle = \frac{\hbar^2}{\mathcal{B}_\lambda}\delta_{\kappa,\lambda}, \qquad (8.65)$$
$$\langle\phi|[\hat{Q}_\kappa, [\hat{H}, \hat{P}_\lambda]]|\phi\rangle = 0.$$

However, in contrast to Equations (8.46)–(8.48), these equations continue to have paired solutions when some of the $\{\omega_\lambda\}$ vanish, as happens when the corresponding $\{\hat{P}_\lambda\}$ operators commute with $\hat{H}$.

Suppose, for example, that $\{\hat{Q}_\lambda, \hat{P}_\lambda\}$ is a set of operators that satisfy Equations (8.64) and (8.65), and that $\{\hat{P}_k := \hbar\hat{J}_k\}$ is a subset for which $\omega_k = 0$. We then have the equations

$$\langle\phi|[\hat{Q}_\lambda, [\hat{H}, \hat{J}_k]]|\phi\rangle = \langle\phi|[\hat{P}_\lambda, [\hat{H}, \hat{J}_k]]|\phi\rangle = 0,$$
$$\langle\phi|[\hat{Q}_\lambda, [\hat{H}, Q_k]]|\phi\rangle = \frac{\hbar^2}{\mathcal{B}_k}\delta_{\lambda,k}, \qquad (8.66)$$

$$\langle\phi|[\hat{Q}_\lambda, \hat{J}_k]|\phi\rangle = i\,\delta_{\lambda,k}. \qquad (8.67)$$

Thus, when $\hat{J}_k$ is an angular momentum, the corresponding moment of inertia is given by

$$\mathcal{B}_k = \frac{\hbar^2}{\langle\phi|[\hat{Q}_k, [\hat{H}, Q_k]]|\phi\rangle}. \qquad (8.68)$$

561

8.2.6 *Moments of inertia for irrotational flow*

Consider, for example, the moments of inertia for the flows generated by the Cartesian components,

$$\hat{Q}_{ij} := \sum_{n=1}^{A} x_{ni} x_{nj}, \tag{8.69}$$

of the symmetric quadrupole tensor of an A-particle nucleus. The orbital angular momenta of a nucleus are similarly expressed by the components,

$$\hbar \hat{L}_{kl} := \sum_{n=1}^{A} \left(x_{nk} p_{nl} - x_{nl} p_{nk} \right), \tag{8.70}$$

of an antisymmetric tensor. It is readily ascertained that the commutation relations $[\hat{Q}_{ii}, \hat{L}_{kl}]$ all vanish. Thus, we consider the subset of quadrupole moments $\hat{Q}_{ij}$ with $i < j$ and angular momenta $\hat{L}_{kl}$ with $k < l$. Then, with the Hartree-Fock state, $|\phi\rangle$, chosen to be one for which the quadrupole moments are diagonal,[11] i.e.,

$$\langle \phi | \hat{Q}_{ij} | \phi \rangle = \delta_{ij} \lambda_i, \tag{8.71}$$

we obtain

$$\langle \phi | [\hat{Q}_{ij}, \hat{L}_{kl}] | \phi \rangle = i \, \delta_{ik} \delta_{jl} \left(\lambda_i - \lambda_j \right), \quad \text{for } i < j, \ k < l. \tag{8.72}$$

Also, if the quadrupole moments commute with every component of the Hamiltonian except for the kinetic energy,

$$\hat{T} := \frac{1}{2M} \sum_{n=1}^{A} \sum_{i=1}^{3} \hat{p}_{ni}^2, \tag{8.73}$$

it is determined (cf. Exercise (6.20)) that

$$\langle \phi | [\hat{Q}_{ij}, [\hat{H}, \hat{Q}_{ij}]] | \phi \rangle = \frac{\hbar^2}{M} (\lambda_i + \lambda_j). \tag{8.74}$$

Thus, when we renormalise the quadrupole moments, so that

$$\langle \phi | [\hat{Q}'_{ij}, \hbar \hat{L}_{kl}] | \phi \rangle = i\hbar \, \delta_{ik} \delta_{jl}, \tag{8.75}$$

by setting $\hat{Q}'_{ij} := \hat{Q}_{ij} / (\lambda_i - \lambda_j)$, we obtain

$$\mathcal{B}_{ij} = \frac{\hbar^2}{\langle \phi | [\hat{Q}'_{ij}, [\hat{H}, \hat{Q}'_{ij}]] | \phi \rangle} = \frac{M (\lambda_i - \lambda_j)^2}{\lambda_i + \lambda_j}, \tag{8.76}$$

---

[11]Note that the quadrupole moments, $\langle \phi | \hat{Q}_{ij} | \phi \rangle$, are zero if the state $|\phi\rangle$ is rotationally invariant. On the other hand, when $|\phi\rangle$ is not rotationally invariant, there is a set of equal-energy Hartree-Fock states obtained by rotating $|\phi\rangle$ through all possible angles. Hence, because the Cartesian quadrupole moments are real and symmetric they can be diagonalised by a rotation. This means that $|\phi\rangle$ can always be chosen such that the Cartesian quadrupole moments are diagonal.

which are interpreted as the moments of inertia for the classical kinetic energy

$$\mathcal{T}_{\text{rot}} := \sum_{i<j} \frac{\hbar^2 \mathcal{L}_{ij}^2}{2\mathcal{B}_{ij}}. \tag{8.77}$$

The mass parameters, $\{\mathcal{B}_{ij}\}$, in this expression are moments of inertia, $\mathscr{I}_i := \mathcal{B}_{ij}$ $(i,j,k$ cyclic), for irrotational flow. In fact, as we now show, any one-body operator, $\hat{Q} = \sum_n Q(\mathbf{r}_n)$, for which $Q(\mathbf{r}_n)$ is a simple function of the particle position coordinates, is an infinitesimal generator of a flow that is irrotational. This is seen from the observation that the mass density and current operators (see Exercise 8.7) for the nucleus are given by

$$\hat{\rho}(\mathbf{r}) := M \sum_{n=1}^{A} \delta(\mathbf{r} - \mathbf{r}_n), \tag{8.78}$$

$$\hat{\mathbf{j}}(\mathbf{r}) := -\frac{\mathrm{i}\hbar}{2} \sum_{n=1}^{A} \left[\nabla_n \delta(\mathbf{r} - \mathbf{r}_n) + (\mathbf{r} - \mathbf{r}_n)\nabla_n \right]. \tag{8.79}$$

Thus, if the current flow $\langle \phi | \hat{\mathbf{j}}(\mathbf{r}) | \phi \rangle$ is zero for the state $|\phi\rangle$ then, for the state

$$|\phi(\alpha)\rangle := e^{\hat{X}(\alpha)} |\phi\rangle \quad \text{with } \hat{X} = \frac{\mathrm{i}}{\hbar} \alpha \hat{Q}, \tag{8.80}$$

it is given to leading order (cf. Exercise 8.8) by

$$\mathbf{j}(\mathbf{r}) = \langle \phi | [\hat{\mathbf{j}}(\mathbf{r}), \hat{X}(\alpha)] | \phi \rangle = \frac{\alpha}{M} \rho(\mathbf{r}) \, \nabla Q(\mathbf{r}), \tag{8.81}$$

where

$$\rho(\mathbf{r}) = \langle \phi | \hat{\rho}(\mathbf{r}) | \phi \rangle. \tag{8.82}$$

When expressed as the product $\mathbf{j}(\mathbf{r}) := \rho(\mathbf{r})\mathbf{v}(\mathbf{r})$, the current flow of Equation (8.81) implies that the velocity field,

$$\mathbf{v}(\mathbf{r}) = \frac{\alpha}{M} \nabla Q(\mathbf{r}), \tag{8.83}$$

associated with this flow satisfies the equation

$$\nabla \times \mathbf{v}(\mathbf{r}) = \frac{\alpha}{M} \nabla \times \nabla Q(\mathbf{r}) = 0. \tag{8.84}$$

Thus, the velocity field has zero vorticity and the flow is said to be *irrotational*.

Although this result is instructive, it is important to recognise that there is no reason to expect that the infinitesimal generators of rotational flows obtained from the solution to Equations (8.65) and (8.64) will be simple functions of the single-particle position coordinates and, hence, be infinitesimal generators of irrotational flows.

Exercises

8.4 Use the Jacobi identity

$$[\hat{X}, [\hat{H}, \hat{Y}]] = [[\hat{X}, \hat{H}], \hat{Y}] + [\hat{H}, [\hat{X}, \hat{Y}]] \tag{8.85}$$

to show that, if $|\phi\rangle$ is a Hartree-Fock state for a Hamiltonian, $\hat{H}$, and $\hat{X}$ and $\hat{Y}$ are one-body operators, then

$$\langle\phi|[\hat{X}, [\hat{H}, \hat{Y}]]|\phi\rangle = \langle\phi|[[\hat{X}, \hat{H}], \hat{Y}]|\phi\rangle. \tag{8.86}$$

8.5 Show that,if $\{(Y(\lambda), Z(\lambda)\}$ are solutions of the TDHF equation, (8.27), with respective energies $\{\hbar\omega_\lambda\}$ that are real, then these solutions satisfy the equation

$$(\hbar\omega_\kappa - \hbar\omega_\lambda)\big(Y^\dagger(\kappa),\ Z^\dagger(\kappa)\big) \begin{pmatrix} Y(\lambda) \\ -Z(\lambda) \end{pmatrix} = 0 \text{ if } \omega_\kappa \neq \omega_\lambda. \tag{8.87}$$

Hence, show that the solutions can be normalised such that they satisfy the orthonormality relationship

$$\big(Y^\dagger(\kappa),\ Z^\dagger(\kappa)\big) \begin{pmatrix} Y(\lambda) \\ -Z(\lambda) \end{pmatrix} = \begin{cases} \delta_{\kappa,\lambda} & \text{if } \omega_\lambda > 0, \\ -\delta_{\kappa,\lambda} & \text{if } \omega_\lambda < 0. \end{cases} \tag{8.88}$$

8.6 Show that if $\hat{Q}_{ij}$ is given by Equation (8.69) and $\hat{T}$ is given by Equation (8.73), then, for $i < j$,

$$[\hat{Q}_{ij}, [\hat{T}, \hat{Q}_{ij}]] = \frac{\hbar^2}{M}(\hat{Q}_{ii} + \hat{Q}_{jj}). \tag{8.89}$$

Hence, derive Equation (8.74).

8.7 Show that if the density distribution for a many-particle system is given by Equation (8.78) and the velocity of a particle is related to its momentum by $\mathbf{v}_n = \mathbf{p}_n/M$, then the classical expression for the current flow, $j(\mathbf{r}) = v(\mathbf{r})\rho(\mathbf{r})$, maps to the quantum-mechanical expression given by Equation (8.79).

8.8 Show that $\big[\delta(\mathbf{r} - \mathbf{r}_m), \sum_n Q(\mathbf{r}_n)\big] = 0$ and, hence, that

$$\Big[\sum_n \nabla_m \delta(\mathbf{r} - \mathbf{r}_m), \sum_n Q(\mathbf{r}_n)\Big] = \nabla Q(\mathbf{r}) \sum_n \delta(\mathbf{r} - \mathbf{r}_n). \tag{8.90}$$

Use this result to derive Equation (8.81).

8.3 Classical mechanics as constrained quantum mechanics

The harmonic oscillator interpretation of the small-amplitude TDHF results relies on the simple relationship between the classical and quantum mechanics of harmonic oscillators. As we now show, the classical mechanics of arbitrary systems can be derived from quantum mechanics much more generally.

In quantum mechanics, a state of a system is described by a wave packet that has precise mean values of position and momentum coordinates at any instant but, unlike idealised classical mechanics,[12] it has spreads in the values of these coordinates that evolve in time according to the principles of wave mechanics. This section shows that, if the width of a quantum-mechanical wave packet is constrained from evolving in time, the resulting equations of motion are classical Hamilton equations.

A map from quantum to classical mechanics for a system with a dynamical group is defined in parallel with that for the simple harmonic oscillator in three basic steps: the first shows that a classical phase space can be identified with a manifold of coherent states; the second equates the values of classical observables to the expectation values of the quantal operators in the corresponding coherent states; the third step shows that quantum mechanics, when constrained to the coherent-state phase space, reduce to classical Hamilton mechanics. The restriction of the quantum mechanics of a system to such a space of coherent states has been described variously as *density dynamics*,[13] and *generalised mean-field dynamics*.

8.3.1 *Phase spaces of coherent states*

We suppose that the Hilbert space, $\mathbb{H}$, for the system under consideration carries a unitary irrep, $\hat{T}$, of a dynamical group G. For a choice of a so-called *fiducial state vector* $|\phi\rangle \in \mathbb{H}$, a system of *coherent-state vectors*[14] is then given by elements of the group orbit

$$\mathcal{M} := \{|\phi(g)\rangle := \hat{T}(g)|\phi\rangle; g \in G\}. \tag{8.91}$$

A corresponding manifold of phase-equivalent coherent states is defined by the set of densities

$$\overline{\mathcal{M}} := \{\rho_\psi; |\psi\rangle \in \mathcal{M}\}, \tag{8.92}$$

where $\rho_\psi(\hat{X}) := \langle\psi|\hat{X}|\psi\rangle$ for $\hat{X}$ a linear operator on $\mathbb{H}$. In order that the manifold, $\overline{\mathcal{M}}$, should have the properties of a classical phase space, all that is needed is a suitable choice of the fiducial state. In particular, for the equations of motion, (8.4), to be well-defined, we require that *the fiducial state should be uniquely determined, to within a phase factor, by the expectation values in this state of the elements of the Lie algebra of G*. If this property is satisfied for the fiducial state, then, as

---

[12]We speak of idealised classical mechanics because in any physical situation, there are necessarily uncertainties in the values of the position and momentum coordinates of any system that are much larger than the minimum uncertainties imposed by quantum mechanics.

[13]Rowe D.J., Vassanji M.G. and Rosensteel G. (1983), *Phys. Rev.* **A28**, 1951.

[14]This definition was given by Perelomov A.M. (1972), *Comm. Math. Phys.* **26**, 222. Variations on this definition have been given by Klauder J.R. (1963), *J. Math. Phys.* **4**, 1058, Gilmore R. (1972), *Ann. Phys. (NY)* **74**, 391, and others. Reviews of coherent-state methods are given in the books of Perelomov A.M. (1986), *Generalized Coherent States and their Applications* (Springer-Verlag, Berlin); Klauder J.R. and Skagerstam B.-S. (1985), *Coherent States; Applications in Physics and Mathematical Physics* (World Scientific, Singapore); and Zhang W.-M., Feng D.H. and Gilmore R. (1990), *Rev. Mod. Phys.* **62**, 867.

shown by Exercise 8.11, it is satisfied by all the coherent states generated from this fiducial state. This ensures that the expectation values of the Hamiltonian and other observables are uniquely defined for every coherent state and that the dynamics on the manifold of phase-equivalent coherent states is determined by the time evolution of the expectation values of elements of the Lie algebra, $\mathbf{g}$, in accordance with Equation (8.4). At the same time, it makes it possible to take advantage of a well-known theorem which states that the restriction of a coherent-state manifold of densities, as defined by Equation (8.92), to the Lie algebra of G, is isomorphic to a classical phase space.[15,16] A natural choice of fiducial state with this property is given by a highest- or a lowest-weight state for the irrep, $\hat{T}$, when there is one.

In accordance with the definition of a group orbit in Section 7.4.2, the manifold $\overline{\mathcal{M}}$ is isomorphic to the factor space G/G_ϕ, where G_ϕ is the subgroup of elements of G that leave the fiducial state, $\rho_\phi \in \overline{\mathcal{M}}$, invariant. This subgroup, known as the *isotropy subgroup* at $\rho_\phi \in \overline{\mathcal{M}}$, contains all elements, $h \in G$, for which

$$\rho_{\phi(g)}(\hat{X}) = \langle \phi | \hat{T}(g^{-1}) \hat{X} \hat{T}(g) | \phi \rangle. \tag{8.93}$$

It then follows that $h \in G$ is an element of G_ϕ if and only if

$$\rho_\phi \big(\hat{T}(h^{-1}) \hat{X} \hat{T}(h) \big) = \rho_\phi(\hat{X}), \tag{8.94}$$

for $\hat{X}$ any linear operator on the Hilbert space. However, if the expectation values of elements of the Lie algebra $\mathbf{g}$ are sufficient to uniquely determine a coherent state, as we have required, it is also sufficient to restrict X in Equation (8.94) to elements of $\mathbf{g}$. Thus, an element $Y \in \mathbf{g}$ belongs to the Lie algebra, $\mathbf{g}_\phi$, of the isotropy subgroup G_ϕ if and only if

$$\rho_\phi([\hat{X}, \hat{Y}]) = 0, \quad \forall X \in \mathbf{g}. \tag{8.95}$$

To label states of the coherent-state manifold, it is useful to define systems of coordinates. Global coordinates generally have singularities, e.g., the latitude and longitude coordinates for the Earth's surface have singularities at the North and South Poles. Thus, it useful to define a coordinate chart for a neighbourhood about any point of the manifold that is free of singularities. Given that all elements of $\mathbf{g}_\phi$ are infinitesimal generators of null displacements of the coherent state ρ_ϕ, it follows that a linearly-independent set, $\{\hat{X}_\nu\}$, of non-null infinitesimal generators of displacement of $\rho_\phi \in \overline{\mathcal{M}}$ are given by a basis for a subspace of the Lie algebra $\mathbf{g}$ that is complementary to the isotropy subalgebra, $\mathbf{g}_\phi$. These displacement operators define a local system of real coordinates, $\{\xi^\mu\}$, for the neighbourhood of the state

---

[15]Marsden J.E. and Ratiu T.S. (1999), *Introduction to Mechanics and Symmetry* (Springer, New York).

[16]Such restrictions of coherent-state manifolds are known as *coadjoint orbits*.

ρ_ϕ for which

$$|\phi(\xi)\rangle := \exp\left[\frac{i}{\hbar}\sum_\mu \xi^\mu \hat{X}_\mu\right]|\phi\rangle. \qquad (8.96)$$

We now show that the manifold of coherent states is a phase space in the sense that, for any state $\rho_\phi \in \overline{\mathcal{M}}$, we can construct a complete set of position and momentum coordinates with infinitesimal generators that satisfy the identities

$$\rho_\phi([\hat{P}_\mu, \hat{Q}_\nu]) = \langle \phi|[\hat{P}_\mu, \hat{Q}_\nu]|\phi\rangle = -i\hbar\delta_{\mu,\nu}. \qquad (8.97)$$

First observe that, for every state $\rho_\phi \in \overline{\mathcal{M}}$, there is an antisymmetric form, ω_ϕ, defined on the infinitesimal generators of displacement by $\omega_\phi(\hat{X}, \hat{Y}) := \frac{i}{\hbar}\rho_\phi([\hat{X}, \hat{Y}])$. A manifold with such a form is said to be *symplectic*, which means that it has the properties of a phase space.[17] For such a form, a given basis, $\{\hat{X}_\nu\}$, of infinitesimal generators at ρ_ϕ, defines a matrix, with elements

$$\omega_{\mu\nu}(\phi) := \frac{i}{\hbar}\rho_\phi([\hat{X}_\mu, \hat{X}_\nu]), \qquad (8.98)$$

that is real, antisymmetric, and invertible. To see that it is invertible, observe that if $\rho_\phi([\hat{X}_\mu, \hat{X}_\nu])$ were zero for all $\hat{X}_\nu$, then $\hat{X}_\mu$ would have to be an element of the isotropy subalgebra at ρ_ϕ and, hence, an infinitesimal generator of a null displacement. Thus, for a suitable choice of basis, the matrix ω assumes the canonical form

$$\omega = \begin{pmatrix} 0 & I \\ -I & 0 \end{pmatrix}. \qquad (8.99)$$

Such a basis of infinitesimal generators satisfies Equation (8.97). Moreover, by comparison with expressions for the simple harmonic oscillator, it is seen that the operators $\{\hat{P}_\mu\}$ and $\{\hat{Q}_\mu\}$ act locally in the neighbourhood of a state $\rho_\phi \in \overline{\mathcal{M}}$ as infinitesimal generators of displacement in locally-defined position and momentum coordinates.

8.3.2 The choice of coherent states

As emphasised above, an essential requirement for a set of coherent states to be a classical phase space for a system is that the fiducial state for the set should be a state in the Hilbert space for the unitary irrep of the chosen dynamical group, G, that is uniquely defined by the expectation values of elements of the Lie algebra, **g**, of G. We shall assume that, when the system is finite (however large), such a state can always be found for some finite-dimensional choice of dynamical group. It is also desirable to choose the fiducial state to be as close as possible to a minimal uncertainty state in order that the corresponding classical Hamiltonian dynamics should

---

[17]Marsden and Ratiu, *op. cit.* Footnote 15 on Page 566.

be as close as possible to the idealised classical mechanics in which uncertainties in the values of physical observables are not considered.

Such a fiducial state is given by a highest- or a lowest-weight state for the irrep of $\mathbf{g}$ when there is such a state. For example, if $\mathbf{g}$ is semi-simple it has a Cartan subalgebra spanned by commuting elements $\{h_\nu\}$ and its complex extension is spanned by elements $\{h_\nu, X_\nu^+, X_\nu^-\}$ which, for example, satisfy the commutation relationships

$$[h_\nu, X_\nu^\pm] = \pm X_\nu^\pm, \quad [X_\nu^+, X_\nu^-] = 2h_\nu. \tag{8.100}$$

A highest-weight state, $|\text{h.wt.}\rangle$, for an irrep of $\mathbf{g}$, if there is one, satisfies the identities

$$\hat{h}_\nu |\text{h.wt.}\rangle = \lambda(\nu)|\text{h.wt.}\rangle, \quad \hat{X}_\nu^+ |\text{h.wt.}\rangle = 0, \tag{8.101}$$

and a lowest-weight state, $|\text{l.wt.}\rangle$, if there is one, satisfies the identities

$$\hat{h}_\nu |\text{l.wt.}\rangle = \lambda'(\nu)|\text{l.wt.}\rangle, \quad \hat{X}_\nu^- |\text{l.wt.}\rangle = 0, \tag{8.102}$$

where λ and λ' are, respectively, highest and lowest weights. In both cases, the extremal state is defined uniquely by the expectation values of elements of the Cartan subalgebra. Moreover such states are minimal uncertainty states as the following shows.

If $|\phi\rangle$ is either a highest- or a lowest-weight state for the irrep of $\mathbf{g}$ on the Hilbert space of the system then, for suitably adjusted values of the constants $\{a_{|}nu\}$ the operators

$$\hat{Q}_\nu = \frac{1}{\sqrt{2}}(\hat{X}_\nu^+ + \hat{X}_\nu^-), \quad \hat{P}_\nu = \frac{i\hbar a_\nu}{\sqrt{2}}(\hat{X}_\nu^+ - \hat{X}_\nu^-), \tag{8.103}$$

satisfy the required identities

$$\langle\phi|[\hat{Q}_\mu, \hat{Q}_\nu]|\phi\rangle = \langle\phi|[\hat{P}_\mu, \hat{P}_\nu]|\phi\rangle = 0, \quad \langle\phi|[\hat{P}_\mu, \hat{Q}_\nu]|\phi\rangle = -i\hbar\delta_{\mu,\nu}. \tag{8.104}$$

It is then follows, cf. Equation 8.3, that these operators satisfy the defining equations

$$\sigma_\phi(\hat{Q}_\nu)\sigma_\phi(\hat{P}_\nu) = \tfrac{1}{2}|\langle\phi|[\hat{Q}_\nu, \hat{P}_\nu]|\phi\rangle|, \tag{8.105}$$

for a minimum-uncertainty state. It is also seen (see Exercises 8.10 and 8.11) that, if the fiducial state is a minimum-uncertainty state, then so is every coherent state generated from this fiducial state. This is a simple consequence of the fact that if $|\phi\rangle$ is a highest- (lowest-) weight state with respect to a given set of raising and lowering operators, $\{\hat{X}_\nu^\pm\}$, then the state $|\phi(g)\rangle := \hat{T}(g)|\phi\rangle$ is a highest- (lowest-)weight states with respect to the transformed raising and lowering operators, $\{\hat{X}_\nu^\pm(g) := \hat{T}(g)\hat{X}_\nu^\pm\hat{T}(g^{-1})\}$.

8.3.3 Classical observables

An observable in quantum mechanics is a Hermitian linear operator on a Hilbert space whereas a classical observable is a real function on the phase space for the system. Thus, an observable, such as the energy or the angular momentum, of a classical system takes a particular value for any given state in the phase space of the system. It is then appropriate to define the classical observable $\mathcal{F}$, corresponding to a quantal observable $\hat{F}$, as the function with values given by

$$\mathcal{F}(\phi) := \langle \phi | \hat{F} | \phi \rangle = \rho_\phi(\hat{F}), \quad \rho_\phi \in \overline{\mathcal{M}}. \tag{8.106}$$

Of particular importance for defining classical Hamilton dynamics on $\overline{\mathcal{M}}$ is the Hamiltonian function

$$\mathcal{H}(\phi) := \langle \phi | \hat{H} | \phi \rangle = \rho_\phi(\hat{H}). \tag{8.107}$$

8.3.4 Hamilton dynamics on $\overline{\mathcal{M}}$

To derive the equation of motion for the time evolution of a state $|\phi(\xi)\rangle$, with $\{\xi^\mu\}$ coordinates defined by Equation (8.96), we start with the equation of motion

$$i\hbar \frac{d}{dt} \langle \phi | \hat{X}_\mu | \phi \rangle = \langle \phi | [\hat{X}_\mu, \hat{H}] | \phi \rangle. \tag{8.108}$$

The left-hand side of this equation has the expansion,

$$\begin{aligned}
\frac{d}{dt} \langle \phi(\xi) | \hat{X}_\mu | \phi(\xi) \rangle \Big|_{\xi=0} &= \sum_\nu \frac{\partial}{\partial \xi^\nu} \langle \phi(\xi) | \hat{X}_\mu | \phi(\xi) \rangle \Big|_{\xi=0} \dot{\xi}^\nu \\
&= \frac{i}{\hbar} \sum_\nu \langle \phi | [\hat{X}_\mu, \hat{X}_\nu] | \phi \rangle \rangle \dot{\xi}^\nu \\
&= \sum_\nu \omega_{\mu\nu}(\phi) \dot{\xi}^\nu.
\end{aligned} \tag{8.109}$$

Thus, because the matrix ω defined by Equation (8.98) is invertible, the time derivatives of the coordinates are given by

$$\dot{\xi}^\mu = \sum_\nu \omega^{\mu\nu} \frac{d}{dt} \langle \phi | \hat{X}_\nu | \phi \rangle = \frac{i}{\hbar} \sum_\nu \omega^{\mu\nu} \langle \phi | [\hat{H}, \hat{X}_\nu] | \phi \rangle, \tag{8.110}$$

where $\{\omega^{\mu\nu}\}$ is the inverse of the matrix $\{\omega_{\mu\nu}\}$, for which

$$\sum_\nu \omega^{\mu\nu} \omega_{\nu\mu'} = \delta_{\mu,\mu'}. \tag{8.111}$$

Hence, from Equation (8.96) and the definition, (8.107), of $\mathcal{H}(\xi) := \langle \phi(\xi) | \hat{H} | \phi(\xi) \rangle$, we obtain

$$\dot{\xi}^\mu = \sum_\nu \omega^{\mu\nu} \frac{\partial \mathcal{H}}{\partial \xi^\nu}. \tag{8.112}$$

The time derivative of any classical observable, $\mathcal{F}$, evaluated as a function of the coordinates for a time-evolving state, is now given by the equation

$$\dot{\mathcal{F}} = \sum_\mu \frac{\partial \mathcal{F}}{\partial \xi^\mu} \dot{\xi}^\mu, \tag{8.113}$$

and can be expressed as a Poisson bracket,

$$\dot{\mathcal{F}} = \left\{ \mathcal{F}, \mathcal{H} \right\} := \sum_{\mu\nu} \frac{\partial \mathcal{F}}{\partial \xi^\mu} \omega^{\mu\nu} \frac{\partial \mathcal{H}}{\partial \xi^\nu}. \tag{8.114}$$

Thus, when constrained to a phase space of coherent states, the equations of quantum mechanics reduce to classical Hamilton equations of motion.

Exercises

8.9 Show that if

$$\hat{Q} = \frac{1}{\sqrt{2}}(\hat{X}_+ + \hat{X}_-), \quad \hat{P} = \frac{i\hbar}{\sqrt{2}}(\hat{X}_+ - \hat{X}_-), \tag{8.115}$$

where $\hat{X}_\pm$ are a pair of raising and lowering operators of a semi-simple Lie algebra, and if $|\phi\rangle$ is either a highest- or a lowest-weight state for a unitary irrep of this Lie algebra, then $\sigma_\phi(\hat{Q})\sigma_\phi(\hat{P}) = \frac{1}{2}\left| \langle\phi|[\hat{Q},\hat{P}]|\phi\rangle \right|$, where

$$\sigma_\phi^2(\hat{A}) := \langle\phi|\hat{A}^2|\phi\rangle - \langle\phi|\hat{A}|\phi\rangle^2. \tag{8.116}$$

8.10 Show that if $\sigma_\phi(\hat{Q})\sigma_\phi(\hat{P}) = \frac{1}{2}\left| \langle\phi|[\hat{Q},\hat{P}]|\phi\rangle \right|$, then

$$\sigma_{\phi(g)}(\hat{Q}(g))\sigma_{\phi(g)}(\hat{P}(g)) = \frac{1}{2}\left| \langle\phi(g)|[\hat{Q}(g),\hat{P}(g)]|\phi(g)\rangle \right|, \tag{8.117}$$

where $|\phi(g)\rangle := \hat{T}(g)|\phi\rangle$ and $\hat{A}(g) := \hat{T}(g)\hat{X}\hat{T}(g^{-1})$, for any $g \in G$.

8.11 With the notations of Exercise 8.10, show that if a state $|\phi\rangle$ in the Hilbert space for the irrep $\hat{T}$ is uniquely determined, to within a phase factor, by the expectation values in this state of the elements of the Lie algebra of G, then each of the coherent states $\{|\phi(g)\rangle; g \in G\}$ is similarly determined.

8.4 Mean-field dynamics in an adiabatic expansion

Of particular interest in nuclear physics are the so-called *adiabatic* degrees of freedom which are those associated with relatively low-energy states, such as low-frequency vibrations and low-angular-velocity rotations. Such adiabatic dynamics are relevant, for example, to the description of low-energy collective states in nuclear structure theory and the description of heavy-ion and fission reactions in nuclear reaction theory.[18] In describing the classical dynamics of adiabatic degrees of freedom, it is reasonable, as a first approximation, to restrict the Hamiltonian

---

[18]for references see Footnote 3 on Page 553.

to quadratic terms in the momentum coordinates. Such an expansion enables a separation of the classical potential and kinetic energies and a replacement of the Hamilton dynamics of mean-field theory by an equivalent Lagrangian dynamics.

8.4.1 *Position and momentum coordinates*

The first concern in implementing an adiabatic expansion is to separate the coordinates of a phase space into subsets of position and momentum coordinates. In an abstract situation, the separation is somewhat arbitrary and it is not clear that it can always be done in a globally consistent way. However, in physical situations there are usually ways to proceed. In particular, it is desirable to have position and momentum observables that are, respectively, even and odd under time reversal. For example, when the $\hat{Q}_\nu$ and $\hat{P}_\nu$ displacement operators can be expressed in terms of local raising and lowering operators, as in Equation (8.103), it is possible to choose the phases of the raising and lowering operators such that they are invariant under time reversal. The $\hat{Q}_\nu$ and $\hat{P}_\nu$ operators are then, respectively, even and odd under time reversal. Moreover, if a particular coherent state $|\phi\rangle$ is even under time reversal, it follows that all the states of the form

$$|\phi(q, p = 0)\rangle := \exp\left[-\frac{i}{\hbar} \sum_\nu q_\nu \hat{P}^\nu \right] |\phi\rangle \qquad (8.118)$$

are also even under time reversal. A submanifold of a phase space in which all momentum coordinates are zero is known as a *Lagrangian submanifold*.

For example, it is possible to construct a basis of single-particle wave functions, $\{\psi_\nu\}$, for use in Hartree-Fock theory, that are all invariant under time reversal. This is possible because, if any single-particle wave function, ψ_ν, is not time-reversal invariant, it can be combined with its time-reverse, denoted by $\psi_{\bar{\nu}}$, to form a pair of time-reversal-invariant wave functions $\psi_\nu + \psi_{\bar{\nu}}$ and $i(\psi_\nu - \psi_{\bar{\nu}})$. We can then describe these single-particle wave functions and any real linear combinations of them as being *real*. Thus, there are well-defined Lagrangian submanifolds of real Slater determinants, for example, that are all invariant under time reversal.

8.4.2 *Potential and kinetic energies*

Given a Lagrangian submanifold, $\mathcal{LM}$, as a subspace of zero-momentum states of the phase space for a classical approximation to the quantum dynamics of a system, it is insightful to consider the expansion of the energy function up to quadratic terms in the momenta,

$$\mathcal{H}(q, p) = \langle \phi(q, p)|\hat{H}|\phi(q, p)\rangle = \mathcal{V}(q) + \tfrac{1}{2} \sum_{\mu\nu} B_{\mu\nu}(q)\, p^\mu p^\nu + O(p^4), \qquad (8.119)$$

where $\mathscr{V}(q)$ is the "potential energy" function,

$$\mathscr{V}(q) := \langle\phi(q, p = 0)|\hat{H}|\phi(q, p = 0)\rangle, \tag{8.120}$$

and we have assumed that the quantal Hamiltonian, $\hat{H}$, is time-reversal invariant so that only terms that are even in the p coordinates appear. The inertial parameters, $B_{\mu\nu}(q)$, of the kinetic energy component of this energy function are derived as follows. First observe that

$$\dot{q}_\mu = \frac{\partial\mathcal{H}}{\partial p_\mu} = \frac{i}{\hbar}\langle\phi(q, p)|[\hat{H}, \hat{Q}_\mu]|\phi(q, p)\rangle = \sum_\nu B_{\mu\nu}(q)\, p^\nu + 0(p^3), \tag{8.121}$$

where

$$B_{\mu\nu}(q) = \frac{\partial^2\mathcal{H}}{\partial p^\mu \partial p^\nu}(q, p = 0) = \frac{1}{\hbar^2}\langle\phi(q)|[\hat{Q}_\mu, [\hat{H}, \hat{Q}_\nu]]|\phi(q)\rangle. \tag{8.122}$$

There are several ways to make use of the above results. For example, one can seek solutions of the classical equations of motion for the Hamiltonian $\mathcal{H}(q, p)$ corresponding to slow-moving (i.e., low-momentum) trajectories that stay close to the valley floors of the potential energy surface and, in this way, identify the important low-energy collective degrees of freedom of the nucleus under consideration. We show in the following that one can also make use of the expression for collective model mass parameters given by Equation (8.122).

8.4.3 *Lagrangian equations of motion*

The classical equations of motion for a Hamiltonian with only quadratic momentum terms are easy to solve, with the help of a computer. When expressed in terms of velocities instead of momenta, the above Hamiltonian, $\mathcal{H}$, assumes the form

$$\mathcal{H}(q, \dot{q}) = \mathscr{V}(q) + \tfrac{1}{2}\sum_{\mu\nu} B^{\mu\nu}(q)\, \dot{q}_\mu \dot{q}_\nu + O(\dot{q}^4), \tag{8.123}$$

where $\{B^{\mu\nu}(q)\}$ is the inverse of the matrix $\{B_{\mu\nu}(q)\}$. The dynamics are then Lagrangian. Recall that the Hamiltonian, expressed in terms of velocities, is related to the Lagrangian by the Legendre transform

$$\mathcal{H} = \sum_\mu \dot{q}_\mu \frac{\partial\mathcal{L}}{\partial\dot{q}_\mu} - \mathcal{L}. \tag{8.124}$$

The Lagrangian and the equations of motion are then given, respectively, by

$$\mathcal{L} = \tfrac{1}{2}\sum_{\mu\nu} B^{\mu\nu}(q)\, \dot{q}_\mu \dot{q}_\nu - \mathscr{V}(q) + O(\dot{q}^4), \tag{8.125}$$

$$\frac{d}{dt}\left(\frac{\partial\mathcal{L}}{\partial\dot{q}_\mu}\right) - \frac{\partial\mathcal{L}}{\partial q_\mu} = 0. \tag{8.126}$$

To find solutions, using a step-by-step integration of Equation (8.126) from a given starting point, it is useful to first make a linear transformation of the $\{q_\mu\}$ coordinates at each step so that the Lagrangian takes the form

$$\mathcal{L} = \tfrac{1}{2} \sum_\mu \dot{q}_\mu^2 - \mathscr{V}'(q) + O(\dot{q}^4). \tag{8.127}$$

The equations of motion then simplify to

$$\ddot{q}_\mu = -\frac{\partial \mathscr{V}'}{\partial q_\mu} + O(\dot{q}^3 \ddot{q}). \tag{8.128}$$

Note that there are infinitely-many solutions to the equations of motion. Thus, before embarking on a solution, it is worthwhile to determine the valleys of the potential energy surface in which the solutions of interest are expected to lie. However, in so doing, it is important to recognise that the manifold is not flat and that the coordinates are curvilinear. Fortunately, this presents no problem because the relevant metric for the space is well-defined as the following shows.

8.4.4 The mass tensor as a Riemannian metric

The $\{B^{\mu\nu}(q)\}$ functions have a physical interpretation as the components of a mass tensor. They also have a geometrical interpretation as components of a Riemannian metric for the Lagrangian manifold, $\mathcal{LM}$. As such, they provide a measure of distance between any two points along a path on $\mathcal{LM}$ by a line integral

$$d_{12} := \int_1^2 ds, \tag{8.129}$$

with

$$ds^2 := \sum_{\mu\nu} B^{\mu\nu}(q)\, dq_\mu\, dq_\nu. \tag{8.130}$$

With this metric, the kinetic energy along any path is given simply by

$$\text{K.E.} = \frac{1}{2} \left(\frac{ds}{dt} \right)^2. \tag{8.131}$$

The metric makes it meaningful to interpret the potential-energy function in topographical terms of hills and valleys, etc., on a Riemannian manifold[19] and, thereby, identify the valleys of the potential-energy surface of primary relevance for a classical interpretation of adiabatic collective dynamics.

---

[19]The interpretation of constrained quantum mechanics in related topographical terms was reviewed in the papers of Rowe D.J. (1982), *Nucl. Phys.* **A391**, 307, and Rowe D.J. and Ryman A. (1982), *J. Math. Phys.* **23**, 732. A review of alternative viewpoints was given by Goeke K., Reinhard P.-G. and Rowe D.J. (1981), *Nucl. Phys.* **A359**, 408.

8.5 Constrained mean-field methods

The irrotational-flow moments of inertia, given by Equation (8.76), deviate substantially from observed values and suggest that, while quadrupole moments may provide useful coordinates for nuclear quadrupole collective models, they are not the relevant infinitesimal generators of collective flows. In fact, there is no reason to suppose that they should be. Thus, one is led to consider the operators that are conjugate to the angular-momenta in the solutions expressed, for example, by Equations (8.66) and (8.67) of the TDHF equations.

8.5.1 *The self-consistent cranking-model*

A conceptually simple alternative[20] to starting with Equations (8.66) and (8.67) is to consider mean-field solutions for a Hamiltonian, $\hat{H}_\sigma := \hat{H} - \sigma \hat{J}_k$, where $\sigma \hat{J}_k$ is a Lagrangian constraint included to induce a non-zero expectation value of the angular momentum, $\hat{J}_k$. When the Hartree-Fock minimum-energy state, $|\phi_\sigma\rangle$, for the Hamiltonian $\hat{H}_\sigma$ has been determined, a moment of inertia is defined by the expansion, to leading order in σ,

$$\langle\phi_\sigma|\hat{H}|\phi_\sigma\rangle := E_0 + \frac{\hbar^2}{2\mathscr{I}_k}\langle\phi_\sigma|\hat{J}_k|\phi_\sigma\rangle^2 + \cdots . \tag{8.132}$$

As we now show, this constrained Hartree-Fock prescription gives exactly the same result as that obtained by solution of the TDHF equations. Note that the self-consistent cranking model, described here, differs from an earlier cranking model of Inglis,[21,22] which uses perturbation theory to determine the ground state of a Hamiltonian, $\hat{H}_\sigma = \hat{H}_0 - \sigma \hat{J}_k$, where $\hat{H}_0$ is a fixed independent-particle model Hamiltonian.

Let $|\phi\rangle$ denote a Hartree-Fock minimum-energy solution for the Hamiltonian $\hat{H}$ and suppose that, for a small value of σ, the minimum-energy solution for $\hat{H}_\sigma$ is given by the state

$$|\phi_\sigma\rangle = \exp\Big[\sum_\lambda \frac{\mathrm{i}}{\hbar}\alpha_\lambda \hat{Q}_\lambda\Big]|\phi\rangle, \tag{8.133}$$

where the operators $\{\hat{Q}_\lambda\}$ are obtained from the solutions to the TDHF equations in the form given by Equations (8.66) and (8.67). From the expansion

$$\langle\phi_\sigma|\hat{H}-\sigma\hat{J}_k|\phi_\sigma\rangle = \langle\phi|\hat{H}|\phi\rangle + \frac{\mathrm{i}\sigma}{\hbar}\sum_\lambda \alpha_\lambda\langle\phi|[\hat{Q}_\lambda,\hat{J}_k]|\phi\rangle$$

$$+ \frac{1}{2\hbar^2}\sum_{\kappa\lambda}\alpha_\kappa\alpha_\lambda\langle\phi|[\hat{Q}_\kappa,[\hat{H},\hat{Q}_\lambda]]|\phi\rangle + \cdots, \tag{8.134}$$

---

[20]Thouless, *op. cit.* Footnote 9 on Page 559.
[21]Inglis D.R. (1954), *Phys. Rev.* **96**, 1059.
[22]Inglis D.R. (1955), *Phys. Rev.* **97**, 701.

the variational condition, $\partial \langle \phi_\sigma | \hat{H} - \sigma \hat{J}_k | \phi_\sigma \rangle / \partial \alpha_\lambda = 0$, then gives

$$i\hbar\sigma\langle\phi|[\hat{Q}_\lambda, \hat{J}_k]|\phi\rangle + \sum_\kappa \alpha_\kappa\langle\phi|[\hat{Q}_\kappa, [\hat{H}, \hat{Q}_\lambda]]|\phi\rangle + \cdots = 0. \qquad (8.135)$$

Recall, cf. Section 8.2.5, that if $\hat{H}$ is rotationally invariant but the variational ground state, $|\phi\rangle$, is not, then $\hat{J}_k$ is a zero-frequency excitation operator of the TDHF equations. We can then make use of Equations (8.66) and (8.67) to obtain

$$i\hbar\sigma\langle\phi|[\hat{Q}_\lambda, \hat{J}_k]|\phi\rangle = -\hbar\sigma\delta_{\lambda,k}, \qquad (8.136)$$

$$\sum_\kappa \alpha_\kappa\langle\phi|[\hat{Q}_\kappa, [\hat{H}, \hat{Q}_\lambda]]|\phi\rangle = \alpha_\lambda \frac{\hbar^2}{\mathcal{B}_\lambda}. \qquad (8.137)$$

It follows from Equation (8.135) that

$$\alpha_\lambda = \delta_{\lambda,k}\frac{\mathcal{B}_k}{\hbar}\sigma + 0(\sigma^2) \qquad (8.138)$$

To leading order in σ, we also have

$$\langle\phi_\sigma|\hbar\hat{J}_k|\phi_\sigma\rangle = -i\sum_\lambda \alpha_\lambda\langle\phi|[\hat{Q}_\lambda, \hat{J}_k]|\phi\rangle = \alpha_k. \qquad (8.139)$$

Thus, from the expansion of the energy expectation value for the state $|\phi_\sigma\rangle$,

$$\langle\phi_\sigma|\hat{H}|\phi_\sigma\rangle = E_0 + \frac{1}{2\hbar^2}\sum_{\kappa\lambda} \alpha_\kappa\alpha_\lambda\langle\phi|[\hat{Q}_\kappa, [\hat{H}, \hat{Q}_\lambda]]|\phi\rangle + \cdots = E_0 + \frac{\alpha_k^2}{2\mathcal{B}_k} + \ldots, \quad (8.140)$$

and, from the definition of the self-consistent-cranking model moment of inertia given by Equation (8.132), we obtain the result (for small σ)

$$\mathcal{I}_i = \mathcal{B}_k. \qquad (8.141)$$

The self-consistent cranking approach can also be used to determine moments of inertia for increasingly larger mean values of the angular momentum, i.e., by a non-perturbative solution of the variational equations, $\partial \langle \phi_\sigma | \hat{H} - \sigma \hat{J}_k | \phi_\sigma \rangle / \partial \alpha_\lambda = 0$. It should be noted, however, that the approach breaks down when the Hartree-Fock state, $|\phi\rangle$, is an eigenstate of $\hat{J}_k$.

8.5.2 *Mapping out energy surfaces*

Mean-field methods can be used to map out the classical potential- and kinetic-energy components of a given quantum-mechanical Hamiltonian. Such energy functions might be used to extend the small amplitude TDHF results to higher order, e.g., by identifying anharmonicities in collective dynamics and in designing effective quantum-mechanical ways of handling them. They can also be used to gain insights into the nature of a low-dimensional model Hamiltonian by viewing it from a classical perspective.

A classical potential-energy function is defined on a Lagrangian submanifold of coherent states by

$$\mathscr{V}(q) := \langle \phi(q)|\hat{H}|\phi(q)\rangle. \tag{8.142}$$

A classical kinetic-energy function, expressed in terms of local momentum coordinates, is similarly defined. In practice, however, one is most interested in energy functions in just a few relevant degrees of freedom. For example, one might be interested in viewing the potential energy as a function of, say, a beta- or a gamma-deformation parameter, in order to learn about the stability of a given mean-field solution against quadrupole deformation or the existence of other low-lying Hartree-Fock solutions with different deformations. The Lagrangian constraint method is then most useful. For example, to determine an energy as a function of beta deformation, one might seek the minimum energy Hartree-Fock solution for the Hamiltonian,

$$\hat{H}(\beta) = \hat{H} - \sigma\hat{Q}_0, \tag{8.143}$$

where $\hat{Q}_0$ is the $M = 0$ component of the quadrupole tensor, $\{\hat{Q}_\mu, \mu = 0, \pm 1, \pm 2\}$, and σ is a Lagrange multiplier with values adjusted such that $\beta = \langle \phi(\sigma)|\hat{Q}_0|\phi(\sigma)\rangle$ takes a range of values of interest. Potential energy functions, defined by $\mathscr{V}(\beta) := \langle \phi(\sigma)|\hat{H}|\phi(\sigma)\rangle$, are illustrated in Figure 7.4. Such calculations may show, for example, a potential energy barrier between two local Hartree-Fock minima along the path defined by $\hat{H}(\beta)$. It should be noted, however, that on a multi-dimensional manifold there will be different paths between two local minimum-energy states some of which will present lesser barriers. Consider, for example, a landscape model for which $\mathscr{V}(q)$ represents the height of the terrain. While a path from one hollow to another may go over a hill, another may go around the hill. It should also be noted that the standard self-consistent-field method of solving the Hartree-Fock equations proceeds in steps by constructing a lower-energy state at each step. Thus, the standard method of finding solutions for a Hamiltonian with a Lagrangian constraining field will not normally converge to a solution for which the curvature of the energy along the path of approach is negative. Other numerical methods for finding solutions must then be used

Procedures for calculating energy functions and using them to determine maximally-decoupled collective submanifolds were widely pursued during the 1970-80 decade.[23] It was shown, as one might expect from a landscape-model perspective, that a maximally-decoupled one-dimensional collective manifold (or collective path) is a path along a valley floor of the potential-energy function. Optimal constraining fields were then defined to derive the energy functions along such valleys. Note that,

---

[23]The search for collective submanifolds was pursued by many following proposals by Holzworth G. and Yukawa T. (1974), *Nucl. Phys.* **A219**, 125 and Rowe D.J. and Basserman R. (1974), *Nucl. Phys.* **A220**, 404. References and a review of the several constrained mean-field methods was given by Rowe D.J. (1982), *Nucl. Phys.* **A391**, 307.

for the small-amplitude TDHF solutions, which describe the harmonic normal-mode vibrations of the system about equilibrium, the lowest-frequency solution, from the perspective of the landscape model, describes an oscillatory motion along the path for which the gradient increases most slowly. Thus, to the extent that the direction of this path is unique, it defines the beginning of a valley through the terrain of the multi-dimensional Hartree-Fock energy function. To map out this valley, one can proceed incrementally by seeking the minimal-energy state, $|\phi(\sigma)\rangle$, for the constrained Hamiltonian, $\hat{H}(\sigma) := \hat{H} - \sigma \hat{Q}_1$, for a range of values of σ, such that, for each σ, $\hat{Q}_1$ is the lowest-frequency time-even solution of Equations (8.66) and (8.67) for the Hamiltonian $\hat{H}(\sigma)$. Similarly, a maximally decoupled two-dimensional submanifold is obtained by a self-consistent determination of the minimal-energy states for the Hamiltonian $\hat{H}(\sigma) := \hat{H} - \sigma_1 \hat{Q}_1 - \sigma_2 \hat{Q}_2$, where $\hat{Q}_1$ and $\hat{Q}_2$ are the two lowest-frequency time-even solutions of Equations (8.66) and (8.67) for the Hamiltonian $\hat{H}(\sigma)$. However, in assessing the quantum-mechanical relevance of a given potential energy surface, it is essential to take due account of the metric given by the inertial mass tensor. For example, for a given kinetic energy, a potential barrier might be essentially impenetrable for one value of the mass parameter and much less for another. As shown above, the inertial mass tensor is well-defined at any point of the space.

Mean-field energy functions have been used to obtain classical interpretations of algebraic-model Hamiltonians,[24,25,26] for which they yield insights into the nature of a phase transition of a model from one dynamical symmetry limit to another.

Exercises

8.12 Consider an interacting boson model,[27] with a u(6) spectrum generating algebra spanned by Hermitian linear combinations of the operators $\{s^\dagger s, d_i^\dagger s.s^\dagger d_i, d_i^\dagger d_j\}$, where $\{s^\dagger, d_i^\dagger, i = 1, \ldots, 5\}$ are the raising operators or a six-dimensional harmonic oscillator and satisfy the boson commutation relations

$$[s, s^\dagger] = 1, \quad [d_i, d_j^\dagger] = \delta_{i,j} \hat{I}, \quad [s, d_i^\dagger] = 0. \tag{8.144}$$

Show that a Cartan decomposition of the complex extension of u(6) is given by the operators

$$\begin{aligned} d_i^\dagger s, \ d_i^\dagger d_j \ (i < j), &\qquad \text{raising operators,} \\ s^\dagger s, \ d_i^\dagger d_i, &\qquad \text{Cartan operators,} \\ s^\dagger d_i, \ d_i^\dagger d_j \ (i > j), &\qquad \text{lowering operators.} \end{aligned} \tag{8.145}$$

Show that a lowest-weight state for the N-boson harmonic-oscillator Hilbert space is given by the state $|\phi_0\rangle := \frac{1}{\sqrt{N!}} (s^\dagger)^N |0\rangle$, where $|0\rangle$ is the zero-boson vacuum state.

[24] Gilmore R. (1979), *J. Math. Phys.* **20**, 891.

[25] Ginocchio N. and Kirson M.W. (1980), *Phys. Rev. Lett.* **44**, 1744.

[26] Dieperink A.E.L., Scholten O. and Iachello F. (1980), *Phys. Rev. Lett.* **44**, 1747.

[27] Iachello F. and Arima A. (1987), *The Interacting Boson Model* (Cambridge University Press, Cambridge).

8.13 Show that a manifold of phase-equivalent coherent states for the above N-boson irrep of u(6) is given by states of the form

$$|\phi(z)\rangle = \exp\left(\sum_i \left(z_i d_i^\dagger s - z_i^* s^\dagger d_i\right)\right)|\phi_0\rangle, \qquad (8.146)$$

where the z_i coefficients are complex numbers and $|\phi(z)\rangle$ is proportional to the unnormalised state

$$|\phi(z)\rangle \propto e^{\hat{X}}|\phi_0\rangle, \qquad (8.147)$$

where $\hat{X} := \sum_i z_i d_i^\dagger s$.

8.14 Show that

$$e^{\hat{X}} s^\dagger e^{-\hat{X}} = s^\dagger + \sum_i z_i d_i^\dagger \qquad (8.148)$$

and that if

$$z_i := \frac{1}{\sqrt{1-\beta^2}} q_i, \qquad (8.149)$$

where q_i is real and $\beta^2 = \sum_i q_i^2 \le 1$, then

$$e^{\hat{X}} s^\dagger e^{-\hat{X}} = \frac{1}{\sqrt{1-\beta^2}} B^\dagger, \qquad (8.150)$$

where

$$B^\dagger := \sqrt{1-\beta^2}\, s^\dagger + \sum_i q_i d_i^\dagger \qquad (8.151)$$

satisfies the commutation relation $[B, B^\dagger] = 1$. From this result, show that the normalised state $|\phi(z)\rangle$ is given by

$$|\phi(z)\rangle = \frac{1}{\sqrt{N!}}\left(B^\dagger\right)^N |0\rangle. \qquad (8.152)$$

8.15 Show that if $\hat{H} = \sum_i \varepsilon d_i^\dagger d_i$, then with z_i real and expressed in terms of q_i as defined above,

$$\mathscr{V}(q) := \langle\phi(z)|\hat{H}|\phi(z)\rangle = N\varepsilon\beta^2. \qquad (8.153)$$

Chapter 9

Concluding remarks

Nuclear structure is a fundamental level of organisation of matter. It is also an extraordinary "window" into the world of many-body quantum mechanics of finite systems. The elucidation of the quantum mechanics of the nucleus is now sixty years old. It received enormous impetus from the foundational models — the Bohr model, the independent-particle shell model, the pairing model, and mean-field models — that were put forward in the early years of its development. The subject of the present volume is the detailing of these foundational models in a modern context, using the full power of dynamical group and algebraic methods. Indeed, we have pursued the deployment of these methods much more than is done in other treatise on nuclear physics.

Remarkably, the basic models of the 1950's have continued to be the foundational models on which subsequent models and theories have been constructed. This has held true during the explosion of new information that has come from the development of a wide variety of accelerators, beam-handling techniques, detection devices, and computer processing facilities. Collective models have been developed in many ways to provide better interpretations of observed phenomena and to take better account of the underlying many-nucleon structure. But, they continue to be phrased in terms of the rotational and vibrational concepts of the basic Bohr model. The many-nucleon shell model has also evolved in numerous sophisticated ways which enable huge calculations to be executed with a variety of nucleonic interactions, including realistic interactions derived from quantum chromodynamics. Nevertheless, the original shell-model technology, based on the use of coupling-schemes and coefficients of fractional parentage, continues to be deployed.

The foundational models provide the basic language and methodology of nuclear structure. In this volume, we have aimed to give this language a deeper meaning and to acquire better insights into its significance. In particular, we have emphasised the remarkable versatility of dynamical group and algebraic methods for providing a unifying perspective on the many models of nuclear structure. Our intent is to set the stage for a forthcoming volume which will focus on unified views of nuclear models.

In presenting the models and theories of nuclear structure, we adopt the perspective that the many observed properties of nuclei are "emergent phenomena" that could not have been predicted from an underlying, first-principles, theory. In the pre-1950 days, it was even considered that the strong short-range repulsive nature of nucleon interactions precluded the existence of a realistic predictive microscopic theory of nuclear structure. Thus, it was only in response to the overwhelming evidence from observed nuclear properties that the nuclear shell model with phenomenological interactions emerged. Subsequently, with the development of effective interaction theories and powerful computers, it became possible to be more optimistic. In fact, nuclear theory is currently remarkably successful at interpreting observed nuclear properties, in terms of models and shell model calculations with phenomenological interactions, and even in making successful predictions of where to observe similar phenomena in other nuclei. Shell model studies are also becoming quite successful in deriving many properties of nuclei with realistic interactions derived from nuclear scattering data. Nevertheless, in spite of all the progress that has been made, the paradigm for the development of nuclear structure physics continues to begin with the systematic study of nuclear properties and how they change with variation of proton and neutron number. Phenomenological models are then proposed to interpret the patterns and explain what is going on in physical terms. These models can then be used to guide the formulation of microscopic description of what is observed in terms of interacting nucleons. In fact, it appears that, once the pattern of a wide body of data is recognised, an interpretation of the underlying phenomenon is often almost self-evident. Rotational bands and nuclear deformation are good examples of this.

The development of models to interpret observed phenomena in physical terms is key not only to the generation of insight into what is going on but also in providing a means to summarise large bodies of data. Moreover, the implications of a model can be explored and new experiments devised to test them; this leads to the accumulation of more critical data and a refined model. If the model proves to be successful in a significant range of situations, the challenge is then to express it in terms of interacting neutrons and protons or, if need be, in terms of sub-nucleon degrees of freedom. If this proves to be unreasonably difficult, it may be appropriate to adjust the model so that it can be given a microscopic interpretation.

The last step is vital because the only way in which we can hope to learn if an underlying microscopic theory exists is to discover if successful model interpretations of the data are compatible with such a theory. In addition to gaining an understanding of nuclear phenomena in this way, one also gains an understanding of the ways nucleons interact and correlate their motions. An essential aid to such an interpretation of a model, that is often overlooked in a desire to 'sell' a model, is to determine the circumstances in which the model breaks down, i.e., determine its domain of validity. After all, every model that is amenable to practical application, has a limited domain of validity. *Knowing the circumstances in which a model fails is every bit as important as knowing those in which it succeeds.* (We note that, at another fundamental level of organisation of matter, the failure of the Standard Model of particles and fields is being sought, avidly.) It is only in the last decade or so that some very serious failures are emerging in the models of nuclear structure. Titles that contain the words "collapse" and "failure" are easily found in recent papers on nuclei thought to be good examples of closed shells and collective behaviour. The road to deeper insight of these unexpected occurrences, and the full complexity of nuclear structure in general is, we believe, to understand the interplay and competition of the simple structures outlined in this volume.

In focusing on the above strategy for the evolution of nuclear structure, it has proved useful to identify models with shell-model coupling schemes. It turns out that many solvable models, which satisfy the desirable property of having a microscopic interpretation, have algebraic structures and associated dynamical symmetries which are the same as those which give rise to solvable sub-models of the nuclear shell model. The recognition of this fact, is another reason for the emphasis in this volume, and the volume that is to follow, on an analysis of nuclear models in algebraic terms. For if a model has a spectrum generating algebra, and if this SGA has an expression as a subalgebra of shell-model observables, then the model has an immediate interpretation as a sub-model of the shell model whose degrees of freedom are restricted to those of the model algebra. One can then try to understand the circumstances in which interactions between nucleons would favour such a restriction, i.e., the circumstances under which the interactions that would break the symmetries of the sub-model are sufficiently small that the sub-model is able to persist as a valid, even if approximate, representation of the phenomena.

The development in this volume of the foundational models, with such a perspective in mind, and the development of the algebraic and group theoretical apparatus to facilitate its pursuit, now provides a powerful framework for understanding nuclear phenomena and their microscopic interpretation which will be pursued in the following volume.

With some confidence, we would say that "We are now at the end of the beginning and are moving to the beginning of the middle: a period that will be measured in decades, where a truly unified view of nuclear structure is within reach." At the experimental end, this will come from the extraordinary power of spectroscopic methods that are now available and the ease of access to samples of single isotopic species of high purity that number in the thousands. At the theoretical end, we anticipate that huge advances will come from the identification of successful phenomenological models with shell-model coupling schemes. The development of the facility to carry out shell-model model calculations in many coupling schemes, will then enable shell-model calculations to be carried out with realistic interactions and truncated model space dictated by model fits to the new data available.

Appendix A

Some basics of angular-momentum theory

This appendix gives a brief review of some important definitions and relationships from the theory of angular momentum, with a particular emphasis on the phase conventions used in coupling and recoupling states and operators. The theory of angular momentum is presented in many standard texts.[1,2,3]

A.1 Angular-momentum basis states

The eigenstates $\{|\alpha jm\rangle; m = -j, \ldots, +j\rangle\}$, of a rotationally invariant Hamiltonian, are naturally labelled by $SU(2) \supset U(1)$ quantum numbers, jm, and a set of extra labels, α, as needed to uniquely identify a state. For each α and j, the states $\{|\alpha jm\rangle\}$ are a basis for an $SU(2)$ irrep, with respect to which the angular-momentum operators $\{\hat{J}_0, \hat{J}_\pm\}$ are infinitesimal generators, which act according to the equations

$$\hat{J}_0|\alpha jm\rangle = m|\alpha jm\rangle, \quad \hat{J}_\pm|\alpha jm\rangle = \sqrt{(j \mp m)(j \pm j + 1)}\,|\alpha j, m \pm 1\rangle. \quad \text{(A.1)}$$

Under a finite $SU(2)$ rotation, the above states transform according to the equation

$$\hat{R}(\Omega)|\alpha jm\rangle = \sum_k |\alpha jk\rangle \mathscr{D}^j_{km}(\Omega), \quad \Omega \in SU(2), \quad \text{(A.2)}$$

where $\mathscr{D}^j_{km}(\Omega)$ is a unitary *rotation matrix*. The properties of the rotation matrices are discussed in Section A.8.

A.2 SU(2) coupling coefficients

If the states $\{|j_1 m_1\rangle\}$ and $\{|j_2 m_2\rangle\}$ are bases for $SU(2)$ irreps of angular momenta j_1 and j_2, respectively, their tensor products $\{|j_1 m_1\rangle \otimes |j_2 m_2\rangle\}$ can be coupled to states

[1] Rose M.E. (1995), *Elementary Theory of Angular Momentum* (Dover Publications, New York; originally published by Wiley, 1957).

[2] Brink D.M. and Satchler G.R. (1994), *Angular Momentum* (Oxford University Press), third edn.

[3] Edmonds A.R. (1996), *Angular Momentum in Quantum Mechanics* (Princeton University Press, Princeton, N.J.), paperback edition.

with good angular momentum quantum numbers, jm, by a unitary transformation,

$$|[j_1 j_2]jm\rangle \equiv \left[|j_1\rangle \otimes |j_2\rangle\right]_{jm} := \sum_{m_1 m_2} |j_1 m_1\rangle \otimes |j_2 m_2\rangle (j_2 m_2\, j_1 m_1|jm), \qquad (A.3)$$

where the overlap matrix elements in this expansion,

$$(j_2 m_2\, j_1 m_1|jm) = \left[\langle j_2 m_2| \otimes \langle j_1 m_1|\right]|[j_1 j_2]JM\rangle, \qquad (A.4)$$

are known as *Clebsch-Gordan coefficients*. With the standard phase conventions of Condon and Shortley[4] these CG coefficients are real.[5] Thus, the inverse unitary transformation is expressed by

$$|j_1 m_1\rangle \otimes |j_2 m_2\rangle = \sum_{jm} \left[|j_1\rangle \otimes |j_2\rangle\right]_{jm} (j_2 m_2\, j_1 m_1|jm). \qquad (A.6)$$

A frequently occurring coefficient that is worth remembering is:

$$(j, -m\, jm|00) = \frac{(-1)^{j+m}}{\sqrt{2j+1}}. \qquad (A.7)$$

The values of some CG coefficients are given in Tables A.1–A.4. Other coefficients can be obtained from published tables[6] or, more usefully, by direct computation.[7]

The CG coefficients obey a number of basic relationships:

---

[4]Condon E.U. and Shortley G.H. (1935), *Theory of Atomic Spectra* (Cambridge University Press, Cambridge).

[5] The CG coefficients, defined with Condon-Shortley phase conventions, are standard and generally accepted. However, authors use them to define coupled tensor products in two different ways. In this book we use the definition, given by Equation (A.3), which we refer to as right-to-left coupling and distinguish this coupling by enclosing the coupled states in square brackets. The alternative left-to-right coupling (denoted using round brackets) is given by

$$\left(|j_1\rangle \otimes |j_2\rangle\right)_{jm} := \sum_{m_1 m_2} |j_1 m_1\rangle \otimes |j_2 m_2\rangle (j_1 m_1\, j_2 m_2|jm) \equiv (-1)^{j_1 + j_2 - j_3}\left[|j_1\rangle \otimes |j_2\rangle\right]_{jm}. \quad (A.5)$$

The choice is, in principle, arbitrary and corresponds to a phase difference arising from the symmetry relation $(j_1 m_1\, j_2 m_2|jm) = (-1)^{j_1 + j_2 - j}(j_2 m_2\, j_1 m_1|jm)$. In fact, there are several instances in which arbitrary choices are encountered in angular-momentum theory, and it appears that (apart from the values of the Clebsch-Gordan and recoupling coefficients) all the widely-used texts on angular-momentum theory differ in some respects in their choices. In the following, we draw attention to the different conventions as they arise. Fortunately, while it is important to be consistent, it is easy to relate the results obtained by different authors provided one knows the conventions they have used. Relationships between results obtained with different phase conventions are discussed in Section A.5.

[6]Appel H. (1968), *Numerical Tables for Angular Correlation Computations in α-, β- and γ-spectroscopy: 3j-, 6j-, 9j-Symbols, F- and Γ-Coefficients*, *Landolt-Börnstein New Series*, Vol. I/3 (Springer-Verlag, Berlin/Heidelberg).

[7]Computer codes for calculating CG coefficients, 6-j and 9-j symbols can be found, for example, in the book Heyde, *op. cit.* Footnote 78 on Page 468. A web-based program is also available from Stevenson P.D. (2002), *Comp. Phys. Comm.* **147**, 853.

Table A.1: $(j_1 m_1 \frac{1}{2} m_2 | jm)$

j	$m_2 = \frac{1}{2}$	$m_2 = -\frac{1}{2}$
$j_1 + \frac{1}{2}$	$\sqrt{\dfrac{j+m}{2j_1+1}}$	$\sqrt{\dfrac{j-m}{2j_1+1}}$
$j_1 - \frac{1}{2}$	$-\sqrt{\dfrac{j-m+1}{2j_1+1}}$	$\sqrt{\dfrac{j+m+1}{2j_1+1}}$

Table A.2: $(j_1 m_1 1 m_2 | jm)$

j	$m_2 = 1$	$m_2 = 0$	$m_2 = -1$
$j_1 + 1$	$\sqrt{\dfrac{(j+m-1)(j+m)}{(2j_1+1)(2j_1+2)}}$	$\sqrt{\dfrac{(j-m)(j+m)}{(2j_1+1)(j_1+1)}}$	$\sqrt{\dfrac{(j-m-1)(j-m)}{(2j_1+1)(2j_1+2)}}$
j_1	$-\sqrt{\dfrac{(j+m)(j-m+1)}{2j_1(j_1+1)}}$	$\dfrac{m}{\sqrt{j_1(j_1+1)}}$	$\sqrt{\dfrac{(j-m)(j+m+1)}{2j_1(j_1+1)}}$
$j_1 - 1$	$\sqrt{\dfrac{(j-m+1)(j-m+2)}{2j_1(2j_1+1)}}$	$-\sqrt{\dfrac{(j-m+1)(j+m+1)}{j_1(2j_1+1)}}$	$\sqrt{\dfrac{(j+m+2)(j+m+1)}{2j_1(2j_1+1)}}$

Table A.3: $(j_1 m_1 \frac{3}{2} m_2 | jm)$

j	$m_2 = \frac{3}{2}$	$m_2 = \frac{1}{2}$
$j_1 + \frac{3}{2}$	$\sqrt{\dfrac{(j+m-2)(j+m-1)(j+m)}{(2j_1+1)(2j_1+2)(2j_1+3)}}$	$\sqrt{\dfrac{3(j+m-1)(j+m)(j-m)}{(2j_1+1)(2j_1+2)(2j_1+3)}}$
$j_1 + \frac{1}{2}$	$-\sqrt{\dfrac{3(j+m-1)(j+m)(j-m+1)}{2j_1(2j_1+1)(2j_1+3)}}$	$-(j-3m+1)\sqrt{\dfrac{j+m}{2j_1(2j_1+1)(2j_1+3)}}$
$j_1 - \frac{1}{2}$	$\sqrt{\dfrac{3(j+m)(j-m+1)(j-m+2)}{(2j_1-1)(2j_1+1)(2j_1+2)}}$	$-(j+3m)\sqrt{\dfrac{j-m+1}{(2j_1-1)(2j_1+1)(2j_1+2)}}$
$j_1 - \frac{3}{2}$	$-\sqrt{\dfrac{(j-m+1)(j-m+2)(j-m+3)}{2j_1(2j_1-1)(2j_1+1)}}$	$\sqrt{\dfrac{3(j+m+1)(j-m+1)(j-m+2)}{2j_1(2j_1-1)(2j_1+1)}}$
j	$m_2 = -\frac{1}{2}$	$m_2 = -\frac{3}{2}$
$j_1 + \frac{3}{2}$	$\sqrt{\dfrac{3(j+m)(j-m-1)(j-m)}{(2j_1+1)(2j_1+2)(2j_1+3)}}$	$\sqrt{\dfrac{(j-m-2)(j-m-1)(j-m)}{(2j_1+1)(2j_1+2)(2j_1+3)}}$
$j_1 + \frac{1}{2}$	$(j+3m+1)\sqrt{\dfrac{j-m}{2j_1(2j_1+1)(2j_1+3)}}$	$\sqrt{\dfrac{3(j+m+1)(j-m-1)(j-m)}{2j_1(2j_1+1)(2j_1+3)}}$
$j_1 - \frac{1}{2}$	$-(j-3m)\sqrt{\dfrac{j+m+1}{(2j_1-1)(2j_1+1)(2j_1+2)}}$	$\sqrt{\dfrac{3(j+m+1)(j+m+2)(j-m)}{(2j_1-1)(2j_1+1)(2j_1+2)}}$
$j_1 - \frac{3}{2}$	$-\sqrt{\dfrac{3(j+m+1)(j+m+2)(j-m+1)}{2j_1(2j_1-1)(2j_1+1)}}$	$\sqrt{\dfrac{(j+m+1)(j+m+2)(j+m+3)}{2j_1(2j_1-1)(2j_1+1)}}$

(i) *Orthogonality properties:*

$$\sum_{j_3, m_3} (j_1 m_1 \, j_2 m_2 | j_3 m_3)(j_1 m_1' \, j_2 m_2' | j_3 m_3) = \delta_{m_1, m_1'} \delta_{m_2, m_2'}, \qquad (A.8)$$

$$\sum_{m_1, m_2} (j_1 m_1 \, j_2 m_2 | j_3 m_3)(j_1 m_1 \, j_2 m_2 | j_3' m_3') = \delta_{j_3, j_3'} \delta_{m_3, m_3'}. \qquad (A.9)$$

Table A.4: $(j_1 m_1\, 2 m_2 | j m)$

j	$m_2 = 2$	$m_2 = 1$
$j_1 + 2$	$\sqrt{\frac{(j+m-3)(j+m-2)(j+m-1)(j+m)}{(2j_1+1)(2j_1+2)(2j_1+3)(2j_1+4)}}$	$\sqrt{\frac{(j-m)(j+m)(j+m-1)(j+m-2)}{(2j_1+1)(j_1+1)(2j_1+3)(j_1+2)}}$
$j_1 + 1$	$-\sqrt{\frac{(j+m-2)(j+m-1)(j+m)(j-m+1)}{2j_1(j_1+1)(j_1+2)(2j_1+1)}}$	$-(j-2m+1)\sqrt{\frac{(j+m)(j+m-1)}{2j_1(j_1+1)(j_1+2)(2j_1+1)}}$
j_1	$\sqrt{\frac{3(j+m-1)(j+m)(j-m+1)(j-m+2)}{(2j_1-1)2j_1(j_1+1)(2j_1+3)}}$	$(1-2m)\sqrt{\frac{3(j-m+1)(j+m)}{(2j_1-1)2j_1(j_1+1)(2j_1+3)}}$
$j_1 - 1$	$-\sqrt{\frac{(j+m)(j-m+1)(j-m+2)(j-m+3)}{2(j_1-1)j_1(j_1+1)(2j_1+1)}}$	$(j+2m)\sqrt{\frac{(j-m+2)(j-m+1)}{2(j_1-1)j_1(j_1+1)(2j_1+1)}}$
$j_1 - 2$	$\sqrt{\frac{(j-m+1)(j-m+2)(j-m+3)(j-m+4)}{(2j_1-2)(2j_1-1)2j_1(2j_1+1)}}$	$-\sqrt{\frac{(j-m+3)(j-m+2)(j-m+1)(j+m+1)}{(j_1-1)(2j_1-1)j_1(2j_1+1)}}$
j	$m_2 = 0$	
$j_1 + 2$	$\sqrt{\frac{3(j-m)(j-m-1)(j+m)(j+m-1)}{(2j_1+1)(2j_1+2)(2j_1+3)(j_1+2)}}$	
$j_1 + 1$	$m\sqrt{\frac{3(j-m)(j+m)}{j_1(j_1+1)(j_1+2)(2j_1+1)}}$	
j_1	$\frac{3m^2-j(j+1)}{\sqrt{(2j_1-1)j_1(j_1+1)(2j_1+3)}}$	
$j_1 - 1$	$-m\sqrt{\frac{3(j-m+1)(j+m+1)}{(j_1-1)j_1(j_1+1)(2j_1+1)}}$	
$j_1 - 2$	$\sqrt{\frac{3(j-m+2)(j-m+1)(j+m+2)(j+m+1)}{(2j_1-2)(2j_1-1)j_1(2j_1+1)}}$	
j	$m_2 = -1$	$m_2 = -2$
$j_1 + 2$	$\sqrt{\frac{(j-m)(j-m-1)(j-m-2)(j+m)}{(j_1+1)(j_1+2)(2j_1+1)(2j_1+3)}}$	$\sqrt{\frac{(j-m-3)(j-m-2)(j-m-1)(j-m)}{(2j_1+1)(2j_1+2)(2j_1+3)(2j_1+4)}}$
$j_1 + 1$	$(j+2m+1)\sqrt{\frac{(j-m)(j-m-1)}{j_1(j_1+2)(2j_1+1)(2j_1+2)}}$	$\sqrt{\frac{(j-m-2)(j-m-1)(j-m)(j+m+1)}{j_1(j_1+1)(2j_1+1)(2j_1+4)}}$
j_1	$(2m+1)\sqrt{\frac{3(j-m)(j+m+1)}{j_1(2j_1-1)(2j_1+2)(2j_1+3)}}$	$\sqrt{\frac{3(j-m-1)(j-m)(j+m+1)(j+m+2)}{j_1(2j_1-1)(2j_1+2)(2j_1+3)}}$
$j_1 - 1$	$-(j-2m)\sqrt{\frac{(j+m+2)(j+m+1)}{j_1(j_1-1)(2j_1+1)(2j_1+2)}}$	$\sqrt{\frac{(j-m)(j+m+1)(j+m+2)(j+m+3)}{j_1(j_1-1)(2j_1+1)(2j_1+2)}}$
$j_1 - 2$	$-\sqrt{\frac{(j-m+1)(j+m+3)(j+m+2)(j+m+1)}{j_1(j_1-1)(2j_1-1)(2j_1+1)}}$	$\sqrt{\frac{(j+m+1)(j+m+2)(j+m+3)(j+m+4)}{2j_1(2j_1-1)(2j_1-2)(2j_1+1)}}$

(ii) *Symmetry relations:*

$$(j_1 m_1\, j_2 m_2 | j_3 m_3) = (-1)^{j_1+j_2-j_3}(j_1, -m_1\, j_2, -m_2 | j_3, -m_3) \qquad (\text{A.10})$$

$$= (-1)^{j_1+j_2-j_3}(j_2 m_2\, j_1 m_1 | j_3 m_3), \qquad (\text{A.11})$$

$$(j_1 m_1\, j_2 m_2 | j_3 m_3) = (-1)^{j_1-m_1}\sqrt{\frac{2j_3+1}{2j_2+1}}\,(j_1 m_1\, j_3, -m_3 | j_2, -m_2) \qquad (\text{A.12})$$

$$= (-1)^{j_2+m_2}\sqrt{\frac{2j_3+1}{2j_1+1}}\,(j_3, -m_3\, j_2 m_2 | j_1, -m_1). \qquad (\text{A.13})$$

(iii) *Wigner 3-j symbols:*

The symmetry relations of SU(2) CG coefficients are summarised succinctly in terms of Wigner's 3-j symbols, defined by

$$\begin{pmatrix} j_1 & j_2 & j_3 \\ m_1 & m_2 & m_3 \end{pmatrix} := \frac{(-1)^{j_1-j_2-m_3}}{\sqrt{2j_3+1}} (j_1 m_1 j_2 m_2 | j_3, -m_3). \qquad (A.14)$$

A 3-j symbol is invariant under even (cyclic) permutations of its columns and is multiplied by $(-1)^{j_1+j_2+j_3}$ under an odd permutation or simultaneous change of sign of m_1, m_2 and m_3.

A.3 Spherical tensor operators

To take advantage of the classification of basis states by their group symmetries, it is desirable to have a similar classification of the operators that act on these states. We are particularly concerned in nuclear physics with the transformations of linear operators under rotations.

A.3.1 *The transformation of linear operators*

Let G denote a group of transformations of a Hilbert space, $\mathbb{H}$, in which a state $|\psi\rangle \in \mathbb{H}$ transforms according to the equation

$$|\psi\rangle \to \hat{U}(g)|\psi\rangle, \quad \forall\, g \in G. \qquad (A.15)$$

If $\hat{W}$ is a linear operator on $\mathbb{H}$, then the transformation

$$\hat{W}|\psi\rangle \to \hat{U}(g)\hat{W}|\psi\rangle = \hat{U}(g)\hat{W}U(g^{-1})\hat{U}(g)|\psi\rangle, \quad g \in G, \qquad (A.16)$$

implies a corresponding transformation of $\hat{W}$, given by

$$\hat{W} \to \hat{U}(g)\hat{W}U(g^{-1}), \quad g \in G. \qquad (A.17)$$

Similarly, the transformation of $\hat{W}$ by an infinitesimal generator $\hat{X}$, of the group, G, is given by

$$\hat{W} \to [\hat{X}, \hat{W}], \quad X \in \mathbf{g}, \qquad (A.18)$$

where $\mathbf{g}$ is the Lie algebra of G (see Exercise A.1). With the observation that

$$\hat{X}\hat{W}|\psi\rangle = [\hat{X}, \hat{W}]|\psi\rangle + \hat{W}\hat{X}|\psi\rangle, \qquad (A.19)$$

this expression is seen to be consistent with the Leibnitz rule which states that a differential operator $\hat{D}$ on a product function fh is given by

$$\hat{D}(fh) = (\hat{D}f)h + f(\hat{D}h) = [\hat{D}, f]h + f[\hat{D}, h]. \qquad (A.20)$$

A.3.2 Irreducible spherical tensors

Spherical tensors are sets of linear operators that transform among themselves under SO(3) or SU(2) rotations in the same way as basis states for SO(3) or SU(2) representations. An *irreducible spherical tensor* is a spherical tensor whose components transform in the same way as basis states for an SO(3) or SU(2) irrep. Thus, whereas a set of states $\{|JM\rangle; M = -J, \ldots, J\}$, which reduce the subgroup chain SU(2) $\supset$ U(1), transform under SU(2) according to the standard equation

$$\hat{R}(\Omega)|JM\rangle = \sum_K |JK\rangle \mathscr{D}^J_{KM}(\Omega), \quad \Omega \in \mathrm{SU}(2), \tag{A.21}$$

the components of a spherical tensor $\hat{W}_J$ transform according to the equation

$$\hat{R}(\Omega)\hat{W}_{JM}\hat{R}(\Omega^{-1}) = \sum_K \hat{W}_{JK}\mathscr{D}^J_{KM}(\Omega), \quad \Omega \in \mathrm{SU}(2). \tag{A.22}$$

A.3.3 The Wigner-Eckart theorem

Expressing linear operators as components of tensors results in efficient ways of computing their matrix elements when the Clebsch-Gordan coefficients for the group are known. The Wigner-Eckart theorem applies to the matrix elements of the irreducible tensors of any group. We consider here its application to matrix elements of SU(2) spherical tensors.

According to the Wigner-Eckart theorem, matrix elements of the components $\{\hat{W}_{J_2 M_2}\}$ of a spherical tensor between SU(2)-coupled states are related by the identity

$$\langle \beta J_3 M_3 | \hat{W}_{J_2 M_2} | \alpha J_1 M_1 \rangle = (J_1 M_1 J_2 M_2 | J_3 M_3) \frac{\langle \beta J_3 \| \hat{W}_{J_2} \| \alpha J_1 \rangle}{\sqrt{2J_3 + 1}}, \tag{A.23}$$

where $\langle \beta J_3 \| \hat{W}_{J_2} \| \alpha J_1 \rangle$ is a so-called *reduced matrix element*.[8] This identity shows that many matrix elements are determined by a much smaller number of reduced matrix elements in combination with Clebsch-Gordan coefficients.

The Wigner-Eckart theorem follows directly from the observation that the overlap of any two states, $\langle \beta J_3 M_3 | \gamma J_3 M_3 \rangle$, where $\{|\beta J_3 M_3\rangle, M_3 = -J_3, \ldots, +J_3\}$

---

[8]Variations on the detailed expression of the Wigner-Eckart theorem, and hence the definition of a reduced matrix element, can be found in the literature. Some authors omit the factor $\sqrt{2J_3 + 1}$ in Equation (A.23), which is included to give the reduced matrix elements extra symmetry. Others add a factor $(-1)^{2J_2}$ to the right-hand side of Equation (A.23); this factor appears to have a historical origin in the way the Wigner-Eckart theorem was expressed by Racah, *op. cit.* Footnote 1 on Page 399, in terms of his so-called *V*-coefficients which were later replaced by Wigner's 3-*j* symbols. The definition given by Equation (A.23) is in common use in nuclear physics, e.g., it is the definition used in the collective model by Bohr A. and Mottelson B.R. (1969), *Nuclear Structure*, Vol. 1 (Benjamin, New York), (republished by World Scientific, Singapore), in the shell model by French J.B. (1966), in *Proceedings of the International School of Physics "Enrico Fermi", Course XXXVI*, edited by C. Bloch (Academic Press, New York), and in quantum mechanics by Messiah A. (1966), *Quantum Mechanics*, Vol. 2 (North Holland, Amsterdam), sixth printing.

and $\{|\gamma J_3 M_3\rangle, M_3 = -J_3, \ldots, +J_3\}$ are orthonormal bases for SU(2) irreps with angular-momentum quantum numbers, J_3, M_3, is independent of M_3 (see Exercise A.2).[9] If we make a coupled expansion of the state $\hat{W}_{J_2 M_2}|\alpha J_1 M_1\rangle$ in parallel with that of Equation (A.6), i.e.,

$$\hat{W}_{J_2 M_2}|\alpha J_1 M_1\rangle = \sum_{J_3 M_3} \left[\hat{W}_{J_2} \otimes |\alpha J_1\rangle\right]_{J_3 M_3} (J_1 M_1 \, J_2 M_2 | J_3 M_3), \qquad (A.24)$$

we obtain the expression

$$\langle \beta J_3 M_3 | \hat{W}_{J_2 M_2} | \alpha J_1 M_1\rangle = (J_1 M_1 J_2 M_2 | J_3 M_3)\langle \beta J_3 M_3 | \left[\hat{W}_{J_2} \otimes |\alpha J_1\rangle\right]_{J_3 M_3}. \qquad (A.25)$$

Thus, the Wigner-Eckart theorem is simply a definition of the M_3-independent reduced matrix:

$$\langle \beta J_3 \| \hat{W}_{J_2} \| \alpha J_1 \rangle := \sqrt{2J_3 + 1} \, \langle \beta J_3 M_3 | \left[\hat{W}_{J_2} \otimes |\alpha J_1\rangle\right]_{J_3 M_3}. \qquad (A.26)$$

For orthonormal basis states, the Wigner-Eckart theorem can be used to express the coupled tensor product, defined by Equation (A.24), in the useful form

$$\left[\hat{W}_{J_2} \otimes |\alpha J_1\rangle\right]_{J_3 M_3} = \sum_{\beta} |\beta J_3 M_3\rangle \frac{\langle \beta J_3 \| \hat{W}_{J_2} \| \alpha J_1\rangle}{\sqrt{2J_3 + 1}}. \qquad (A.27)$$

A.4 Recoupling coefficients

Systematic techniques for coupling and recoupling spherical tensors are provided by diagrammatic representations of coupled tensor products.[10] For example, the right-to-left coupled tensor product

$$\left[\hat{W}_{J_2} \otimes |\alpha J_1\rangle\right]_{J_3 M_3} := \sum_{M_1 M_2} \hat{W}_{J_2 M_2}|\alpha J_1 M_1\rangle(J_1 M_1 \, J_2 M_2 | J_3 M_3), \qquad (A.28)$$

has the diagrammatic representation:

$$\left[\hat{W}_{J_2} \otimes |\alpha J_1\rangle\right]_{J_3} := J_3 \quad \overbrace{\diagdown}^{\hat{W}_{J_2}} |\alpha J_1\rangle \, . \qquad (A.29)$$

---

[9]This is a particular case of a general result. An overlap of two states, $\langle \alpha \lambda' \nu' | \beta \lambda \nu \rangle$, where λ' and λ label irreps of a compact Lie group, ν' and ν label irreps of a subgroup, and α, β, are multiplicity indices, is zero if $\lambda' \neq \lambda$ and/or $\nu' \neq \nu$. Moreover, the overlaps $\langle \alpha \lambda \nu | \beta \lambda \nu \rangle$ take common values for all ν, whenever ν labels orthonormal basis states for the irrep λ which transform in the same way for every occurrence of the irrep λ.

[10]French J.B. (1966), in *Proceedings of the International School of Physics "Enrico Fermi", Course XXXVI*, edited by C. Bloch (Academic Press, New York).

A.4.1 *Racah coefficients*

There are two natural choices for the coupling of three angular-momentum states to good combined angular momentum: $\left|\left[j_3 \otimes [j_2 \otimes j_1]_{J_{12}}\right]_J\right\rangle$ and $\left|\left[[j_3 \otimes j_2]_{J_{23}} \otimes j_1\right]_J\right\rangle$. The states defined in these alternative ways are related by a unitary transformation:

$$\left|\left[j_3 \otimes [j_2 \otimes j_1]_{J_{12}}\right]_J\right\rangle = \sum_{J_{23}} U(j_1 j_2 J j_3 : J_{12} J_{23}) \left|\left[[j_3 \otimes j_2]_{J_{23}} \otimes j_1\right]_J\right\rangle, \quad \text{(A.30)}$$

$$\left|\left[[j_3 \otimes j_2]_{J_{23}} \otimes j_1\right]_J\right\rangle = \sum_{J_{12}} U(j_1 j_2 J j_3 : J_{12} J_{23}) \left|\left[j_3 \otimes [j_2 \otimes j_1]_{J_{12}}\right]_J\right\rangle, \quad \text{(A.31)}$$

where the U coefficients are so-called *Racah U-coefficients*. These identities are expressed in the more memorable graphical form by

$$\quad \text{(A.32)}$$

$$\quad \text{(A.33)}$$

The Racah U-coefficients are useful because they are the coefficients of a unitary transformation. However, it is common practice to express recoupling results in terms of Racah's W-coefficients, which have simpler symmetry relations, or in terms of the so-called 6-j symbols, whose symmetries are still more evident. These recoupling coefficients are related by the identities:

$$U(j_1 j_2 J j_3 : J_{12} J_{23}) = \sqrt{(2J_{12} + 1)(2J_{23} + 1)}\, W(j_1 j_2 J j_3 : J_{12} J_{23}) \quad \text{(A.34)}$$

$$= (-1)^{j_1 + j_2 + j_3 + J} \sqrt{(2J_{12} + 1)(2J_{23} + 1)}$$

$$\times \begin{Bmatrix} j_1 & j_2 & J_{12} \\ j_3 & J & J_{23} \end{Bmatrix}. \quad \text{(A.35)}$$

The 6-j symbols are invariant under permutations of columns and under the simultaneous exchange of upper and lower entries in any two columns.

Useful special cases of these relations are given by the identities

$$U(aabb : 0c) = U(abab : c0) = (-1)^{a+b-c} \sqrt{\frac{2c+1}{(2a+1)(2b+1)}}, \quad \text{(A.36)}$$

$$\begin{Bmatrix} a & b & c \\ b & a & 0 \end{Bmatrix} = (-1)^{a+b+c} \left[(2a+1)(2b+1)\right]^{-1/2}. \quad \text{(A.37)}$$

A.4.2 9-j symbols

When there are four coupled systems as frequently occurs, e.g., when a pair of coupled operators, $[\hat{T}_L \otimes \hat{U}_S]_{JM}$, operates on a coupled product state, $|[L_i \otimes S_i]J_iM_i\rangle$, one may want to consider the coupled states $[[\hat{T}_L \otimes \hat{U}_S]_J \otimes |[L_i \otimes S_i]J_i\rangle]_{J_fM_f}$. Moreover, if it should happen that the tensor $\hat{T}_L$ operates only on the first and the tensor $\hat{U}_S$ operates only on the second of the coupled states, one will want to express the state $[[\hat{T}_L \otimes \hat{U}_S]_J \otimes |[L_i \otimes S_i]J_i\rangle]_{J_fM_f}$ in terms of states in the re-ordered form $[[\hat{T}_L \otimes |L_i\rangle]_{L_f} \otimes [\hat{U}_S \otimes |S_i\rangle]_{S_f}]_{J_fM_f}$. The 9-$j$ coefficients are defined for this purpose.

The recoupling is given by the unitary transformation

$$
[[\hat{T}_L \otimes \hat{U}_S]_J \otimes |[L_i \otimes S_i]J_i\rangle]_{J_fM_f} := \sum_{L_fS_f} \overline{\begin{Bmatrix} S_i & L_i & J_i \\ S & L & J \\ S_f & L_f & J_f \end{Bmatrix}} [[\hat{T}_L \otimes |L_i\rangle]_{L_f} \otimes [\hat{U}_S \otimes |S_i\rangle]_{S_f}]_{J_fM_f},
$$

$$(A.38)$$

where the matrix array denotes a so-called *unitary 9-j symbol*. This identity can also be expressed in the more memorable graphical form

$$(A.39)$$

The unitary (barred) 9-j symbols are related to the standard (unbarred) 9-j symbols, which are non-unitary but have more symmetry, by

$$
\overline{\begin{Bmatrix} S_i & L_i & J_i \\ S & L & J \\ S_f & L_f & J_f \end{Bmatrix}} = [(2J_i+1)(2J+1)(2S_f+1)(2L_f+1)]^{1/2} \begin{Bmatrix} S_i & L_i & J_i \\ S & L & J \\ S_f & L_f & J_f \end{Bmatrix}. \quad (A.40)
$$

The standard 9-j symbol is invariant under an even permutation of rows or columns or a transposition (i.e., a reflection in either diagonal). An odd permutation of rows or columns is equivalent to multiplying it by $(-1)^\Sigma$, where Σ is the sum of the entries in the symbol.

A useful special case is given by

$$
\begin{Bmatrix} j_1 & j_2 & J_{12} \\ j_3 & j_4 & J_{12} \\ J_{13} & J_{13} & 0 \end{Bmatrix} = \frac{(-1)^{j_2+j_3+J_{12}+J_{13}}}{[(2J_{12}+1)(2J_{13}+1)]^{1/2}} \begin{Bmatrix} j_1 & j_2 & J_{12} \\ j_4 & j_3 & J_{13} \end{Bmatrix} \quad (A.41)
$$

$$
= \frac{(-1)^{j_1+j_4-J_{12}-J_{13}}}{(2J_{12}+1)(2J_{13}+1)} U(j_1j_2j_3j_4 : J_{12}J_{13}). \quad (A.42)
$$

A.5   Examples of recoupling expressions

We derive here some expressions for typical matrix elements that arise, for example, in applications of the shell model, first for the right-to-left coupling convention used in this book and then for the alternative left-to-right convention. It will be seen that there is a need to keep track of fewer phase factors with right-to-left coupling.

A.5.1   *Reduced matrix elements for right-to-left coupling*

Example (i):

Suppose that $\hat{T}_L$ is a spherical tensor which acts only on the first factor of an orthonormal system of coupled tensor-product states of the form $|[\alpha L_i S]JM\rangle :=$ $[|\alpha L_i\rangle \otimes |S\rangle]_{JM}$. Such a situation is encountered, for example, in the evaluation of matrix elements of operators that act only on the orbital wave functions in a tensor product space of coupled spin-orbit states. The reduced matrix elements of $\hat{T}_L$ are given by

$$\langle[\alpha L_f S]J_f\|\hat{T}_L\|[\beta L_i S]J_i\rangle = \sqrt{\frac{2J_f+1}{2L_f+1}}\, U(SL_iJ_fL:J_iL_f)\langle\alpha L_f\|\hat{T}_L\|\beta L_i\rangle. \quad \text{(A.43)}$$

This result follows by direct substitution of the expressions given in the preceeding sections. From Equation (A.26), the required reduced matrix element is expressed as the M_f-independent overlap,

$$\langle[\alpha L_f S]J_f\|\hat{T}_L\|[\beta L_i S]J_i\rangle = \sqrt{2J_f+1}\,\langle[\alpha L_f S]J_fM_f|[\hat{T}_L \otimes |[\beta L_i S]J_i\rangle]_{J_fM_f}. \quad \text{(A.44)}$$

Equation (A.30), is then used to recouple $[\hat{T}_L \otimes |[\beta L_i S]J_i\rangle]_{J_fM_f}$ to the form

$$[\hat{T}_L \otimes [|\beta L_i\rangle \otimes |S\rangle]_{J_i}]_{J_fM_f} = \sum_{L_f} U(SL_iJ_fL:J_iL_f)\,[[\hat{T}_L \otimes |\beta L_i\rangle]_{L_f} \otimes |S\rangle]_{J_fM_f}, \quad \text{(A.45)}$$

and Equation (A.27) is used to make the substitution

$$[\hat{T}_L \otimes |\beta L_i\rangle]_{L_f} = \sum_{\alpha} |\alpha L_f\rangle \frac{\langle\alpha L_f\|\hat{T}_L\|\beta L_i\rangle}{\sqrt{2L_f+1}}. \quad \text{(A.46)}$$

Example (ii):

If spherical tensors $\hat{A}_{J_1}$ and $\hat{B}_{J_2}$ both act on an orthonormal system of states of the form $|\alpha JM\rangle$, then

$$\langle\alpha J_f\|[\hat{A}_{J_1} \otimes \hat{B}_{J_2}]_J\|\beta J_i\rangle$$
$$= \sum_{\gamma J'} U(J_iJ_2J_fJ_1:J'J)\, \frac{\langle\alpha J_f\|\hat{A}_{J_1}\|\gamma J'\rangle\langle\gamma J'\|\hat{B}_{J_2}\|\beta J_i\rangle}{\sqrt{(2J'+1)}} \quad \text{(A.47)}$$

and

$$\langle \alpha J \| [\hat{A}_{J_1} \otimes \hat{B}_{J_1}]_0 \| \beta J \rangle = \sum_{\gamma J'} (-1)^{J+J_1-J'} \frac{\langle \alpha J \| \hat{A}_{J_1} \| \gamma J' \rangle \langle \gamma J' \| \hat{B}_{J_1} \| \beta J \rangle}{\sqrt{(2J+1)(2J_1+1)}}. \quad (A.48)$$

To derive the first equation, start with the Racah recoupling identity

$$\left[[\hat{A}_{J_1} \otimes \hat{B}_{J_2}]_J \otimes |\beta J_i\rangle \right]_{J_f} = \sum_{J'} U(J_i J_2 J_f J_1 : J' J) \left[\hat{A}_{J_1} \otimes [\hat{B}_{J_2} \otimes |\beta J_i\rangle]_{J'} \right]_{J_f}. \quad (A.49)$$

Then use the Wigner-Eckart expression, as given by Equation (A.27), twice in sequence to express the right-hand side of this equation in terms of reduced matrix elements:

$$\sum_{J'} U(J_i J_2 J_f J_1 : J' J) \left[\hat{A}_{J_1} \otimes [\hat{B}_{J_2} \otimes |\beta J_i\rangle]_{J'} \right]_{J_f M_f}$$

$$= \sum_{J'\gamma} U(J_i J_2 J_f J_1 : J' J) [\hat{A}_{J_1} \otimes |\gamma J'\rangle]_{J_f M_f} \frac{\langle \gamma J' \| \hat{B}_{J_2} \| \beta J_i \rangle}{\sqrt{2J'+1}}$$

$$= \sum_{J'\alpha\gamma} U(J_i J_2 J_f J_1 : J' J) |\alpha J_f M_f\rangle \frac{\langle \alpha J_f \| \hat{A}_{J_1} \| \gamma J' \rangle \langle \gamma J' \| \hat{B}_{J_2} \| \beta J_i \rangle}{\sqrt{(2J_f+1)(2J'+1)}}. \quad (A.50)$$

Equation (A.26) then gives the desired result. The special $J = 0$ case is obtained with the Racah coefficient given by Equation (A.36).

Example (iii):

If $\hat{T}_L$ is a spherical tensor that acts only on the first factor and $\hat{U}_S$ is a spherical tensor that acts only the second factor of an orthonormal system of coupled tensor-product states of the form $|[L_i S] JM\rangle := [|L_i\rangle \otimes |S\rangle]_{JM}$, then

$$\langle [L_f S_f] J_f \| [\hat{T}_L \otimes \hat{U}_S]_J \| [L_i S_i] J_i \rangle = \sqrt{(2J_f+1)(2J+1)(2J_i+1)}$$

$$\times \begin{Bmatrix} S_i & L_i & J_i \\ S & L & J \\ S_f & L_f & J_f \end{Bmatrix} \langle L_f \| \hat{T}_L \| L_i \rangle \langle S_f \| \hat{U}_S \| S_i \rangle, \quad (A.51)$$

and

$$\langle [L_f S_f] J_i \| [\hat{T}_L \otimes \hat{U}_L]_0 \| [L_i S_i] J_i \rangle$$

$$= \frac{(-1)^{L_i + S_f - J_i - L}}{2L+1} U(L_i S_i L_f S_f : J_i L) \langle L_f \| \hat{T}_L \| L_i \rangle \langle S_f \| \hat{U}_L \| S_i \rangle. \quad (A.52)$$

These equations are derived by using Equation (A.46) to express the right-hand side of Equation (A.38) in the form

$$\left[[\hat{T}_L \otimes |L_i\rangle]_{L_f} \otimes [\hat{U}_S \otimes |S_i\rangle]_{S_f}\right]_{J_f} = \sum_{\alpha\beta} [|\alpha L_f\rangle \otimes |\beta S_f\rangle]_{J_f} \frac{\langle\alpha L_f\|\hat{T}_L\|L_i\rangle\langle\beta S_f\|\hat{U}_S\|S_i\rangle}{\sqrt{(2L_f+1)(2S_f+1)}}$$

(A.53)

and thereby obtain the identity

$$\left[[\hat{T}_L \otimes \hat{U}_S]_J \otimes |[L_i \otimes S_i]_{J_i}\rangle\right]_{J_f M_f}$$

(A.54)

$$= \sum_{\alpha L_f \beta S_f} \begin{Bmatrix} S_i & L_i & J_i \\ S & L & J \\ S_f & L_f & J_f \end{Bmatrix} \frac{\langle\alpha L_f\|\hat{T}_L\|L_i\rangle\langle\beta S_f\|\hat{U}_S\|S_i\rangle}{\sqrt{(2L_f+1)(2S_f+1)}} [|\alpha L_f\rangle \otimes |\beta S_f\rangle]_{J_f}.$$

Then, using the expression of the Wigner-Eckart theorem, as given by Equation (A.27), one obtains the reduced matrix elements

$$\langle[L_f \otimes S_f]_{J_f}\|[\hat{T}_L \otimes \hat{U}_S]_J\|[L_i \otimes S_i]_{J_i}\rangle$$

(A.55)

$$= \begin{Bmatrix} S_i & L_i & J_i \\ S & L & J \\ S_f & L_f & J_f \end{Bmatrix} \sqrt{\frac{2J_f+1}{(2L_f+1)(2S_f+1)}} \langle L_f\|\hat{T}_L\|L_i\rangle\langle S_f\|\hat{U}_S\|S_i\rangle,$$

which becomes Equation (A.51), when expressed in terms of a standard (non-unitary) 9-j symbol. For the special, $J = 0$ case,

$$\langle[L_f \otimes S_f]_J\|[\hat{T}_L \otimes \hat{U}_L]_0\|[L_i \otimes S_i]_J\rangle$$

$$= (2J+1) \begin{Bmatrix} L_i & S_i & J \\ L_f & S_f & J \\ L & L & 0 \end{Bmatrix} \langle L_f\|\hat{T}_L\|L_i\rangle\langle S_f\|\hat{U}_S\|S_i\rangle$$

(A.56)

simplifies, by use of Equation (A.42), to Equation (A.52).

A.5.2 *Reduced matrix elements for left-to-right coupling*

In the left-to-right (round-bracket) coupling convention, coupled spherical tensors are defined and expressed graphically by

$$\left(\hat{W}_J \otimes |\beta J_i\rangle\right)_{J_f M_f} := \sum_{M_i M} \hat{W}_{JM} \otimes |\beta J_i M_i\rangle \, (JM\, J_i M_i | J_f M_f)$$

$$= \quad \overset{\hat{W}_J}{\underset{J_f}{\diagdown}} \quad |\beta J_i\rangle\,,$$

(A.57)

where the left-to-right arrow distinguishes this coupling from the right-to-left (square-bracket) coupling to which it is related by the simple identity (cf. Footnote 5 on Page 584)

$$\left(\hat{W}_J \otimes |\beta J_i\rangle\right)_{J_f M_f} = (-1)^{J_i + J - J_f} \left[\hat{W}_J \otimes |\beta J_i\rangle\right]_{J_f M_f}. \tag{A.58}$$

Thus, because of the phase factor, the left-to-right coupled products have expansion in terms of the above-defined reduced matrix elements given by

$$\left(\hat{W}_J \otimes |\beta J_i\rangle\right)_{J_f M_f} = (-1)^{J + J_i - J_f} \sum_\beta |\beta J_3 M_3\rangle \frac{\langle \beta J_3 \|\hat{W}_{J_2}\| \alpha J_1\rangle}{\sqrt{2J_3 + 1}}. \tag{A.59}$$

As we now show, simply adding these phase factors, directly reproduces the following expressions derived by French.[11]

Example (i):

Replacing the order of coupling,

$$|[LS]JM\rangle \to |(LS)JM\rangle = (-1)^{L+S-J}|[LS]JM\rangle, \tag{A.60}$$

for both the initial and final states of Equation (A.43) gives

$$\langle (L_f S)J_f \|\hat{T}_L\| (L_i S)J_i\rangle = (-1)^{L_f - J_f - L_i + J_i}$$

$$\times \sqrt{\frac{2J_f + 1}{2L_f + 1}} \, U(SL_i J_f L : J_i L_f)\langle \alpha L_f \|\hat{T}_L\| \beta L_i\rangle$$

$$= \sqrt{\frac{2J_i + 1}{2L_i + 1}} \, U(SL_f J_i L : J_f L_i)\langle \alpha L_f \|\hat{T}_L\| \beta L_i\rangle. \tag{A.61}$$

Example (ii):

With the substitution

$$\left[\hat{A}_{J_1} \otimes \hat{B}_{J_2}\right]_J \to \left(\hat{A}_{J_1} \otimes \hat{B}_{J_2}\right)_J = (-1)^{J_1 + J_2 - J}\left[\hat{A}_{J_1} \otimes \hat{B}_{J_2}\right]_J, \tag{A.62}$$

Equation (A.47) becomes

$$\langle \alpha J_f \|\left(\hat{A}_{J_1} \otimes \hat{B}_{J_2}\right)_J\| \beta J_i\rangle \tag{A.63}$$

$$= (-1)^{J_1 + J_2 - J} \sum_{\gamma J'} U(J_i J_2 J_f J_1 : J' J) \frac{\langle \alpha J_f \|\hat{A}_{J_1}\| \gamma J'\rangle \langle \gamma J' \|\hat{B}_{J_2}\| \beta J_i\rangle}{\sqrt{(2J' + 1)}};$$

and (A.48) is unchanged

$$\langle \alpha J \|\left(\hat{A}_{J_1} \otimes \hat{B}_{J_1}\right)_0\| \beta J\rangle = \sum_{\gamma J'} (-1)^{J - J_1 - J'} \frac{\langle \alpha J \|\hat{A}_{J_1}\| \gamma J'\rangle \langle \gamma J' \|\hat{B}_{J_1}\| \beta J\rangle}{\sqrt{(2J + 1)(2J_1 + 1)}}. \tag{A.64}$$

---

[11] French, *op. cit.* Footnote 10 on Page 589.

Example (iii):

Changing the order of the three coupled products in Equation (A.51) adds a phase factor $(-1)^\Sigma$, where Σ is the sum of the entries in the 9-j symbol, and is equivalent to interchanging two columns of the 9-j symbol. Thus, with left-to-right coupling, Equation (A.51) becomes

$$\langle (L_f S_f) J_f \| (\hat{T}_L \otimes \hat{U}_S)_J \| (L_i S_i) J_i \rangle = \sqrt{(2J_f + 1)(2J + 1)(2J_i + 1)}$$
$$\times \begin{Bmatrix} L_i & S_i & J_i \\ L & S & J \\ L_f & S_f & J_f \end{Bmatrix} \langle L_f \| \hat{T}_L \| L_i \rangle \langle S_f \| \hat{U}_S \| S_i \rangle, \qquad (A.65)$$

(see also Exercise A.7) and

$$\langle (L_f S_f) J_i \| [\hat{T}_L \otimes \hat{U}_L]_0 \| (L_i S_i) J_i \rangle$$
$$= \frac{(-1)^{S_i + L_f - J_i - L}}{2L + 1} U(S_i L_i S_f L_f : JL) \langle L_f \| \hat{T}_L \| L_i \rangle \langle S_f \| \hat{U}_L \| S_i \rangle. \quad (A.66)$$

A.6 Covariant and contravariant tensors

There are two kinds of tensor: covariant and contravariant. The distinction between the two is particularly relevant in the quantum theory of angular momentum. For example, if a nucleon creation operator, $a_{jm}^\dagger$, transforms under su(2) in the same way as a ket vector, $|jm\rangle$, of angular momentum j, we say that it is a component of a covariant tensor and label it with lower indices. Conversely, an operator a^{jm}, which annihilates a nucleon in an angular-momentum state $|jm\rangle$, transforms under su(2) in the same way as the bra vector $\langle jm|$. We then say that it is a component of a contravariant tensor and label it with upper indices. Equivalently, we might speak of such an annihilation operator as a spherical tensor of angular momentum j with contravariant components, $\{a^{jm}\}$, and covariant components, $\{a_{jm}\}$, as we now define and explain.

To determine the way in which a contravariant tensor transforms under su(2), the appropriate starting point is the observation that, by definition, an su(2) invariant combination of covariant and contravariant tensors of the same angular momentum is given by summing over repeated indices, thereby defining a *contraction*,

$$\hat{V}_J \cdot W^J := \sum_M \hat{V}_{JM} \hat{W}^{JM}. \qquad (A.67)$$

For this contraction to be an su(2) invariant scalar, we require that

$$\hat{V}_J \cdot \hat{W}^J = \hat{R}(\Omega) \hat{V}_J \cdot \hat{W}^J \hat{R}(\Omega^{-1}) = \sum_M \hat{R}(\Omega) \hat{V}_{JM} \hat{R}(\Omega^{-1}) \, \hat{R}(\Omega) \hat{W}^{JM} \hat{R}(\Omega^{-1}). \quad (A.68)$$

According to the definition, the components of a (covariant) spherical tensor transform according to the equation

$$\hat{R}(\Omega)\hat{V}_{JM}\hat{R}(\Omega^{-1}) = \sum_{K}\hat{V}_{JK}\mathscr{D}^{J}_{KM}(\Omega). \tag{A.69}$$

It follows that the components of a contravariant tensor must transform according to the equation

$$\hat{R}(\Omega)\hat{W}^{JM}\hat{R}(\Omega^{-1}) = \sum_{K}\mathscr{D}^{J}_{MK}(\Omega^{-1})\hat{W}^{JK} = \sum_{K}\hat{W}^{JK}\mathscr{D}^{J\,*}_{KM}(\Omega), \tag{A.70}$$

i.e., according to the complex conjugate of the representation J.

We also know that to form an su(2)-invariant combination of two covariant tensors, we have only to couple them to angular momentum zero, viz.

$$[\hat{V}_J \otimes \hat{W}_{J'}]_0 = \delta_{J,J'} \sum_{M=-J}^{+J} (J,-M\,JM|00)\hat{V}_{JM}\hat{W}_{J,-M}$$

$$= \delta_{J,J'} \sum_{M=-J}^{+J} \frac{(-1)^{J+M}}{\sqrt{2J+1}}\hat{V}_{JM}\hat{W}_{J,-M}. \tag{A.71}$$

This relationship indicates that the indices of a covariant tensor, $\hat{W}_J \equiv \{\hat{W}_{JM}\}$, can be raised to define a contravariant tensor with components

$$\hat{W}^{JM} := (-1)^{J+M}\hat{W}_{J,-M}, \tag{A.72}$$

where the phase $(-1)^J$ is not essential but is conveniently included when $2J$ is odd so that $(-1)^{J+M}$ is always real and equal to ± 1.

For example, the contraction of the covariant tensor $\{a^{\dagger}_{jm}\}$ of fermion creation operators with the contravariant tensor $\{a^{jm}\}$ of fermion annihilation operators is the number operator $\hat{n}_j = \sum_m a^{\dagger}_{jm}a^{jm}$, which has the useful expression

$$\hat{n}_j = \sum_{m}(-1)^{j+m}a^{\dagger}_{jm}a_{j,-m} = \sqrt{2j+1}\,[a^{\dagger}_j \otimes a_j]_0. \tag{A.73}$$

As an example of a situation in which the arbitrary factor $(-1)^J$ is naturally omitted, consider two vector operators, $\hat{\mathbf{V}}$ and $\hat{\mathbf{W}}$, with Cartesian components given, respectively, by $\{\hat{V}_x, \hat{V}_y, \hat{V}_z\}$ and $\{\hat{W}_x, \hat{W}_y, \hat{W}_z\}$. The scalar (dot) product of $\hat{\mathbf{V}}$ and $\hat{\mathbf{W}}$ is defined in the usual way by

$$\hat{\mathbf{V}} \cdot \hat{\mathbf{W}} = \hat{V}_x\hat{W}_x + \hat{V}_y\hat{W}_y + \hat{V}_z\hat{W}_z. \tag{A.74}$$

However, in a spherical SO(3) basis, labelled by $\hat{L}_0 \equiv \hat{J}_0$-angular-momentum eigenvalues, a vector $\hat{\mathbf{V}}$ has an integer angular momentum $L = 1$ and (covariant) components $\{\hat{V}_{1m}, m = 0, \pm 1\}$ with

$$\hat{V}_{10} = \hat{V}_z, \quad \hat{V}_{1,\pm1} = \mp\frac{1}{\sqrt{2}}\left(\hat{V}_x \pm \mathrm{i}\hat{V}_y\right). \tag{A.75}$$

To agree with the standard Cartesian dot product, the scalar product is then appropriately expressed in the form

$$\hat{\mathbf{V}} \cdot \hat{\mathbf{W}} = \sum_{m=-1}^{+1} (-1)^m \hat{V}_{1m} \hat{W}_{1,-m} = \sum_m \hat{V}_{1m} \hat{W}^{1m}, \tag{A.76}$$

with $\hat{W}^{1m} = (-1)^m \hat{W}_{1,-m}$.

It should be noted that, for SO(3) and SU(2), the complex conjugate of an irrep is not a new irrep; it is the same irrep but expressed in terms of a complex conjugated basis. Thus, every spherical tensor has covariant and contravariant components. It also follows that, because the Cartesian components of a spherical tensor, e.g., the quadrupole tensor $\{Q_{ij} = \sum_{n=1}^A x_{ni} x_{nj}; 1 \le i, j \le 3\}$, are real, the SO(3) transformation matrices are also real in a Cartesian basis. Thus, in a Cartesian basis, there is no need to distinguish between covariant and contravariant components of a spherical tensor; they are equal, e.g., $Q_{ij} = Q^{ij}$.

A.7 Hermitian adjoints, covariant adjoints, and their reduced matrix elements

The Hermitian adjoint, $\hat{X}^\dagger$, of an operator $\hat{X}$ on a Hilbert space $\mathbb{H}$ is defined by the equation

$$\langle \psi | \hat{X}^\dagger | \varphi \rangle := \langle \hat{X}\psi | \varphi \rangle, \tag{A.77}$$

for all $|\psi\rangle$ in the domain of $\hat{X}$. An operator $\hat{X}$ is said to be Hermitian if $\hat{X}^\dagger \equiv \hat{X}$.

A.7.1 *Covariant adjoints*

The definition (A.77) applies to the components of spherical tensor operators. However, as noted above for the fermion creation operators, the Hermitian adjoints of covariant spherical tensors are contravariant tensors. For, whereas the components $\{\hat{X}_{JM}\}$ of a spherical tensor $\hat{X}_J$ transform under an element $\Omega \in SU(2)$ according to the equation

$$\hat{X}_{JM} \to \sum_K \hat{X}_{JK} \mathscr{D}_{KM}^J(\Omega), \tag{A.78}$$

their Hermitian adjoints transform according to

$$(\hat{X}_{JM})^\dagger \to \left[\sum_K \hat{X}_{JK} \mathscr{D}_{KM}^J(\Omega) \right]^\dagger = \sum_K (\hat{X}_{JK})^\dagger \mathscr{D}_{KM}^{J*}(\Omega). \tag{A.79}$$

Thus, from Equation (A.70), the Hermitian adjoint of a covariant spherical tensor is a contravariant spherical tensor, $\hat{Z}^J := (\hat{X}_J)^\dagger$, with components

$$\hat{Z}^{JM} = (\hat{X}_{JM})^\dagger. \tag{A.80}$$

Because angular-momentum coupling theory is designed for use with covariant tensors, it is useful to lower the indices of the (contravariant) Hermitian adjoint of a spherical tensor, $\hat{X}_J$, to define a *covariant adjoint* tensor,

$$\hat{X}^{\ddagger}_{JM} := (-1)^{J-M}\hat{Z}^{J,-M} = (-1)^{J-M}\left(\hat{X}_{J,-M}\right)^{\dagger}. \qquad (A.81)$$

Such tensors have been used by, for example, French[12] and Edmonds.[13]

A.7.2 *Covariant and contravariant fermion operators*

In accordance with the above definition, (A.81), the covariant annihilation operators $\{a_{jm}\}$, defined by

$$a_{jm} := (-1)^{j-m}a^{j,-m} = (-1)^{j-m}\left(a^{\dagger}_{j,-m}\right)^{\dagger}, \qquad (A.82)$$

are the components of the covariant adjoint of the creation tensor operator $a^{\dagger}_j$.

For several reasons, notably because of the importance of time-reversal in pairing theory, it is often useful to express the components of tensors with respect to a contravariant basis (which with an appropriate phase choice is also a time-reversed basis) in covariant form. Thus, for example, we define the fermion operators

$$a^{\dagger}_{j\bar{m}} := (-1)^{j+m}a^{\dagger}_{j,-m}, \quad a_{j\bar{m}} := (-1)^{j+m}a_{j,-m} = a^{jm}. \qquad (A.83)$$

This notation lets us express the scalar (dot) product of pairs of fermion creation and annihilation operators in the simple form

$$a^{\dagger}_j \cdot a^{\dagger}_j = \sum_m a^{\dagger}_{jm}a^{\dagger}_{j\bar{m}}, \quad a_j \cdot a_j = \sum_m a_{j\bar{m}}a_{jm}. \qquad (A.84)$$

The fermion anticommutation relations (for which $2j$ is odd)

$$\{a^{jm}, a^{\dagger}_{jn}\} = \delta_{m,n}, \qquad (A.85)$$

have the expression

$$\{a_{j\bar{m}}, a^{\dagger}_{jn}\} = \delta_{m,n}, \quad \{a_{jm}, a^{\dagger}_{j\bar{n}}\} = -\delta_{m,n}. \qquad (A.86)$$

It is important to note, however, that

$$a^{\dagger}_{j\bar{\bar{m}}} = (-1)^{2j}a^{\dagger}_{jm} = -a^{\dagger}_{jm}, \quad a_{j\bar{\bar{m}}} = -a_{jm}. \qquad (A.87)$$

Thus, taking the time-reverse of a single-fermion state twice, reverses it sign.

---

[12]French, *op. cit.* Footnote 10 on Page 589.
[13]Edmonds, *op. cit.* Footnote 3 on Page 583.

A.7.3 Reduced matrix elements of covariant adjoint tensors

The above definition of a covariant adjoint tensor implies that

$$\langle \alpha J_3 M_3 | \hat{X}^{\dagger}_{J_2 M_2} | \beta J_1 M_1 \rangle = (-1)^{J_2 - M_2} \langle \alpha J_3 M_3 | (\hat{X}_{J_2, -M_2})^{\dagger} | \beta J_1 M_1 \rangle$$
$$= (-1)^{J_2 - M_2} \langle \beta J_1 M_1 | \hat{X}_{J_2, -M_2} | \alpha J_3 M_3 \rangle^*. \quad (A.88)$$

Thus, from the Wigner-Eckart theorem and the Clebsch-Gordan coefficient symmetry relation

$$(J_1 M_1 \, J_2 M_2 | J_3 M_3) = (-1)^{J_3 - J_1 - M_2} \sqrt{\frac{2J_3 + 1}{2J_1 + 1}} \, (J_3 M_3 \, J_2, -M_2 | J_1 M_1), \quad (A.89)$$

we obtain the identity

$$\langle \alpha J_3 \| \hat{X}^{\dagger}_{J_2} \| \beta J_1 \rangle = (-1)^{J_1 + J_2 - J_3} \langle \beta J_1 \| \hat{X}_{J_2} \| \alpha J_3 \rangle^*. \quad (A.90)$$

For example, in the important case of the single-fermion operators, the matrix elements are related by

$$\langle \alpha J_3 \| a_j \| \beta J_1 \rangle = (-1)^{J_3 - J_1 - j} \langle \beta J_1 \| a^{\dagger}_j \| \alpha J_3 \rangle^*. \quad (A.91)$$

A.8 Rotation matrices

Rotation matrices $\{ \mathscr{D}^J_{KM}(\Omega) \}$ are the unitary representation matrices, relative to the standard orthonormal basis, of an element $\Omega \in \mathrm{SU}(2)$ as they appear in the equation

$$\hat{R}(\Omega) | JM \rangle = \sum_K | JK \rangle \mathscr{D}^J_{KM}(\Omega), \quad \forall \, \Omega \in \mathrm{SU}(2). \quad (A.92)$$

Thus, $\mathscr{D}^J_{KM}(\Omega)$ is the matrix element

$$\mathscr{D}^J_{KM}(\Omega) = \langle JK | \hat{R}(\Omega) | JM \rangle, \quad (A.93)$$

of a rotation operator. In a parameterisation of a rotation by Euler angles, (α, β, γ), it is expressed in the form

$$\mathscr{D}^J_{KM}(\alpha, \beta, \gamma) = \langle JK | e^{-\mathrm{i}\alpha \hat{J}_z} e^{-\mathrm{i}\beta \hat{J}_y} e^{-\mathrm{i}\gamma \hat{J}_z} | JM \rangle = e^{-\mathrm{i}\alpha K} d^J_{KM}(\beta) e^{-\mathrm{i}\gamma M}, \quad (A.94)$$

where the angular-momentum operators act in accordance with an su(2) irrep, defined by Equation (A.1), and

$$d^J_{KM}(\beta) = \langle JK | e^{-\mathrm{i}\beta \hat{J}_y} | JM \rangle, \quad (A.95)$$

is a so-called *reduced rotation matrix*. The latter matrix has explicit expression

$$d^J_{KM}(\beta) = \sum_N (-1)^N \frac{[(J+K)!(J-K)!(J+M)!(J-M)!]^{1/2}}{(J+K-N)!N!(J-M-N)!(M-K+N)!}$$
$$\times (\cos \beta/2)^{2J+K-M-2N}(\sin \beta/2)^{M-K+2N}, \qquad (A.96)$$

in which the summation is over all integer values of N for which the arguments of the factorials are non-negative.

A.8.1   *Some properties of rotation matrices*

(i) Because they are the matrices of a unitary representation of SU(2), rotation matrices satisfy the identities

$$\mathscr{D}^J_{KM}(-\gamma, -\beta, -\alpha) = \mathscr{D}^{J*}_{MK}(\alpha, \beta, \gamma),$$
$$\sum_K \mathscr{D}^{J*}_{KM}(\alpha, \beta, \gamma)\mathscr{D}^J_{KM'}(\alpha, \beta, \gamma) = \delta_{MM'}, \qquad (A.97)$$
$$\sum_M \mathscr{D}^{J*}_{KM}(\alpha, \beta, \gamma)\mathscr{D}^J_{K'M}(\alpha, \beta, \gamma) = \delta_{KK'}.$$

(ii) The reduced rotation matrices are real and therefore, using (i),

$$d^J_{KM}(-\beta) = d^J_{MK}(\beta). \qquad (A.98)$$

(iii) From the expression (A.96), it follows that

$$d^J_{-M-K}(\beta) = d^J_{KM}(\beta) = (-1)^{K-M}d^J_{MK}(\beta) \qquad (A.99)$$

and, hence, that

$$\mathscr{D}^J_{KM}(\alpha, \beta, \gamma) = (-1)^{K-M}\mathscr{D}^{J*}_{-K-M}(\alpha, \beta, \gamma). \qquad (A.100)$$

(iv) Some useful values of the matrices are given by

$$d^J_{KM}(\pi) = \delta_{M,-K}(-1)^{J+K},$$
$$d^J_{KM}(-\pi) = \delta_{M,-K}(-1)^{J-K}. \qquad (A.101)$$

(v) Rotation matrices can be combined according to the equations

$$\mathscr{D}^{J_1}_{K_1M_1}(\Omega)\mathscr{D}^{J_2}_{K_2M_2}(\Omega)$$
$$= \sum_{JKM}(J_2K_2\,J_1K_1|JK)(J_2M_2\,J_1M_1|JM)\mathscr{D}^J_{KM}(\Omega)$$
$$= \sum_{JKM}(J_1K_1\,J_2K_2|JK)(J_1M_1\,J_2M_2|JM)\mathscr{D}^J_{KM}(\Omega). \qquad (A.102)$$

(vi) The rotation matrices, with $J = L$ integer, are related to the spherical harmonics by

$$Y_{LM}(\beta, \alpha) = \sqrt{\frac{2L+1}{4\pi}}\,\mathscr{D}_{M0}^{L\,*}(\alpha, \beta, \gamma) = \sqrt{\frac{2L+1}{4\pi}}\,\mathscr{D}_{0M}^{L}(-\gamma, -\beta, -\alpha).$$

(A.103)

A.8.2   The Wigner $\mathscr{D}$ functions

The Wigner $\mathscr{D}$ functions are invaluable in nuclear physics because they provide basis wave functions for the rotor model and other models, such as the SU(3) model. They have their origins in the Peter-Weyl theorem, a theorem which provides the foundations of character theory, Fourier analysis, and harmonic analysis in general.

Let $\{\hat{T}^{(\lambda)}\}$ denote the irreps of a group G, and let $\{T_{\mu\nu}^{(\lambda)}(g)\}$, for $g \in G$, denote the matrix elements of these irreps relative to an orthonormal basis for the irrep. Then, each irrep defines a set of so-called *matrix coefficient functions*, $\{T_{\mu\nu}^{(\lambda)}\}$, on G whose values are the matrix elements for the irrep $\hat{T}^{(\lambda)}$.

Theorem A.1 (Peter-Weyl) *If $\{\hat{T}^{(\lambda)}\}$ is a complete set of inequivalent unitary irreps of a finite or compact Lie group G and $\{T_{\mu\nu}^{(\lambda)}\}$ is a corresponding set of matrix coefficient functions on G, then the re-normalised functions $\{\sqrt{d_\lambda}T_{\mu\nu}^{(\lambda)}\}$, where d_λ is the dimension of the irrep $\hat{T}^{(\lambda)}$, is an orthonormal basis for the Hilbert space, $\mathcal{L}^2(G)$, of square-integrable functions on G with respect to the G-invariant volume element.*[14,15]

A direct application of this theorem gives an interpretation of the matrix co-efficient functions for SU(2) as wave functions, known as Wigner $\mathscr{D}$ functions.[16] According to the theorem, these functions have orthonormality relations given by

$$\frac{2J+1}{8\pi^2}\int_{SU(2)} \mathscr{D}_{KM}^{J\,*}(\Omega)\mathscr{D}_{K'M'}^{J'}(\Omega)\,d\Omega_{SU(2)} = \delta_{JJ'}\delta_{KK'}\delta_{MM'},$$

(A.104)

---

[14]The Peter-Weyl theorem is proved in many texts on group theory, e.g., Barut A.O. and Raczka R. (1988), *Theory of Group Representations and Applications* (World Scientific, Singapore).

[15]A volume element on a group, G, is said to be right-invariant if, for any function f defined on the group for which the integral $\int_G f(g)\,dv(g)$ converges,

$$\int_G f(g)\,dv(g\alpha) = \int_G f(g)\,dv(g), \quad \forall\,\alpha \in G.$$

A volume element on a group is similarly said to be left-invariant if, for such a function, f,

$$\int_G f(g)\,dv(\alpha g) = \int_G f(g)\,dv(g), \quad \forall\,\alpha \in G.$$

A volume element that is both left- and right-invariant is simply described as an invariant volume element or an *invariant measure* for the group. It is also called a *Haar measure*.

[16]Wigner E.P. (1959), *Group Theory and Its Applications to the Quantum Mechanics of Atomic Spectra* (Academic Press, New York).

where $d\Omega_{\mathrm{SU}(2)}$ is the SU(2)-invariant volume element. (Note that the factor $8\pi^2$, which is conventionally included, could be omitted by absorbing it into the definition of the volume element.) Moreover, the theorem also shows that any square-integrable function, f, of SU(2) rotations has an expansion, in parallel with a Fourier plane-wave expansion, given by

$$f(\Omega) = \sum_{KJM} f_{KM}^{J} \mathscr{D}_{KM}^{J}(\Omega), \qquad (\mathrm{A.105})$$

where

$$f_{KM}^{J} := \frac{2J+1}{8\pi^2} \int_{SU(2)} \mathscr{D}_{KM}^{J*}(\Omega) f(\Omega) \, d\Omega_{\mathrm{SU}(2)}. \qquad (\mathrm{A.106})$$

The above properties also apply to functions on the group SO(3) with the understanding that the values of J are then restricted to integer values. When applied to SO(3), the volume element is the familiar SO(3)-invariant volume element, expressed in terms of Euler angles by

$$d\Omega_{\mathrm{SO}(3)} = \sin\beta \, d\alpha \, d\beta \, d\gamma, \qquad (\mathrm{A.107})$$

and an integral of a function f over SO(3) is defined by

$$\int_{\mathrm{SO}(3)} f(\Omega) \, d\Omega_{\mathrm{SO}(3)} := \int_0^{2\pi} d\alpha \int_0^{2\pi} d\gamma \int_0^{\pi} \sin\beta \, d\beta \, f(\alpha,\beta,\gamma), \qquad (\mathrm{A.108})$$

so that

$$\int_{\mathrm{SO}(3)} d\Omega_{\mathrm{SO}(3)} = 8\pi^2. \qquad (\mathrm{A.109})$$

However, for SU(2), the range of each integral is doubled to maintain the orthogonality properties for half-odd integer values of J.[17] To retain the normalisation given by Equation (A.104), an integral of a function f over SU(2) is then

$$\int_{\mathrm{SU}(2)} f(\Omega) \, d\Omega_{\mathrm{SU}(2)} := \tfrac{1}{8} \int_0^{4\pi} d\alpha \int_0^{4\pi} d\gamma \int_0^{2\pi} \sin\beta \, d\beta \, f(\alpha,\beta,\gamma) \qquad (\mathrm{A.110})$$

and the corresponding SU(2) volume element is adjusted to

$$d\Omega_{\mathrm{SU}(2)} = \tfrac{1}{8} \sin\beta \, d\alpha \, d\beta \, d\gamma. \qquad (\mathrm{A.111})$$

---

[17]Because SU(2) is the double cover of SO(3), it is strictly only necessary to double the integral of the β range. However, the integral is simplified by doubling the α and γ ranges also so that, for example, $\int_0^{4\pi} e^{\mathrm{i}(K-K')\alpha} \, d\alpha = 4\pi\delta_{K,K'}$.

Exercises

A.1 Show that, if an operator $\hat{X}$ maps another operator $\hat{W}$ to the new operator $[\hat{X}, \hat{W}]$, the map defined by $e^{\hat{X}}$ leads to the Baker-Campbell-Hausdorff identity,

$$e^{\hat{X}} \hat{W} e^{-\hat{X}} = \hat{W} + [\hat{X}, \hat{W}] + \frac{1}{2}[\hat{X}, [\hat{X}, \hat{W}]] + \frac{1}{3!}[\hat{X}, [\hat{X}, [\hat{X}, \hat{W}]]] + \dots. \quad (A.112)$$

A.2 Use Equation (A.107) to show that

$$\langle \alpha JM | \beta JM \rangle = \frac{1}{8\pi^2} \int_{SO(3)} \langle \alpha JM | \hat{R}(\Omega^{-1}) \hat{R}(\Omega) | \beta JM \rangle \, d\Omega. \quad (A.113)$$

Then use Equations (A.92) and (A.104) to show that

$$\langle \alpha JM | \beta JM \rangle = \frac{1}{2J+1} \sum_K \langle \alpha JK | \beta JK \rangle \quad (A.114)$$

and, hence, that $\langle \alpha JM | \beta JM \rangle$ has an M-independent value.

A.3 By considering $\hat{J}$ to be an SU(2) angular momentum tensor with components $\{\hat{J}_M, M = 0, \pm 1\}$, use the identity $\hat{J}_0 | \alpha jm \rangle = m | \alpha jm \rangle$ and the known Clebsch-Gordan coefficient $(jm \, 10 | jm) = m/\sqrt{j(j+1)}$ to derive the reduced matrix element

$$\langle \beta j' \| \hat{J} \| \alpha j \rangle = \delta_{\alpha\beta} \delta_{j',j} \sqrt{j(j+1)(2j+1)}. \quad (A.115)$$

A.4 With coupled commutation relations defined for spherical tensors by

$$[\hat{A}_{J_1}, \hat{B}_{J_2}]_{JM} = \sum_{M_1 M_2} (J_2 M_2 \, J_1 M_1 | JM) [\hat{A}_{J_1 M_1}, \hat{B}_{J_2 M_2}], \quad (A.116)$$

show that

$$[\hat{J}, \hat{J}]_L = \delta_{L,1} \sqrt{2} \, \hat{J}. \quad (A.117)$$

A.5 Show that the coupled anticommutator of fermion operators is given by

$$\{a_j, a_j^\dagger\}_{JM} = -\sqrt{2j+1} \, \delta_{J,0} \delta_{M,0}. \quad (A.118)$$

A.6 Two spherical harmonics with a common argument satisfy the relationship

$$Y_{l_2 m_2}(\theta, \varphi) Y_{l_1 m_1}(\theta, \varphi) = \sum_{l_3 m_3} \left[\frac{(2l_1 + 1)(2l_2 + 1)}{4\pi(2l_3 + 1)} \right]^{\frac{1}{2}} (l_1 0 \, l_2 0 | l_3 0)$$
$$\times (l_1 m_1 \, l_2 m_2 | l_3 m_3) Y_{l_3 m_3}(\theta, \varphi). \quad (A.119)$$

Show that this expression is summarised by the reduced matrix element

$$\langle l_3 \| \hat{Y}_{l_2} \| l_1 \rangle = \left[\frac{(2l_1 + 1)(2l_2 + 1)}{4\pi} \right]^{\frac{1}{2}} (l_1 0 \, l_2 0 | l_3 0), \quad (A.120)$$

where $\hat{Y}_l$ is a spherical tensor with components defined by the equation

$$\hat{Y}_{lm} Y_{l_1 m_1}(\theta, \varphi) := Y_{lm}(\theta, \varphi) Y_{l_1 m_1}(\theta, \varphi). \quad (A.121)$$

A.7 Use the symmetry relation for $9j$ symbols given in Section A.4.2 to show that

$$\begin{Bmatrix} S_i & L_i & J_i \\ S & L & J \\ S_f & L_f & J_f \end{Bmatrix} = \begin{Bmatrix} L_i & S_i & J_i \\ L_f & S_f & J_f \\ L & S & J \end{Bmatrix}. \qquad (A.122)$$

A.8 Show that the Dirac delta function in SO(3) orientation angles has an expansion

$$\delta(\Omega - \Omega') = \sum_{L \geq 0} \frac{2L+1}{8\pi^2} \mathscr{D}^{L*}_{KM}(\Omega) \mathscr{D}^{L}_{KM}(\Omega'), \qquad (A.123)$$

where the sum is over non-negative integers.

Appendix B

Useful relationships for electromagnetic properties of nuclei

B.1 General expressions

Electric and magnetic moment operators are defined in the long wavelength limit[1] by

$$\mathfrak{M}(E\lambda; \mu) := \sum_i g_i^{(l)} r_i^\lambda Y_{\lambda\mu}(\theta_i, \varphi_i), \tag{B.1}$$

$$\mathfrak{M}(M\lambda; \mu) := \frac{e\hbar}{2Mc} \sum_i \left[\nabla_i r_i^\lambda Y_{\lambda\mu}(\theta_i, \varphi_i)\right] \cdot \left[g_i^{(l)} \frac{2}{\lambda+1} \mathbf{L}_i + g_i^{(s)} \mathbf{s}_i\right], \tag{B.2}$$

where $g_i^{(l)} = 1$ and $g^{(s)} = 5.586$ for a proton and $g_i^{(l)} = 0$ and $g^{(s)} = -3.862$ for a neutron, λ is the multipolarity, and M is the nucleon mass.

Reduced Eλ or Mλ transition rates are evaluated from the equation

$$B(\lambda; i \to f) := \frac{1}{2J_i + 1} |\langle J_f || \mathfrak{M}(\lambda) || J_i \rangle|^2. \tag{B.3}$$

Transition rates are then expressed in terms of these reduced rates by

$$T(\lambda; i \to f) = \frac{8\pi}{\hbar} \frac{\lambda+1}{\lambda[(2\lambda+1)!!]^2} k^{2\lambda+1} B(\lambda; i \to f), \tag{B.4}$$

where $k = E_\gamma/\hbar c$ and E_γ is the energy of the emitted gamma ray.

Reduced $B(E\lambda)$ and $B(M\lambda)$ transition rates are related to experimentally measurable quantities by

$$B(E\lambda; i \to f) = \frac{\mathcal{F}_\lambda^{(E)}}{(E_\gamma)^{2\lambda+1} T_{1/2}^\gamma(E\lambda) \ln 2} e^2 \cdot b^\lambda \tag{B.5}$$

---

[1]The long wavelength limit makes the approximation $j_\lambda(kr) \approx (kr)^\lambda/(2\lambda+1)!!$ for $kr \ll 1$ in the expansion of a plane wave function, e^{ikr}. This approximation, known as Siegert's theorem, makes it possible to simplify the expressions for photon transition matrix elements by use of the charge continuity equation and thereby avoid involving less reliable matrix elements for the nuclear current operators.

$$B(M\lambda; i \to f) = \frac{\mathcal{F}_\lambda^{(M)}}{(E_\gamma)^{2\lambda+1} T_{1/2}^\gamma(M\lambda) \ln 2} e^2 . \mu_N^{\lambda-1}, \qquad (B.6)$$

where the gamma-ray energy, E_γ, is expressed in MeV, the partial half-life with respect to electric or magnetic λ-multipole γ-ray emission, $T_{1/2}^\gamma(E\lambda) \ln 2$ or $T_{1/2}^\gamma(M\lambda)$, in ps (pico-seconds), for the $i \to f$ gamma decay, and $\mathcal{F}_\lambda$ is given in Table B.1.

Table B.1: Values of the parameters $\mathcal{F}_\lambda^{(E)}$ and $\mathcal{F}_\lambda^{(E)}$ appearing in Equations (B.5) and (B.6).

λ	1	2	3	4	5
$\mathcal{F}_\lambda^{(E)}$	$6.290\,E{-}6$	$8.162\,E{-}2$	$1.753\,E{+}3$	$5.896\,E{+}7$	$2.894\,E{+}12$
$\mathcal{F}_\lambda^{(M)}$	$5.689\,E{-}2$	$7.382\,E{+}2$	$1.586\,E{+}7$	$5.332\,E{+}11$	$2.617\,E{+}16$

Note that $T_{1/2}^\gamma(E\lambda) \ln 2$ and $T_{1/2}^\gamma(M\lambda)$ are obtained by dividing the corresponding observed (total) half lives, $T_{1/2}^{\text{obs}}$, by the gamma-decay branch fraction, e.g., when there is a mixed E2/M1 multipolarity, the ratio

$$\delta := \frac{\langle J_f || \mathfrak{M}(E2) || J_i \rangle}{\langle J_f || \mathfrak{M}(M1) || J_i \rangle} \qquad (B.7)$$

gives $T_{1/2}^\gamma(E2) = (1 + 1/\delta^2) T_{1/2}^\gamma$. For $2_1^+ \to 0_1^+$, E2 transitions, $T_{1/2}^\gamma(E2) = (1+\alpha) T_{1/2}^{\text{obs}}$, where α is the total internal conversion coefficient.

The electric quadrupole moment, $Q^{(e)}(\alpha I)$, of a state, $|\alpha I M\rangle$, of angular momentum I is defined by

$$Q^{(e)}(\alpha I) := \sqrt{\frac{16\pi}{5}} \langle \alpha I, M = I | \mathfrak{M}(E2; 0) | I, M = I \rangle. \qquad (B.8)$$

The magnetic dipole moment, $\mu(\alpha I)$, of a state, $|\alpha I M\rangle$, of angular momentum I is defined by

$$\mu(\alpha I) := \sqrt{\frac{4\pi}{3}} \langle \alpha I, M = I | \mathfrak{M}(M1; 0) | I, M = I \rangle. \qquad (B.9)$$

B.2   Weisskopf units

Transition rates are frequently expressed in Weisskopf (single-particle) units,[2,3] W.u. The Weisskopf units for electric and magnetic transition rates are given, respectively, by

$$B_{\text{W.u.}}(E\lambda) := \frac{e^2}{4\pi} \left(\frac{3}{\lambda+3} \right)^2 R^{2\lambda}, \qquad (B.10)$$

[2]Weisskopf V.F. (1951), *Phys. Rev.* **83**, 1073.

[3]Wilkinson D.H. (1960), in *Nuclear Spectroscopy*, edited by F. Ajzenberg-Selove (Academic Press, New York), Part B.

and

$$B_{\mathrm{W.u.}}(\mathrm{M}\lambda) := \frac{10}{\pi}\left(\frac{e\hbar}{2mc}\right)^2\left(\frac{3}{\lambda+3}\right)^2 R^{2\lambda-2}, \tag{B.11}$$

where R, the nuclear radius, is conventionally assigned the value $R = 1.2A^{\frac{1}{3}}$ fm. For example, for an E2 transition the Weisskopf unit is

$$B_{\mathrm{W.u.}}(\mathrm{E2}) = 5.940 \times 10^{-6} A^{4/3}\, e^2 b^2. \tag{B.12}$$

It should be clearly understood that, while the model used by Weisskopf in esti-mating single-particle transition rates gives reduced transition rates that depend on the initial and final angular momenta of the states involved, the so-called Weisskopf unit, in which reduced transition rates is expressed, is independent of these angular momenta.

B.3 Electric quadrupole moments and transitions in the axially-symmetric rotor model

The following relationships are included for convenience because they are in common use. Some of them are based on unified models that will be discussed in Volume 2. They are important because they are the equations by which data are reduced for comparison with theory.

The basic rotor model makes the underlying assumption that, while the states of a rotational band have different rotational wave functions, they all share a common intrinsic wave function, φ. Thus, it is useful to characterise the intrinsic shape of a rotor by its intrinsic quadrupole moments. For an axially-symmetric rotor, an intrinsic electric quadrupole moment is defined by

$$\overline{Q}_0^{(e)} := \sqrt{\frac{16\pi}{5}}\,\langle\varphi|\mathfrak{M}(\mathrm{E2},0)|\varphi\rangle. \tag{B.13}$$

If a given rotational band has a lowest angular-momentum state with component $K \neq 1$, then the states of angular momentum I of this band have electric quadrupole moments given by

$$Q_0^{(e)}(KI) = \frac{3K^2 - I(I+1)}{(I+1)(2I+3)}\,\overline{Q}_0^{(e)}. \tag{B.14}$$

In particular, for a $K = 0$ band,

$$Q_0^{(e)}(0I) = -\frac{I}{2I+3}\,\overline{Q}_0^{(e)}. \tag{B.15}$$

(When $K = 1$ there is, generally, an additional small term.)

Electric quadrupole transition rates between the states of an axially-symmetric rotor model can likewise be expressed in terms of the intrinsic quadrupole moments:

for $K \neq 1$,

$$\langle KI_f \| \mathfrak{M}(\text{E2}) \| KI_i \rangle = \sqrt{\frac{5}{16\pi}} \, (2I_i + 1)^{\frac{1}{2}} \, (I_i K \, 20 | I_f K) \, \overline{Q}_0^{(e)}. \tag{B.16}$$

(Again, when $K = 1$ there is, generally, an additional small term.) Thus, for a $K = 0$ band

$$B(\text{E2}; I \to I - 2) = \frac{15I(I-1)}{32\pi(2I-1)(2I+1)} \, \left(\overline{Q}_0^{(e)}\right)^2. \tag{B.17}$$

We also obtain the useful relationships

$$B(\text{E2}; 2 \to 0) = \frac{1}{16\pi} \, \left(\overline{Q}_0^{(e)}\right)^2, \tag{B.18}$$

$$B(\text{E2}; I \to I - 2) = \frac{15I(I-1)}{2(2I-1)(2I+1)} \, B(\text{E2}; 2 \to 0). \tag{B.19}$$

For historic reasons, intrinsic quadrupole moments are often expressed in terms of the surface shape parameters of the liquid-drop model. These parameters are defined in an expansion of the nuclear surface radius (presumed to be well defined in the liquid-drop model)

$$R(\theta, \varphi) := R_0 \Big[1 + \sum_{\lambda\mu} \alpha_{\lambda\mu}^* Y_{\lambda\mu}(\theta, \varphi) + \dots \Big]. \tag{B.20}$$

In this model, the electric quadrupole moment of the nucleus is given to first order by

$$\overline{Q}_0^{(e)} = \frac{3}{\sqrt{5\pi}} Z e R_0^2 \, \alpha_{20} + 0(\alpha^2) \tag{B.21}$$

and, from Equation (B.18),

$$\alpha_{20} \approx \frac{4\pi}{3 Z R_0^2 e} \sqrt{5 B(\text{E2}; 2 \to 0)}, \tag{B.22}$$

where R_0 is conventionally given the value $1.2 A^{\frac{1}{3}}$ fm.

Fundamental physical constants and conversion factors

Speed of light in vacuum	c	$= 299\,792\,458 \times 10^8$ m s^{-1}
Elementary charge	e	$= 1.602\,176\,49 \times 10^{-19}$ C
	$e^2/4\pi\epsilon_0$	$= 1.439\,964\,45 \times 10^{-9}$ eV m
Planck constant/2π	$\hbar$	$= 1.054\,571\,63 \times 10^{-34}$ J s
		$= 6.582\,1190 \times 10^{-22}$ MeV s
Fine-structure constant	$e^2/4\pi\epsilon_0\hbar c$	$= 1/137.035\,999\,68$
Nuclear magneton	$\mu_N = e\hbar/2m_p c$	$= 5.050\,783\,2 \times 10^{-27}$ J T^{-1}
(Unified) atomic mass unit	u	$= 931.494\,03$ MeV
Electric constant	ϵ_0	$= 8.854\,187\,817\,10^{-12}$ F m^{-1}
Electron mass	$m_e c^2$	$= 0.510\,998\,91$ MeV
Neutron mass	$m_n c^2$	$= 939.565\,35$ MeV
Proton mass	$m_p c^2$	$= 938.272\,01$ MeV

$$\hbar c \quad = 1.973\,269\,63 \times 10^{-7} \text{ eV m} = 1.973\,269\,63 \text{ MeV fm}$$

$$1 \text{ MeV} = 1.602\,176\,49 \times 10^{-13} \text{ J} = 1.602\,176\,49 \times 10^{-6} \text{ erg}$$

$$1 \text{ fm} \quad = 10^{-15} \text{ m} = 10^{-13} \text{ cm}$$

$$1 \text{ b} \quad = 10^{-28} \text{ m}^2 = 10^{-24} \text{ cm}^2$$

Bibliography

Abramowitz M. and Stegun I.A. (1968), *Handbook of Mathematical Functions* (Dover Publications, New York), fifth printing.

Ajzenberg-Selove F. (1988), *Nucl. Phys.* **A490**, 1.

Ajzenberg-Selove F. (1991), *Nucl. Phys.* **A523**, 1.

Alhassid Y. and Leviatan A. (1992), *J. Phys. A: Math. Gen.* **25**, L1265.

Alkhazov G.D. *et al.* (1988), *Nucl. Phys.* **A477**, 37.

Allaart K. *et al.* (1988), *Phys. Repts.* **169**, 209.

Anderson E. *et al.* (2008), *Phys. Rev.* **C77**, 037001.

Anderson P.W. (1958), *Phys. Rev.* **112**, 1900.

Anderson P.W. (1972), *Science* **177**, 393.

Anderson R.E. (1979), *Phys. Rev.* **C19**, 2138.

Andreozzi F. (1996), *Phys. Rev.* **C54**, 684.

Angeli I. (2004), *At. Data Nucl. Data Tables* **87**, 185.

Anselment M. *et al.* (1986), *Phys. Rev.* **C34**, 1052.

Anthony M.S. and Pape A. (1984), *Phys. Rev.* **C30**, 1286.

Appel H. (1968), *Numerical Tables for Angular Correlation Computations in α-, β- and γ-spectroscopy: 3j-, 6j-, 9j-Symbols, F- and Γ-Coefficients, Landolt-Börnstein New Series*, Vol. I/3 (Springer-Verlag, Berlin/Heidelberg).

Arfken G. (1985), *Mathematical Methods for Physicists* (Academic Press, San Diego).

Arima A. and Kawarada H. (1964), *J. Phys. Soc. Japan* **19**, 1768.

Asorey M., Falceto F. and Sierra G. (2002), *Nucl. Phys.* **B622**, 593.

Audi G., Wapstra A.H. and Thibault C. (2003), *Nucl. Phys.* **A729**, 337.

Auerbach N. (1983), *Phys. Repts.* **98**, 273.

Babu S. and Brown G.E. (1973), *Ann. Phys. (NY)* **78**, 1.

Bacher R.F. and Goudsmit S. (1934), *Phys. Rev.* **46**, 948.

Baer H.W. *et al.* (1991), *Phys. Rev.* **C43**, 1458.

Bahri C. and Draayer J.P. (1994), *Comp. Phys. Comm.* **83**, 59.

Bahri C. and Rowe D.J. (2000), *Nucl. Phys.* **A662**, 125.

Bakri M.M. (1967), *Nucl. Phys.* **A96**, 115.

Bang J. and Krumlinde J. (1970), *Nucl. Phys.* **A141**, 18.

Bar-Touv J. and Kelson I. (1965), *Phys. Rev.* **138**, B1035.

Baranger M. (1961), *Phys. Rev.* **122**, 992.

Bardeen J., Cooper L.N. and Schrieffer J.R. (1957), *Phys. Rev.* **108**, 1175.

Barnes P.D. *et al.* (1972), *Nucl. Phys.* **A195**, 146.

Barrett B.R., ed. (1975), *Effective Interactions and Operators in Nuclei, Lecture Notes in Physics*, Vol. 40 (Springer-Verlag, Berlin).

Barrett B.R., Hewitt R.G.L. and McCarthy R.J. (1971), *Phys. Rev.* **C3**, 1137.

Barrett B.R. and Kirson M.W. (1973), *Adv. Nucl. Phys.* **6**, 219.

Bartlett S.D. and Rowe D.J. (2003), *J. Phys. A: Math. Gen.* **36**, 1683.

Bartlett S.D., Rowe D.J. and Repka J. (2002), *J. Phys. A: Math. Gen.* **35**, 5599, 5625.

Barut A.O. and Raczka R. (1988), *Theory of Group Representations and Applications* (World Scientific, Singapore).

Baxter R.J. (1982), *Exactly solved models in statistical mechanics* (Academic Press, London).

Bayman B.F. (1960), *Nucl. Phys.* **15**, 33.

Bayman B.F. (1960), *Some Lectures on Groups and their Applications to Spectroscopy* (Nordita, Copenhagen).

Bayman B.F. and Lande A. (1966), *Nucl. Phys.* **77**, 1.

Belyaev S.T. (1959), *Mat. Fys. Medd. Dan. Vid. Selsk.* **31** (11).

Bender M., Heenen P.-H. and Reinhard P.-G. (2003), *Rev. Mod. Phys.* **75**, 121.

Bender M. *et al.* (1999), *Phys. Rev.* **C60**, 034304.

Bentley M.A. *et al.* (1991), *J. Phys. G: Nucl. Part. Phys.* **17**, 481.

Bentley M.A. *et al.* (2006), *Phys. Rev.* **C73**, 024304.

Berger J.-F., Girod M. and Gogny D. (1984), *Nucl. Phys.* **A428**, 23c.

Bertsch G.F. (1965), *Nucl. Phys.* **74**, 234.

Bertsch G.F. (1972), *The Practitioner's Shell Model* (North-Holland, Amsterdam).

Bertsch G.F. (2007), *J. of Phys: Conf. Series* **78**, 012005.

Bertsch G.F. and Mekjian A. (1972), *Ann. Rev. Nucl. Sci* **22**, 25.

Bès D.R. (1959), *Nucl. Phys.* **10**, 373.

Bès D.R. and Broglia R.A. (1966), *Nucl. Phys.* **80**, 289.

Bès D.R. and Sorensen R.A. (1969), *Adv. Nucl. Phys.* **2**, 129.

Bethe H.A. (1931), *Z. Phys.* **71**, 205.

Bethe H.A. and Goldstone J. (1957), *Proc. Roy. Soc. London* **A238**, 551.

Beyer R.T., ed. (1949), *Foundations of Physics* (Dover Publications, New York).

Bhaduri R.K. (1988), *Models of the Nucleon from Quarks to Soliton* (Addison-Wesley, Redwood City, California).

Bishop R.F. *et al.* (1993), *J. Phys. G: Nucl. Part. Phys.* **19**, 1163.

Bjerregaard J.H. *et al.* (1968), *Nucl. Phys.* **A110**, 1.

Bjerregaard J.H. *et al.* (1969), *Nucl. Phys.* **A131**, 481.

Blankert P.J. (1979), Ph.D. thesis, Free University, Amersterdam.

Bloch C. (1958), *Nucl. Phys.* **6**, 329.

Bloch C. and Horowitz J. (1958), *Nucl. Phys.* **8**, 91.

Blomqvist J. (1984), in *International Review of Nuclear Physics*, Vol. 2 (World Scientific, Singapore), pp. 1–32.

Bogner S.K., Furnstahl R.J. and Perry R.J. (2007), *Phys. Rev.* **C75**, 061001(R).

Bogner S.K., Kuo T.T.S. and Coraggio L. (2001), *Nucl. Phys.* **A684**, 432c.

Bogner S.K., Kuo T.T.S. and Schwenk A. (2003), *Phys. Repts.* **386**, 1.

Bogner S.K. *et al.* (2002), *Phys. Rev.* **C65**, 051301(R).

Bogner S.K. *et al.* (2008), *Nucl. Phys.* **A801**, 21.

Bogolyubov N.N. (1958), *Nuovo Cimento* **7**, 794.

Bogolyubov N.N. (1959), *Sov. Phys.–Uspekhi* **2**, 236.

Boguta J. and Bodmer A.R. (1977), *Nucl. Phys.* **A292**, 413.

Bohm A., Ne'eman Y. and Barut A.O., eds. (1988), *Dynamical Groups and Spectrum Generating Algebras*, Vol. 1 & 2 (World Scientific, Singapore).

Bohr A. (1952), *Mat. Fys. Medd. Dan. Vid. Selsk.* **26** (14).

Bohr A. (1976), *Rev. Mod. Phys.* **48**, 365.

Bohr A. and Mottelson B.R. (1953), *Mat. Fys. Medd. Dan. Vid. Selsk.* **27** (16).

Bohr A. and Mottelson B.R. (1969), *Nuclear Structure*, Vol. 1 (Benjamin, New York), (republished by World Scientific, Singapore).

Bohr A. and Mottelson B.R. (1975), *Nuclear Structure*, Vol. 2 (Benjamin, Reading, Mass.), (republished by World Scientific, Singapore).

Bohr A., Mottelson B.R. and Pines D. (1958), *Phys. Rev.* **110**, 936.

Bohr N. (1936), *Nature* **137**, 344.

Bohr N. and Kalckar F. (1937), *Mat. Fys. Medd. Dan. Vid. Selsk* **14** (10).

Bohr N. and Wheeler J.A. (1939), *Phys. Rev.* **56**, 426.

Born M., Heisenberg W. and Jordan P. (1926), *Z. Phys.* **35**, 557, (reprinted in van der Waerden B.L., ed. (1968), *Sources of Quantum Mechanics* (Dover Publications, New York)).

Brandow B.H. (1967), *Rev. Mod. Phys.* **39**, 771.

Brink D.M. and Broglia R.A. (2005), *Nuclear superfluidity: pairing in finite systems* (Cambridge University Press, Cambridge).

Brink D.M. and Satchler G.R. (1994), *Angular Momentum* (Oxford University Press), third edn.

Britz J., Pape A. and Anthony M.S. (1998), *At. Data Nucl. Data Tables* **69**, 125.

Brody T.A. and Moshinsky M. (1960), *Tables of Transformation Brackets* (Monografias del Instituto de Fisica, Universidad Nacional Autonoma de Mexico, Mexico).

Broglia R.A., Hansen O. and Riedel C. (1973), in *Adv. Nucl. Phys.*, Vol. 6, edited by M. Baranger and E. Vogt (Plenum, New York), p. 287.

Bromley D.A., ed. (1985), *Treatise on Heavy-Ion Science*, Vol. 3 (Plenum Press, New York and London).

Brown B.A. (2001), *Prog. Part. Nucl. Phys.* **47**, 517.

Brown B.A. (2002), *Nucl. Phys.* **A704**, 11c.

Brown B.A. and Richter W.A. (2006), *Phys. Rev.* **C74**, 0343150.

Brown B.A. and Wildenthal B.H. (1988), *Ann. Rev. Nucl. Part. Sci.* **38**, 29.

Brown G.E. (1967), *Unified Theory of Nuclear Models and Forces* (North-Holland, Amsterdam).

Brown G.E. and Green A.M. (1965), *Phys. Lett.* **15**, 168.

Brown G.E. and Green A.M. (1966), *Nucl. Phys.* **75**, 401.

Brueckner K.A. (1955), *Phys. Rev.* **97**, 1353.

Brussaard P.J. and Glaudemans P.W.M. (1977), *Shell-Model Applications in Nuclear Spectroscopy* (North-Holland, Amsterdam).

Buchinger F. *et al.* (1990), *Phys. Rev.* **C41**, 2883.

Buck B. and Merchant A.C. (1996), *Nucl. Phys.* **A600**, 387.

Butler S.T. (1951), *Proc. Roy. Soc. London* **A208**, 559.

Butler S.T. (1957), *Nuclear Stripping Reactions* (J. Wiley, New York).

Cambiaggio M.C., Rivas A.M.F. and Saraceno M. (1997), *Nucl. Phys.* **A624**, 157.

Caprio M.A., Cejnar P. and Iachello F. (2008), *Ann. Phys. (NY)* **323**, 1106.

Caprio M.A., Rowe D.J. and Welsh T.A. (2009), *Comp. Phys. Comm.* **180**, 1150–1163.

Carvalho M.J. and D'Agostino S. (2001), *Comp. Phys. Comm.* **141**, 282.

Carvalho M.J. and Rowe D.J. (1997), *Nucl. Phys.* **A618**, 65.

Casimir H. (1931), *Proc. Roy. Acad. Amsterdam* **34**, 844.

Caurier E. *et al.* (2005), *Rev. Mod. Phys.* **77**, 427.

Celenza L.S., Harindranath A. and Shakin C.M. (1985), *Phys. Rev.* **C32**, 2173.

Celenza L.S., Pong W.S. and Shakin C.M. (1983), *Phys. Rev.* **C27**, 1799.

Chabanat E. *et al.* (1998), *Nucl. Phys.* **A635**, 231, (erratum, *ibid.* **A643**, 441).

Chacón E. and Moshinsky M. (1977), *J. Math. Phys.* **18**, 870.

Chacón E., Moshinsky M. and Sharp R.T. (1976), *J. Math. Phys.* **17**, 668.

Chen H., Brownstein J.R. and Rowe D.J. (1990), *Phys. Rev.* **C42**, 1422.

Chen H., Song T. and Rowe D.J. (1995), *Nucl. Phys.* **A582**, 181.

Čížek J. and Paldus J. (1977), *Int. J. Quantum Chem.* **12**, 875.

Coester F. and Kümmel H. (1960), *Nucl. Phys.* **17**, 477.

Cohen B.L. and Price R.E. (1961), *Phys. Rev.* **121**, 1441.

Cohen S. and Kurath D. (1965), *Nucl. Phys.* **73**, 1.

Cohen S. and Kurath D. (1967), *Nucl. Phys.* **A101**, 1.

Coleman A.J. (1965), *J. Math. Phys.* **6**, 1425.

Condon E.U. and Shortley G.H. (1935), *Theory of Atomic Spectra* (Cambridge University Press, Cambridge).

Cooke T.H. and Wood J.L. (2002), *Amer. J. Phys.* **70**, 945.

Coon S.A. and Han H.K. (2001), *Few-body Systems* **30**, 131.

Cooper L.N. (1956), *Phys. Rev.* **104**, 1189.

Coraggio L. *et al.* (2002), *Phys. Rev.* **C66**, 021303(R).

Coraggio L. *et al.* (2007), *Phys. Rev.* **C76**, 061303(R).

Coraggio L. *et al.* (2009), *Prog. Part. Nucl. Phys.* **62**, 135.

Corminbouef F. *et al.* (2000), *Phys. Rev.* **C63**, 014305.

Covello A. *et al.* (2007), *Prog. Part. Nucl. Phys.* **59**, 401.

Czosnyka T. *et al.* (1986), *Nucl. Phys.* **A458**, 123.

da Providencia J. and Shakin C.M. (1964), *Ann. Phys.* (*NY*) **30**, 95.

Davidson P.M. (1932), *Proc. Roy. Soc. London* **135**, 459.

De Baerdemacker S., Heyde K. and Hellemans V. (2007), *J. Phys. A: Math. Gen.* **40**, 2733.

de Shalit A. and Goldhaber M. (1953), *Phys. Rev.* **92**, 1211.

de Shalit A. and Talmi I. (1963), *Nuclear Shell Theory* (Academic Press, New York).

Dean D. *et al.* (2004), *Prog. Part. Nucl. Phys.* **53**, 419.

Dean D.J. and Hjorth-Jensen M. (2003), *Rev. Mod. Phys.* **75**, 607.

Decharge J. and Gogny D. (1980), *Phys. Rev.* **C21**, 1568.

Des Cloizeaux J. (1960), *Nucl. Phys.* **20**, 321.

Dieperink A.E.L., Scholten O. and Iachello F. (1980), *Phys. Rev. Lett.* **44**, 1747.

Dietrich K., Mang H.J. and Pradal J.H. (1964), *Phys. Rev.* **135**, B22.

Dirac P.A.M. (1930), *Cambridge Philos. Soc* **26**, 376.

Dobaczewski J. *et al.* (2000), *Phys. Rev.* **C62**, 014311.

Domínguez F., Esebbag C. and Dukelsky J. (2006), *J. Phys. A: Math. Gen.* **39**, 11349.

Dossat C. *et al.* (2007), *Nucl. Phys.* **A792**, 18.

Dothan Y. (1970), *Phys. Rev.* **D2**, 2944.

Draayer J.P., Weeks K.J. and Rosensteel G. (1984), *Nucl. Phys.* **A413**, 215.

Draayer J.P. *et al.* (1989), *Comp. Phys. Comm.* **56**, 279.

Drut J., Furnstahl R. and Platter L. (2009), arXiv:0906.1463v2 [nucl-th].

Dukelsky J., Esebbag C. and Schuck P. (2001), *Phys. Rev. Lett.* **87**, 066403.

Dukelsky J., Pittel S. and Sierra G. (2004), *Rev. Mod. Phys.* **76**, 643.

Dukelsky J. and Sierra G. (1999), *Phys. Rev. Lett.* **83**, 172.

Dukelsky J. *et al.* (2006), *Phys. Rev. Lett.* **96**, 072503.

Dumitrescu T.S. and Hamamoto I. (1982), *Nucl. Phys.* **A383**, 205.

Dussel G.G. *et al.* (2007), *Phys. Rev.* **C76**, 011302(R).

Dyson F.J. (1956), *Phys. Rev.* **102**, 1217.

Dytrych T. *et al.* (2007), *Phys. Rev. Lett.* **98**, 162503.

Dytrych T. *et al.* (2008), *J. Phys. G: Nucl. Part. Phys.* **35**, 123101.

Edmonds A.R. (1996), *Angular Momentum in Quantum Mechanics* (Princeton University Press, Princeton, N.J.), paperback edition.

Edmonds A.R. and Flowers B.H. (1952), *Proc. Roy. Soc. London* **A214**, 515.

Eisberg R.M. and Porter C.E. (1961), *Rev. Mod. Phys.* **33**, 190.

Eisenberg J.M. and Greiner W. (1987), *Nuclear Models* (North Holland, Amsterdam), third edn.

El Samra N. and King R.C. (1979), *J. Phys. A: Math. Gen.* **12**, 2317.

Ellegaard C., Patnaik B. and Barnes P.D. (1970), *Phys. Rev.* **C2**, 2450.

Elliott J.P. (1958), *Proc. Roy. Soc. London* **A245**, 128.

Elliott J.P. (1958), *Proc. Roy. Soc. London* **A245**, 562.

Elliott J.P., Evans J.A. and Park P. (1986), *Phys. Lett.* **B169**, 309.

Elliott J.P. and Flowers B.H. (1955), *Proc. Roy. Soc. London* **A229**, 536.

Elliott J.P. and Lane A.M. (1957), in *Handbuch der Physik*, Vol. 39, edited by S. Flügge (Springer-Verlag, Berlin).

Elliott J.P. *et al.* (1968), *Nucl. Phys.* **A121**, 241.

Ellis P.J. and Osnes E. (1977), *Rev. Mod. Phys.* **49**, 777.

Emch G.G. (1984), *Mathematical and Conceptual Foundations of 20th Century Physics* (North-Holland, Amsterdam).

Endt P.M. (1979), *At. Data Nucl. Data Tables* **23**, 3.

Endt P.M. (1990), *Nucl. Phys.* **A521**, 1.

Entem D.R. and Machleidt R. (2002), *Phys. Lett.* **B524**, 93.

Entem D.R. and Machleidt R. (2003), *Phys. Rev.* **C68**, 041001(R).

Epelbaum E., Hammer H.W. and Meissner U.G. (2008) ArXiv:0811.1338v1 [nucl-th].

Epelbaum E. *et al.* (2002), *Phys. Rev.* **C66**, 064001.

Fadeev L.D. (1960), *Zh. Eksperim. i Teor. Fiz.* **39**, 1459, (English translation: Fadeev L.D. (1961), *Soviet Phys. JETP* **12**, 1014).

Fahlander C. *et al.* (1992), *Nucl. Phys.* **A541**, 157.

Fahlander C. *et al.* (1996), *Phys. Lett.* **B388**, 475.

Fahlander C. *et al.* (2001), *Phys. Rev.* **C63**, 021307(R).

Feldmeier H. and Schnack J. (1997), *Prog. Part. Nucl. Phys.* **39**, 393.

Feldmeier H. *et al.* (1998), *Nucl. Phys.* **A632**, 61.

Fermi E. (1932), *Rev. Mod. Phys.* **4**, 87.

Feshbach H. (1962), *Ann. Phys. (NY)* **19**, 287.

Feynman R.P. (1972), *Statistical Mechanics* (Benjamin, Reading, Mass.).

Fielding H.W. *et al.* (1977), *Nucl. Phys.* **A281**, 389.

Fierz M. (1939), *Helv. Phys. Acta* **12**, 3.

Fleming D.G. (1970), *Nucl. Phys.* **A157**, 1.

Fleming D.G. (1982), *Can. J. Phys.* **60**, 428.

Flowers B.H. (1952), *Proc. Roy. Soc. London* **A212**, 248.

Flowers B.H. and Szpikowski S. (1964), *Proc. Phys. Soc.* **84**, 673.

Flowers B.H. and Szpikowski S. (1964), *Proc. Phys. Soc.* **84**, 193.

Fock V.A. (1930), *Z. Phys.* **61**, 126.

Frame J.S., Robinson G. de B. and Thrall R.M. (1954), *Can. J. Math.* **6**, 316.

French J.B. (1960), *Nucl. Phys.* **15**, 393.

French J.B. (1966), in *Proceedings of the International School of Physics "Enrico Fermi", Course XXXVI*, edited by C. Bloch (Academic Press, New York).

French J.B. *et al.* (1969), in *Advances in Nuclear Physics, Vol. 3*, edited by M. Baranger and E. Vogt (Plenum, New York), p. 193.

Frois B. and Papanicolas C. (1987), *Ann. Rev. Nucl. Part. Sci.* **37**, 133.

Fujii S., Okamoto R. and Suzuki K. (2004), *Phys. Rev.* **C69**, 034328.

Fujii S., Okamoto R. and Suzuki K. (2009), *Phys. Rev. Lett.* **103**, 182501.

Fultz S.C. *et al.* (1962), *Phys. Rev.* **127**, 1273.

Gallagher Jr. C.J. and Soloviev V.G. (1962), *Mat. Fys. Skr. Dan. Vid. Selsk.* **2** (2).

Gambhir Y.K., Rimini A. and Weber T. (1969), *Phys. Rev.* **188**, 1573.

Gambhir Y.K., Rimini A. and Weber T. (1973), *Phys. Rev.* **C7**, 1454.

Gammel J. and Thaler R. (1957), *Phys. Rev.* **107**, 1337.

Garrett P.E. (2001), *J. Phys. G: Nucl. Part. Phys.* **27**, R1.

Garrett P.E. *et al.* (1977), *Phys. Lett.* **B400**, 250.

Garrett P.E. *et al.* (1997), *Phys. Rev. Lett.* **78**, 4545.

Garrett P.E. *et al.* (2001), *Phys. Rev. Lett.* **87**, 132502.

Garrett P.E. *et al.* (2007), *Phys. Rev.* **C75**, 054310.

Gaudin M. (1976), *J. Phys. (Paris)* **37**, 1087.

Geiger H. (1910), *Proc. Roy. Soc. London* **A83**, 492.

Geiger H. and Marsden E. (1909), *Proc. Roy. Soc. London* **A82**, 495.

Gelbart S. (1979), *Proc. Sympos. Pure Math.* **33**, 287.

Gel'fand I.M. and Tsetlin M.A. (1950), *Dokl. Akad. Nauk. USSR* **71**, 825.

Genilloud L. *et al.* (2001), *Nucl. Phys.* **A683**, 287.

Geyer H.B. and Hahne F.J.W. (1980), *Phys. Lett.* **B90**, 6.

Gilmore R. (1972), *Ann. Phys. (NY)* **74**, 391.

Gilmore R. (1979), *J. Math. Phys.* **20**, 891.

Ginocchio N. and Kirson M.W. (1980), *Phys. Rev. Lett.* **44**, 1744.

Glazek S.D. and Wilson K.G. (1993), *Phys. Rev.* **D48**, 5863.

Glazek S.D. and Wilson K.G. (1994), *Phys. Rev.* **D49**, 4214.

Gneuss G. and Greiner W. (1971), *Nucl. Phys.* **A171**, 449.

Goeke K. and Reinhard P.-G., eds. (1982), *Proc. Int. Symposium, held at Bad Honnef, Germany, June 7–11, 1982, Lecture Notes in Physics*, Vol. 171 (Springer-Verlag, Berlin).

Goeke K., Reinhard P.-G. and Rowe D.J. (1981), *Nucl. Phys.* **A359**, 408.

Goeke K. *et al.* (1983), *Prog. Theor. Phys. Suppl.* **74**, 33.

Goeppert-Mayer M. (1949), *Phys. Rev.* **75**, 1969.

Gogny D. (1973), in *Proc. Int. Conf. on Nucl. Phys. (Munich)*, edited by J. de Boer and H.J. Mang (North-Holland, Amsterdam).

Goldstone J. (1957), *Proc. Roy. Soc. London* **A239**, 267.

Górska M. *et al.* (1998), *Phys. Rev.* **C58**, 108.

Grzywacz R. *et al.* (1995), *Phys. Lett.* **B429**, 247.

Guardiola R. *et al.* (1996), *Nucl. Phys.* **A609**, 218.

Guazzoni P. (2004), *Phys. Rev.* **C69**, 024619.

Hahn B., Ravenhall D.G. and Hofstadter R. (1956), *Phys. Rev.* **101**, 1131.

Hamada T. and Johnston I.D. (1962), *Nucl. Phys.* **34**, 382.

Hamermesh M. (1962), *Group theory and its applications to physical problems* (Addison-Wesley).

Hartree D.R. (1928), *Proc. Camb. Phil. Soc.* **24**, 89.

Hasegawa M. and Tazaki S. (1987), *Phys. Rev.* **C35**, 1508.

Hasegawa M. and Tazaki S. (1993), *Phys. Rev.* **C47**, 188.

Haxel O., Jensen J.H.D. and Suess H.E. (1949), *Phys. Rev.* **75**, 1766.

Hayes A.C., Navrátil P. and Vary J.P. (2003), *Phys. Rev. Lett.* **91**, 012502.

Hecht K.T. (1965), *Nucl. Phys.* **63**, 177.

Hecht K.T. (1987), *Nucl. Phys.* **A475**, 276.

Hecht K.T. (1987), *The Vector Coherent State Method and Its Application to Problems of Higher Symmetries, Lecture Notes in Physics*, Vol. 290 (Springer-Verlag, Berlin).

Hecht K.T. (1993), *J. Phys. A: Math. Gen.* **26**, 329.

Hecht K.T. (1994), *J. Phys. A: Math. Gen.* **27**, 3445.

Hecht K.T. (2000), *Quantum Mechanics* (Springer, New York).

Hecht K.T. and Elliott J.P. (1985), *Nucl. Phys.* **A438**, 29.

Hecht K.T. and Pang S.C. (1969), *J. Math. Phys.* **10**, 1571.

Heisenberg J.H. and Mihaila B. (1999), *Phys. Rev.* **C59**, 1440.

Heisenberg W. (1932), *Z. Phys.* **77**, 1.

Helmers K. (1961), *Nucl. Phys.* **23**, 594.

Herman R. and Hofstadter R. (1960), *High Energy Electron Scattering Tables*, Stanford Univ. Press, Stanford, California.

Hess P.O., Maruhn J.A. and Greiner W. (1981), *J. Phys. G: Nucl. Phys.* **7**, 737.

Hess P.O. *et al.* (1980), *Z. Phys.* **A296**, 147.

Heyde K. *et al.* (1983), *Phys. Repts.* **102**, 291.

Heyde K. *et al.* (1987), *Nucl. Phys.* **A466**, 189.

Heyde K.L.G. (1990), *The Nuclear Shell Model* (Springer-Verlag, Berlin).

Hicks N.J. (1971), *Notes on Differential Geometry* (Van Nostrand, London).

Högaasen-Feldman J. (1961), *Nucl. Phys.* **28**, 258.

Hohenberg P. and Kohn W. (1964), *Phys. Rev.* **136**, B864.

Holzworth G. and Yukawa T. (1974), *Nucl. Phys.* **A219**, 125.

Honma M., Mizusaki T. and Otsuka T. (1995), *Phys. Rev. Lett.* **75**, 1284.

Hotta A. *et al.* (1987), *Phys. Rev.* **C36**, 2212.

Howe R. (1979), *Proc. Sympos. Pure Math.* **33**, 275.

Howe R. (1985), *Lect. Appl. Math.* **21**, 179.

Howe R. (1989), *Trans. Amer. Math. Soc.* **313**, 539.

Hubbard L.B. (1971), *At. Data Nucl. Data Tables* **9**, 85.

Hund F. (1937), *Z. Phys.* **105**, 202.

Iachello F. and Arima A. (1987), *The Interacting Boson Model* (Cambridge University Press, Cambridge).

Ichimura M. (1964), *Prog. Theor. Phys.* **32**, 757.

Igo G., Barnes P.D. and Flynn E.R. (1970), *Phys. Rev. Lett.* **24**, 470.

Infeld L. and Hull T.E. (1951), *Rev. Mod. Phys.* **23**, 21.

Inglis D.R. (1954), *Phys. Rev.* **96**, 1059.

Inglis D.R. (1955), *Phys. Rev.* **97**, 701.

Irvine J.M. (1972), *Nuclear Structure Theory* (Pergamon Press, Oxford).

Ishii N., Aoki S. and Hatsuda T. (2007), *Phys. Rev. Lett.* **99**, 022001.

Jahn H.A. (1951), *Proc. Roy. Soc. London* **A205**, 192.

Jahn H.A. and van Wieringen H. (1951), *Proc. Roy. Soc. London* **A209**, 502.

Jänecke J. (1969), in *Isospin in Nuclear Physics*, edited by D.H. Wilkinson (North Holland, Amsterdam), chap. 8.

Jänecke J. *et al.* (1991), *Nucl. Phys.* **A526**, 1.

Jarrio M., Wood J.L. and Rowe D.J. (1991), *Nucl. Phys.* **A528**, 409.

Ji X. and Wildenthal B.H. (1988), *Phys. Rev.* **C37**, 1256.

Ji X. and Wildenthal B.H. (1988), *Phys. Rev.* **C38**, 2849.

Jonsson N.-G. *et al.* (1981), *Nucl. Phys.* **A371**, 333.

Jordan P., von Neumann J. and Wigner E.P. (1934), *Annals of Math.* **35**, 29.

Jordan P. and Wigner E.P. (1928), *Z. Phys.* **47**, 631, (reprinted in Schwinger J., ed. (1958), *Selected Papers on Quantum Electrodynamics* (Dover Publications, New York)).

Judd B.R. (1968), in *Group Theory and its Applications*, edited by M. Loebl (Academic Press, New York), p. 183.

Juutinen S. *et al.* (1996), *Phys. Lett.* **B386**, 80.

Juutinen S. *et al.* (1997), *Nucl. Phys.* **A617**, 74.

Kadi M. *et al.* (2003), *Phys. Rev.* **C68**, 031306(R).

Kamada H. and Glöckle W. (1992), *Nucl. Phys.* **A548**, 205.

Kashiwara M. and Vergne M. (1978), *Invent. Math.* **44**, 1.

Kavka A.E. *et al.* (1995), *Nucl. Phys.* **A593**, 177.

Kelson I. and Levinson C.A. (1964), *Phys. Rev.* **134**, B269.

Kerman A.K. (1961), *Ann. Phys. (NY)* **12**, 300.

Kerman A.K. and Koonin S.E. (1976), *Ann. Phys. (NY)* **100**, 332.

Kibédi T. and Spear R.H. (2002), *At. Data Nucl. Data Tables* **80**, 35.

King R.C. (1971), *J. Math. Phys.* **12**, 1588.

King R.C. (1975), *J. Phys. A: Math. Gen.* **8**, 429.

Kirillov A.A. (1976), *Elements of the Theory of Representations* (Springer-Verlag, Berlin).

Kirson M.W. (1971), *Ann. Phys. (NY)* **66**, 624.

Kirson M.W. (1974), *Ann. Phys. (NY)* **82**, 345.

Kisslinger L.S. and Sorensen R.A. (1963), *Rev. Mod. Phys.* **35**, 853.

Klauder J.R. (1963), *J. Math. Phys.* **4**, 1058.

Klauder J.R. and Skagerstam B.-S. (1985), *Coherent States; Applications in Physics and Mathematical Physics* (World Scientific, Singapore).

Klein A. and Marshalek E.R. (1991), *Rev. Mod. Phys.* **63**, 375.

Kohn W. and Sham L. (1965), *Phys. Rev.* **140**, A1133.

Koonin S.E., Dean D.J. and Langanke K. (1997), *Phys. Repts.* **278**, 2.

Koopman T.A. (1933), *Physica* **1**, 104.

Kostant B. (1970), in *Group Representations in Mathematics and Physics, Lecture*

Notes in Physics, Vol. 6, edited by V. Bargmann (Springer-Verlag, Berlin).

Kota V.K.B. and Castilho Alcarás J.A. (2006), *Nucl. Phys.* **A764**, 181.

Kotliński B. *et al.* (1990), *Nucl. Phys.* **A517**, 365.

Kowalski K. *et al.* (2004), *Phys. Rev. Lett.* **92**, 132501.

Kramers H.A. (1930), *Proc. Koninkl. Ned. Akad. Wetenschop.* **33**, 959.

Kramers P. and Saraceno M. (1981), *Geometry of the Time-Dependent Variational Principle in Quantum Mechanics, Lecture Notes in Physics*, Vol. 140 (Springer-Verlag, Berlin).

Krane K.S. (1988), *Introductory Nuclear Physics* (Wiley, New York).

Krenciglowa E.M. and Kuo T.T.S. (1974), *Nucl. Phys.* **A235**, 171.

Kuhn T.S. (1996), *The Structure of Scientific Revolutions* (University of Chicago Press, Chicago), third edn.

Kuo T.T.S. (1974), *Ann. Rev. Nucl. Sci.* **24**, 101.

Kuo T.T.S. and Brown G.E. (1966), *Nucl. Phys.* **85**, 40.

Kuo T.T.S. and Brown G.E. (1968), *Nucl. Phys.* **A114**, 241.

Kuo T.T.S., Lee S.Y. and Ratcliff K.F. (1971), *Nucl. Phys.* **A176**, 65.

Kuo T.T.S. and Osnes E. (1990), *Folded-Diagram Theory of the Effective Interaction in Nuclei, Atoms and Molecules, Lecture Notes in Physics*, Vol. 364 (Springer-Verlag, Berlin).

Lacombe M. *et al.* (1980), *Phys. Rev.* **C21**, 861.

Lalazissis G.A., König J. and Ring P. (1997), *Phys. Rev.* **C55**, 540.

Lalazissis G.A., Raman S. and Ring P. (1999), *At. Data Nucl. Data Tables* **71**, 1.

Lamm I.-L. (1969), *Nucl. Phys.* **A125**, 504.

Landau L.D. and Lifshitz E.M. (1980), *Statistical Physics* (Butterworth-Heinemann, Oxford), third edn.

Lande A. (1965), *Ann. Phys. (NY)* **31**, 525.

Lane A.M. (1964), *Nuclear Theory* (Benjamin, New York and Amsterdam).

Langevin M. *et al.* (1985), *Phys. Lett.* **B150**, 71.

Lassila K.E. *et al.* (1962), *Phys. Rev.* **126**, 881.

Lattes C.M.G., Occhialini G.P.S. and Powell C.F. (1947), *Nature* **160**, 453.

Lauritsen T. *et al.* (2002), *Phys. Rev. Lett.* **88**, 042501.

Lawson R.D. (1980), *Theory of the Shell Model* (Oxford University Press).

Lawson R.D. (1981), *Z. Phys.* **A303**, 51.

Lawson R.D. and Macfarlane M.H. (1965), *Nucl. Phys.* **66**, 80.

Lee J., Teang M.B. and Lynch W.G. (2007), *Phys. Rev.* **C75**, 064320.

Lee S.Y. and Suzuki K. (1980), *Phys. Lett.* **91B**, 173.

Lehmann H. *et al.* (1996), *Phys. Lett.* **B387**, 259.

Lerma S. *et al.* (2007), *Phys. Rev. Lett.* **99**, 032501.

Leviatan A. (1996), *Phys. Rev. Lett.* **77**, 818.

Levinson C.A. (1963), *Phys. Rev.* **132**, 2184.

Lievens P. *et al.* (1991), *Phys. Lett.* **B256**, 141.

Lievens P. *et al.* (1992), *Phys. Rev.* **C46**, 797.

Links J. *et al.* (2002), *J. Phys. A: Math. Gen.* **35**, 6459.

Lipkin H.J. (1960), *Ann. Phys. (NY)* **9**, 272.

Lippman B.A. and Schwinger J. (1950), *Phys. Rev.* **79**, 469.

Littlewood D.E. (1950), *The Theory of Group Characters and Matrix Representations of Groups* (Oxford University Press), second edn.

Littlewood D.E. (1958), *Can. J. Math.* **10**, 17.

Littlewood D.E. and Richardson A.R. (1934), *Philos. Trans. Roy. Soc. London* **A233**, 99.

Lorazo B. (1969), *Phys. Lett.* **B29**, 150.

Lorazo B. (1970), *Nucl. Phys.* **A153**, 255.

Louck J.D. and Galbraith H.W. (1972), *Rev. Mod. Phys.* **44**, 540.

Lozeva R.L. *et al.* (2008), *Phys. Rev.* **C77**, 064313.

Macdonald I.G. (1979), *Symmetric Functions and Hall Polynomials* (Oxford University Press).

Macfarlane M.H. (1966), *Lectures in Theoretical Physics*, Vol. **8c** (University of Colorado Press, Boulder).

Macfarlane M.H. and French J.B. (1960), *Rev. Mod. Phys.* **32**, 567.

Machleidt R. (2001), *Phys. Rev.* **C63**, 024001.

Machleidt R. and Šlaus I. (2001), *J. Phys. G: Nucl. Part. Phys.* **27**, R69.

Mackey G.W. (1976), *The Theory of Unitary Group Representations* (University of Chicago Press, Chicago).

Marsden D.C.J. *et al.* (2002), *Phys. Rev.* **C66**, 044007.

Marsden J.E. and Ratiu T.S. (1999), *Introduction to Mechanics and Symmetry* (Springer, New York).

Mårtensson-Pendrill A.-M. *et al.* (1992), *Phys. Rev.* **A45**, 4675.

Martínez-Pinedo G. *et al.* (1997), *Phys. Rev.* **C55**, 187.

Mayer M.G. (1950), *Phys. Rev.* **78**, 16.

Mayer M.G. and Jensen J.H.D. (1955), *Elementary Theory of Nuclear Shell Structure* (Wiley, New York).

McNeill J.H. *et al.* (1989), *Phys. Rev. Lett.* **63**, 860.

Messiah A. (1966), *Quantum Mechanics*, Vol. 2 (North Holland, Amsterdam), sixth printing.

Messiah A. (1966), *Quantum Mechanics*, Vol. 1 (North Holland, Amsterdam), fourth printing.

Metag V., Habs D. and Specht H.J. (1980), *Phys. Repts.* **65**, 1.

Meyer-ter-Vehn J. (1975), *Nucl. Phys.* **A249**, 111.

Miller G.A., Nefkens B.M.K. and Šlaus I. (1990), *Phys. Repts.* **194**, 1.

Mizusaki T. *et al.* (1999), *Phys. Rev.* **C59**, R1846.

Möller P. *et al.* (1995), *At. Data Nucl. Data Tables* **59**, 185.

Morinaga H. (1956), *Phys. Rev.* **101**, 254.

Morita T. (1963), *Prog. Theor. Phys.* **29**, 351.

Moshinsky M. (1959), *Nucl. Phys.* **13**, 104.

Moshinsky M. and Quesne C. (1970), *J. Math. Phys.* **11**, 1631.

Moshinsky M. and Quesne C. (1971), *J. Math. Phys.* **12**, 1772.

Moszkowski S.A. and Scott B.L. (1960), *Ann. Phys. (NY)* **11**, 65.

Moszkowski S.A. and Scott B.L. (1961), *Ann. Phys. (NY)* **14**, 107.

Mottelson B.R. (1958), in *The Many-Body Problem* (Dunod, Paris), p. 283.

Mottelson B.R. (1976), *Rev. Mod. Phys.* **48**, 375.

Mottelson B.R. and Valatin J.G. (1960), *Phys. Rev. Lett.* **5**, 511.

Murnaghan F.D. (1938), *The Theory of Group Representations* (The Johns Hopkins Press, Baltimore).

Nadjakov E.G., Marinova K.P. and Gangrsky Y.P. (1994), *At. Data Nucl. Data Tables* **56**, 133.

Navratil P., Geyer H.B. and Kuo T. (1993), *Phys. Lett.* **B315**, 165.

Navrátil P. and Ormand W.E. (2002), *Phys. Rev. Lett.* **88**, 152502.

Navrátil P. and Ormand W.E. (2003), *Phys. Rev.* **C68**, 034305.

Navrátil P., Vary J.P. and Barrett B.R. (2000), *Phys. Rev.* **C62**, 054311.

Navrátil P., Vary J.P. and Barrett B.R. (2000), *Phys. Rev. Lett.* **84**, 5728.

Navrátil P. *et al.* (2001), *Phys. Rev. Lett.* **87**, 172502.

Nazarewicz W. (1993), *Int. J. Mod. Phys.* **E2**, 51.

Neddermeyer S.H. and Anderson C.D. (1937), *Phys. Rev.* **51**, 884.

Negele J.W. (1970), *Phys. Rev.* **C1**, 1260.

Newell M.J. (1951), *Proc. Roy. Irish Acad.* **A54**, 153.

Nilsson S.G. (1955), *Mat. Fys. Medd. Dan. Vid. Selsk.* **29** (16).

Nishiyama S. and Fukutome H. (1992), *J. Phys. G: Nucl. Part. Phys.* **18**, 317.

Nogami Y. (1964), *Phys. Rev.* **134**, B313.

Nogga A., Kamada H. and Glöckle W. (2000), *Phys. Rev. Lett.* **85**, 944.

Okubo S. (1954), *Prog. Theor. Phys.* **12**, 603.

Ortiz J.V., Weiner B. and Öhrn Y. (1981), *Int. J. Quantum Chem.* **S15**, 113.

Otsuka T., Tanihata I. and Sakurai H., eds. (2002), *Proceedings of The International Symposium Shell Model 2000, Nucl. Phys.*, Vol. A704.

Otsuka T. *et al.* (2001), *Prog. Part. Nucl. Phys.* **47**, 319.

Ottaviani P.L. and Savoia M. (1969), *Phys. Rev.* **187**, 1306.

Ottaviani P.L. and Savoia M. (1970), *Nuovo Cimento* **A67**, 630.

Otten E.W. (1989), in *Treatise on Heavy-Ion Science, Nuclei Far from Stability*, Vol. 8, edited by D.A. Bromley (Plenum Press, New York), p. 517.

Pape A., Anthony M.S. and Georgiadis A. (1988), *Phys. Rev.* **C38**, 1952.

Parikh J.C. (1965), *Nucl. Phys.* **63**, 214.

Park P. *et al.* (1984), *Nucl. Phys.* **A414**, 93.

Patera J. and Sharp R.T. (1980), *J. Phys. A: Math. Gen.* **13**, 397.

Pauli W. (1933), *Handbuch der Physik Bd.*, Vol. XXIV/1 (Springer-Verlag, Berlin).

Pauli W. (1940), *Phys. Rev.* **58**, 716.

Peierls R.E. and Yoccoz J. (1957), *Proc. Phys. Soc.* **70**, 381.

Perelomov A.M. (1972), *Comm. Math. Phys.* **26**, 222.

Perelomov A.M. (1986), *Generalized Coherent States and their Applications* (Springer-Verlag, Berlin).

Peterson R.J. *et al.* (1991), *Phys. Rev.* **C44**, 136.

Pieper S.C., Varga K. and Wiringa R.B. (2002), *Phys. Rev.* **C66**, 044310.

Pieper S.C. *et al.* (2001), *Phys. Rev.* **C64**, 014001.

Pudliner B.S. *et al.* (1997), *Phys. Rev.* **C56**, 1720.

Putterman S.J. (1974), *Superfluid Hydrodynamics* (North-Holland, Amsterdam).

Racah G. (1942), *Phys. Rev.* **62**, 438.

Racah G. (1943), *Phys. Rev.* **63**, 367.

Racah G. and Talmi I. (1952), *Physica* **18**, 1097.

Raghavan P. (1989), *At. Data Nucl. Data Tables* **42**, 189.

Rainwater J. (1950), *Phys. Rev.* **79**, 432.

Rainwater J. (1976), *Rev. Mod. Phys.* **48**, 385.

Raman S., Nestor Jr. C.W. and Tikkanen P. (2001), *At. Data Nucl. Data Tables* **78**, 1.

Raman S., Walkiewicz T.A. and Behrens H. (1975), *At. Data Nucl. Data Tables* **16**, 451.

Redmond P.J. (1954), *Proc. Roy. Soc. London* **A222**, 84.

Reid R.V. (1968), *Ann. Phys. (NY)* **50**, 411.

Reinhard P.-G. and Flocard H. (1995), *Nucl. Phys.* **A584**, 467.

Richardson R.W. (1963), *Phys. Lett.* **3**, 277.

Richardson R.W. (1963), *Phys. Lett.* **5**, 82.

Richardson R.W. and Sherman N. (1964), *Nucl. Phys.* **52**, 221.

Richardson R.W. and Sherman N. (1964), *Nucl. Phys.* **52**, 253.

Ring P. and Schuck P. (1980), *The Nuclear Many-Body Problem* (Springer-Verlag, New York).

Robinson G. de B. (1958), *Can. Math. Bull.* **1**, 21.

Robinson G. de B., ed. (1977), *The Collected Papers of Alfred Young* (University of Toronto Press).

Rohoziński S.G., Srebrny J. and Horbaczewska K. (1974), *Z. Phys.* **268**, 401.

Rombouts S., Van Neck D. and Dukelsky J. (2004), *Phys. Rev.* **C69**, 061303(R).

Rose M.E. (1995), *Elementary Theory of Angular Momentum* (Dover Publications, New York; originally published by Wiley, 1957).

Rosensteel G. and Rowe D.J. (1977), *Phys. Rev. Lett.* **38**, 10.

Rosensteel G. and Rowe D.J. (1980), *Ann. Phys. (NY)* **126**, 343.

Rosensteel G. and Rowe D.J. (2003), *Phys. Rev.* **C67**, 014303.

Rosensteel G. and Rowe D.J. (2007), *Nucl. Phys.* **A797**, 94.

Roth R. *et al.* (2004), *Nucl. Phys.* **A745**, 3.

Rowe D.J. (1966), *Nucl. Phys.* **80**, 209.

Rowe D.J. (1968), *Rev. Mod. Phys.* **40**, 153.

Rowe D.J. (1970), *Nuclear Collective Motion: Models and Theory* (Methuen, London).

Rowe D.J. (1982), *Nucl. Phys.* **A391**, 307.

Rowe D.J. (1984), *J. Math. Phys.* **25**, 2662.

Rowe D.J. (1985), *Rep. Prog. Phys.* **48**, 1419.

Rowe D.J. (2001), *Nucl. Phys.* **A691**, 691.

Rowe D.J. (2004), *Nucl. Phys.* **A735**, 372.

Rowe D.J. (2004), in *Int. Conf. on Computational and Group-Theoretical Methods in Nuclear Physics*, edited by J. Escher *et al.* (World Scientific, Singapore).

Rowe D.J. (2005), *J. Phys. A: Math. Gen.* **38**, 10181.

Rowe D.J. and Bahri C. (1998), *J. Phys. A: Math. Gen.* **31**, 4947.

Rowe D.J. and Basserman R. (1974), *Nucl. Phys.* **A220**, 404.

Rowe D.J. and Basserman R. (1976), *Can. J. Phys.* **54**, 1941.

Rowe D.J. and Carvalho M.J. (2007), *J. Phys. A: Math. Gen.* **40**, 471.

Rowe D.J. and Repka J. (1991), *J. Math. Phys.* **32**, 2614.

Rowe D.J. and Rosensteel G. (1976), *Int. J. Theor. Phys.* **15**, 501.

Rowe D.J., Rosensteel G. and Carr R. (1984), *J. Phys. A: Math. Gen.* **17**, L399.

Rowe D.J., Rosensteel G. and Gilmore R. (1985), *J. Math. Phys.* **26**, 2787.

Rowe D.J. and Ryman A. (1982), *J. Math. Phys.* **23**, 732.

Rowe D.J., Ryman A. and Rosensteel G. (1980), *Phys. Rev.* **A22**, 2362.

Rowe D.J., Song T. and Chen H. (1991), *Phys. Rev.* **C44**, R598.

Rowe D.J., Thiamova G. and Wood J.L. (2006), *Phys. Rev. Lett.* **97**, 202501.

Rowe D.J. and Turner P.S. (2005), *Nucl. Phys.* **A753**, 94.

Rowe D.J., Turner P.S. and Repka J. (2004), *J. Math. Phys.* **45**, 2761.

Rowe D.J., Vassanji M.G. and Rosensteel G. (1983), *Phys. Rev.* **A28**, 1951.

Rowe D.J., Welsh T.A. and Caprio M.A. (2009), *Phys. Rev.* **C79**, 054304.

Rowe D.J. *et al.* (preprint).

Rutherford E. (1911), *Phil. Mag.* **21**, 669.

Saint-Laurent M.G. *et al.* (1987), *Phys. Rev. Lett.* **59**, 33.

Samyn M. *et al.* (2002), *Nucl. Phys.* **A700**, 142.

Schrödinger E. (1940-41), *Proc. Roy. Irish Acad.* **A46**, 9.

Schrödinger E. (1940-41), *Proc. Roy. Irish Acad.* **A46**, 183.

Schrödinger E. (1941), *Proc. Roy. Irish Acad.* **A47**, 53.

Schucan T.H. and Weidenmüller H.A. (1972), *Ann. Phys. (NY)* **73**, 108.

Schucan T.H. and Weidenmüller H.A. (1973), *Ann. Phys. (NY)* **76**, 483.

Schur I. (1901), *Uber eine Klasse von Matrizen, die sich einer gegebenen Matrix zuordnen lassen*, Ph.D. thesis, Berlin.

Schwinger J., ed. (1958), *Selected Papers on Quantum Electrodynamics* (Dover Publications, New York).

Schwinger J. (1965), in *Quantum Theory of Angular Momentum*, edited by L.C. Biedenharn and H. Van Dam (Academic Press, New York), p. 229.

Shera E.B. *et al.* (1976), *Phys. Rev.* **C14**, 731.

Shirokov A.M. *et al.* (2004), *Phys. Rev.* **C70**, 044005.

Shirokov A.M. *et al.* (2005), *J. Phys. G: Nucl. Part. Phys.* **31**, S1283.

Shlomo S. (1972), *Nucl. Phys.* **A184**, 545.

Shlomo S. (1978), *Rep. Prog. Phys.* **41**, 957.

Shlomo S. and Talmi I. (1972), *Nucl. Phys.* **A198**, 81.

Shoup R. *et al.* (1969), *Nucl. Phys.* **A135**, 689.

Sierra G. (2000), *Nucl. Phys.* **B572**, 517.

Sierra G. *et al.* (2000), *Phys. Rev.* **B61**, R11890.

Sinatkas J. *et al.* (1992), *J. Phys. G: Nucl. Part. Phys.* **18**, 1377.

Skensved P. *et al.* (1981), *Nucl. Phys.* **A366**, 125.

Skyrme T.H.R. (1957), *Proc. Phys. Soc.* **A70**, 433.

Skyrme T.H.R. (1958/9), *Nucl. Phys.* **9**, 615.

Skyrme T.H.R. (1958/9), *Nucl. Phys.* **9**, 635.

Smith M.B. *et al.* (2003), *Phys. Rev.* **C68**, 031302(R).

Soloviev V.G. (1961), *Mat. Fys. Skr. Dan. Vid. Selsk.* **1** (11).

Souriau J.-M. (1966), *Comm. Math. Phys.* **1**, 374.

Spreng W. *et al.* (1983), *Phys. Rev. Lett.* **51**, 1522.

Srebrny J. *et al.* (2006), *Nucl. Phys.* **A766**, 25.

Stachel J. *et al.* (1982), *Nucl. Phys.* **A383**, 429.

Stachel J. *et al.* (1984), *Nucl. Phys.* **A419**, 589.

Stapp H.P., Ypsilantis T.J. and Metropolis N. (1957), *Phys. Rev.* **105**, 302.

Stetcu I. *et al.* (2005), *Phys. Rev.* **C71**, 044325.

Stevenson P.D. (2002), *Comp. Phys. Comm.* **147**, 853.

Stoks V.G.J. *et al.* (1994), *Phys. Rev.* **C49**, 2950.

Stone N.J. (2005), *At. Data Nucl. Data Tables* **90**, 75.

Street J.C. and Stevenson E.C. (1937), *Phys. Rev.* **52**, 1003.

Strutinski V.M. (1967), *Nucl. Phys.* **A95**, 420.

Suzuki K. and Lee S.Y. (1980), *Prog. Theor. Phys.* **64**, 2091.

Suzuki K. and Okamoto R. (1986), *Prog. Theor. Phys.* **75**, 1388.

Suzuki K. and Okamoto R. (1986), *Prog. Theor. Phys.* **76**, 127.

Suzuki K. and Okamoto R. (1994), *Prog. Theor. Phys.* **92**, 1045.

Suzuki K. and Okamoto R. (1995), *Prog. Theor. Phys.* **93**, 905.

Suzuki T. and Matsuyanagi K. (1976), *Prog. Theor. Phys.* **56**, 1156.

Svensson L.E. (1989), Ph.D. thesis, Univ. of Uppsala.

Svensson L.E. *et al.* (1995), *Nucl. Phys.* **A584**, 547.

Sviratcheva K.D. *et al.* (2009), Talk available from the INT workshop website http://www.int.washington.edu/talks/WorkShops/int_09_1/.

Talmi I. (1952), *Helv. Phys. Acta* **25**, 185.

Talmi I. (1971), *Nucl. Phys.* **A172**, 1.

Talmi I. (1975), *Phys. Lett.* **B55**, 255.

Talmi I. (1993), *Simple Models of Complex Nuclei* (Harwood Academic Publishers, Chur, Switzerland).

Talmi I. and Unna I. (1960), *Nucl. Phys.* **19**, 225.

Talmi I. and Unna I. (1962), *Nucl. Phys.* **30**, 280.

Thouless D.J. (1960), *Nucl. Phys.* **21**, 225.

Thouless D.J. (1961), *Nucl. Phys.* **22**, 78.

Tickle R. and Bardwick J. (1971), *Phys. Lett.* **B36**, 32.

Tilley D.R., Weller H.R. and Cheves C.M. (1993), *Nucl. Phys.* **A564**, 1.

Tjøm P.O. and Elbek B. (1968), *Nucl. Phys.* **A107**, 385.

Towner I.S. and Hardy J.C. (1969), *At. Data Nucl. Data Tables* **6**, 153.

Trainor L.E.H. (1952), *Phys. Rev.* **85**, 962.

Troltenier D., Maruhn J.A. and Hess P.O. (1991), in *Computational Nuclear Physics 1*, edited by K. Langanke, J.A. Maruhn and S.E. Koonin (Springer, Berlin), p. 105.

Turner P.S. and Rowe D.J. (2005), *Nucl. Phys.* **A756**, 333.

Tuttle III W.K. *et al.* (1976), *Phys. Rev.* **C13**, 1036.

Uhlenbeck G.E. and Goudsmit S. (1925), *Naturwissenschaften* **47**, 953.

Ungrin J. *et al.* (1971), *Mat. Fys. Medd. Dan. Vid. Selsk.* **38** (8).

Uusitalo J. *et al.* (1998), *Phys. Rev.* **C57**, 2259.

Valatin J.G. (1958), *Nuovo Cimento* **7**, 843.

van der Waerden B.L., ed. (1968), *Sources of Quantum Mechanics* (Dover Publications, New York).

Vary J.P. and Plastino A. (1983), *Phys. Rev.* **C28**, 2494.

Vautherin D. and Brink D.M. (1972), *Phys. Rev.* **C5**, 626.

von Delft J. and Poghossian R. (2002), *Phys. Rev.* **B66**, 134502.

von Delft J. and Ralph D.C. (2001), *Phys. Repts.* **345**, 61.

Wada Y., Takano F. and Fukuda N. (1958), *Prog. Theor. Phys.* **19**, 597.

Wahlborn S. (1962), *Nucl. Phys.* **37**, 554.

Walecka J.D. (1974), *Ann. Phys. (NY)* **83**, 491.

Watanable H. (1964), *Prog. Theor. Phys.* **32**, 106.

Wegner F. (1994), *Ann. Phys. (Leipzig)* **3**, 77.

Weisskopf V.F. (1951), *Phys. Rev.* **83**, 1073.

Welsh T.A. (2008), (accessible from http://www.physics.utoronto.ca/~rowe/group.html).

Wendt K. *et al.* (1988), *Z. Phys.* **A329**, 407.

Wesseling J. *et al.* (1991), *Nucl. Phys.* **A535**, 285.

Weyl H. (1950), *The Theory of Groups and Quantum Mechanics* (Dover Publications, New York), translation of *Gruppentheorie und Quantenmechanik*, published in 1931.

Weyl H. (1953), *Classical Groups, their Invariants and Representations* (Princeton University Press, Princeton, N.J.), second edn.

Whitehead R.R. *et al.* (1977), *Adv. Nucl. Phys.* **9**, 123.

Wick G.C. (1950), *Phys. Rev.* **80**, 268.

Wigner E.P. (1937), *Phys. Rev.* **51**, 106.

Wigner E.P. (1957), in *Proc. of the Robert A. Welch Conferences on Chemical*

Research, Vol. 1, edited by W.O. Milligan (Robert A. Welch Foundation, Houston, Texas), p. 67.

Wigner E.P. (1959), *Group Theory and Its Applications to the Quantum Mechanics of Atomic Spectra* (Academic Press, New York).

Wilets L. and Jean M. (1956), *Phys. Rev.* **102**, 788.

Wilkinson D.H. (1960), in *Nuclear Spectroscopy*, edited by F. Ajzenberg-Selove (Academic Press, New York), Part B.

Wilkinson D.H., ed. (1969), *Isospin in Nuclear Physics* (North Holland, Amsterdam).

Williams S.A. and Pursey D.L. (1968), *J. Math. Phys.* **9**, 1230.

Wiringa R.B., Stoks V.G.J. and Schiavilla R. (1995), *Phys. Rev.* **C51**, 38.

Wiringa R.B. *et al.* (2000), *Phys. Rev.* **C62**, 014001.

Wollersheim H.J. *et al.* (1993), *Nucl. Phys.* **A556**, 261.

Wood J.L. *et al.* (1992), *Phys. Repts.* **215**, 101.

Woods R.D. and Saxon D.S. (1954), *Phys. Rev.* **95**, 577.

Wu C.Y. *et al.* (1991), *Nucl. Phys.* **A533**, 359.

Wybourne B.G., SCHUR, An interaction program for calculating properties of Lie groups and symmetric functions, distributed by S. Christensen `http://smc.vnet.net/Christensen.html`.

Wybourne B.G. (1970), *Symmetry Principles and Atomic Spectroscopy* (Wiley, New York).

Wybourne B.G. (1974), *Classical Groups for Physicists* (Wiley, New York).

Yakubovsky O.A. (1967), *Sov. J. Nucl. Phys.* **5**, 937.

Yennie D.R., Ravenhall D.G. and Wilson R.N. (1954), *Phys. Rev.* **95**, 500.

Yoshida S. (1961), *Phys. Rev.* **123**, 2122.

Yoshida S. (1962), *Nucl. Phys.* **33**, 685.

Yukawa H. (1935), *Proc. Phys. Math. Soc. (Japan)* **17**, 48, (reprinted in Beyer R.T., ed. (1949), *Foundations of Physics* (Dover Publications, New York)).

Zamick L. and Escuderos A. (2005), *Phys. Rev.* **C71**, 054308.

Zamick L. and Escuderos A. (2006), *Ann. Phys. (NY)* **321**, 987.

Zhang W.-M., Feng D.H. and Gilmore R. (1990), *Rev. Mod. Phys.* **62**, 867.

Author index

Subject index